MATH 6

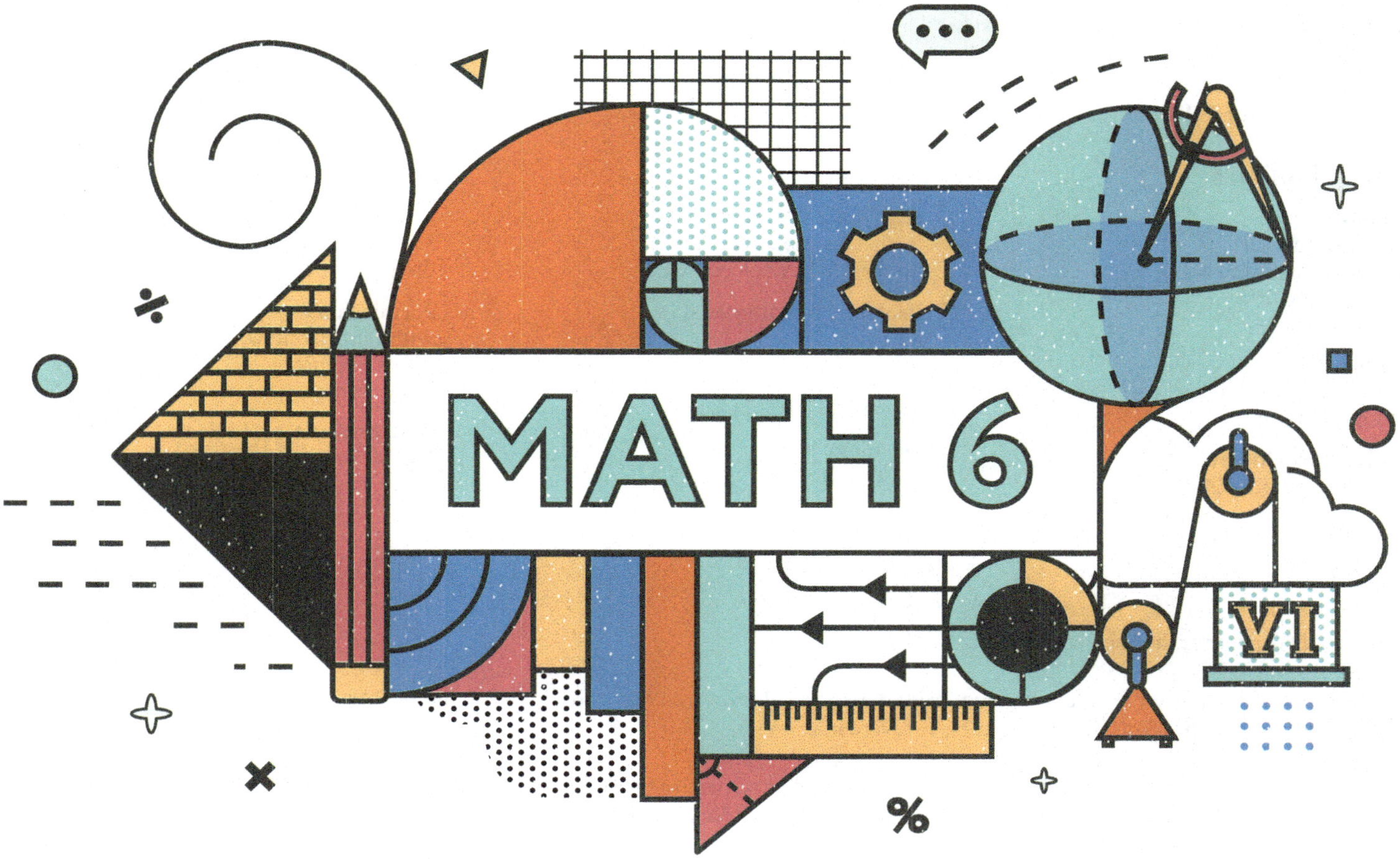

Fourth Edition

bju press®

Greenville, South Carolina

Part 2

Note: The fact that materials produced by other publishers may be referred to in this volume does not constitute an endorsement of the content or theological position of materials produced by such publishers. Any references and ancillary materials are listed as an aid to the student or the teacher and in an attempt to maintain the accepted academic standards of the publishing industry.

Math 6 Teacher Edition
Part 2
Fourth Edition

Writers
Charlene McCall
Lindsey Dickinson, MEd
Kathleen Hynicka
Rita Lovely

Editors
Natalie Bonczek
Emily N. House

Biblical Worldview
Tyler Trometer, MDiv

Academic Integrity
Jeff Heath, EdD

Academic Oversight
Rachel Santopietro, MEd
Michael Winningham, MEd

Project Coordinator
Kyla J. Smith

Production Designer
Maribeth Hayes

Designer
Kathryn Ratje

Cover Designer
Michael Asire
Brenna Short

Permissions
Ruth Bartholomew
Carrie Hanna
Stacy Stone

Post-Production Liaison
Peggy Hargis

Photo credits appear at the back of the book.

The text for this book is set in Adobe Minion Pro, Adobe Myriad Pro, Arial MT, Circe by Paratype, Helvetica, MathematicalPi, Nimbus Sans by URW, Times, and Verdana.

All trademarks are the registered and unregistered marks of their respective owners. BJU Press is in no way affiliated with these companies. No rights are granted by BJU Press to use such marks, whether by implication, estoppel, or otherwise.

BJU Press grants the original purchaser a limited license to copy the reproducible pages in this book for use in his classroom. Further reproduction or distribution shall be a violation of this license. Copies may not be sold.

© 2022 BJU Press
Greenville, South Carolina 29609
First Edition © 1981 BJU Press
Second Edition © 1995 BJU Press
Third Edition © 2011 BJU Press

Printed in the United States of America
All rights reserved

ISBN 978-1-64626-031-7 (two parts)

15 14 13 12 11 10 9 8 7 6

CONTENTS

Part 1

BIBLICAL WORLDVIEW SHAPING

IN *MATH 6*, FOURTH EDITION

What is math all about? Is it just numbers and symbols? Math is not just about these characters that represent math. Math is about numbers in space in God's creation. Using the four biblical worldview themes below, students will begin learning to view mathematics biblically.

MATH 6, Fourth Edition, answers the questions posed below to help students begin to think about math the way that God intends.

Knowledge about Math

God shows us what the world is like in the Scriptures and in creation. Math is a human activity that enables people to explore God's creation. Some people want to think that math can give them the most complete and reliable knowledge of the world around them.

- How did God make math possible for humans?
- How does Scripture inform a Christian's view of the reliability of math?

Ch 1	Ch 5	Ch 9	Ch 14
E(3), R	E(5), EV	E, EV(3)	E(2), A, F

KEY
(for tables)

R: Recall biblical teaching.

E: Explain biblical teaching.

EV: Evaluate controversial concepts.

F: Formulate a biblical understanding of a controversial concept.

A: Apply a biblical understanding to life.

The number in parentheses indicates the number of times a specific application of a theme occurs in the chapter.

Modeling with Math

Mathematical modeling is a human way of understanding and representing the world God made. Sometimes, however, people put too much confidence in math models. Some people even believe that math models are the only really reliable way to find the truth.

- What do math models require in order to be useful?
- How should a Christian respond to someone who claims that math is completely objective?

Ch 2	Ch 7	Ch 10	Ch 17
R(3), E	E(3)	R, E, EV	E(3), F

Service with Math

People uniquely bear the image of God and therefore they possess the ability to use math to serve others. Math enables people to serve others as God intended. However, because all people are sinful, they naturally learn and use math in selfish ways and for distorted purposes.

- How can a Christian use math to serve others?
- What must always guide the way a Christian uses math?

Ch 3	Ch 8	Ch 12	Ch 15
E(4)	E(4)	E(2), A	E(3), F, A

Design in Math

As we use math to solve problems, we discover that our world has been carefully designed. God intends for us to praise Him for His good and wise design of creation. But many people argue that the appearance of design seen in mathematical patterns resulted from natural processes, not God.

- How should a Christian explain the mathematical order found in Creation?
- What role does the Bible play in helping us understand God's design of creation?

Ch 4	Ch 6	Ch 11	Ch 13	Ch 16
R, E(2)	E(3)	R, E, EV, F	R, E(2), EV, A	R, E, F

Above are the biblical worldview themes that are important for sixth grade math students to know. Early in the course the students will recall and explain these themes. However, as these themes are repeated, the students will evaluate ideas within them, formulate a biblical understanding of them, and apply what they have learned about these themes to real-world situations. High levels of internalization are expected wherever the students are required to apply their learning.

BUILDING ACADEMIC RIGOR

WITH *Math 6*, FOURTH EDITION

Desired Learning Outcomes
What do I want students to learn?

- Number sense
- Understanding of ratios and proportional relationships
- Ability to manipulate algebraic expressions to solve equations
- Ability to analyze data sets
- Mathematical literacy and fluency
- Twenty-first-century skills: collaboration, creativity, critical thinking, problem solving, technology literacy
- Understanding of the biblical worldview themes of knowledge, modeling, service, and design, as they relate to mathematics
- Math practices including perseverance in problem solving, reasoning abstractly and quantitatively, constructing arguments and critiquing others' reasoning, using models, using appropriate math tools, encouraging precision, looking for structure and patterns, and developing the ability to employ consistent reasoning

Teaching and Learning Supports
How will I help students learn?

- Model processes and procedures for students to apply to their own problem solving.
- Provide opportunities for students to practice mental math.
- Use cumulative reviews to strengthen retention of recent content.
- Guide students in standardized test practice.
- Play mathematics games to strengthen student understanding of new content.
- Guide visual analyses of charts and graphs.
- Lead number talks to encourage students to think creatively about problem solving.
- Help students grasp new content by using mathematical vocabulary in context.
- Respond to varying learning levels with differentiated instruction.
- Develop twenty-first-century skills using collaborative STEM projects.
- Assign collaborative activities in speaking, listening, and writing.
- Redefine educational experiences through strategic use of technology.
- Pique interest in worldview and mathematical concepts using essential questions.
- Encourage class discussions to determine misconceptions and to answer biblical worldview questions.

Evidence of Understanding
How do I evaluate student learning?

- Completing frequent and varied preassessments and formative assessments
- Answering questions during class discussions and activities
- Making and checking predictions
- Consistent, accurate reasoning
- Independent, self-monitored learning
- Perseverance in problem solving
- Written and oral communication

A PANORAMA OF ACADEMIC RIGOR

Academic rigor is the educational experience that engages students in content appropriate to their academic level and helps them learn to analyze, evaluate, and ultimately create.

The Teacher's Role

The teacher is the person who crafts this learning environment. The teacher creates a learning community with high educational expectations, ignites interest and passion to help students persevere in challenging educational tasks, and inspires them by providing models of learning. The teacher instructs growing learners, supporting, critiquing, and praising their efforts along the way. While laying a foundation for lifelong learning, the teacher helps students transfer their academic knowledge, understanding, and skills to life with the purpose of shaping students' worldview.

The Learning Experience

The textbook opens doors to the world of the learning experience. It captures student interest and significantly contributes to the educational environment. Educational and technological resources enable the teacher to develop knowledge, understanding, and skills in a scaffolded sequence to build a foundation that students can transfer to life. Teachers use educational tools to craft relevant, authentic learning experiences that develop creativity and problem-solving skills in a variety of contexts. These kinds of learning experiences allow students to take responsibility for their learning in the present and discover the joy of serving God with their vocational knowledge and skills in the future.

The Learning Environment

Learning happens in a context. Physical and social environments create this context. For students to be stimulated and engaged and to perform at high levels, they need an environment that connects with and shapes their interests and values. Students need a flexible environment that adapts to remedy their deficits and capitalize on their proficiencies. They need an interactive environment that intrinsically motivates them by giving them structure, freedom, and choices in learning. They need a safe environment that invites questioning, risk-taking, and honest discussion. Most importantly, students need positive relationships with mentors and peers and one-on-one time with caring, responsible instructors and parents. This kind of environment does not happen by accident; an educational climate like this must be crafted.

TECHNOLOGY SOLUTIONS

Why Educational Technology?

BJU Press has always maintained that the teacher is the key for learning in the classroom. The teacher knows best what the students require so they can accomplish educational objectives. The teacher guides and directs learning, using various methods to accomplish his or her purposes.

Today, teachers are expected to use technology to implement and supplement their methods. Of course, educational technology has always been around—it includes the alphabet, which allows us to visualize sounds and words, various devices used to write, and surfaces to write on. But when we think of technology in the twenty-first century, we picture digital devices, projections, and online interactivity. BJU Press's offerings are growing with the demand for educational technology in Christian schools. The purpose is not to have a shiny new gadget or method. The purpose is to equip today's teachers to fulfill the goal of Christian education: "to develop redeemed man in the image of God" (BJU Press, *Handbook of Christian Education* [2017], 33).

What Does BJU Press Offer?
Teacher Tools Online

Teacher Tools Online from BJU Press is the premier resource for teachers using BJU Press materials in schools. This powerful catalog of digital resources encourages teachers in their efforts to expand their lessons and enhances their teaching strategies for greater depth. Available resources include

- editable PowerPoint® presentations
- lesson plan overviews
- various media such as videos, artwork, and web links
- instructional aids
- ExamView® test databases
- eTexts (student and teacher editions)
- Curriculum Trak (curriculum maps of BJU Press materials)
- professional development (opportunity to earn CEUs)
- ShopTalk Community (teacher collaboration)

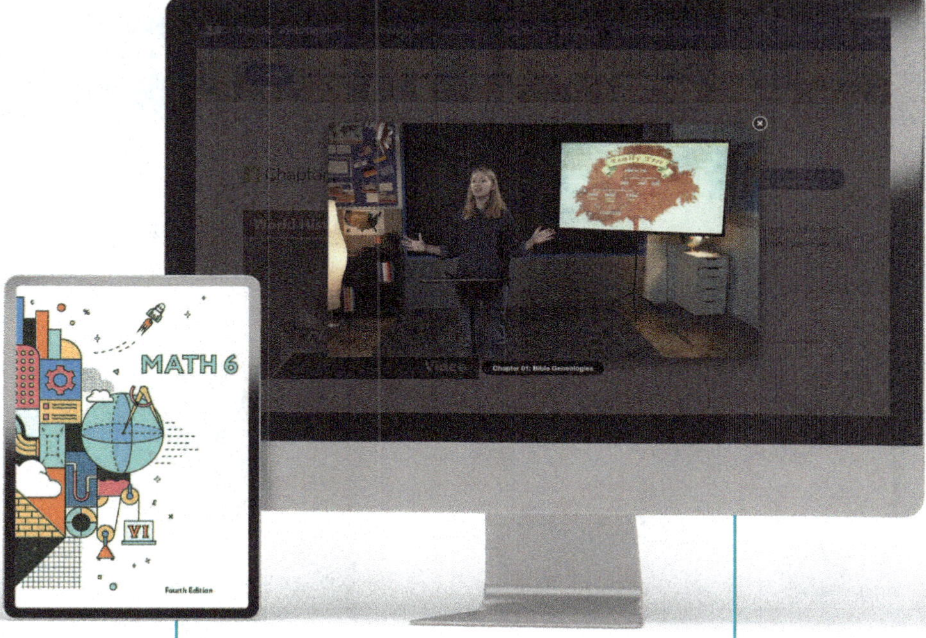

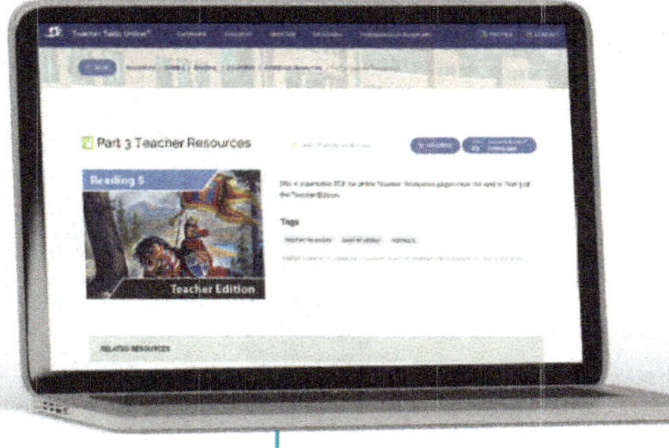

Teacher Tools Online provides practical aid as you plan and teach.

eTexts allow convenient access at home and school for both students and teachers.

Video classes provide instruction for the students you proctor.

Homeschool Hub

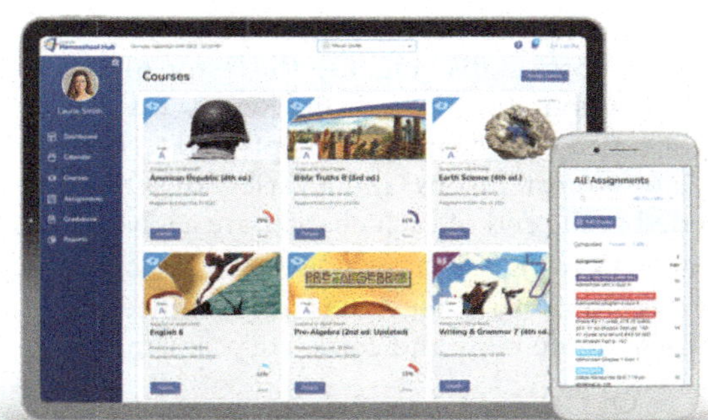

Homeschool Hub provides homeschool parents and facilitators with access to select digital instructional aids that support the lessons. Homeschool Hub also provides a digital scheduler and gradebook.

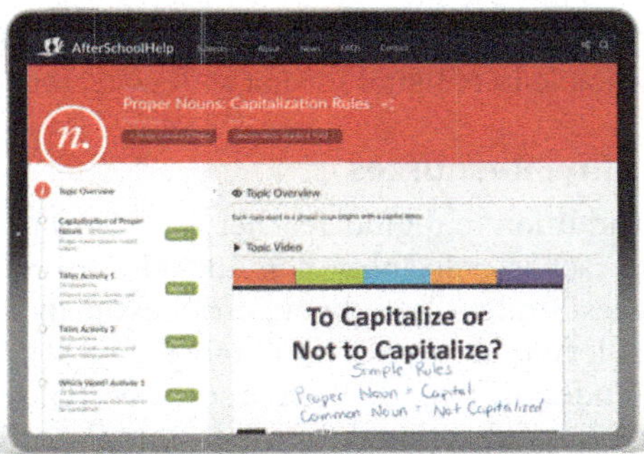

AfterSchoolHelp

AfterSchoolHelp allows your students to freely access tutorials and extra practice for select BJU Press math and English courses. The materials on AfterSchoolHelp align with the chapters of BJU Press textbooks. It also is the home of audio files for Spanish 1 and Spanish 2.

How to Access Technology Offered by BJU Press

TeacherToolsOnline.com may be accessed by Christian schools that purchase a one-time, transferrable license.

HomeschoolHub.com offers free access to homeschoolers using BJU Press materials.

AfterSchoolHelp.com offers free access to anyone who wants to use its resources.

INSTRUCTIONAL MATERIALS

Student Materials

Student Edition

The Student Edition provides two pages of explanation and practice problems to reinforce skills taught in the lesson and in earlier lessons. In addition, it provides Daily Review exercises for previously taught concepts. Key math concepts and terms are listed on the first page of most lessons, and a variety of activities are provided. Included at the end of each chapter are a Chapter Review and a Cumulative Review.

The Student Edition is nonconsumable and is not designed to be written in. Students should copy and complete problems on their own paper.

Assessments

The assessments packet includes a test and one or more quizzes for each of the seventeen chapters.

Teacher Materials

Teacher Edition

This Teacher Edition provides for 180 teaching days divided into seventeen chapters.

Each chapter is introduced by a chapter opener, which provides the following: chapter objectives, a list of skills taught, and a chapter essential question.

Each lesson includes reduced Student Edition and Daily Review pages with answer overprint, and the steps used to solve word problems and other mathematical procedures on the Student Edition pages are provided in the Solutions section. Each chapter concludes with a Chapter Review lesson and a Cumulative Review lesson.

Teacher Resources

Pages intended for use as instructional aids during lessons are found in the back of this Teacher Edition. These pages are also available in a digital format at TeacherToolsOnline.com.

Online Resources

In addition to digital Teacher Resources pages, online resources at TeacherToolsOnline.com include Teaching Tips, Application Pages, student Handbook, preassessment checklists, Fact Reviews, Exploring Ideas pages, and more. The Solutions section also includes the long-division process, partial-products multiplication, graphs, and optional drawings used for solving problems on Student Edition pages. Speed drills are available online at AfterSchoolHelp.com.

Assessments Answer Key

The assessments packet includes an answer key for each of the tests and quizzes.

THE TEACHING CYCLE IN *MATH 6*

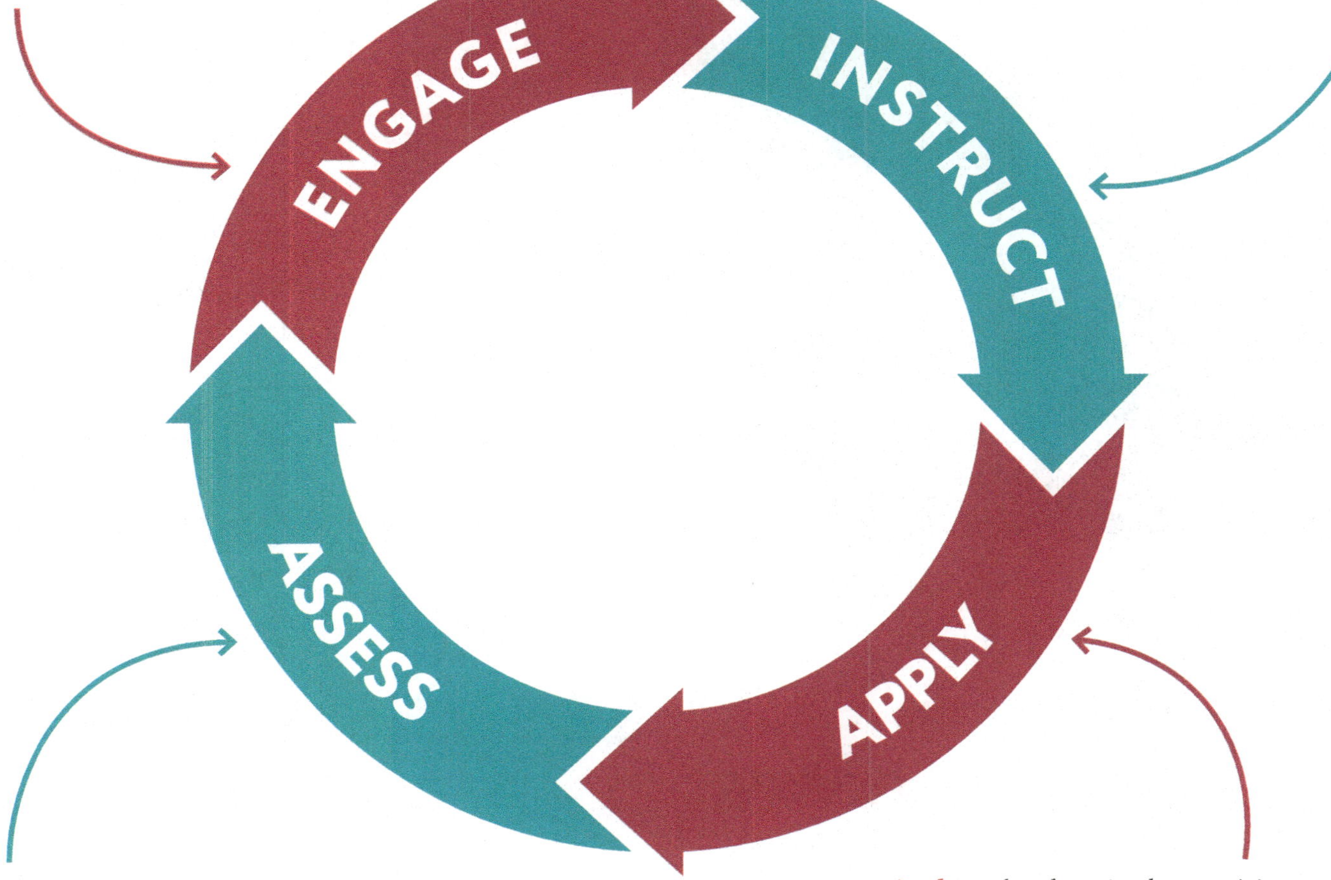

Engage students by capturing their attention, activating prior knowledge, and motivating them to connect with new content.

- Think-Pair-Share
- Brainstorm
- Think-Pair-Write
- Turn and Talk
- Ticket in the Door

Instruct students by using direct, indirect, and interactive strategies to expand and extend their knowledge and skills.

- Discussions
- Modeling
- Analysis
- Mental math
- Graphic organizers
- Guided practice
- Number sense
- Guided discovery

Assess student understanding by using a variety of tools to systematically evaluate knowledge, skills, attitudes, and beliefs in order to improve student learning.

- Preassessments
- Formative assessments
- Summative assessments
- Reviews
- STEM rubrics

Apply student learning by practicing knowledge and skills and connecting them to real life.

- Games
- Math Talks
- Guess and Check
- Demonstrations
- Activities

LESSON FEATURES

Preparation

Visit TeacherToolsOnline.com for additional STEM resources on bridge design.

Locate pictures of bridges of varying design, including historic, unusual (very long, innovatively designed), and artistically beautiful bridges, to display during the lesson. Include pictures of the bridges mentioned but not pictured on Student Edition page 265, if available. A list of suggested bridges to display is provided at the end of this lesson.

To save time during the lesson, prepare the *Graph Paper* pages by folding under one of the pages along the top of the grid. Tape it to the bottom of the grid on the other page to provide a $17\frac{3}{4}$ in. continuous grid that the students can use for drawing their part of a bridge plan.

Display the *Engineering Design Process* page and refer to it as you proceed through the steps of the STEM project.

Ask

Bridge design

- Guide a **visual analysis** of bridges to help the students connect bridge design with bridge efficiency in the bridge they will build.

- Direct the students to read the career link about Mr. Rosales at the bottom of Student Edition page 265 silently.
  What are Mr. Rosales's bridge designs known for? their beauty and how they fit into their surroundings
  Besides beauty, what other qualities of a bridge do you think are important to a bridge designer? sample answers: reliability (strength and safety), cost of the materials, length required, how well it will function in the setting in which it is needed, availability of materials, durability (how long it will last)

Objectives point out the skills taught in the lesson.

Discussion of real-world math problems helps students relate math to **biblical worldview truths**.

The **Materials** section lists items that are used in the lesson.

A variety of **activities** allows the students to practice analytical thinking and see math at work in real-world contexts.

Instructional strategies provide the means for presenting educational content.

LESSON 131

- Distribute a copy of the *Bridge & Truss Designs* page to each student and display your copy. Discuss the drawing and description of each type of bridge. Point out the 3 common truss designs.

  Point out that the Liberty Bridge designed by Mr. Rosales, shown in the inset picture on Student Edition page 265, is an innovatively designed cable-stayed bridge.

- Display the pictures of bridges that you located, guiding the students as they identify the structural design of each bridge by comparing it with the *Bridge & Truss Designs* page.

- Point out that because of the different ways different bridge designs handle the forces or load put on them, certain designs are better suited to certain settings.

  Why do you think a bridge designer might choose a suspension bridge over a beam bridge for a longer span? A beam bridge must rest on pillars beneath its deck, which might not be practical or even possible for a long distance. A suspension bridge is supported by cables above the deck connected to towers which help bear the load. A suspension bridge can typically stretch a longer distance than a beam bridge.

  Point out that many bridges combine different structures in their design. For example, suspension bridges sometimes also use trusses under the deck for support.

The problem

- Use **direct instruction** to help the students identify the problem they need to solve.

- Write the word "efficiency" for display. Explain that a bridge's efficiency is a measurement of how well the bridge performs. Point out that the bridges the students build for the STEM project will be tested for efficiency.

  The suggested requirements for the project may be adjusted as desired.

6. Collaborate to divide the construction process among your group members. Record the responsibilities of each person. *Answers will vary.*

7. Using graph paper, draw a full-size model of the section of the bridge that you are responsible to build. *Answers will vary.*

Create to implement your solution.

8. Build a bridge according to the specifications of your project.

Test and improve your solution.

9. Weigh the bridge and record its weight in grams. Round to the nearest tenth. *Answers will vary.*

10. Test the bridge and calculate its efficiency.
    a. Record the weight of the load the bridge held until failure. Round to the nearest whole number. *Answers will vary.*
    b. Convert the weight of the load from pounds to kilograms. (2.2 lb = 1 kg) Round to the nearest whole number. Rename kilograms to grams for the final answer. (1 kg = 1,000 g)
    c. Calculate the bridge's efficiency, using a ratio. Round to the nearest whole number.

11. How could I improve my design?

Reflect.

12. Why can I build a sturdy bridge? *The reliability of ratios in the design of God's world makes building sturdy bridges possible (Hebrews 1:3).*

10. b. *Divide the weight of the load (lb) by 2.2 to convert to kilograms (round to the nearest whole number); multiply the number of kilograms by 1,000 to rename to grams.*

    c. *Answers will vary.*

    $$\text{bridge efficiency} = \frac{\text{mass of load}}{\text{mass of bridge}}$$

11. *sample answers: I could change my truss design, add more cross bracing, strengthen my beams, or use fewer sticks for greater efficiency.*

294   Chapter 13

- Explain to the students that they will work with their group to design and build a bridge made of only 100 or fewer popsicle sticks and all-purpose glue. The bridge must span a 12 in. opening but be no longer than 15 in. The bridge must be able to support a load placed on top of it so that it can be tested to failure (until it breaks and no longer supports the load) to determine its efficiency.

  What problem do you need to solve? how to design and build the most efficient bridge from the specified materials that will span a 12 in. opening

  Direct the students to answer problems 1–2 on Student Edition page 293.

Imagine

Bridge efficiency

- Guide a **discussion** to help the students identify an efficient design for their bridge. Emphasize the importance of good planning in bridge design. Refer to Student Edition page 265.

  What type of planning does Mr. Rosales do when designing a bridge? He makes a lot of 3D drawings and models and computer drawings.

- Explain that building the strongest bridge possible with the least amount of materials makes it efficient. A bridge's efficiency can be calculated mathematically by comparing the mass of the load the bridge can bear to the bridge's mass.

REVIEW FEATURES

Review with the chapter sequence

The chapter sequence of core math topics provides review. This sequence allows time for the teacher to review concepts, and reteach if necessary, before the student moves to a higher level in that same math topic.

Review daily with practice activities

Practice activities are drill activities that usually require more memory than understanding. These include measurement equivalents and fact memorization. Math facts should be practiced for at least 5–10 minutes daily. Fact Reviews are provided at TeacherToolsOnline.com.

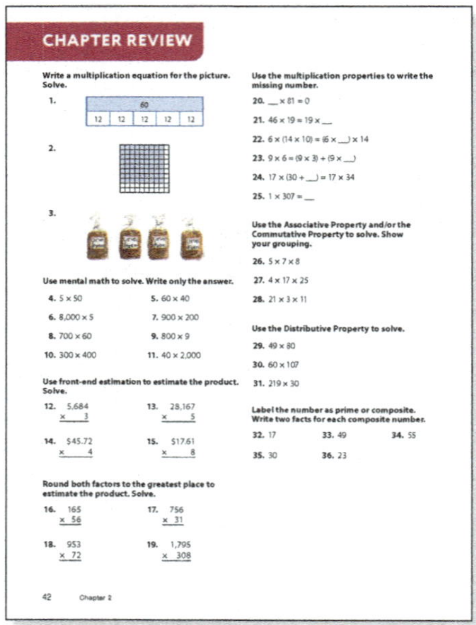

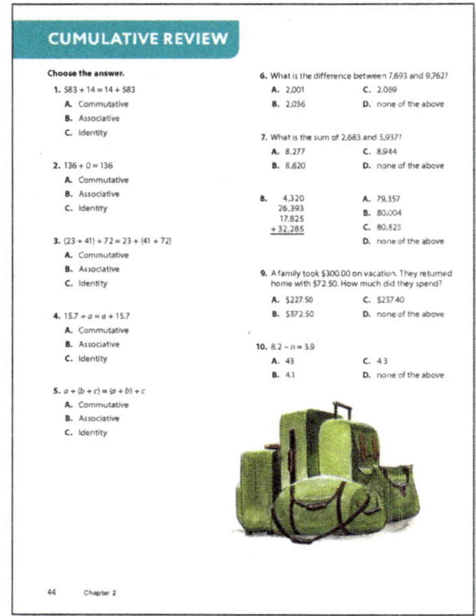

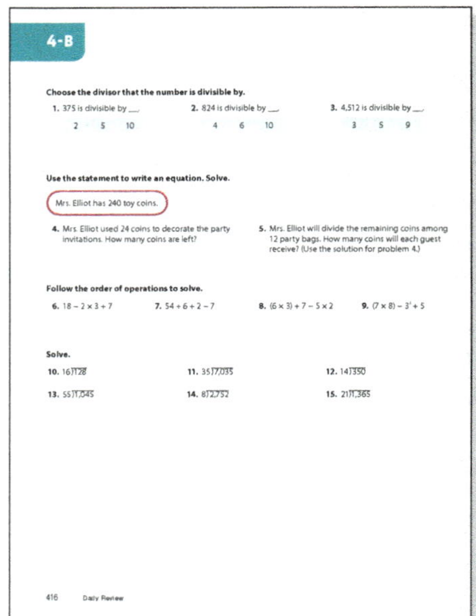

Review the chapter with a Chapter Review

A Chapter Review is included at the end of each chapter. During the Chapter Review lesson, the teacher reviews the main concepts of the chapter. The two Student Edition pages allow the teacher to evaluate what the student has learned; these pages along with the Daily Review exercises can be used as a study guide for the Chapter Test.

Review preceding chapters with a Cumulative Review

A Cumulative Review follows the Chapter Review at the end of Chapters 2–17. These Student Edition pages include skills from several earlier chapters and can be used to review math concepts or to identify concepts the student has forgotten or not yet mastered. This review enables the teacher to evaluate the specific needs of students in order to reteach these students individually or in small groups. A fifth-grade review is at the end of Chapter 1.

Review with Daily Reviews

A section of Daily Review exercises is included for each chapter of the Student Edition. These exercises review a concept taught or practiced in an earlier chapter or in fifth grade. These may be used for additional review, reteaching of concepts, and evaluation.

NEW TO THIS EDITION

STEM

Four chapters (5, 9, 13, and 15) feature special lessons that emphasize science, technology, engineering, and math (STEM). Each STEM lesson is intended to pique students' interest as they collaborate to solve a real-world problem through inquiry, active learning, and creativity by following an engineering design process. STEM lessons may be used at any time following the lesson in which they are introduced and are excellent springboards to further student investigation on related topics. Encourage the students to make this a time of learning and experimenting rather than one of producing a polished product. There is much to be learned from the process of trying out ideas.

Essential Questions

Each chapter and lesson have an essential question, which is posed near the beginning of the chapter or lesson and answered at a later point. The sample answers provided can be used to guide the students to answer the essential question on the Student Edition pages.

Math Talks

Most chapters include a Math Talk, a short, open-ended question designed to foster conversations and critical thinking. During this activity, students should share and expand their ideas, listen to others, and deepen their understanding of relevant problems.

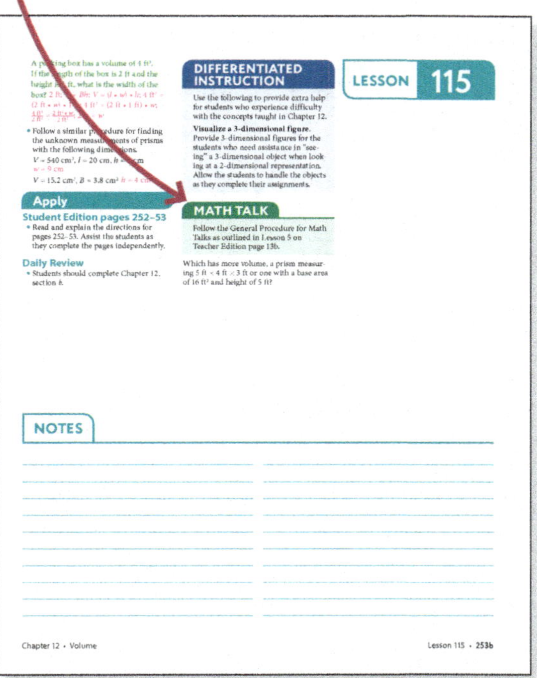

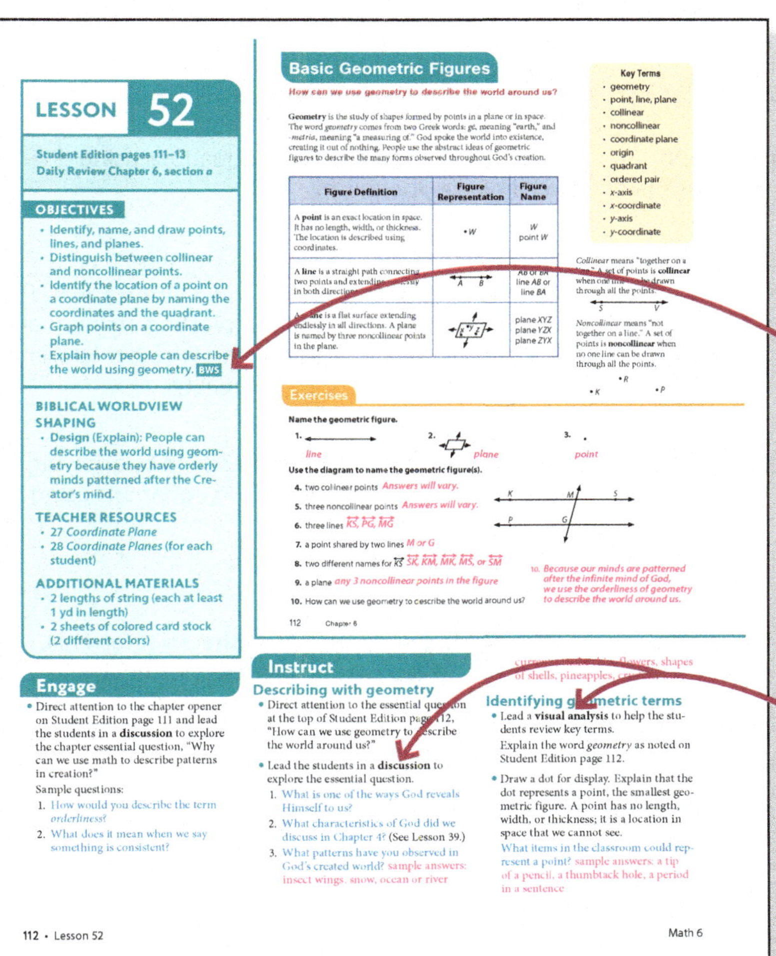

Biblical Worldview Shaping

Biblical worldview themes are specifically highlighted in certain lesson objectives (indicated by **BWS**). These sections will help students learn to apply a biblical worldview of mathematics to real-world problems.

Strategies

Effective teaching strategies are offered in bold for each different topic in the flow of the lesson. These may be used to provide variety and to sustain interest within and between the lessons.

Lesson Plan Overview

Chapter 10: Equations

Day	Lesson	Student Edition Pages	Teacher Edition Pages	Teacher Resources & Additional Materials	Topics, Skills & Biblical Worldview Shaping
94	94	203–5	203–5b		**Topic:** Expressions **Skills:** writing numerical or algebraic expressions, evaluating expressions using substitution
95	95	206–7	206–7b		**Topic:** Equations **Skills:** writing an equation with two equal expressions, determining the unknown in a word problem and writing it as a variable **BWS:** Modeling
96	96	208–9	208–9b	Teacher Resources • 52 *Variable Cards N* • 53 *Variable Cards X* Additional Materials • 18 round counters	**Topic:** Simplifying Expressions **Skills:** computing sums and products that are multiples of 10, simplifying algebraic expressions by applying the Commutative and Associative Properties
97	97	210–11	210–11b	Additional Materials • 1 *n* card and 1 *x* card from the patterns provided on the *Variable Cards* pages • 30 round counters	**Topic:** Addition & Subtraction Equations **Skills:** solving addition and subtraction equations using inverse operations, checking addition and subtraction equations using substitution **BWS:** Modeling
98	98	212–13	212–13b	Additional Materials • 5 *n* cards from the pattern provided on the *Variable Cards N* page • 15 round counters	**Topic:** Multiplication Equations **Skills:** solving multiplication equations using division (inverse operation)
99–100	99	214–15	214–15b		**Topic:** Multiplication & Division Equations **Skills:** solving multiplication and division equations using inverse operations, plotting an inequality on a number line, determining whether a given number is a solution to an inequality
101	100	216–17	216–17b	Additional Materials • geographic compass or picture of one Assessments • Chapter 10 Quiz 1	**Topic:** Equivalent Expressions **Skills:** finding equivalent expressions by applying the Distributive Property, solving equations using inverse operations **BWS:** Modeling
102	101	218–19	218–19b	Teacher Resources • 54 *Graphing an Equation*	**Topic:** *Distance = Rate × Time* **Skills:** solving problems using the $d = rt$ formula, completing a table using the $d = rt$ formula **BWS:** Modeling
103	102	220–21	220–21b		**Topic:** Chapter Review
104	103	222–24	222–24		**Topic:** Test & Cumulative Review

CHAPTER 10

CHAPTER OBJECTIVES

- Write, relate, and simplify algebraic expressions.
- Simplify algebraic expressions by applying mathematical properties to find equivalent expressions.
- Solve and check equations using inverse operations.
- Write an equation with two equal expressions or with a variable.
- Solve problems using the $d = rt$ formula to complete a table and create a line graph.
- Evaluate mathematical models in light of reality.

10   EQUATIONS

Do math models represent reality?

To determine whether your students are prepared for the concepts taught in this chapter, you may wish to consult the preassessment checklist found on TeacherToolsOnline.com.

Make fact practice, both oral and written, part of your daily math routine to help the students with mastery.

Visit AfterSchoolHelp.com for math practice resources, or visit TeacherToolsOnline.com for additional resources to enhance the lessons.

Student Edition pages 203–5
Daily Review Chapter 10, section *a*

OBJECTIVES

- Write a numerical or an algebraic expression for a word phrase.
- Write algebraic expressions with more than one operation.
- Evaluate an expression using substitution.

Engage

- Direct attention to the chapter opener on Student Edition page 203 and lead the students in a **discussion** to explore the chapter essential question, "Do math models represent reality?"

  You may remember studying models in Chapters 2 and 7. What is a model? A model is a simplified explanation or representation of something in the world.

  What is meant by a model representing reality? It comes as close as possible to real life. It accurately represents God's world.

- Point out that in this chapter the students will be exploring how well a math model represents reality.

Instruct

Writing an expression for a word phrase

- Guide a **discussion** to connect previous content to new information to help the students write an expression.

- Read aloud the following word problem.

  Five hikers were hiking on the nature trail. As they proceeded up the path, 2 other hikers caught up with them and joined their group.

  How many hikers were on the trail at first? 5

  Which word phrase tells you what happened to the original 5 hikers? They continued on the path, and 2 other hikers joined the group.

  What operation sign and number can you write to show that 2 hikers joined the group of 5? + 2

Expressions

How can a comma help me write a word phrase for an expression?

An **expression** is a mathematical phrase made up of numbers, operation signs, and sometimes variables. A **variable** is a letter used to represent an unknown value. An expression can be written by interpreting a word phrase.

two added to thirty: $30 + 2$
a number divided by 3: $n \div 3$

An **algebraic expression** always contains a variable. When a number is multiplied by a variable, the number is written before the variable and is called the **coefficient** of the variable. A multiplication sign is *not* needed.

the product of a number and 12: $12n$
$12n = 12 \times n$
12 is the coefficient of the variable n.

Evaluate algebraic expressions by **substituting** a value for the variable. Calculate to find a numerical value for the expression.

Key Terms
- expression
- variable
- algebraic expression
- coefficient
- substitution

Home Zone offers a discount to customers who purchase 4 or more gallons of paint. Mr. Lehman purchased 4 gal of paint for his music studio. Write an algebraic expression to show the cost of paint for Mr. Lehman. Find the cost if the discounted price is $13.00 for each gallon.

$$4g = \underline{\quad}$$
$g = \$13.00$
$4(\$13.00) = \$52.00$
The cost of 4 gal of paint is $52.00.

Exercises

Complete the word phrase for the expression. *Answers may vary.*

1. 15 + 32: __ added to __ *32; 15*

2. $14\frac{1}{2} \div 4 + 3$: __ added to __ divided by __ *3; $14\frac{1}{2}$; 4*

3. 6c: __ times __ *6; a number*

4. 8x − 2: __ less than __ times __ *2; 8; a number*

Write the numerical expression for the word phrase.

5. 52 less than 96 *96 − 52*

6. 8 multiplied by 2 *2 × 8*

7. the product of 4 and 6 *4 × 6*

8. 0.9 more than 1.06 *1.06 + 0.9*

9. 12 decreased by 1.7 *12 − 1.7*

10. the quotient of 32 and 4 *32 ÷ 4*

11. 3 to the second power added to 9 *$9 + 3^2$*

12. 17.3 increased by three-tenths *17.3 + 0.3*

Write the algebraic expression for the word phrase.

13. a number divided by 7 *n ÷ 7*

14. 1.5 less than n *n − 1.5*

15. 9 times a number *9n*

16. n divided by 8, decreased by 4 *n ÷ 8 − 4*

17. 0.2 less than a number *n − 0.2*

18. 2 less than 3 times n *3n − 2*

19. 16.8 more than n *n + 16.8*

20. the sum of 6 and a number, divided by 3 *(6 + n) ÷ 3*

21. $\frac{3}{4}$ of a number *$\frac{3}{4}n$*

22. $\frac{2}{3}$ of a number divided by 12 *$\frac{2}{3} \times (n \div 12)$ or $(\frac{2}{3} \times n) \div 12$*

- Write "5 + 2" for display. Explain that expressions are mathematical phrases made up of numbers, operation signs, and sometimes variables. Numerical expressions are mathematical phrases such as 5 + 2; the phrases contain numbers and operations. Recognizing a mathematical operation in written form can help them to translate a word phrase into a numerical expression.

  What other word phrases can you think of that indicate the addition of 2 to 5? sample answers: 5 increased by 2; 2 added to 5; the sum of 5 and 2; and 2 more than 5

- Read aloud the following word problem.

  After the 7 hikers had walked 1 mi, the path became very steep, so 3 hikers left the group and went back to the trailhead.

  How many hikers were walking together? 7

  Which word phrase tells you what happened to the 7 hikers? 3 hikers left.

  What operation sign and number can you write to show that 3 hikers left the group? − 3

- Write "7 − 3" for display.

  What other word phrases can you think of that indicate the subtraction of 3 from 7? sample answers: 7 minus 3; 7 decreased by 3; the difference of 7 and 3; 3 subtracted from 7; and 3 less than 7

Write the algebraic expression. Identify what the variable represents in the expression.

23. The pet shop sold 3 of the parakeets. *p − 3; p represents the number of parakeets.*

24. Jillian grew 3 in. *s + 3; s represents the original height of Jillian.*

25. Cameron read $2\frac{1}{2}$ times as many books as Joseph.

26. Matthew scored 10 fewer points than Alexa on the math test. *x − 10; x represents Alexa's math test score.*

27. The apples were divided into 8 bags. *d ÷ 8; d represents the number of apples.*

28. Mrs. Sullivan bought 3 dozen eggs at the farmers' market. By the time she arrived home, some of the eggs had broken. *(3 × 12) − n; n represents the number of eggs broken.*

Evaluate the expression. Let $x = 3$.

29. $2x − 5$
(2 × 3) − 5 = 1

30. $15 ÷ x + 4$
15 ÷ 3 + 4 = 9

31. $x + 7 × 3$
3 + 7 × 3 = 24

32. $12 − x + 2$
12 − 3 + 2 = 11

25. *$2\frac{1}{2}$ • b; b represents the number of books read by Joseph.*

Practice & Application

33. List the factors for 36 and 72. What is the greatest common factor? *GCF = 36*

34. Write an equation showing that 42 is a multiple of 6. *7 × 6 = 42*

35. Rename $\frac{5}{6}$ and $\frac{7}{8}$, using a common denominator. Write a comparison statement by using >. *$\frac{5}{6} = \frac{20}{24}$; $\frac{7}{8} = \frac{21}{24}$, $\frac{21}{24} > \frac{20}{24}$ or $\frac{7}{8} > \frac{5}{6}$*

36. Explain how the fractions $\frac{6}{16}$, $\frac{9}{24}$, and $\frac{12}{32}$ are related to $\frac{3}{8}$.

37. Write the next three numbers in the pattern: 24, 48, 96. *192, 384, 768*

38. Draw and name two quadrilaterals with no right angles. *Answers will vary.*

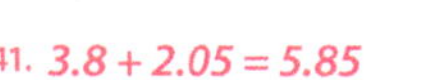

trapezoid   parallelogram   rhombus

41. *3.8 + 2.05 = 5.85   5.85 − 2.05 = 3.8
2.05 + 3.8 = 5.85   5.85 − 3.8 = 2.05*

39. How many cookies are in $3\frac{1}{2}$ dozen? *(3 × 12) + 6 = 42 cookies*

40. Evaluate $7(b) + 3.5$ if $b = 6$. *7(6) + 3.5 = 42 + 3.5 = 45.5*

41. Write two related addition equations and two related subtraction equations, using 3.8, 2.05, and 5.85.

42. Write $\frac{3}{4}$ as a decimal. *$\frac{3}{4} = 0.75$*

43. Explain why it is important to use the order of operations to solve the problem $3x + 24 ÷ 3 − 2$ if $x = 10$.

44. How can a comma help me write a word phrase for an expression? *A comma can help me clarify the meaning of a word phrase. Example: I can use a comma in a word phrase to indicate addition needs to happen before division.*

43. *(3 × 10) + (24 ÷ 3) − 2 = 30 + 8 − 2 = 36; Following the order of operations is the only way to get the correct answer.*

Lesson 94   205

What expression can you write for this situation? $3 × h$, $3 • h$, or $3h$; Since the number of hikers in the morning is unknown, the variable h can be used to represent the unknown number.

- Write "$3h$" for display. Remind the students that $3h$ is another way to write "3 times h." Explain that a number multiplied with a variable is a *coefficient*; 3 is the coefficient of the variable h. Point out that the known factor (the coefficient) precedes the variable, even for a word phrase such as "the product of h and 3," and that a multiplication sign is not needed.

- Point out that $h ÷ 2$, $\frac{h}{2}$, $3 × h$, $3 • h$, and $3h$ are all algebraic expressions because they include a variable.

- Guide the students in writing a numerical or an algebraic expression for each of the following word phrases. Point out that since addition and multiplication are commutative, the order of the addends or factors may vary. However, the numbers must be written in the correct order when the operation is subtraction or division because subtraction and division are not commutative. If the numbers are not written in the correct order, the expression will not be representative of the word phrase; its value will differ from the word phrase.

the quotient of 12 and 4 *12 ÷ 4 or $\frac{12}{4}$*
2 to the third power minus 3 *$2^3 − 3$*
16 less than 30 *30 − 16*
the sum of a number and 5 *n + 5 or 5 + n*
6 times a number *6x*
the sum of 5 and a number, divided by 2 *(5 + n) ÷ 2*
the product of 8 and 7 *8 × 7 or 7 × 8*
a number divided by 6 *n ÷ 6 or $\frac{n}{6}$*
a number decreased by 4 *a − 4*

- Read aloud the following word problem.

Glenn saw a large group of hikers walking together on the nature trail. When the trail became narrow, the guide divided the hikers into groups of 2.

How many hikers did Glenn see? It was a large group; the number of hikers is unknown.

Which word phrase tells you what happened to the large group of hikers? divided the hikers into groups of 2

- Remind students that any letter can be used as a variable. Point out that using an initial letter such as h to represent "hikers" can help them remember what the variable represents. Explain that when

a mathematical phrase includes a variable, the phrase is an *algebraic expression*.

What algebraic expression can you write to show that an unknown number of hikers was divided into groups of 2? $h ÷ 2$ or $\frac{h}{2}$

- Write "$h ÷ 2$" for display.

What other word phrases can you think of that indicate division by 2? sample answers: the quotient of a number and 2; grouped in pairs; cut in half; shared by 2 people

- Read aloud the following word problem.

The trail guide led 3 times more hikers on the afternoon hike than he led on the morning hike.

Writing algebraic expressions

- **Model** use of the Problem-Solving Plan to help the students write algebraic expressions.

- Explain that word phrases may indicate more than one operation. They must read the words carefully to be sure that they perform the order of operations correctly.

10-A

Read aloud the following word problem.

Mrs. Donner delivered a container of cookies to the church for a youth activity. Alisa brought 3 dozen more cookies. How many cookies did Mrs. Donner and Alisa bring to the church?

What is the question asking you to find? how many cookies Mrs. Donner and Alisa brought to the church

What information is given? A container of cookies and 3 dozen more cookies were brought to the church.

What algebraic expression can you write to show an unknown number of cookies in the container and 3 dozen more cookies? $c + 3 \times 12$; "C" represents the unknown number of cookies in the container, "more" indicates addition, and "3 dozen" can be interpreted as 3×12.

Some students may mentally calculate the 3 dozen cookies to be 36 cookies; therefore, accept $c + 36$ as a correct expression, but continue the lesson using $c + 3 \times 12$.

Write "$c + 3 \times 12$." The order of operations must be considered when writing a multistep expression.

What do you do first to solve $c + 3 \times 12$? I multiply 3×12; there are no parentheses or exponents, and multiplication and division are performed before addition and subtraction.

Remind the students that parentheses can be placed around the first step in a multistep problem to indicate which operation should be performed first. However, because of the order of operations, parentheses are not always necessary.

Write "$c + (3 \times 12)$." Point out that with or without parentheses, students multiply before they add.

Can you find how many cookies there are? No; the number of cookies in the container is not known.

Use mental math to solve.

1. $34.7 \div 10$ *3.47* **2.** $67.83 \div 100$ *0.6783* **3.** $821.3 \div 1,000$ *0.8213*

Rename the fraction, using a power of 10 as the denominator. Write the fraction as a decimal.

4. $\frac{3}{4}$ *$\frac{75}{100}$; 0.75* **5.** $\frac{1}{2}$ *$\frac{5}{10}$; 0.5* **6.** $\frac{1}{4}$ *$\frac{25}{100}$; 0.25* **7.** $\frac{1}{5}$ *$\frac{2}{10}$; 0.2*

Solve.

8. $5\overline{)16.25}$ *3.25* **9.** $1.5\overline{)5.79}$ *3.86* **10.** $0.21\overline{)4.641}$ *22.1* **11.** $6\overline{)\$39.54}$ *\$6.59*

12. $\begin{aligned} \$4{,}128.45 \\ + \$2{,}397.15 \\ \hline \textit{\$6{,}525.60} \end{aligned}$ **13.** $\begin{aligned} 395.1 \\ \times \quad 4 \\ \hline \textit{1{,}580.4} \end{aligned}$ **14.** $\begin{aligned} 158 \\ \times \quad 25 \\ \hline \textit{3{,}950} \end{aligned}$ **15.** $2.5 - 1.860$ *0.64* **16.** $54.3 \div 6$ *9.05*

Evaluating using substitution

Use **direct instruction** to help the students evaluate an expression.

Explain that although students cannot know for sure how many cookies are in the container, they can use substitution to assign a value to the variable c and evaluate the expression (i.e., determine a possible total number of cookies).

What will be the total number of cookies if there are 27 cookies in the container or $c = 27$? 63; $27 + (3 \times 12) = 27 + 36 = 63$

Guide the students as they find the total number of cookies if $c = 12$, $c = 30$, and $c = 36$, using an input/output table.

c	$c + (3 \times 12)$
12	48
27	63
30	66
36	72

- Write "the sum of 8 and a number divided by 2."

  What do the words "the sum" tell you? There will be 2 addends.

  What expressions can you write for this word phrase? $(8 + n) \div 2$ and $8 + (n \div 2)$.

  Write both expressions. Point out that either expression is correct. The word phrase can be interpreted to mean addends 8 and n or addends 8 and $(n \div 2)$.

  Explain that a comma can clarify the meaning of a word phrase. Insert a comma after the word *number* in the word phrase: the sum of 8 and a number, divided by 2. Point out that the comma separates the phrase to indicate that addition is the first operation and division is the second operation. Because of the order of operations, parentheses must be placed around $8 + n$ to show that the addition is performed first: $(8 + n) \div 2$.

- Direct attention to the essential question at the top of Student Edition page 204, "How can a comma help me write a word phrase for an expression?" A comma can help me clarify the meaning of a word phrase. For example, I can use a comma in a word phrase to indicate addition needs to happen before division.

- Guide the students as they write algebraic expressions for the following word phrases.

  2 times a number, divided by 4 $2n \div 4$ or $(2 \times n) \div 4$

  a number increased by 7, multiplied by 3 $(n + 7) \times 3$ or $3 \times (n + 7)$

  n divided by 5, decreased by 2 $(n \div 5) - 2$ or $n \div 5 - 2$

LESSON **94**

Apply

Student Edition pages 204–5

- Read and explain the directions for pages 204–5. Assist the students as they complete the pages independently.

Daily Review

- Students should complete Chapter 10, section *a*.

NOTES

Student Edition pages 206–7
Daily Review Chapter 10, section _b_

OBJECTIVES

- Write an equation with two equal expressions.
- Determine the unknown in a word problem and write it as a variable in an equation.
- Explain the meaning of a mathematically true equation. **BWS**
- Evaluate and relate expressions by using >, <, or =.

BIBLICAL WORLDVIEW SHAPING

- **Modeling (Recall):** True equations are consistent and match real life.

Engage

- Direct attention to the essential question at the top of Student Edition page 206, "What does it mean for an equation to be mathematically true?"

- Lead the students in a **Mind Warm-up** to explore the essential question. Write the following equations for display:

$6 \times 3 = 15$

$6 \times 3 = 15$

$6 \times 3 = 15$

In this example, is the answer in each equation the same, or consistent? yes
Is the answer in each equation accurate? no
Could you test this model using real objects? yes

- Explain that the students will find out in this lesson what is meant by the phrase "mathematically true."

Instruct

Writing an equation

- Guide a **discussion** to help the students write an equation.

Equations

What does it mean for an equation to be mathematically true?

An **equation** is a mathematical sentence stating that two expressions are equal. An equation can be written with or without variables.

25 added to 75 is 100.	$75 + 25 = 100$
Two less than y is 12.	$y - 2 = 12$
Dividing c by 4 equals 9.	$c \div 4 = 9$

Key Terms
- equation

Exercises

Complete the sentence for the equation.

1. $\frac{30}{y} = 15$: _30_ divided by _y_ equals _15_.

2. $3x + 8 = 20$: The sum of _3_ times _x_ and 8 is _20_.

3. $n + 7 = 25$: A number increased by _7_ equals _25_.

4. $6p - 4 = 32$: _4_ less than _6_ times _p_ equals _32_.

5. $9s = 63$: The product of _9_ and _s_ is _63_.

6. $\frac{m}{2} - 2 = 3$: The quotient of _m_ and _2_ decreased by _2_ equals _3_.

Write an equation for the sentence.

7. 5 less than n is 17. _$n - 5 = 17$_

8. 3 times s equals 27. _$3s = 27$_

9. 8 less than 6 times r equals 4. _$6r - 8 = 4$_

10. A number decreased by 2 is 27. _$z - 2 = 27$_

11. A number divided by 7 is 6. _$\frac{x}{7} = 6$_

12. 9 added to n equals 21. _$n + 9 = 21$_

Evaluate the expression. Let $s = 8$. Write a comparison sentence by using >, <, or =.

13. $2 + s \underline{\le} 15 - 3$ _$10 < 12$_

14. $\frac{24}{s} \underline{\le} 2^2$ _$3 < 4$_

15. $4s \underline{\ge} 30$ _$32 > 30$_

16. $12 - s \underline{=} 2 \cdot 2$ _$4 = 4$_

17. $\frac{s}{4} \underline{\ge} 2 - 1$ _$2 > 1$_

18. $s^2 \underline{=} 60 + 4$ _$64 = 64$_

Complete the table, using the given values to evaluate the expression.

19.

x	$\frac{x}{4} - 3$
12	0
16	1
20	2
24	3

20.

x	$5x + 2$
2	12
3	17
5	27
7	37

21.

x	$6 + x^2$
2	10
3	15
4	22
5	31

- Write "10" for display. Below the 10 write "$9\frac{3}{4} + \frac{1}{4}$"; "$12.5 - 2.5$"; "$2 \times 5$"; and "$\frac{30}{3}$." What do you notice about these numerical expressions? Each expression represents the value 10.

- Direct the students to write the number 24 and then to write in one minute as many numerical expressions for 24 as possible. Encourage them to use all four operations and various types of numbers (fractions and decimals). Allow the students to share their expressions. sample answers: $18 + 6$; $26 - 2$; 12×2; $48 \div 2$; $20\frac{1}{2} + 3\frac{1}{2}$; $22.5 + 1.5$

How many different expressions for 24 do you think there are? There is an infinite number of expressions.

- Explain that any two expressions for 24 can be written with an equal sign (=) between them because each expression has a value of 24; they are equivalent.

- Choose pairs of students to write equations for display. Instruct each student in a pair to write one of his expressions for 24 so that the expressions are side by side, and then to write an equal sign (=) between the expressions to complete the equation.

Write an addition equation and a subtraction equation for each word problem. Use a variable for the unknown value.

22. Nine pieces of pizza were left after the family ate 7 pieces. How many pieces of pizza did the family begin with? *9 + 7 = p; p − 7 = 9*

23. The basketball team won 8 of its 12 basketball games. How many games were lost? *8 + g = 12; 12 − g = 8 or 12 − 8 = g*

Evaluate the expression. Let *m* = 6.

24. $30 \div m - 3$
30 ÷ 6 − 3 = 5 − 3 = 2

25. $m + 8 \times 3$
6 + 8 × 3 = 6 + 24 = 30

26. $2(m + 1) \div 2$
2(6 + 1) ÷ 2 = 2(7) ÷ 2 = 14 ÷ 2 = 7

Write the algebraic expression for the word phrase.

27. 4 times a number divided by 3 *4n ÷ 3 or 4(n ÷ 3)*

28. *x* divided by 8, decreased by 2 $\frac{x}{8} - 2$

Practice & Application

The Hagan twins are planning a deep-sea fishing trip. The ticket cost is $130 per person. The Hagan twins figured the cost of the fishing trip and the cost of getting to the dock to be $280.

29. Write an equation, using *c* for the unknown cost of getting to the boat dock. Solve to find the value of *c*. *$280 − (2 x $130) = c; $280 − $260 = $20*

30. Write an equation to show the cost of the deep-sea fishing trip for the boys if they split the total cost evenly. *$280 ÷ 2 = $140*

31. If Mr. Hagan goes with the boys on the trip, what will be the total cost of the trip? *(3 × $130) + $20 = $390 + $20 = $410*

32. An expression is a mathematical phrase made up of numbers and operation signs but not a relation sign; an equation is a mathematical sentence stating that 2 expressions are equal. Both expressions and equations can use variables.

32. Explain the difference between an expression and an equation.

33. What does it mean for an equation to be mathematically true? *It is accurate and consistent.*

What is an equation? Answers may vary, but an equation is a mathematical sentence that states that two expressions are equal.

• Write for display: "A number divided by 6 equals 2." Direct attention to the phrase "equals 2."

What does the phrase "equals 2" tell you about the phrase before the word "equals"? The phrase before the word "equals" has a value of 2.

Draw a simple picture of a balanced scale. Explain that the equal sign in an equation is like the fulcrum of a balanced scale, and the expressions on each side of the equal sign are like the equal weights on each side of a balanced scale.

What word phrase represents the value on the left side of the equation? a number divided by 6

How can you mathematically write "a number divided by 6 equals 2"? $n \div 6 = 2$ or $\frac{n}{6} = 2$

Write "$\frac{n}{6} = 2$" above the pictured scale so that the equal sign is directly above the fulcrum.

• Write for display: "12 decreased by *a* is 7."

What verb in this word phrase is similar to "equals"? is

What algebraic expression can you write for 12 decreased by *a*? $12 - a$

Choose a student to write for display the algebraic equation for 12 decreased by *a* is 7. $12 - a = 7$

• Follow a similar procedure for the following sentences. Remind the students that the words "equals" and "is" indicate where the equal sign should be written in the equation.

The product of 9 and *a* equals 54. $9a = 54$

Dividing *x* by 7 is 4. $\frac{x}{7} = 4$

The sum of 73 and *b* equals 97. $73 + b = 97$

A number decreased by 2 to the power of 3 is 11. $n - 2^3 = 11$

Determining the unknown in a word problem

• **Model** using a written statement to help the students determine the unknown in a word problem.

• Read aloud the following word problem.

Last week, Mitchell earned $115 for mowing lawns. That is $35 more than he earned for mowing lawns this week. Write an equation to find how much money Mitchell earned for mowing this week. $115 = w + $35

What is the word problem asking you to find? how much money Mitchell earned this week

What statement tells you the amount of money Mitchell earned this week? That is $35 more than he earned for mowing lawns this week.

• Write for display: "That is $35 more than he earned this week." Write the mathematical interpretation below the written statement as students answer the following questions.

That	is	$35 more than he earned this week.
$115	=	*w* + 35.

What word in the statement indicates an equal sign in the equation for the word problem? is

What does the word "that" refer to? the $115 Mitchell earned last week

Point out that just as the word "that" comes before the word "is" in the written statement, the $115 can be written before (to the left of) the equal sign (=) in the mathematical statement: $115 =.

LESSON 95

What is $115 equal to? **$35 more than Mitchell earned this week**

What algebraic expression can you write to the right of the equal sign to show that the $115 Mitchell earned last week is $35 more than he earned this week? **$w + $35**

W, or any other variable given by the students, represents the unknown amount that Mitchell earned this week.

- **Demonstrate** using a part-whole model to help the students determine the unknown in a word problem.

- Read aloud the following word problem.

  Three children from the sixth-grade class remained at school after 12 children from the class went home on the school bus. Write an equation to find how many students are in the class.

- Draw 2 part-whole models side by side for display. Label the sections of the first model "whole," "part," and "part" as shown. Explain to the students that their understanding of part-whole models can help them write an equation when the words in a problem are unclear. Read the word problem aloud again.

  What represents the whole in the word problem? **the number of students in the class**

  Write "*s* = students in the class" for display. Then write *s* in the whole section of the second model.

  What represents the parts in the problem? **the 3 students who were still at school and the 12 students who went home**

  Write "3" and "12" in the part sections of the second model.

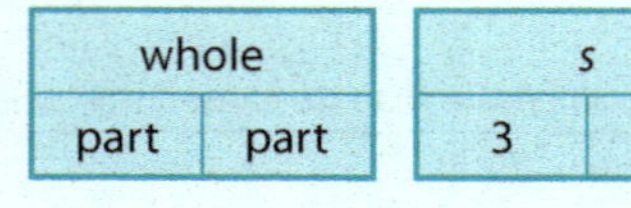

whole		*s*	
part	part	3	12

Using the part-whole model, what equation can you write to solve for *s*, the number of students in the class? **sample answers: $s = 3 + 12$; $s = 12 + 3$; $s - 12 = 3$**

- Instruct the students to write an equation for each of the following word problems as you read the word problem aloud several times. Encourage them to sketch a part-whole model or a picture if needed.

The youth chamber orchestra has 23 members. Ten of the members are boys. Write an equation to find how many girls are in the orchestra. **sample answers: $23 - 10 = g$; $g + 10 = 23$**

Cal divided some apples into 5 bags. When he was finished dividing the apples, there were 8 apples in each bag. Write an equation to find how many total apples he put in all the bags. **sample answers: $a \div 5 = 8$; $5 \times 8 = a$**

At halftime of the basketball game, one team's score was 48. The team had tripled the score it had at the end of the first quarter. Write an equation to find the score at the end of the first quarter. **sample answers: $3s = 48$; $\frac{48}{3} = s$**

Explaining the phrase "mathematically true"

- Guide a **discussion** to help the students answer the essential question.

- Redirect the students' attention to the question about the orchestra.

  In the equation $23 - 10 = g$ or $g + 10 = 23$, how could you tell if the equation is really true mathematically? How could you test this equation? **sample answer: I could separate the orchestra members into boys and girls and count to make sure my answer matches the answer for the equation.**

  Point out that if students get the same correct answer each time, the equation is both accurate and consistent.

Make a factor tree for the number. Write the prime factorization for the number in exponent form. *Factorization may vary.*

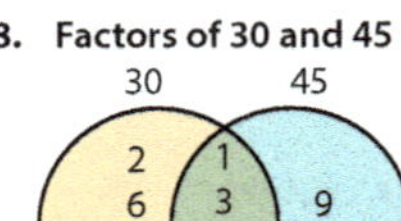

1. 81 3^4 $3 \cdot 3 \cdot 3 \cdot 3$
2. 56 $2^3 \cdot 7$ $7 \cdot 2 \cdot 2 \cdot 2$

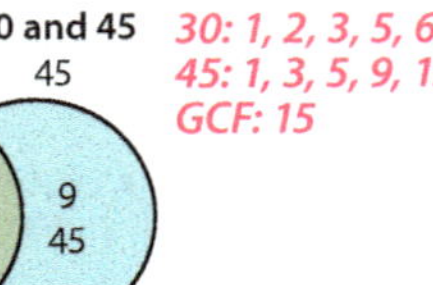

3. 64 2^6 $2 \cdot 2 \cdot 2 \cdot 2 \cdot 2 \cdot 2$

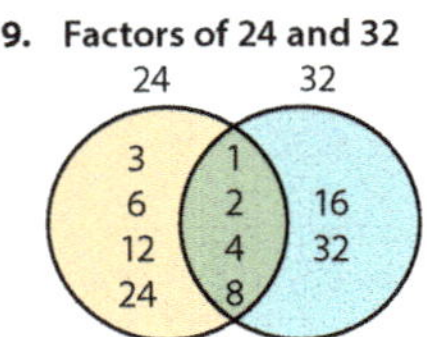

4. 75 $3 \cdot 5^2$ $5 \cdot 5 \cdot 3$

Find the greatest common factor (GCF) by listing the factors of each number.

5. 12 and 18   *GCF: 6*
   12: 1, 2, 3, 4, 6, 12
   18: 1, 2, 3, 6, 9, 18

6. 21 and 35   *GCF: 7*
   21: 1, 3, 7, 21
   35: 1, 5, 7, 35

7. 36 and 48   *GCF: 12*
   36: 1, 2, 3, 4, 6, 9, 12, 18, 36
   48: 1, 2, 3, 4, 6, 8, 12, 16, 24, 48

Use the Venn diagram to list the factors. Find the GCF.

8. Factors of 30 and 45   *30: 1, 2, 3, 5, 6, 10, 15*
   45: 1, 3, 5, 9, 15
   GCF: 15

9. Factors of 24 and 32   *24: 1, 2, 3, 4, 6, 8, 12*
   32: 1, 2, 4, 8, 16
   GCF: 8

Use the GCF to rename the fractions in lowest terms.

10. $\frac{12}{18}$ $\frac{2}{3}$
11. $\frac{21}{35}$ $\frac{3}{5}$
12. $\frac{36}{48}$ $\frac{3}{4}$
13. $\frac{30}{45}$ $\frac{2}{3}$
14. $\frac{24}{32}$ $\frac{3}{4}$

Daily Review   461

What does it mean for an equation to be mathematically true? It is accurate and consistent. The model matches real life.

Evaluating & relating expressions

- **Ask questions** to help the students relate expressions by using >, <, or =.

- Explain that expressions are not always equal in value; therefore, other mathematical symbols such as a greater-than sign (>) or a less-than sign (<) are used to make a statement true.

- Write "$a + 15$ ___ $19 - 3$" for display.
  If $a = 2$, what is the value of $a + 15$? 17
  What is the value of $19 - 3$? 16
  What mathematical symbol shows the relationship between the two expressions? the greater-than sign; 17 > 16
  Complete the comparison sentence.

- Repeat the procedure for the following comparison sentences, using $a = 6$.
  $\frac{48}{a} = 2^3$; 8 = 8
  $a - 2 = \frac{12}{3}$; 4 = 4
  $11 + 3 < 3a$; 14 < 18
  $7a + 3 > 6 \times 7$; 45 > 42

Apply

Student Edition pages 206–7

- Read and explain the directions for pages 206–7. Assist the students as they complete the pages independently.

Daily Review

- Students should complete Chapter 10, section b.

DIFFERENTIATED INSTRUCTION

Use the following to provide extra help for students who experience difficulty with the concepts taught in Chapter 10.

Isolate the variable to solve an equation.
Prepare symbol cards (+, −, ×, ÷, =), number cards, and variable cards using 3×5 cards. Use the cards to display an equation such as $n \times 3 = 27$. Ask the students to identify the operation that is being performed on the variable in the equation. multiplication; N is being multiplied by 3. Ask the students to tell the inverse operation they should use to isolate the variable. division

Direct the students to place cards in the equation so that both sides of the equation are divided by 3: $n \times 3 \div 3 = 27 \div 3$.

Remind them that any operation that is performed on one side of the equation must also be performed on the other side to keep the equation balanced (the same value on both sides of the equation). Instruct the students to remove the cards for $\times 3 \div 3$ from the left side of the equation. Remind them that since $3 \div 3 = 1$, dividing the left side of the equation by 3 isolates the variable because $n \times 1 = n$ (Identity Property of Multiplication). Direct the students to perform the same operation ($\div 3$) on the right side of the equation, simplifying the numerical expression by replacing $27 \div 3$ with the card that shows the answer. 9 Repeat the procedure, varying the numbers and operation in each equation.

Student Edition pages 208–9
Daily Review Chapter 10, section c

OBJECTIVES

- Compute sums and products that are multiples of 10 using the Commutative and Associative Properties.
- Simplify algebraic expressions by applying the Commutative and Associative Properties.

TEACHER RESOURCES

- 52 *Variable Cards N*
- 53 *Variable Cards X*

ADDITIONAL MATERIALS

- 18 round counters

Preparation

- Prepare 6 *n* cards and 15 *x* cards for demonstration using the patterns provided on the *Variable Cards* pages.

Engage

- Direct attention to the essential question at the top of Student Edition page 208, "How do the Commutative and Associative Properties help me simplify expressions?"

- Guide the students in a **review** to explore the essential question.

  1. Which property of addition states that the order of the addends can be changed without changing the sum? the Commutative Property of Addition

  2. Which property states that the grouping of addends can be changed without changing the sum? the Associative Property of Addition

Instruct

Computing with multiples of 10

- **Model** using properties to help the students compute sums and products.

Simplifying Expressions

How do the Commutative and Associative Properties help me simplify expressions?

Key Terms
- simplify expressions
- like terms

Simplify expressions by combining **like terms**. The Commutative and Associative Properties of Addition or Multiplication allow you to rewrite the expression to reorder and group the like terms together.

The Commutative Property allows you to change the order of addends.

$n + 6 + n =$
$n + n + 6 = 2n + 6$

The Associative Property allows you to group like terms.

$4 + x + 3 =$
$x + 4 + 3 = x + 7$

Repeated addends can be combined using multiplication.

$x + x + x + x = 4x$

$2x + 2x + 2x + 2x =$
$4(2x) = 8x$

Exercises

Write the missing number or variable. Name the property used.

1. $(p + \underline{}) + r = p + (q + r)$
   q; Associative Property
2. $45 + (29 + 80) = (45 + 29) + \underline{}$
   80; Associative Property
3. $4 \cdot 9 = \underline{} \cdot 4$
   9; Commutative Property
4. $2a + (a + 8) = (2a + \underline{}) + 8$
   a; Associative Property
5. $m + n = n + \underline{}$
   m; Commutative Property
6. $(5 + 7x) + 8x = 5 + (7x + \underline{})$
   8x; Associative Property
7. $5 \cdot (7 \cdot 4) = (\underline{} \cdot 7) \cdot 4$
   5; Associative Property
8. $3s \times 4 \times 7s = \underline{} \times 3s \times 7s$
   4; Commutative Property

Choose the simplified expression for the model.

9. $4y + 9 + 5$
   (4y + 14)
   $y + y + y + y + 14$

10. $6 + a + 2a + 2$
    $8 + a + 2a$
    (8 + 3a)

11. *(2(2c) + 8)*
    $2c + 2c + 8$
    $c + c + c + c + 8$

12. $x + x + 6$
    (2x + 6)
    $x + x + 3 + 3$

Simplify the expression.

13. $6 + 4x + 2$ *4x + 8*

14. $x + 4x$ *5x*

15. $2x + 3x$ *5x*

16. $x + 5 + x$ *2x + 5*

17. $3 + x + 6$ *x + 9*

18. $x + 2x + 3$ *3x + 3*

19. $2x + 1 + x$ *3x + 1*

20. $5x + 2x + 3$ *7x + 3*

21. $4 \cdot 2 \cdot x$ *8x*

Order of terms may vary.

208   Chapter 10

- Write "85 + 25 __ 25 + 85" for display.

  What do you notice about the addends in the expressions? The addends are the same, but their order is different.

  Invite a student to write the sum below each of the expressions. 110

  Choose another student to write the correct symbol between the expressions. =

  Explain that 85 + 25 and 25 + 85 are equivalent expressions; they have the same value.

  Did the order of the addends change the sum? no

- Write for display "a + b = b + a."

  What is different about the addends in this equation? They are letters or variables rather than numbers.

  What property is shown using the variables? the Commutative Property of Addition; The order of the addends was changed.

- Write "(25 + 10) + 45 __ 25 + (10 + 45)" for display.

  What do the parentheses in the equation tell you? I should add the numbers within the parentheses first.

  Choose a student to write the sum below each of the expressions and then write the correct symbol between the expressions. 80; =

Simplify the expression. Show each step used to combine the like terms. Name the property used in each step. *Order of terms may vary.*

22. $8x + (2 + 4x)$ *$12x + 2$* **23.** $3(7x)$ *$21x$* **24.** $(6 + 5x) + 7$ *$5x + 13$*

25. $5 \cdot x$ *$5x$* **26.** $8 + (2 + x)$ *$x + 10$* **27.** $(x \cdot 5) \cdot 6$ *$30x$*

28. $2 + (3 + x)$ *$x + 5$* **29.** $x \cdot 8 \cdot 9$ *$72x$* **30.** $12(2x)$ *$24x$*

Write an equation for the sentence.

31. 5 added to y equals 18. *$y + 5 = 18$* **32.** A number divided by 3 is 7. *$n \div 3 = 7$*

33. 12 decreased by 7 is 5. *$12 - 7 = 5$* **34.** 58 more than n is 134. *$n + 58 = 134$*

35. The product of 9 and n is 36. *$9n = 36$* **36.** The sum of b and 8 equals 25. *$b + 8 = 25$*

Evaluate the expression. Let $y = 4$.

37. $40 \div y \times 3$
$40 \div 4 \times 3 =$
$10 \times 3 = 30$

38. $y + 7 - 6$
$4 + 7 - 6 =$
$11 - 6 = 5$

39. $(36 - y) \div 4$
$(36 - 4) \div 4 =$
$32 \div 4 = 8$

40. $2 + y - 3$
$2 + 4 - 3 =$
$6 - 3 = 3$

Practice & Application

41. Write an addition equation and a multiplication equation for the part-whole model.

15.2			
3.8	3.8	3.8	3.8

$3.8 + 3.8 + 3.8 + 3.8 = 15.2$; $4 \times 3.8 = 15.2$

42. The diameter of one of the largest Ferris wheels in the world is 492 ft. What is the length of any radius? *$492 \div 2 = 246$ ft*

Use the equation $x \div 45 = \$40$ for problems 43–45.

43. Solve to find the value of x. Write the related multiplication equation to solve.
$x = 45 \times \$40$; $x = \$1,800$

44. If x represents the amount paid to the amusement park for entrance fees for 45 students, what does the quotient represent? *the cost for each student to get into the park*

45. Each student paid the entrance fee and took an additional \$10 for gas and \$50 for food and games. How much money did each person take?
$\$40 + \$10 + \$50 = \$100$

46. Use the information in problem 45 to write a word problem showing how Jake spent his food and game money at the amusement park. *Answers will vary.*

47. How do the Commutative and Associative Properties help me simplify expressions? *The Commutative and Associative Properties help me rewrite an expression to organize the like terms.*

The Singapore Flyer is 165 m tall. The passengers ride in air-conditioned capsules that take about 37 min to make one full rotation on the wheel.

- Point out that the Commutative Properties involve changing the order of the addends or the factors, and the Associative Properties involve regrouping the addends or the factors.

- Write the following expressions for display. Guide the students as they use the Commutative and the Associative Properties to make sums and products that are multiples of 10. Remind the students that multiples of 10 can be easily added and multiplied. State the value of each expression.

  The order of the addends and the factors may vary.

  $(8 + 23) + 2$ *$(8 + 2) + 23 = 10 + 23 = 33$*
  $(8 + 25) + (75 + 20)$ *$(75 + 25) + (20 + 8) = 100 + 28 = 128$*
  $(7 \times 5) \times 6$ *$7 \times (5 \times 6) = 7 \times 30 = 210$*
  $(5 \times 25) \times (2 \times 3)$ *$(5 \times 2) \times (25 \times 3) = 10 \times 75 = 750$*

Simplifying algebraic expressions

- **Demonstrate** combining like terms to help the students simplify expressions.

- Write for display "$5n + 3 + n$."

- Illustrate the expression with the n cards and counters as shown below. Explain that the variable n, without a coefficient (multiplier) preceding it, is the same as $1n$.

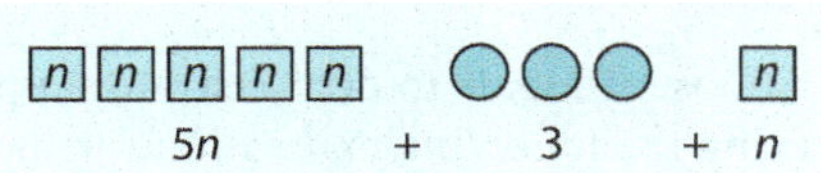

Which two addends do you think are similar? sample answer: $5n$ and n are similar because they both have the variable n.

Explain that $5n$ and n are like terms; they can be combined to make 1 term or addend.

If you added from left to right, would you be adding like terms? No; $5n$ and 3 are beside each other in the expression, so unlike terms would be grouped together.

How can you reorder the addends so that the like terms are side by side? $5n + n + 3$; The Commutative Property of Addition allows me to reorder the addends without changing the value of the expression.

Did the grouping of addends change the sum? no

- Choose a student to write for display an example of the Associative Property of Addition using variables. sample answer: $(a + b) + c = a + (b + c)$

- Write "$12 \times 10 = 120$" and "$10 \times 12 = 120$" for display.

  What do you notice about these multiplication equations? The factors and the products are the same, but the order of the factors is different.

  What property states that the order of the factors can be changed without changing the product? the Commutative Property of Multiplication

- Choose a student to write for display an example of the Commutative Property of Multiplication using variables. sample answer: $a \times b = b \times a$

  What does the Associative Property of Multiplication state? The grouping of the factors may be changed without changing the product.

- Choose a student to write for display an example of the Associative Property of Multiplication using numbers. Choose another student to write an example using variables. sample answers: $(10 \times 11) \times 5 = 10 \times (11 \times 5)$; $a \times (b \times c) = (a \times b) \times c$

LESSON 96

- Invite a student to reorder the 3 counters and the one n card. Write "$5n + n + 3$" below the illustration.

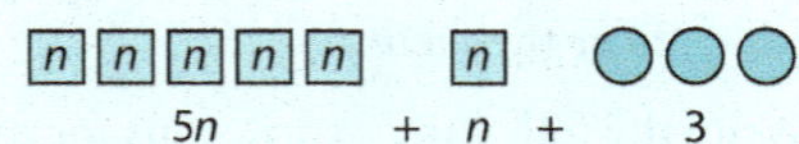

How can you regroup the addends to combine addends of like terms? I can group $5n$ and n together to make $6n$; the Associative Property of Addition allows me to regroup the addends without changing the value of the expression.

Draw parentheses around $5n + n$.

- Choose another student to combine the like terms (combine the n cards). Explain that combining like terms is referred to as simplifying the expression. Point out that both the Commutative and the Associative Properties of Addition were used to organize the expression so that they could combine the like terms.

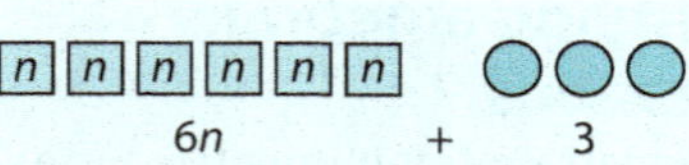

What is the simplified expression? $6n + 3$
Write "$6n + 3$."

Why do you think the expression is not simplified as $9n$? $9n$ represents $9 \times n$ or 9 addends of n and there are only 6 n's in the original expression; therefore, the simplified expression of 6 n's and 3 counters is correct.

- Choose students to demonstrate simplifying the following expressions, using the x cards and counters if needed. Instruct each student to explain the property or properties he uses as he simplifies the expression. Remind the students that applying the Commutative and the Associative Properties to an expression does not change its value.

$x + 8 + 5x$　$6x + 8$

$8 + 7x + 2$　$7x + 10$

$3x + 12 + 6x$　$9x + 12$

$15 + 3 + 9x$　$9x + 18$

$6x + 10 + x$　$7x + 10$

Use mental math to solve.

1. 3×40　*120*
2. 100×5.76　*576*
3. $85 \div 100$　*0.85*
4. 30×40　*1,200*
5. $1,000 \times 3.187$　*3,187*
6. $29.7 \div 10$　*2.97*
7. 300×40　*12,000*
8. $217 \div 10$　*21.7*
9. $0.835 \div 10$　*0.0835*
10. 10×32.1　*321*
11. $385 \div 100$　*3.85*
12. $87.32 \div 100$　*0.8732*

Solve.

13.	14.	15.	16.	17.	18.
23	$20.00	137.50	4.50	382	401
47	− $15.37	21.83	− 0.372	× 175	× 342
52	*$4.63*	+ 0.98	*4.128*	*66,850*	*137,142*
+ 89		*160.31*			
211					

Solve. Round to the nearest hundredth.

19. $178 \div 24$　*7.416 ≈ 7.42*

20. $4,065 \div 31$　*131.129 ≈ 131.13*

462　Daily Review

- Write "$(36 + x) + 19$" for display.

How do you think you can simplify this expression? I can use the Commutative Property to change the order of the addends in the parentheses and then use the Associative Property to regroup the addends so that 36 and 19 are grouped together.

Choose students to apply the properties and simplify the expression. sample answer: $(x + 36) + 19$; $x + (36 + 19)$; $x + 55$

Although the variable is normally written first in a simplified addition or multiplication expression, at this introductory level of instruction, accept answers in which the variable is written in another position (e.g., accept $2x + 6$ written as $6 + 2x$).

- Write "5(3*x*)" for display. Remind the students that multiplication can be indicated as $5 \times 3x$, $5 \cdot 3x$, or by parentheses around the second factor. Choose a student to read the expression. 5 times 3*x*

  Choose a student to illustrate the expression. 5 sets of 3*x*

  When you combine 5 sets of 3*x*, what is the product? 15*x*

- Direct attention to 5(3*x*) written for display. Point out that the Associative Property of Multiplication can be applied to the expression, simplifying it without using manipulatives: $5(3x) = 5 \cdot (3 \cdot x) = (5 \cdot 3) \cdot x = 15 \cdot x = 15x$.

- Guide the students as they simplify the following expressions. Discuss applying the properties as needed.

  $x + 8 + 2x + 7$ 3*x* + 15

  $(x \cdot 5) \cdot 12$ 60*x*

  $8(2x)$ 16*x*

  $9x + (11 + 4x)$ 13*x* + 11

  How do the Commutative and Associative Properties help you simplify expressions? The Commutative and Associative Properties help me to rewrite an expression to organize the like terms.

Apply

Student Edition pages 208–9

- Read and explain the directions for pages 208–9. Assist the students as they complete the pages independently.

Daily Review

- Students should complete Chapter 10, section *c*.

MATH TALK

Follow the General Procedure for Math Talks as outlined in Lesson 5 on Teacher Edition page 13b.

Make It Equal

Example:

9	2	20	9

$9 + 9 = 20 - 2$

$9 + 2 = 20 - 9$

20	4	10	2

$20 \times 2 = 4 \times 10$

$20 \div 10 = 4 \div 2$

NOTES

LESSON 97

Student Edition pages 210–11
Daily Review Chapter 10, section *d*

OBJECTIVES

- Solve addition and subtraction equations using inverse operations.
- Check addition and subtraction equations using substitution.
- Explain how some people's mathematical models contradict reality. **BWS**

BIBLICAL WORLDVIEW SHAPING

- **Modeling (Explain): People may use wrong assumptions to make models that contradict reality.**

ADDITIONAL MATERIALS

- 1 *n* card and 1 *x* card from the patterns provided on the *Variable Cards* pages (52 and 53)
- 30 round counters

Engage

- Direct attention to the essential question at the top of Student Edition page 210, "Why do some people's mathematical models contradict reality?"
- Direct the students to do a **Quick Write** to jot down an initial answer to the essential question.

Instruct

Solving equations using inverse operations

- **Model** an equation to help the students solve addition and subtraction equations.
- Write "7 + 5 __ 12" for display.
  What math symbol can you write to make the number sentence true? an equal sign; Both expressions have a value of 12.
  Write the equal sign.
- Draw an equal sign for display. Place a set of 7 counters and a set of 5 counters

---

Why do some people's mathematical models contradict reality?

An equation with a variable is solved by finding the value of the variable. The value must make the equation true to be called a **solution**.

To solve, isolate the variable on one side of the equals sign by using the **inverse operation**. The inverse operation of addition is subtraction. Likewise, the inverse operation of subtraction is addition. Keep the equation **balanced** by performing the same operation on both sides of the equation.

To check, substitute (replace) the variable with its numerical value and evaluate the original equation.

<table>
<tr><th>$n + 6 = 10$</th><th>$n - 3 = 8$</th></tr>
<tr><td>

$n + 6 = 10$
$n + 6 - 6 = 10 - 6$
$n = 4$
$4 + 6 = 10$

1. Identify the inverse operation. The inverse operation of addition is subtraction.
2. **Subtract 6** from both sides of the equation to isolate *n* on the left side.
3. Substitute 4 for *n* in the original equation to check.

</td><td>

$n - 3 = 8$
$n - 3 + 3 = 8 + 3$
$n = 11$
$11 - 3 = 8$

1. Identify the inverse operation. The inverse operation of subtraction is addition.
2. **Add 3** to both sides of the equation to isolate *n* on the left side.
3. Substitute 11 for *n* in the original equation to check.

</td></tr>
</table>

The equation mat shows the result of subtracting 6 from both sides of the equation.

$n + 6 = 10$

$6 - 6 = 0 \quad 10 - 6 = 4$

The number lines show the result of adding 3 to both sides of the equation.

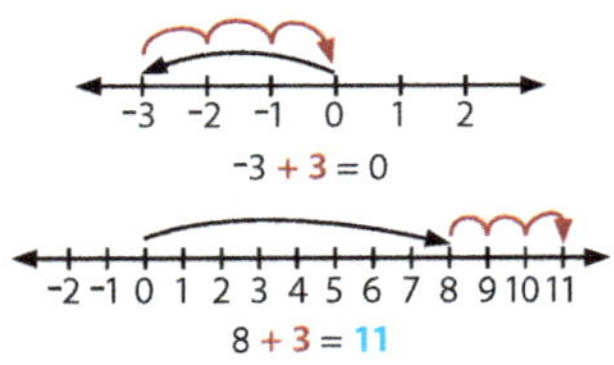

$-3 + 3 = 0$

$8 + 3 = 11$

Exercises

Solve the equation by using the inverse operation. Check the solution.

1. $n + 5 = 21$ *n = 16*
2. $d + 45 = 90$ *d = 45*
3. $d + 43 + 17 = 85$ *d = 25*
4. $x + 12 = 40$ *x = 28*
5. $16 + f = 35$ *f = 19*
6. $s + 14 - 3 + 0.5 = 20$ *s = 8.5*
7. $c - 6 = 17$ *c = 23*
8. $s - 39 = 61$ *s = 100*
9. $3.8 + 16 + b = 29$ *b = 9.2*
10. $a - 4 = 36$ *a = 40*
11. $24 + n = 100$ *n = 76*
12. $\frac{3}{4} + \frac{1}{2} + f = 1\frac{1}{2}$ *f = $\frac{1}{4}$*

210   Chapter 10

---

to the left of the equal sign and a set of 12 counters to the right of the equal sign. Remind the students that an equation is like a balanced scale; any operation that is performed on one side of the equation must also be performed on the other side to keep the expressions equal (the equation balanced).

- Remove the 5 counters that are to the left of the equal sign.
  Are the expressions still equal? No; 5 counters were removed from only one side of the equation; 7 does not equal 12.
  Write "7 + 5 − 5 ≠ 12" for display and then write "7 ≠ 12" below it, aligning the not-equal signs.

What must you do to balance the values on both sides of the equal sign? I must also remove 5 of the counters that are to the right of the equal sign.

Remove 5 of the counters from the right of the equal sign.

Are the expressions equal? Yes; the same amount (5 counters) was removed from both sides of the equation; 7 equals 7.

Write "7 + 5 − 5 = 12 − 5" and then write "7 = 7" below it, aligning the equal signs. Continue to display the equation.

- Repeat the procedure, using 7 + 5 = 12. Allow students to remove or add an equal number of counters on both sides of the equal sign. Guide the students in

State whether the given value is the solution to the equation. If an equation has an incorrect given value, solve to find the correct value of the variable.

13. $a - 8 = 6$ $a = 14$ *yes*

14. $n + 0.8 = 1.7$ $n = 0.8$ *no; $n = 0.9$*

15. $x + 13 = 40$ $x = 17$ *no; $x = 27$*

16. $a - 1\frac{1}{2} = 6\frac{1}{2}$ $a = 8$ *yes*

17. $f - 17 = 9$ $f = 2.6$ *no; $f = 26$*

18. $b + 17 + 3.5 = 40.7$ $b = 20.2$ *yes*

Evaluate the expression. Let $x = 3$.

19. $15 \div x - 4$
$15 \div 3 - 4 = 5 - 4 = 1$

20. $(x + 15) \div 2$
$(3 + 15) \div 2 = 18 \div 2 = 9$

21. $3^2 + 7 - x$
$9 + 7 - 3 = 16 - 3 = 13$

22. $4 + x - 2$
$4 + 3 - 2 = 7 - 2 = 5$

23. $x(2.3)$
$3(2.3) = 6.9$

24. $x(2^3) - 6$
$3(8) - 6; 24 - 6 = 18$

25. $(18 - x) \div 5$
$(18 - 3) \div 5 = 15 \div 5 = 3$

26. $2x + 3x$
$2(3) + 3(3) = 6 + 9 = 15$

27. $\sqrt{49} + 6x$
$7 + 6(3) = 7 + 18 = 25$

Write an equation for the sentence. Solve the equation by using the inverse operation. Check the solution.

28. 5 more than n equals 12.
$n + 5 = 12; n = 7$

29. 6 subtracted from a number is 12.
$n - 6 = 12; n = 18$

30. 8 less than n is 3.
$n - 8 = 3; n = 11$

31. A number increased by 3 equals 16.
$n + 3 = 16; n = 13$

32. The sum of 10 and a number is 17.
$10 + n = 17; n = 7$

33. A number decreased by 7 is 9.
$n - 7 = 9; n = 16$

34. The difference of a number and 2 equals 5.
$n - 2 = 5; n = 7$

35. 7 and a number equals 10.
$7 + n = 10; n = 3$

Practice & Application

36. Find the product of 113 and 609, using only two partial products. *68,817*

37. Write equations to show that 468 is divisible by 2, 3, 4, and 6. *$468 \div 2 = 234; 468 \div 3 = 156; 468 \div 4 = 117; 468 \div 6 = 78$*

38. Without dividing, explain how you know that 317 is *not* divisible by 5. *There is not a 5 or a 0 in the ones place.*

39. Use front-end estimation to estimate the sum of 398,640 and 954,207.
$390{,}000 + 950{,}000 = 1{,}340{,}000$

40. What is the value of x in $x + 5 = 14$?
$x + 5 - 5 = 14 - 5; x = 9$

41. What is the value of n in $17.3 + 16.8 + n = 35$?
$34.1 - 34.1 + n = 35 - 34.1; n = 0.9$

42. What whole number is equivalent to $\frac{51}{3}$? *17*

43. Three triangles represent $\frac{1}{2}$ of a set. How many triangles are in the whole set? *6 triangles*

44. Draw a number line to show the sum for $^-9 + 13$.

45. Round each factor to the greatest place to estimate the product of 17.8 and 21.03.
$20 \times 20 = 400$

46. Explain how the inverse operation helps you find the value of n in $n + 3\frac{1}{2} = 6\frac{3}{4}$. Solve.

47. Why do some people's mathematical models contradict reality? *People may use incorrect assumptions to determine what is true.*

44.

46. *Subtraction is used to cancel the value that is added to the variable. Keep the equation balanced by subtracting the same value from the other side of the equation.*
$n + 3\frac{1}{2} - 3\frac{1}{2} = 6\frac{3}{4} - 3\frac{1}{2}; n = 3\frac{1}{4}$

Lesson 97 211

writing equations for the operations that are performed (e.g., remove 7 counters from each side $7 + 5 - 7 = 12 - 7; 5 = 5$; add 3 counters to both sides $7 + 5 + 3 = 12 + 3; 15 = 15$).

Direct attention to $7 + 5 - 5 = 12 - 5$ and $7 + 5 - 7 = 12 - 7$. Subtracting from both sides of the equation an amount that is equal to one of the addends will cancel out the value of that addend on one side of the equation, making the value of that expression the other addend (e.g., $7 + 5 - 5 = 7 + 0 = 7$ [Identity Property of Addition]).

- Write for display: "The sum of a number and 5 equals 8."

What algebraic equation can you write for this sentence? $n + 5 = 8$; "Equals" tells me that an equation is needed, and "sum" tells me to add n and 5.

- Write "$n + 5 = 8$" for display. Explain that to solve an equation with a variable, they need to find the value of the variable. The value of the variable must make the number sentence true; it is the solution to the equation.

Point out that the goal when solving an equation with an unknown value is to isolate the variable so that it stands alone on one side of the equation, giving them the value of n (e.g., $n =$ __ or __ $= n$).

LESSON 97

How can you illustrate $n + 5$? I can place 1 n card and 5 counters to the left of a written equal sign.

How can you illustrate the sum 8? I can place 8 counters to the right of the equal sign.

- Invite a student to picture the equation.

How do you think you can isolate the variable n? I need to remove the 5 counters that are to the left of the equal sign.

If you remove the 5 counters that are to the left of the equal sign, what must you do to the right of the equal sign? I must remove 5 counters.

Choose a student to remove 5 counters from both sides of the equal sign.

What is the value of n? $n = 3$

- Direct attention to $n + 5 = 8$ that was written for display. Explain that to solve the equation, they first need to identify the operation that is being performed on the variable.

What operation is being performed on the variable in the equation? 5 is being added to n.

What operation must you perform to isolate n? I must subtract 5 from the left side of the equation because subtraction is the inverse operation of addition.

If you subtract 5 from the left side of the equation, what must you do to the right side? subtract 5

- Write "$n + 5 - 5 = 8 - 5$" below $n + 5 = 8$. Point out that subtracting 5 from the left side of the equation cancels out the addend 5 ($5 - 5 = 0$). The value of the equation does not change because 5 is being subtracted from both sides of the equation. Write "$n + 0 = 3$" below $n + 5 - 5 = 8 - 5$.

What does the Identity Property of Addition tell you about adding zero to any number? The sum will be the other addend.

What is the value of n? $n = 3$

Write "$n = 3$" below $n + 0 = 3$.

Checking equations

- Guide a **discussion** to help the students check addition and subtraction equations.

LESSON 97

How could you check the solution to this equation? I can substitute 3 for the variable (*n*) in the original equation.

What does 3 + 5 equal? 8

Is this solution correct? yes

- Write "4 + *x* = 15" and invite a student to solve it for display, using the *x* card and the counters.

  What operation is being performed on the variable? 4 is being added to *x*.

- Write "*x* + 4 = 15" for display. The Commutative Property of Addition allows students to change the order of the addends. Explain that applying the Commutative Property clearly shows that 4 is being added to *x*; it places the plus sign (+) directly in front of the addend 4, making it easier to determine the inverse operation.

  What inverse operation can be performed to isolate the variable? subtract 4

  Direct the students to solve either of the equations. Give guidance as needed.

  $4 + x = 15$ $x + 4 = 15$
  $4 - 4 + x = 15 - 4$ $x + 4 - 4 = 15 - 4$
  $x = 11$ $x = 11$

  How can you check the solution? I can substitute 11 for the variable (*x*) in the original equation.

  What is 4 + 11? 15

  Is the solution correct? yes

- Follow a similar procedure for the following equations. Use counters if needed.

  $7 = 2 + n$ $6 + n = 10$ $x + 7 = 9$
  $n = 5$ $n = 4$ $x = 2$

- Write "*x* − 4 = 2" for display.

  What operation is being performed in this equation? 4 is being subtracted from *x*.

  What inverse operation can be performed to isolate the variable? add 4

  If you add 4 to the left side of the equation, what must you do to the right side? I must add 4 to the right side of the equation to keep the equation balanced.

  Write "*x* − 4 + 4 = 2 + 4" below *x* − 4 = 2. Point out that adding 4 to the left side of the equation cancels out the subtrahend 4 (i.e., *x* − 4 + 4 = *x*). The value of the equation does not change because 4 is being added to both sides of the equation.

Write *ones*, *thousands*, *millions*, or *billions* to name the underlined period.

1. 2̲3̲7̲,910,845
millions

2. 819,061,2̲4̲3̲,755
thousands

3. 4̲,603,754,103
billions

4. 1,399,0̲5̲7
ones

Round to the greatest place.

5. 89,371
90,000

6. 1,430,995
1,000,000

7. 7,510,249,631
8,000,000,000

8. 349,275,670
300,000,000

Use the numbers in the box to write the answer.

| 320,941,855 | 39,850,274 | 321,801,327 | 41,273,089 |

9. Write the numbers from least to greatest.

10. Which number has a 3 in the ten millions place? *39,850,274*

11. Which numbers have a 1 in the one millions place? *321,801,327 and 41,273,089*

12. Which number is even? *39,850,274*

13. Which number equals 300,000,000 + 20,000,000 + 1,000,000 + 800,000 + 1,000 + 300 + 20 + 7? *321,801,327*

14. Which number equals 39 millions, 850 thousands, and 274 ones? *39,850,274*

15. Which numbers round to 300,000,000? *320,941,855 and 321,801,327*

16. Which numbers round to 40,000,000? *39,850,274 and 41,273,089*

17. Which numbers have the estimated sum of 80,000,000? *39,850,274 and 41,273,089*

18. Which number is divisible by 5? *320,941,855*

9. *39,850,274; 41,273,089; 320,941,855; 321,801,327*

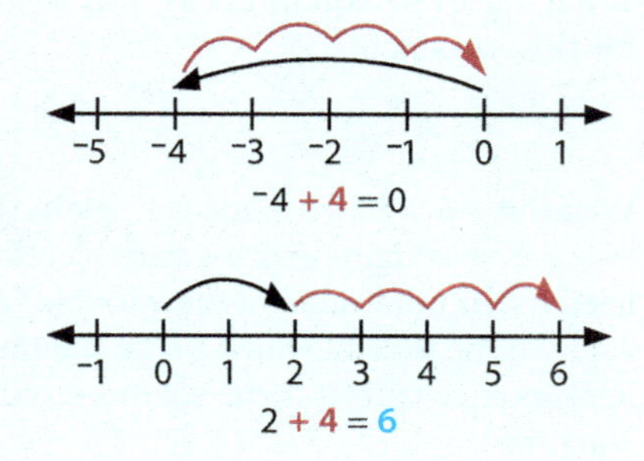

Write "*x* − 0 = 6" below *x* − 4 + 4 = 2 + 4.

What does the Zero Principle of Subtraction state? When zero is subtracted from any number, the answer is that number.

What is the value of *x*? *x* = 6

Write "*x* = 6."

Is the solution correct? Yes; when I substitute 6 for the variable (*x*) in the original equation, the equation is true; 6 − 4 = 2.

- Guide the students as they solve the following equations.

  $b + 12 = 18$ $j - 7 = 21$
  $b = 6$ $j = 28$

  $a + 3 + 7 = 12$ $c - 5 = 14$
  $a = 2$ $c = 19$

- Guide the students as they solve the following equations by first simplifying the expression to the left of the equal sign and then applying the inverse operation to find the value of the variable. Point

---

out that solving equations with decimals or fractions is similar to solving equations with whole numbers.

$9.4 - 2 + d = 13$
$d = 5.6$

$f + \frac{2}{3} + \frac{5}{6} = 5\frac{1}{2}$
$f = 4$

$t + 4 + 2.7 - 1.9 = 14.6$
$t = 9.8$

$m - 3\frac{3}{4} = 2\frac{1}{4}$
$m = 6$

Explaining how models can contradict reality

- Use **critical thinking questions** to help the students answer the essential question.

  You have learned it is important that mathematical models match real life. What could happen if someone failed to find out if his model matched reality? sample answer: The model would not be true to the object or idea that it was representing.

- Read the following scenario aloud.

  Ava learns that a baby grows $\frac{1}{2}$" to 1" per month and gains 5–7 oz per week from birth to 6 months of age. She makes a model to represent that growth and, based on the model, predicts that the child at age 11 will weigh at least 200 lb and stand about 10 ft tall!

Does this model match a child's actual growth? no

What assumption did Ava make, or what wrong idea did she accept as true? that children grow at the same rate as an infant throughout childhood

- Point out that Ava made an assumption that infant growth continued up through age 11 at the same rate. Actually, newborn babies grow at a faster rate. Growth slows down a little after 6 months and again after 1 year.

  What went wrong with Ava's model? It was based on a wrong idea, or assumption.

- Explain that some people check to be sure their model matches real life—what God reveals in nature. If a model is based on a wrong assumption, it is not a good model. The student who doesn't check that the model matches reality may believe the model and look to it as a source of truth.

  Why do some people's mathematical models contradict reality? People may use incorrect assumptions to determine what is true.

LESSON 97

Apply

Student Edition pages 210–11
- Read and explain the directions for pages 210–11. Assist the students as they complete the pages independently.

Daily Review
- Students should complete Chapter 10, section *d*.

NOTES

LESSON 98

Student Edition pages 212–13
Daily Review Chapter 10, section e

OBJECTIVES

- Solve multiplication equations using division.
- Solve a word problem with a variable by writing an equation.

ADDITIONAL MATERIALS

- 5 *n* cards from the pattern provided on the *Variable Cards N* page
- 15 round counters

Engage

- Direct attention to the essential question at the top of Student Edition page 212, "How does using the inverse operation help me solve real-world problems?"
- Direct the students to do a **Think-Pair-Share** to explore the essential question.

Instruct

Solving multiplication equations

- **Model** equations with manipulatives to help the students solve multiplication equations.
- Write for display: "The product of *n* and 3 equals 15."

  What words indicate a mathematical operation or symbol? "Product" indicates multiplication (×), and "equals" indicates an equal sign (=).

  Choose students to write multiplication equations for the sentence. sample answers: $3 \times n = 15$; $3 \cdot n = 15$; $n \times 3 = 15$; $n \cdot 3 = 15$; $3n = 15$

  Direct attention to $3n = 15$. (Write the equation if it was not written by a student.) Remind the students that when the multiplication is written as $3n$, the coefficient (multiplier) is written in front of the variable.

  How do you know that $3 \cdot n$, $n \times 3$, and $3n$ all have the same value? They all represent multiplication, and the Commutative Property of Multiplication states that the order of the factors can be changed without changing the product.

- Remind the students that an equation is a mathematical sentence that states that two expressions are equal; the expression to the left of the equal sign and the expression to the right of the equal sign have the same value. To solve $3n = 15$, students must isolate the variable (*n*) on one side of the equal sign to find the number that makes the sentence true when it is substituted for the variable. The value of *n* is the solution to the equation.

  How can you illustrate $3n = 15$? I can place 3 *n* cards to the left of a written equal sign and 15 counters to the right.

- Direct a student to picture the equation for display.

  What operation is being performed on the variable in the equation? *N* is being multiplied by 3.

  How do you think you can isolate the variable *n*? I can divide the 3 *n* cards by 3; division is the inverse operation of multiplication.

  Demonstrate dividing the 3 *n* cards into 3 groups.

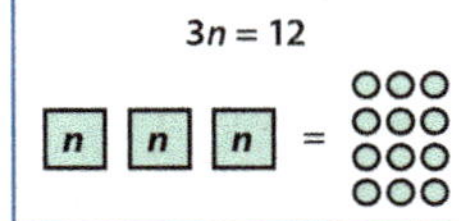

Multiplication Equations

How does using the inverse operation help me solve real-world problems?

Solving Multiplication Equations

$3n = 12$

1. Identify the operation that is being performed on the variable: *n* is multiplied by 3.
2. Apply the inverse operation to both sides to isolate the variable: divide both sides of the equation by 3.
3. Check the solution by substituting the value of the variable into the original equation.

$3n = 3 \cdot n$

$\dfrac{\cancel{3}n}{\cancel{3}} = \dfrac{12}{3}$

$n = 4$

$3 \cdot 4 = 12$

Exercises

Solve. Check the solution. *Answers are shown using cancellation.*

1. $a \times 5 = 30$ *a = 6*
2. $9p = 54$ *p = 6*
3. $c \cdot 7 = 49$ *c = 7*
4. $2m = 18$ *m = 9*
5. $8 \times p = 64$ *p = 8*
6. $d \times 2^2 = 48$ *d = 12*
7. $f \cdot 3 = 24$ *f = 8*
8. $4n = 32$ *n = 8*
9. $a \cdot 0.9 = 3.6$ *a = 4*
10. $12 \cdot b = 36$ *b = 3*
11. $7n = 56$ *n = 8*
12. $1.8 \times a = 36$ *a = 20*

Write an equation for the sentence. Solve.

13. The product of *a* and 7 equals 21. *a = 3*
14. *y* times 9 is 54. *y = 6*
15. *b* multiplied by 8 equals 48. *b = 6*
16. 3 times a number is 36. *n = 12*
17. The product of *x* and 7 equals 35. *x = 5*

Evaluate the expressions. Let $a = 4$. Write a comparison sentence by using >, <, or =.

18. $a + 8 \geq 20 \div 2$ *12 > 10*
19. $\dfrac{32}{a} = a \cdot 2$ *8 = 8*
20. $16 - a \leq 15 - 2$ *12 < 13*
21. $9a \geq 5 \cdot 7$ *36 > 35*
22. $\dfrac{a}{2} \leq 2 \cdot 2$ *2 < 4*
23. $28 \div a \leq 2^3$ *7 < 8*

Solve. *Steps to solve may vary.*

24. $8 = x + 2$ *x = 6*
25. $16 = 9 + x$ *x = 7*
26. $3 \cdot m = 30$ *m = 10*
27. $x - 5 = 8$ *x = 13*
28. $a - 9 = 25$ *a = 34*
29. $b - 3 = 12$ *b = 15*
30. $b \times 8 = 48$ *b = 6*
31. $a + 9 = 20$ *a = 11*
32. $x + 2 = 25$ *x = 23*
33. $5m = 40$ *m = 8*
34. $6n = 36$ *n = 6*
35. $24 = 4x$ *x = 6*

212 Chapter 10

---

36. Write an equation showing that Lyla is 6 yr younger than her brother. *l = b − 6*

37. Write an equation showing that Liam's science test score was 6 points more than Zoe's. *l = z + 6*

38. Write an equation to show 58 boys divided evenly among 4 baseball teams. Explain your answer.

39. What terms can be combined in the expression $2s + \frac{1}{4} + \frac{2}{3}$? *$\frac{1}{4}$ and $\frac{2}{3}$*

40. If $a = 4$ and $b = 6$, is $\frac{24}{a} = b$ a true statement? Explain your answer. *yes; $\frac{24}{4} = 6$*

41. Write and solve the multiplication equation that can be used to find the value of c in $c \div 2.3 = 17$. *c = 17 × 2.3; c = 39.1*

42. Find the product of $\frac{12}{25}$ and $\frac{5}{8}$ in lowest terms.

43. Find the least common denominator for $\frac{1}{8}$ and $\frac{1}{12}$. What is the sum of these fractions in lowest terms? *LCD = 24; $\frac{3}{24} + \frac{2}{24} = \frac{5}{24}$*

44. Write an equation to show that factors r and s equal 36. *r × s = 36*

45. Substitute numbers for the variables in the equation for problem 44. Explain why 6 cannot be a factor in this problem.

46. A student solved the following equations incorrectly. Find the mistakes and correct them. Explain how to solve the problem correctly.

$$4x = 36 \qquad\qquad 8m = 24$$
$$\frac{4x}{2} = \frac{36}{2} \qquad\qquad \frac{8m}{8} = \frac{24}{4}$$
$$2x = 18 \qquad\qquad m = 6$$

47. How does using the inverse operation help me solve real-world problems? *The inverse operation helps me isolate and find the value of a variable in a multiplication equation.*

38. *58 is not evenly divisible by 4; there is a remainder of 2. All 4 teams will have 14 boys. Two of the teams will have an extra player.*

42. $\frac{12}{25} \times \frac{5}{8} = \frac{3}{10}$
Answer is shown using cancellation.

45. *9 × 4 = 36; 6 cannot represent both r and s in the same equation.*

46.
$$\frac{4x}{4} = \frac{36}{4} \qquad\qquad \frac{8m}{8} = \frac{24}{8}$$
$$x = 9 \qquad\qquad m = 3$$

Since x is multiplied by 4, both sides of the equation must be divided by 4. *Since m is multiplied by 8, both sides of the equation must be divided by 8.*

Lesson 98   213

Choose a student to demonstrate solving the equation. 3(5) = 15

Is $n = 5$ the correct solution? Yes; 3(5) = 15 is a true mathematical statement.

- Follow a similar procedure with the following equations. Direct them to solve the equations and check the solutions.

$$5n = 10 \qquad\qquad 2n = 8$$
$$\frac{5n}{5} = \frac{10}{5} \qquad\qquad \frac{2n}{2} = \frac{8}{2}$$
$$n = 2 \qquad\qquad n = 4$$

Solving a word problem with a variable

- **Demonstrate** translating word phrases into numerical values to help the students solve word problems.

- Read aloud the following word problem.

Mrs. Johnson is 3 times as old as her daughter, Kathryn. If Mrs. Johnson is 36 years old, how old is Kathryn? 12 years old

What is the question asking you to find? Kathryn's age

What information are you given? Mrs. Johnson is 3 times as old as Kathryn, and Mrs. Johnson is 36 years old.

Write for display these two sentences: "Mrs. Johnson is 3 times Kathryn's age" and "Mrs. Johnson is 36 years old."

What word could be written as an equal sign in each of these sentences? is

Erase "is" in each of the sentences and write an equal sign (=) in its place: Mrs. Johnson = 3 times Kathryn's age and Mrs. Johnson = 36 years old.

Which of these word phrases expresses the same value? 3 times Kathryn's age and 36 years old

What value do both phrases express? Mrs. Johnson's age

- Explain that since "3 times Kathryn's age" and "36 years old" both express Mrs. Johnson's age, students can write them as equivalent expressions. Write "3 times Kathryn's age = 36 years old" for display.

How could you write "3 times Kathryn's age" as an algebraic expression? sample answer: 3k; Kathryn's age is unknown and can be written using a variable.

How many n cards are in each of the 3 sets? 1

- Remind the students that an equation is similar to a balanced scale. Any operation that is performed on one side of the equation must also be performed on the other side to keep the equation balanced.

Since you divided the n cards to the left of the equal sign by 3, what must you do to the counters to the right of the equal sign? I must divide them by 3 or divide them into 3 equal groups.

Invite a student to divide the 15 counters into 3 equal groups.

How many counters are in each of the 3 groups? 5

Explain that each n card on the mat is equivalent to 1 set of 5 counters, showing that $n = 5$.

- Direct attention to $3n = 15$ that was written for display and demonstrate solving the equation without using manipulatives: $\frac{3n}{3} = \frac{15}{3}$, $n = 5$. The variable (n) is being multiplied by 3; therefore, they must use division (the inverse operation) and divide the value on each side of the equation by 3.

How can you check the solution? I can substitute 5 for the variable (n) in the original equation.

Accept any variable, but remind the students that using a variable that is related to the information can be helpful when solving a word problem.

- Write "$3k = 36$" for display. Direct the students to write the equation, solve it, and then check their solution.

  How old is Kathryn? 12 years old

  How do you know that 12 years old is the correct solution? When 12 is substituted for k in $3k = 36$, the mathematical statement is true: $3(12) = 36$.

  How does using the inverse operation help you to solve real-world problems? The inverse operation helps me to isolate and find the value of a variable in a multiplication equation.

- Direct the students to solve the following equations. Remind them to identify the operation that is being performed on the variable and then apply the inverse operation. Give guidance as needed.

  $6a = 24$ $3 \times c = 21$ $4m = 36$
  $a = 4$ $c = 7$ $m = 9$

  $54 = 9b$ $6.3 = 0.7 \cdot h$
  $b = 6$ $h = 9$

Apply

Student Edition pages 212–13
- Read and explain the directions for pages 212–13. Assist the students as they complete the pages independently.

Daily Review
- Students should complete Chapter 10, section e.

Find the volume of the figure.

1. 

$\underline{7} \times \underline{4} \times \underline{2} = \underline{}$ units^3 *56*
$l \quad w \quad h$

2. 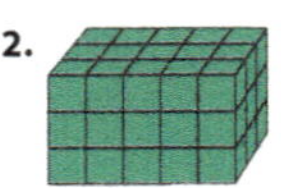

$\underline{5} \times \underline{3} \times \underline{3} = \underline{}$ units^3 *45*
$l \quad w \quad h$

3. 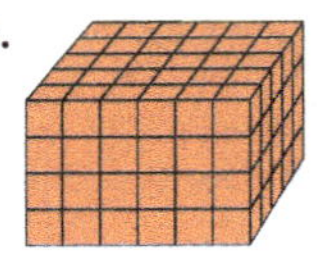

$\underline{6} \times \underline{5} \times \underline{4} = \underline{}$ units^3 *120*
$l \quad w \quad h$

Write a multiplication equation to find the area of the figure.

4. 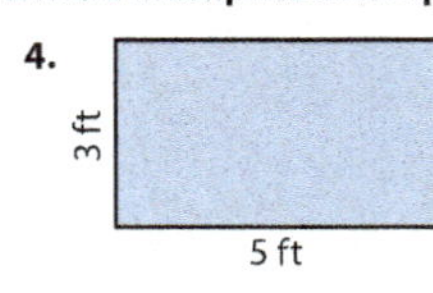
3 ft / 5 ft
__ ft^2
3 × 5 = 15

5. 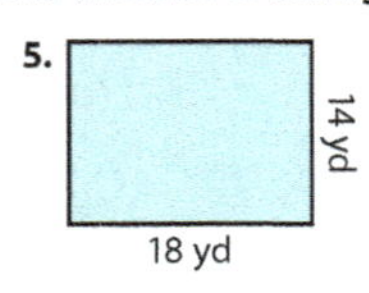
14 yd / 18 yd
__ yd^2
18 × 14 = 252

6. 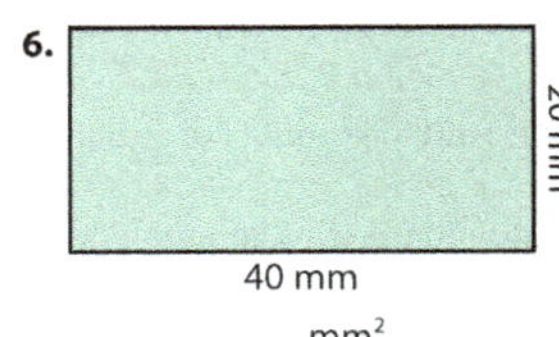
20 mm / 40 mm
__ mm^2
40 × 20 = 800

Find the perimeter of the figure.

7. 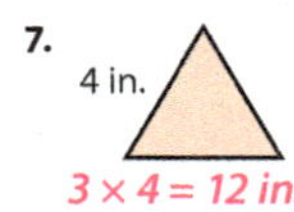
4 in.
3 × 4 = 12 in.

8. 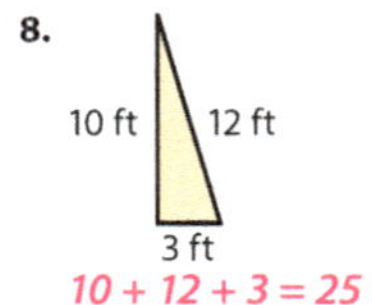
10 ft / 12 ft / 3 ft
10 + 12 + 3 = 25 ft

9. 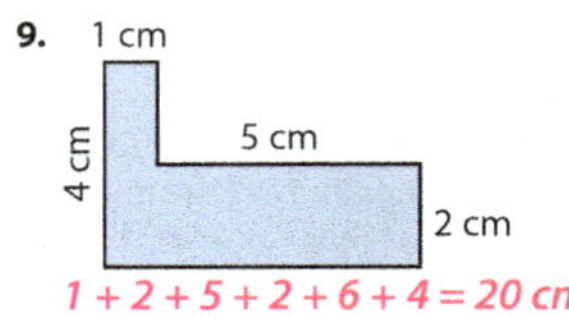
1 cm / 4 cm / 5 cm / 2 cm
1 + 2 + 5 + 2 + 6 + 4 = 20 cm

LESSON 99

Student Edition pages 214–15
Daily Review Chapter 10, section *f*

OBJECTIVES

- Solve multiplication and division equations by using inverse operations.
- Solve a word problem with a variable by writing an equation.
- Plot an inequality on a number line.
- Determine whether a given number is a solution to an inequality.

Lesson 99 is designed to be a 2-day lesson. Student Edition page 214 may be assigned on Day 1.

Engage

- Direct attention to the essential question at the top of Student Edition page 214, "How does a number line help me picture an inequality?"

- Guide the students in a **discussion** to contrast the words *equality* and *inequality*.

  What does it mean when two mathematical expressions are equal or in a state of equality? They have the same value.

  What does it mean when the two mathematical expressions show a state of inequality? They are not equal; one is greater than the other.

Instruct

Solving equations using inverse operations

- Use **guided discovery** to help the students use inverse operations to solve equations.

- Write for display: "The quotient of a number and 4 equals 12."

  What is an equation? a mathematical statement in which two expressions are equal

  Can the sentence be written as an equation? Yes; "equals" indicates an equal sign, and "quotient" indicates that division is being performed.

Multiplication & Division Equations

How does a number line help me picture an inequality?

Key Terms
- inequality

Multiplication and division are inverse operations.

$$36 \div 4 \times 4 = 36 \qquad 8 \cdot 2 \div 2 = 8$$

$$\frac{36}{4} \times 4 = 36 \qquad \frac{8 \cdot 2}{2} = 8$$

Isolate the variable on one side of the equals sign by using the inverse operation. An equation is much like a balanced scale: when you perform an operation on the left side of the equation, you must perform the same operation on the right side of the equation.

$5x = 35$

The inverse of multiplying by 5 is dividing by 5.

$$5x = 35$$
$$\frac{5x}{5} = \frac{35}{5}$$
$$x = 7$$
$$5 \cdot 7 = 35$$

$n \div 8 = 4$

The inverse of dividing by 8 is multiplying by 8.

$$n \div 8 = 4$$
$$\frac{n}{8} \cdot 8 = 4 \cdot 8$$
$$n = 32$$
$$\frac{32}{8} = 4$$

Exercises

Solve. Check the solution.

1. $n \div 7 = 63$ *n = 441*
2. $\frac{m}{4} = 3$ *m = 12*
3. $x \div 8 = 14.5$ *x = 116*
4. $3 \cdot x = 15$ *x = 5*
5. $n \cdot \frac{1}{5} = 9$ *n = 45*
6. $r \div 0.3 = 57.5$ *r = 17.25*
7. $\frac{2}{3} \cdot x = 15$ *x = 22$\frac{1}{2}$*
8. $\frac{x}{9} = 18$ *x = 162*
9. $4 \cdot c = 32$ *c = 8*
10. $\frac{1}{2}p = 45$ *p = 90*
11. $3n = 78.12$ *n = 26.04*
12. $p \cdot \frac{3}{5} = 5$ *p = 8$\frac{1}{3}$*

Write an equation for the sentence. Solve the equation by using the inverse operation. Check the solution.

13. The quotient of a number divided by 12 is 3.
    n ÷ 12 = 3 or $\frac{n}{12}$ = 3; n = 36
14. The quotient of *y* and 7 equals 2.
    y ÷ 7 = 2 or $\frac{y}{7}$ = 2; y = 14
15. The product of *x* and 7 is 35.
    x • 7 = 35 or 7x = 35; x = 5
16. A number divided by 20 equals 3.
    n ÷ 20 = 3 or $\frac{n}{20}$ = 3; n = 60
17. A number divided by 5 equals 11.
    n ÷ 5 = 11 or $\frac{n}{5}$ = 11; n = 55
18. Twice a number is 18.
    2n = 18 or 2 • n = 18; n = 9
19. 6 times a number is 54.
    6n = 54 or 6 • n = 54; n = 9
20. The product of a number and 5 is 10.
    n • 5 = 10 or 5n = 10; n = 2

Complete the table, using the given values to evaluate the expression.

21.

x	$2x + 3$
11	*25*
13	*29*
15	*33*
17	*37*

22.

x	$x^2 + 5$
8	*69*
10	*105*
12	*149*
14	*201*

23.

x	$4(x)$
25	*100*
35	*140*
45	*180*
55	*220*

What are the two equal expressions in the sentence? the quotient of a number and 4; 12

How do you represent an unknown number in an equation? I can use a variable to represent the unknown value.

- Choose a student to write the equation for display. Choose another student to write the equation in fraction form.
  $x \div 4 = 12; \frac{x}{4} = 12$

  What operation is being performed on the variable? *X* is being divided by 4.

  How can you find the value of *x*? Since multiplication is the inverse operation of division, I can multiply both sides of the equation by 4 to isolate the variable.

Why must you multiply both sides of the equation by 4? to keep the values on both sides of the equation equal (keep the equation balanced)

- Explain each step as you demonstrate solving both equations as shown. For the equation written in fraction form, explain that $\frac{4x}{4}$ is the same as $(\frac{4}{4})x$, $1x$, or x.

You may remind the students that the whole number 4 can be written as the improper fraction $\frac{4}{1}$ ($\frac{x}{4} \cdot \frac{4}{1} = 12 \cdot 4$) and then demonstrate solving the equation.

An **inequality** is a mathematical sentence in which two expressions are not equal. The greater than (>) and less than (<) symbols are used to express an inequality. A number line shows all solutions for the inequality.

$$x > 2$$

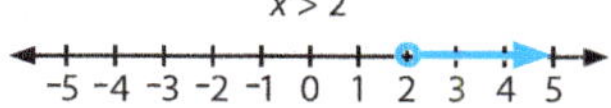

The number line shows that x is any value *greater than* 2.

The open circle on the number line indicates that the circled number is *not* included in the solution.

$$x < 3 - 1$$

The number line shows that x is any value *less than* 2.

Draw a number line to illustrate the inequality. *Figures may vary.*

24. $x < 3$

25. $b > {}^-2$

26. $y > 15 \div 3$

27. $c < 5 - 4$

State whether the given value is a solution to the inequality.

28. $x < 12$ if $x = 10$ *yes*

29. $y > 3$ if $y = 2$ *no*

30. $n < {}^-2$ if $n = {}^-1$ *no*

31. $b > 10$ if $b = 3^2$ *no*

32. $m < 5$ if $m = \frac{19}{4}$ *yes*

33. $s > 7$ if $s = \sqrt{9}$ *no*

Practice & Application *Equations may vary.*

34. There will be 23 people at Josiah and Jacob's birthday party. How many packages of hot dogs will need to be purchased if 2 hot dogs are cooked for each person at the party? There are 10 hot dogs in each package. *$(23 \times 2) \div 10 = 46 \div 10 = 4.6$; 5 packages*

35. There are 8 hot dog buns in each package. How many packages of buns will need to be purchased for the number of hot dogs cooked? (Use the solution from problem 34.) *$46 \div 8 = 5.75$; 6 packages*

36. Two ice-cream cakes were made for the party. Each cake was cut into 12 equal pieces. There were 8 pieces of cake left at the end of the day. What fraction of the cake was left? *$2 \times 12 = 24$; $\frac{8}{24} = \frac{1}{3}$ of the cake*

37. Four hundred water balloons were filled. Jacob figured that each non-adult guest would have 26 balloons to throw. If 8 of the guests were adults, how many extra water balloons were there? (Use the information from problem 34.) *$23 - 8 = 15$ non-adult guests; $400 - (15 \times 26) = 400 - 390 = 10$ extra water balloons*

38. Write an equation for the water balloon purchase. Use the variable t for the unknown amount. Solve to find the amount of tax paid.
2 packages of balloons: $2.94 each
2 balloon launchers: $11.99 each
shipping: $5.00
tax: ___
total: $36.65

39. How does a number line help me picture an inequality? *A number line helps me picture all solutions for the inequality.*

38. *$(2 \times \$2.94) + (2 \times \$11.99) + \$5.00 + t = \$36.65$*
$\$5.88 + \$23.98 + \$5.00 + t = \$36.65$
$\$34.86 + t = \$36.65$
$\$36.65 - \$34.86 = \$1.79$
$t = \$1.79$

$x \div 4 = 12$ $\qquad$ $\frac{x}{4} = 12$

$x \div 4 \times 4 = 12 \times 4$ $\qquad$ $\frac{x}{4} \cdot 4 = 12 \cdot 4$

$x = 48$ $\qquad$ (or $\frac{4x}{4} = 12 \cdot 4$)

$\qquad\qquad\qquad\qquad\quad x = 48$

Is $x = 48$ the correct solution? Yes; when 48 is substituted for x in the original equation, the mathematical statement is true; $48 \div 4 = 12$ and $\frac{48}{4} = 12$.

Point out that writing each step in the solution helps them to accurately solve an equation and makes it easier to identify possible errors.

- Follow a similar procedure for the following equations. Point out that solving equations with fractions or decimals is similar to solving equations with whole numbers. While solving $f \cdot \frac{3}{4} = 7$, discuss with the students that they can use what they know about division (any number divided by itself is 1; $\frac{3}{4} \div \frac{3}{4} = 1$) or multiply by the reciprocal ($\frac{3}{4} \cdot \frac{4}{3} = 1$).

$f \cdot \frac{3}{4} = 7$ $\qquad\qquad$ $f \cdot \frac{3}{4} = 7$

$f \cdot \frac{3}{4} \div \frac{3}{4} = 7 \div \frac{3}{4}$ $\qquad$ $f \cdot \frac{3}{4} \div \frac{3}{4} = 7 \div \frac{3}{4}$

$f = 7 \cdot \frac{4}{3}$ $\qquad\qquad$ $f \cdot \frac{3}{4} \cdot \frac{4}{3} = 7 \cdot \frac{4}{3}$

$f = \frac{28}{3} = 9\frac{1}{3}$ $\qquad$ $f = \frac{28}{3} = 9\frac{1}{3}$

$d \div 0.4 = 23.2$ $\qquad\qquad$ $d \div 0.4 = 23.2$

$d \div 0.4 \cdot 0.4 = 23.2 \cdot 0.4$ $\qquad$ $\frac{d}{0.4} \cdot 0.4 = 23.2 \cdot 0.4$

$d = 23.2 \cdot 0.4$ $\qquad\qquad$ $d = 9.28$

$d = 9.28$

- Write the following equations for display and direct the students to solve the problems. Remind them to first identify the operation that is being performed on the variable and then apply the inverse operation. Give guidance as needed.

$\frac{x}{6} = 7$ $\qquad$ $n \div 5 = 4$ $\qquad$ $3y = 2.7$
$x = 42$ $\qquad$ $n = 20$ $\qquad$ $y = 0.9$

$7 = t \cdot \frac{2}{5}$ $\qquad$ $p \times 8 = 32$ $\qquad$ $a \div \frac{4}{5} = 10$
$t = 17\frac{1}{2}$ $\qquad$ $p = 4$ $\qquad$ $a = 8$

Solving a word problem with a variable

- Guide a **discussion** to help the students solve a problem with a variable.

- Read aloud the following word problem.

Nicole and Jane are making a photo album of the 165 vacation pictures they took. Nicole took twice as many pictures as Jane. How many vacation pictures did Nicole and Jane each take?

What word sentence can you write to find the total number of pictures taken? The number of pictures Nicole took and the number of pictures Jane took equal 165 pictures.

Write the sentence for display.

What information do you need in order to find how many pictures Nicole took? I need the number of pictures that Jane took.

How can you represent the unknown number of pictures Jane took? I can use the variable j; a variable is used to represent an unknown value, and j is the first letter in Jane's name.

Write "j = number of pictures Jane took" for display. Point out that although an uppercase j begins Jane's name, variables are written using a lowercase letter.

If j represents the number of pictures Jane took, what algebraic expression could represent the number of pictures Nicole took? *$2j$; Nicole took twice as many pictures as Jane.*

LESSON 99

Write "$2j$ = number of pictures Nicole took" for display.

- Guide the students as they write a mathematical equation for the word sentence: $2j + j = 165$.

  How can you simplify the equation? $3j = 165$

- Choose a student to solve the equation and to find the number of pictures that Nicole took while the other students solve the problems on paper. $j = 55$, Jane took 55 pictures; and $2j = 110$, Nicole took 110 pictures.

  How can you check the answers to the word problem? I can substitute 55 for the variable (j) in the original equation.

  $(2 \times 55) + 55 = 165$; therefore, $j = 55$ pictures and $2j = 110$ pictures are correct answers.

You may end the lesson here and teach the remainder of the lesson and use Student Edition page 215 on Day 2.

Plotting an inequality on a number line

- **Model** plotting values on a number line to help the students picture an inequality.

- Write "inequality" for display and the symbols > and <. Explain that an inequality is a mathematical sentence in which two expressions are not equal; it can be expressed using a greater-than (>) or a less-than (<) symbol. Point out that inequalities have more than one solution.

- Write "$x > 3$." Ask students to provide fraction, decimal, and whole-number solutions such as $3\frac{1}{2}$, 4.2, and 75.

  Explain that there are an infinite number of values greater than 3; therefore, it is impossible to name all of the values for x that will make the number sentence true. The best way to show the solutions for an inequality is to graph them on a number line.

- Draw for display a number line for 0–10.

  How do you show $x = 3$ on a number line? I can plot a point at 3.

  Where on the number line are numbers greater than 3 located? to the right of 3

Solve. Shade the picture to illustrate the answer.

1. 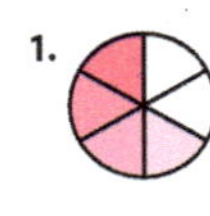 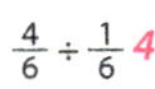
$\frac{4}{6} \div \frac{1}{6}$ **4**

2. 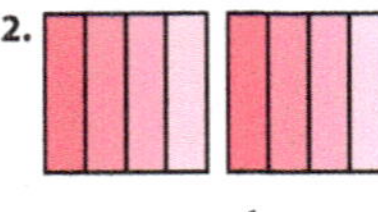
$2 \div \frac{1}{4}$ **8**

3. 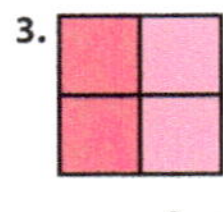
$1 \div \frac{2}{4}$ **2**

Solve. Write the answer in lowest terms. *Answer is shown using cancellation.*

4. $8 \div \frac{1}{2}\ \frac{8}{1} \times \frac{2}{1} = 16$

5. $2\frac{1}{9} \div 3\ \frac{19}{9} \times \frac{1}{3} = \frac{19}{27}$

6. $\frac{4}{6} \div \frac{1}{3}\ \frac{4}{6} \times \frac{3}{1} = \frac{4}{2} = 2$

7. $\frac{6}{12} \div \frac{2}{3}\ \frac{6}{12} \times \frac{3}{2} = \frac{3}{4}$

8. $\frac{3}{4} \div \frac{1}{8}\ \frac{3}{4} \times \frac{8}{1} = \frac{6}{1} = 6$

9. $\frac{4}{5} \div \frac{1}{5}\ \frac{4}{5} \times \frac{5}{1} = \frac{4}{1} = 4$

10. $\frac{5}{6} \div \frac{2}{8}\ \frac{5}{6} \times \frac{8}{2} = \frac{20}{6} = $
$3\frac{2}{6} = 3\frac{1}{3}$

11. $\frac{3}{4} \div 8\ \frac{3}{4} \times \frac{1}{8} = \frac{3}{32}$

Use the data to answer the question.

12. Noah prepared half of the trail mix recipe. How many cups of mix did he make?

$\underline{1\frac{1}{2}}$ c + $\underline{\frac{3}{4}}$ c + $\underline{1\frac{1}{8}}$ c = $\underline{3\frac{3}{8}}$ c
cereal   raisins   candy   total

13. Mom doubled the trail mix recipe to take to the church fellowship. How many cups of mix did she make?

$\underline{6}$ c + $\underline{3}$ c + $\underline{4\frac{1}{2}}$ c = $\underline{13\frac{1}{2}}$ c
cereal   raisins   candy   total

Trail Mix Recipe
3 c of cereal
$1\frac{1}{2}$ c of raisins
$2\frac{1}{4}$ c of candy

Daily Review   465

Draw an open (unshaded) circle to mark 3 on the number line and shade the part of the number line to the right of 3. Explain that since the number sentence tells them that x is greater than 3, x is not equal to 3; therefore, 3 is not a solution to the inequality. The open circle shows that 3 is not included in the solution. The shaded part of the number line shows that any value greater than 3 is included in the solution, whether the value is slightly or significantly greater than 3.

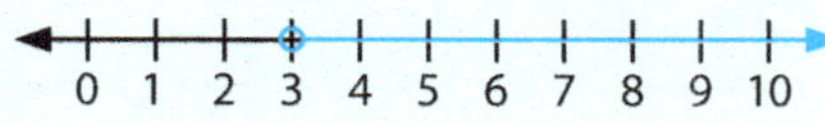

Determining a solution to an inequality

- Conduct a **practice exercise** to help the students determine whether a given number is a solution to an inequality.

- Write the following numbers for display and guide the students as they identify which of the following values are included in the set of solutions for $x > 3$.

The magenta numbers indicate the values that are included in the set of solutions.

6	-2	11	3.0
3.00001	$\frac{7}{1}$	$\frac{12}{4}$	$\sqrt{9}$
2.999999	$\sqrt{25}$	$\frac{72}{8}$	3.3

- Draw 3 number lines for ⁻5–5. Choose students to graph the solutions for the following inequalities.

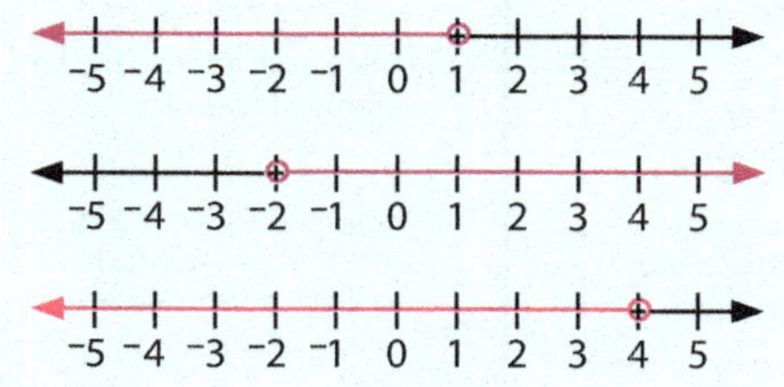

$y < 1$

$x > ⁻2$

$n < 4$

How does a number line help me picture an inequality? A number line helps me picture all solutions for an inequality.

- Write the following inequalities for display. Choose students to tell whether the given value is a solution to the inequality and have them explain their answer.

$x > 7$ if $x = 6$ No; 6 is less than 7.

$y < 4$ if $y = 2$ Yes; 2 is less than 4.

$n > ⁻1$ if $n = ⁻3$ No; ⁻3 is less than ⁻1.

$x < 7$ if $x = 2^3$ No; 2^3 equals 8, and 8 is greater than 7.

Apply

Student Edition pages 214–15

- Read and explain the directions for pages 214–15. Assist the students as they complete the pages independently.

Daily Review

- Students should complete Chapter 10, section *f*.

NOTES

Student Edition pages 216–17
Daily Review Chapter 10, section g

OBJECTIVES

- Find equivalent expressions by applying the Distributive Property.
- Explain the meaning of consistency when writing equivalent expressions. **BWS**
- Solve equations using inverse operations.

BIBLICAL WORLDVIEW SHAPING

- **Modeling (Explain):** Consistency with equivalent expressions refers to the ability to get the same answer by following procedures.

ADDITIONAL MATERIALS

- a geographic compass or a picture of one

ASSESSMENTS

- Chapter 10 Quiz 1

Engage

- Direct attention to the essential question at the top of Student Edition page 216, "What does it mean to get consistent answers when we write equivalent expressions?"

- Direct the students to do a **Quick Write** to explain the statement "I *consistently* do my best" to prepare the students to answer the essential question.

Instruct

Finding equivalent expressions

- **Ask questions** to activate prior knowledge to help the students apply the Distributive Property.

- Write for display "$a + b = b + a$" and "$x \cdot y = y \cdot x$." Explain that properties are important when studying algebraic expressions and equations.

Equivalent Expressions

What does it mean to get consistent answers when we write equivalent expressions?

The Associative and Commutative Properties of Addition or Multiplication are used to simplify expressions by reordering (Commutative Property) and regrouping (Associative Property) like terms.

$4 + (n + 5) =$
$4 + (5 + n) =$
$(4 + 5) + n = 9 + n$

$6 \cdot (n \cdot 9) =$
$6 \cdot (9 \cdot n) =$
$(6 \cdot 9) \cdot n = 54n$

The Distributive Property of Multiplication over Addition can also be used to simplify expressions. Or it can be used to find an equivalent expression by finding a common factor in the terms.

Simplifying an Expression

$3(2n + 4)$

$3(2n + 4) =$
$3(2n) + 3(4) =$
$6n + 12$

1. Determine the number of sets of each addend. There are 3 sets of each addend.
2. Multiply each addend by the multiplier, 3.
3. Write the simplified expression.

Key Terms
- Addition & Multiplication Properties

Finding an Equivalent Expression

$9 + 15b$

1. Find the common factor of the addends: 3.
   $9 + 15b =$
2. Divide each addend by the common factor, 3.
   $\frac{9}{3} + \frac{15b}{3} =$
3. Group the addends and write the common factor outside the parentheses.
   $3(3 + 5b)$

Exercises

Apply the Distributive Property to write an equivalent expression. *Order of terms may vary.*

1. $9(2 + x)$ $9(2) + 9(x) = 18 + 9x$
2. $5(n + 21)$ $5(n) + 5(21) = 5n + 105$
3. $7(3 + 4y)$ $7(3) + 7(4y) = 21 + 28y$
4. $2(5x + 3.5)$ $2(5x) + 2(3.5) = 10x + 7$
5. $16(5 + 3a)$ $16(5) + 16(3a) = 80 + 48a$
6. $3(3x + 0.8)$ $3(3x) + 3(0.8) = 9x + 2.4$
7. $4(y + \frac{1}{8})$ $4(y) + 4(\frac{1}{8}) = 4y + \frac{1}{2}$
8. $6(n + \frac{1}{4})$ $6(n) + 6(\frac{1}{4}) = 6n + 1\frac{1}{2}$

Choose the equivalent expression.

9.	$6x + 18$	$3(2x + 6)$	$3(x + 9)$	$2(3x) + 2(6)$
10.	$4(7 + 3a)$	$11 + 7a$	$28 + 12a$	$\frac{7}{4} + \frac{3}{4}a$
11.	$12(y) + 3$	$12 + 3y$	$15y$	$12y + 3$

Choose the inverse operation that would be used to solve the equation.

12.	$5 + y = 25$	addition	subtraction	multiplication
13.	$\frac{x}{7} = 7$	multiplication	division	addition
14.	$3y = 48$	addition	multiplication	division

216 Chapter 10

What properties do these equations represent? the Commutative Property of Addition and the Commutative Property of Multiplication

What does the Commutative Property of Addition allow you to do? change the order of the addends without changing the sum

What does the Commutative Property of Multiplication allow you to do? change the order of the factors without changing the product

- Write for display $(a + b) + c = a + (b + c)$ and $(a \times b) \times c = a \times (b \times c)$.

What properties do these equations represent? the Associative Property of Addition and the Associative Property of Multiplication

What does the Associative Property of Addition allow you to do? change the grouping of the addends without changing the sum

What does the Associative Property of Multiplication allow you to do? change the grouping of the factors without changing the product

- Write "$(5 + 2x) + 3$" for display.

How can you apply the Commutative and the Associative Properties of Addition to simplify the expression? I can apply the Commutative Property to change the order of the addends within the parentheses: $(2x + 5) + 3$. Then I can apply the Associative Property to change the grouping of the addends: $2x + (5 + 3)$.

State whether the given value is the solution to the equation. *Steps to solve may vary.*

15. $r - 42 = 59$ if $r = 17$ *no;*
$17 - 42 = ^-25$

16. $3.5 + w = 17.9$ if $w = 14.4$ *yes;*
$3.5 + 14.4 = 17.9$

17. $16.08n = 1{,}608$ if $n = 100$
yes; $16.08 \times 100 = 1{,}608$

18. $156 \div h = 12$ if $h = 12$ *no;*
$156 \div 12 = 13$

19. $\frac{1}{8}x = 6$ if $x = 50$ *no;*
$\frac{1}{8}(50) = \frac{50}{8} = 6.25$

20. $\frac{x}{3} = 13$ if $x = 39$
yes; $\frac{39}{3} = 13$

Solve. Check the solution.

21. $a + 5 = 33$ *a = 28*

22. $8x = 480$ *x = 60*

23. $1.5w = 30$ *w = 20*

24. $x - 1.2 = 10$ *x = 11.2*

25. $y - 43 = 129$ *y = 172*

26. $3.8p = 64.6$ *p = 17*

27. $\frac{a}{12} = 3$ *a = 36*

28. $\frac{3}{4}x = 6$ *x = 8*

29. $\frac{x}{9} = 4$ *x = 36*

30. $8x = 1$ *x = $\frac{1}{8}$*

31. $2x = 14.8$ *x = 7.4*

32. $a + 1.7 = 1.9$ *a = 0.2*

33. $n - 16 = 140$ *n = 156*

34. $x - 6 = 1.4$ *x = 7.4*

35. $x \div 12 = 62$ *x = 744*

State whether the given value is a solution to the inequality.

36. $y > 3$ if $y = 2$ *no*

37. $x < 2$ if $x = ^-1$ *yes*

38. $a > 7$ if $a = 3^2$ *yes*

Solve. *Equations may vary.*

39. How many pizzas will be ordered for the Sunday school picnic if each pizza is cut into 8 slices and 200 slices are needed? *$200 \div 8 = p$; p = 25 pizzas*

40. On Monday Forrest found 9 more insects for his science project. Now he can make a display of all 20 insects. How many insects did he have before Monday? *$x + 9 = 20$; x = 11 insects*

Practice & Application

41. Simplify the expression: $(2n + 4) + (3n + 6)$.
$2n + 3n + 4 + 6 = 5n + 10$

42. What operation cancels out addition?
subtraction

43. What is the product of $3n$ if $n = 36$?
$3 \cdot 36 = 108$

44. If $\frac{1}{2}n = 118$, what is the value of n?
n = 236

45. Write an expression for a number plus 3, multiplied by 6. *$(n + 3) \times 6$*

46. $y < $ *2*

-5 -4 -3 -2 -1 0 1 2 3 4 5

47. Identify as true or false: $3(8 + 12) = 24 + 36$.
true; $3(8) + 3(12) = 24 + 36$
$24 + 36 = 24 + 36$

48. *$30 \times \frac{1}{5} = 6$; $\frac{1}{3} \times 39 = 13$; $\frac{1}{3} \times 39$ has a greater value.*

48. Which expression has a greater value:
$30 \times \frac{1}{5}$ or $\frac{1}{3} \times 39$?

49. Identify the steps shown as the Associative, the Commutative, and/or the Distributive Property. Simplify the expression.

$3x + 4 + 2x$	$n + (3 + 5n)$	$7(y + 3y)$
$3x + 2x + 4$	$n + (5n + 3)$	$(7 \cdot y) + (7 \cdot 3y)$
	$(n + 5n) + 3$	

50. What does it mean to get consistent answers when we write equivalent expressions? *It means that we get the same answer each time if we follow the correct procedures.*

49. *Commutative; 5x + 4*
Commutative, Associative; 6n + 3
Distributive; 7y + 21y = 28y

Write the equations as they are given.

What is the expression in simplified form? $2x + 8$

- Write for display "Distributive Property of Multiplication over Addition" and "$a(b + c) = ab + ac$." Remind the students that the Distributive Property involves both multiplication and addition. Explain that this property is also important when studying algebraic expressions and equations.

What is different about the expression to the right of the equal sign? Each addend is being multiplied by the multiplier, a.

Write "$2(3 + 5) = $ ___" for display. Direct the students to apply the Distributive Property and to write an equivalent expression. $(2 \times 3) + (2 \times 5)$

Write the expression in the blank.

- Choose students to give the values for each side of the equation.

How do you know that the expressions are equivalent? Both expressions have a value of 16.

- Write "$4(3 + x) = $" for display.

How can you apply the Distributive Property to write an equivalent expression? I can multiply each addend by the multiplier, 4.

Choose a student to apply the Distributive Property and to write the equivalent expression below $4(3 + x) =$. $(4 \cdot 3) + (4 \cdot x)$

LESSON 100

How can you simplify $(4 \cdot 3) + (4 \cdot x)$ to write an equivalent expression? I can multiply $4 \cdot 3$ and multiply $4 \cdot x$, and then I can add the products.

Choose a student to simplify the expression and to write the equivalent expression below $(4 \times 3) + (4 \times x)$. $12 + 4x$

Because of the Commutative Property, $12 + 4x$ can also be written as $4x + 12$.

- Write "$4b + 24$" for display. Explain that the students can also find an equivalent expression by factoring a common factor out of each term. Point out that factoring out the greatest common factor (GCF) will usually make it easier to find an equivalent expression.

What are the common factors of 4 and 24? 2 and 4

What is the greatest common factor? 4

What times 4 equals $4b$? $1b$

What number times 4 equals 24? 6

Write "$4(b + 6)$" to the right of $4b + 24$, leaving a space between the 2 expressions. Explain that when a common factor is factored out of the addends, it is written outside the parentheses to show that it is now a factor in the expression. Remind the students that when the coefficient of a variable is 1, the 1 does not need to be written (Identity Property of Multiplication).

Is $4(b + 6)$ equal to $4b + 24$? Yes; b multiplied by 4 equals $4b$, and 6 multiplied by 4 equals 24.

Complete the equation: $4b + 24 = 4(b + 6)$.

- Write the following expressions for display. Instruct the students to apply the Distributive Property and to write the equivalent expressions. Encourage them to use the GCF when factoring out common factors. Give guidance as needed. Continue to display the answers.

$3(x + 2)$ $3x + 6$ $5(4 + y)$ $20 + 5y$

$4(n + 1)$ $4n + 4$ $18 + 6a$ $6(3 + a)$

$14x + 21$ $7(2x + 3)$ $9x + 24$ $3(3x + 8)$

- Write "$a + a + a$" for display.

How could you write the repeated addends using multiplication? $3a$; The coefficient 3 tells how many times the variable a is multiplied.

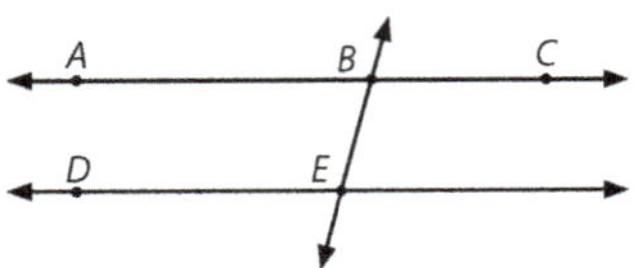

LESSON 100

- Write "$x \cdot x \cdot x$" for display.

  How could you write the repeated factors using exponents? x^3; The exponent tells how many times the factors are repeated.

- Write the following expressions and direct the students to write the equivalent expressions.

  $y + y + y + y$ *4y* $x \cdot x$ *x^2*

Understanding consistency

- Guide the students to make an **analogy** to help them distinguish between consistency and biblical truth.

- Redirect attention to the equation $5(4 + y)$ $= 20 + 5y$ previously written for display. Point out that $20 + 5y$ is an equivalent expression to $5(4 + y)$.

  When we use the Distributive Property to write an equivalent expression, do you think we would get the same answer if we repeat the procedure? *yes*

- Point out that when we get the same answer each time we repeat the math procedure, we say that our answer is consistent. But getting a consistent answer does not mean that the answer is biblically true.

- Display the compass or picture of one.

- Explain that a map made by someone using a broken compass, for example, will deceive people. It may be consistent because it always gives the same confusing directions, but it is still wrong because it doesn't match God's unchangeable truth—reality.

- Remind the students that each answer or conclusion a model provides must be what God reveals to us in nature or in Scripture.

  When we use the Distributive Property to write equivalent expressions, what does it mean that our answers are consistent? *It means that we get the same answer each time if we follow the correct procedures.*

Use the diagram to name the geometric figure.

1. two collinear points *sample answers: A and C; B and E; D and E*
2. three noncollinear points *sample answers: D, B, and A; B, E, and C*
3. three lines
   $\overleftrightarrow{AC}$, $\overleftrightarrow{BE}$, and $\overleftrightarrow{DE}$
4. a point shared by two lines *E or B*
5. two different names for $\overleftrightarrow{AC}$
   $\overleftrightarrow{CA}$, $\overleftrightarrow{AB}$, $\overleftrightarrow{BC}$

Write *hexagon, octagon, pentagon, quadrilateral,* or *triangle* to classify the polygon.

6. 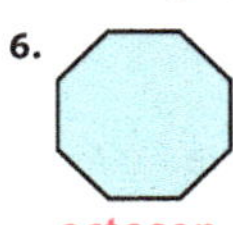   *octagon*
7. 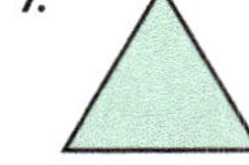   *triangle*
8. *quadrilateral*
9. 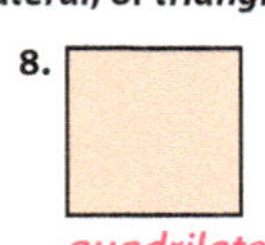   *pentagon*
10. 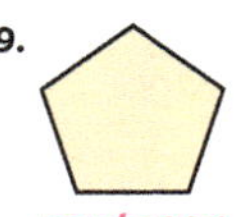   *hexagon*

Write *equilateral, isosceles,* or *scalene* to classify the triangle.

11.    *scalene*
12. 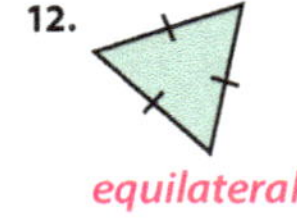   *equilateral*
13. 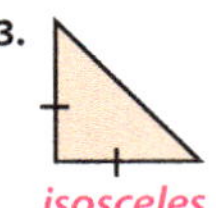   *isosceles*

Solving equations using inverse operations

- Guide a **review** to help the students solve equations using inverse operations.

- Write "$m - 3 = 17$" for display.

  How can you solve the equation? *I can add 3 to both sides of the equation.*

  Why is it necessary to apply the inverse operation to both sides of the equation? *I apply the inverse operation to both sides of the equation to keep the expressions equal or balanced while isolating the variable.*

- Choose a student to demonstrate solving the equation. Remind the students that the value of the variable (m) is the solution to the equation. *$m - 3 + 3 = 17 + 3$; $m = 20$*

  Is 20 the correct solution? *Yes; when 20 is substituted for m in the original equation ($m - 3 = 17$), the mathematical statement is true.*

- Write the following equations for display and direct the students to solve them, showing their work for each solution. Choose students to write their solutions for display and to explain them.

$9 + a = 12$
$a + 9 - 9 = 12 - 9$
$a = 3$

$y - 5 = 13$
$y - 5 + 5 = 13 + 5$
$y = 18$

$5x = 35$
$\frac{5x}{5} = \frac{35}{5}$
$x = 7$

$n \div 4 = 8$
$n \div 4 \times 4 = 8 \times 4$
$n = 32$

$6 \cdot x = 48$
$6 \cdot \frac{x}{6} = \frac{48}{6}$
$x = 8$

$\frac{y}{7} = 3$
$\frac{y}{7} \cdot 7 = 3 \cdot 7$
$y = 21$

Apply

Student Edition pages 216–17

- Read and explain the directions for pages 216–17. Assist the students as they complete the pages independently.

Daily Review

- Students should complete Chapter 10, section g.

Assess

Quiz 1

- Use the **summative assessment** to evaluate the students' progress at this point in the chapter.

NOTES

Student Edition pages 218–19
Daily Review Chapter 10, section *h*

OBJECTIVES

- Solve problems using the $d = rt$ formula.
- Evaluate mathematical models in light of reality. **BWS**
- Complete a table using the formula $d = rt$.
- Create a line graph relating to the formula $d = rt$.

BIBLICAL WORLDVIEW SHAPING

- **Modeling (Evaluate):** Math models based on God's revealed truth represent reality well.

TEACHER RESOURCES

- **54** *Graphing an Equation* (for the teacher and for each student)

Engage

- Direct the students to do a **Turn and Talk** to help prepare them for the content of this lesson.

  If you walked 2 mi with your friends in $\frac{1}{2}$ hr, how fast did you walk?

Instruct

Solving distance problems

- **Ask questions** to guide the students in applying the distance formula.

- Write the words *distance*, *rate*, and *time* for display. Point out that rate often refers to speed.

  How would you define these words? Distance is how far someone travels, rate is how fast someone travels, and time is how long it takes to travel a distance.

  Write for display "*distance = rate × time*." Explain that a formula is a method of calculating information in a way that works every time.

- Write "$d = r \times t$" below distance = rate × time. Point out that d represents distance, r represents rate, and t represents time.

  How can you find the distance if you are given the rate and time? I can multiply the rate times the time.

- Read aloud the following word problem.

  If a car travels at a rate of 55 mph for 2 hr, how far will it travel? 110 mi

  How can you find how far the car will travel? The distance can be found by multiplying the rate (speed) times the time, using the formula $d = r \times t$.

  What equation can you write? $d = 55 \times 2$

  Write the equation for display and choose a student to solve it without labeling the answer. $d = 110$

  What label should you write for the answer? The label should be miles

  because "how far" indicates a distance, and the rate is given in miles per hour.

  Label the answer: $d = 110$ mi.

  What distance will the car travel? 110 mi

- Follow a similar procedure for the following word problems. Point out that the word "speed" and the phrase "88 km/hr" both represent rate.

  How far will a car travel in 3 hr if its average speed is 65 mph? $d = 65 \times 3$; $d = 195$ mi

  A car traveled 88 km/hr for 2 hr. How far did it travel? $d = 88 \times 2$; $d = 176$ km

- Explain that you can also use the formula $d = r \times t$ to find the time it takes to

---

Distance = Rate × Time

Do math models represent reality?

Scientists and mathematicians have discovered many formulas. A formula is a mathematical statement that is used to solve a problem. A formula allows you to substitute known information into an equation to solve for an unknown part.

$$\textbf{distance} = \textbf{rate} \times \textbf{time} \text{ or } \textbf{d} = \textbf{r} \times \textbf{t}$$

d (distance): how far   *r* (rate of speed): how fast   *t* (time): how long

When finding the rate of speed, two different units (distance and time) are being compared. The label will use both units of measurement.

$$r = \frac{d}{t} = \frac{100 \text{ mi}}{2 \text{ hr}} = 50 \text{ miles per hour (mph)}$$

1. Read the question carefully to find the unknown.
2. Substitute the known information for the variables in the formula.
3. Solve for the unknown variable.
4. Label your answer appropriately.

Use related facts to solve for the unknown.

product = factor × factor
product ÷ factor = factor

How many miles would a truck travel in 6 hr at an average speed of 55 mph?	What is the average speed if a truck traveled 330 mi in 6 hr?	How many hours would it take a truck to travel 330 mi at an average speed of 55 mph?
$d = r \times t$ $\quad d = 55 \times 6$ $d = \underline{\quad}$ $\quad d = 330$ $r = 55$ mph $t = 6$ hr	$d = r \times t$ $\quad 330 = r \times 6$ $d = 330$ mi $\quad 330 \div 6 = 55$ $r = \underline{\quad}$ $\qquad r = 55$ $t = 6$ hr	$d = r \times t$ $\quad 330 = 55 \times t$ $d = 330$ mi $\quad 330 \div 55 = 6$ $r = 55$ mph $\qquad t = 6$ $t = \underline{\quad}$
The truck would travel **330 mi**.	The truck would travel **55 mph**.	It would take **6 hr**.

Exercises

Solve. *Equations may vary.*

1. In 1929 a Zeppelin airship flew 11,247 km from Germany to Japan in just under 102 hr. At what speed was it traveling? *approximately* ***110 km/hr***

2. The Carson family drove at an average speed of 85 km per hour for 4 hr. How many kilometers did they travel? ***340 km***

3. A family took a bicycle trip through the Roosevelt National Forest on their vacation. They cycled a total of 96 mi in 4 days. What is the average number of miles they cycled each day? ***24 mi/day***

4. Grant and his friend Derek biked 33 mi at an average speed of 11 mph. About how long did it take to bike that distance? ***3 hr***

5. If an airship flew for 52 hr at an average speed of 93 km per hour, how many kilometers would it travel? ***4,836 km***

6. Austin's remote-control car can travel at a speed of 22 yd per minute. How far could the car travel in 15 min? ***330 yd***

7. A train traveled 380 mi at an average speed of 95 mph. How long did it take to travel the 380 mi? ***4 hr***

8. Use the information from problem 6 to find how long it would take the remote-control car to travel 1 mi if it continued at the same rate. ***80 min or 1 hr 20 min***

Key Terms

- *distance = rate × time*
- $d = r \times t$

If you travel a distance of 1,500 mi at a speed of 100 mph, how many hours will you have traveled? **15 hr**

Draw a diagram to show the relationship of the *number of miles traveled* to the *amount of time traveled.*

100 mi	100 mi	100 mi	100 mi	100 mi	100 mi	100 mi	100 mi	100 mi	100 mi	100 mi	100 mi	100 mi	100 mi	100 mi
1 hr	1 hr	1 hr	1 hr	1 hr	1 hr	1 hr	1 hr	1 hr	1 hr	1 hr	1 hr	1 hr	1 hr	1 hr

1. Use the formula for distance to find the unknown time.
2. Substitute the known information into the formula to make an equation.
3. Use the inverse operation to find the unknown part.

$$d = r \times t \qquad\qquad 1{,}500 = 100 \times t$$

$$t = \underline{\quad}$$
$$r = 100 \text{ mph}$$
$$d = 1{,}500 \text{ mi}$$

$$\frac{1{,}500}{100} = \frac{\overset{1}{100}t}{\underset{1}{100}}$$

$$15 = t$$

Travel time will be **15 hr**.

Find the number of hours it will take to travel a distance of 1,500 mi at each given rate of speed.

9.

rate (mph)	100	200	300	400	500	600
time (hours)	15	*7.5*	*5*	*3.75*	*3*	*2.5*

Use the table in problem 9 to complete the line graph.

10.

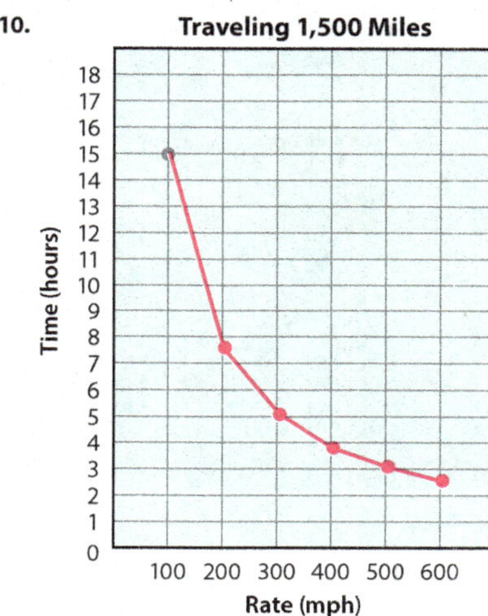

11. Make a statement to summarize your findings from problems 9–10.

12. Do math models represent reality?
Math models based on God's revealed truth in nature represent reality well.

11. *As the rate of speed increases, the time required to travel a given distance decreases.*

How can you solve this equation? Since division is the inverse operation of multiplication, I can divide both sides of the equation by 20 to isolate the variable (t).

Choose a student to solve the equation without labeling the answer. $\frac{400}{20} = \frac{20t}{20}$; $t = 20$

What label should you write for the answer? The label should be days because "how long did it take" indicates an amount of time and the rate is given in "miles per day."

Label the answer: $t = 20$ days.

How long would it have taken for Joseph, Mary, and Jesus to travel from Bethlehem to Egypt? 20 days

- Point out that you can also use the formula $d = r \times t$ to find the rate at which someone or something traveled when given the distance and the time.

Henri Giffard built an airship in 1852. If the airship flew 17 mi in 3.4 hr, how many miles an hour did it travel? 5 mph

What information is given in this problem? The distance traveled was 17 mi, and the time it took to travel the distance was 3.4 hr.

- Direct the students to rewrite $d = r \times t$, substituting the given information for the appropriate variables.

What equation did you write? $17 = r \times 3.4$ or $17 = 3.4r$

Write "17 = 3.4r" for display. Point out that the Commutative Property of Multiplication allows them to write $r \times 3.4$ as $3.4r$.

How can you solve this equation? I can divide both sides of the equation by 3.4 to isolate the variable (r).

Choose a student to solve the equation without labeling the answer. $\frac{17}{3.4} = \frac{3.4r}{3.4}$; $r = 5$

- Explain that when finding the rate, two different units of measure are being compared in the division, and the division bar can be read *per* (e.g., $\frac{17 \text{ mi}}{3.4 \text{ hr}}$ can be read "17 mi per 3.4 hr"). Therefore, both units of measure are used in the label (e.g., miles per hour [mph] or kilometers per hour [km/hr]).

---

travel a given distance when given the rate (speed).

The distance that Joseph, Mary, and the young child, Jesus, traveled from Bethlehem to Egypt may have been about 400 mi. If they had traveled 20 mi per day, how long would it have taken them to travel from Bethlehem to Egypt? 20 days

- Direct attention to the formula $d = r \times t$.

What is the possible distance that Joseph, Mary, and Jesus traveled? 400 mi

Write "$d = 400$ mi" for display.

At what rate would Joseph, Mary, and Jesus have traveled? 20 mi per day

Write "$r = 20$ mi per day."

What is the word problem asking you to find? how long it may have taken Joseph, Mary, and Jesus to travel from Bethlehem to Egypt

Which variable in the formula will also be the variable representing the unknown value in the equation for solving the word problem? In the formula, t represents time, and I need to find the amount of time it may have taken for Joseph, Mary, and Jesus to travel from Bethlehem to Egypt.

- Choose a student to rewrite the formula substituting the known information for d (distance) and r (rate). $400 = 20 \times t$, $400 = 20 \cdot t$, or $400 = 20t$

LESSON 101

What label should you write for the answer? *miles per hour*

Label the answer: "$r = 5$ mph." Explain that *mph* is the abbreviation for "miles per hour."

At what rate did the airship travel? *5 mph*

- Direct the students to solve the problem posed by the Turn and Talk at the beginning of the lesson. *$d = r \times t$; $2 = r \times \frac{1}{2}$ or $2 = \frac{1}{2}r$; $r = 4$; I walked 4 mph.*

Evaluating models

- Direct attention to Student Edition page 218 and guide a **discussion** to answer the chapter essential question.

- Direct attention back to the airplane illustration on the chapter opener on Student Edition page 203.

  Is the model for the duration of the New York to London flight always representative of reality? Explain. *No; the estimated flight time is an average. If there is a significant headwind or tailwind, the time could be slower or faster.*

- Remind the students that models that reflect God's truth as revealed in nature are true to reality. Models made without acknowledging biblical truth may be consistent but may not align with truth revealed in nature or the Bible. Such models would not be good models because they do not represent reality well. They may be based on the wrong assumptions.

  Do math models represent reality? *Math models based on God's revealed truth in nature represent reality well.*

Completing a table & a line graph using the formula $d = r \times t$

- **Model** writing equations to help the students complete a table using a formula.

- Distribute a *Graphing an Equation* page to each student and display your copy. Review the formula displayed above the table.

- Explain that a log or a table can be used to calculate distance, rate, or time. Then the data can be graphed.

Use the circle graph to answer the question.

Land Owned by the US Government

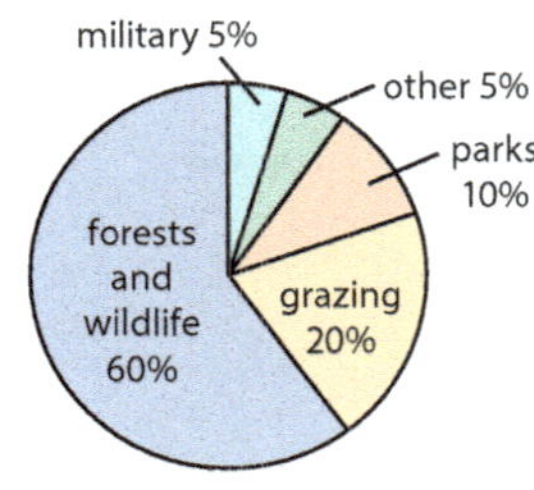

1. What is the sum of the percents shown on this graph? *100%*

2. Which category shows the greatest percentage of land owned by the federal government? *forests and wildlife*

3. What percentage of land owned by the government is used for grazing and parks? *20% + 10% = 30%*

4. Which two categories together make up about one-fourth of federal land? *military (or other) and grazing*

Solve.

5. $\begin{array}{r} 8,374 \\ 6,985 \\ +\,4,876 \\ \hline \mathit{20,235} \end{array}$ 6. $\begin{array}{r} 45,799 \\ +\,86,964 \\ \hline \mathit{132,763} \end{array}$

7. $\begin{array}{r} 900,000 \\ -\,318,974 \\ \hline \mathit{581,026} \end{array}$ 8. $\begin{array}{r} 60,005 \\ -\,32,057 \\ \hline \mathit{27,948} \end{array}$

Solve. Round the decimal quotient to the nearest hundredth.

9. $84\overline{)420}$ *5*

10. $56\overline{)1,975}$ *35.267 ≈ 35.27*

Point out that the rates in the table are given in miles per hour. By completing the table, students will find out how much time it would take for them to travel 200 mi at each of the given rates.

What equation can you write to find how much time it would take to travel 200 mi at the rate of 10 mph? *$200 = 10 \times t$*

- Write "$200 = 10t$" for display. Choose a student to solve the equation and label the answer. *$\frac{200}{10} = \frac{10t}{10}$; $t = 20$ hr*

  What does $t = 20$ hr mean? *Traveling a distance of 200 mi at a rate of 10 mph would take 20 hr.*

  Direct the students to write "20" in the table as you write it for display.

- Follow a similar procedure to guide the students in completing the table. Instruct them to round decimal answers to the nearest tenth of an hour. *10; 6.7; 5; 4; 3.3; 2.9; 2.5; 2.2; 2*

- Model plotting and connecting points to help the students complete a line graph.

- Direct attention to the grid on the page. Point out that the numbers along the left side represent time in hours, and the numbers along the bottom represent rate (speed) in miles per hour.

  What information is given first in the table? *10 mph and 20 hr*

- Explain that each rate and its corresponding time are coordinates that can be graphed similar to graphing coordinates on a coordinate plane. Instruct the students to locate the point where the rate of 10 mph and the time of 20 hr intersect and to draw a dot at the point of intersection. Demonstrate.

- Guide the students in plotting the points for the rest of the information given in the table and in connecting the points.

  What type of graph did you make? a line graph

  What do you think the line graph illustrates? sample answer: The curved line shows that as the rate of speed increases, the time required to travel the 200 mi decreases.

Apply

Student Edition pages 218–19

- Read and explain the directions for pages 218–19. Assist the students as they complete the pages independently.

Daily Review

- Students should complete Chapter 10, section *h*.

NOTES

CHAPTER REVIEW

OBJECTIVES

- Write algebraic expressions and equations.
- Simplify algebraic expressions using addition and multiplication properties: Commutative Property, Associative Property, and Distributive Property.
- Evaluate expressions using substitution.
- Solve equations using inverse operations.
- Write an equation with a variable to solve a word problem.
- Determine whether a given value is a solution to the inequality.

The Chapter Review offers an opportunity for students to discuss the concepts they have learned in the chapter. They may work collaboratively or independently as you review concepts. Circulate among the students, giving individual help as needed. Students who demonstrate proficiency with the discussion, the modeling, and the Student Edition pages are ready for the Chapter Test. Students who encounter difficulties with the review concepts would benefit from additional coaching and practice before testing.

Writing algebraic expressions & equations

- Write for display: "Joseph lost 2 of his pens."

How many pens did Joseph have to begin with? The amount is unknown.

How can you represent an unknown value? A variable can be used to write an unknown value.

When a mathematical expression includes a variable, is it an algebraic or a numerical expression? an algebraic expression; Algebraic expressions include a variable rather than just numbers as in a numerical expression.

If the letter p represents the number of pens Joseph had before losing 2, what algebraic expression can you write to express the number of pens Joseph has now? $p - 2$

Why is $p - 2$ an algebraic expression rather than an algebraic equation? Algebraic expressions are made up of numbers, operation signs, and variables, and an algebraic equation requires an equal sign to show that two algebraic expressions have the same value.

What words would indicate that an equal sign is needed when writing a word phrase or sentence mathematically? "is" or "equals"

How do you know what to write on each side of the equal sign in an algebraic equation? The word "is" or "equals" separates the two equal word phrases in a sentence, indicating what should be written on each side of the equal sign.

- Write the following word phrases and sentences for display. Choose students to tell whether the phrases and sentences represent an expression or an equation. Instruct each student to write the algebraic expression or equation for the word phrase or sentence. When writing

---

CHAPTER REVIEW

Write the algebraic expression. Identify what the variable represents in the expression.

1. Eli read 3 times as many books as Nolan.

2. Brynn broke 4 water glasses. *g − 4; g represents the total number of glasses.*

3. The florist divided the roses among 5 vases. *r ÷ 5; r represents the total number of roses.*

4. The product of a number and 5. *5n*

5. Two less than 20 times n. *20n − 2*

6. Sixty more than n, divided by 3. *(n + 60) ÷ 3*

Evaluate the expression. Let $n = 4$.

7. $n + 7 - 3$ 8. $5n ÷ 2$

9. $3 + (n ÷ 2)$ 10. $3n + 5$

11. $(2.7 \cdot n) - 3$ 12. $100 - 12n$

Simplify the expression. *Order of terms may vary.*

13. $3(6x)$ *18x*   14. $8x + (3 + 7x)$

15. $3 + x + 6$ 16. $(6 + 2x) + 3$

17. $y + y + y$ *3y*   18. $2(8x)$ *16x*

Simplify the expression by using the Distributive Property. *Order of terms may vary.*

19. $3(8 + 2a)$ *3(8) + 3(2a) = 24 + 6a*

20. $6(n + 2)$ *6(n) + 6(2) = 6n + 12*

21. $5(7x + 5.1)$ *5(7x) + 5(5.1) = 35x + 25.5*

1. *3b; b represents the number of books Nolan read.*

7. *4 + 7 − 3 = 11 − 3 = 8*   8. *(5 · 4) ÷ 2 = 20 ÷ 2 = 10*

9. *3 + (4 ÷ 2) = 3 + 2 = 5*   10. *(3 · 4) + 5 = 12 + 5 = 17*

11. *(2.7 · 4) − 3 = 10.8 − 3 = 7.8*   12. *100 − (12 · 4) = 100 − 48 = 52*

Complete the table, using the given values to evaluate the expression.

22.

b	$2b + 5$
6	*17*
12	*29*
25	*55*
49	*103*

23.

x	$\frac{x}{3} - 2$
9	*1*
15	*3*
21	*5*
33	*9*

24.

n	$n^2 + 9$
5	*34*
9	*90*
12	*153*
20	*409*

Write an equation for the sentence.

25. 8 less than a number is 14. *n − 8 = 14*

26. The quotient of a number divided by 2 equals 9. *n ÷ 2 = 9*

27. The sum of 2 times a number and 8 is 14. *2n + 8 = 14*

14. *8x + 7x + 3 = 15x + 3*   15. *x + 3 + 6 = x + 9*

16. *2x + 6 + 3 = 2x + 9*

220   Chapter 10

State whether the given value is the solution to the equation.

28. $n - 12 = 5$ if $n = 15$ *no*

29. $\frac{x}{6} = 7$ if $x = 42$ *yes*

30. $8a = 56$ if $a = 7$ *yes*

31. $9 + n = 14$ if $n = 4$ *no*

32. $m \div 8 = 2$ if $m = 16$ *yes*

33. $5 \cdot a = 50$ if $a = 10$ *yes*

Solve. Write the inverse operation used to solve. Check the solution.

34. $\frac{x}{4} = 9$ **35.** $n - 8 = 61$

36. $4a = 56$ **37.** $n + 3 = 32$

38. $x - 1.6 = 1.4$ **39.** $y \div 5 = 25$

40. $b + 5 = 48$ **41.** $7n = 85.4$

State whether the given value is a solution to the inequality.

42. $y < {}^-1$ if $y = {}^-2$ *yes*

43. $x > 2$ if $x = 1$ *no*

44. $y < 5$ if $y = 2$ *yes*

34. *x = 36; multiplication*

35. *n = 69; addition*

36. *a = 14; division*

37. *n = 29; subtraction*

38. *x = 3; addition*

39. *y = 125; multiplication*

40. *b = 43; subtraction*

41. *n = 12.2; division*

45.
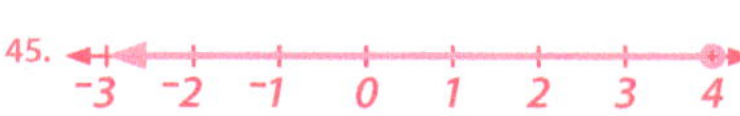

Draw a number line to illustrate the inequality.

45. $b < 4$ **46.** $w > 1$

Write an equation with a variable. Solve.

47. One-third of the rancher's cattle are calves. If he puts 2 calves in each of his 20 stalls, how many head of cattle does the rancher have?
Equations may vary.

46.

47. $\frac{1}{3}x = 2 \cdot 20$

    $3 \cdot \frac{1}{3}x = 2 \cdot 20 \cdot 3$

       x = 120 head of cattle

---

a multiplication expression, such as three times a number, the coefficient (multiplier) precedes the variable ($3n$).

a number divided by 6 *expression;* $n \div 6$ *or* $\frac{n}{6}$

the sum of a number and 2 *expression;* $n + 2$

the product of a number and 3 *expression;* $3n$

a number decreased by 4 *expression;* $n - 4$

9 more than a number is 14. *equation;* $n + 9 = 14$

5 less than a number equals 12. *equation;* $n - 5 = 12$

2 times a number equals 8. *equation;* $2n = 8$

A number divided by 3 is 4. *equation;* $n \div 3 = 4$ *or* $\frac{n}{3} = 4$

- Write "2 times a number, divided by 3."

  How many operations are indicated in the word phrase? 2 operations; "Times" indicates multiplication, and "divided by" indicates division.

  Direct the students to write an algebraic expression for the word phrase. $2n \div 3$ *or* $\frac{2n}{3}$

- Instruct the students to write algebraic expressions for the following word phrases. Remind them that using parentheses is helpful when writing expressions with more than 1 operation.

---

a number decreased by 7, multiplied by 3
$(n - 7) \times 3;\ 3 \times (n - 7);$ *or* $3(n - 7)$

a number divided by 5, increased by 2
$(n \div 5) + 2;\ n \div 5 + 2;$ *or* $\frac{n}{5} + 2$

Simplifying algebraic expressions

- Write $3(2n)$ for display.

  What property can you use to simplify the expression? I can use the Associative Property of Multiplication because the grouping of the factors can be changed without changing the value of the expression.

  Choose a student to apply the Associative Property to $3(2n)$ and then simplify the expression. $(3 \times 2) \times n;\ 6n$

- Follow a similar procedure for the following expressions. Like terms are combined when simplifying an expression such as $8x + 2 + 5x$.

  $8x + 2 + 5x$ *Commutative Property of Addition; The order of the addends can be changed without changing the value of the expression;* $8x + 5x + 2;\ 13x + 2.$

  $4(3x + 2)$ *Distributive Property; Each addend can be multiplied by 4 without changing the value of the expression;* $(4 \times 3x) + (4 \times 2);\ 12x + 8.$

- Write "$s + s + s + s$" for display.

  What is the simplified expression for $s + s + s + s$? 4s; Multiplication is repeated addition.

- Direct the students to simplify the following expressions. Choose students to write the answers for display and discuss the properties that were used to simplify each expression. Point out that both the Commutative and the Associative Properties of Addition were used to simplify $8a + (5 + 6a)$.

  $y + y$ *2y*

  $4(3y)$ *12y*

  $6 + 2y + 3$ *2y + 9*

  $8a + (5 + 6a)$ *14a + 5*

  $2(5x + 7)$ *10x + 14*

  $4(3 + y)$ *12 + 4y*

Evaluating expressions

- Write the following values and expressions for display. Choose students to rewrite the expressions, substituting the given value for the variable, and to evaluate the expressions.

$a = 4$

$(20 + a) \div 3$
$(20 + 4) \div 3$
$24 \div 3 = 8$

$b = 3$

$5b - 9.7$
$(5 \times 3) - 9.7$
$15 - 9.7 = 5.3$

$c = 2$

$34 - (14 \div c)$
$34 - (14 \div 2)$
$34 - 7 = 27$

Solving equations using inverse operations

- Write "$n + 9 = 14$"; "$m - 5 = 12$"; "$2a = 8$"; and "$\frac{x}{3} = 4$" for display.

- How can you solve these algebraic equations? I can identify what operation is being performed on the variable and then apply the inverse operation to both sides of the equation.

  Why is it necessary to apply the inverse operation to both sides of the equation? to keep the expressions equal or balanced while isolating the variable

- Choose students to demonstrate solving the equations by applying the inverse operation. Ask each student to name the operation in the equation and the inverse operation used to find the solution. $n = 5$, subtraction; $m = 17$, addition; $a = 4$, division; $x = 12$, multiplication

  How can you check each of the solutions? I can substitute the solution for the variable in the original equation. If the equation is a true mathematical statement, the solution is correct.

- Write the following equations for display. Choose students to solve the equations and check the solutions while the other students find the solutions and check them.

$a + 7 = 15 \quad n - 3 = 7 \quad 8a = 16 \quad \frac{x}{7} = 3$
$a = 8 \qquad\quad n = 10 \qquad a = 2 \qquad x = 21$

Writing an equation with a variable

- Read aloud the following word problem.

  Samantha has 48 coins in her collection. This amount is 16 coins fewer than Mia has in her collection. How many coins does Mia have? 64 coins

- Direct the students to write what they need to find and the given information. Reread the problem as needed. the number of coins that Mia has; Samantha has 48 coins, and Samantha has 16 fewer coins than Mia.

- Write for display these two sentences: "Samantha has 48 coins" and "Samantha has 16 fewer coins than Mia."

  Guide the students in using the information in the sentences to write an algebraic equation. (See Lesson 99.) $48 = m - 16$

- Direct the students to solve the equation. $48 + 16 = m - 16 + 16; m = 64$

  How many coins does Mia have? 64 coins

  How do you know that 64 coins is the correct answer? 64 substituted for m in the original equation is a true mathematical statement; $48 = 64 - 16$.

- Guide the students in determining what information in the multistep word problem on Student Edition page 221 should be represented using a variable. Then guide them in solving the word problem. $x = 120$ head of cattle

Determining a solution to an inequality

  What is an inequality? a mathematical sentence in which two expressions are not equal

  Inequalities have an infinite number of solutions.

- Choose students to draw number lines to illustrate the solutions for $x > 6$, $y < {}^-1$, and $x > 3$. The number lines will vary, but the given value should be graphed using an open (unshaded) circle, and the part of the number line showing the solutions should be shaded. (See Lesson 99.)

- Write the following inequalities. Choose students to tell whether the given value is a solution to the inequality and have them explain their answer.

  $x > 2$ if $x = 7$ Yes; 7 is greater than 2.
  $y < {}^-2$ if $y = 1$ No; 1 is greater than ${}^-2$.
  $n < 5$ if $n = 6$ No; 6 is greater than 5.
  $x > {}^-1$ if $x = 1$ Yes; 1 is greater than ${}^-1$.

Student Edition pages 220–21

- Read and explain the directions for pages 220–21. Assist the students as they complete the pages independently.

LESSON 103

Student Edition pages 222–24

CHAPTER 10 TEST

CUMULATIVE REVIEW

CONCEPT REVIEW

- Reading and interpreting a bar graph (Lesson 41)
- Identifying plane figures when given the characteristics (Lessons 56, 58)
- Determining the perimeter of a figure (Grade 5)
- Solving a fraction word problem (Chapter 4)
- Adding, multiplying, and dividing fractions (Chapters 5, 7, 8)
- Finding a missing factor (Grade 5)
- Identifying the reciprocal of a fraction (Lesson 75)
- Multiplying and dividing decimals (Chapters 7, 9)
- Determining the value of an expression (Lessons 94–95)
- Determining the lowest term of a fraction or a mixed number (Lesson 35)

- To prepare the students for the format of achievement tests, instruct them to work on a separate sheet of paper, if necessary, and to mark their answers on the *Cumulative Review Answer Sheet*.

Student Edition pages 222–24

The Cumulative Review provides additional practice of previously learned concepts. These pages may be completed during this lesson or anytime after this lesson, since they require limited or no teaching.

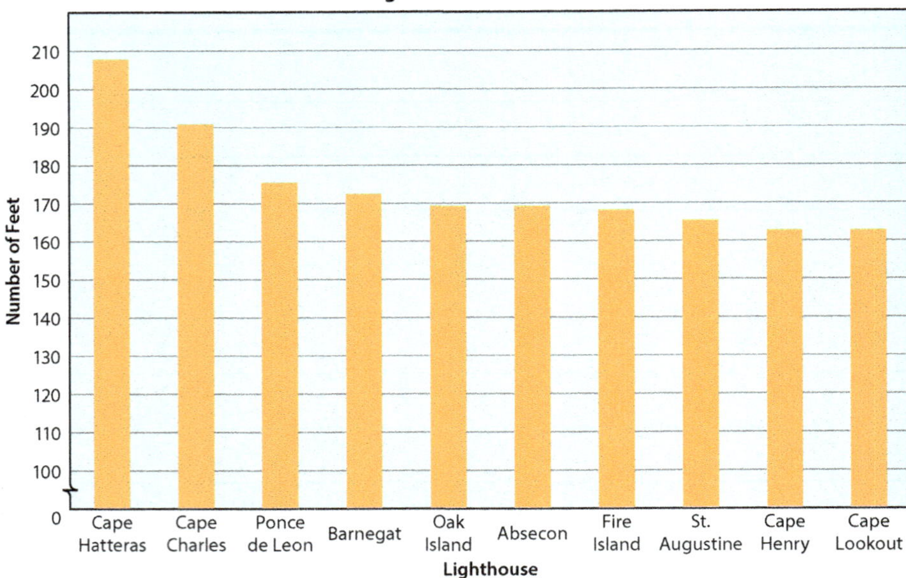

Use the bar graph to find the answer.

1. This bar graph displays which of the following?
   - **A.** the amount of material used to construct each lighthouse
   - **B.** the height of 10 lighthouses
   - **C.** the number of tourists that visit lighthouses
   - **D.** the number of ships saved by each lighthouse

2. Estimate the difference in height between the tallest and shortest lighthouses on the graph.
   - **A.** 10 ft
   - **B.** 20 ft
   - **C.** 30 ft
   - **D.** 40 ft

3. Which expression shows the height of the Ponce de Leon lighthouse?
   - **A.** 1.75×10^1 ft
   - **B.** 1.75×10^2 ft
   - **C.** 17×10^2 ft
   - **D.** 1.7×10^3 ft

4. According to the graph, which of the following is true about Cape Henry?
   - **A.** It is approximately 180 ft tall.
   - **B.** It is taller than the Oak Island lighthouse.
   - **C.** It is the same height as the Cape Lookout lighthouse.
   - **D.** It is the tallest lighthouse in the United States.

5. About how much taller is Barnegat than Absecon?
   - **A.** less than 10 ft
   - **B.** about 10 ft
   - **C.** more than 20 ft
   - **D.** about 20 ft

6. Which lighthouse measures 165 ft tall?
   - **A.** Barnegat
   - **B.** Ponce de Leon
   - **C.** Cape Charles
   - **D.** St. Augustine

222   Chapter 10

Choose the answer.

7. The Cape Henry lighthouse tower has an octagonal shape. How many sides does the lighthouse have?

   A. 4   **C.** 6

   B. 5   **D.** 8

8. What specific name is given to a quadrilateral with 4 congruent sides and 4 right angles?

   A. rhombus   **C.** square

   B. rectangle   **D.** parallelogram

9.

26.8 ft

17.3 ft

What is the perimeter of the figure?

   A. 44.1 ft   **C.** 100.3 ft

   B. 88.2 ft   **D.** 122.4 ft

10. What shapes are created when a diagonal line is drawn in a rectangle?

   A. congruent equilateral triangles

   B. congruent right triangles

   C. congruent squares

   D. similar rectangles

11. Three pizzas were ordered for family night. Each pizza had 8 slices. Three-fourths of the pizza was eaten. How many slices of pizza were left?

   A. 6   **C.** 10

   B. 8   **D.** 12

12. Which statement is true?

   A. $1\frac{1}{2} = \frac{3}{4} + \frac{3}{4}$ **C.** $1\frac{1}{2} = \frac{3}{4} \times \frac{3}{4}$

   B. $1\frac{1}{2} = \frac{3}{4} - \frac{3}{4}$ **D.** $1\frac{1}{2} = \frac{3}{4} \div \frac{3}{4}$

13. $7 \times n = 469$

   A. $n = 64$ **C.** $n = 66$

   B. $n = 65$ **D.** $n = 67$

14. $7\overline{)64}$

   A. $8\frac{6}{7}$ **C.** $9\frac{5}{7}$

   B. $9\frac{1}{7}$ **D.** $10\frac{1}{7}$

Cumulative Review   223

Choose the answer.

15. $3\frac{7}{8} + 4\frac{2}{3}$

   A. $7\frac{9}{11}$ **C.** $8\frac{13}{24}$

   B. $8\frac{1}{8}$ **D.** 9

16. $\frac{3}{9} \times \underline{\quad} = 1$

   A. $\frac{3}{9}$ **C.** $\frac{9}{3}$

   B. $\frac{3}{3}$ **D.** $\frac{9}{9}$

17. $\frac{1}{3} \div \frac{1}{2}$

   A. $\frac{2}{3}$ **C.** $1\frac{1}{2}$

   B. $\frac{1}{6}$ **D.** 2

18. 1.63×100

   A. 0.0163 **C.** 16.3

   B. 0.163 **D.** 163

19. $1.5\overline{)4.05}$

   A. 0.27 **C.** 27

   B. 2.7 **D.** 270

20. $3^2 + 2^3$

   A. 12 **C.** 36

   B. 17 **D.** 72

21. $3 + 7 \times 4 - 8$

   A. 23 **C.** 45

   B. 30 **D.** 80

22. $4 \times (7.8 - 3.3) \div 2$

   A. 4 **C.** 9.8

   B. 9 **D.** 12

23. Rename $17\frac{15}{10}$ to lowest terms.

   A. $17\frac{1}{2}$ **C.** $18\frac{3}{4}$

   B. $18\frac{1}{2}$ **D.** 19

24. Which fraction is in lowest terms?

   A. $\frac{4}{10}$ **C.** $\frac{7}{14}$

   B. $\frac{6}{8}$ **D.** $\frac{9}{16}$

25. What is the sum of $\frac{3}{4}$, $\frac{5}{8}$, and $\frac{1}{2}$?

   A. 1 **C.** $1\frac{7}{8}$

   B. $1\frac{1}{2}$ **D.** $2\frac{1}{4}$

Lesson Plan Overview

<table>
<tr><td colspan="6">Chapter 11: Perimeter & Area</td></tr>
<tr><th>Day</th><th>Lesson</th><th>Student Edition Pages</th><th>Teacher Edition Pages</th><th>Teacher Resources & Additional Materials</th><th>Topics, Skills & Biblical Worldview Shaping</th></tr>
<tr>
<td>105</td><td>104</td><td>225–27</td><td>225–27b</td>
<td>Additional Materials
• ruler</td>
<td>Topic: Perimeter
Skills: calculating the unknown length of a side of a polygon, solving an algebraic expression to find the perimeter of a rectangle</td>
</tr>
<tr>
<td>106</td><td>105</td><td>228–30</td><td>228–30a</td>
<td>Teacher Resources
• 55 Finding the Circumference
Additional Materials
• cylindrical objects
• rulers
• calculators
• 12 in. lengths of string</td>
<td>Topic: Circumference
Skills: calculating the circumference of a circle using a formula, calculating the diameter of a circle given the circumference
BWS: Design</td>
</tr>
<tr>
<td>107</td><td>106</td><td>231–33</td><td>231–33a</td>
<td>Teacher Resources
• 56 Area: Rectangles, Squares & Parallelograms
Additional Materials
• scissors
Assessments
• Chapter 11 Quiz 1</td>
<td>Topic: Area of Parallelograms
Skills: calculating the areas of parallelograms using a formula, calculating the area of a complex figure</td>
</tr>
<tr>
<td>108</td><td>107</td><td>234–35</td><td>234–35b</td>
<td>Teacher Resources
• 57 Area: Triangles
Additional Materials
• rulers
• calculators</td>
<td>Topic: Area of Triangles
Skills: calculating the area of triangles using a formula, calculating the unknown height or base of a triangle
BWS: Design</td>
</tr>
<tr>
<td>109</td><td>108</td><td>236–37</td><td>236–37b</td>
<td>Teacher Resources
• 58 Area: Circles
Additional Materials
• rulers
• calculators</td>
<td>Topic: Area of Circles
Skills: calculating the area of a circle using a formula, estimating the area of a circle</td>
</tr>
<tr>
<td>110</td><td>109</td><td>238–39</td><td>238–39b</td>
<td>Teacher Resources
• 59 Surface Area: Triangular Prism
Additional Materials
• classroom set of 3-dimensional figures or real-world objects
• cereal boxes
• rulers
• transparent tape
Assessments
• Chapter 11 Quiz 2</td>
<td>Topic: Surface Area of Prisms
Skills: calculating the surface area of prisms using formulas, constructing a triangular prism
BWS: Design</td>
</tr>
</table>

Day	Lesson	Student Edition Pages	Teacher Edition Pages	Teacher Resources & Additional Materials	Topics, Skills & Biblical Worldview Shaping
111	110	240–41	240–41b	**Teacher Resources** • 60 *Surface Area* **Additional Materials** • small oatmeal container • 2 sheets of 12" × 18" construction paper • rulers • transparent tape • cylinders • sheets of construction paper • calculators	**Topic:** Surface Area of Cylinders **Skills:** calculating the surface area of a cylinder using formulas, constructing a cylinder net
112	111	242–43	242–43b	**Teacher Resources** • 11 *Graph Paper* • 61 *Floor Plan Activity* • 62 *Floor Plan Grid* **Additional Materials** • samples of floor plans • rulers • calculators • scissors • sheets of 9" × 12" construction paper • glue sticks	**Topic:** Fixed Areas **Skills:** calculating the area and perimeter of a rectangle, calculating the area of a complex figure, relating geometry to real-world situations **BWS:** Design
113	112	244–45	244–45b	**Teacher Resources** • 60 *Surface Area* • 63 *Geometry Review I* • 64 *Geometry Review II* • 65 *Nets* **Additional Materials** • calculators	**Topic:** Chapter Review
114	113	246–48	246–48		**Topic:** Test & Cumulative Review

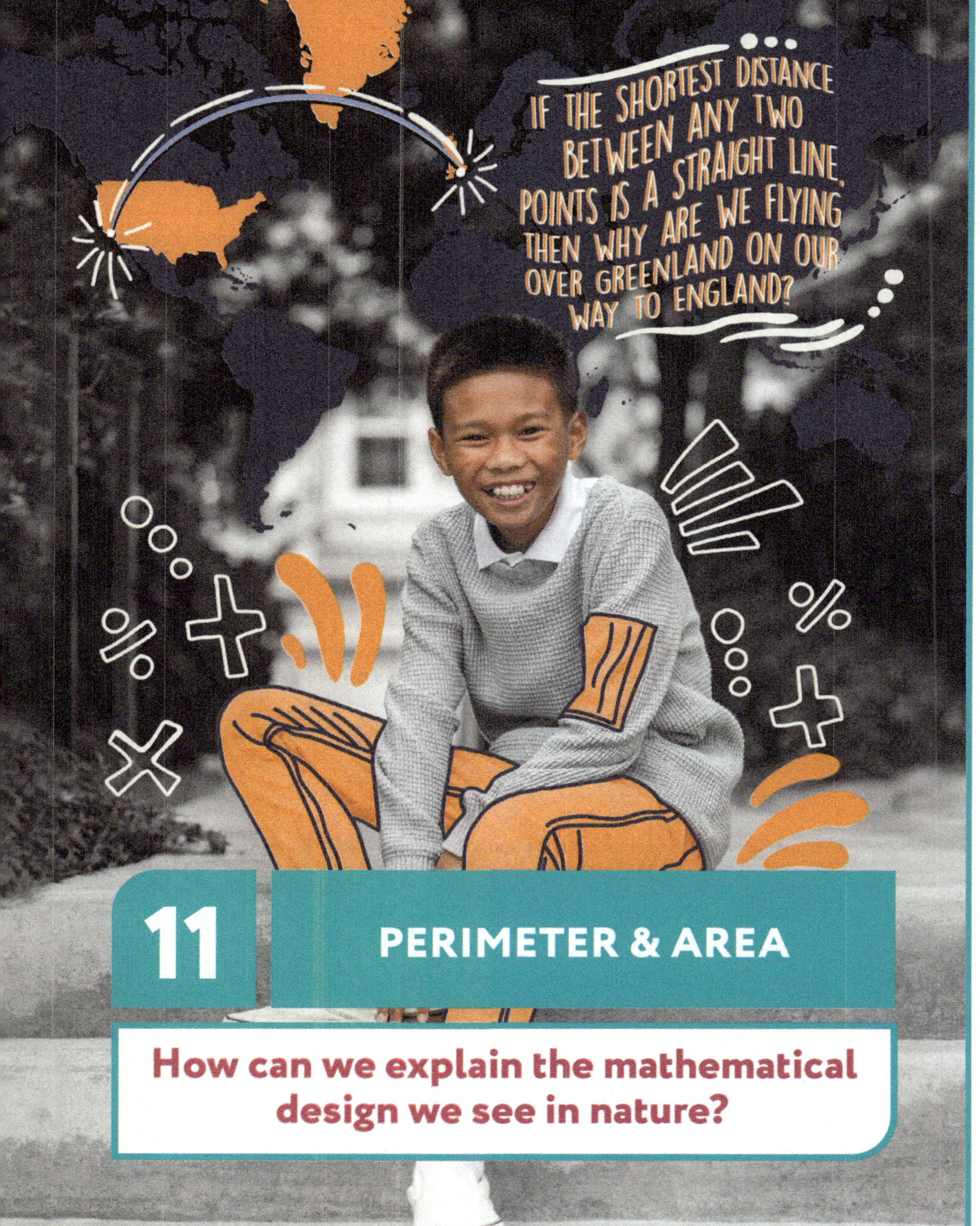

CHAPTER OBJECTIVES

- Calculate the perimeter and area of a polygon by using a formula.
- Calculate the diameter, circumference, and area of a circle.
- Calculate the surface area of 3-dimensional figures by using formulas.
- Relate geometry to real-world situations.
- Defend the claim that God is the source of mathematical design in nature.

11 PERIMETER & AREA

How can we explain the mathematical design we see in nature?

To determine whether your students are prepared for the concepts taught in this chapter, you may wish to consult the preassessment checklist found on TeacherToolsOnline.com.

Make fact practice, both oral and written, part of your daily math routine to help the students with mastery.

Visit AfterSchoolHelp.com for math practice resources, or visit TeacherToolsOnline.com for additional resources to enhance the lessons.

Student Edition pages 225–27
Daily Review Chapter 11, section *a*

OBJECTIVES

- Calculate the perimeter of a polygon by using a formula.
- Calculate the unknown length of a side of a polygon.
- Find the perimeter of a rectangle by solving an algebraic expression.

ADDITIONAL MATERIALS

- a ruler

To reinforce the students' understanding of finding perimeter and area, encourage them throughout the chapter to first write the appropriate formula and then to substitute the number of sides and/or the measurement(s) into the formula.

Engage

- Direct attention to the chapter opener on Student Edition page 225 and lead the students in a **discussion** to explore the chapter essential question, "How can we explain the mathematical design we see in nature?"

  1. Is there a relationship between the diameter and the circumference of a circle?
  2. Do you think the relationship is the same for every circle, no matter how large?
  3. How is the earth like a circle? Does it have a circumference?

Instruct

Calculating perimeter

- Use **guided discovery** to help the students calculate the perimeter of an object.
- Remind the students that a polygon is a closed figure made up of line segments. Choose students to draw a polygon for display and to tell its number of sides.

Perimeter

How can I determine the amount of fence I need for my dog kennel?

Perimeter is the distance around a geometric figure and is represented by *P*.

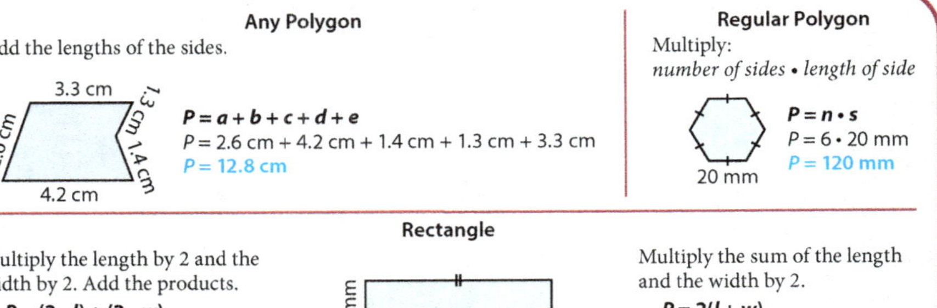

Any Polygon
Add the lengths of the sides.

$P = a + b + c + d + e$
$P = 2.6\ cm + 4.2\ cm + 1.4\ cm + 1.3\ cm + 3.3\ cm$
$P = 12.8\ cm$

Regular Polygon
Multiply:
number of sides • length of side

$P = n \cdot s$
$P = 6 \cdot 20\ mm$
$P = 120\ mm$

Rectangle
Multiply the length by 2 and the width by 2. Add the products.
$P = (2 \cdot l) + (2 \cdot w)$
$P = (2 \cdot 56\ mm) + (2 \cdot 19\ mm)$
$P = 112\ mm + 38\ mm$
$P = 150\ mm$

Multiply the sum of the length and the width by 2.
$P = 2(l + w)$
$P = 2(56\ mm + 19\ mm)$
$P = 2 \cdot 75\ mm$
$P = 150\ mm$

Key Terms
- perimeter
- $P = n \cdot s$
- $P = 2(l + w)$

Exercises

Measure the length of each side to the nearest centimeter. Find the perimeter.

1.   2.   3.

Write an equation to find the perimeter of the figure. Use multiplication if possible. *Equations may vary.*

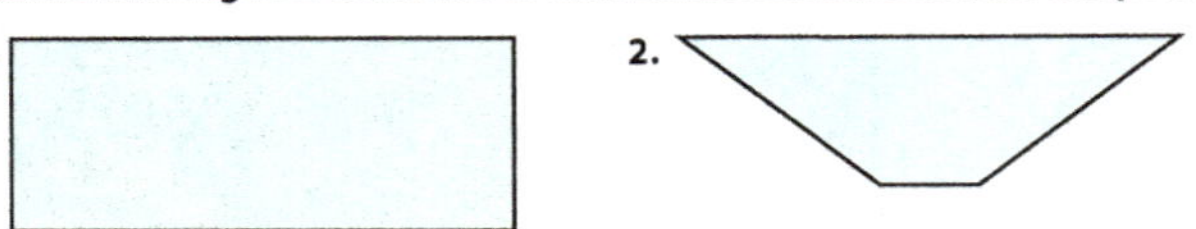

4. $P = 1\frac{1}{2}\ cm + 2\frac{1}{2}\ cm + 1\ cm + 2\ cm$
$P = 7\ cm$

5. $P = 8\ m + 7\ m + 2\ m + 1\ m + 1.5\ m + 5\ m$
$P = 24.5\ m$

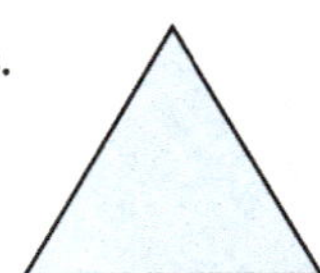

6. $P = 3\ in. + 2.5\ in. + 3\ in. + 2.5\ in. + 6\ in. + 5\ in.$
$P = 22\ in.$

7. $P = 5\ cm + 10\ cm + 6\ cm + 3\ cm + 3\ cm + 2\ cm$
$P = 29\ cm$

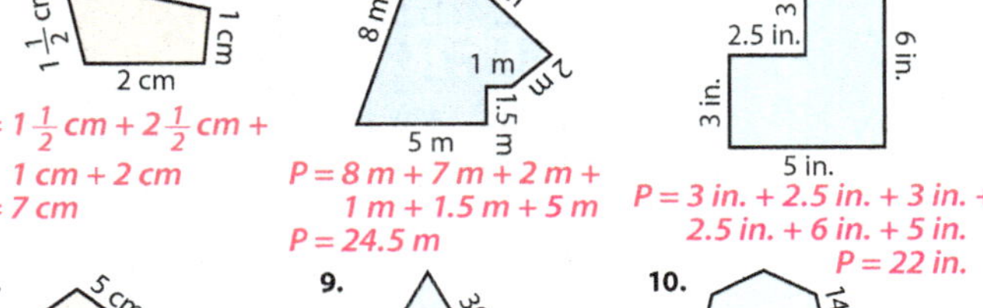
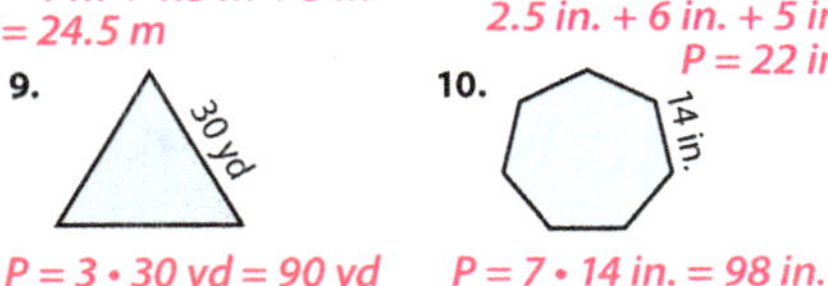
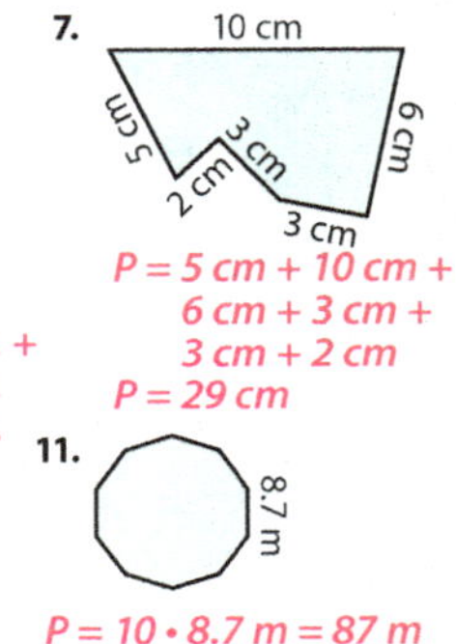

8. $P = 5 \cdot 5\ cm = 25\ cm$

9. $P = 3 \cdot 30\ yd = 90\ yd$

10. $P = 7 \cdot 14\ in. = 98\ in.$

11. $P = 10 \cdot 8.7\ m = 87\ m$

226   Chapter 11

- Direct each student to draw a polygon and to write "Park" inside of it. Tell them that the polygon represents an imaginary city park around which they can jog with their dog 1 time each day.

  If 1 cm on your drawing represents 1 mi, how can you find the distance that you jog around the entire park each day? I can use a ruler to measure the length of each side and then add the lengths together.

- Explain that perimeter is the distance around a geometric figure. Direct each student to find the perimeter of his park. Lead the students to conclude that their answers should be in miles because each centimeter of the drawing represents 1 mi.

- Allow volunteers to tell the distances they can jog around the park each day.

- Write for display "P = the sum of the lengths of the sides." Explain that this formula can be used to find the perimeter of any polygon.

- Direct each student to draw a quadrilateral having 4 right angles and all 4 sides measuring 6 cm.

  What type of quadrilateral did you draw? a square

  Draw the square for display; write "6 cm" along each side.

  Using the formula, what equation can you write to find the perimeter of the square? $P = 6\ cm + 6\ cm + 6\ cm + 6\ cm$

Write an equation to find the perimeter of the rectangle. *Equations may vary.*

12. 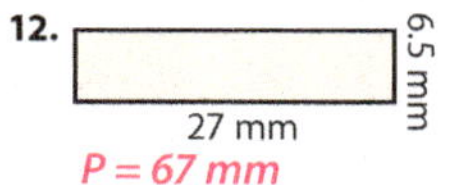 6.5 mm
27 mm
P = 67 mm

13. 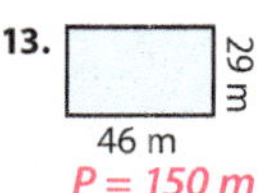 29 m
46 m
P = 150 m

14. 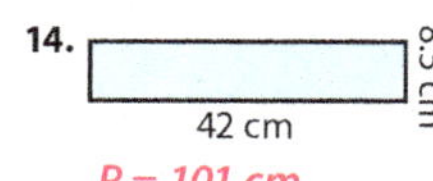 8.5 cm
42 cm
P = 101 cm

15. 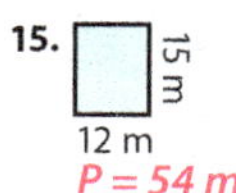 15 m
12 m
P = 54 m

<table>
<tr><td colspan="2">

Using the Perimeter to Find an Unknown Side Length

Find the sum of the known sides. Subtract the sum from the perimeter to find the measurement of the unknown side.

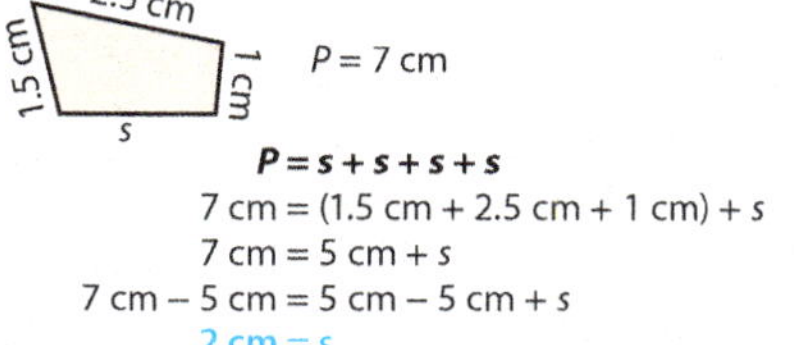
2.5 cm
1.5 cm
1 cm
s
$P = 7$ cm

$$P = s + s + s + s$$
$$7 \text{ cm} = (1.5 \text{ cm} + 2.5 \text{ cm} + 1 \text{ cm}) + s$$
$$7 \text{ cm} = 5 \text{ cm} + s$$
$$7 \text{ cm} - 5 \text{ cm} = 5 \text{ cm} - 5 \text{ cm} + s$$
$$2 \text{ cm} = s$$

The length of the unknown side is 2 cm.

</td><td>

Using Algebraic Expressions to Find the Perimeter

Find the perimeter of the rectangle if the width is 5 cm.

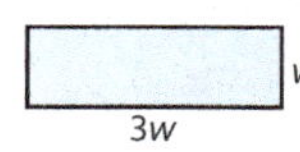
3w
w
The length of the rectangle is 3 times the width (3w).

$$P = (2 \cdot l) + (2 \cdot w) \qquad l = 3 \cdot 5 \text{ cm} = 15 \text{ cm}$$
$$P = (2 \cdot 15 \text{ cm}) + (2 \cdot 5 \text{ cm})$$
$$P = 30 \text{ cm} + 10 \text{ cm}$$
$$P = 40 \text{ cm}$$

The perimeter of the rectangle is 40 cm.

</td></tr>
</table>

Use the given perimeter to find the length of the unknown side.

16. 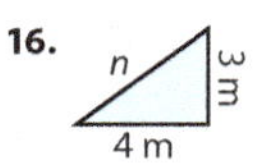 n, 3 m, 4 m
$P = 12$ m   *n = 5 m*

17. 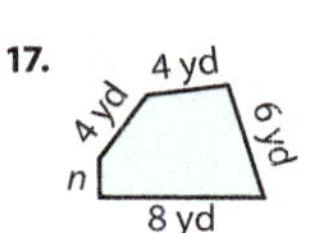 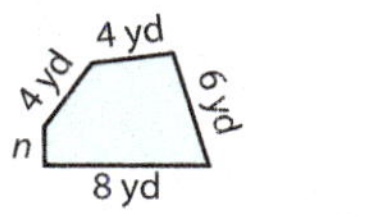 4 yd, 4 yd, 6 yd, n, 8 yd
$P = 24$ yd   *n = 2 yd*

Find the perimeter of the rectangle if w = 4 cm.

18. 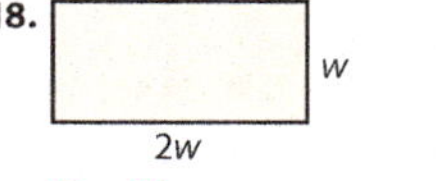 2w, w
P = 24 cm

19. 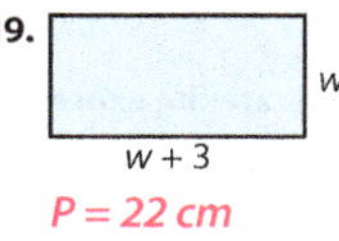 w + 3, w
P = 22 cm

Solve. *Equations may vary.*

20. Dad plans to place molding around the ceiling of the living room. If the length of the living room is 18 ft and the width is 12 ft, how many feet of molding will he use?
(2 • 18 ft) + (2 • 12 ft) = 60 ft

21. Mr. Franklin is fencing in a dog kennel in his backyard. The kennel dimensions are 12 ft, 20 ft, 10 ft, and 14 ft. How many feet of fencing does he need? *12 ft + 20 ft + 10 ft + 14 ft = 56 ft*

22. How can I determine the amount of fence I need for my dog kennel?
22. I can measure the length of each side and add the measurements to find the perimeter of my dog kennel.

Practice & Application   *Equations may vary.*

23. What is the perimeter of the field?
P = 300 yd
24. What is the perimeter of the parking lot?
P = 140 yd
25. What is the perimeter of the high school building?
P = 210 yd
26. What is the perimeter of the elementary school building? *P = 180 yd*

27. What is the perimeter of the entire school property?
P = 550 yd

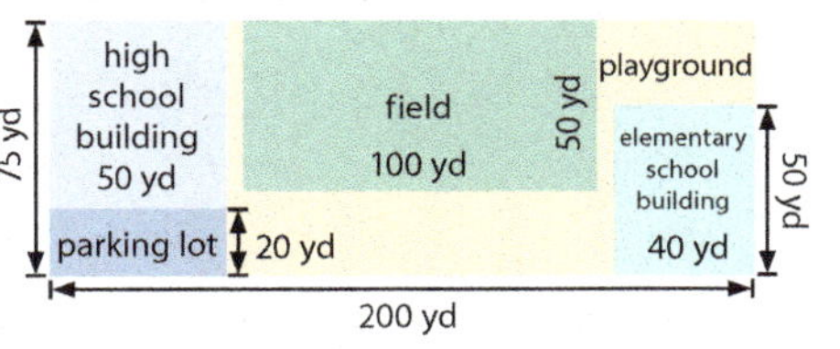

What is the perimeter of the square?
24 cm

- Refer again to the square drawn for display.

What multiplication equation can you write to find the perimeter of the square? 4×6 cm $= 24$ cm; Since all 4 sides have the same length, the length of 1 side can be multiplied by 4 to find the perimeter (the total distance around the square).

Since all the sides of a square are congruent, what multiplication formula can you use for finding the perimeter of a square? $P = 4 \times s$ or $P = 4s$

Write the formulas for display.

Choose a student to use the multiplication formula to find the perimeter of a square in which each side measures 7 in. $P = 4 \times 7$ in.; $P = 28$ in.

- Direct a student to draw a regular hexagon for display.

What formula can you use to find the perimeter of a regular hexagon? $P = 6 \times s$ or $P = 6s$

LESSON 104

Choose a student to use the multiplication formula to find the perimeter of a regular hexagon in which each side measures 7 cm. $P = 6 \times 7$ cm; $P = 42$ cm

- Follow a similar procedure for finding the perimeter of the following polygons.
a regular pentagon: 4 in. sides $P = 5s$; $P = 5 \times 4$ in.; $P = 20$ in.
a regular octagon: 5 in. sides $P = 8s$; $P = 8 \times 5$ in.; $P = 40$ in.

- Direct each student to draw another quadrilateral having 4 right angles, 1 pair of congruent opposite sides measuring 6 cm, and the other pair of congruent opposite sides measuring 10 cm. What type of quadrilateral did you draw? a rectangle

- Draw the rectangle for display and write the measurements (6 cm and 10 cm) along the sides. Lead the students to conclude that two multistep equations could be written to find the perimeter of the rectangle. Write both equations for display.
$P = (2 \times 10 \text{ cm}) + (2 \times 6 \text{ cm})$
$P = 2(10 \text{ cm} + 6 \text{ cm})$
Why can both of these equations be used to find the perimeter of the rectangle? Both equations represent the same value. The Distributive Property allows me to write $2(10 \text{ cm} + 6 \text{ cm})$ as $(2 \times 10 \text{ cm}) + (2 \times 6 \text{ cm})$.

- Choose a student to solve $P = (2 \times 10 \text{ cm}) + (2 \times 6 \text{ cm})$, and choose another student to solve $P = 2(10 \text{ cm} + 6 \text{ cm})$. 32 cm

- Guide the students in writing a formula for each equation: $P = (2 \cdot l) + (2 \cdot w)$ and $P = 2(l + w)$.
Write the equations for display.
Is this equation a true mathematical statement? Yes; both expressions represent the same value, and the Distributive Property allows me to write the expression either way.

Calculating an unknown side length

- **Model** writing an equation to help the students find an unknown side length.

- Draw an isosceles triangle for display. Write "5 m" along two sides and *s* along the third side. Write "*P* = 14 m" beside the triangle.

  How can you find the length of the third side of the triangle? I can subtract the sum of the lengths of the known sides from the perimeter (14 m); the difference is the length of the third side.

- Write "*P* = *s* + *s* + *s*" for display. Point out that *s* represents each side.

  Choose a student to rewrite the equation using the known measurements. 14 m = 5 m + 5 m + *s*

  Write parentheses around 5 m + 5 m. Lead the students to conclude that although the parentheses are not needed to solve the equation, they are helpful in isolating the unknown measurement.

  Choose a student to solve the equation. 14 m = 10 m + *s*; 14 m − 10 m = 10 m − 10 m + *s*; 4 m = *s*

- Follow a similar procedure to find the measurement of the unknown side of an irregular quadrilateral with a perimeter of 22 cm.

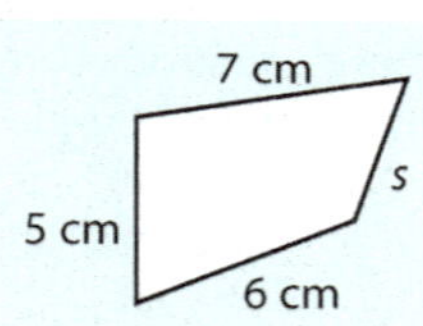

  22 cm = (7 cm + 5 cm + 6 cm) + *s*; 22 cm = 18 cm + *s*; 22 cm − 18 cm = 18 cm − 18 cm + *s*; 4 cm = *s*

- Draw a square for display. Write "*s*" along one side.

  If the perimeter of the square is 28 in., what equation can you write to find the unknown length of the side? Solve. 28 in. = 4 × *s*; 28 in. ÷ 4 = 4 ÷ 4 × *s*; 7 in. = *s*

- Draw a rectangle for display. Write "4 in." as the length and *w* as the width.

  If the perimeter of this rectangle is 20 in., what equation can you write to find the measure of the width? Solve. 20 in. = (2 × 4 in.) + (2 × *w*); 20 in. = 8 in. + (2 × *w*); 20 in. − 8 in. = 8 in. − 8 in. + (2 × *w*); 12 in. = 2 × *w*; 12 in. ÷ 2 = 2 ÷ 2 × *w*; 6 in. = *w*

Write the numerical expression for the word phrase. Solve.

1. 15 take away 2 *15 − 2 = 13*
2. the sum of 14 and 16 *14 + 16 = 30*
3. 1 more than a dozen *12 + 1 = 13*
4. one-half of ten $\frac{1}{2} \times \frac{10}{1} = \frac{5}{1} = 5$
5. the product of 4 and 5 *4 × 5 = 20*
6. seven times three *7 × 3 = 21*
7. 6 to the second power $6^2 = 36$
8. the difference between 3 and 8 *8 − 3 = 5*

Write an algebraic expression for the word phrase.

9. 4 times a number *4n*
10. a number divided by 10 $n \div 10$ or $\frac{n}{10}$
11. $\frac{1}{2}$ of a number $\frac{1}{2}n$
12. 20 more than a number *n + 20*
13. 6 less than a number *n − 6*
14. a number to the second power n^2

Evaluate the expression. Let *n* = 2. Write a comparison sentence by using >, <, or =.

15. $7 + 5 \geq n \cdot 5$
16. $\frac{18}{n} \leq 9 + 9$
17. $3n = 4 + 2$

Complete the table, using the given values to evaluate the expression.

18.

x	x + 3
7	*10*
11	*14*

19.

a	a · 4
3	*12*
6	*24*

20.

n	12 ÷ n
3	*4*
6	*2*

Solving an algebraic expression

- **Demonstrate** a procedure to help the students solve a word problem.

- Draw a rectangle for display. Write "*w*" along the width and "*2w*" along the length.

- Read aloud the following word problem.

  Mr. Alexander plans to put up fencing for a rectangular dog kennel. The length of the kennel will be twice the width. If the width of the kennel will be 10 ft, how much fencing will Mr. Alexander need for the kennel? 60 ft

  How wide will the kennel be? 10 ft Write "*w* = 10 ft."

  If the length of the kennel is twice as long as the width, what would the length be? 20 ft (*l* = 2 × 10 ft)

- Direct attention to the essential question at the top of Student Edition page 226.

  How can I determine the amount of fence I need for my dog kennel? I can measure the length of each side and add the measurements to find the perimeter of my dog kennel.

  What is another way this problem could be solved? by using the perimeter formula

- Write for display "$l = 20$ ft" and "$P = (2 \times l) + (2 \times w)$." Choose a student to use the length and width measurements to write the equation and solve the problem.
  $P = (2 \cdot 20 \text{ ft}) + (2 \cdot 10 \text{ ft}); P = 40 \text{ ft} + 20 \text{ ft}; P = 60 \text{ ft}$

- Erase "$2w$" from the rectangle and write the length "$w + 5$."
  If the length of the kennel will be 5 ft longer than the width, what will the length be? 15 ft ($l = 10 \text{ ft} + 5 \text{ ft}$)

- Follow a similar procedure to guide the students in finding how much fencing will be needed. $P = (2 \cdot 15 \text{ ft}) + (2 \cdot 10 \text{ ft}); P = 30 \text{ ft} + 20 \text{ ft}; P = 50 \text{ ft}$

Apply

Student Edition pages 226–27
- Read and explain the directions for pages 226–27. Assist the students as they complete the pages independently.

Daily Review
- Students should complete Chapter 11, section *a*.

NOTES

LESSON 105

Student Edition pages 228–30
Daily Review Chapter 11, section *b*

OBJECTIVES

- Relate the circumference of a circle to its diameter.
- Calculate the circumference of a circle using a formula.
- Calculate the diameter of a circle given the circumference.
- Recall that God has revealed Himself through nature. **BWS**

BIBLICAL WORLDVIEW SHAPING

- **Design (Recall):** God has revealed Himself in the consistent relationships found in circles.

TEACHER RESOURCES

- 55 *Finding the Circumference* (for the teacher and for each student)

ADDITIONAL MATERIALS

- a cylindrical object, such as a can or mug (for each pair of students) (Provide objects varying in size.)
- a ruler (for the teacher and for each student)
- a calculator (for each student)
- a 12 in. length of string (for each pair of students)

Circumference

How does pi demonstrate the Creator's design?

The **circumference** (C) of a circle is a little more than 3 times its diameter (d). The ratio $\frac{C}{d}$ has a value of **π (pi)**. Pi is a non-repeating and non-terminating decimal with an approximate value of **3.14** or $\frac{22}{7}$. Use the approximate value of pi to find an unknown circumference or diameter.

Key Terms
- circumference
- π (pi) ≈ 3.14
- $C = \pi d$
- $C = 2\pi r$

Finding the Circumference Given the Diameter
$$C = \pi d$$

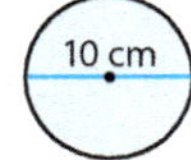

$C = \pi d$
$C = 3.14 \times 10$
$C = 31.4$ cm

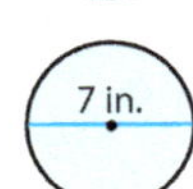

$C = \pi d$
$C = \frac{22}{7} \times 7$
$C = \frac{22}{7} \times \frac{7}{1}$
$C = 22$ in.

Finding the Circumference Given the Radius
$$C = 2\pi r$$

$C = 2\pi r$
$C = 2 \times 3.14 \times 5$
$C = 31.4$ m

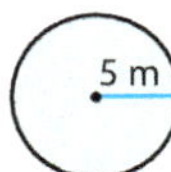

$C = 2\pi r$
$C = 2 \times \frac{22}{7} \times 3\frac{1}{2}$
$C = 2 \times \frac{22}{7} \times \frac{7}{2}$
$C = 22$ in.

Finding the Diameter Given the Circumference

Since $C = \pi d$, then $\frac{C}{\pi} = d$.

$\frac{C}{\pi} = d$

$\frac{28.26}{3.14} = d$

$C = 28.26$

$d = 9$ cm

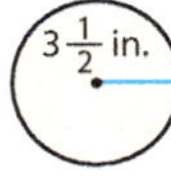

Exercises

Find the circumference of the circle, using 3.14 for π.

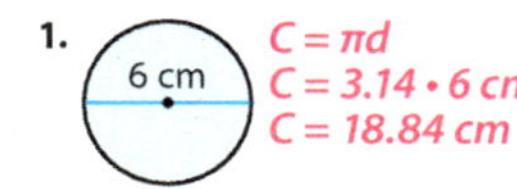

1. $C = \pi d$
   $C = 3.14 \cdot 6$ cm
   $C = 18.84$ cm

2. $C = 2\pi r$
   $C = 2 \cdot 3.14 \cdot 4.5$ m
   $C = 28.26$ m

3. $C = 2\pi r$
   $C = 2 \cdot 3.14 \cdot 10$ m
   $C = 62.8$ m

Find the circumference of the circle, using $\frac{22}{7}$ for π.

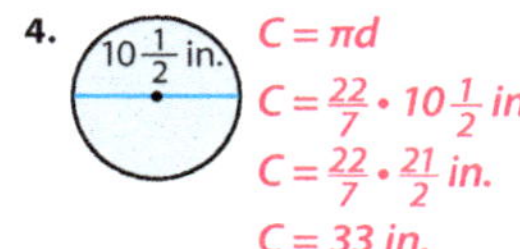

4. $C = \pi d$
   $C = \frac{22}{7} \cdot 10\frac{1}{2}$ in.
   $C = \frac{22}{7} \cdot \frac{21}{2}$ in.
   $C = 33$ in.

5. $C = 2\pi r$
   $C = 2 \cdot \frac{22}{7} \cdot 14$ ft
   $C = 88$ ft

6. $C = \pi d$
   $C = \frac{22}{7} \cdot \frac{7}{8}$ yd
   $C = \frac{11}{4}$ yd
   $C \approx 2.75$ yd
   or $2\frac{3}{4}$ yd

Preparation

Label the cylindrical objects by writing different numbers on pieces of masking tape, placing one label on each object.

Rather than using cylindrical objects, you may make different sized cylinders by rolling construction paper into cylinders and taping the sides.

Allow the students to use calculators throughout this chapter so that they can focus on finding the perimeter and area of the various figures rather than on doing lengthy calculations.

Engage

- Direct attention to the essential question at the top of Student Edition page 228, "How does pi demonstrate the Creator's design?"
- Direct the students to **brainstorm** to explore the essential question.

Instruct

Relating circumference to diameter

- Use a **collaborative learning activity** to help the students find the diameter/circumference relationship.

- Distribute a copy of the *Finding the Circumference* page to each student along with a ruler and a calculator. Pair the students and distribute the lengths of string and cylindrical objects to each pair.

  What is the distance around a geometric figure called? the perimeter

- Explain that circumference is the perimeter or distance around a circle.

- Display the *Finding the Circumference* page and explain instructions 1–3. Demonstrate finding the circumference: place the string around a cylindrical object to measure the circumference, and then measure that length of string to the

DID YOU KNOW?

The Bible indicates that the circumference of a circle is about three times greater than the diameter.

And he made a molten sea, ten cubits from the one brim to the other… and a line of thirty cubits did compass it round about.

1 Kings 7:23

Find the diameter, using the circumference.

7. If the circumference is 23.55 cm, then the diameter is ___ cm.

8. If the circumference is 47.1 cm, then the diameter is ___ cm.

7. $\frac{C}{\pi} = d$
$d = \frac{23.55}{3.14}$ cm
$d = 7.5$ cm

8. $\frac{C}{\pi} = d$
$d = \frac{47.1}{3.14}$ cm
$d = 15$ cm

Find the circumference or the perimeter of the figure. Use 3.14 for π. Round decimals to the nearest hundredth.

9. 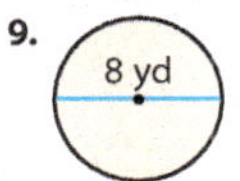 8 yd
$C = 25.12$ yd

10.  3.5 m
$C = 10.99$ m

11. 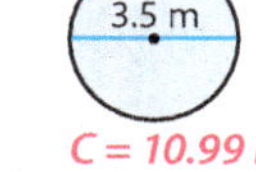 2.7 cm
$C = 16.956$ cm ≈ 16.96 cm

12. 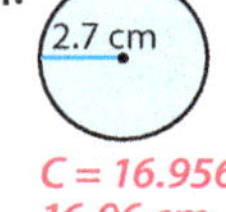 $1\frac{3}{4}$ ft
$C = 5.495$ ft ≈ 5.50 ft or 5.5 ft

13. 13 m
$C = 81.64$ m

14. 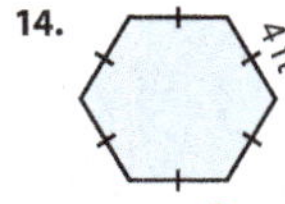 4 ft
$P = 24$ ft

15. 6 cm, 2 cm
$P = 16$ cm

16. 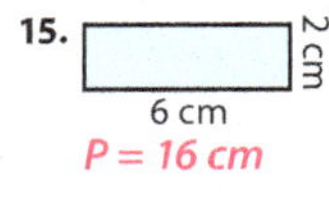 15 m, 25 m, 20 m, 15 m, 20 m, 30 m
$P = 125$ m

Find the perimeter of the figure. (*Hint:* Find the circumference of the whole circle to determine the distance around the half circle. Use 3.14 for π.)

17. 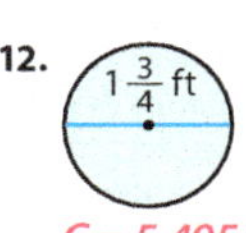 10 m, 10 m
$3 \cdot 10 = 30$ m
$\frac{1}{2}(3.14 \cdot 10) = 15.7$ m
$30 + 15.7 = 45.7$ m

18. 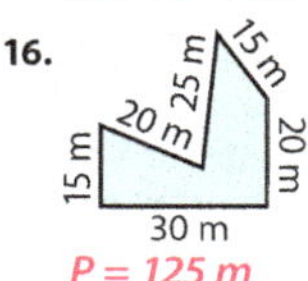 6 m, 6 m
$3 \cdot 6 = 18$ m
$\frac{1}{2}(3.14 \cdot 6) = 9.42$ m
$18 + 9.42 = 27.42$ m

---

circle's circumference to its diameter. Pi is a non-repeating, non-terminating decimal. The approximate values 3.14 and $\frac{22}{7}$ are used for calculating the circumference of a circle.

Calculating the circumference of a circle

- Use **critical thinking questions** to help the students calculate the circumference of a circle.

  If the circumference of a circle is about 3.14 or $\frac{22}{7}$ times greater than the diameter of a circle, how could you calculate the circumference of a circle when you know the diameter? I could multiply the diameter by 3.14 or $\frac{22}{7}$ to find the circumference.

- Direct attention to the formula for finding circumference on the displayed page. Lead the students to conclude that C represents circumference and d represents diameter.

- Direct attention to instruction 5 and direct the students to multiply the diameter by 3.14 to find each circumference. Guide a discussion about how the calculated circumferences compare to the measured circumferences.

- Direct the students to use their calculators to divide 22 by 7.

- Guide the students in calculating the circumference of the circles at the bottom of the page using $\frac{22}{7}$ for pi. Point out that using $\frac{22}{7}$ to find the circumference is especially helpful when working with fractions and whole numbers that are compatible with the denominator 7.

  $C = \frac{22}{7} \times \frac{28}{1}$ in.
  $C = 88$ in.

  $C = \frac{22}{7} \times \frac{7}{1}$ ft
  $C = 22$ ft

  $C = \frac{22}{7} \times 17\frac{1}{2}$ yd
  $C = \frac{22}{7} \times \frac{35}{2}$ yd
  $C = 55$ yd

- Draw a circle for display. Draw a radius and write "21 yd" as its measure.

---

nearest millimeter. Explain that measuring around the curved surface anywhere on a cylindrical object will give you the circumference of the circular bases. Demonstrate measuring the object's diameter using a ruler.

- Direct pairs of students to measure the circumference and the diameter of their object to the nearest millimeter and to record the measurements. Instruct the pairs to exchange objects.

- Repeat the procedure so that each pair measures at least 4 different objects.

- Direct attention to instruction 4. Instruct each student to divide the circumference by the diameter for each object he measured and to record his answer.

  What common relationship do you see between the circumference and diameter of a circle? The circumference is a little more than 3 times the diameter.

- Explain that the relationship between a circle's circumference and its diameter is a God-ordained constant. The circumference of every circle is a little more than 3 times greater than its diameter.

- Write "π ≈ 3.14 or $\frac{22}{7}$" for display. Explain that mathematicians use a Greek letter, *pi*, to represent the relationship (ratio) of a

How do you think you can find the circumference of the circle when given only the measurement of the radius? Since the radius is half the length of the diameter, I multiply the radius by 2 to find the diameter and then multiply the diameter by pi to find the circumference.

- Write "$C = 2\pi r$" for display. Guide the students in using the formula to find the circumference of the circle using $\frac{22}{7}$ for pi. $C = 2 \times \frac{22}{7} \times \frac{21}{1}$ yd; $C = 132$ yd Remind them that the coefficient (a number) precedes the variable.

- Follow a similar procedure for finding the circumference of circles with the following radii.
  radius = 1.4 cm; $\pi \approx 3.14$ $C = 2 \times 3.14 \times 1.4$ cm; $C = 8.792$ cm

  radius = $\frac{3}{4}$ in.; $\pi \approx \frac{22}{7}$ $C = 2 \times \frac{22}{7} \times \frac{3}{4}$ in.; $C = 4\frac{5}{7}$ in.

- Read aloud the following word problem.
  Mrs. Hagan makes and decorates baskets to sell at a craft fair. She made a circular basket that has a diameter of 27 cm. Mrs. Hagan has 75 cm of ribbon left over from her last basket. Is this enough ribbon to go around the rim of the basket? No; $C = 3.14 \times 27$ cm; $C = 84.78$ cm; 84.78 cm of ribbon is needed.

  What is the question asking you to determine? whether 75 cm of ribbon is enough to go around a basket with a diameter of 27 cm

  What equation can you use to solve the problem? $C = 3.14 \times 27$ cm; Since the circumference is also the amount of ribbon needed on the basket, I can multiply pi by the diameter of the basket to find its circumference.

- Choose a student to write the equation for display and solve it while the other students solve it independently. $C = 84.78$ cm

- Follow a similar procedure for the following word problem. Lead the students to conclude that it is a multistep problem; they must combine the circumference of the circular cake and the perimeter of the rectangular cake to find the length of fondant that is needed to decorate both cakes.

---

19. Molly made a circular braided rug. The radius of the rug is 4 ft. What is the circumference of the rug? *C = 25.12 ft*

20. Mom is hemming a circular tablecloth. How many inches will she hem if the diameter of the tablecloth measures 60 in.? *C = 188.4 in.*

21. Mr. Byers is placing edging around a rectangular fishpond. The length of the fishpond is 10 ft and the width is 5 ft. If the edging costs $4.50 per foot, how much will Mr. Byers spend? *P = 30 ft; 30 • $4.50 = $135.00*

22. The Ferris wheel at the Texas State Fair has a diameter of 212 ft. What is the circumference of the Ferris wheel? *C = 665.68 ft*

23. Noah wants to know how far his bicycle wheel travels when it goes all the way around one time. The wheel is 20 in. across. How far does his bicycle wheel travel in one turn? *C = 62.8 in.*

24. Mrs. Kwan is putting wallpaper border up in her bedroom. The length of the bedroom is 12 ft and the width is 8 ft. Each roll of border is 15 ft. How many rolls will Mrs. Kwan need? *P = 40 ft; 40 ÷ 15 = 2.6̄ rolls; 3 rolls*

25. How does pi demonstrate the Creator's design? *The fact that the pi relationship is the same in all circles shows the orderliness and consistency in God's creation.*

MEET THE MATHEMATICIAN

Archimedes (287–212 BC) was the greatest mathematician of the ancient world. He lived in Greece over 200 years before Jesus was born. Archimedes developed the first law of hydrostatics and worked out the approximate value of π. Putting mathematics into action, Archimedes was also an inventor. His military inventions included a catapult, a device for dropping heavy objects on ships to sink them, and a machine that picked up ships in the harbor and turned them over! Because of these inventions, the city of Syracuse was able to fend off the Romans for over two years. Archimedes' tomb was engraved with a sphere inside a cylinder to memorialize his great contributions to the study of geometry.

Computers have calculated π to over a trillion decimal places but have not arrived at its exact value. We may never know the exact value of π, but we do know that it is constant for every circle.

230   Chapter 11

Evie is decorating 2 cakes: a circular cake that is 8 in. in diameter and a rectangular cake that is 9 in. × 13 in. She needs to place a fondant rope around the bottom of each cake. What is the length of fondant needed for both cakes? $C = 3.14 \times 8$ in. $= 25.12$ in.; $P = (2 \times 9 \text{ in.}) + (2 \times 13 \text{ in.}) = 44$ in.; 44 in. + 25.12 in. = 69.12 in.

Calculating the diameter of a circle

- Guide the students in a **practice exercise** to find the diameter when given the circumference.

How do you think you find the diameter of a circle when given the circumference? Since the product of the diameter and pi equals the circumference ($C = \pi d$), the circumference divided by pi equals the diameter.

- Lead the students to write the formula for finding the diameter when the circumference is known. $\frac{C}{\pi} = \frac{\pi d}{\pi}$; $\frac{C}{\pi} = d$

- Guide the students in finding the diameter of circles with the following circumferences.

  circumference = 37.68 m; $\pi \approx 3.14$
  $\frac{37.68 \text{ m}}{3.14} = d$; $d = 12$ m

  circumference = 14.13 m; $\pi \approx 3.14$
  $\frac{14.13 \text{ m}}{3.14} = d$; $d = 4.5$ m

- Guide the students in solving the following word problem. Direct the students to round the answer to the nearest meter.

Solve.

1. $3.47 + $1.62 **$5.09**	**2.** 45,816 + 21,437 **67,253**	**3.** 86,045 + 19,057 **105,102**	**4.** 832 + 659 **1,491**

5. 371 422 + 870 **1,663**	**6.** 419 27 + 132 **578**	**7.** 15 32 18 + 604 **669**	**8.** 38 + 44 **82**

9. $0.78 $2.52 $0.07 + $1.18 **$4.55**	**10.** 517,053 + 13,267 **530,320**	**11.** 60,984 + 321,786 **382,770**	**12.** 417,035 + 562,809 **979,844**

13. 2.135 + 41.03
43.165

14. $39.76 + $124.01
$163.77

15. $2\frac{1}{2} + 1\frac{3}{4}$
$2\frac{2}{4} + 1\frac{3}{4} = 3\frac{5}{4} = 4\frac{1}{4}$

16. 0.278 + 1.93
2.208

A circular swimming pool has a circumference of 25 m. Approximately how far does a swimmer have to swim across the pool to get from one side to the point exactly opposite his starting location?
25 m ÷ 3.14 = d; d = 7.96 m; 8 m

How does pi demonstrate the Creator's design? The fact that the pi relationship is the same in all circles shows the orderliness and consistency in God's creation.

God's design in nature
- Guide the students in a **formative assessment** to answer the essential question.
- Remind the students that the value of pi (3.14) is the same for all circles regardless of the size. This consistency is evidence of design in nature.

Apply

Student Edition pages 228–30
- Read and explain the directions for pages 228–30. Assist the students as they complete the pages independently.

Daily Review
- Students should complete Chapter 11, section *b*.

LESSON 106

Student Edition pages 231–33
Daily Review Chapter 11, section c

OBJECTIVES

- Calculate the area of rectangles, squares, and parallelograms using a formula.
- Calculate the area of a complex figure.
- Calculate the area of parallelograms.
- Calculate the unknown side (length or width) of a rectangle or a square.

TEACHER RESOURCES

- 56 *Area: Rectangles, Squares & Parallelograms* (for the teacher and for each student)

ADDITIONAL MATERIALS

- scissors (for each student)

ASSESSMENTS

- Chapter 11 Quiz 1

Engage

- Direct attention to the top of Student Edition page 231 and assign a **bell ringer** to help the students explore the essential question, "What am I measuring when I find the perimeter and area of a figure?"

  Read the following phrases and ask whether they describe perimeter or area.

  the frame around a picture perimeter

  the floor of a bedroom area

  the space within a football field area

  the fence around a swimming pool perimeter

Instruct

Calculating the area of rectangles

- Guide a **discussion** to help the students use a formula to find the area of rectangles.

Area of Rectangles, Squares & Parallelograms

What am I measuring when I find the perimeter and area of a figure?

Area is the space within a region. The area of a region is the number of square units needed to cover its surface.

Area of a Rectangle

$A = l \cdot w$
$A = 8 \cdot 5$
$A = 40 \text{ units}^2$

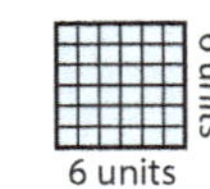

Area of a Square

$A = l \cdot w \text{ or } A = s^2$
$A = 6 \times 6 \text{ or } 6^2$
$A = 36 \text{ units}^2$

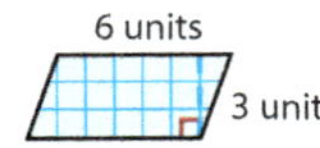

Area of a Parallelogram

When a parallelogram does not have right angles, its area is still determined by multiplying the length of its base (*b*) and the length of its height (*h*). The height is the length of a line segment that is perpendicular to both bases; it is the shortest distance between the bases.

$A = b \cdot h$
$A = 6 \cdot 3$
$A = 18 \text{ units}^2$

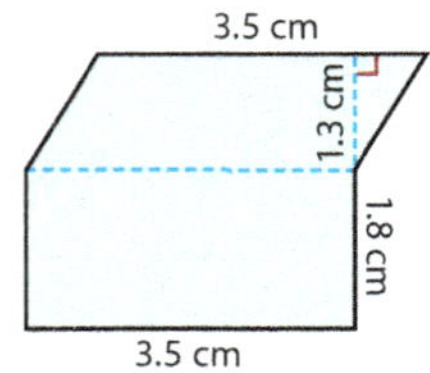

Area of a Complex Figure

The area of an irregular polygon is determined by dividing the figure into smaller regions, finding the area of each smaller region, and then adding the areas.

$A = (l \cdot w) + (b \cdot h)$
$A = (3.5 \cdot 1.8) + (3.5 \cdot 1.3)$
$A = 6.3 + 4.55$
$A = 10.85 \text{ cm}^2$

Using the Area to Find an Unknown Side Length

Divide the given area by the measurement of the known side.

$n = A \div s$
$n = 42 \div 7$
$n = 6 \text{ in.}$

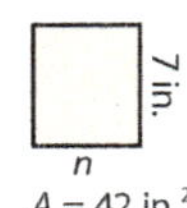

Key Terms

- area
- $A = l \cdot w$
- $A = b \cdot h$
- $A = l \cdot w \text{ or } A = s^2$
- $A = (l \cdot w) + (b \cdot h)$

- Remind students that area is the space within a region (figure) and is measured in square units. Point out that a small raised 2 is written with the label in the answer to indicate the square units.

- Distribute a copy of the *Area: Rectangles, Squares & Parallelograms* page to each student and display your copy.

  How can you find the number of square units in the rectangle at the top of the page? I can count the square units or multiply the length times the width.

- Write the formula for display. $A = l \times w$

  Instruct the students to write the equation to find the area of the rectangle and to write the solution on the blanks.
  $A = 11 \text{ units} \times 5 \text{ units}; A = 55 \text{ units}^2$

Choose a student to count the square units to verify the area.

Can the formula $A = l \times w$ be used to find the area of the square? Yes; a square is a type of rectangle.

- Instruct the students to write the equation to find the area of the square and the solution. $A = 9 \text{ units} \times 9 \text{ units}; A = 81 \text{ units}^2$

  When finding the area of a square, what do you notice about the equation? Since the side lengths are the same, I multiply a number by itself.

  How could you rewrite this equation using an exponent? $A = 9^2$ or $A = (9 \text{ units})^2$

Use the formula $A = l \cdot w$ to find the area of the figure.

1. 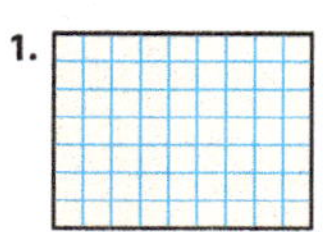
$A = 9 \cdot 7 = 63 \text{ units}^2$

2. 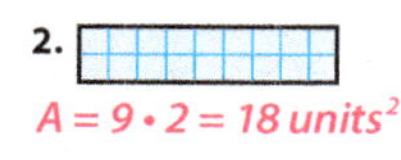
$A = 9 \cdot 2 = 18 \text{ units}^2$

3. 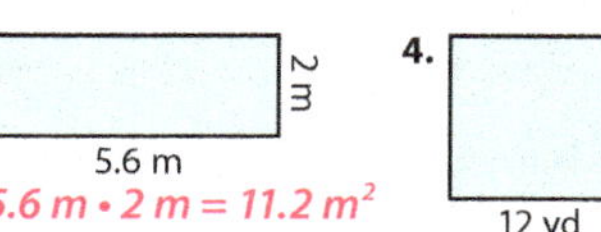
5.6 m · 2 m
$A = 5.6 \text{ m} \cdot 2 \text{ m} = 11.2 \text{ m}^2$

4. 12 yd · 12 yd
$A = 12 \text{ yd} \cdot 12 \text{ yd} = 144 \text{ yd}^2$

Find the area of the figure.

5. 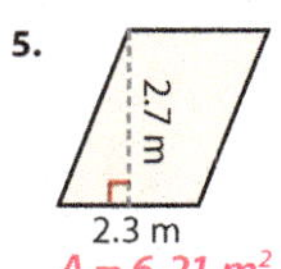
2.7 m; 2.3 m
$A = 6.21 \text{ m}^2$

6. 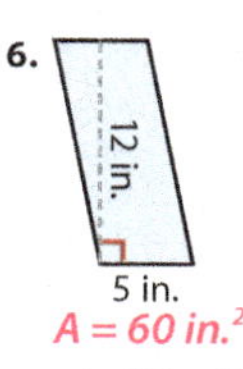
12 in.; 5 in.
$A = 60 \text{ in.}^2$

7. 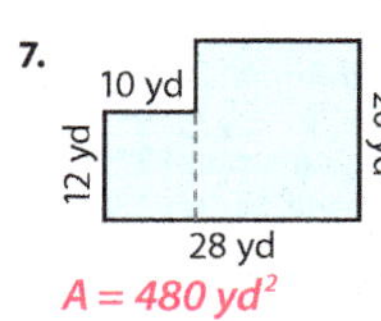
10 yd; 12 yd; 20 yd; 28 yd
$A = 480 \text{ yd}^2$

8. 2 cm; 2 cm; 2 cm; 3 cm; 1 cm; 2 cm; 1 cm
$A = 7 \text{ cm}^2$

Find the unknown measurement of the figure, using the given area.

9. 5 ft; n
$A = 25 \text{ ft}^2$ *5 ft*

10. n; 17 yd
$A = 136 \text{ yd}^2$ *8 yd*

11. 5.6 m; n
$A = 70 \text{ m}^2$ *12.5 m*

12. 10 cm; n
$A = 140 \text{ cm}^2$ *14 cm*

Find the area of the figure.

13. 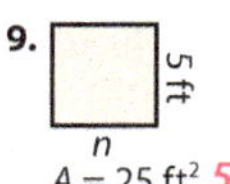
15 cm; 35 cm
$A = 525 \text{ cm}^2$

14. 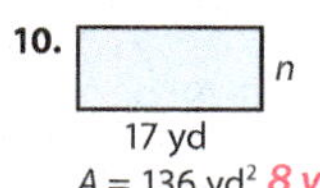
4 m; 5 m
$A = 20 \text{ m}^2$

15. 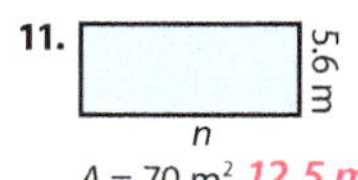
1.9 yd
$A = 3.61 \text{ yd}^2$

16. 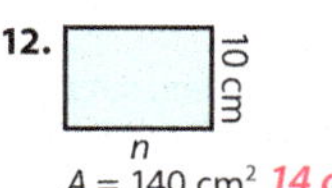
10 ft; 18 ft
$A = 180 \text{ ft}^2$

17. 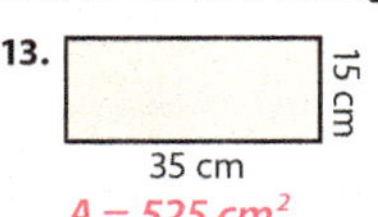
$3\frac{1}{2}$ in.
$A = 12\frac{1}{4} \text{ in.}^2$

18. 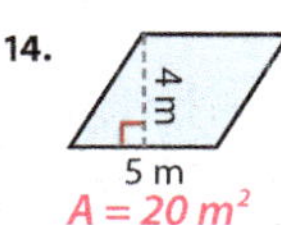
3.2 m; 4 m
$A = 12.8 \text{ m}^2$

19. 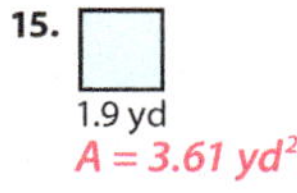
5 cm; 6 cm; 11 cm; 10 cm
$A = 85 \text{ cm}^2$

20. 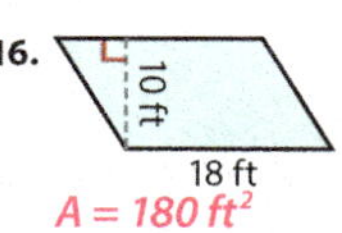
3 m; 7 m; 3 m; 6 m; 9 m; 13 m
$A = 96 \text{ m}^2$

Use the given dimensions to find the area.

21. $l = 4$ in.
$w = 8$ in.
$A =$ __ *32 in.²*

22. $b = 15$ ft
$h = 24$ ft
$A =$ __ *360 ft²*

23. $b = 4.5$ m
$h = 7.5$ m
$A =$ __ *33.75 m²*

Use the given measurements to find the unknown measurement.

24. $l = 7$ yd
$w =$ __ *12 yd*
$A = 84 \text{ yd}^2$

25. $l =$ __ *9 m*
$w = 5.5$ m
$A = 49.5 \text{ m}^2$

26. $l =$ __ *10 ft*
$w = 10$ ft
$A = 100 \text{ ft}^2$

27. What am I measuring when I find the perimeter and area of a figure?
For perimeter, I am measuring the distance around a figure. For area, I am measuring the space within a region.

- Guide the students in drawing a line in the irregular polygon to form an 11×5 rectangle and a 9×9 square.

  What expression can you use to find the area of the left rectangle? 11 units × 5 units

  Write the expression for display.

  What expression can you use to find the area of the square? 9 units × 9 units or (9 units)²

  Write either expression beside 11 units × 5 units; leave space between the expressions.

  Write parentheses around each multiplication expression, a plus sign (+) between them, and "$A =$" to the left to show how the entire area is calculated: $A = $ (11 units × 5 units) + (9 units × 9 units) or $A = $ (11 units × 5 units) + (9 units)².

  Choose a student to solve the problem. $A = 136 \text{ units}^2$

- Guide the students in erasing the line drawn on the figure and then in drawing a new line to show a 20×5 rectangle and a 9×4 rectangle.

  What equation could you use to find the total area if you divided the irregular polygon into 2 rectangles? $A = $ (20 units × 5 units) + (9 units × 4 units); $A = 136 \text{ units}^2$

  Did the area change when you partitioned the complex figure a different way? No; although a complex figure can be partitioned in different ways, the total area is the same when all the areas are added.

- Guide the students in solving the following word problem. Allow them to draw pictures if necessary.

  Isaac's dad fenced in the backyard and the adjoining side yard for a play area. The backyard is 25 ft long and 35 ft wide. The side yard is 10 ft long and 15 ft wide. What is the total area of the fenced-in yard? $A = $ (25 ft × 35 ft) + (10 ft × 15 ft); $A = 1{,}025 \text{ ft}^2$

Calculating the area of parallelograms

- **Model** reshaping a parallelogram to help the students calculate area.

- Direct the students to count the square units in the parallelogram on the page.

How could you write the formula for finding the area of a square using an exponent? $A = s^2$

- Guide the students in solving the following word problems, using the formulas for finding the area of a rectangle and a square.

  Mrs. Barnes's kitchen floor is rectangular, measuring 12 ft long and 9 ft wide. How many square-foot tiles does she need to cover the floor? $A = 12 \text{ ft} \times 9 \text{ ft}$; $A = 108 \text{ ft}^2$; 108 square-foot tiles

  Mr. Hamblin poured a concrete patio in the backyard. The patio is square, measuring 15 ft on each side. What is the area of the patio? $A = 15 \text{ ft} \times 15 \text{ ft}$ or $A = (15 \text{ ft})^2$; $A = 225 \text{ ft}^2$

- Guide the students in distinguishing perimeter and area.

  What am I measuring when I find the perimeter and area of a figure? For perimeter, I am measuring the distance around a figure. For area, I am measuring the space within a region.

Calculating the area of a complex figure

- Use **guided discovery** to help the students find the area of a complex figure.

  How could you find the area of the irregular polygon on the page? sample answers: count the square units; partition the polygon into parts (1 rectangle and 1 square or 2 rectangles), calculate the area of each part, and then add the areas

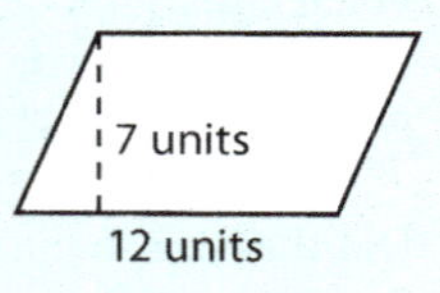

How many square units did you count? sample answer: It is difficult to count because there are so many partial squares.

- Direct the students to cut the bottom section from the page and then to cut out the parallelogram. Instruct them to cut the parallelogram into 2 pieces along the dashed vertical line.

  What shape can you make if you slide the triangular piece to the opposite side of the parallelogram? a rectangle

  Point out that reshaping the parallelogram formed perpendicular sides, making it easier to see its length and width.

  What formula can you use to find the area of the parallelogram now? $A = l \times w$

  What equation and solution can you write for the area of the parallelogram? $A = 9$ units $\times 4$ units; $A = 36$ units^2

- Draw a parallelogram for display. Explain that it is not necessary to reshape a parallelogram to know its length and width. Students can identify the width, or height, that is perpendicular to the length, or base.

  Write "12 units" along the base of the parallelogram. Draw a line segment perpendicular to the bases to represent the height inside the parallelogram and write "7 units" beside it. Explain that the height is the shortest distance between both bases (top and bottom edges) and is a line segment that is perpendicular to both bases. They multiply the base (length) times the height (width) to find the area of a parallelogram.

- Write "$A = b \times h$" for display. Guide the students in using the formula to find the area of the parallelogram. $A = 12$ units $\times 7$ units; $A = 84$ units^2

- Follow a similar procedure for parallelograms having these dimensions.
  base = 6 m; height = 3.1 m $A = 6$ m $\times 3.1$ m; $A = 18.6$ m^2

  base = 5.6 cm; height = 3 cm $A = 5.6$ cm $\times 3$ cm; $A = 16.8$ cm^2

28. Luke built a table with a square top. The tabletop measures 30 in. on each side. How many 1 in.^{2} tiles are needed to cover the whole tabletop? *A = 30 in. • 30 in. = 900 1 in.^{2} tiles*

29. Owen's family has a rectangular-shaped swimming pool in their backyard. The swimming pool is 30 ft long and 15 ft wide. What is the area of the swimming pool? *A = 30 ft • 15 ft = 450 ft^2*

30. Owen's backyard has the shape of a rectangle. It is 75 ft long and 50 ft wide. What is the area of the backyard that is *not* part of the swimming pool? (Use the information in problem 29.) *A = (75 ft x 50 ft) − 450 ft^2 = 3,300 ft^2*

31. Mr. Martinez is placing edging around 2 trees in the front yard. The diameter of each circle will be 3 ft. What is the total amount of edging he needs for these trees? *C = (3.14 • 3 ft) • 2 = 18.84 ft*

32. Mrs. Anderson is putting a frame around an octagonal mirror. Each side of the mirror is 12 in. How much molding will Mrs. Anderson use to frame the mirror? *P = 8 • 12 in. = 96 in.*

33. Explain where to draw a line segment in a parallelogram to measure the height of the figure. *perpendicular to the two bases*

34. Why is area measured in square units and perimeter measured in units?

34. *Perimeter is a 1-dimensional measurement—length (in., cm, ft). Area is a 2-dimensional measurement—length × width or base × height (in. × in. = in.^{2}; cm × cm = cm^2; ft × ft = ft^2).*

Calculating an unknown side length or width

- Use **critical thinking questions** to help the students calculate an unknown side length.

- Read aloud the following word problem.

  The men laid 12 yd^2 of carpet in the bedroom. If the length of the bedroom is 3 yd, what is the width of the room? 4 yd

  What information is given in the word problem? The area of the bedroom is 12 yd^2 and its length is 3 yd.

- Choose a student to draw a picture to represent the bedroom. Instruct him to write "n" along the unknown side. Write "$A = l \times w$" below the figure.

  How can understanding the formula for finding area help you solve the word problem? Since the product of the length (3 yd) and width (n yd) equals the area (12 yd^2), the area divided by the length equals the width.

- Explain that since multiplication and division are inverse operations, $w = A \div l$ (the inverse of $A = l \times w$) can be used to solve the word problem.

Solve.

1. $67.48
− $17.70
$49.78

2. 37,604
− 28,442
9,162

3. 525,004
− 317,423
207,581

4. 719,604
− 385,260
334,344

5. 8,042
− 5,609
2,433

6. 45,697
− 13,806
31,891

7. 200,345
− 124,670
75,675

8. 63,089
− 20,428
42,661

9. 6,839
− 3,860
2,979

10. 747,222
− 648,203
99,019

11. 832,587
− 604,388
228,199

12. 783,054
− 332,867
450,187

13. $5.00 − $1.32
$3.68

14. 14.03 − 2.5
11.53

15. $6\frac{1}{8} - 3\frac{1}{2}$

$6\frac{1}{8} - 3\frac{4}{8} = 5\frac{9}{8} - 3\frac{4}{8} = 2\frac{5}{8}$

16. 89 − 15.75
73.25

17. 2,000 − 1,947
53

18. $13 − $1.98
$11.02

Apply

Student Edition pages 231–33
- Read and explain the directions for pages 231–33. Assist the students as they complete the pages independently.

Daily Review
- Students should complete Chapter 11, section *c*.

Assess

Quiz 1
- Use the **summative assessment** to evaluate the students' progress at this point in the chapter.

- Choose a student to write for display and solve the equation needed to find the width of the bedroom.
  $w = 12$ yd^2 ÷ 3 yd; $w = 4$ yd

  Explain that the formula $n = A ÷ s$ (the unknown side = the area ÷ the known side) can be used to calculate the unknown side. Write the formula for display.

- Follow a similar procedure for the following problems.

  The men placed 192 ft^2 of vinyl flooring in the kitchen. If the width of the kitchen is 12 ft, what is the length of the kitchen?
  $n = A ÷ s$; $n = 192$ ft^2 ÷ 12 ft; $n = 16$ ft

  The baby's playpen has an area of 9 ft^2. If the length of the playpen is 3 ft, what is the width of the playpen? $n = A ÷ s$; $n = 9$ ft^2 ÷ 3 ft; $n = 3$ ft

Student Edition pages 234–35
Daily Review Chapter 11, section *d*

OBJECTIVES
- Calculate the area of triangles using a formula.
- Relate the area of triangles to God's design. **BWS**
- Calculate the area of a complex figure.
- Calculate the unknown height or base of a triangle.

BIBLICAL WORLDVIEW SHAPING
- **Design (Explain): The same formula can be used to find all triangle areas because God designed an orderly world.**

TEACHER RESOURCES
- 57 *Area: Triangles* (for the teacher and for each student)

ADDITIONAL MATERIALS
- a ruler (for the teacher and for each student)
- a calculator (for each student)

Engage
- Direct attention to the essential question at the top of Student Edition page 234, "Why is the formula for the area of a triangle always the same?"
- Direct the students to **brainstorm** to explore the essential question.

Instruct
Calculating the area of triangles
- Use **guided discovery** to help the students find the area of a triangle.
- Distribute an *Area: Triangles* page to each student and display your copy.
  What is the formula for finding the area of a rectangle or square? $A = l \cdot w$

What is the formula for finding the area of a parallelogram? $A = b \cdot h$

Remind the students that a rectangle and a square can also be classified as a parallelogram: a quadrilateral whose opposite sides are parallel. Explain that since rectangles and squares are quadrilaterals that have a width that is perpendicular to their length, the width can also be called the height and the length can also be called the base.

Could you use the formula $A = b \cdot h$ to find the area of a square and a rectangle? Yes; a rectangle and a square both have sides perpendicular to each other, so they have a base and a height.

- Write "$A = b \cdot h$" for display and direct the students to write on the *Area: Triangles* page the equation and solution for the area of the rectangle. $A = 12 \text{ units} \times 6 \text{ units}; A = 72 \text{ units}^2$

- Instruct the students to use a ruler and draw a diagonal from 1 vertex to another to divide the rectangle in half. Demonstrate.

What 2 geometric figures were formed? 2 triangles

What fraction of the whole rectangle is each triangle? $\frac{1}{2}$; A rectangle is 2 congruent triangles.

Area of Triangles

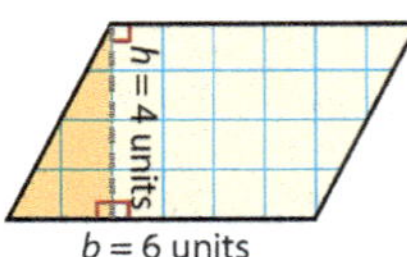

Why is the formula for the area of a triangle always the same?

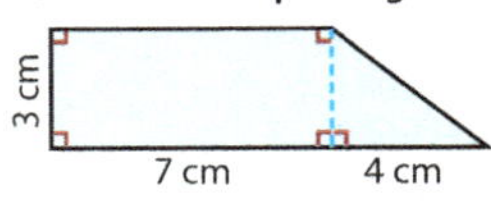

Area of a Parallelogram
$A = b \cdot h$

Area of a Rectangle
$A = l \cdot w$

To change a parallelogram to a rectangle, you can remove a triangle from one side of the parallelogram and connect it to the other side.

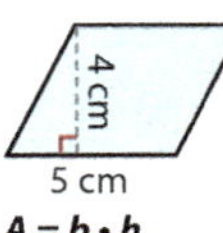

$A = b \cdot h$
$A = 6 \cdot 4$
$A = 24 \text{ units}^2$

Area of a Complex Figure

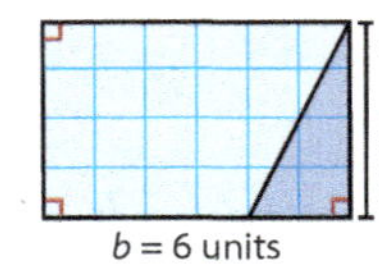

$A = (7 \cdot 3) + \frac{1}{2}(4 \cdot 3)$
$A = 21 + \frac{1}{2}(12)$
$A = 21 + 6$
$A = 27 \text{ cm}^2$

Area of a Triangle

A diagonal divides a parallelogram into two congruent triangles. The area of each triangle is $\frac{1}{2}$ of the area of the parallelogram.

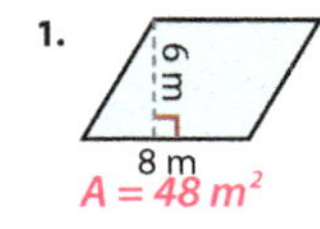

$A = b \cdot h$
$A = 5 \cdot 4$
$A = 20 \text{ cm}^2$

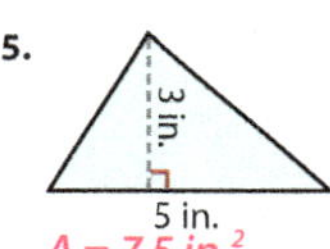

$A = \frac{1}{2}(b \cdot h)$
$A = \frac{1}{2}(5 \cdot 4)$
$A = \frac{1}{2}(20)$
$A = 10 \text{ cm}^2$

Each triangle has an area of 10 cm².

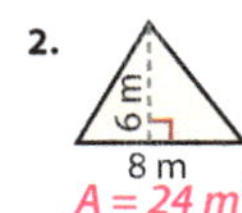

$A = l \cdot w$
$A = 6 \cdot 4$
$A = 24 \text{ cm}^2$

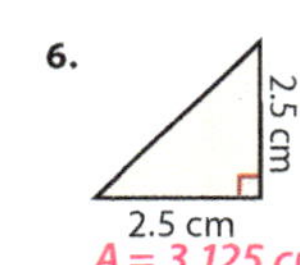

$A = \frac{1}{2}(b \cdot h)$
$A = \frac{1}{2}(6 \cdot 4)$
$A = \frac{1}{2}(24)$
$A = 12 \text{ cm}^2$

Each triangle has an area of 12 cm².

Exercises

Find the area of the figure.

1.
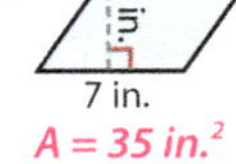

$A = 48 \text{ m}^2$

2.

$A = 24 \text{ m}^2$

3.
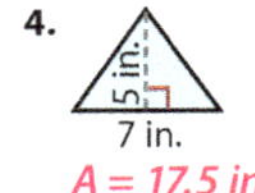

$A = 35 \text{ in.}^2$

4.

$A = 17.5 \text{ in.}^2$

5.
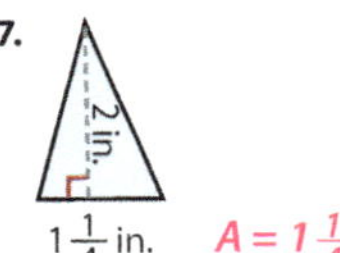

$A = 7.5 \text{ in.}^2$

6.

$A = 3.125 \text{ cm}^2$

7.

$A = 1\frac{1}{4} \text{ in.}^2$

8. Why is the formula for the area of a triangle always the same?
The formula is the same for all triangles because God designed an orderly world.

234   Chapter 11

Find the perimeter and the area of the complex figure.

9. 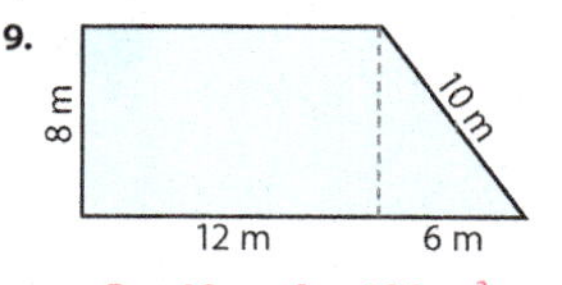
8 m / 12 m / 6 m / 10 m
P = 48 m; A = 120 m²

10. 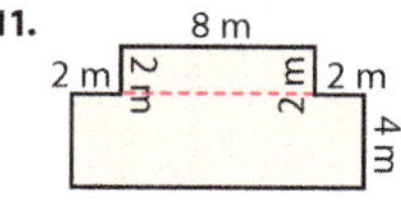
2 ft / 5 ft / 9 ft / 12 ft
P = 42 ft; A = 68 ft²

11. 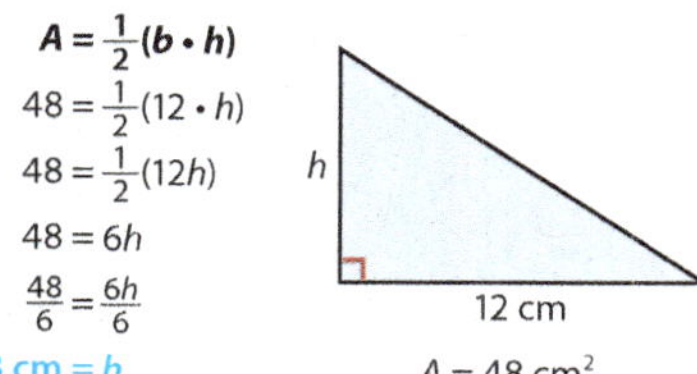
8 m / 2 m / 2 m / 2 m / 2 m / 4 m
P = 36 m; A = 64 m²

Finding an Unknown Measurement

Substitute the known measurements into the formula and then solve for the unknown.

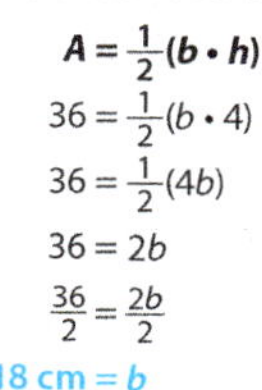

$A = \frac{1}{2}(b \cdot h)$

$48 = \frac{1}{2}(12 \cdot h)$

$48 = \frac{1}{2}(12h)$

$48 = 6h$

$\frac{48}{6} = \frac{6h}{6}$

8 cm = h

h / 12 cm / *A = 48 cm²*

$A = \frac{1}{2}(b \cdot h)$

$36 = \frac{1}{2}(b \cdot 4)$

$36 = \frac{1}{2}(4b)$

$36 = 2b$

$\frac{36}{2} = \frac{2b}{2}$

18 cm = b

4 cm / *b* / *A = 36 cm²*

Find the unknown measurement of the triangle.

12. $A = 30$ ft²
$b = 10$ ft
$h = \_\_$ *6 ft*

13. $A = 15$ m²
$b = \_\_$ *3 m*
$h = 10$ m

14. $A = 24$ yd²
$b = 6$ yd
$h = \_\_$ *8 yd*

Estimate the area of the figure. Each square on the grid represents 1 cm².

15. 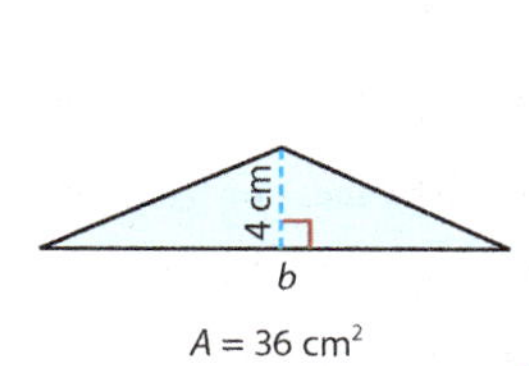
A = 42 cm²

16. 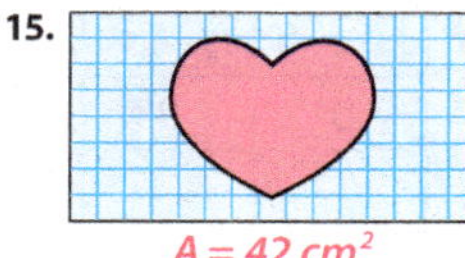
A = 24 cm²

17. 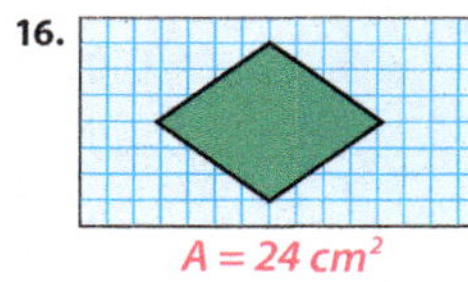
A = 48 cm²

Solve. Round to the nearest whole number if needed. *Equations may vary.*

18. The "Welcome Home, Soldiers!" banner over the church entrance is a rectangle with a length that is 2 times the width. The width of the banner is 6 ft. What is the area of the banner?
A = (2 • 6 ft) • 6 ft = 72 ft²

19. A builder plans to place ceramic tile in a bathroom that is 10 ft long and 6 ft wide. The tiles are 12 in. by 12 in. How many tiles will the builder need to cover the entire floor?
A = 10 ft • 6 ft = 60 ft²; 60 ÷ 1 = 60 tiles

20. Ani made a circular clay pot with a diameter of 8.6 in. She wants to glue beads around the rim of the pot. How many beads will she use if she glues 1 in. beads around the rim?
C = 3.14 × 8.6 = 27.004; 27 beads

21. Mr. Jeffrey is placing a small fence around his garden to keep the rabbits out. His square garden is 15 ft on each side. How much fencing does he need to go around the garden?
P = 4 ft • 15 ft = 60 ft

22. Parker is making felt pennants for the school soccer team. Each pennant will be 18 in. long and have a height of 8 in. How many square inches of felt does he need for each triangular pennant? *A = $\frac{1}{2}$(18 in. • 8 in.) = 72 in.²*

23. On Saturday Colin painted a wall in his bedroom. The wall is 12 ft long and 8 ft high. What is the area of the wall? *A = 12 ft • 8 ft = 96 ft²*

What is the area of each triangle? 36 units²; The area of each triangle is $\frac{1}{2}$ the area of the rectangle.

Since every rectangle can be partitioned to form two congruent triangles, what formula can you write to find the area of each triangle? Explain. $A = \frac{1}{2}(l \times w)$ or $A = \frac{1}{2}(b \times h)$; I can find the area of the rectangle and multiply the area by $\frac{1}{2}$ or divide it by 2.

- Write "$A = \frac{1}{2}(b \cdot h)$" for display. Explain that the area of any triangle is $\frac{1}{2}$ the area of its related rectangle (parallelogram).

Guide the students in using the formula to write the equation to find the area of one of the triangles: $A = \frac{1}{2}(12 \times 6)$.

Instruct them to solve the equation. $A = \frac{1}{2}(72$ units²$)$; $A = 36$ units² Discuss the solution as needed and instruct the students to write the equation and solution for the area of the triangle. $A = \frac{1}{2}(12$ units × 6 units$)$; $A = 36$ units²

- Follow a similar procedure to guide the students in finding the area of the square and then the area of the triangles that are formed by drawing a diagonal inside the square. $A = 8$ units × 8 units, $A = 64$ units²; $A = \frac{1}{2}(8$ units × 8 units$)$, $A = 32$ units²

- Direct attention to the parallelogram.

Does a parallelogram have perpendicular sides? no

What must you know in order to calculate the area of a parallelogram? I must identify the base and the height of the parallelogram.

What represents the height in a parallelogram? the length of a line segment that is perpendicular to both bases

Follow a procedure similar to the one used for the rectangle and the square as you guide the students in finding the area of the parallelogram and then the area of the triangles that are formed by drawing a diagonal inside the parallelogram. $A = 11$ units × 5 units, $A = 55$ units²; $A = \frac{1}{2}(11$ units × 5 units$)$, $A = 27.5$ units²

- Draw for display a right triangle, an acute triangle, and an obtuse triangle similar to the ones below. Direct the students to draw similar triangles on the back of the *Area: Triangles* page.

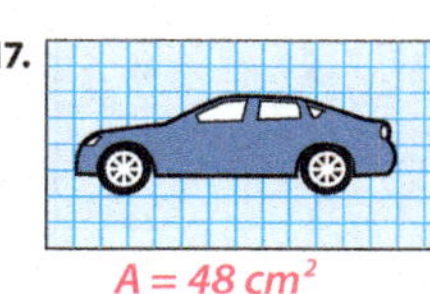

- Instruct the students to measure the base and height of their right triangles to the nearest millimeter and to write each measurement along the line segment. Point out that in a right triangle the perpendicular sides are the base and the height.

What formula can you use to find the area of your right triangle? $\frac{1}{2}(b \cdot h)$

Guide the students in writing and solving the equation to find the area of their right triangles. Answers will vary.

- Direct attention to the acute triangle drawn for display.

How do you think you can identify the height of an acute triangle? I can draw a line segment from 1 of the vertices to the opposite side; the line segment must be perpendicular to the opposite side.

Demonstrate drawing a line segment to represent the height of your acute triangle.

Instruct each student to first draw the line segment representing the height of his acute triangle and then to measure the base and the height to the nearest millimeter and write the measurements. Then direct the students to write and solve the equation to calculate the area of their triangle. Answers will vary.

LESSON 107

• Explain that the height of an obtuse triangle may be located outside the triangle. Demonstrate drawing the height of the obtuse triangle drawn for display.

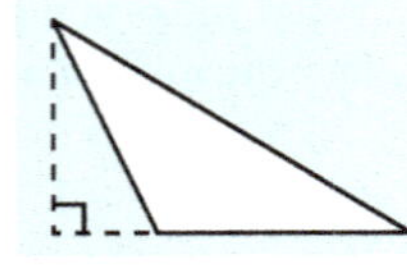

Follow a similar procedure to that of the other triangles to write an equation to find the area and solve it.

Recognizing God's design

• Guide the students in a **review** of their previous brainstorming activity. Encourage them to share some of their discussion answers to the essential question.

Why is the formula for the area of a triangle always the same? The formula is the same for all triangles because God designed an orderly world.

Calculating the area of a complex figure

• Guide the students in a **discussion** to help them find the area of a complex figure.

• Direct attention to the complex figure on the *Area: Triangles* page.

How can you find the area of this irregular polygon? sample answers: count the square units; partition the polygon into 1 square and 1 triangle, calculate the two areas, and then add the areas

• Direct the students to write and solve an equation to find the area of the square and then to write and solve an equation to find the area of the triangle. $A = 8$ units × 8 units, $A = 64$ units²; $A = \frac{1}{2}(8$ units × 6 units$)$, $A = 24$ units²

What equation can you write to find the total area of the polygon? $A = (8$ units × 8 units$) + \frac{1}{2}(8$ units × 6 units$)$

Write the equation for display.

Instruct the students to solve the equation and then to write the solution on the page. $A = 88$ units²

Write the missing number or variable. Name the property used.

1. $(5 \cdot 3) \cdot 4 = 5 \cdot (\underline{3} \cdot 4)$
Associative Property

2. $a + b = \underline{b} + a$
Commutative Property

3. $3 + 2a = 2a + \underline{3}$
Commutative Property

Simplify the expression.

4. $x + 5x$ *6x*

5. $x + 8 + x$ *8 + 2x*

6. $x \cdot 4 \cdot 5$ *20x*

Solve the equation by using the inverse operation.

7. $a + 10 = 25$
$a = \underline{\quad}$ *15*

8. $3 \cdot n = 18$
$n = \underline{\quad}$ *6*

9. $12 - x = 7$
$x = \underline{\quad}$ *5*

10. $\frac{x}{3} = 9$
$x = \underline{\quad}$ *27*

11. $8n = 32$
$n = \underline{\quad}$ *4*

12. $15 \div c = 3$
$c = \underline{\quad}$ *5*

Complete the table.

13.

x	4x
5	20
7	28
10	40

14.

x	x²
2	4
4	16
6	36

15.

x	3x − 1
3	8
5	14
7	20

Daily Review   471

Calculating the unknown height or base of a triangle

• Guide a **problem-solving exercise** to help the students calculate the unknown triangle height or base.

• Read aloud the following word problem.

The sixth-grade classes are making pennants for the game. Each pennant has an area of 45 in.². If each pennant is 10 in. long, what is the height of each pennant? 9 in.

• Write "$A = \frac{1}{2}(b \cdot h)$" for display. Explain that if you know the area of a triangle and the length of its base, you can find the measurement of its height.

What number can you substitute for A? 45; The area of each pennant is 45 in.².
What number can you substitute for b? 10; Each pennant is 10 in. long.
Write "$45 = \frac{1}{2}(10h)$" for display.
What is $\frac{1}{2}$ of 10h? 5h

Write "$45 = 5h$" below $45 = \frac{1}{2}(10h)$.
What is the height of each banner? 9 in.; I divide both sides of the equation by 5 to solve for h.

Why do you label the answer *inches*? Base and height are 1-dimensional units used to measure perimeter. The base is given in inches, so the height is also measured in inches.

- Follow a similar procedure for the following word problem. Lead the students to conclude that they can use the Commutative Property of Multiplication to change the order of the factors that represent the base and the height so that the coefficient is written in front of the variable (b): $A = \frac{1}{2}(b \cdot h)$; $6 = \frac{1}{2}(b \cdot 4)$; $6 = \frac{1}{2}(4b)$.

  Grandpa is making a triangular table. The table has an area of 6 ft^2. If the height of the triangular tabletop is 4 ft, what is the measurement of the base of the tabletop? $6 = \frac{1}{2}(4b)$; $6 = 2b$; $3 = b$; base = 3 ft

- Direct the students to find the unknown height or base of other triangles using these measurements. Give guidance as needed.

  Area = 20 ft^2
  base = 4 ft
  height = 10 ft

  Area = 25 ft^2
  base = 10 ft
  height = 5 ft

  Area = 36 ft^2
  base = 12 ft
  height = 6 ft

Apply

Student Edition pages 234–35

- Read and explain the directions for pages 234–35. Assist the students as they complete the pages independently.

Daily Review

- Students should complete Chapter 11, section *d*.

NOTES

Student Edition pages 236–37
Daily Review Chapter 11, section e

OBJECTIVES

- Calculate the area of a circle using a formula.
- Estimate the area of a circle.
- Solve an area word problem and interpret the solution.

TEACHER RESOURCES

- 58 *Area: Circles* (for the teacher and for each student)

ADDITIONAL MATERIALS

- a ruler (for each student)
- a calculator (for each student)

Engage

- Direct attention to the essential question at the top of Student Edition page 236, "How can I relate the area of a circle to the formula $A = b \cdot h$?"

- Lead the students in a **guided discovery** activity to explore the essential question.

Distribute a copy of the *Area: Circles* page to each student. Direct the students to find the approximate area of the top circle by counting the whole and partial square units. sample answer: 200 units

Why is it difficult to find the area of a circle by counting units? A circle contains many partial squares.

Instruct

Calculating the area of a circle

- Display the *Area: Circles* page and **demonstrate** each step as you guide the students in determining the formula for finding the area of a circle. (Refer to the figure pictured at the top of page 236 in the Student Edition.)

  1. Cut out both circles on the page. Lead the students to conclude that the circles are congruent. Set aside the circle containing the grid.
  2. Fold the other circle in half along the edge of the shaded area.

Area of Circles

How can I relate the area of a circle to the formula
$A = b \cdot h$?

The **area of a circle** can be found by using the formula for the area of a parallelogram ($A = b \cdot h$) and the formula for the circumference of a circle ($C = \pi d$).

Key Terms
- area of a circle
- $A = \pi r^2$

Determining the Formula for the Area of a Circle

Divide a circle into 8 congruent wedges. Shade each half circle.

Since $C = \pi d$, then $C = 2\pi r$, or $\frac{1}{2}C = \pi r$.

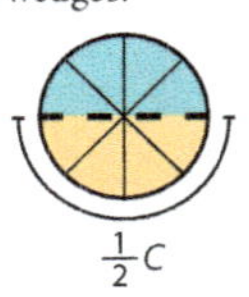

Arrange the wedges to form a figure that is similar to a parallelogram.

- The base of the figure is half the circumference of the circle.
- The height of the figure is the radius of the circle.

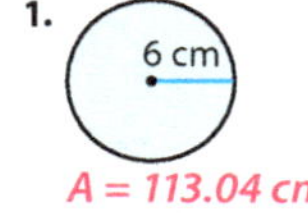

$A = b \cdot h$
$A = (\frac{1}{2}C)(r)$
$A = \frac{1}{2}(2\pi r)r$
$A = (\frac{1}{2} \cdot 2)\pi(r \cdot r)$
$A = \pi r^2$

The area of a circle is π times the radius squared: $A = \pi r^2$.

Finding the Area of a Circle

Substitute the length of the radius for r and 3.14 for π.

$r = 7$ ft
$A = \pi r^2$
$A = 3.14(7^2)$
$A = 3.14(49)$
$A = 153.86$ ft^2

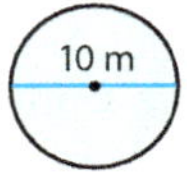

Remember that the radius is half the length of the diameter.

If $d = 10$ m, then $r = 5$ m.

$A = \pi r^2$
$A = 3.14(5^2)$
$A = 3.14(25)$
$A = 78.5$ m^2

Estimating the Area of a Circle

To estimate the area of a circle, round π to 3.

$A = \pi r^2$
$A = 3(4^2)$
$A = 3(16)$
$A = 48$ ft^2
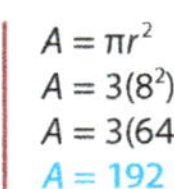

$A = \pi r^2$
$A = 3(8^2)$
$A = 3(64)$
$A = 192$ m^2
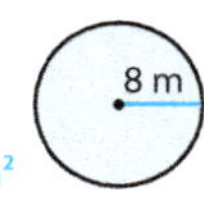

Exercises

Find the area of the circle. Use 3.14 for π.

1. 6 cm
$A = 113.04$ cm^2

2. 9 yd
$A = 254.34$ yd^2

3. 42 in.
$A = 1,384.74$ in.^{2}

4. 20 m
$A = 314$ m^2

Estimate the area of the circle. Round π to 3.

5. 15 m
$A = 675$ m^2

6. 4 cm
$A = 12$ cm^2

7. 6 ft
$A = 108$ ft^2

8. 26 ft
$A = 507$ ft^2

236 Chapter 11

---

3. Fold the circle in half again: fourths.
4. Fold the circle in half one more time: eighths.
5. Unfold the circle and cut out the eighths along the fold lines.
6. Place the white wedges pointing up in a row, with the bottom edges touching.
7. Place the shaded wedges between the white wedges, pointing down.

You may have students glue the wedges onto a sheet of paper and cut out the figure along its outer edges.

What geometric figure do the wedges resemble when they are arranged like this? The wedges form a figure that resembles a parallelogram.

- Direct the students to compare the parallelogram-type figure to the whole circle. Remind them that both of the circles were congruent.

What formula do you use to find the area of a parallelogram? $A = b \cdot h$

Write the formula for display.

How can I relate the area of a circle to the formula $b \cdot h$? I can divide the circle into 8 congruent wedges and arrange them to form a figure similar to that of a parallelogram.

Write the number that is needed to find the area of the circle.

9. 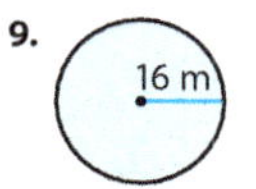
$A = 3.14 \cdot$ ___ 16^2

10. 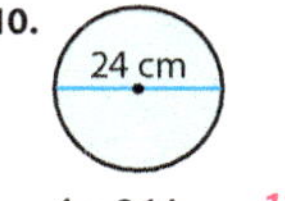
$A = 3.14 \cdot$ ___ 12^2

11. 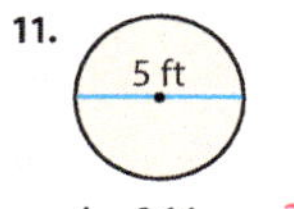
$A = 3.14 \cdot$ ___ 2.5^2

12. 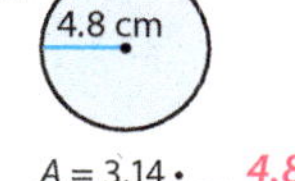
$A = 3.14 \cdot$ ___ 4.8^2

13. 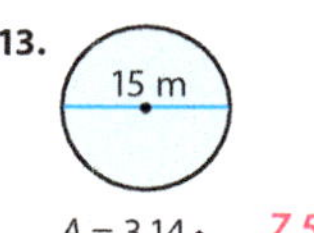
$A = 3.14 \cdot$ ___ 7.5^2

14. 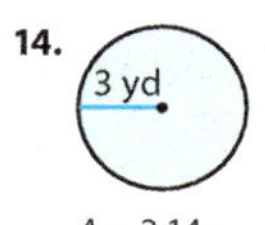
$A = 3.14 \cdot$ ___ 3^2

Find the area and the circumference of the circle.

15. 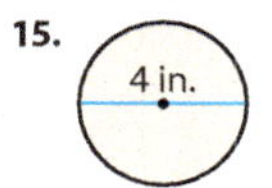
A = 12.56 in.²; C = 12.56 in.

16.
A = 28.26 ft²; C = 18.84 ft

17. 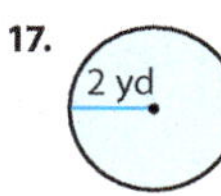
A = 12.56 yd²; C = 12.56 yd

18.
A = 3.14 yd²; C = 6.28 yd

19. 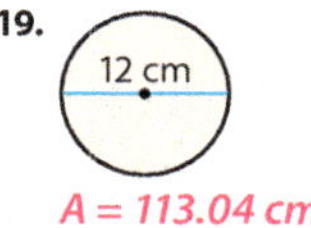
*A = 113.04 cm²;
C = 37.68 cm*

20. 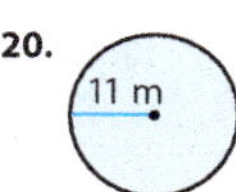
A = 379.94 m²; C = 69.08 m

Find the area of the orange-shaded region.

21. 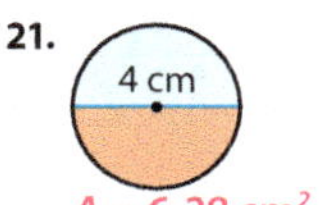
A = 6.28 cm²

22. 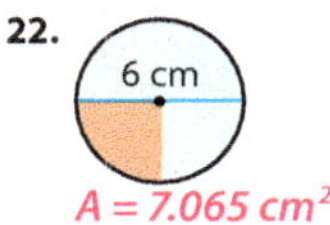
A = 7.065 cm²

23. 10 ft | 10 ft
A = 21.5 ft²

Practice & Application *Equations may vary.*

24. Julia designed a circular flower bed. The radius of the flower bed is 8 ft. What is the area of the flower bed? *A = 200.96 ft²*

25. Mrs. Davenport found a round tablecloth with an area of 11 ft². Will this tablecloth cover her table that has a diameter of 4 ft? *A = 12.56 ft²; no*

26. Mr. King built a circular sandbox for the playground. The radius of the sandbox is 6 ft. What is the area of the sandbox? *A = 113.04 ft²*

27. How can I relate the area of a circle to the formula $A = b \cdot h$? *I can divide the circle into 8 congruent wedges and arrange them to form a figure similar to that of a parallelogram.*

Do you think the height of your curved "parallelogram" is closer in length to the radius or the diameter of the whole circle? the radius

Instruct the students to slide the whole circle underneath the "parallelogram" to see that the height is equal to the radius. Direct attention to the base of the "parallelogram."

How many of the 8 edges of the circle make up the length of the base? 4; One-half of the edges make up the length of the base.

Remind the students that the circumference of a circle is the measurement of the edge of a whole circle. Lead the students to conclude that half of the circumference of the circle makes up the length of the base of the "parallelogram."

- Direct attention to $A = b \cdot h$ written for display.

Since the length of the base is $\frac{1}{2}$ of the circumference, what can we substitute for b in the formula? $\frac{1}{2}C$

Since the height is equal to the measurement of the radius, what can we substitute for h in the formula? r

- Write "$A = \frac{1}{2}C \cdot r$" for display.

What formula do you use to find the circumference of a circle? πd or $2\pi r$

LESSON 108

Show students that when $2\pi r$ is substituted into the formula for circumference, it becomes $A = \frac{1}{2}(2\pi r)r$. Explain that this formula can be written in a simpler form. Since $\frac{1}{2} \times 2 = 1$ and the radius multiplied by itself is r-squared, the formula for the area of a circle can be written $A = \pi r^2$.

$$A = (\tfrac{1}{2}C)(r)$$
$$A = \tfrac{1}{2}(2\pi r)(r)$$
$$A = (\tfrac{1}{2} \cdot \tfrac{2}{1})\pi(r \cdot r)$$
$$A = \pi r^2$$

- Direct attention to the whole circle that was cut from the *Area: Circles* page. Guide the students in using the formula to find the area of the whole circle. $A = \pi r^2$; $A = 3.14 \times (8 \text{ units})^2$; $A = 3.14 \times 64 \text{ units}^2$; $A = 200.96 \text{ units}^2$

- Draw for display two circles, each with one of the following radii. Write each measurement along the corresponding radius. Guide the students in using $A = \pi r^2$ to find the area of each circle. Continue to display the circles.

radius = 10 m
$A = 3.14 \times (10 \text{ m})^2$
$A = 3.14 \times 100 \text{ m}^2$
$A = 314 \text{ m}^2$

radius = 4 cm
$A = 3.14 \times (4 \text{ cm})^2$
$A = 3.14 \times 16 \text{ cm}^2$
$A = 50.24 \text{ cm}^2$

How can you determine the radius if you only know the measurement of the diameter? I can divide the diameter by 2 or multiply it by $\frac{1}{2}$; the radius is half of the diameter.

- Draw for display a circle with a diameter of 10 cm.

What is the length of the radius? 5 cm; The radius is $\frac{1}{2}$ of the diameter; $10 \div 2 = 5$ cm or $\frac{1}{2} \times 10 = 5$ cm.

Guide the students in using the formula to find the area. $A = 3.14 \times (5 \text{ cm})^2$; $A = 3.14 \times 25 \text{ cm}^2$; $A = 78.5 \text{ cm}^2$ Continue to display the circles.

Estimating the area of a circle

- Use a **practice exercise** to help the students estimate the area of a circle.

LESSON 108

- Explain that when an exact answer is not needed, it is helpful to estimate the area of a circle by rounding pi to the nearest whole number.

  How could you estimate the area of a circle? I can round pi (3.14) to the nearest whole number (3) and then multiply by the radius squared.

- Display the whole circle again. Remind the students that the radius is 8 units long and the area is 200.96 units². Choose a student to write for display and solve an equation to estimate the area of the circle. $A = 3 \times (8 \text{ units})^2$; $A = 3 \times 64 \text{ units}^2$; $A = 192 \text{ units}^2$

- Guide the students in estimating the area of the circles that were previously drawn for display. Compare the estimated areas to the calculated areas.

  radius = 10 m $A = 300 \text{ m}^2$

  radius = 4 cm $A = 48 \text{ cm}^2$

  diameter = 10 cm $r = 5 \text{ cm}$; $A = 75 \text{ cm}^2$

Solving an area word problem

- Guide the students in **drawing a picture** to help them solve word problems.

- Guide the students in first estimating and then solving the following word problems. While solving the last two problems, remind the students that the Commutative and Associative Properties of Multiplication allow them to reorder and regroup the factors.

  Mrs. Arnold purchased a circular rug to be placed under the kitchen table. The radius of the rug is 3 ft. What is the area of the rug? $A = 3.14(3^2)$; $A = 3.14(9)$; $A = 28.26 \text{ ft}^2$

  Mrs. Arnold also purchased a smaller rug to be placed at the doorway in the kitchen. This rug is a half-circle with a radius of 2 ft. What is the area of the rug? $A = \frac{1}{2}(3.14 \times 2^2)$; $A = (\frac{1}{2} \times 2^2) \times 3.14$; $A = (\frac{1}{2} \times 4) \times 3.14$; $A = 2 \times 3.14$; $A = 6.28 \text{ ft}^2$

The school has a circular flower bed around the flag pole. The radius of this circle is 6 ft. The sixth-grade class planted roses in $\frac{1}{4}$ of the flower bed. What is the area of the section that has roses planted by the sixth-grade class? $A = \frac{1}{4}(3.14 \times 6^2)$; $A = (\frac{1}{4} \times 6^2) \times 3.14$; $A = (\frac{1}{4} \times 36) \times 3.14$; $A = 9 \times 3.14$; $A = 28.26 \text{ ft}^2$

Apply

Student Edition pages 236–37

- Read and explain the directions for pages 236–37. Assist the students as they complete the pages independently.

Daily Review

- Students should complete Chapter 11, section e.

Solve.

1.	2.	3.	4.	5.
324 × 12 = 3,888	835 × 15 = 12,525	1,280 × 21 = 26,880	238 × 34 = 8,092	507 × 42 = 21,294

6.	7.	8.	9.	10.
450 × 312 = 140,400	513 × 142 = 72,846	831 × 123 = 102,213	452 × 171 = 77,292	324 × 214 = 69,336

11.	12.	13.	14.	15.
12,475 × 20 = 249,500	$15.75 × 4 = $63.00	0.03 × 0.21 = 0.0063	2.53 × 0.04 = 0.1012	$21.48 × 5 = $107.40

16. $3 \times \$1.75$ *$5.25*   **17.** 2.4×3.7 *8.88*   **18.** $8\frac{1}{2} \times 2\frac{1}{3}$ *$19\frac{5}{6}$*   **19.** $\frac{3}{5} \cdot 3\frac{9}{5} = 1\frac{4}{5}$ **20.** $\frac{3}{4} \cdot \frac{2}{3}$ *$\frac{6}{12} = \frac{1}{2}$*

472   Daily Review

Use the following to provide extra help for students who experience difficulty with the concepts taught in Chapter 11.

Identify the net of a 3-dimensional figure.

Provide the students with the polyhedrons that were constructed in Lesson 62, or construct new polyhedrons using the faces on the *Polyhedrons* page. Ask the students to select and identify one of the polyhedrons (e.g., a square pyramid). Explain that they can think of the base of the square pyramid as the center of its net with one side of each triangular face connected to each side of the square base. Direct the students to sketch a prediction of what the figure's net would look like. Then guide them in cutting the square pyramid so that it opens flat to form its net. Guide the students in comparing their sketched net to the net of the actual figure.

Repeat the activity using the other polyhedrons from Lesson 62. Remind the students that prisms have 2 congruent bases.

(*Note:* You may provide the student with objects such as a small cereal box, a Toblerone™ chocolate box, or a small oatmeal container rather than use the polyhedrons from Lesson 62.)

NOTES

OBJECTIVES

- Identify a 3-dimensional figure from its net.
- Calculate the surface area of rectangular, square, and triangular prisms using formulas.
- Construct a triangular prism.
- Evaluate the claim that order and consistency in nature explain the world apart from God. **BWS**

BIBLICAL WORLDVIEW SHAPING

- **Design (Evaluate):** People explain order and consistency in nature apart from God by rejecting God's Word and attributing patterns to chance.

TEACHER RESOURCES

- 59 *Surface Area: Triangular Prism* (for the teacher and for each student)

ADDITIONAL MATERIALS

- a classroom set of 3-dimensional figures or real-world objects
- a cereal box (rectangular prism) (for the teacher and for each 2 or 3 students)
- a ruler (for the teacher and for each student)
- transparent tape (for each student)

ASSESSMENTS

- Chapter 11 Quiz 2

Engage

- Direct attention to the essential question at the top of Student Edition page 238, "How do people who don't believe in God explain order and consistency?" Lead the students in a **discussion** to explore the essential question.

Do you think that people that do not believe in God can still observe orderly patterns in nature? **yes**

Surface Area of Prisms

How do people who don't believe in God explain order and consistency?

A **net** is a flat pattern of 2-dimensional surfaces that can be shaped into a 3-dimensional figure.

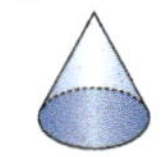 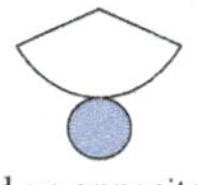

A **cone** has 1 base and an opposite vertex.

A **pyramid** is a type of cone that has a polygon instead of a circle as a base. A pyramid is named for the shape of its base. All other faces of a pyramid are triangles.

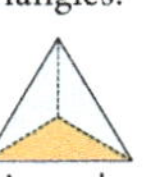  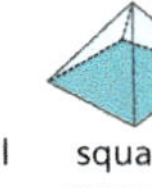

triangular pyramid | pentagonal pyramid | square pyramid

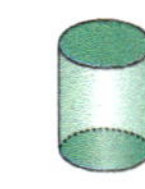 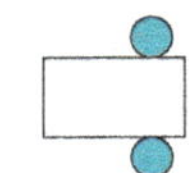

A **cylinder** has 2 congruent circular bases that are parallel.

A **prism** is a type of cylinder that has 2 congruent polygon bases instead of circular bases. A prism is named for the shape of its bases. All other faces of a prism are parallelograms.

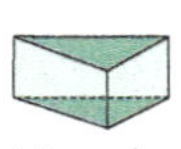 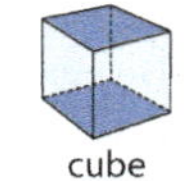 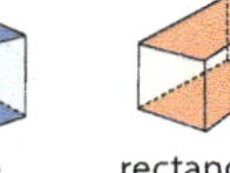

triangular prism | cube (square prism) | rectangular prism

Key Terms
- net
- cone
- pyramid
- cylinder
- prism
- surface area
- rectangular prism
- triangular prism

Exercises

Write the name of the figure that the net will make. The bases of the nets are shaded.

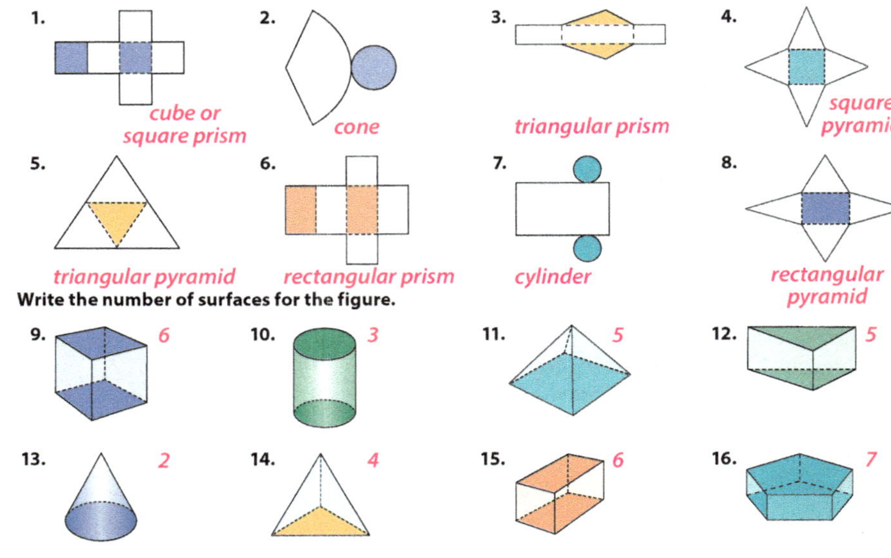

Write the number of surfaces for the figure.

9. 6
10. 3
11. 5
12. 5
13. 2
14. 4
15. 6
16. 7

238   Chapter 11

How do you think they explain the consistency and design they see? **sample answer: They may claim a belief in Mother Nature or a higher intelligence without naming God, or they may believe patterns resulted by chance.**

Challenge the students to think about this question as they progress through the lesson.

Instruct

Identifying a 3-dimensional figure

- Use **visual aids** to help the students identify 3-dimensional figures.

What is the difference between 2-dimensional and 3-dimensional figures? **2-dimensional figures lie in the same plane and have length and width; 3-dimensional figures lie in more than one plane and have length, width, and height.**

- Display your examples of a cone and a cylinder. Remind the students that conical figures are 3-dimensional figures with 1 base, similar to a cone. Cylindrical figures are 3-dimensional figures with 2 congruent bases, similar to a cylinder. The 2 bases may vary in size when compared to other cylindrical figures, but within a figure the 2 bases must be congruent.

Finding the Surface Area of Prisms

The **surface area** of a 3-dimensional figure is the sum of the areas of all its surfaces.

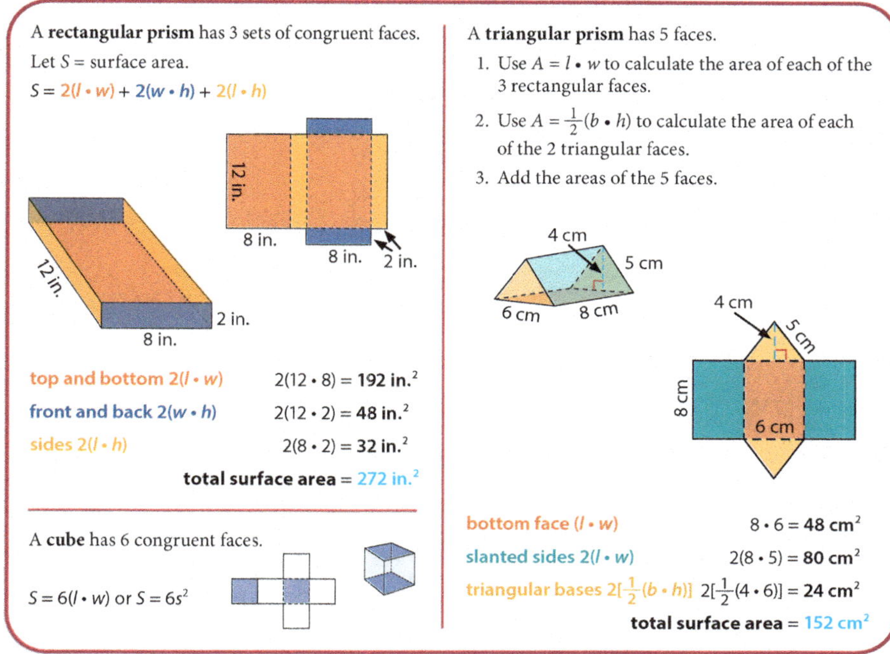

A **rectangular prism** has 3 sets of congruent faces.
Let S = surface area.

$$S = 2(l \cdot w) + 2(w \cdot h) + 2(l \cdot h)$$

top and bottom $2(l \cdot w)$	$2(12 \cdot 8) =$ **192 in.^{2}**
front and back $2(w \cdot h)$	$2(12 \cdot 2) =$ **48 in.^{2}**
sides $2(l \cdot h)$	$2(8 \cdot 2) =$ **32 in.^{2}**
total surface area =	**272 in.^{2}**

A **cube** has 6 congruent faces.

$$S = 6(l \cdot w) \text{ or } S = 6s^2$$

A **triangular prism** has 5 faces.

1. Use $A = l \cdot w$ to calculate the area of each of the 3 rectangular faces.
2. Use $A = \frac{1}{2}(b \cdot h)$ to calculate the area of each of the 2 triangular faces.
3. Add the areas of the 5 faces.

bottom face $(l \cdot w)$	$8 \cdot 6 =$ **48 cm^2**
slanted sides $2(l \cdot w)$	$2(8 \cdot 5) =$ **80 cm^2**
triangular bases $2[\frac{1}{2}(b \cdot h)]$	$2[\frac{1}{2}(4 \cdot 6)] =$ **24 cm^2**
total surface area =	**152 cm^2**

Find the surface area of the prism.

17.

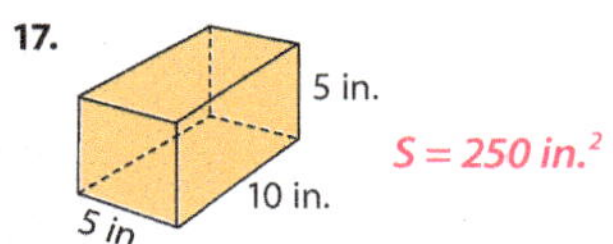

$S = 250$ in.^{2}

18.

$S = 150$ in.^{2}

19.

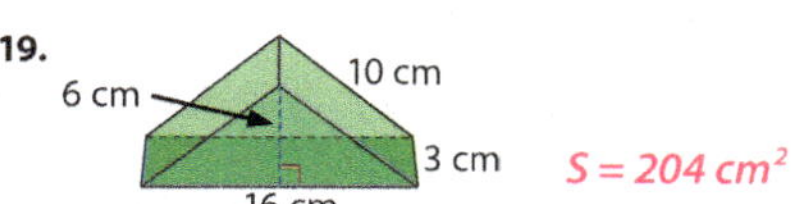

$S = 204$ cm^2

20.

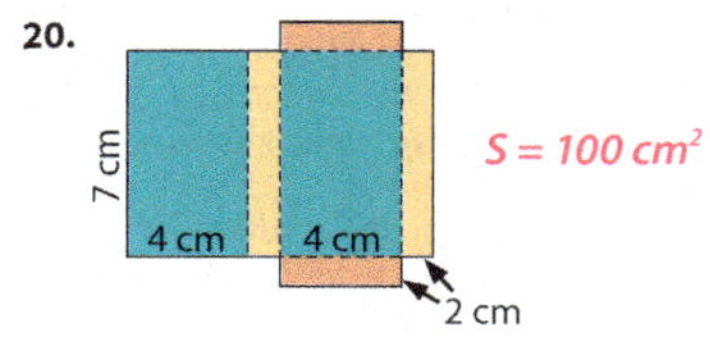

$S = 100$ cm^2

21. Explain why a cube and a rectangular prism have 6 faces, but a triangular prism has only 5 faces. *A rectangle has 4 sides plus the 2 bases, giving the prism 6 faces; a triangle has 3 sides plus the 2 bases, giving the prism 5 faces.*

22. How do people who don't believe in God explain order and consistency? *They reject the Word of God and think patterns in nature happen by chance.*

Calculating the surface area of prisms; constructing a triangular prism

- Use **collaborative learning** to help the students calculate surface area.

- Arrange the students in groups of 2 or 3 and distribute a cereal box to each group.

  How do you find the area of one face on your box? I multiply the face's length times its width.

  Explain that the surface area of a 3-dimensional figure is the sum of the areas of all its surfaces (faces).

  How do you think you can find the surface area of your box? I must first find the area of each of the 6 faces and then add all of the areas.

  Do you need to measure the length and width of all 6 faces to find the areas? No; the opposite faces in a rectangular prism are congruent, so I need to measure and find the area of only 3 different rectangles (1 in each of the pairs of congruent faces).

- Direct each group of students to develop a plan for finding the surface area of their rectangular prism. Allow a student from each group to share the group's ideas. Conclude that there are 3 pairs of congruent faces in any rectangular prism: adding the congruent top and bottom areas, the congruent front and back areas, and the congruent side areas together will give them the surface area of the rectangular prism.

  Write for display "Surface Area of a rectangular prism = (Area of top and bottom) + (Area of front and back) + (Area of sides)."

- Direct each group of students to measure the length and the width of each face of their cereal box to the nearest inch, calculate the area of each face, and add the 6 areas to find the surface area of the box (rectangular prism). Discuss each group's findings.

Display the other 3-dimensional objects. Choose students to identify each object, tell the number and shape of the bases, and then classify the object as cylindrical or conical.

- Display the cereal box. Lead the students to conclude that it is a rectangular prism.

  How many faces does a rectangular prism have? 6

  What do you notice about the faces of a rectangular prism? There are 3 pairs of congruent rectangular faces.

  Cut the cereal box along its edges so that it opens flat to form a net with 6 rectangular faces, including 2 rectangular bases.

Explain that a *net* is the 2-dimensional pattern of a 3-dimensional figure. The net for a rectangular prism includes 6 rectangles.

- Direct attention to the nets pictured on Student Edition page 238. Lead the students to conclude which illustrations are pyramid nets and which are prism nets. Point out the polygon base in the net for each pyramid and explain that pyramids and prisms are named for their bases. Also point out that the net for each prism has 2 congruent bases and that all of the side faces are parallelograms (rectangles).

LESSON 109

Are the length and width dimensions you used to determine the area of each face the same length and width dimensions of the prism? Explain. *Yes and no; the top and bottom faces ($l \times w$) are the same length and width measurements of the prism, the side faces are the width and height of the prism ($w \times h$), and the front and back faces are the length and height of the prism ($l \times h$).*

What one mathematical formula could you write to find the surface area of a rectangular prism, using its length, width, and height measurements? *$S = 2(l \times w) + 2(w \times h) + 2(l \times h)$*

Write "$S = 2(l \times w) + 2(w \times h) + 2(l \times h)$" below the formula that was written in word form. Explain that memorizing this formula is not necessary if you can determine which dimensions of the prism are used as the length and width of the faces.

- Guide the students in solving the following word problem. Allow them to refer to the formula and to draw a picture if needed.

When finding surface area, allow the students to solve one equation or several equations as shown on Student Edition page 239.

Molly is wrapping a birthday present in a rectangular box. The box is 2 ft long, 3 ft wide, and 1 ft high. What is the surface area of the box that Molly needs to cover? *$2(2 \text{ ft} \times 3 \text{ ft}) + 2(3 \text{ ft} \times 1 \text{ ft}) + 2(2 \text{ ft} \times 1 \text{ ft}) = 22 \text{ ft}^2$*

- Display your example of a cube. Remind the students that a cube is a square prism.
  How many faces does a cube have? *6*

  Do you need to measure the length and width of all 6 faces of a cube to find their areas? *No; all the faces of a cube are congruent squares, so I need to measure only 1 side of 1 face, find the area of 1 face, and multiply the area by 6.*

  Write for display, "Surface Area of a cube = 6(Area of 1 side)." Lead the students to conclude that "Area of 1 side" is ($l \times w$) or ($s \times s$). Write "$S = 6(l \bullet w)$" or "$S = 6s^2$" below the formula word form.

Use the table and the graphs to answer the question.

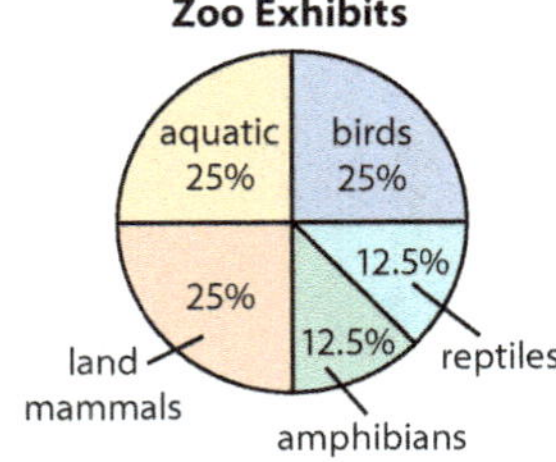

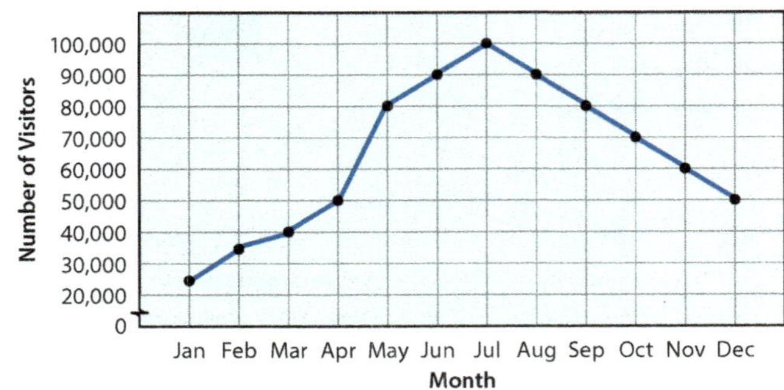

1. Which graph or table shows a change in the number of visitors over time? *the Visitors in 2021 line graph*
2. Which graph or table tells how many people visited the zoo in 2021? *the Visitors in 2021 line graph*
3. Which graph or table compares parts to a whole? *the Zoo Exhibits circle graph*
4. The Zoo Exhibits circle graph represents 800 zoo animals. How many animals are land mammals? How many are reptiles? *200 land mammals; 100 reptiles*
5. Which graph or table gives basic information about zoo admission costs? *the Zoo Admission table*
6. The Anderson family bought tickets to spend a day at the zoo. How much money did they spend on tickets for Mr. and Mrs. Anderson, 3 school-age boys, and Grandma Larson? *$(2 \times \$11.00) + (4 \times \$8.00) = \$54.00$*

Daily Review   473

- Choose a student to measure 1 side of the cube to the nearest inch, centimeter, or millimeter. Direct the students to find the surface area.

- Instruct the students to solve the following word problems. Give guidance as needed.

Mrs. Grant is making a pillow for her granddaughter by covering a foam cube. Each side of the cube is 12 in. Mrs. Grant needs to calculate the surface area of the foam cube in order to determine how much fabric to purchase. What is the surface area of the foam cube? *$S = 6(12 \text{ in.} \times 12 \text{ in.})$ or $S = 6(12 \text{ in.})^2$; $S = 864 \text{ in.}^2$*

Mrs. Grant purchased 1 yd of fabric that is 45 in. wide. Will she have enough fabric to cover the foam cube for her granddaughter's pillow? *yes; 36 in. × 45 in. = 1,620 in.²*

- Distribute a copy of the *Surface Area: Triangular Prism* page to each student and display your copy. Remind the students that a net can be used to construct a 3-dimensional figure.

  What does the net show you about the 3-dimensional figure called a triangular prism? *The net shows that the triangular prism has 5 faces. Two of the faces are congruent triangular bases; they show* that the figure is a triangular prism and give the prism its name. The other 3 faces are rectangles; two of the rectangular faces are congruent and the third rectangular face is 1 unit wider than the other two rectangular faces.

Do you need to find the area of each base? No; I can find the area of one of the congruent triangular bases and multiply it by 2.

What formula can you use to find the area of one triangular base? $A = \frac{1}{2}(b \cdot h)$

Write the formula for display.

What equation can you use to find the area of both triangular bases? $A = 2[\frac{1}{2}(6 \times 4)]$

Write the equation for display.

Do you need to find the area of each rectangular face? No; I can find the area of one of the congruent faces and multiply it by 2 to find the area of the two congruent faces, and then I can find the area of the other face.

- Direct the students to find the area of the 2 triangular bases and the 3 rectangular faces. Give guidance as needed.

2 triangular bases $A = 2[\frac{1}{2}(6\ \text{units} \times 4\ \text{units})]$; $A = 2(\frac{1}{2} \times 24\ \text{units}^2)$; $A = 2(12\ \text{units}^2)$; $A = 24\ \text{units}^2$

2 rectangular faces $A = 2(l \cdot w)$; $A = 2(5\ \text{units} \times 12\ \text{units})$; $A = 2(60\ \text{units}^2)$; $A = 120\ \text{units}^2$

1 rectangular face $A = l \cdot w$; $A = 6\ \text{units} \times 12\ \text{units}$; $A = 72\ \text{units}^2$

How can you find the surface area of the triangular prism? I can add the areas of all the faces.

What is the surface area? $24 + 120 + 72 = 216\ \text{units}^2$

- Instruct the students to cut out the net, fold it along the lines shared by 2 faces, and tape it together.

Order in nature without God?

- Guide a **verse analysis** to help the students interpret a Bible verse in order to evaluate a claim.

- Read aloud Romans 1:20.

According to this verse, God's eternal power and being have been clearly seen since when? creation

Since people can see the evidence of God through the created world, can they claim that they do not know about Him? No; the Bible says that they do not have an excuse.

- Read aloud Isaiah 42:5.

What does this verse tell us about God? He created the heaven and the earth; He created people and gave them breath.

- Point out that unbelievers may think the design we see happened by chance apart from God. Remind the students that God has revealed characteristics of Himself through nature. He has created our world so that we can recognize order and consistency.

How do you know that order comes from God? God shows us in creation and tells us in His Word.

How do people who don't believe in God explain order and consistency? They reject the Word of God and think patterns in nature happen by chance.

Apply

Student Edition pages 238–39
- Read and explain the directions for pages 238–39. Assist the students as they complete the pages independently.

Daily Review
- Students should complete Chapter 11, section *f*.

Assess

Quiz 2
- Use the **summative assessment** to evaluate the students' progress at this point in the chapter.

NOTES

LESSON 110

Student Edition pages 240–41
Daily Review Chapter 11, section *g*

OBJECTIVES

- Calculate the surface area of rectangular, square, and triangular prisms using formulas.
- Calculate the surface area of a cylinder using formulas.
- Construct a cylinder net.

TEACHER RESOURCES

- 60 *Surface Area*

ADDITIONAL MATERIALS

- a small oatmeal container
- 2 sheets of 12" × 18" construction paper (to cover the surface of the oatmeal container)
- a ruler (for the teacher and for each student)
- transparent tape (for the teacher and for each student)
- a cylinder (a can or a potato chip container with a top) (for each student)
- 1–2 sheets of construction paper (to cover the surface of the cylinder) (for each student)
- a calculator (for each student)

Engage

- Direct attention to the essential question at the top of Student Edition page 240, "How can finding the area of a rectangle help me find the surface area of a cylinder?"

Display a rectangle (sheet of paper) and direct the students to **brainstorm** to explore the essential question. sample answer: I need to see that the curved surface of a cylinder is actually a rectangle.

Instruct

Calculating the surface area of prisms

- Guide a **discussion** to help the students find the surface area of prisms.

Surface Area of Cylinders

How can finding the area of a rectangle help me find the surface area of a cylinder?

A **cylinder** has 2 congruent circular bases and 1 curved surface.

1. Calculate the area of one circular base, using $A = \pi r^2$. Multiply by 2 to find the area of both circular bases.
2. Calculate the area of the curved surface. The curved surface is a rectangle when lying flat. Use $A = l \cdot w$. The width of the rectangle is the height of the cylinder. The length of the rectangle is the circumference of the circular bases.

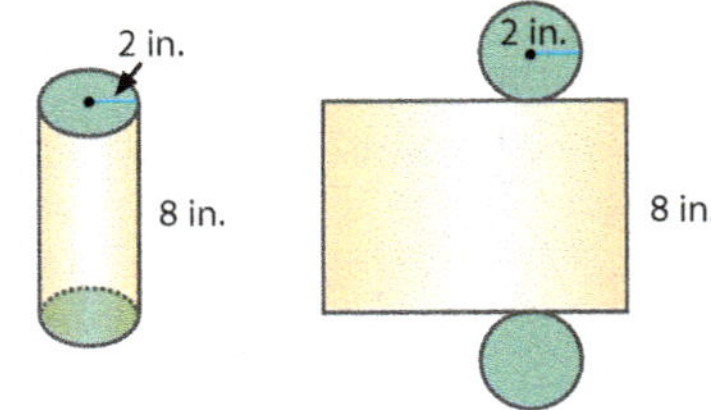

circular bases $2(\pi r^2)$ $2(3.14 \cdot 2^2) = $ **25.12 in.²**
curved surface ($l \cdot w$) $(3.14 \cdot 4)8 = $ **100.48 in.²**

total surface area = 125.60 in.²

Exercises

Find the surface area of the cylinder.

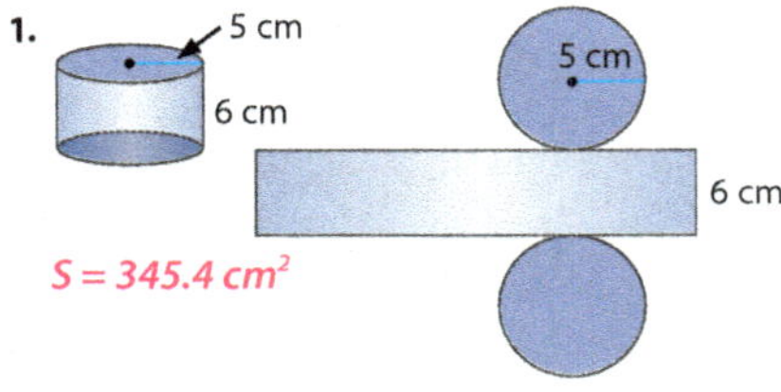

S = 345.4 cm²

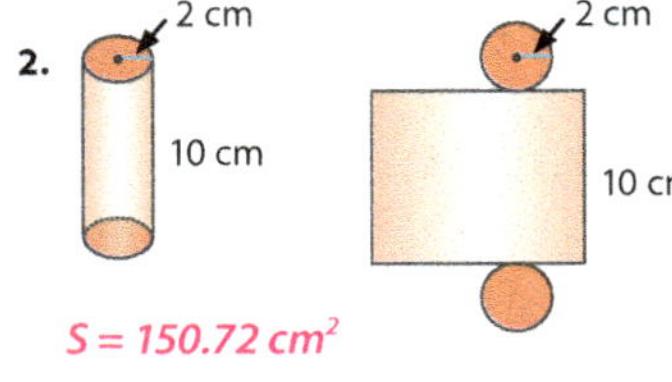

S = 150.72 cm²

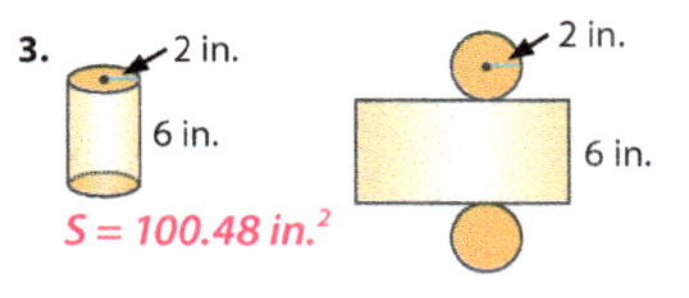

S = 100.48 in.²

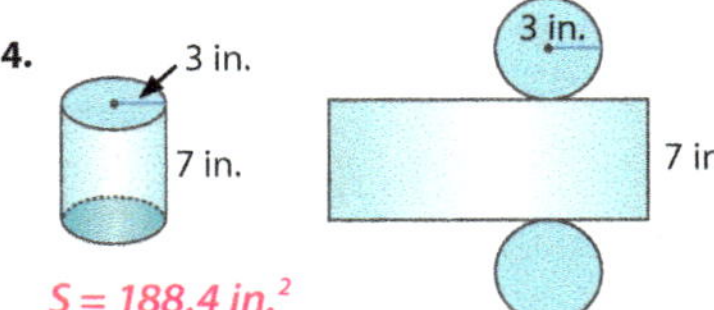

S = 188.4 in.²

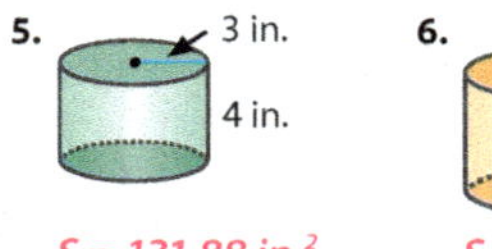

S = 131.88 in.²   *S = 226.08 cm²*

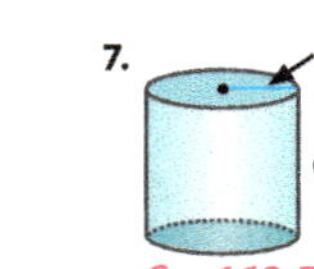

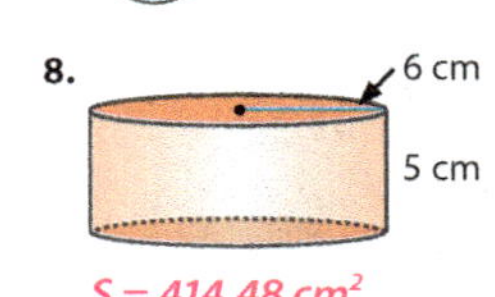

S = 169.56 in.²   *S = 414.48 cm²*

Write the answer.

9. The bases of the cylinders in problem 5 and problem 7 are congruent, but the heights are not. How will the nets differ? *The rectangle in the net for problem 7 will be wider (taller).*

10. The height of the cylinders in problem 6 and problem 8 is the same, but the bases are not congruent. How will the nets differ? *The rectangle in the net for problem 8 will be longer.*

What is surface area? the number of square units it takes to cover the surface of a 3-dimensional figure

How do you find the surface area of a figure? I find the area of each of the figure's faces and add all the areas.

- Read aloud the following word problem.

Beth made several scratching toys for her pet cats by covering the surfaces of some wooden blocks with carpet. She made one toy by using a wooden block that had 6 square faces and was 2 ft high. How much carpet did Beth use to cover all the surfaces of the block? 24 ft²

What is the shape of the wooden block Beth covered with carpet? a cube or a square prism; Each face is a square measuring 2 ft × 2 ft.

- Display the *Surface Area* page. Invite a student to write the height (2 ft), length (2 ft), and width (2 ft) dimensions along the edges of the cube for display.

How can you find the surface area of a cube? Since all 6 faces are congruent squares, I can first find the area of 1 square and then multiply the area of 1 square by 6.

What mathematical formula could you write to help you find the surface area of any cube? $S = 6(l \times w)$ or $S = 6s^2$

Write both formulas for display.

Guide the students in finding the surface area of the cube using either formula.

Find the surface area of the prism.

11. 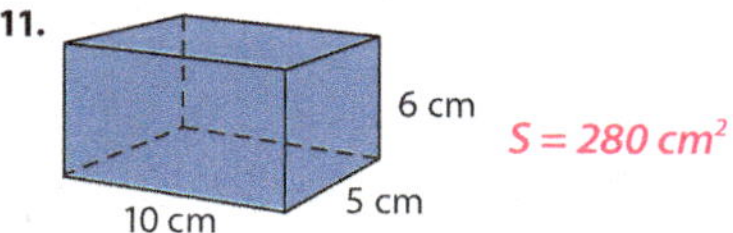
$S = 280 \text{ cm}^2$

12. 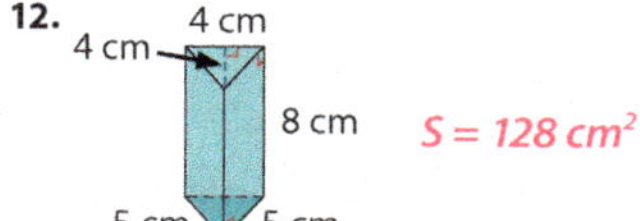
$S = 128 \text{ cm}^2$

13. 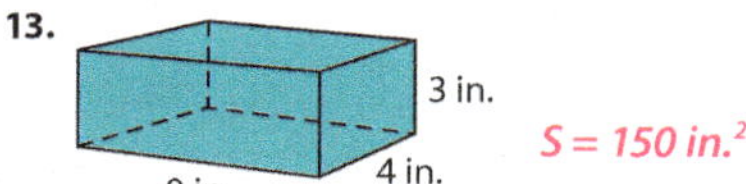
$S = 150 \text{ in.}^2$

14. 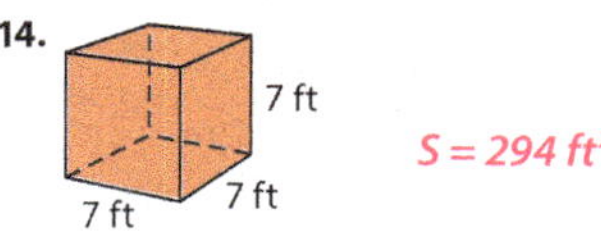
$S = 294 \text{ ft}^2$

15. 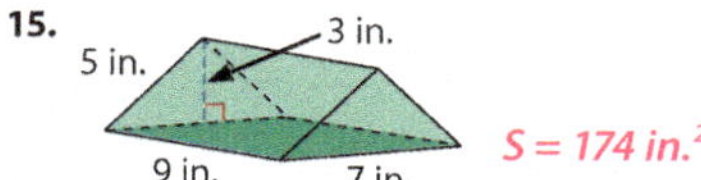
$S = 174 \text{ in.}^2$

16. 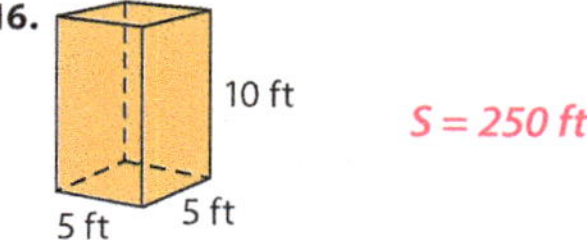
$S = 250 \text{ ft}^2$

Solve. Draw a model if needed.

17. Emma plans to cover a box with colorful adhesive paper. The box is 8 in. long, 6 in. wide, and 5 in. high. Will 225 in.² of adhesive paper cover the box?
2(8 • 6) + 2(8 • 5) + 2(6 • 5) = 236 in.²; no

18. Grace made a pillow in the shape of an octagon. Each side is 12 in. long. Will 100 in. of fringe be enough to go around the entire pillow?
8 • 12 = 96 in.; yes

19. Mr. Watkins painted a dodge ball circle on the playground. The circle has a diameter of 18 ft. What is the circumference of the circle?
3.14 • 18 = 56.52 ft

20. Pedro wants to paint 3 walls in his bedroom. His bedroom is 10 ft long and 10 ft wide, and the ceiling is 10 ft high. How much surface area is he going to paint? *3(10 • 10) = 300 ft²*

21. Why would you measure the perimeter of a rectangle in centimeters and the area of that same rectangle in square centimeters?

22. Can a cylinder have circular bases with different areas? Why or why not? *No, a cylinder is defined as having 2 congruent circular bases and 1 curved surface.*

23. How can finding the area of a rectangle help me find the area of a cylinder?

DID YOU KNOW?

Solomon had the interior of the temple in Jerusalem overlaid with gold. Workers had to figure the area of each surface to be covered in order to prepare the gold overlays.

So Solomon overlaid the house within with pure gold: and he made a partition by the chains of gold before the oracle; and he overlaid it with gold.

1 Kings 6:21

23. *The curved surface of a cylinder is a rectangle when lying flat. I can calculate the area of a rectangle to help find the surface area of a cylinder.*

Guide the students in identifying the 6 rectangular faces (3 pairs of congruent faces) and their length and width dimensions. Then use the $l \times w$ formula to guide the students in finding the area of each face.

top and bottom $A = 2 \text{ ft} \times 1 \text{ ft}; A = 2 \text{ ft}^2$
front and back $A = 2 \text{ ft} \times 1.5 \text{ ft}; A = 3 \text{ ft}^2$
sides $A = 1 \text{ ft} \times 1.5 \text{ ft}; A = 1.5 \text{ ft}^2$

What equation can you write to find the surface area for the rectangular prism? $S = 2(2 \text{ ft}^2) + 2(3 \text{ ft}^2) + 2(1.5 \text{ ft}^2)$; or $S = 2 \text{ ft}^2 + 2 \text{ ft}^2 + 3 \text{ ft}^2 + 3 \text{ ft}^2 + 1.5 \text{ ft}^2 + 1.5 \text{ ft}^2; S = 13 \text{ ft}^2$

- Read aloud the following word problem.

Beth made the third scratching toy from a wooden block with 5 faces. Each of the 3 rectangular sides has a length of 2 ft and a width of 1.5 ft. The 2 triangular faces have a base of 2 ft and a height of 1.75 ft. How much carpet did Beth use to cover all the surfaces of this block? 12.5 ft²

What is the shape of this scratching toy? a triangular prism; 3 sides (faces) are rectangles, and the top and bottom faces (bases) are triangles.

Lead the students to conclude that the 3 rectangular sides are congruent and the 2 triangular bases are congruent.

What formula can help you to find the area of a rectangle? $A = l \times w$

What formula can help you to find the area of a triangle? $A = \frac{1}{2}(b \times h)$

- Choose a student to identify and write the length and width dimensions of the 3 congruent rectangular sides (2 ft and 1.5 ft) for display. Use the $l \times w$ formula to find the area of each face. 3 ft² Next, choose another student to identify and write the base and height measurements of the 2 congruent triangles (2 ft and 1.75 ft) for display. Use the formula $\frac{1}{2}(b \times h)$ to find the area of each triangle. 1.75 ft²

What equation can you write to find the surface area for the triangular prism? $S = 3(3 \text{ ft}^2) + 2(1.75 \text{ ft}^2), S = 9 \text{ ft}^2 + 3.5 \text{ ft}^2$; or $S = 3 \text{ ft}^2 + 3 \text{ ft}^2 + 3 \text{ ft}^2 + 1.75 \text{ ft}^2 + 1.75 \text{ ft}^2$; $S = 12.5 \text{ ft}^2$

Choose a student to write the solution for display. $S = 6(2 \text{ ft} \times 2 \text{ ft}), S = 24 \text{ ft}^2$; or $S = 6 \times (2 \text{ ft})^2, S = 24 \text{ ft}^2$

What is the surface area of the cube-shaped scratching toy? 24 ft²

- Read aloud the following word problem.

To make the second scratching toy, Beth used a wooden block with 6 rectangular sides. The block has a length of 2 ft, a width of 1 ft, and a height of 1.5 ft. How much carpet did Beth need to cover all the surfaces of this block? 13 ft²

What is the shape of the wooden block that Beth used to make this scratching toy? a rectangular prism; The length, width, and height measurements are all different, so there are 3 pairs of opposite, congruent faces that are rectangles.

- Invite a student to write the length, width, and height dimensions along the edges of the rectangular prism for display.

How can you find the surface area of a rectangular prism? First, I need to identify each face and its dimensions, and find the area of each side or the area of 1 of each pair of congruent sides and multiply the area by 2; then I can add the areas together to find the surface area.

What formula can help you to find the area of a rectangle? $A = l \times w$

LESSON 110

11-G

Calculating the surface area of a cylinder; constructing a cylinder net

- Use **visual aids** to help the students find the surface area of a cylinder.

- Display the oatmeal container.
  How do you think you can find the surface area of a cylinder? I can find the sum of the area of all the surfaces.
  How many bases does a cylinder have? 2
  What is the shape of the base? a circle
  What formula is used to find the area of a circle? $A = \pi r^2$
  Write the formula for display.
  What information is needed to use this formula? The measurement of a radius or a diameter is needed.

- Trace the circular bases of the oatmeal container onto 1 sheet of 12" × 18" construction paper and cut them out. Choose a student to measure the circle to the nearest inch to find the diameter.
  What is the radius? Answers will vary based on the size of the circle. The radius is half of the diameter.

  Guide the students in using the equation $A = 3.14 \times r^2$ to find the area of one of the circles. Lead the students to conclude that to find the area of both circular faces, they will need to multiply the expression "$3.14 \times r^2$" by 2. Guide them in finding the area of both circular bases using the equation $A = 2(3.14 \times r^2)$. Answers will vary.

  What kinds of items cover a curved surface similar to the curved surface of a cylinder? sample answers: paper towels, wrapping paper, tape, soup can labels
  What shape do you think the curved surface of the cylinder is when you lay it flat? rectangular

- Cut the other sheet of construction paper to match the height and distance around the cylinder. Lay it out flat to show the rectangle.
  What formula can you use to find the area of a rectangle? $A = l \cdot w$
  Write the formula for display.

Solve. Annex 0s if needed. Round decimal answers to the nearest hundredth.

1. $8\overline{)72}$ → 9
2. $9\overline{)54}$ → 6
3. $7\overline{)56}$ → 8
4. $8\overline{)64}$ → 8
5. $6\overline{)42}$ → 7
6. $5\overline{)60}$ → 12

7. 21 ÷ 7 = 3
8. 32 ÷ 8 = 4
9. 81 ÷ 9 = 9
10. 50 ÷ 10 = 5
11. 49 ÷ 7 = 7
12. 36 ÷ 12 = 3

13. $7\overline{)154}$ → 22
14. $9\overline{)8,362}$ → 929.111 ≈ 929.11
15. $6\overline{)4,032}$ → 672
16. $5\overline{)\$15.35}$ → $3.07
17. $4\overline{)4.2}$ → 1.05

18. $21\overline{)3,407}$ → 162.238 ≈ 162.24
19. $132\overline{)13,465}$ → 102.007 ≈ 102.01
20. $4.1\overline{)1,484.2}$ → 362
21. $231\overline{)23,573}$ → 102.047 ≈ 102.05

- Choose a student to place the cut piece of construction paper on the cylinder without taping it. Tape the top and bottom bases (circles) to the curved surface to make a net. Remove the paper.

- Display the net and point out that the length of the rectangle is equal to the circumference of the circle. Conclude that they can use the formula for finding the circumference of a circle, πd or $2\pi r$, to find the length of the rectangle. Choose a student to calculate the length of the rectangle. Answers will vary.

  Explain that the width of the rectangle is equal to the height of the cylinder. Choose a student to measure the height of the cylinder to the nearest inch. Guide the students in finding the area of the curved surface (rectangle) by multiplying the calculated length by the width. Answers will vary.

  Guide the students in adding the areas of the circular bases and the curved surface (rectangle) to find the surface area of the cylinder. Answers will vary.

- Distribute the cylinders and construction paper. Direct each student to construct a net for his cylinder and to determine its surface area to the nearest inch or centimeter. Give guidance as needed.

- Read aloud the following word problem.

The final scratching toy Beth made for her cats was 4 ft high. It had 2 circular bases and 1 curved surface. Each circle had a radius of 0.5 ft. How much carpet did Beth need to cover all of the surfaces of this toy? 14.13 ft^2

What is the shape of this scratching toy? a cylinder; It has 2 circular bases and 1 curved surface.

- Write the dimensions of the cylinder pictured on the *Surface Area* page and guide the students in finding the surface area. Remind them that the length of the curved surface is the circumference of the cylinder's base, and that the width is the height of the cylinder.

2 circular bases $A = 2[3.14 \times (0.5 \text{ ft})^2]$; $A = 1.57 \text{ ft}^2$

1 curved surface $C = 3.14 \times 2(0.5 \text{ ft})$; $C = 3.14 \text{ ft}$; $A = 3.14 \text{ ft} \times 4 \text{ ft}$; $A = 12.56 \text{ ft}^2$

total surface area $1.57 \text{ ft}^2 + 12.56 \text{ ft}^2 = 14.13 \text{ ft}^2$

How can finding the area of a rectangle help me find the surface area of a cylinder? The curved surface of a cylinder is a rectangle when lying flat. I can calculate the area of a rectangle to help find the surface area of a cylinder.

Apply

Student Edition pages 240–41

- Read and explain the directions for pages 240–41. Assist the students as they complete the pages independently.

Daily Review

- Students should complete Chapter 11, section *g*.

NOTES

OBJECTIVES

- Calculate the area and perimeter of a rectangle.
- Draw rectangles having the same area with different perimeters.
- Create a basic floor plan from a fixed area.
- Calculate the area of a complex figure.
- Defend the claim that God is the source of the mathematical design in nature. **BWS**

BIBLICAL WORLDVIEW SHAPING

- **Design (Formulate):** Mathematical design is a result of the Creator-God of the Bible rather than a result of chance.

TEACHER RESOURCES

- 11 *Graph Paper* (for the teacher and for each student)
- 61 *Floor Plan Activity* (for the teacher and for each pair of students)
- 62 *Floor Plan Grid* (for each pair of students, extra copies as needed)

ADDITIONAL MATERIALS

- samples of floor plans
- a ruler (for each student)
- a calculator (for each student)
- scissors (for each student)
- a sheet of 9" × 12" construction paper (for each pair of students)
- glue stick

Floor plans can be found on websites, in home decorating magazines, and in brochures provided by builders of local housing developments. You may make a model floor plan using the *Floor Plan Grid* page to help students better understand the floor plan they will make.

Fixed Areas

How can we explain the mathematical design we see in nature?

Perimeter is the distance around a geometric figure. **Area** is the space within a region. Geometric figures can have the same area but different perimeters.

Area can be found by counting the number of square units needed to cover the surface.

The perimeter of any polygon can be found by adding the lengths of the sides.

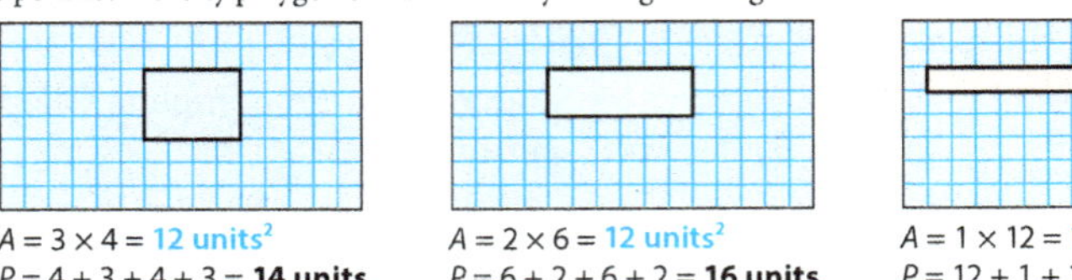

$A = 3 \times 4 = $ 12 units^2
$P = 4 + 3 + 4 + 3 = $ **14 units**

$A = 2 \times 6 = $ 12 units^2
$P = 6 + 2 + 6 + 2 = $ **16 units**

$A = 1 \times 12 = $ 12 units^2
$P = 12 + 1 + 12 + 1 = $ **26 units**

The formula for the area of a rectangle is length (l) times width (w). $A = l \cdot w$

The formula for the perimeter of a rectangle is 2 times the length plus 2 times the width. $P = (2 \cdot l) + (2 \cdot w)$

The three figures below all have the area of 18 cm^2, but each figure has different dimensions. Find factor pairs of 18 to use for the dimensions of each figure.

factor pairs of 18: 1×18 2×9 3×6

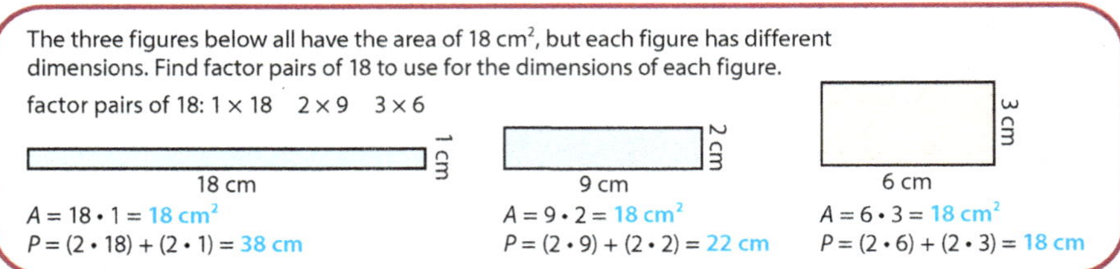

$A = 18 \cdot 1 = $ 18 cm^2
$P = (2 \cdot 18) + (2 \cdot 1) = $ 38 cm

$A = 9 \cdot 2 = $ 18 cm^2
$P = (2 \cdot 9) + (2 \cdot 2) = $ 22 cm

$A = 6 \cdot 3 = $ 18 cm^2
$P = (2 \cdot 6) + (2 \cdot 3) = $ 18 cm

Key Terms

- perimeter
- $P = (2 \cdot l) + (2 \cdot w)$
- area
- $A = l \cdot w$

Exercises

Write factor pairs for the given area. Use factor pairs for the dimensions of a figure to find the perimeter. Draw diagrams if needed. *Answers may vary.*

1. 6 m^2 **2.** 10 in.^{2} **3.** 16 ft^2

Choose the two perimeter equations for the given area.

4. 16 cm^2 $P = (2 \cdot 8) + (2 \cdot 2)$ $P = (2 \cdot 4) + (2 \cdot 4)$ $P = (2 \cdot 16) + (2 \cdot 16)$

5. 21 cm^2 $P = (2 \cdot 1) + (2 \cdot 21)$ $P = (2 \cdot 11) + (2 \cdot 10)$ $P = (2 \cdot 3) + (2 \cdot 7)$

6. 28 cm^2 $P = (2 \cdot 4) + (2 \cdot 7)$ $P = (2 \cdot 3) + (2 \cdot 7)$ $P = (2 \cdot 14) + (2 \cdot 2)$

7. How can 2 rectangles have the same area but different perimeters? *2 rectangles can be different sizes (have different perimeters) but still have the same number of square units.*

8. How can we explain the mathematical design we see in nature? *Mathematical design is a result of the Creator-God of the Bible.*

242   Chapter 11

Engage

- Guide the students in a **review** to help them understand the relationship between perimeter and area.

  What is perimeter? the distance around a geometric figure

  What is area? the space within a region

Instruct

Calculating area; recognizing perimeter can vary for a fixed area

- Direct the students to **brainstorm** to help them form as many different rectangles as possible with a prescribed area.

- Distribute a copy of the *Graph Paper* page and display your copy.

  How many different rectangles do you think you can make with an area of 12 square units? Answers will vary.

- Direct each student to draw on the graph paper a rectangle with an area of 12 square units. Choose several students to describe their rectangles by giving the dimensions. Draw for display on the *Graph Paper* page 3 different rectangles that are congruent to the ones described by the students.

- Lead the students to conclude that there are only 3 different factor pairs that make up rectangles with an area of 12 square units: 1 unit × 12 units or

9.

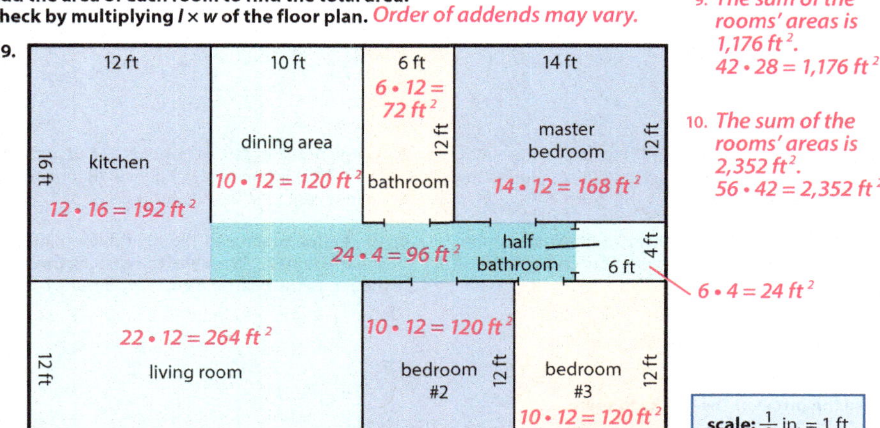

9. The sum of the rooms' areas is 1,176 ft². 42 • 28 = 1,176 ft²

10. The sum of the rooms' areas is 2,352 ft². 56 • 42 = 2,352 ft²

10.

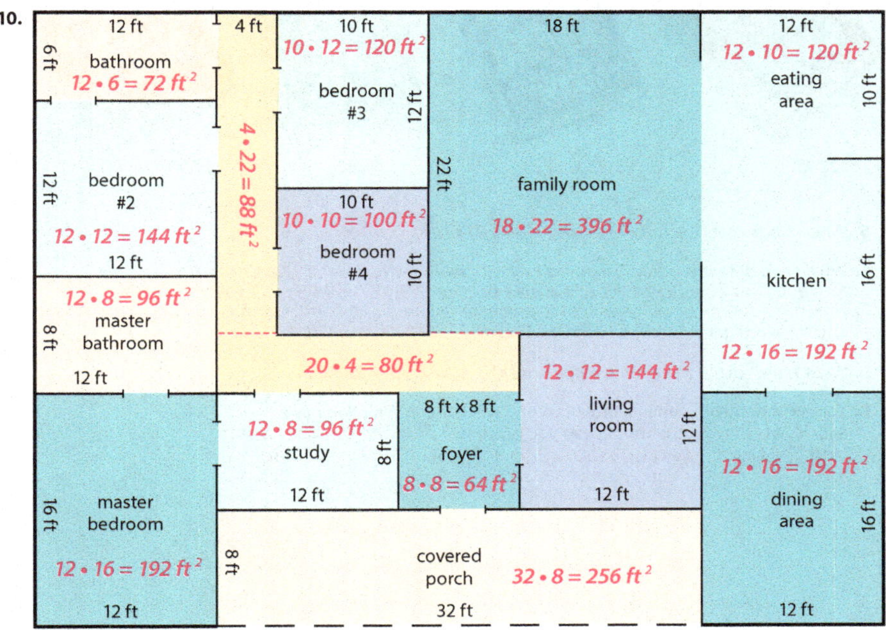

12 units × 1 unit; 2 units × 6 units or 6 units × 2 units; 3 units × 4 units or 4 units × 3 units.

Write "*A* = 12 units²" inside each rectangle. Point out that 12 square units can be referred to as a fixed area; the area remained the same even though the perimeters changed.

- Guide the students as they calculate the perimeter of each displayed rectangle.

1 × 12 rectangle $P = (2 • 1) + (2 • 12) = 26$ units

2 × 6 rectangle $P = (2 • 2) + (2 • 6) = 16$ units

3 × 4 rectangle $P = (2 • 3) + (2 • 4) = 14$ units

What do you notice about the perimeter and the area of the rectangles? Rectangles that are equal in area can have perimeters that are not equal.

- Direct each student to find the perimeter of his rectangle and to write the area and perimeter inside the rectangle (e.g., *A* = 12 units², *P* = 26 units).

- Repeat the procedure for rectangles with an area of 24 square units.

1 × 24 or 24 × 1; $P = (2 • 1) + (2 • 24) = 50$ units

2 × 12 or 12 × 2; $P = (2 • 2) + (2 • 12) = 28$ units

LESSON 111

$3 × 8$ or $8 × 3$; $P = (2 • 3) + (2 • 8) = 22$ units

$4 × 6$ or $6 × 4$; $P = (2 • 4) + (2 • 6) = 20$ units

Why can the perimeters be different when rectangles have the same fixed area? The same number of square units can be arranged in different ways, changing the dimensions which determine the perimeter of a given region.

Creating a basic floor plan

- Direct the students to do a **Think-Pair-Sketch** to help them create a floor plan.

Have you ever walked into a friend's house or apartment and noticed that the rooms are not in the same location as in your house?

- Display the floor plans. Explain that houses, apartments, offices, and other buildings are designed by architects. There are many ways to design a floor plan for a fixed area. A floor plan is a scaled drawing which is later used in creating blueprints for the builder. Blueprints give more detailed information about the placement of cabinets, electrical outlets, windows, and so on.

- Arrange the students in pairs. Distribute the *Floor Plan Activity* page to each pair of students and display your copy. Read aloud and explain the requirements of the project.

- Distribute a copy of the *Floor Plan Grid* page and a sheet of construction paper to each pair of students. Explain that there are 1,500 squares on the page, and each square represents 1 ft² in the house.

- Read aloud and discuss the procedures for creating the floor plan. Encourage the students to discuss and sketch possible layouts for their house before cutting apart the *Floor Plan Grid* page. Direct them to cut out the squares for each room or hallway, and to arrange the rooms on the construction paper as they cut out each room. Provide assistance as needed. You may check the floor plans before the students glue the rooms onto the construction paper.

LESSON 111

Calculating the area of a complex figure

- Assist the students in a **guided discovery** to help them find the area of a complex figure.

  What is (or should be) the area of the floor plan of your house? 1,500 ft²; All 1,500 squares ("square feet") of the *Floor Plan Grid* were used to make the floor plans.

  How do you calculate the area of a complex figure? I can partition (divide) the complex figure into smaller figures, find the area of each smaller figure, and add all the areas.

  How can you calculate the area of your floor plan? I can find the area of each room or hallway, and then add all the areas.

- Instruct the students to find the area of each room and hallway on their floor plan, and then find the sum of all these areas. Challenge them to write one equation for the area of their floor plan.

Relating geometry to real-world situations

> The following enrichment activities are beneficial in helping the students apply math to real-world situations. The activities may be completed anytime during the school year.

- Assign a **formative assessment** as you instruct the students to calculate the outside perimeter of their floor plan.

- Guide a discussion about the extra expense that comes from having a larger outside perimeter due to more siding or bricks being needed for the exterior walls.

- Display the students' floor plans. Allow the students in your class or invite the students in other classes to vote for their favorite floor plan. Guide the class as they use sidewalk chalk to draw on the parking lot a life-size model of the winning floor plan.

- Instruct each student to measure some or all of the rooms in his home and compare the measurements with the measurements of the corresponding room on his floor plan.

- Place the floor plan on posterboard. Cover the floor of each room with carpet or paper tiles. Calculate the cost of all the floor coverings based on prices found online or in local store advertisements.

- Make a 3-dimensional model from the floor plan. Place the floor plan on posterboard. Use 3 × 5 cards or card stock to form walls. Calculate the surface area of the walls. Calculate the cost of painting the walls based on prices found online or in local store advertisements.

- Invite an architect or builder to the class to discuss how he uses math for his career.

Mathematical design by God

- After reviewing the key ideas in this lesson, direct the students to do a **Think-Pair-Share** to answer the chapter essential question.

- Key Ideas:
  1. The relationship of pi in all circles is the same because God created an orderly world.
  2. The consistency seen in the formulas for the area of triangles and rectangles shows God's orderliness and design.

Solve.

1. Michelle purchased a 5.07 oz tube of oil paint for $5.10. What was the cost per ounce? (Round to the nearest cent.) *$1.01 per ounce*

2. A large bottle of soft drink holds 67.6 oz and costs $1.39. What is the price per ounce? (Round to the nearest cent.) *$0.02 per ounce*

3. A car traveled 158.75 mi in 2.5 hr. What was the average speed in miles per hour? *63.5 mph*

4. Mrs. Patton purchased 12.5 lb of chicken on sale. She spent $11.13. What was the cost per pound? (Round to the nearest cent.) *$0.89 per pound*

Use the prices of the books to solve.

$22.95 $13.98 $9.95 $19.99

5. Which book costs the most? *The Big Book of Brain Games*

6. Which two different books could you buy with twenty-five dollars? *The Challenge Sudoku and The Quest Word Games books; $13.98 + $9.95 = $23.93*

7. How much money would you need to purchase the puzzle and riddle book and the word game book? *$19.99 + $9.95 = $29.94*

8. What is the cost of three brain game books? *3 × $22.95 = $68.85*

9. You want to buy the brain game book and two other books. You have $50.00. Which two other books can you purchase? *The Challenge Sudoku and The Quest Word Games; $22.95 + $13.98 + $9.95 = $46.88*

Daily Review   475

---

3. Some people who don't believe in God think that design in nature happened by chance. God's Word tells us God created and designed all things.

Additional information

- You may refer to the chapter opener on Student Edition page 225 and discuss great circle routes. Point out that formulas are useful because God made an orderly world. Just as circumference is the distance around a regular circle, the earth is a sphere and its circumference is a "great circle."

- Display an orange to represent the shape of the earth. Cut the orange crossways in half. The halves represent the earth's hemispheres. The circle you see on the face of the cut sphere is the great circle.

- Because the earth is not represented well on a flat map, the shortest distance between two cities may follow a route over the arctic region. That is why flights between some cities in the US and Europe pass over Greenland. Because of God's orderly design of the earth, pilots can rely on great circle distance to economize on fuel and reach destinations more efficiently.

- For further ideas you may search the internet using the key words "Videos of Great Circle Route."

Apply

Student Edition pages 242–43

- Read and explain the directions for pages 242–43. Assist the students as they complete the pages independently.

Daily Review

- Students should complete Chapter 11, section *h*.

MATH TALK

Follow the General Procedure for Math Talks as outlined in Lesson 5 on Teacher Edition page 13b.

- Draw each set of 3 figures for display and direct the students to build the next train with the rule.

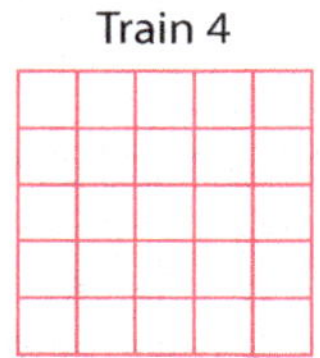

- Challenge the students to relate the train number to the number of unit squares.

NOTES

CHAPTER REVIEW

OBJECTIVES

- Calculate the perimeter and area of polygons and complex figures.
- Calculate the circumference and the area of a circle.
- Calculate the unknown side (length or width) of a rectangle or a square.
- Name the 3-dimensional figure that can be formed from a net.
- Calculate the surface area of rectangular, square, and triangular prisms.
- Calculate the surface area of a cylinder.

TEACHER RESOURCES

- 60 *Surface Area*
- 63 *Geometry Review I*
- 64 *Geometry Review II*
- 65 *Nets*

ADDITIONAL MATERIALS

- a calculator (for each student)

CHAPTER REVIEW

Find the circumference or the perimeter of the figure. Use 3.14 for π. Round decimals to the nearest hundredth.

1. 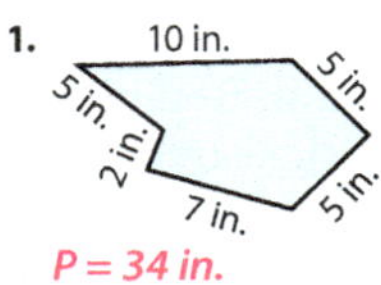

P = 34 in.

2. 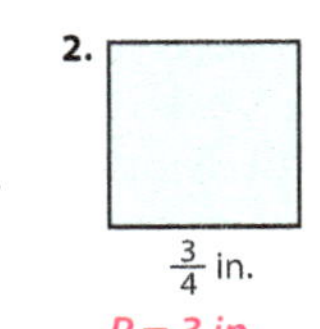

P = 3 in.

3. 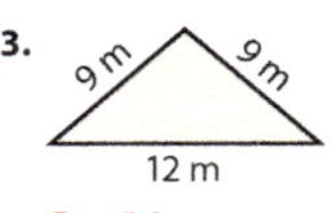

P = 30 m

4. 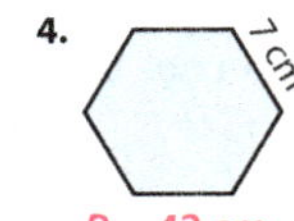

P = 42 cm

5. 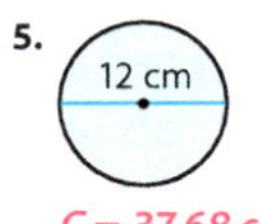

C = 37.68 cm

6. 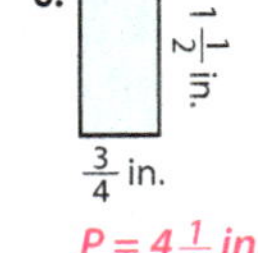

P = 4$\frac{1}{2}$ in.

7. 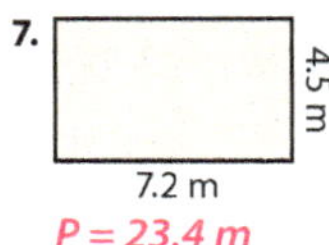

P = 23.4 m

8. 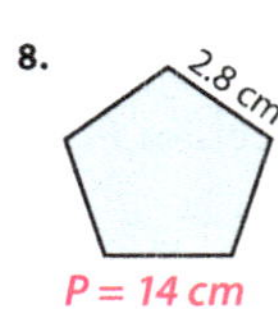

P = 14 cm

9. 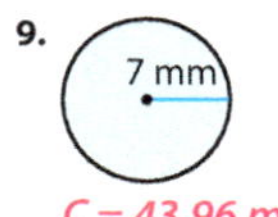

C = 43.96 mm

Find the area of the figure.

10. 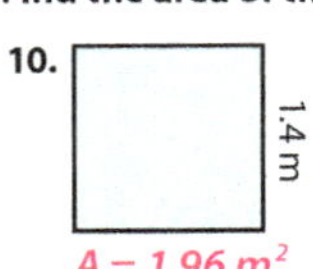

A = 1.96 m²

11. 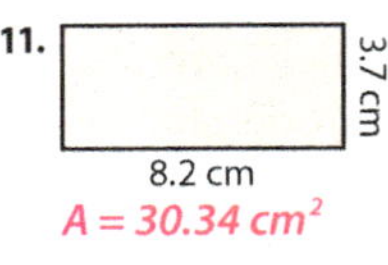

A = 30.34 cm²

12. 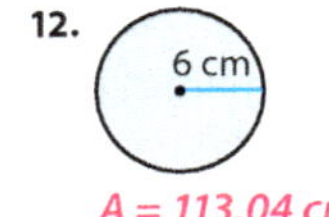

A = 113.04 cm²

13. 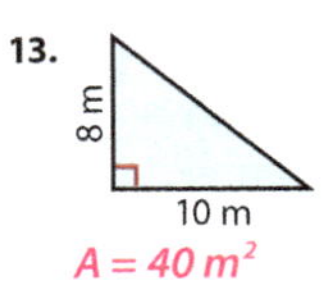

A = 40 m²

14. 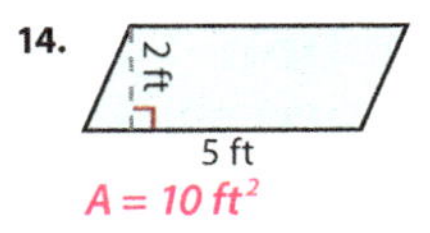

A = 10 ft²

15. 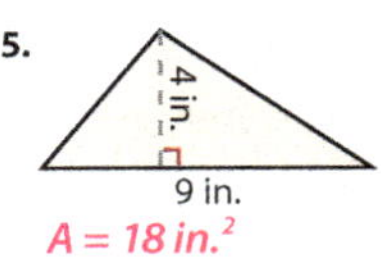

A = 18 in.²

Find the perimeter and the area of the complex figure.

16. 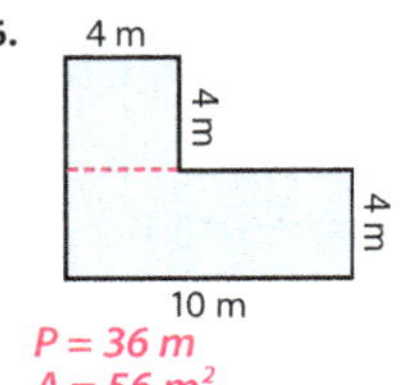

P = 36 m
A = 56 m²

17. 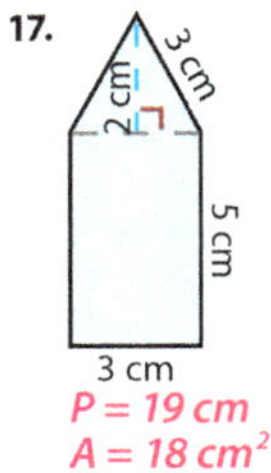

P = 19 cm
A = 18 cm²

18. 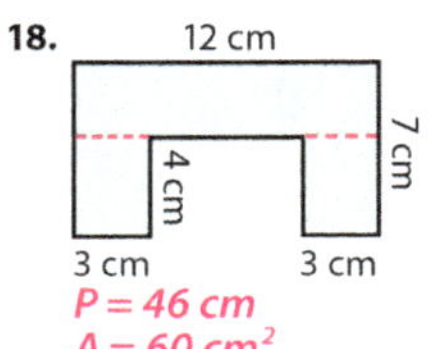

P = 46 cm
A = 60 cm²

19. 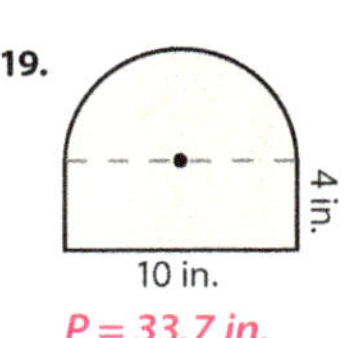

P = 33.7 in.
A = 79.25 in.²

244   Chapter 11

The Chapter Review offers an opportunity for students to discuss the concepts they have learned in the chapter. They may work collaboratively or independently as you review concepts. Circulate among the students, giving individual help as needed. Students who demonstrate proficiency with the discussion, the modeling, and the Student Edition pages are ready for the Chapter Test. Students who encounter difficulties with the review concepts would benefit from additional coaching and practice before testing.

Calculating perimeter, circumference & area

- Display the *Geometry Review I* page and direct attention to the square. Assist the students to recall the formulas used to calculate the perimeter and the area of a square. $P = 4 \cdot s$ or $P = s + s + s + s$; $A = s^2$ Review the formulas as needed.

  Direct the students to find the perimeter and the area of the square. $P = 4 \cdot 7$ m, $P = 28$ m; $A = (7 \text{ m})^2$, $A = 49$ m²

- Follow a similar procedure for all of the figures on both this page and the *Geometry Review II* page. If additional review of any formula is needed, choose a student to provide different dimensions for the figure and guide the students as they

calculate the new perimeter or circumference and area.

rectangle
$P = (2 \cdot l) + (2 \cdot w)$
$P = (2 \cdot 9.6 \text{ cm}) + (2 \cdot 4 \text{ cm})$; $P = 27.2$ cm
$A = l \cdot w$
$A = 9.6 \text{ cm} \cdot 4 \text{ cm}$; $A = 38.4$ cm²

circle
$C = 2\pi r$
$C = 2 \times 3.14 \times 6 \text{ cm}$; $C = 37.68$ cm
$A = \pi r^2$
$A = 3.14(6 \text{ cm})^2$; $A = 113.04$ cm²

parallelogram
$P = s + s + s + s$
$P = 12 \text{ in.} + 10 \text{ in.} + 12 \text{ in.} + 10 \text{ in.} = 44$ in.
$A = b \cdot h$
$A = 12 \text{ in.} \cdot 8 \text{ in.}$; $A = 96$ in.²

Use the area or the perimeter given to find the unknown measurement of the figure.

20. 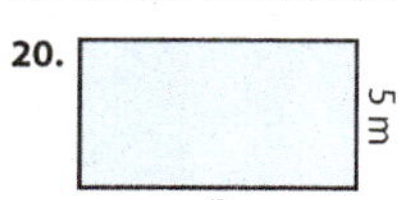

$A = 45$ m^2 **$n = 9$ m**

21. 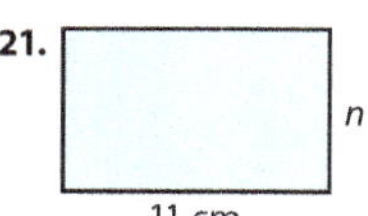

$A = 77$ cm^2 **$n = 7$ cm**

22. 

$P = 20$ in. **$n = 5$ in.**

Write the name of the figure that the net will make. The bases of the nets are shaded.

23. 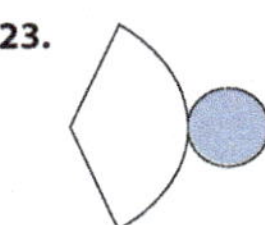

cone

24.

rectangular prism

25. 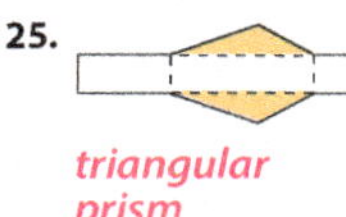

triangular prism

26. 

triangular pyramid

27. 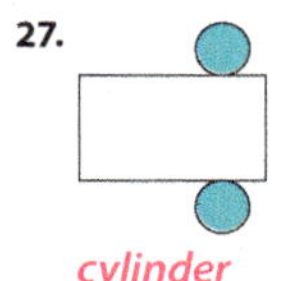

cylinder

28. 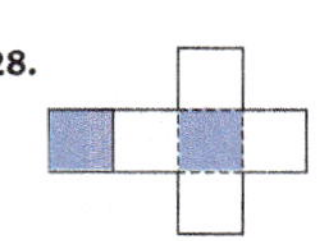

cube or square prism

Find the surface area of the figure.

29. 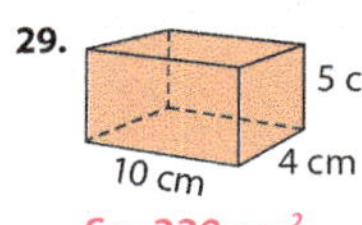

$S = 220$ cm^2

30. 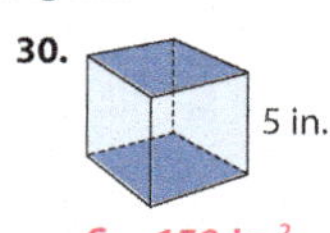

$S = 150$ in.^{2}

31. 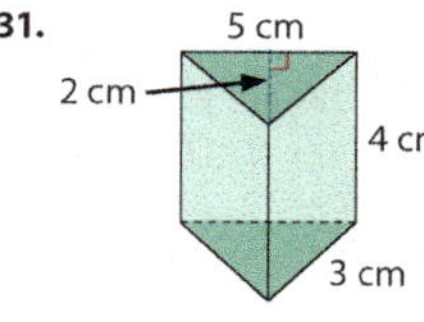

$S = 54$ cm^2

32. 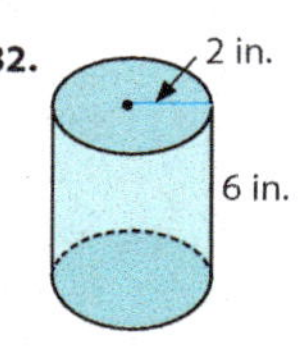

$S = 100.48$ in.^{2}

Solve. *Equations may vary.*

33. Mr. Franz is placing molding around the ceiling in the family room. If the length of the room is 12 ft and the width is 14 ft, how much molding will he use? **$(2 \times 12) + (2 \times 14) = 52$ ft**

34. Mr. Franz is laying new carpet in the family room. If the length of the room is 12 ft and the width is 14 ft, how much carpet will he use? **$12 \times 14 = 168$ ft^2**

35. Mr. Franz built a wooden cube. Mrs. Franz is going to cover the cube with carpet. The faces on the cube are 2 ft by 2 ft. How much carpet will she need to cover all the sides? **$6(2 \times 2) = 24$ ft^2**

36. Mrs. Franz had some leftover carpet, so she asked her husband to build a wooden cylinder about the same size as the wooden cube. The height of the cylinder is 2 ft and the radius of each circular base is 1 ft. How much carpet will she need to cover the cylinder?

37. Mrs. Franz is making a circular wall hanging for the family room. If the radius of the wall hanging is 3 ft, what is the area of the wall of the wall hanging? **$3.14 \times 3^2 = 28.26$ ft^2**

38. Mrs. Franz decided to attach fringe around the outside edge of the circular wall hanging. How much fringe will she need? **$2(3.14 \times 3) = 18.84$ ft**

36. **$2(3.14 \cdot 1^2) = 6.28$ ft^2; $(3.14 \cdot 2) \cdot 2 = 12.56$ ft^2; $6.28 + 12.56 = 18.84$ ft^2**

Calculating the unknown side length

- Read aloud the following word problem.

  Mr. Lawrence planted a vegetable garden that has an area of 3,000 ft^2. The garden is rectangular and has a width of 100 ft. What is the length of the garden? 30 ft

  What formula could you use to find the area of the garden if both the length and width were known? $A = l \cdot w$

  How can you find the length of the unknown side since you know the area and the measurement of the other? I can divide the area by the known measure; $l = A \div w$ or $n = A \div s$.

- Direct the students to write an equation for the word problem and solve it. Encourage them to draw pictures if needed. $n = 3{,}000$ ft$^2 \div 100$ ft; $n = 30$ ft

- Use the formula $P = 4 \times s$ for the following word problem. Discuss that since students know the perimeter of the square kitchen, they can find the length of the unknown side by dividing by 4 because all four sides of the kitchen are equal in length.

- Read aloud the following word problem.

  Mrs. Stevens is putting up wallpaper border near the ceiling in her square-shaped kitchen. The perimeter of her kitchen is 28 yd. What is the length of each wall? 28 yd $= 4 \times s$; 28 yd $\div 4 = 4 \div 4 \times s$; $s = 7$ yd

- Draw a rectangle for display. Write "5 ft" along the length and write n along the width.

  If the perimeter of the rectangle is 22 feet, how can you find the measurement of the unknown width? I can subtract the measurement of the two known sides ($2 \cdot 5$ ft) from the perimeter (22 ft) and divide the difference (12) by 2.

  Direct the students to find the measurement of the unknown width. 22 ft $= (2 \cdot 5$ ft$) + (2 \cdot n)$; 22 ft $- 10$ ft $= 10$ ft $- 10$ ft $+ 2n$; 12 ft $= 2n$; $\frac{12\text{ ft}}{2} = \frac{2n}{2}$; 6 ft $= n$

right triangle

$P = s + s + s$
$P = 3$ ft $+ 5$ ft $+ 4$ ft; $P = 12$ ft
$A = \frac{1}{2}(b \cdot h)$
$A = \frac{1}{2}(4$ ft $\cdot 3$ ft$)$; $A = 6$ ft^2

obtuse triangle

$P = s + s + s$
$P = 5$ in. $+ 5$ in. $+ 8$ in.; $P = 18$ in.
$A = \frac{1}{2}(b \cdot h)$
$A = \frac{1}{2}(8$ in. $\cdot 3$ in.$)$; $A = 12$ in.^{2}

hexagon

$P = n \cdot s$
$P = 6 \cdot 5$ m; $P = 30$ m
$A = (l \cdot w) + 2[\frac{1}{2}(b \cdot h)]$
$A = (5$ m $\cdot 8$ m$) + 2[\frac{1}{2}(8$ m $\cdot 3$ m$)]$;
$A = 64$ m^2

complex figure

$P = s + s + s + s + s$
$P = 8$ cm $+ 5$ cm $+ 5$ cm $+ 8$ cm $+ 12$ cm;
$P = 38$ cm
$A = (b \cdot h) + \frac{1}{2}(b \cdot h)$
$A = (12$ cm $\cdot 4$ cm$) + \frac{1}{2}(12$ cm $\cdot 4$ cm$)$;
$A = 72$ cm^2

Naming a figure from its net

- Display the *Nets* page.

  What is a net? the flat or 2-dimensional pattern of a 3-dimensional figure

  How many bases do conical figures have? 1 base

  How many bases do cylindrical figures have? 2 bases

- Choose students to identify each net by name and tell whether it is a conical or a cylindrical figure. Write the name of each figure below the net.

  You may point out that prisms and cylinders are part of the larger category of cylindrical figures and that pyramids and cones are part of the category of conical figures.

  1. cylinder; cylindrical figure
  2. cone; conical figure
  3. triangular prism; cylindrical figure
  4. rectangular prism; cylindrical figure
  5. square prism or cube; cylindrical figure
  6. square pyramid; conical figure
  7. triangular pyramid; conical figure
  8. rectangular pyramid; conical figure

Calculating surface area of prisms & a cylinder

- Display the *Surface Area* page. Write the following dimensions along the edges of the rectangular prism: length = 4 ft, width = 2 ft, and height = 3 ft.

- Assist the students to conclude that they can find the surface area of a rectangular prism by finding the area of 1 face in each pair of congruent faces, multiplying each area by 2, and adding the areas; or they can add the areas of all 6 faces.

  Allow the students to solve more than 1 equation to find the surface area of each figure if needed.

  What formula could you use to find the surface area of a rectangular prism? $S = 2(l \bullet w) + 2(w \bullet h) + 2(l \bullet h)$

  Write the formula for display and review it as needed.

- Direct the students to find the surface area of the rectangular prism. $S = 2(4 \text{ ft} \times 2 \text{ ft}) + 2(2 \text{ ft} \times 3 \text{ ft}) + 2(4 \text{ ft} \times 3 \text{ ft}) = 52 \text{ ft}^2$; or $S = 8 \text{ ft} + 8 \text{ ft} + 6 \text{ ft} + 6 \text{ ft} + 12 \text{ ft} + 12 \text{ ft} = 52 \text{ ft}^2$

- Follow a similar procedure for the other figures on the page.

  cube—one side = 5 in.
  I can multiply the area of 1 face by 6: $S = 6(l \bullet w)$ or $S = 6s^2$; $S = 6(5 \text{ in.} \times 5 \text{ in.})$ or $S = 6(5 \text{ in.})^2$; $S = 150 \text{ in.}^2$.

  triangular prism—2 rectangular faces: length = 5 ft, width = 3 ft; 1 rectangular face: length = 6 ft, width = 3 ft; 2 triangular faces: base = 6 ft, height = 4 ft
  I can multiply the area of 1 of the 2 congruent rectangular faces by 2, find the area of the non-congruent rectangular face, multiply the area of 1 of the triangular bases by 2, and add the areas: $S = 2(l \bullet w) + (l \bullet w) + 2[\frac{1}{2}(b \bullet h)]$; $S = 2(5 \text{ ft} \times 3 \text{ ft}) + (6 \text{ ft} \times 3 \text{ ft}) + 2[\frac{1}{2}(6 \text{ ft} \times 4 \text{ ft})] = 72 \text{ ft}^2$.

  cylinder—radius = 1 ft; height = 3 ft
  Since the circular bases are congruent, I can multiply the area of 1 of the bases by 2 and add that area to the area of the curved surface.

  Remind students that the length of the curved surface is the circumference of the cylinder's base, and the width is the height of the cylinder. $S = 2(\pi r^2) + (2\pi r \bullet h)$; $S = 2[3.14 \times (1 \text{ ft})^2] + (2 \times 3.14 \times 1 \text{ ft}) \times 3 \text{ ft}$; $S = 25.12 \text{ ft}^2$

Student Edition pages 244–45

- Read and explain the directions for pages 244–45. Assist the students as they complete the pages independently.

LESSON 113

Student Edition pages 246–48

CHAPTER 11 TEST

CUMULATIVE REVIEW

CONCEPT REVIEW

- **Identifying prime and composite numbers (Lesson 12)**
- **Solving problems using the guess-and-check strategy (Lesson 46)**
- **Identifying equivalent expressions (Lesson 100)**
- **Estimating products and sums (Lessons 2, 5, 15, 16)**
- **Identifying all factors of a number (Lessons 31–32)**
- **Adding and multiplying decimals (Lesson 5, 68)**
- **Determining the value of a variable (Lessons 95–100)**
- **Reading and interpreting a pictograph (Lesson 13)**
- **Reading and interpreting a Venn diagram (Lesson 88)**

- To prepare the students for the format of achievement tests, instruct them to work on a separate sheet of paper, if necessary, and to mark their answers on the *Cumulative Review Answer Sheet*.

Student Edition pages 246–48

The Cumulative Review provides additional practice of previously learned concepts. These pages may be completed during this lesson or anytime after this lesson, since they require limited or no teaching.

Choose the answer.

1. What two prime numbers are between 20 and 30?
   - **A.** 21 and 23
   - **B.** 23 and 29
   - **C.** 25 and 27
   - **D.** none of the above

2. Two addends have a sum of 30. The second addend is 2 times the first addend.
   - **A.** 14 + 16
   - **B.** 10 + 20
   - **C.** 5 + 25
   - **D.** all of the above

3. $300 =$ ___
   - **A.** 3×10^2
   - **B.** 2×30
   - **C.** 30×100
   - **D.** none of the above

4. $400 =$ ___
   - **A.** $1,000 \div 2.5$
   - **B.** $8,000 \div 20$
   - **C.** $2\frac{2}{3} \times 150$
   - **D.** all of the above

5. $(15 \times 200) \div 60 + 90 =$ ___
   - **A.** 20
   - **B.** 100
   - **C.** 140
   - **D.** none of the above

6. Estimate the product of 726×398.
   - **A.** 21,000
   - **B.** 28,000
   - **C.** 200,000
   - **D.** 280,000

7. Use front-end estimation for $189,786 + 346,398$.
   - **A.** 300,000
   - **B.** 520,000
   - **C.** 600,000
   - **D.** 720,000

8. Estimate the inventory of 169,387 nails to the nearest one thousand.
   - **A.** 169,000
   - **B.** 170,000
   - **C.** 200,000
   - **D.** 201,000

9. Which list shows all the factors of 72?
   - **A.** 8, 9
   - **B.** 2, 3, 6, 8, 9
   - **C.** 1, 2, 3, 8, 9, 12
   - **D.** 1, 2, 3, 4, 6, 8, 9, 12, 18, 24, 36, 72

10. $\frac{3}{4} + \frac{5}{6} =$ ___
    - **A.** The estimated sum is 2.
    - **B.** The estimated sum is 1.
    - **C.** The sum is less than 1.
    - **D.** The sum is greater than 2.

Use the number boxes to find the answer.

 0.987   0.087   0.7

11. Choose the numbers ordered from least to greatest.

- **A.** 0.7, 0.087, 0.987
- **B.** 0.087, 0.7, 0.987
- **C.** 0.987, 0.7, 0.087
- **D.** 0.987, 0.087, 0.7

12. Choose the sum of the numbers.

- **A.** 1.081
- **B.** 1.774
- **C.** 2.557
- **D.** 25.57

13.  $\times$ ▦ $= n$

- **A.** $n = 609$
- **B.** $n = 6.09$
- **C.** $n = 0.609$
- **D.** $n = 0.0609$

14. (▦ $+$ ▦) $\times$ ▦ $= n$

- **A.** $n = 0.07$
- **B.** $n = 0.7518$
- **C.** $n = 7.518$
- **D.** $n = 8$

15. ▦ $+ n =$ ▦

- **A.** $n = 0.9$
- **B.** $n = 0.09$
- **C.** $n = 0.009$
- **D.** $n = 9$

Choose the answer.

16. $x > 2$

- **A.** $x = 16$
- **B.** $x = 2.3$
- **C.** $x = \frac{15}{3}$
- **D.** all of the above

17. $x + 10 - 3 = 29.8$

- **A.** $x = 19.8$
- **B.** $x = 20.5$
- **C.** $x = 22.8$
- **D.** all of the above

18. $17(n) = 68$

- **A.** $n = 2$
- **B.** $n = 3$
- **C.** $n = 4$
- **D.** $n = 5$

19. $\frac{n}{8} = 7$

- **A.** $n = 48$
- **B.** $n = 56$
- **C.** $n = 64$
- **D.** $n = 77$

20. Write an equivalent expression for 8×9, using prime numbers and exponents.

- **A.** $2^3 \times 3^2$
- **B.** $2^2 \times 3^2$
- **C.** $2^2 \times 3^3$
- **D.** none of the above

Use the pictograph to find the answer.

February	
weekdays	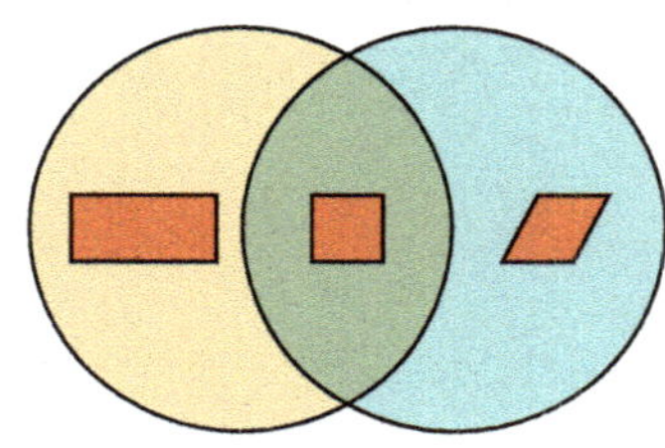
weekend days	
special days	
national holidays	
school holidays	

Key
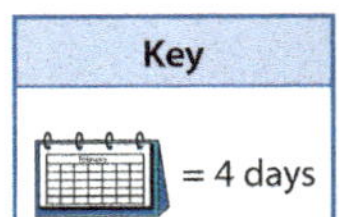 = 4 days

21. Which two lines of the pictograph give the total number of days in February?

   A. weekdays + special days

   B. national holidays + weekdays

   C. weekdays + weekend days

   D. none of the above

22. How many special days are in February?

   A. 3

   B. 4

   C. 5

   D. 6

23. How many more weekdays are there than weekend days?

   A. 2 times as many weekdays

   B. $2\frac{1}{2}$ times as many weekdays

   C. 15 more weekdays

   D. 20 more weekdays

24. If school closes for national holidays and school holidays, how many vacation days will there be?

   A. 1 + 2 = 3 days

   B. 4 + 2 = 6 days

   C. 4 + 1 = 5 days

   D. none of the above

Use the Venn diagram to find the true statement.

25. **Quadrilaterals**

   A. A square can be classified as a rectangle and a rhombus.

   B. A rectangle is a rhombus.

   C. A rhombus is not a quadrilateral.

   D. A rectangle, a square, and a rhombus are not related at all.

Lesson Plan Overview

<table>
<tr><th colspan="6">Chapter 12: Volume</th></tr>
<tr>
<th>Day</th>
<th>Lesson</th>
<th>Student Edition Pages</th>
<th>Teacher Edition Pages</th>
<th>Teacher Resources & Additional Materials</th>
<th>Topics, Skills & Biblical Worldview Shaping</th>
</tr>
<tr>
<td>115</td>
<td>114</td>
<td>249–51</td>
<td>249–51b</td>
<td>Teacher Resources
• 66 Shipping Volume
Additional Materials
• blank sheets of $8\frac{1}{2}$" × 11" paper
• a clear storage container with lid
• cube-shaped blocks
• a rectangular prism-shaped block
• a ruler</td>
<td>Topic: Volume of Rectangular Prisms
Skills: finding the volume of a rectangular prism by using a model or a formula
BWS: Service</td>
</tr>
<tr>
<td>116</td>
<td>115</td>
<td>252–53</td>
<td>252–53b</td>
<td>Additional Materials
• blank sheets of $8\frac{1}{2}$" × 11" paper
• cube-shaped blocks</td>
<td>Topic: Volume of Cubes
Skills: finding the volume of a cube by using a model or a formula, calculating an unknown measurement, solving a word problem</td>
</tr>
<tr>
<td>117</td>
<td>116</td>
<td>254–55</td>
<td>254–55b</td>
<td>Teacher Resources
• 67 Triangular Prisms & Cylinders
Additional Materials
• blank sheets of $8\frac{1}{2}$" × 11" paper
• cube-shaped blocks
• calculators
Assessments
• Chapter 12 Quiz 1</td>
<td>Topic: Volume of Other 3D Figures
Skills: finding the volume of an irregular prism by using a model, calculating the volume of a triangular prism or a cylinder
BWS: Service</td>
</tr>
<tr>
<td>118–19</td>
<td>117</td>
<td>256–57</td>
<td>256–57b</td>
<td>Teacher Resources
• 68 Fixed Volume
Additional Materials
• cube-shaped blocks
• sheets of 9" × 12" construction paper
• rulers
• transparent tape
• box lids or rectangular pans
• 1 c measuring cups
• rice, unpopped popcorn, or dried beans
• calculators</td>
<td>Topic: Fixed Volumes & Fixed Lateral Surfaces
Skills: constructing prisms with the same volume but different surface areas, constructing cylindrical figures with the same lateral surface area but different volumes</td>
</tr>
<tr>
<td>120</td>
<td>118</td>
<td>258–59</td>
<td>258–59b</td>
<td>Teacher Resources
• 69 Volume Review
• 70 Volume Word Problems
Additional Materials
• copies of area drawings
• cube-shaped blocks
• calculators</td>
<td>Topic: Chapter Review
BWS: Service</td>
</tr>
<tr>
<td>121</td>
<td>119</td>
<td>260–63</td>
<td>260–63</td>
<td></td>
<td>Topic: Test & Cumulative Review</td>
</tr>
</table>

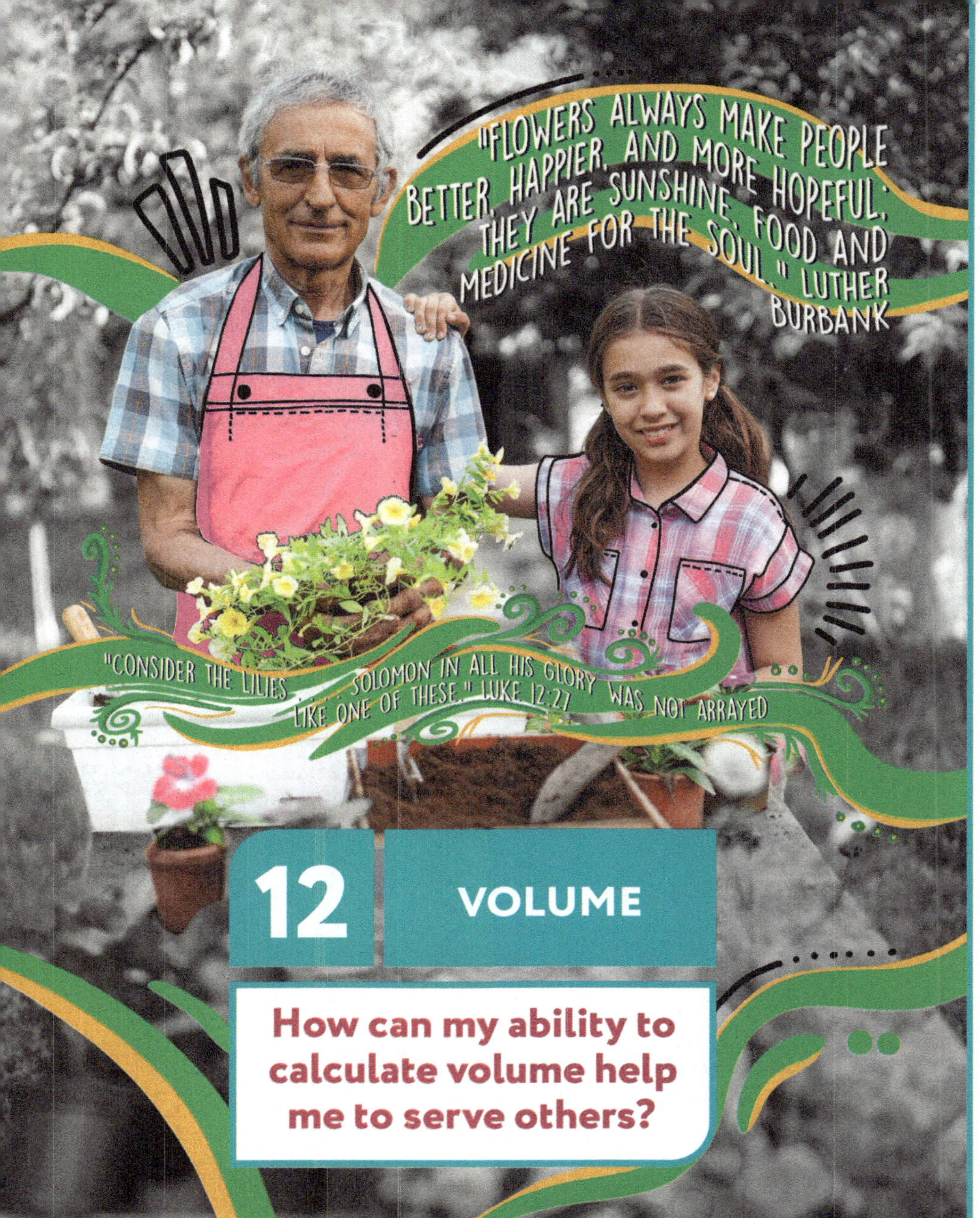

CHAPTER OBJECTIVES

- Find the volume of prisms and cylinders.
- Calculate the unknown measurement of prisms.
- Compare surface area and volume of various prisms.
- Use the ability to calculate volume to serve others.

To determine whether your students are prepared for the concepts taught in this chapter, you may wish to consult the preassessment checklist found on TeacherToolsOnline.com.

Make fact practice, both oral and written, part of your daily math routine to help the students with mastery.

Visit AfterSchoolHelp.com for math practice resources, or visit TeacherToolsOnline.com for additional resources to enhance the lessons.

Student Edition pages 249–51
Daily Review Chapter 12, section *a*

OBJECTIVES

- Find the volume of a rectangular prism by using a model.
- Calculate the volume of a rectangular prism by using a formula.
- Explain how finding the volume of a container can make people's lives better. **BWS**

BIBLICAL WORLDVIEW SHAPING

- **Service (Explain):** Finding the volume of a container serves people by enabling efficient delivery.

TEACHER RESOURCES

- 66 *Shipping Volume* (for the teacher and for each student)

ADDITIONAL MATERIALS

- 3 blank sheets of $8\frac{1}{2}$" × 11" paper
- a clear storage container with lid (shoebox size)
- cube-shaped blocks (to fill the clear container and build a 60 unit^3 tower)
- a rectangular prism-shaped block
- a ruler

Preparation

Prepare each of the following area drawings on a separate sheet of blank $8\frac{1}{2}$" × 11" paper. Make the squares in each drawing the same size as the cubes used in the lesson. Retain the drawings for use in Lesson 118.

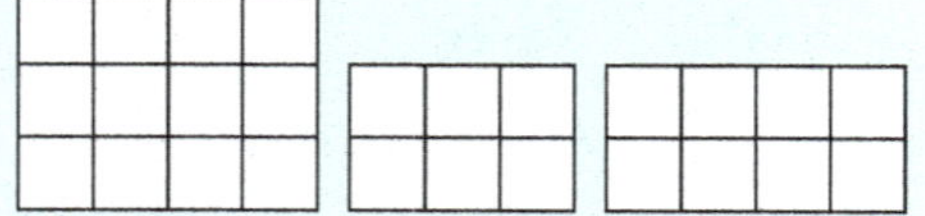

If enough cubes are available, you may prepare copies of the area drawings so that groups of students can build the towers as you demonstrate.

Volume of Rectangular Prisms

How can finding the volume of a container help a shipper make people's lives better?

The area of a figure is the number of square units a flat, 2-dimensional space covers. The **volume** of a figure is the number of cubic units the figure contains. Volume builds on the area of a figure. To calculate the volume of a rectangular prism, multiply the area of the **base** ($B = l \times w$) by the number of cubic unit layers (**height**). The formula is $V = Bh$.

Area is measured using square units. **units^2**

Volume is measured using cubic units. **units^3**

1 square cm or cm^2

1 cm × 1 cm = 1 cm^2

1 cubic cm or cm^3

1 cm × 1 cm × 1 cm = 1 cm^3

The volume of any prism can be found using the volume formula. Because prism bases are parallel and congruent, opposite bases have the same area. Substitute $l \times w$ for B (base).

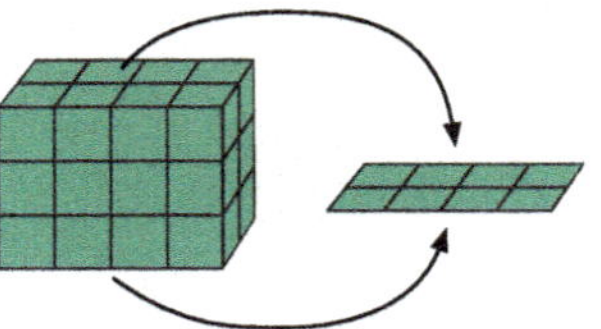

$V = Bh$
$V = (l \times w) \times h$
$V = (4 \text{ rows of } 2 \text{ cubes}) \times 3 \text{ layers}$
$V = 8 \text{ cubes} \times 3 \text{ layers}$
$V = 24 \text{ cubes or } 24 \text{ units}^3$

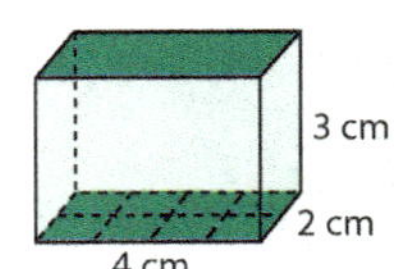

$V = Bh$
$V = (l \times w) \times h$
$V = (4 \times 2) \times 3$
$V = 8 \times 3$
$V = 24 \text{ cm}^3$

Exercises

Write an equation to find the volume of the model. Label the answer with units^3.

1. $V = 56 \text{ units}^3$
2. $V = 45 \text{ units}^3$
3. $V = 120 \text{ units}^3$
4. $V = 24 \text{ units}^3$
5. $V = 40 \text{ units}^3$
6. $V = 42 \text{ units}^3$

Find the area of the base: $B = l \times w$. Find the volume of the rectangular prism that could be built using centimeter cubes for the given height.

7. 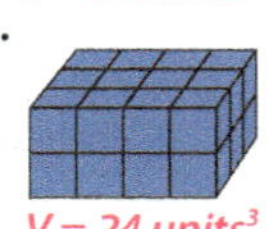 $V = 18 \text{ cm}^3$
$h = 3 \text{ cm}$

8. 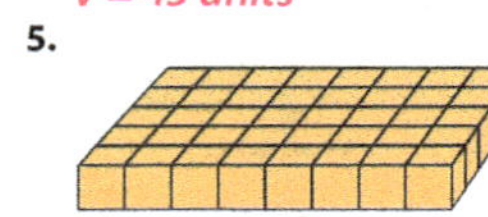 $V = 8 \text{ cm}^3$
$h = 2 \text{ cm}$

9. 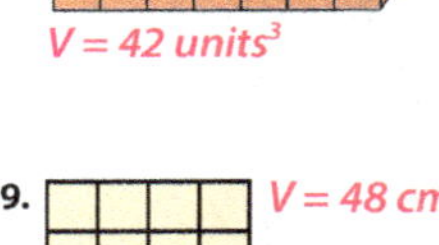 $V = 48 \text{ cm}^3$
$h = 4 \text{ cm}$

250   Chapter 12

Engage

- Assign a **Quick Write** (1–5 min) to explore the chapter essential question on Student Edition page 249, "How can my ability to calculate volume help me to serve others?" Direct the students to give as many examples of serving others by calculating volume as they can think of during the time allowed.

- Direct them to share their answers with a nearby partner. Allow each pair to share an example with the class.

- Read aloud the essential question at the top of Student Edition page 250, "How can finding the volume of a container help a shipper make people's lives better?" Explain that later in the lesson they will discuss how shippers use volume calculations to provide a valuable service that affects their life every day.

Instruct

Finding volume by using a model

- Use a **3D model** to guide the students as they find the volume of a rectangular prism.

- Display the clear storage container.

  What 3-dimensional figure is this container? a rectangular prism

---

Write an equation to find the volume. Round decimals to the nearest tenth.

10. 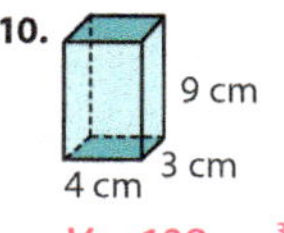
V = 108 cm³

11. 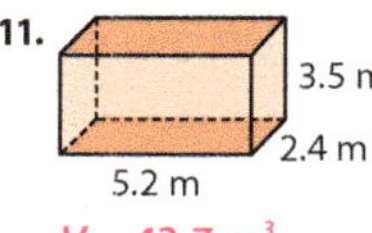
V = 43.7 m³

12. 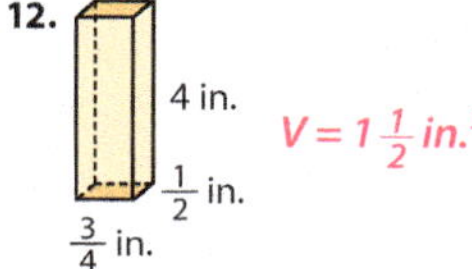
$V = 1\frac{1}{2}$ *in.³*

13. 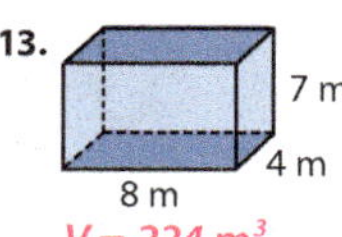
V = 224 m³

14.
V = 88.2 cm³

15. 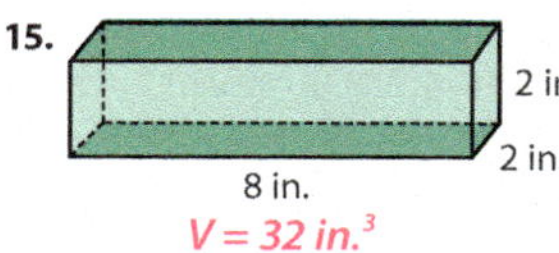
V = 32 in.³

Find the volume of a prism with the given dimensions.

16. $l = 5$ in., $w = 7$ in., $h = 2$ in.
V = 70 in.³

17. $l = 12$ cm, $w = 8$ cm, $h = 10$ cm
V = 960 cm³

18. $l = \frac{3}{8}$ in., $w = \frac{2}{3}$ in., $h = \frac{1}{2}$ in.
$V = \frac{1}{8}$ *in.³*

19. $B = 15$ ft², $h = 2$ ft
V = 30 ft³

20. $B = 25$ m², $h = 8$ m
V = 200 m³

21. $B = 46$ cm², $h = 5.2$ cm
V = 239.2 cm³

Practice & Application

Equations may vary.

24. *yes; (2 • 9 in.) + (2 • 12 in.) = 42 in. needed;*
4 × 12 = 48 in. available

On Memorial Day weekend, Carmen and her family attended a hot air balloon festival.

29. *sample answer: Knowing the volume of a container helps a shipper load his freight efficiently so he can provide better service to his customers.*

22. Carmen selected a box in which she will store her souvenirs. The box is 24 in. long, 18 in. wide, and 12 in. high. What is the volume of the box?
5,184 in.³

23. Carmen wants to cover the outside of her box with decorative contact paper. What is the least amount of contact paper Carmen will need? (Use the information from problem 22.) *1,872 in.²*

24. Mr. Fields has 4 ft of molding to make a rectangular frame for a photo he took at the festival. Does he have enough to make a 9 in. by 12 in. frame? Explain your answer.

25. Mr. Fields purchased a 9 in. by 12 in. piece of glass for the frame. What is the area of the glass?
9 in. × 12 in. = 108 in.²

26. Mrs. Fields wants to move the shadow box displaying her new balloon figurines. How much space does a shadow box that measures 2 ft long, 1 ft wide, and $\frac{1}{2}$ ft high take up?
2 ft × 1 ft × $\frac{1}{2}$ ft = 1 ft³

28. *L = 27.25 ft ÷ 4 ft = 6.81 ≈ 6 skid units*
W = 8.33 ft ÷ 4 ft = 2.08 ≈ 2 skid units
H = 8.92 ft ÷ 4 ft = 2.23 ≈ 2 skid units

27. Identify whether perimeter, area, or volume would be used to find the following. Explain your answer.
   a. the amount of air to fill a balloon
   b. the amount of material needed to make a balloon
   c. the amount of space in a basket
   d. the amount of cushioned edging to go around a basket

28. An enclosed shipping trailer's inside measurements are 27.25 ft × 8.33 ft × 8.92 ft. How many 4 ft × 4 ft × 4 ft skid units³ (skid cubes) will fit in the trailer?

29. How can finding the volume of a container help a shipper make people's lives better?

27. a. *volume; finding the cubic units within a figure*
   b. *area; finding the square units in a figure*
   c. *volume; finding the amount of cubic space within a figure*
   d. *perimeter; finding the distance around a figure*

V = Bh
V = (l × w) × h
V = (6 × 2) × 2
V = 12 × 2 = 24 skid units³

What do you notice about the number of cubic units needed to cover the bottom of the container and the number of square units for the bottom face? The number of cubic units is the same as the number of square units.

How many layers of cubes do you estimate are needed to fill the container? Answers will vary.

- Place a second layer of cubes in the container.

How many cubic units are in the second layer? the same number of cubic units that is in the first layer

How many cubic units are now in the container? Answers will vary; the number of cubic units is 2 times the number in the first layer.

- Repeat the procedure until the container is filled.

How many cubes were needed to fill the container? Answers will vary.

What is the volume of the box in cubic units? Answers will vary.

Write the volume for display (e.g., 28 units³). Point out that the exponent 3 indicates cubic units.

Restate that volume is the measurement of a 3-dimensional object's capacity or the amount the object can hold.

- Display the area drawing that shows 3 rows of 4 square units.

What is the area of this figure? 12 square units; $A = l × w$; $A = 3 × 4 = 12$ units²

If you are building a tower with blocks on this base plan, how many blocks will be needed for the first floor? 12 blocks; Each square unit in the base plan represents 1 face of each block (cubic unit) needed for the first floor.

- Choose a student to use cubes to build the first floor of the tower (3 rows of 4 cubes).

What is the volume of the tower since it is 1 cube high? 12 cubic units; A total of 12 cubes makes up 1 floor with a base that is a 3 × 4 area.

- Draw for display a table similar to the one provided and complete the first row using the data for the first floor of the tower.

What is a prism? It is a cylindrical figure having 2 congruent and parallel bases that are polygons; all of its other faces are parallelograms.

How can you calculate the surface area of this container? I can add the areas of all the faces, or I can add the congruent top and bottom areas, the congruent front and back areas, and the side areas using the formula $S = 2(l × w) + 2(w × h) + 2(l × h)$.

- Remind the students that perimeter and area are attributes of 2-dimensional figures. Surface area is an attribute of 3-dimensional figures. Explain that another attribute of 3-dimensional figures is volume, the capacity of a 3-dimensional object or the amount that the object can hold. Write "volume" for display.

- Cover the interior bottom of the container with cubes. Point out that, by looking through the bottom of the container, the students can see the number of square units that form the bottom face of the prism.

How many square units form the bottom face of the container? Answers will vary based on the container's size.

How many cubes were needed to make one layer in the bottom of the container? Answers will vary.

Explain that when they use cubes to find the volume of a 3-dimensional object, each cube is 1 cubic unit.

LESSON 114

- Repeat the procedure until the tower has 5 floors (layers). Guide the students as they conclude that the volume of the tower after each floor is built is the area of the base multiplied by the number of layers or the height.

Area of Base (units²)	Number of Layers—Height (unit)	Volume (units³)
12 units²	1 unit	12 units³
12 units²	2 units	24 units³
12 units²	3 units	36 units³
12 units²	4 units	48 units³
12 units²	5 units	60 units³

- Display the area drawing that shows 2 rows of 3 square units. Follow a similar procedure for building a tower. Guide the students as they multiply the area of the base by the height after each floor is built to calculate the volume of the tower at that height.

- Display the area drawing that shows 2 rows of 4 square units.

  What is the area of this base? 8 units²

  If a tower with a height of 5 units was built on this base, what would its volume be? 40 units³; The volume equals the area of the base multiplied by the number of units high; 8 units² × 5 units = 40 units³.

Calculating volume by using a formula

- Guide a **discussion** to help the students use a formula to calculate the volume of a rectangular prism.

- Display the rectangular prism-shaped block.

  What shape is the base of this prism? a rectangle

  How can you find the volume of the prism? I can multiply the area of the base times the height of the prism.

  Explain that when finding the volume of a solid figure, they are finding the amount of space the object fills rather than how much the object can hold.

Evaluate the expression. Let $n = 6$.

1. $(1.3 \cdot n) - 4$ *$(1.3 \cdot 6) - 4 =$*
   $7.8 - 4 = 3.8$

2. $75 - 7n$ *$75 - (7 \cdot 6) =$*
   $75 - 42 = 33$

3. $5n \div 2$ *$(5 \cdot 6) \div 2 =$*
   $30 \div 2 = 15$

Simplify the expression.

4. $4(3x)$ *$12x$*

5. $7(n + 4)$ *$7n + 28$*

6. $8y + (3y + 4)$ *$11y + 4$*

Write the algebraic expression for the sentence.

7. The fence is 7 times longer than the gate. *$7g$*

8. Sarah ran 2 mi more than Abby. *$m + 2$*

9. David popped 5 balloons. *$b - 5$*

10. Josh is 3 yr older than Aaron. *$a + 3$*

Solve.

11. $4a = 64$ *$a = 16$*

12. $k + 7 = 48$ *$k = 41$*

13. $\frac{x}{7} = 56$ *$x = 392$*

14. $b - 6.4 = 1.8$ *$b = 8.2$*

15. $a \div 16 = 4$ *$a = 64$*

16. $20r = 400$ *$r = 20$*

476 Daily Review

- Write for display "V = area of the base (B) × height of the prism (h)" and write "$V = Bh$" below it.

  What does the h in the formula for volume represent? the height of the prism

  What does the uppercase B represent? the area of the base

- Choose a student to measure the length and the width of the prism's base to the nearest centimeter and then calculate the area of the base using $A = l \times w$.

  Choose another student to measure the height to the nearest centimeter.

Guide the students as they find the volume of the solid figure by substituting the calculated area of the base and the measured height into the formula $V = Bh$.

How do you think you can find the volume of the prism using one equation? Since the B in the formula represents the area of the base, I can multiply the length of the base times its width and then multiply by the height.

- Write "$V = (l \times w) \times h$" below $V = Bh$. Guide the students as they substitute the prism's dimensions into $V = (l \times w) \times h$ and solve it.

- Guide the students as they use the formula $V = Bh$ to calculate the volume of prisms with the following dimensions.

  $l = 2.5$ m, $w = 1.5$ m, $h = 3$ m $V = Bh$; $V = (l \times w) \times h$; $V = (2.5$ m $\times 1.5$ m$) \times 3$ m; $V = 3.75$ m$^2 \times 3$ m; $V = 11.25$ m^3

  $B = 20$ in.^{2}, $h = 6$ in. $V = Bh$; $V = 20$ in.$^2 \times 6$ in.; $V = 120$ in.^{3}

- Guide the students as they solve the following word problems by substituting the dimensions into the appropriate formula.

- Read aloud the following word problems.

  Kyle has an aquarium that is 40 cm long, 30 cm wide, and 20 cm high. What is the volume of Kyle's aquarium? $V = Bh$; $V = (l \times w) \times h$; $V = (40$ cm $\times 30$ cm$) \times 20$ cm; $V = 24{,}000$ cm^3

  One of the fish tanks at the zoo is 3.2 m long, 2.6 m wide, and 1.5 m high. What is the volume of the aquarium tank? $V = Bh$; $V = (l \times w) \times h$; $V = (3.2$ m $\times 2.6$ m$) \times 1.5$ m; $V = 12.48$ m^3

  At the zoo, the parrots are kept in a wire cage that is 3 yd long, 3 yd wide, and 4 yd high. What is the volume of the parrot cage? $V = Bh$; $V = (l \times w) \times h$; $V = (3$ yd $\times 3$ yd$) \times 4$ yd; $V = 36$ yd^3

  The gift shop at the zoo is 25 ft long and 30 ft wide. What area does the gift shop cover? $A = l \times w$; $A = 25$ ft $\times 30$ ft; $A = 750$ ft^2

Making people's lives better by finding volume

- Guide a **discussion** to help the students answer the essential question at the top of Student Edition page 250.

- Distribute a *Shipping Volume* page to each student and display a copy.

  Explain that shipping companies transport most of the items that the students purchase in stores or online, providing a valuable service for customers and employing many workers.

- Instruct the students to read the information and examine the picture of a trailer and freight at the top of the *Shipping Volume* page.

  How does calculating the volume of his trailer (container) help a shipper make his customers' lives better? It helps him load freight more efficiently and provide better service to his customers.

  What geometric figure does an enclosed shipping trailer remind you of? a rectangular prism

- Point out the equivalency, 4 ft = 1 skid unit. Explain that they will use this to convert the dimensions of the trailer from feet to skid units in order to calculate the trailer's volume in cubic skid units (skid units^3).

- Guide the students as they complete the problem at the bottom of the page. They may use a calculator.

  Check the answers together and direct them to correct any wrong answers. Explain that they will use the completed page as a guide to help them answer problems 28–29 on the Student Edition page.

Apply

Student Edition pages 250–51

- Read and explain the directions for pages 250–51. Assist the students as they complete the pages independently.

Daily Review

- Students should complete Chapter 12, section *a*.

NOTES

Student Edition pages 252–53
Daily Review Chapter 12, section *b*

OBJECTIVES

- Find the volume of a cube by using a model or a formula.
- Calculate the unknown measurement of a rectangular prism.
- Solve a word problem and interpret the solution.

ADDITIONAL MATERIALS

- 3 blank sheets of $8\frac{1}{2}" \times 11"$ paper
- cube-shaped blocks (to build a 64 unit^3 tower)

Preparation

Prepare each area drawing on a separate sheet of blank paper. Make the squares in each drawing the same size as the cubes used in the lesson. Retain the drawings for use in Lesson 118.

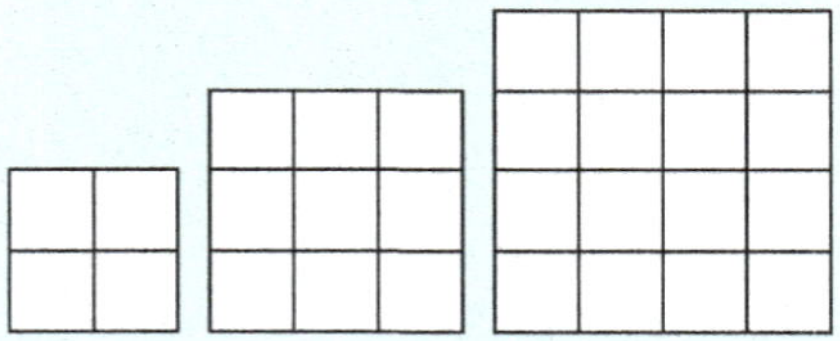

If enough cubes are available, you may prepare copies of the area drawings so that groups of students can build the towers as you demonstrate.

Engage

- Direct the students to **brainstorm** to explore the essential question on Student Edition page 252, "Why can I use the formula $V = s^3$ to find the volume of a cube?"

Instruct

Finding volume by using a model or a formula

- Use a **3D model** to guide the students as they find the volume of a cube.

Volume of Cubes

Why can I use the formula $V = s^3$ to find the volume of a cube?

A cube is a special rectangular prism in which all 6 faces are congruent squares and all sides measure the same.

The formula for volume, $V = Bh$, can be modified for a cube.

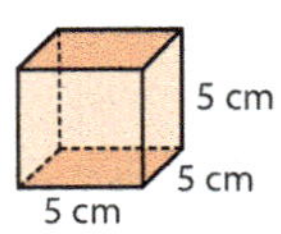

$$\text{volume of a cube} = (\text{side} \times \text{side}) \times \text{side}$$
$$V = s^3$$

$$
\begin{array}{ll}
V = (s \times s) \times s & V = s^3 \\
V = (5 \times 5) \times 5 & V = 5^3 \\
V = 25 \times 5 \quad \text{or} & V = 5 \times 5 \times 5 \\
V = 125 \text{ cm}^3 & V = 125 \text{ cm}^3
\end{array}
$$

Key Terms
- volume of a cube
- $V = s^3$

Exercises

Write an equation to find the volume of the cube. Round decimals to the nearest tenth. *Formula used may vary.*

1. 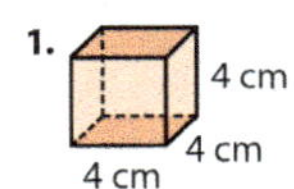 4 cm, 4 cm, 4 cm $V = 64 \text{ cm}^3$

2.  10 in., 10 in., 10 in. $V = 1{,}000 \text{ in.}^3$

3. 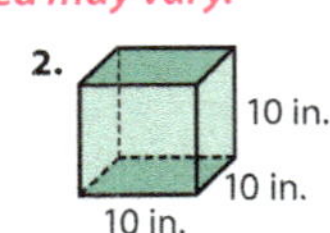 3.2 m, 3.2 m, 3.2 m $V = 32.8 \text{ m}^3$

4. 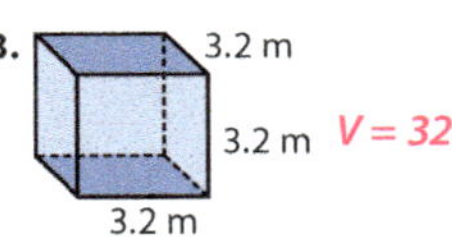 6 ft, 6 ft, 6 ft $V = 216 \text{ ft}^3$

5. $1\frac{1}{2}$ yd, $1\frac{1}{2}$ yd, $1\frac{1}{2}$ yd $V = 3\frac{3}{8} \text{ yd}^3$

6. 4.5 m, 4.5 m, 4.5 m $V = 91.1 \text{ m}^3$

Find the volume of a prism with the given dimensions. *Equations may vary.*

7. rectangular prism: $l = 5$ in., $w = 2$ in., $h = 4$ in. *(5 in. $\times$ 2 in.) $\times$ 4 in. = 40 in.^{3}*

8. rectangular prism: $B = 12$ ft^2, $h = 3$ ft *12 ft^2 $\times$ 3 ft = 36 ft^3*

9. cube (square prism): $s = 7$ m *(7 m)^3 or (7 m $\times$ 7 m) $\times$ 7 m = 343 m^3*

10. cube (square prism): $B = 100$ cm^2, $h = 10$ cm *100 cm^2 $\times$ 10 cm = 1,000 cm^3*

Find the area of the base: $B = l \times w$. Find the volume of the prism that could be built using centimeter cubes for the given height.

11. 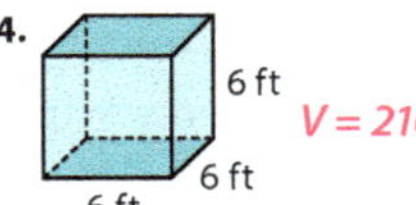 $V = 8 \text{ cm}^3$ $h = 2$ cm

12. 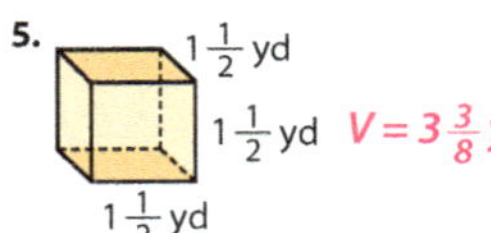 $V = 27 \text{ cm}^3$ $h = 3$ cm

13. 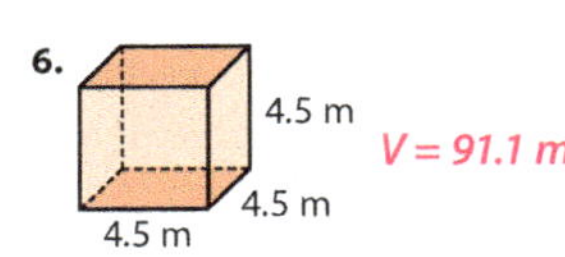 $V = 60 \text{ cm}^3$ $h = 5$ cm

---

What is a prism? It is a cylindrical figure having 2 congruent and parallel bases that are polygons; all of its other faces are parallelograms.

- Display a cube-shaped block.

What is this 3-dimensional figure? a cube; All 6 faces are congruent squares.

Remind the students that a square is a rectangle whose sides are all the same length, so a cube is a special type of rectangular prism whose sides (length, width, and height) are all the same length.

Since a square is a special type of rectangle, a prism with 2 square bases may be called a square prism or a rectangular prism. Not all square prisms are cubes.

- Write the terms "perimeter," "area," "surface area," and "volume" for display.

Which terms are attributes of 2-dimensional figures? perimeter; area

Which are attributes of 3-dimensional figures? surface area; volume

Choose students to give an example of a 2- or 3-dimensional figure to support their answer. sample answer: A rectangle has length and width that I can use to calculate its perimeter and area. A rectangular prism has length, width, and height that I can use to calculate its surface area and volume.

- Display the area drawing of 4 square units.

Finding an Unknown Dimension

The formula for volume of a prism is $V = Bh$. Since the bases of the figure are rectangles, use $V = (l \cdot w) \cdot h$.

When the volume is given and any two of the volume dimensions are known, you can find the unknown third dimension of a figure.

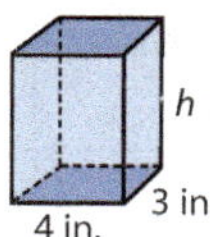

$V = Bh$ $84 \text{ in.}^3 = (4 \cdot 3) \cdot h$
$V = (l \cdot w) \cdot h$ $84 \text{ in.}^3 = 12 \cdot h$
$V = 84 \text{ in.}^3$ $\dfrac{84 \text{ in.}^3}{12} = \dfrac{12h}{12}$
$l = 4 \text{ in.}$ $7 \text{ in.} = h$
$w = 3 \text{ in.}$
$h = \underline{\ \ } \text{ in.}$

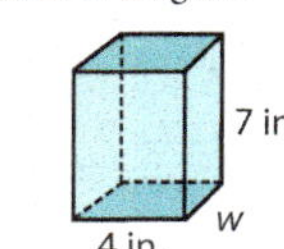

$V = Bh$ $84 \text{ in.}^3 = (4 \cdot w) \cdot 7$
$V = (l \cdot w) \cdot h$ $84 \text{ in.}^3 = 4 \cdot w \cdot 7$
$V = 84 \text{ in.}^3$ $84 \text{ in.}^3 = 4 \cdot 7 \cdot w$
$l = 4 \text{ in.}$ $84 \text{ in.}^3 = 28 \cdot w$
$w = \underline{\ \ } \text{ in.}$ $\dfrac{84 \text{ in.}^3}{28} = \dfrac{28w}{28}$
$h = 7 \text{ in.}$ $3 \text{ in.} = w$

What is the measure of the length, the width, and the height of a cube whose volume is 27 units³?

$V \text{ (of a cube)} = s^3$
$27 = s \cdot s \cdot s$

If $s = 2$: $2 \cdot 2 \cdot 2 = 8$
If $s = 3$: $3 \cdot 3 \cdot 3 = 27$

Each side is 3 units.

Find the unknown measurement of the rectangular prism.

14. $V = 24 \text{ ft}^3$ *h = 4 ft*

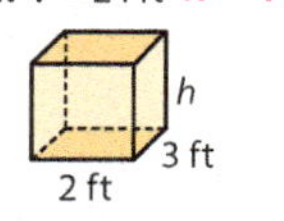

15. $V = 248 \text{ m}^3$ *l = 10 m*

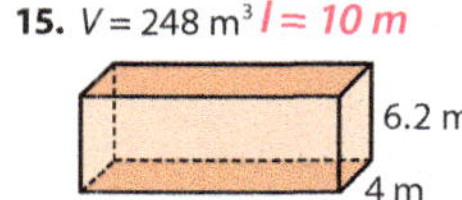

16. $V = 144 \text{ cm}^3$ *w = 3 cm*

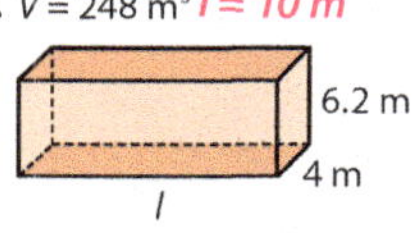

17. $l = 5 \text{ in.}, w = 3 \text{ in.}, h = \underline{\ \ } \text{ in.}, V = 75 \text{ in.}^3$ *h = 5*

18. $l = 2 \text{ cm}, w = \underline{\ \ } \text{ cm}, h = 4 \text{ cm}, V = 48 \text{ cm}^3$ *w = 6*

19. $B = 6 \text{ in.}^2, h = \underline{\ \ } \text{ in.}, V = 90 \text{ in.}^3$ *h = 15*

20. $B = 5.7 \text{ cm}^2, h = \underline{\ \ } \text{ cm}, V = 51.3 \text{ cm}^3$ *h = 9*

Practice & Application   *Equations may vary.*

Mr. Cole is making a small rectangular pond in his backyard. The pond will be 7 m long, 5 m wide, and 1 m deep.

21. What is the volume of the pond?
7 m × 5 m × 1 m = 35 m³

22. One cubic meter can hold about 264 gal of water. Approximately how many gallons of water will the pond hold? *35 × 264 gal = 9,240 gal*

23. Mr. Cole is placing stones along the 4 sides of the pond. What is the distance around the pond's edge? *(2 × 7 m) + (2 × 5 m) = 24 m*

24. Mrs. Cole wants to purchase a tarp to cover the pond during the winter. What is the area of the pond? *7 m × 5 m = 35 m²*

25. Find the surface area and the volume of the prisms.

26. *Since the length, width, and height of a cube are equal lengths, I can multiply the measurement of 1 side 3 times.*

26. Why can I use the formula $V = s^3$ to find the volume of a cube?

What is the area of this figure? 4 square units; $A = l \times w$ or $A = s^2$; $A = 2 \times 2 = 4 \text{ units}^2$ or $A = 2^2 = 4 \text{ units}^2$

If you are building a tower with blocks on this base plan, how many blocks will be needed for the first floor? 4 blocks; Each square unit in the base plan represents 1 face of each block (cubic unit) needed for the first floor.

- Choose a student to use cubes to build the first floor of the tower (2 rows of 2 cubes).

What is the volume of the tower since it is 1 cube high? 4 cubic units; A total of 4 cubes make up 1 floor with a base that is a 2 × 2 area.

- Draw for display a table similar to the one provided. Complete the first row using the data for the first floor of the tower.

- Repeat the procedure for a second floor of the tower.

What is the volume of the tower since it is 2 cubes high? 8 cubic units

- Point out that the volume of the tower after each floor is built is the area of the base multiplied by the number of layers or the height. The length, width, and height of the cube-shaped tower are equal.

Area of Base (units²)	Number of Layers—Height (unit)	Volume (units³)
4 units²	1 unit	4 units³
4 units²	2 units	8 units³

- Display the area drawing of 9 square units.

To make a cube-shaped tower from this base, how many units high must the figure be? 3 units high; The length, width, and height of a cube must be equal.

- Choose a student to place the blocks, layer by layer, on the area drawing to build the 3 × 3 × 3 tower.

What is the volume of this tower? Explain. 27 cubic units; possible explanations: Each 3 × 3 layer contains 9 cubes and there are 3 layers; volume is equal to the area of the base (9) multiplied by the height (3); 9 × 3 = 27.

What formula can be used to find the volume of all prisms? $V = Bh$

Write "$V = Bh$" for display.

- Guide the students as they use a formula to write an equation to find the volume of the tower. Remind them that the factors in parentheses are the length and width, which are multiplied to find the area of the base. $V = (3 \times 3) \times 3$

How can you rewrite this equation using exponents? $V = 3^3$

- Write the new equation for display. Choose a student to give the answer. $V = 27 \text{ units}^3$

How could you modify the formula $V = Bh$ to find the volume of a cube? $V = (s \times s) \times s$ and $V = s^3$

Direct the students' attention to the essential question at the top of Student Edition page 252.

What about the lengths of a cube's sides makes it possible for you to use the formula $V = s^3$ to find the volume of a cube? Since the length, width, and height of a cube are equal lengths, I can multiply the measurement of 1 side 3 times.

Write "$V = (s \times s) \times s$" and "$V = s^3$" for display.

What does $(s \times s)$ in the formula represent? the area of the cube's base

LESSON 115

- Display the area drawing of 16 square units and choose a student to build a cube-shaped tower on the base. Guide the students as they use both formulas to find the volume of the tower.

  $V = (s \times s) \times s$ $V = (4 \times 4) \times 4$; $V = 16 \times 4$; $V = 64$ units^3

  $V = s^3$ $V = 4^3$; $V = 4 \times 4 \times 4$; $V = 64$ units^3

- Guide the students as they use the formulas to calculate the volume of cubes having the following dimensions. While solving the second problem, explain that when a factor is labeled, such as *6 cm*, the exponent must be written outside the parentheses to indicate "6 centimeters cubed" rather than "6 cubic centimeters."

  $l = \frac{1}{2}$ in., $w = \frac{1}{2}$ in., $h = \frac{1}{2}$ in. $V = (s \times s) \times s$; $V = (\frac{1}{2}$ in. $\times \frac{1}{2}$ in.$) \times \frac{1}{2}$ in.; $V = \frac{1}{4}$ in.$^2 \times \frac{1}{2}$ in.; $V = \frac{1}{8}$ in.^{3}

  $s = 6$ cm $V = s^3$; $V = (6$ cm$)^3$; $V = 6$ cm $\times$ 6 cm $\times$ 6 cm; $V = 216$ cm^3

  $B = 16$ m^2, $h = 4$ m $V = Bh$; $V = 16$ m$^2 \times 4$ m; $V = 64$ m^3

Solving a word problem by calculating an unknown measurement

- Guide the students to use the **Problem-Solving Plan** to find the unknown measurement of a rectangular prism.

- Read the following word problem.

  A gift box for earrings has a volume of 24 in.^{3}. If the area of the rectangular base is 12 in.^{2}, what is the height of the gift box? 2 in.

  What is the question asking you to find? the unknown height of the gift box

  What information is given? the total volume of the box (24 in.^{3}) and the area of the rectangular base (12 in.^{2})

  What type of figure is the gift box? a rectangular prism; It has a rectangular base.

  How can you find the unknown height of the gift box? I can think of the formula $V = Bh$; since the volume of the gift box is 24 in.^{3} and the area of its rectangular base is 12 in.^{2}, I can divide the volume by the area of the base to find the height.

Complete the table.

1.

meter	millimeter
1	1,000
4	4,000
2	2,000
9	9,000

2.

gram	kilogram
1,000	1
3,000	3
4,000	4
5,000	5

3.

milliliter	liter
1,000	1
5,000	5
7,000	7
8,000	8

Write a comparison sentence by using >, <, or =.

4. 3 m $\geq$ 300 mm

5. 8,000 g $=$ 8 kg

6. 2,859 mL $\leq$ 4 L

Choose the best measurement.

7.

Capacity	
a bottle of water	
10 mL	(1 L)
a mug of cocoa	
(250 mL)	25 L
water in a bathtub	
150 mL	(150 L)

8.

Mass	
a dog	
(20 kg)	20 g
four jellybeans	
4 kg	(4 g)
a chocolate chip cookie	
1 kg	(10 g)

9.

Temperature	
swimming in the ocean	
(30°C)	70°C
normal body temperature	
(37°C)	98°C
boiling water	
0°C	(100°C)

- Guide the students as they first substitute the given values into the volume formula, and then solve the equation to find the unknown dimension of the prism (the value of the variable).

  $V = Bh$; 24 in.^{3} = (12 in.^{2})h; $\frac{24 \text{ in.}^3}{12 \text{ in.}^2} = \frac{(12 \text{ in.}^2)h}{12 \text{ in.}^2}$; 2 in. = h

  How could you check your answer? I could substitute my answer into the volume formula equation to see if it makes a true statement.

- Follow a similar procedure for the following word problems. Assist the students as they conclude that when they multiply the width of a rectangular prism times the height or multiply the length times the height, they are finding the area of 1 of the faces and area is measured in square units.

- Read aloud the following word problems.

  A gift box for a sweater has a volume of 648 in.^{3}. If the width of the box is 12 in. and the height is 3 in., what is the length of the box? 18 in.; $V = Bh$; $V = (l \cdot w) \cdot h$; 648 in.^{3} = $(l \cdot 12$ in.$) \cdot 3$ in.; 648 in.^{3} = $l \cdot (12$ in. $\cdot 3$ in.$)$; 648 in.^{3} = $l \cdot 36$ in.^{2}; $\frac{648 \text{ in.}^3}{36 \text{ in.}^2} = \frac{l \cdot 36 \text{ in.}^2}{36 \text{ in.}^2}$; 18 in. = l

A packing box has a volume of 4 ft³. If the length of the box is 2 ft and the height is 1 ft, what is the width of the box? 2 ft; $V = Bh$; $V = (l \cdot w) \cdot h$; 4 ft³ = (2 ft $\cdot$ w) $\cdot$ 1 ft; 4 ft³ = (2 ft $\cdot$ 1 ft) $\cdot$ w; $\frac{4\ ft^3}{2\ ft^2} = \frac{2\ ft^2 \cdot w}{2\ ft^2}$; 2 ft = w

- Follow a similar procedure for finding the unknown measurements of prisms with the following dimensions.

  $V = 540$ cm³, $l = 20$ cm, $h = 3$ cm
  $w = 9$ cm

  $V = 15.2$ cm³, $B = 3.8$ cm² $h = 4$ cm

Apply

Student Edition pages 252–53
- Read and explain the directions for pages 252–53. Assist the students as they complete the pages independently.

Daily Review
- Students should complete Chapter 12, section b.

DIFFERENTIATED INSTRUCTION

Use the following to provide extra help for students who experience difficulty with the concepts taught in Chapter 12.

Visualize a 3-dimensional figure.
Provide 3-dimensional figures for the students who need assistance in "seeing" a 3-dimensional object when looking at a 2-dimensional representation. Allow the students to handle the objects as they complete their assignments.

MATH TALK

Follow the General Procedure for Math Talks as outlined in Lesson 5 on Teacher Edition page 13b.

Which has more volume, a prism measuring 5 ft × 4 ft × 3 ft or one with a base area of 16 ft² and height of 5 ft?

NOTES

Student Edition pages 254–55
Daily Review Chapter 12, section c

OBJECTIVES

- Find the volume of an irregular prism by using a model.
- Calculate the volume of a triangular prism by using a formula.
- Calculate the volume of a cylinder by using a formula.
- Explain how finding the volume of a 3D figure can protect others.
  BWS

BIBLICAL WORLDVIEW SHAPING

- **Service (Explain):** Finding the volume of a figure protects others by enabling proper care of pool water.

TEACHER RESOURCES

- *67 Triangular Prisms & Cylinders*

ADDITIONAL MATERIALS

- 2 blank sheets of $8\frac{1}{2}"\times 11"$ paper
- cube-shaped blocks (to build a 25 unit³ tower)
- a calculator (for each student)

ASSESSMENTS

- Chapter 12 Quiz 1

Volume of Other 3D Figures

How can finding the volume of a 3D figure protect others?

Key Terms
- volume of an irregular prism
- volume of a triangular prism
- volume of a cylinder

Volume of an Irregular Prism

1. Count the square units in the base to find the area of the base.

   $B = 6$ units²

2. Substitute the area of the base for B in the volume formula.

   $V = Bh$
   $V = (6)h$

3. Solve.

   $V = (6)(3)$
   $V = 18$ units³

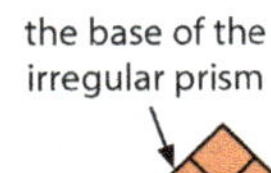
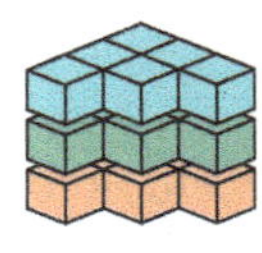

Volume of a Triangular Prism

1. Find the area of the triangular base.

   $B = \frac{1}{2}bh$
   $B = \frac{1}{2}(2 \cdot 3)$
   $B = \frac{1}{2}(6)$
   $B = 3$ units²

2. Substitute the area of the base for B in the volume formula.

   $V = Bh$ or $(\frac{1}{2}bh_1)h_2$
   $V = (3)h$

3. Solve.

   $V = (3)(6)$
   $V = 18$ units³

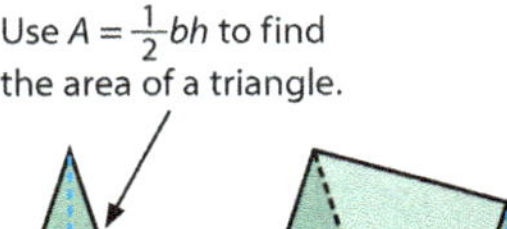

Volume of a Cylinder

1. Find the area of the circular base, given the radius. (Use $r = \frac{1}{2}d$ if a diameter is given.)

   $B = \pi r^2$
   $B = (3.14)(6^2)$
   $B = (3.14)(36)$
   $B = 113.04$ cm²

2. Substitute the area of the circular base for B in the volume formula.

   $V = Bh$ or $(\pi r^2)h$
   $V = (113.04)h$

3. Solve.

   $V = (113.04)(10)$
   $V = 1,130.4$ cm³

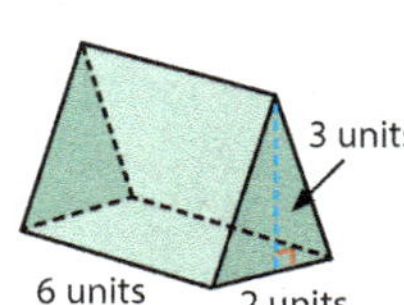

Exercises

Find the volume of the triangular prism. *Equations may vary.*

1. 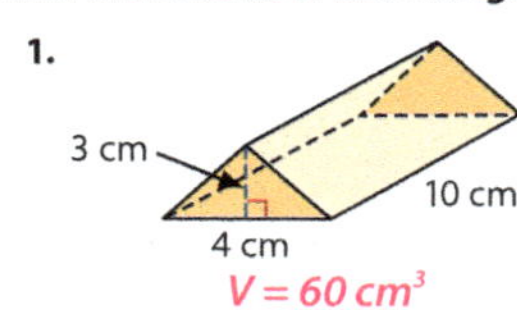

2. 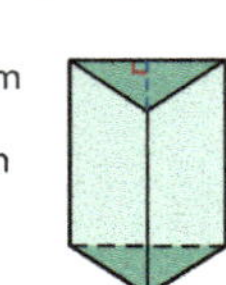

3. 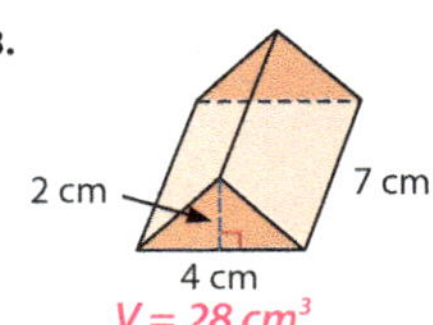

254   Chapter 12

Preparation

Prepare each area drawing on a separate sheet of blank paper. Make the squares in each drawing the same size as the cubes used in the lesson. Retain the drawings for use in Lesson 118.

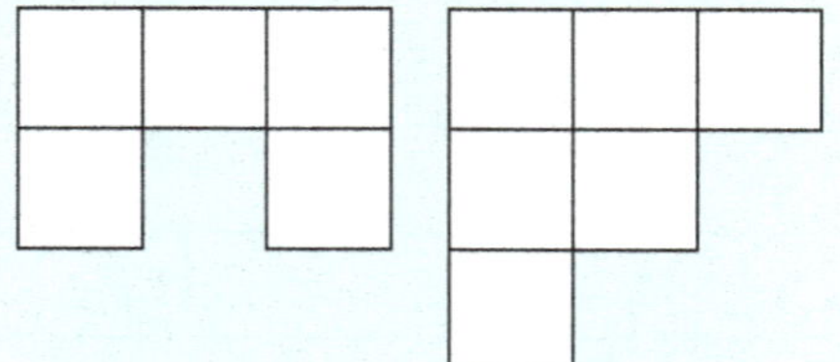

If enough cubes are available, you may prepare copies of the area drawings so that groups of students can build the towers as you demonstrate.

Allow the students to use calculators throughout the remainder of this chapter so that they can focus on finding the volume of the various figures rather than on doing lengthy calculations.

Engage

- Direct the students to **brainstorm** to explore the essential question on Student Edition page 254, "How can finding the volume of a 3D figure protect others?"

Instruct

Volume of an irregular prism

- Use a **3D model** to guide the students as they find the volume of an irregular prism.
- Display the area drawing of 5 square units.

Find the volume of the cylinder. Round decimals to the nearest tenth. *Equations may vary.*

4. 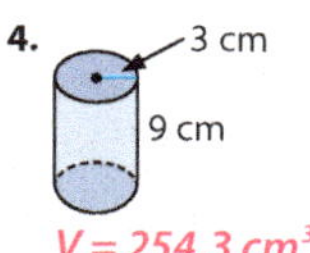

$V = 254.3$ cm³

5. 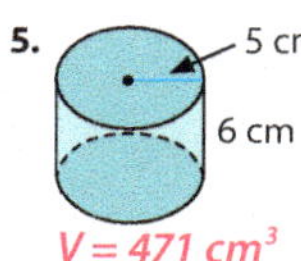

$V = 471$ cm³

6. 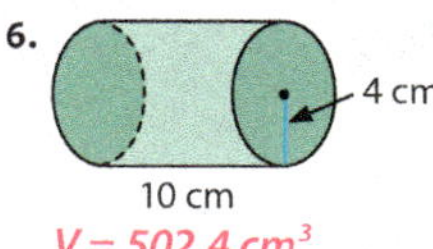

$V = 502.4$ cm³

Count the squares to find the area of the base. Find the volume of the prism that could be built using centimeter cubes for the given height.

7.  $h = 2$ cm
$V = 10$ cm³

8. 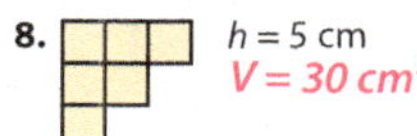 $h = 5$ cm
$V = 30$ cm³

9. 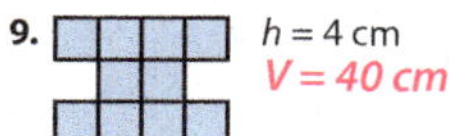 $h = 4$ cm
$V = 40$ cm³

Find the volume of a figure with the given dimensions. *Equations may vary.*

10. rectangular prism: $l = 7$ in., $w = 3$ in., $h = 4$ in.
7 in. × 3 in. × 4 in. = 84 in.³

11. square prism: $s = 2$ m
(2 m)³ = 8 m³

12. rectangular prism: $B = 8$ ft², $h = 5$ ft
8 ft² × 5 ft = 40 ft³

13. square prism: $B = 16$ cm², $h = 4$ cm
16 cm² × 4 cm = 64 cm³

14. triangular prism: $B = 12$ ft², $h = 3$ ft
12 ft² × 3 ft = 36 ft³

15. cylinder: $r = 2$ m, $h = 5$ m
3.14 × (2 m)² × 5 m = 62.8 m³

16. triangular prism: $b = 5$ in., $h = 2$ in., h (prism) = 4 in.
½ (5 in. × 2 in.) × 4 in. = 20 in.³

17. cylinder: $B = 78.5$ cm², $h = 4$ cm
78.5 cm² × 4 cm = 314 cm³

Find the unknown measurement of the rectangular prism.

18. 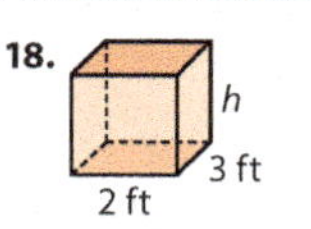

$V = 36$ ft³ *h = 6 ft*

19. 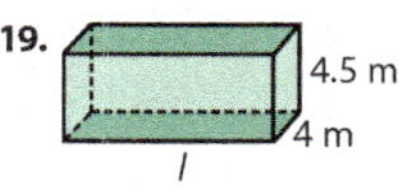

$V = 162$ m³ *l = 9 m*

20. 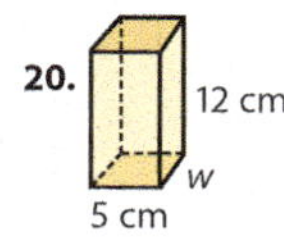

$V = 240$ cm³ *w = 4 cm*

Practice & Application   *Equations may vary.*

21. *½ (6 in. × 7 in.) × 8 in. = 168 in.³*

21. Samuel built a birdhouse in the shape of a triangular prism. Use the given dimensions to find the volume of the birdhouse.

22. Thomas filled a rectangular planter with potting soil. His planter is 2 ft long, $1\frac{1}{2}$ ft wide, and $\frac{1}{2}$ ft high. How much potting soil did it take to fill his planter?
2 ft × $\frac{3}{2}$ ft × $\frac{1}{2}$ ft = $\frac{6}{4}$ ft³ = $1\frac{1}{2}$ ft³

23. Rebecca filled a cylindrical planter with potting soil. Her planter is 1 ft tall and has a diameter of 2 ft. How much potting soil did it take to fill her planter? Remember $r = \frac{1}{2}d$. *3.14 × (1 ft)² × 1 ft = 3.14 ft³*

24. Maya just purchased an above-ground swimming pool. The swimming pool is 4 ft high and has a diameter of 18 ft. What is the volume of her new swimming pool? Remember $r = \frac{1}{2}d$.
3.14 × (9 ft)² × 4 ft = 1,017.36 ft³

25. One cubic foot can hold about 7.5 gal of water. How many gallons of water are needed to fill Maya's swimming pool? (Use the information from problem 24.)
1,017.36 × 7.5 gal = 7,630.2 gal

26. Maya needs 0.00013 oz of chlorine per gallon of water to disinfect her pool. How much chlorine does she need to add? Round your answer to the nearest ounce. *7,630.2 × 0.00013 oz = 0.991926 oz ≈ 1 oz*

27. How did finding the volume of a 3D figure help Maya protect others?

27. It helped her determine how much chlorine she should safely add to the pool water so that those swimming in it could be protected from bacteria or from too much chlorine that could make them sick.

If you are building a tower with blocks using this irregular base plan, how many blocks will be needed for the first floor?
5 blocks; I can count the number of square units in the figure to know how many cubes are needed to cover the first floor.

What is the area of this irregular figure?
5 square units

Choose a student to place the blocks for the base on the area drawing.

- Draw for display a table similar to the one provided and complete the first row using the data for the first floor of the tower.

- Repeat the procedure until the tower has 5 layers. Guide the students to conclude that the volume of the tower after each floor is built is the area of the base multiplied by the number of layers or the height.

Area of Base (units²)	Number of Layers—Height (unit)	Volume (units³)
5 units²	1 unit	5 units³
5 units²	2 units	10 units³
5 units²	3 units	15 units³
5 units²	4 units	20 units³
5 units²	5 units	25 units³

LESSON 116

- Display the area drawing of 6 square units.

  What formula can be used to find the volume of all prisms? $V = Bh$
  Write the formula for display.
  How can you use $V = Bh$ to find the volume of a tower that is 3 units high built on this base? Since the area of the base is 6 square units, I can multiply the area of the base (6 units²) by the height (3 units).
  What is the volume of the tower? 18 units³

- Choose a student to place the blocks, layer by layer, on the area drawing to build the tower.
  Is this 3D figure a prism? Yes; it has 2 congruent and parallel bases, and its other faces are parallelograms.

- Choose a student to write and solve the equation to find the volume of the prism.
  $V = 6$ units² × 3 units; $V = 18$ units³

- Guide the students as they use the appropriate formulas to solve the following volume word problems.

Encourage the students to draw pictures for the word problems if needed.

Ben attended classes to learn how to make ice sculptures. During the first week, Ben learned to make geometric figures. On the first day, he made a cube. Each edge of the cube measured 1.5 ft. To the nearest tenth of a foot, what was the volume of the ice sculpture? $V = s³$ (or $V = s × s × s$); $V = (1.5$ ft$)³$; $V = 1.5$ ft × 1.5 ft × 1.5 ft; $V = 3.375$ ft³ ≈ 3.4 ft³

On the second day of class, Ben made an ice sculpture in the shape of a rectangular prism. The sculpture measured 2 ft long, 1.5 ft wide, and 3 ft high. What was the volume of this ice sculpture? $V = Bh$; $V = (l × w) × h$; $V = (2$ ft × 1.5 ft$) × 3$ ft; $V = 3$ ft² × 3 ft; $V = 9$ ft³

Volume of a triangular prism

- Guide the students as they use a **formula** to calculate the volume of a triangular prism.

LESSON 116

- Display the *Triangular Prisms & Cylinders* page. Explain that although the formula $V = Bh$ is used to find the volume of any prism, the base of the prism determines how to find B (area of the base).
  What is the shape of the congruent parallel bases of the first prism? triangle
  What formula do you use to calculate the area of a triangular base? $A = \frac{1}{2}bh$

- Write for display "$B = \frac{1}{2}bh$." Review what each letter in the formula represents. B = area of the base; b = length of the base; h = height of the base
  Why do you think the formula uses both an uppercase B and a lowercase b? They represent different things. The uppercase B represents the area of the base (square units), and the lowercase b represents the length of the base (linear units).

- Guide the students as they use the formula to find the area of the triangular base. Remind them that the height of a triangle forms a right angle where it intersects the triangle's base. $B = \frac{1}{2}(2 \text{ in.} \times 3 \text{ in.}); B = \frac{1}{2}(6 \text{ in.}^2); B = 3 \text{ in.}^2$

- Guide the students as they use $V = Bh$ to find the volume of the triangular prism. $V = (3 \text{ in.}^2)(5 \text{ in.}); V = 15 \text{ in.}^3$

- Follow a similar procedure for finding the volume of the second triangular prism. $B = \frac{1}{2}bh; B = \frac{1}{2}(4 \text{ in.} \times 5 \text{ in.}); B = \frac{1}{2}(20 \text{ in.}^2); B = 10 \text{ in.}^2$
  $V = Bh; V = (10 \text{ in.}^2)(6 \text{ in.}); V = 60 \text{ in.}^3$

- Explain that they can also calculate the area of the triangular base of the prism while calculating the volume of the triangular prism. Write "$V = (\frac{1}{2}bh_1)h_2$" for display. Point out that when calculating the area of the triangle, they use the base and height of the triangle. When multiplying the area of the triangular base (B) by the height (h), they use the height of the prism. The subscripts 1 and 2 help them to distinguish between the two heights: h_1 is the height of the triangle and h_2 is the height of the prism.

- Guide the students as they find the volume of the second triangular prism using $V = (\frac{1}{2}bh_1)h_2$. $V = (\frac{1}{2} \times 4 \text{ in.} \times 5 \text{ in.})6 \text{ in.}; V = (10 \text{ in.}^2)6 \text{ in.}; V = 60 \text{ in.}^3$

- Guide the students as they use $V = (\frac{1}{2}bh_1)h_2$ to solve the following word problem.
  On the third day, Ben made an ice sculpture in the shape of a triangular prism. The triangular base had a length of 12 in. and a height of 10 in. The prism had a height of 15 in. What was the volume of this ice sculpture? $V = (\frac{1}{2}bh_1)h_2; V = (\frac{1}{2} \times 12 \text{ in.} \times 10 \text{ in.})15 \text{ in.}; V = (60 \text{ in.}^2)15 \text{ in.} = 900 \text{ in.}^3$

Volume of a cylinder

- Guide the students as they use a **formula** to calculate the volume of a cylinder.

- Direct attention to the first cylinder on the *Triangular Prisms & Cylinders* page.
  What is the shape of the cylinder's base? a circle
  What formula do you use to calculate the area of a circle? $A = \pi r^2$
  What other letter could you substitute for A to indicate the area of a circular base? I can use an uppercase B.

- Write "$B = \pi r^2$" for display. Remind the students that they can use 3.14 (or $\frac{22}{7}$) as the value of pi.

Use the graph to answer the question.

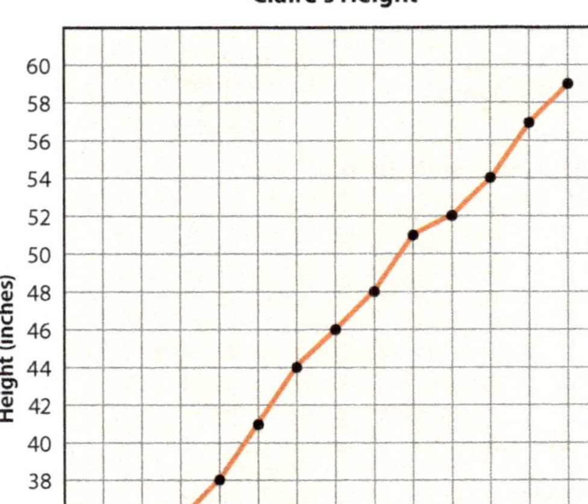

Mrs. West recorded Claire's height on each birthday. Claire took the measurements and put them in a graph form.

1. What kind of graph did Claire make? *a line graph*

2. Why does the line increase rather than decrease? *because Claire grew taller each year*

3. How tall was Claire at age 1? *30 in.*

4. How many inches taller was Claire at age 5 than at age 1? *41 − 30 = 11 in.*

5. Between which two years did Claire grow only 1 in. taller? *9–10*

6. How tall was Claire at age 13? *59 in.*

7. How many inches did Claire gain between ages 6 and 7? *2 in.*

478   Daily Review

Guide the students as they use the formula to find the area of the base. Remind them that when a factor is labeled, such as 4 in., the exponent must be written outside the parentheses so that the factor is "4 inches squared" rather than "4 square inches." $B = (3.14)(4 \text{ in.})^2$; $B = 3.14 \times 16 \text{ in.}^2$; $B = 50.24 \text{ in.}^2$

Since you know the area of the base, what formula do you think you can use to find the volume of the cylinder? $V = Bh$

- Write "$V = Bh$" for display and guide the students as they use the formula to find the volume of the cylinder. Round the answer to the nearest tenth. $V = (50.24 \text{ in.}^2)3 \text{ in.}$; $V = 150.72 \text{ in.}^3 \approx 150.7 \text{ in.}^3$

- Follow a similar procedure for the second cylinder. Assist the students to conclude that half of the diameter (6 cm) is 3 cm, the measure of the radius.

  $B = \pi r^2$; $B = 3.14 \cdot (3 \text{ cm})^2$; $B = 3.14 \times 9 \text{ cm}^2$; $B = 28.26 \text{ cm}^2$

  $V = Bh$; $V = 28.26 \text{ cm}^2 \times 5 \text{ cm}$; $V = 141.3 \text{ cm}^3$

- Explain that they can also calculate the area of the circular base of the cylinder while calculating the volume of the cylinder. Write "$V = (\pi r^2)h$" for display and guide the students as they find the volume of the second cylinder. $V = (\pi r^2)h$; $V = 3.14 \cdot (3 \text{ cm})^2 \cdot 5 \text{ cm}$; $V = (3.14 \times 9 \text{ cm}^2) \cdot 5 \text{ cm}$; $V = 28.26 \text{ cm}^2 \cdot 5 \text{ cm}$; $V = 141.3 \text{ cm}^3$

- Guide the students as they use $V = (\pi r^2)h$ to solve the following word problem.

The last ice sculpture Ben made was in the shape of a cylinder. The diameter of the base was 12 in. and the height of the cylinder was 24 in. To the nearest inch, what was the volume of this sculpture? $V = (\pi r^2)h$; $V = 3.14 \cdot (6 \text{ in.})^2 \cdot 24 \text{ in.}$; $V = (3.14 \times 36 \text{ in.}^2) \cdot 24 \text{ in.}$; $V = 113.04 \text{ in.}^2 \cdot 24 \text{ in.}$; $V = 2{,}712.96 \text{ in.}^3 \approx 2{,}713 \text{ in.}^3$

Protecting others by finding volume

- Guide a **discussion** to prepare the students to answer the essential question at the top of Student Edition page 254.

- Direct attention to problem 26 on Student Edition page 255. Explain that pools must be treated with a disinfecting agent such as salt or chlorine to kill bacteria that can make people sick and to kill algae that can make a pool dirty.

  Too much disinfectant, however, can cause eye and skin irritation, breathing difficulties, or even nausea or dizziness. Read aloud the final phrase of Galatians 5:13.

  How does God call us to treat others? by love to serve one another

  How would you show love to those swimming in your pool if you were in charge of adding chlorine to the pool water? I would calculate to find a safe amount of chlorine to add to the pool water so those swimming in it would be protected from bacteria or from too much chlorine that could make them sick.

- Explain that the students will use the information from the answer to problem 24 to complete problems 25–27.

Apply

Student Edition pages 254–55
- Read and explain the directions for pages 254–55. Assist the students as they complete the pages independently.

Daily Review
- Students should complete Chapter 12, section c.

Assess

Quiz 1
- Use the **summative assessment** to evaluate the students' progress at this point in the chapter.

NOTES

Student Edition pages 256–57
Daily Review Chapter 12, section *d*

OBJECTIVES

- Construct prisms having the same volume but different surface areas.
- Construct cylindrical figures having the same lateral surface area but different volumes.

TEACHER RESOURCES

- 68 *Fixed Volume* (for the teacher and for each group)

ADDITIONAL MATERIALS

- for the teacher and for each group: 18 cube-shaped blocks; 4 sheets of 9" × 12" construction paper; a ruler; transparent tape; a box lid or a rectangular pan; a 1 c measuring cup; rice, unpopped popcorn, or dried beans (approximately 6 c)
- a calculator (for each student)

Lesson 117 is designed to be a 2-day lesson. Student Edition page 256 may be assigned on Day 1.

Preparation

Cut a 1" × 12" strip off 2 of the 4 sheets of construction paper for each group so that the 2 sheets measure 8" × 12".

Engage

- Conduct a **bell ringer** to help the students review volume formulas for 3D figures. Direct the students to number 1–4 on a sheet of paper. Display the names of the following figures in a numbered list.
  1. cube
  2. triangular prism
  3. cylinder
  4. rectangular prism

Display the volume formulas one at a time.

a. $V = (lw)h$
b. $V = (\pi r^2)h$
c. $V = s^3$
d. $V = (\frac{1}{2}bh_1)h_2$

Direct the students to record the letter of the formula beside the number of the correct figure. Check their answers together. 1. c; 2. d; 3. b; 4. a

Instruct

Prisms with the same volume but different surface areas

- Guide the students in a **collaborative activity** to help them compare prisms having the same volume but different dimensions and different surface areas.

  What is surface area? the sum of the areas of all the surfaces of a 3-dimensional figure

  What is volume? the number of cubic units that a 3-dimensional figure holds or the amount of space that the figure fills

- Display the *Fixed Volume* page. Arrange the students into groups of 3 or 4 and distribute to each group a copy of the *Fixed Volume* page and 18 cubes.

  How many different rectangular prisms do you think you can make with 8 cubes? Answers will vary.

- Direct the students to make as many 8 cubic unit rectangular prisms as they

---

Fixed Volumes & Fixed Lateral Surfaces

Why can two cylinders have the same lateral surface area but different volumes?

Geometric figures can have the same volume but different shapes. Volume is the number of cubic units an object holds. Surface area is the sum of the areas of all the surfaces.

Shapes with a Volume of 8 cm³

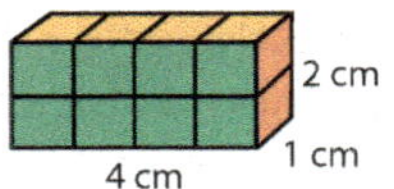

$V = l \cdot w \cdot h$
$V = 8 \cdot 1 \cdot 1$
$V = 8$ cm³

8 cm, 1 cm, 1 cm

Surface Area

top and bottom	$2 \cdot (8 \cdot 1) = $ **16 cm²**
front and back	$2 \cdot (8 \cdot 1) = $ **16 cm²**
left and right sides	$2 \cdot (1 \cdot 1) = $ **2 cm²**
	total surface area = 34 cm²

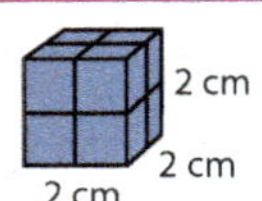

$V = l \cdot w \cdot h$
$V = 4 \cdot 1 \cdot 2$
$V = 8$ cm³

4 cm, 2 cm, 1 cm

top and bottom	$2 \cdot (4 \cdot 1) = $ **8 cm²**
front and back	$2 \cdot (4 \cdot 2) = $ **16 cm²**
left and right sides	$2 \cdot (1 \cdot 2) = $ **4 cm²**
	total surface area = 28 cm²

$V = l \cdot w \cdot h$
$V = 2 \cdot 2 \cdot 2$
$V = 8$ cm³

2 cm, 2 cm, 2 cm

all sides	$6 \cdot (2 \cdot 2) = $ **24 cm²**
	total surface area = 24 cm²

Exercises

Order of dimensions may vary within a row.

The volume and the length of a rectangular prism are given. Use the given dimensions to find the unknown values for the width and the height. Then find the surface area, using the 3 dimensions.

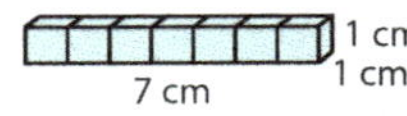

	Volume (cm³)	Length (cm)	Width (cm)	Height (cm)	Surface Area (cm²)
1.	7 cm³	7 cm	1 cm	1 cm	30 cm²
2.	10 cm³	10 cm	1 cm	1 cm	42 cm²
		5 cm	2 cm	1 cm	34 cm²
3.	12 cm³	12 cm	1 cm	1 cm	50 cm²
		6 cm	2 cm	1 cm	40 cm²
		4 cm	3 cm	1 cm	38 cm²
		3 cm	2 cm	2 cm	32 cm²
4.	16 cm³	16 cm	1 cm	1 cm	66 cm²
		8 cm	2 cm	1 cm	52 cm²
		4 cm	2 cm	2 cm	40 cm²
5.	24 cm³	24 cm	1 cm	1 cm	98 cm²
		12 cm	2 cm	1 cm	76 cm²
		8 cm	3 cm	1 cm	70 cm²
		6 cm	4 cm	1 cm	68 cm²
		6 cm	2 cm	2 cm	56 cm²
		4 cm	3 cm	2 cm	52 cm²

To find **lateral surface area**, find the surface area of all faces except the bases. Two geometric figures with different base areas and different heights can have the same lateral surface area but different volumes. Let LS = lateral surface area.

Lateral Surface Area of a Cylinder

LS = circumference $(2\pi r) \times$ height (h); V = Base $(B) \times$ height (h)

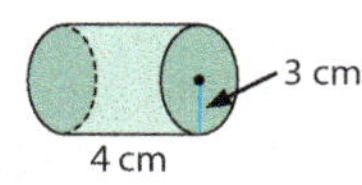

$LS = (2 \cdot 3.14 \cdot 1.6) \cdot 5 \approx 50 \text{ cm}^2$
$V = (3.14 \cdot 1.6^2) \cdot 5 \approx 40 \text{ cm}^3$

$LS = (2 \cdot 3.14 \cdot 0.8) \cdot 10 \approx 50 \text{ cm}^2$
$V = (3.14 \cdot 0.8^2) \cdot 10 \approx 20 \text{ cm}^3$

Find the lateral surface area and the volume of a cylinder, using the given radius and height. Draw the shape if needed. Round the answer to the nearest unit.

	Lateral Surface Area (cm²)	Radius (cm)	Height (cm)	Volume (cm³)
6.	*75 cm²*	3 cm	4 cm	*113 cm³*
		1 cm	12 cm	*38 cm³*
7.	*132 cm²*	1.5 cm	14 cm	*99 cm³*
		1 cm	21 cm	*66 cm³*

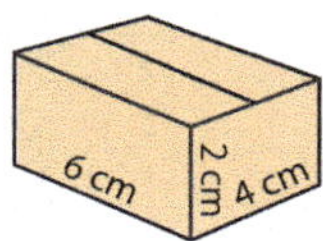

Find the lateral surface area (LS = area of side faces) and the volume ($V = [l \cdot w] \cdot h$) of a rectangular prism, using the given dimensions.

	Lateral Surface Area (cm²)	Length (cm)	Width (cm)	Height (cm)	Volume (cm³)
8.	*40 cm²*	6 cm	4 cm	2 cm	*48 cm³*
		5 cm	5 cm	2 cm	*50 cm³*

9.

A toy manufacturer makes wooden blocks that are 1 in. cubes and packages them in sets. A box with the smallest surface area is the least expensive to produce. Find the length, width, and height for a box that gives the least surface area for each set of blocks.

Order of dimensions may vary.
a small set of 16 blocks
l = 4 in., w = 2 in., h = 2 in.
a medium set of 24 blocks
l = 4 in., w = 3 in., h = 2 in.
a large set of 36 blocks
l = 4 in., w = 3 in., h = 3 in.

10. Why can two cylinders have the same lateral surface area but different volumes?
The curved surface of each cylinder creates different dimensions for the cylinder's circular base and height, which determine the volume.

What are the dimensions of the tower that is still standing upright? *l = 1 unit, w = 1 unit, h = 8 units*

What are the dimensions of the tower that is lying sideways? *l = 8 units, w = 1 unit, h = 1 unit*

What are the dimensions of the tower that is lying backwards? *l = 1 unit, w = 8 units, h = 1 unit*

Point out that congruent figures have the same dimensions (e.g., 1 unit × 1 unit × 8 units). How the figure is positioned or looked at determines whether a dimension is referred to as the length, the width, or the height.

- Direct the groups to follow a similar procedure to make as many rectangular prisms as they can with the following volumes. Direct them to record the volume, dimensions, and surface area of each prism on the *Fixed Volume* page. Accept any configuration of the following dimensions.

 V = 2 cubic units *$2 \times 1 \times 1$, $S = 10$*

 V = 4 cubic units *$4 \times 1 \times 1$, $S = 18$; $2 \times 2 \times 1$, $S = 16$*

 V = 6 cubic units *$6 \times 1 \times 1$, $S = 26$; $2 \times 3 \times 1$, $S = 22$*

 V = 18 cubic units *$18 \times 1 \times 1$, $S = 74$; $9 \times 2 \times 1$, $S = 58$; $6 \times 3 \times 1$, $S = 54$*

You may end the lesson here and teach the remainder of the lesson and use Student Edition page 257 on Day 2.

Cylindrical figures with the same lateral surface area but different volumes

- Guide the students in a **collaborative activity** to help them compare cylindrical figures having the same lateral surface area but different dimensions and different volume.

- Distribute to each group of students the following items: 4 sheets of construction paper (two 9" × 12", two 8" × 12"), a ruler, and tape. Direct the students to measure one of the larger sheets of construction paper to the nearest inch and calculate the area. *A = 9 in. × 12 in.; A = 108 in.²*

---

can. Direct them to record the volume, dimensions, and surface area of each prism on the page. *possible configurations: $V = 8$, $l = 8$, $w = 1$, $h = 1$, $S = 34$; $V = 8$, $l = 4$, $w = 2$, $h = 1$, $S = 28$; $V = 8$, $l = 2$, $w = 2$, $h = 2$, $S = 24$*

- Choose students to share their group's findings and record the dimensions on the displayed page.

 What do you notice about the volume and the surface area of your rectangular prisms? The prisms are equal in volume but have surface areas that are not equal.

 Why can the surface areas be different when rectangular prisms have the same fixed volume? The same number of cubic units can be arranged differently, changing the dimensions which determine the surface area of the figure.

- Choose three students to each use 8 cubes to build a tower that has a base of 1 square unit. Direct one of the students to lay down his tower sideways and another student to lay down his tower backwards. Write the dimensions as the students answer the following questions.

 What is the volume of each of these rectangular prisms? 8 cubic units

 What do you know about the surface areas of the prisms? The surface areas are the same because the prisms are congruent.

LESSON 117

- Demonstrate how to make a tube that is 12 in. tall by rolling one 9" × 12" sheet of construction paper lengthwise, overlapping the ends slightly, and taping the seam at the top, middle, and bottom. Make a 9 in. tall tube using the other 9" × 12" sheet of paper. Direct each group of students to do the same with the 2 larger sheets of construction paper.

- Write "lateral surface area" for display. Explain that the lateral surface area of a 3-dimensional figure is the sum of the surface areas excluding the area of the base(s). A cylinder's lateral surface area is the area of only the curved surface, without the top and bottom circular bases—similar to the 2 paper tubes.

  What is the approximate lateral surface area of the taller tube? 108 in.²; When the curved surface is flat, it is the same shape and size as the rectangular sheet of paper.

  What is the approximate lateral surface area of the shorter tube? 108 in.²; It has approximately the same area as the taller tube (108 in.²) because both tubes were made from congruent rectangular sheets of paper.

  Since the lateral surface area of a cylinder is the area of its curved surface, how can you mathematically find the lateral surface area of each tube? Since the length of the curved surface of the cylinder is equal to the circumference of its base, I can use the same formula that is used to find the surface area of the curved surface of a cylinder; $(\pi d)h$ or $(2\pi r)h$.

- Write "$LS = (\pi d)h$" for display. Point out that LS represents lateral surface area.

  Direct the students to measure the diameter of each tube to the nearest inch. 3 in. for the taller tube; 4 in. for the shorter tube

  Guide the students as they use the formula to calculate the lateral surface area of each tube to the nearest inch. Point out that the answers will be approximate because the diameters were measured to the nearest inch.

  taller tube $LS = (\pi d)h$; $LS \approx (3.14 \cdot 3$ in.$)12$ in.; $LS \approx 9.42$ in. $\cdot 12$ in.; $LS \approx 113.04$ in.² ≈ 113 in.²

  shorter tube $LS = (\pi d)h$; $LS \approx (3.14 \cdot 4$ in.$)9$ in.; $LS \approx 12.56$ in. $\cdot 9$ in.; $LS \approx 113.04$ in.² ≈ 113 in.²

- Distribute to each group of students a box lid or pan, a 1 c measuring cup, and rice. Stand both tubes you constructed upright inside the lid or pan. Direct the students to do the same.

  Do you think these tubes have the same volume? Accept all answers.

- As you demonstrate, direct the students to place the taller tube inside the shorter tube.

- Demonstrate using the measuring cup to fill the entire taller tube with rice. Gently hold the taller tube in place so that the rice doesn't come out of the bottom as it is poured in from the top.

  Point out that the amount of rice in the taller tube is equal to the volume of the tube.

  As you demonstrate, direct the students to remove the taller tube to let the rice fill the shorter tube. Instruct them to gently flatten the rice if necessary to keep the top level.

  Do both tubes have the same volume? No; the rice that filled the taller tube does not fill the shorter tube.

  Point out that the rice that completely filled the taller tube fills only about $\frac{3}{4}$ of the shorter tube.

Solve.

1. 4.5
    × 6.7
    30.15

2. 5)$23.40
    $4.68

3. $6,932.37
    − $5,331.97
    $1,600.40

4. 38.472
    5.391
    + 2.0
    45.863

5. 7.18
    × 2.9
    20.822

6. 47)5,076
    108

7. 20,320
    − 14,410
    5,910

8. $169.95
    $139.49
    + $39.99
    $349.43

9. 442
    × 71
    31,382

10. 31)7,626
    246

11. $9,875
    − $5,769
    $4,106

12. 31,998 + 543,477
    575,475

13. 975
    × 48
    46,800

14. 206)5,150
    25

15. 469.549
    − 203.895
    265.654

16. 6,003 + 6,422
    12,425

Math 6

How can you calculate the volume of each tube? I can measure the diameter of the circular base, calculate the area of the base (πr^2), and then multiply the area by the height.

- Write for display the formula for finding the volume of a cylinder, "$V = (\pi r^2)h$." Guide the students as they calculate the volume of each tube. Round the answers to the nearest tenth.

  taller tube $V = (\pi r^2)h$; $V = 3.14 \cdot (1.5 \text{ in.})^2 \cdot 12$ in.; $V = (3.14 \cdot 2.25 \text{ in.}^2) \cdot 12$ in.; $V = 7.065 \text{ in.}^2 \cdot 12 \text{ in.}; V = 84.78 \text{ in.}^3 \approx 84.8 \text{ in.}^3$

  shorter tube $V = (\pi r^2)h$; $V = 3.14 \cdot (2 \text{ in.})^2 \cdot 9$ in.; $V = (3.14 \cdot 4 \text{ in.}^2) \cdot 9$ in.; $V = 12.56 \text{ in.}^2 \cdot 9 \text{ in.}; V = 113.04 \text{ in.}^3 \approx 113 \text{ in.}^3$

- Direct the students' attention to the essential question at the top of Student Edition page 256, "Why can two cylinders have the same lateral surface area but different volumes?" The curved surface of each cylinder creates different dimensions for the cylinder's circular base and height, which determine the volume of the figure.

- Follow a similar procedure with rectangular prisms. Construct a tall and a short prism by folding one 8" × 12" sheet of construction paper lengthwise into fourths and folding the other sheet of paper widthwise into fourths. Tape the seams of each prism at the top, middle, and bottom.

  Explain that the lateral surface area of a prism is found by adding all the areas of its side faces (without the 2 bases). Remind the students that they find the volume of a prism by multiplying its length, width, and height.

  lateral surface area of both prisms based on the paper size 8 in. × 12 in. = 96 in.²

  taller prism $LS = (2 \text{ in.} \times 12 \text{ in.}) + (2 \text{ in.} \times 12 \text{ in.}) + (2 \text{ in.} \times 12 \text{ in.}) + (2 \text{ in.} \times 12 \text{ in.}); LS = 24 \text{ in.}^2 + 24 \text{ in.}^2 + 24 \text{ in.}^2 + 24 \text{ in.}^2; LS = 96 \text{ in.}^2$ or $LS = 4 \times 24 \text{ in.}^2 = 96 \text{ in.}^2$ $V = (2 \text{ in.} \times 2 \text{ in.})12 \text{ in.}; V = 4 \text{ in.}^2 \times 12 \text{ in.}; V = 48 \text{ in.}^3$

  shorter prism $LS = (3 \text{ in.} \times 8 \text{ in.}) + (3 \text{ in.} \times 8 \text{ in.}) + (3 \text{ in.} \times 8 \text{ in.}) + (3 \text{ in.} \times 8 \text{ in.}); LS = 24 \text{ in.}^2 + 24 \text{ in.}^2 + 24 \text{ in.}^2 + 24 \text{ in.}^2; LS = 96 \text{ in.}^2$ or $LS = 4 \times 24 \text{ in.}^2 = 96 \text{ in.}^2$ $V = (3 \text{ in.} \times 3 \text{ in.})8 \text{ in.}; V = 9 \text{ in.}^2 \times 8 \text{ in.}; V = 72 \text{ in.}^3$

Apply

Student Edition pages 256–57
- Read and explain the directions for pages 256–57. Assist the students as they complete the pages independently.

Daily Review
- Students should complete Chapter 12, section *d*.

NOTES

CHAPTER REVIEW

OBJECTIVES

- Find the volume of a rectangular prism, a cube, and an irregular prism using models.
- Calculate the unknown measurement of a rectangular prism.
- Calculate the volume of a rectangular prism, a cube, a triangular prism, and a cylinder by using formulas.
- Relate volume to real-world situations.
- Use the ability to calculate volume to serve others. **BWS**

BIBLICAL WORLDVIEW SHAPING

- **Service (Apply):** Calculating volume helps others by saving time, money, and effort.

TEACHER RESOURCES

- 69 *Volume Review*
- 70 *Volume Word Problems*

ADDITIONAL MATERIALS

- copies of the area drawings (from Lessons 114–16 preparation)
- cube-shaped blocks (to build towers of varied heights)
- a calculator (for each student)

CHAPTER REVIEW

Use the given formulas to write the formula for calculating the volume of the figure.

$$V = (lw)h$$
$$V = (\pi r^2)h$$
$$V = s^3$$
$$V = (\tfrac{1}{2}bh_1)h_2$$

1. volume of a cube $V = s^3$

2. volume of a triangular prism $V = (\tfrac{1}{2}bh_1)h_2$

3. volume of a cylinder $V = (\pi r^2)h$

4. volume of a rectangular prism $V = (lw)h$

Find the area of the base. Find the volume of the prism that could be built using centimeter cubes for the given height.

5.
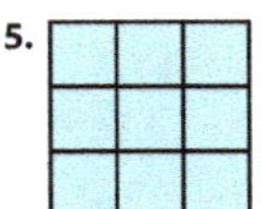
$h = 3$ cm $V = 27$ cm³

6.
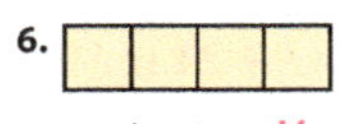
$h = 5$ cm $V = 20$ cm³

7.
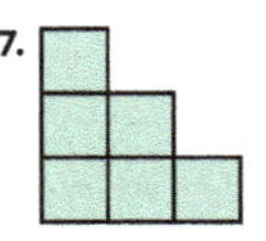
$h = 2$ cm $V = 12$ cm³

Find the volume of the figure. Round decimals to the nearest tenth. *Equations may vary.*

8.
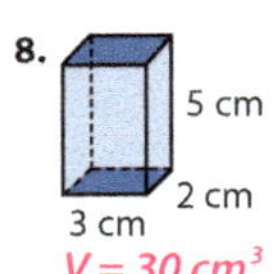

$V = 30$ cm³

9.
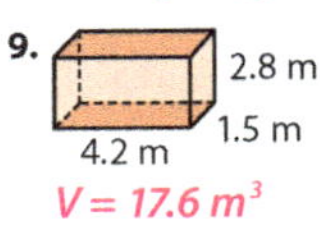

$V = 17.6$ m³

10.
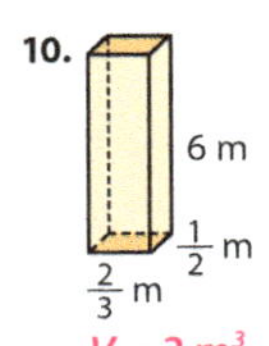

$V = 2$ m³

11.
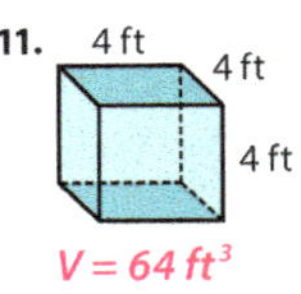

$V = 64$ ft³

12.
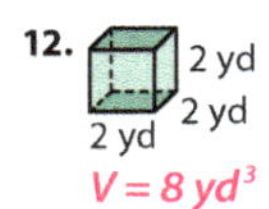

$V = 8$ yd³

13.
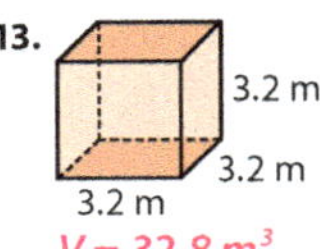

$V = 32.8$ m³

14.
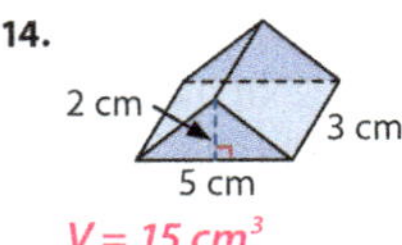

$V = 15$ cm³

15.
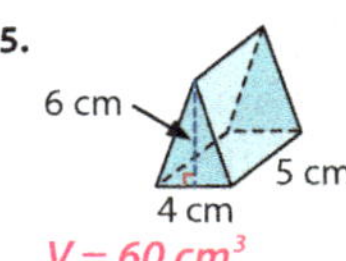

$V = 60$ cm³

16.
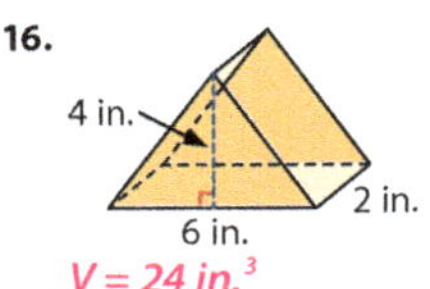

$V = 24$ in.³

17.
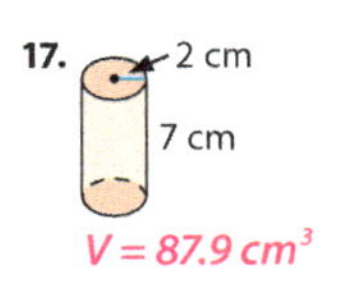

$V = 87.9$ cm³

18.
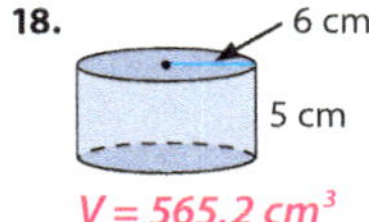

$V = 565.2$ cm³

19.
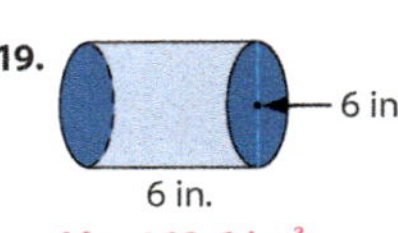

$V = 169.6$ in.³

The Chapter Review offers an opportunity for students to discuss the concepts they have learned in the chapter. They may work collaboratively or independently as you review concepts. Circulate among the students, giving individual help as needed. Students who demonstrate proficiency with the discussion, the modeling, and the Student Edition pages are ready for the Chapter Test. Students who encounter difficulties with the review concepts would benefit from additional coaching and practice before testing.

Finding volume by using a model

- Display several of the area drawings, one at a time, to review finding the volume of a rectangular prism, a cube, and an irregular prism by using a model.

Choose a student to tell the area of the base and build a tower on it with the height of your choice. Choose another student to tell the volume of the tower and to explain his answer.

Find the volume of a figure with the given dimensions. Round decimals to the nearest tenth. *Equations may vary.*

20. rectangular prism: l = 5.3 in., w = 4 in., h = 6 in.
    V = 5.3 in. × 4 in. × 6 in.; V = 127.2 in.³
21. cube: s = 10 m *V = (10 m)³; V = 1,000 m³*

22. triangular prism: b = 4 in., h = 3 in., h (prism) = 5 in. *$V = \frac{1}{2}$ (4 in. × 3 in.) × 5 in.; V = 30 in.³*
23. cylinder: r = 3 m, h = 2 m
    V = 3.14 × (3 m)² × 2 m; V = 56.5 m³

Find the unknown measurement of the rectangular prism.

24. V = 36 ft³ 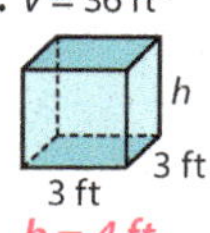
    h = 4 ft
25. V = 70 m³ 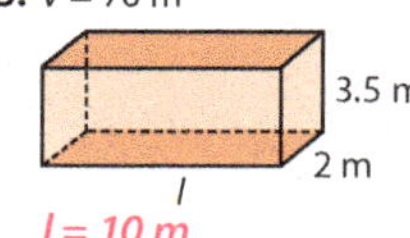
    l = 10 m

26. V = 216 cm³ 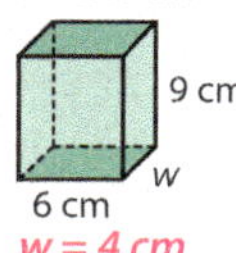
    w = 4 cm

27. l = 5 in., w = 2 in., h = __ in., V = 80 in.³
    h = 8 in.

28. l = 6 cm, w = __ cm, h = 4 cm, V = 144 cm³
    w = 6 cm

34. *sample answer: I could serve my family by calculating how many bags of compost would be needed for our garden without wasting time, money, or resources.*
    Convert ft to in.
    12 ft = 144 in.; 8 ft = 96 in.
    Find the volume of compost needed.
    V = (lw)h
    V = (144 in. × 96 in.) × $\frac{1}{4}$ in.
    V = 13,824 in.² × $\frac{1}{4}$ in. = 3,456 in.³
    There are 1,728 in.³ per ft³ (see #32).
    3,456 in.³ ÷ 1,728 in.³ = 2 ft³
    2 bags needed

Equations may vary.
Solve. Round decimals to the nearest tenth.

29. The Fernandez family is building raised garden beds to grow vegetables. The largest bed is 9 ft long, 4 ft wide, and 1.5 ft high. How many cubic feet of soil will it take to fill the largest bed? *54 ft³*
30. The smallest bed the Fernandez family will build is 4 ft long, 4 ft wide, and 1.5 ft high. How many cubic feet of soil will it take to fill the smallest bed? *24 ft³*
31. Mr. Fernandez wants to use rainwater to water his garden. He purchased a rain barrel that is 37 in. tall and has a 20 in. diameter. What is the volume of the rain barrel? *11,618 in.³*

32. There are 1,728 in.³ in 1 ft³. What is the volume of the rain barrel in cubic feet? *11,618 in.³ ÷ 1,728 in.³ = 6.7 ft³*
33. Since 1 ft³ can hold about 7.5 gal of water, approximately how many gallons of water will the rain barrel hold? *6.7 • 7.5 gal = 50.3 gal*

34. You want to add a $\frac{1}{4}$ in. layer of compost to the soil of your family's 12 ft × 8 ft garden. Compost bags contain 1 ft³ each. Tell how you could serve your family by calculating volume to care for your garden. Show your equations and results.

Calculating an unknown measurement

- Draw a rectangular prism for display.
  What formula do you use to calculate the volume of a rectangular prism?
  $V = Bh$ or $V = (l × w) × h$
  Write "$V = Bh$" and "$V = (lw)h$" for display.

- Direct the students to find the unknown measurement by substituting the given values into the volume formula. Review the solution as needed.
  V = 360 cm³, l = 15 cm, h = 6 cm
  360 cm³ = (15 cm × w)6 cm; 360 cm³ = (15 cm × 6 cm)w; 360 cm³ = 90 cm² × w; $\frac{360 \text{ cm}^3}{90 \text{ cm}^2} = \frac{90 \text{ cm}^2 × w}{90 \text{ cm}^2}$; 4 cm = w

V = 31.5 cm³, B = 4.5 cm²
31.5 cm³ = 4.5 cm² × h; $\frac{31.5 \text{ cm}^3}{4.5 \text{ cm}^2} = \frac{4.5 \text{ cm}^2 × h}{4.5 \text{ cm}^2}$; 7 cm = h

- Repeat the procedure, allowing students to give values for the volume and 2 dimensions.

Using a formula

- Display the *Volume Review* page. Direct the students to calculate the volume of each figure. Discuss the solutions and review the formulas for finding volume.

- Alternately, use the *Volume Review* page as a tic-tac-toe board and arrange the students into 2 teams, Team X and Team O,

to play a game of tic-tac-toe using the following procedure.

- Explain that you will choose 1 member from a team to select a section on the tic-tac-toe board. Then each member of that team will use the appropriate formula to solve the problem at his desk. Direct the students to round decimal answers to the nearest tenth. Explain that you will give 1 point to the team for each team member who gets the answer correct.

- Begin with a student from Team O. If the student correctly finds the volume of the figure in the selected section, write an O in that section. Also award 1 point for every member of Team O who has the correct answer.

If the student answers incorrectly, the section is available for a student from Team X to give the answer, but still award 1 point to Team O for each team member who has the correct answer.

Continue the game by calling on Team X to choose a section of the board. Alternate teams until the game has been completed. Award 5 bonus points to the first team to achieve 3 correct answers in a row on the board (vertically, horizontally, or diagonally), and then add up the points earned. The team with the most points wins the game. You may play the game again by varying the dimensions of each figure.

1. rectangular prism *$V = (lw)h$; (12 in. • 4 in.)5 in. = 240 in.³*
2. cylinder *$V = (\pi r^2)h$; (3.14)(3 cm)² • 9 cm ≈ 254.3 cm³*
3. cube *$V = s^3$; (8 in.)³ = 512 in.³*
4. triangular prism *$V = (\frac{1}{2}bh_1)h_2$; $\frac{1}{2}$(4 in. × 4 in.) × 8 in. = 64 in.³*
5. cylinder *$V = (\pi r^2)h$; (3.14)(2.5 m)² • 5 m ≈ 98.1 m³*
6. triangular prism *$V = (\frac{1}{2}bh_1)h_2$; $\frac{1}{2}$(6 in. × 5 in.) × 12 in. = 180 in.³*
7. cylinder *$V = (\pi r^2)h$; (3.14)(2 cm)² • 6 cm ≈ 75.4 cm³*
8. rectangular prism *$V = (lw)h$; (4.5 cm × 3.5 cm)9 cm ≈ 141.8 cm³*
9. triangular prism *$V = (\frac{1}{2}bh_1)h_2$; $\frac{1}{2}$(4 in. • 5 in.) • 10 in. = 100 in.³*

Volume in real life

- Display the *Volume Word Problems* page and read aloud the first word problem.

  Discuss the type of container described and the formula needed to find the answer. Choose a student to write the formula for display. Direct each student to write and solve the equation for the word problem. Discuss the answer.
  1. cube; $V = s^3$; $V = (2 \text{ ft})^3$; $V = 8 \text{ ft}^3$

- Follow a similar procedure for problems 2–4. Instruct the students to round the decimal answers to the nearest tenth.
  2. cylinder; $V = (\pi r^2)h$; $V = (3.14)(1 \text{ ft})^2 \times 2 \text{ ft}$; $V \approx 6.3 \text{ ft}^3$
  3. rectangular prism; $V = (lw)h$; $V = (4 \text{ ft} \times 2 \text{ ft}) \times 1.5 \text{ ft}$; $V = 12 \text{ ft}^3$
  4. triangular prism; $V = (\frac{1}{2}bh_1)h_2$; $V = \frac{1}{2}(4 \text{ ft} \times 4 \text{ ft}) \times 2 \text{ ft} = 16 \text{ ft}^3$

Serving others by calculating volume

- Guide a discussion of the chapter essential question on Student Edition page 249, "How can my ability to calculate volume help me to serve others?"

- Direct attention to the picture on Student Edition page 249. Point out the Bible verse and the Luther Burbank quote.

  How do you think growing flowers might be related to serving others by calculating volume? sample answer: A gardener might calculate the volume of his flower pots or planters so he knows how much potting soil he needs to grow flowers. People are often encouraged by the beauty of flowers and reminded of God's glory and His promises (Luke 12:28).

  Have you ever given flowers to someone as a gift? Answers will vary.

- Direct attention to the gardening problems on page 259 in the Student Edition.

  What are some ways home gardening serves others? sample answers: Gardening provides healthy exercise; teaches the value of patience, hard work, and dependence on God; saves money; and produces fresh, healthy food for the gardener to eat and to share with others.

- Direct attention to problem 34 as another example of serving others by calculating volume.

  Explain that compost is decomposed organic matter that adds valuable nutrients to the soil. It also helps soil retain water and air, which helps plants grow better.

  Compost can be made from things such as grass clippings and dead leaves from a person's yard, or from fruit and vegetable peels, coffee grounds, and eggshells from their kitchen.

- Explain that gardeners who do not have time to make their own compost can purchase it in bags.

  What things might you want to know before purchasing compost for your garden? sample answers: what kind to buy, how much I need, how much compost a bag contains, how much a bag costs, where I can purchase it, how much each bag weighs, how I will transport and spread it

  How could knowing those things help you save resources, such as time, money, or effort? sample answers: Knowing how much compost I need before I purchase it could save the time, expense, and effort of making a second trip if I needed more. Placing an accurate online order the first time could save me the time, shipping expense, and effort of placing a second order. If I failed to calculate and purchased too much, it might be a wasteful purchase.

- Point out to the students that they can use the equivalency 1,728 in.^{3} = 1 ft^3 given in problem 32 to help them solve problem 34.

- Assist the students as needed as they perform the calculations and then as they consider how calculating volume can help them serve their family.

- Optional: Direct the students to calculate the volume of mulch required to meet your school's landscaping needs.

Student Edition pages 258–59

- Read and explain the directions for pages 258–59. Assist the students as they complete the pages independently.

LESSON 119

Student Edition pages 260–63

CHAPTER 12 TEST

CUMULATIVE REVIEW

CONCEPT REVIEW

- Adding, subtracting, multiplying, and dividing whole numbers, decimals, and fractions (Chapters 1–3, 5, 7–9)
- Renaming a fraction as a decimal (Lesson 37)
- Identifying the ordered pair for a point on a coordinate plane (Lesson 52)
- Determining the radius and the diameter of a circle (Lesson 61)
- Identifying the formula used to find the area of a circle (Lesson 108)
- Identifying geometric figures: a parallelogram, parallel lines, perpendicular lines (Lessons 53, 58)
- Classifying angles as acute, obtuse, or right (Lesson 54)
- Reading and interpreting a line plot (Grade 5)

Choose the answer.

1.
$$36,903$$
$$14,772$$
$$+ \, 21,187$$

A. 71,682   **C.** 72,862
B. 72,753   **D.** 72,992

2. $3 \times 26 \times 16 = $ ___
A. 1,221   **C.** 1,336
B. 1,248   **D.** 1,428

3.
$$17.34$$
$$- \ 8.9286$$

A. 8.4114   **C.** 9.4114
B. 8.6286   **D.** 9.4214

4. $3\frac{3}{16} + 7\frac{1}{4} = $ ___
A. $10\frac{7}{16}$ **C.** $10\frac{3}{16}$
B. $10\frac{1}{4}$ **D.** 11

5. $10 - 3\frac{6}{7} = $ ___
A. $7\frac{3}{7}$ **C.** $6\frac{4}{7}$
B. 7   **D.** $6\frac{1}{7}$

6. Round to the nearest hundredth.
$$112\overline{)4,318}$$
A. 38.55   **C.** 40.51
B. 39.52   **D.** 45.01

7. $\frac{3}{4} \div \frac{1}{2} = $ ___
A. 1   **C.** 2
B. $1\frac{1}{2}$ **D.** $2\frac{1}{2}$

8. $\frac{4}{9} \times \frac{11}{12} = $ ___
A. $\frac{1}{3}$ **C.** $\frac{11}{108}$
B. $\frac{5}{7}$ **D.** $\frac{11}{27}$

9. Rename $\frac{5}{6}$ as a decimal. Round to the nearest thousandth.
A. 0.822   **C.** 0.932
B. 0.833   **D.** 0.983

10. Estimate the difference.
$$\frac{8}{9} - \frac{3}{4} = $$ ___
A. 0   **C.** 1
B. $\frac{1}{2}$ **D.** $1\frac{1}{2}$

- To prepare the students for the format of achievement tests, instruct them to work on a separate sheet of paper, if necessary, and to mark their answers on the *Cumulative Review Answer Sheet*.

Student Edition pages 260–63

The Cumulative Review provides additional practice of previously learned concepts. These pages may be completed during this lesson or anytime after this lesson, since they require limited or no teaching.

Use the coordinate plane to find the answer.

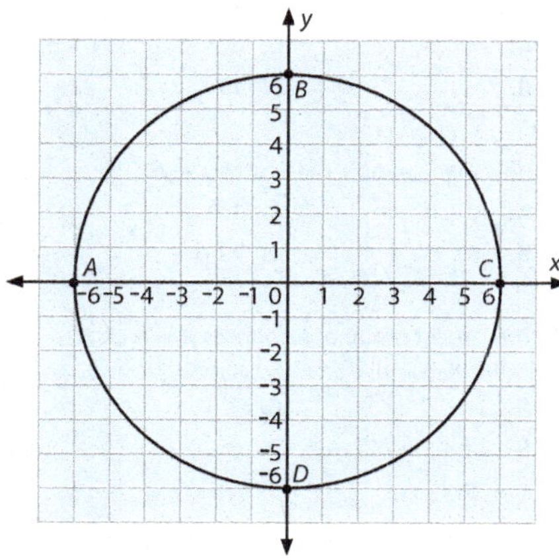

11. What are the coordinates for point *C*?

A. (0, 6) **C.** (0, −6)

B. (−6, 0) **D.** (6, 0)

12. The diameter of the circle is ___ units.

A. 6 **C.** 18

B. 12 **D.** 24

13. The radius of the circle is ___ units.

A. 6 **C.** 18

B. 12 **D.** 24

14. Which formula could be used to find the area of the circle?

A. $A = l \times w$ **C.** $A = \pi r^2$

B. $A = \frac{1}{2}bh$ **D.** $A = bh$

Use the figure to find the answer.

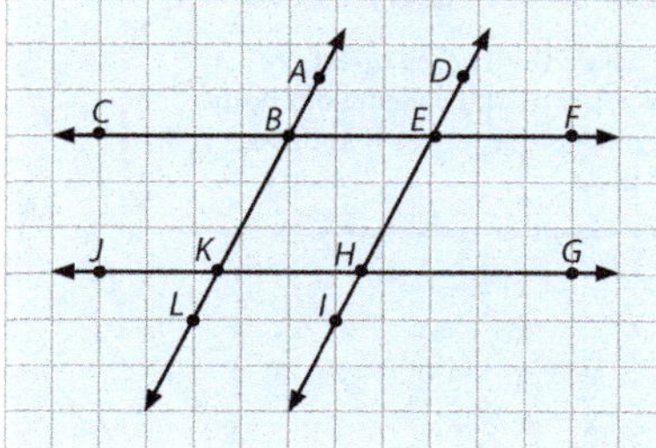

15. What figure is enclosed by the lines?

A. square **C.** parallelogram

B. trapezoid **D.** hexagon

16. Describe the relationship between $\overleftrightarrow{AL}$ and $\overleftrightarrow{DI}$.

A. intersecting lines **C.** perpendicular lines

B. parallel lines **D.** none of the above

17. Describe $\angle BKH$.

A. acute **C.** right

B. obtuse **D.** all of the above

18. Choose the true statement.

A. $\angle GHI = 90°$ **C.** $\angle GHI < 90°$

B. $\angle GHI > 90°$ **D.** $\angle GHI > 180°$

19. Which line is perpendicular to $\overleftrightarrow{CF}$?

A. $\overleftrightarrow{AL}$ **C.** $\overleftrightarrow{DI}$

B. $\overleftrightarrow{JG}$ **D.** none of the above

Cumulative Review 261

Use the line plot to find the answer.

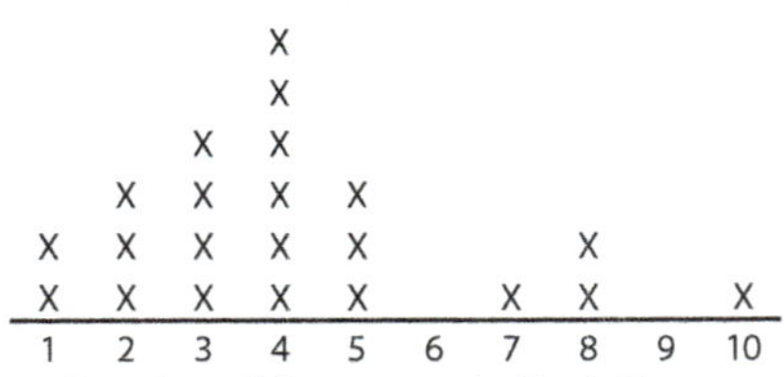


```
                  X
                  X
            X     X
            X  X  X  X
      X  X  X  X     X
      X  X  X  X     X  X     X
      1  2  3  4  5  6  7  8  9  10
```
Number of Occupants in Each Home

20. According to the survey, how many people lived alone?

A. 1   **C.** 6

B. 2   **D.** 10

21. How many people did Scott survey?

A. 10   **C.** 18

B. 15   **D.** 22

22. The largest group of responses was for how many occupants?

A. 2   **C.** 4

B. 3   **D.** 5

23. How many homes had 10 occupants?

A. 1   **C.** 3

B. 2   **D.** 4

24. Can you tell from this graph how many homes have school-age children?

A. yes

B. no

262   Chapter 12

2020 US CENSUS
Age of Family Members

The US Constitution mandates that the Bureau of the Census take a survey to count the US population once every ten years. The purpose of the census is to decide how many seats in the House of Representatives each state receives.

The census also benefits local communities. It lets the federal government know where to disperse grants and supports. The census helps local governments develop natural disaster response plans. It also helps them figure out how many public safety officers the community needs. The census helps add jobs to the community by showing business owners where to create new businesses.

Lesson Plan Overview

Chapter 13: Ratios, Proportions & Percents

Day	Lesson	Student Edition Pages	Teacher Edition Pages	Teacher Resources & Additional Materials	Topics, Skills & Biblical Worldview Shaping
122	120	264, 266–67	264, 266–67b	Additional Materials • pennies, nickels, dimes, and quarters	**Topic:** Ratios & Rates **Skills:** writing a ratio in three forms, finding equivalent ratios, determining the unit rate **BWS:** Design
123	121	265, 268–69	265, 268–69b	Teacher Resources • 72 *Pictured Ratios* • 73 *Ratio Tables*	**Topic:** Ratio Tables **Skills:** completing a ratio table, solving problems using ratio tables
124	122	270–71	270–71b	Additional Materials • black and red counters	**Topic:** Solving Proportions **Skills:** determining whether two ratios are proportional, solving for a missing term in a proportion **BWS:** Design
125	123	272–74	272–74a	Teacher Resources • 11 *Graph Paper* • 74 *Missing Measurements* Additional Materials • rulers or straight edges	**Topic:** Similar Figures **Skills:** finding the unknown measure in similar figures, using indirect measurement to find the unknown measure
126	124	275–77	275–77a	Additional Materials • modeling clay • a map • rulers • calculators Assessments • Chapter 13 Quiz 1	**Topic:** Scale **Skills:** finding actual measurements using a scale and a scale drawing, map, or model; determining the unknown measure on a scale drawing **BWS:** Design
127	125	278–79	278–79b	Teacher Resources • 11 *Graph Paper* • 75 *Percent* Additional Materials • calculators	**Topic:** Percent **Skills:** expressing percents as ratios, decimals, and fractions; expressing decimals and fractions as percents
128	126	280–81	280–81b	Teacher Resources • 76 *Percent Models: Finding the Part* Additional Materials • calculators	**Topic:** Finding the Percent of a Number **Skills:** finding the percent of a number by using an equation, a model, and a proportion; solving percent word problems
129	127	282–84	282–84a	Teacher Resources • 77 *Percent Models: Finding the Whole* Additional Materials • calculators Assessments • Chapter 13 Quiz 2	**Topic:** Finding the Unknown Whole **Skills:** finding the unknown whole in a percent problem by using a model, an equation, and a proportion; solving percent word problems

Day	Lesson	Student Edition Pages	Teacher Edition Pages	Teacher Resources & Additional Materials	Topics, Skills & Biblical Worldview Shaping
130	128	285–87	285–87a	Additional Materials • calculators	**Topic:** Speed, Distance & Time **Skills:** calculating the distance, the rate of speed, and the time; finding an equivalent rate by using a proportion **BWS:** Design
131	129	288–89	288–89b	Teacher Resources • 78 *Circle Graph: Elements in the Earth's Crust* Additional Materials • calculators	**Topic:** Chapter 13 Review
132	130	290–92	290–92		**Topic:** Test & Cumulative Review
133	131	265, 293–94	265, 293–94a	Teacher Resources • 11 *Graph Paper* • 25 *Engineering Design Process* • 79 *Bridge & Truss Designs* Additional Materials • See Lesson 131 list.	**Topic:** STEM: Building a Bridge **Skills:** identifying and researching a problem, suggesting solutions, collaborating on a design
134	132	293–94	294b	Teacher Resources • 80 *STEM Rubric: Building a Bridge* Additional Materials • See Lesson 132 list.	**Topic:** STEM: Building a Bridge **Skills:** building a bridge prototype according to specifications
135	133	293–94	294c–d	Additional Materials • See Lesson 133 list.	**Topic:** STEM: Building a Bridge **Skills:** testing a design, analyzing a design with math, solving a problem by using a ratio **BWS:** Design

CHAPTER 13

CHAPTER OBJECTIVES

- Write ratios to describe comparisons and to determine the unit rate.
- Determine whether two ratios are proportional.
- Find the unknown measure or term in a proportion.
- Solve problems by using the relationship between ratios, decimals, fractions, and percents.
- Model proportions using counters and tables.
- Evaluate the use of proportions in solving real-world problems.
- STEM: Apply learned design and engineering practices to appreciate God's design in creation.

To determine whether your students are prepared for the concepts taught in this chapter, you may wish to consult the preassessment checklist found on TeacherToolsOnline.com.

See Lessons 131–33 for the materials needed for this chapter's STEM project. The STEM lessons require assembling materials in advance and are most effective if they have been previewed before teaching.

Send home the *Parent Letter* page to inform parents of the items needed for Chapter 14.

Make fact practice, both oral and written, part of your daily math routine to help the students with mastery.

Visit AfterSchoolHelp.com for math practice resources, or visit TeacherToolsOnline.com for additional resources to enhance the lessons.

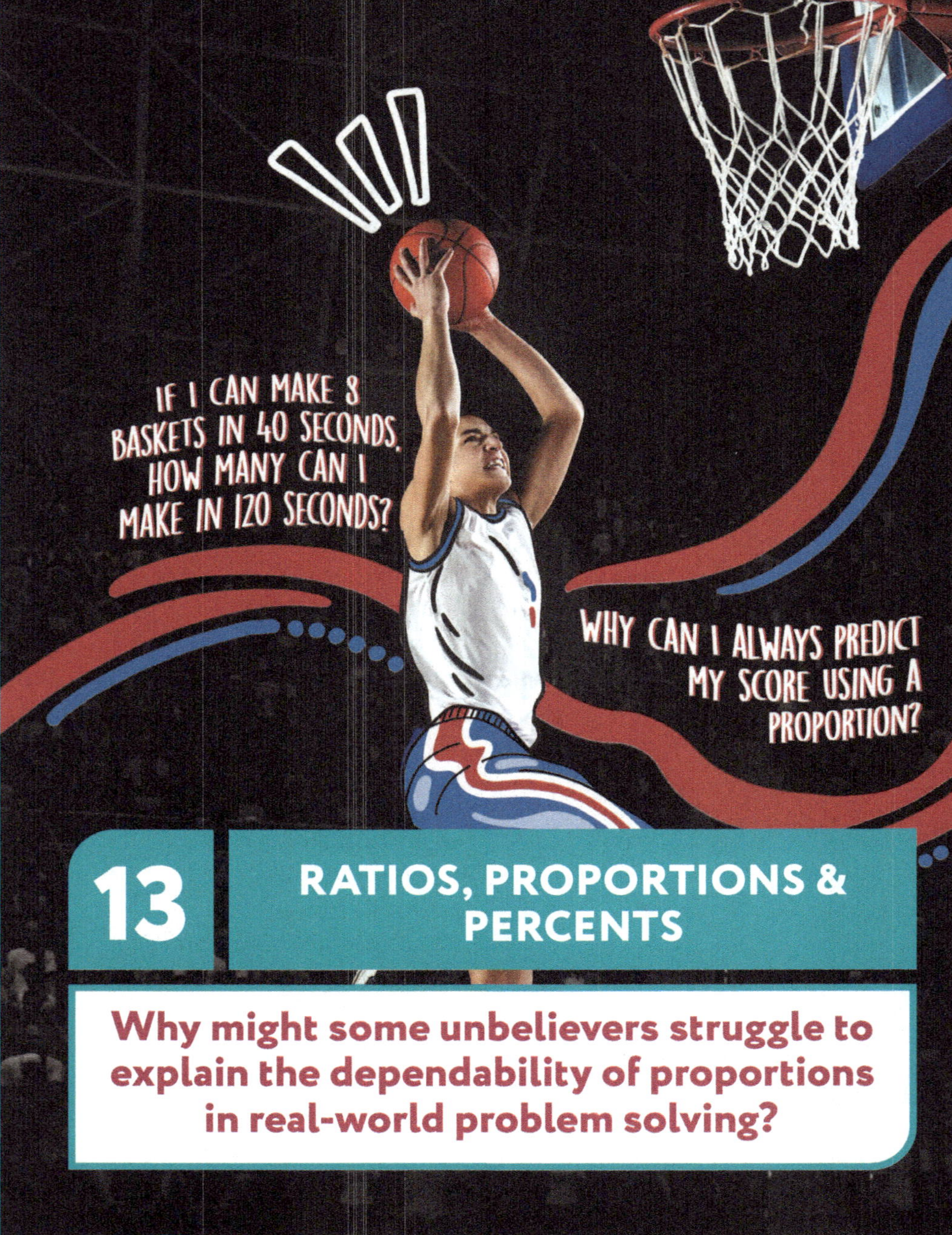

13 RATIOS, PROPORTIONS & PERCENTS

Why might some unbelievers struggle to explain the dependability of proportions in real-world problem solving?

Math 6

BUILDING A BRIDGE

Why can I build a sturdy bridge?

Spanning not only a river canyon but also the centuries, the ancient Incan rope suspension **bridge at Q'eswachaka**, Peru, is still in use.

As one of the millions of sturdy traffic bridges, the striking **Golden Gate Bridge** in San Francisco (pictured) is a frequently photographed suspension bridge.

The **Helix "DNA" Bridge**, a pedestrian bridge in Singapore with a remarkable curved tubular truss design, exhibits beauty and interest.

Miguel Rosales is an internationally recognized bridge architect and the designer of the stunning Liberty Bridge in Greenville, South Carolina. Relying on his strong engineering background, Rosales designs innovative bridges that are beautiful yet functional. In an interview with PBS, Rosales said, "I'm very interested in how a bridge fits in its context . . . [and in] the proportions and the cultural value of the structure. . . . That's why I make a lot of 3-D drawings and models and computer drawings, to try to really see how the bridge is going to appear when it's built."

Liberty Bridge, Greenville, SC

Miguel Rosales
bridge designer

LESSON 120

Student Edition pages 264, 266–67
Daily Review Chapter 13, section _a_

OBJECTIVES

- Write a ratio in word, ratio, and fraction form.
- Write ratios to describe part-to-part, part-to-whole, and whole-to-part comparisons.
- Find equivalent ratios.
- Determine the unit rate.
- Find an equivalent ratio by using the unit rate.
- Recall that God has given us His Word so that we can understand the design of creation. **BWS**

BIBLICAL WORLDVIEW SHAPING

- **Design (Recall):** Rates show orderliness in that they are constant and dependable.

ADDITIONAL MATERIALS

- 4 pennies
- 6 nickels
- 8 dimes
- 12 quarters

Engage

- Direct attention to the chapter opener on Student Edition page 264 and lead the students in a **discussion** to explore the chapter essential question, "Why might some unbelievers struggle to explain the dependability of proportions in real-world problem solving?"

Include the idea that proportions are reliable because they work every time. Question how unbelievers can explain this dependability apart from God designing the world.

Instruct

Writing ratios in 3 forms

- Use a **visual representation** to illustrate the 3 forms of ratios.

- Display 4 pennies and 6 nickels and write "pennies to nickels" for display.

Ratios & Rates

What evidence of design is found in ratios and rates?

A **ratio** is a mathematical comparison of two quantities. The **terms** in the ratio represent the items being compared. A ratio is read using the word _to_. A ratio can be written with _to_, with a colon, or as a fraction.

> There are 4 girls to every 6 boys that play in the county soccer league. The ratio of girls to boys can be written three ways: 4 to 6, 4:6, or $\frac{4}{6}$.

Quantities can be compared in different ways; therefore, the order of the terms must match the order of the quantities being compared in the written statement.

Quantities	Ratio	Word Form	Ratio Form	Fraction Form
part to part	girls to boys	4 to 6	4:6	$\frac{4}{6}$
part to whole	girls to players	4 to 10	4:10	$\frac{4}{10}$
whole to part	players to girls	10 to 4	10:4	$\frac{10}{4}$

Equivalent ratios can be found by multiplying or dividing both terms of the ratio by the same nonzero number (a form of 1). It is similar to renaming a fraction into higher or lower terms. To simplify a ratio, rename it to lowest terms.

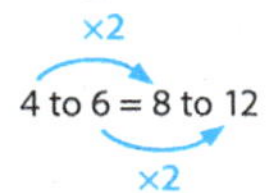 $\times 2$ — 4 to 6 = 8 to 12 — $\times 2$

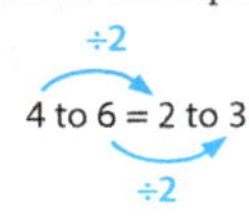 $\div 2$ — 4 to 6 = 2 to 3 — $\div 2$

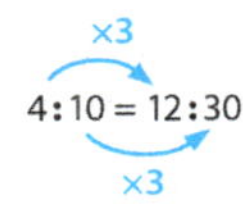 $\times 3$ — 4:10 = 12:30 — $\times 3$

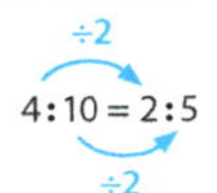 $\div 2$ — 4:10 = 2:5 — $\div 2$

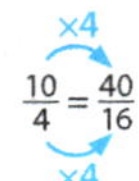 $\times 4$ — $\frac{10}{4} = \frac{40}{16}$ — $\times 4$

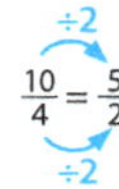 $\div 2$ — $\frac{10}{4} = \frac{5}{2}$ — $\div 2$

Exercises

Use the table to write the ratio.

Number of Volcanoes Per State	
State	**Volcanoes**
Washington	5
Oregon	5
California	3

1. volcanoes in Washington to volcanoes in California (ratio form) *5 : 3*
2. volcanoes in California to volcanoes in Oregon (word form) *3 to 5*
3. volcanoes in Washington to total volcanoes (fraction form) *$\frac{5}{13}$*
4. volcanoes in Washington and Oregon to total volcanoes (ratio form) *10 : 13*

Write an equivalent ratio in higher terms. *Answers will vary.*

5. 3 to 4 *6 to 8*   6. 4:5 *12 : 15*   7. $\frac{7}{15}$ *$\frac{28}{60}$*

Write an equivalent ratio in lower terms. *Answers will vary.*

8. 12 to 20 *3 to 5*   9. 27:45 *3 : 5*   10. $\frac{150}{10}$ *$\frac{15}{1}$*

Find the missing term that completes the equivalent ratio.

11. $\frac{6}{9} = \frac{n}{3}$ *n = 2*   12. $\frac{5}{8} = \frac{n}{64}$ *n = 40*   13. $\frac{3}{4} = \frac{n}{100}$ *n = 75*   14. $\frac{21}{30} = \frac{7}{n}$ *n = 10*   15. $\frac{5}{6} = \frac{30}{n}$ *n = 36*   16. $\frac{24}{42} = \frac{4}{n}$ *n = 7*

Write the ratio as a fraction in lowest terms.

17. 6 parakeets to 36 dogs *$\frac{6}{36} = \frac{1}{6}$*
18. 18 right-handed students to 4 left-handed students *$\frac{18}{4} = \frac{9}{2}$*
19. 2 c of sugar to 10 c of water *$\frac{2}{10} = \frac{1}{5}$*
20. 15 red candies to 10 green candies *$\frac{15}{10} = \frac{3}{2}$*

266   Chapter 13

How many pennies are displayed? 4
How many nickels are displayed? 6
If you were to substitute the number of each coin for its name in the written statement, how would it read? 4 to 6

Write "ratio" for display and write the comparison "4 to 6" below it. Explain that a ratio is a comparison of two quantities and is read using the word _to_. This ratio compares the number of pennies _to_ the number of nickels.

Explain that a ratio can be written in word form, ratio form, or fraction form. The ratio 4 to 6 is written in word form because the word _to_ is written. Write "word form" beside 4 to 6.

- Write "4:6" for display and label it "ratio form." Explain that the ratio form is written with a colon.

- Write "$\frac{4}{6}$" for display and label it "fraction form." Explain that a ratio can also be written in fraction form.

- Point out that all 3 forms represent the same ratio. Similar to the terms of a fraction, the numbers of a ratio are referred to as terms: the first number (4) is referred to as the first term and the second number (6) is referred to as the second term.

What ratio compares the number of pennies to the number of coins displayed? 4 to 10

Write a ratio in fraction form.

The school's soccer team won 7 games and lost 5 games during the fall soccer season.

21. wins to games played $\frac{7}{12}$ 22. wins to losses $\frac{7}{5}$ 23. games played to losses $\frac{12}{5}$

A **rate** is a special ratio comparing two quantities that have different measuring units. The **unit rate** tells how many of a quantity there are *per* one unit of another quantity.

Gasoline is purchased by a *per-gallon rate*.
$3.00 per 1 gallon = $3/gal

A babysitter is paid by a *per-hour rate*.
$6.50 per 1 hour = $6.50/hr

Calories are reported in a *per-serving rate*.
180 calories per 1 serving = 180 calories/serving

Speed is calculated as *miles per hour*.
60 miles per 1 hour = 60 mph or 60 mi/hr

To find the unit rate, rename the ratio using a denominator of 1.

Mom spent $11.00 for 4 packages of cookies. What is the unit rate (cost per package)?

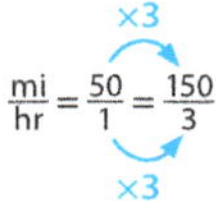

$$\frac{\text{cost}}{\text{package}} = \frac{11}{4} = \frac{2.75}{1}$$

The cookies cost $2.75 per package.

Multiply the terms of the unit rate to find an equivalent ratio.

Each package contains 12 cookies. How many cookies are in 4 packages?

$$\frac{\text{cookies}}{\text{packages}} = \frac{12}{1} = \frac{48}{4}$$

There are 48 cookies in 4 packages.

What distance will a car travel in 3 hr at an average speed of 50 mph?

$$\frac{\text{mi}}{\text{hr}} = \frac{50}{1} = \frac{150}{3}$$

The car will travel 150 mi in 3 hr.

Find the unit rate.

24. Kevin drove 480 mi on 16 gal of gasoline. *30 mi/gal*

25. Elizabeth earned $48 in 8 hr. *$6/hr*

26. Mr. Monroe drove 2,250 mi in 3 days. *750 mi/day*

27. Juliet read 30 pages in 60 min. *0.5 page/min or $\frac{1}{2}$ page/min*

28. Mom paid $3.16 for 4 lb of apples. *$0.79/lb*

29. The office assistant can type 165 words in 3 min. *55 words/min*

Use the unit rate to find the answer.

30. cost of 5 gal at $3.15/gal *$15.75*

31. amount earned for 4 hr at $7/hr *$28*

32. distance traveled after 6 hr at 60 mph *360 mi*

33. distance traveled after 9 days at 230 mi/day *2,070 mi*

34. distance traveled after 0.5 hr at 7 km/hr *3.5 km*

35. amount paid for 3.5 hr at $10/hr *$35*

Solve.

36. Eva is traveling by train. She has traveled 180 mi in 2 hr. At this rate, her trip will take 6 hr. How far will she travel in all? *540 mi*

37. Madison is traveling by plane. She has traveled 420 mi in 2 hr. At this rate, her trip will take 4 hr. How far will she travel in all? *840 mi*

38. Bethany drove 300 mi on 12 gal of gasoline. At this rate, how many gallons of gasoline will she need to travel 1,000 mi? *40 gal*

39. What evidence of design is found in ratios and rates? *Rates show orderliness in that they are constant and dependable.*

Choose a student to write the ratio in the three forms, naming each form as he writes the ratio. 4 to 6, word form; 4:6, ratio form; $\frac{4}{6}$, fraction form

- Choose a student to display a second set (row) of 4 dimes and 6 quarters. Write for display "4 to 6 is like 8 to __," "4:6 is like 8:__," and "$\frac{4}{6}$ is like $\frac{8}{n}$."
What number would complete each of these ratios to make the statement true? 12
Write "12" to complete each ratio.
How do you think you can find an equivalent ratio in higher terms? I can multiply each term by the same nonzero number.
How do you think you can find an equivalent ratio in lower terms? I can divide each term by the same nonzero divisor.
Do you think the ratio 4:6 is written in lowest terms? No; 4 and 6 have a common factor of 2.
Choose a student to write an equivalent ratio in lower terms. 2:3
Divide the original set of 4 dimes to 6 quarters into two equal sets.
Is the ratio 2:3 written in lowest terms? Yes; 2 and 3 have no common factor other than 1.

- Follow a similar procedure for a whole-to-part comparison of the original set (coins to quarters). 10 to 6, 10:6, $\frac{10}{6}$; sample answers: higher terms—20 to 12, 30:18, $\frac{100}{60}$; lower terms—5:3

- Guide the students as they complete the following equivalent ratios.
10 to 7 = __ to 49 70   5:12 = __:60 25
$\frac{15}{45} = \frac{3}{n}$ 9 $\frac{16}{4} = \frac{8}{n}$ 2

- Point out that when students multiply each term of a ratio by the same factor or divide each term of a ratio by the same divisor, they are multiplying or dividing the entire ratio by a value of 1 (e.g., $\times \frac{4}{4}$ or $\div \frac{3}{3}$). Remind the students that the Identity Property of Multiplication states that when 1 is a factor, the product is the other factor. When a ratio written in fraction form is multiplied by a fraction name for 1, the same ratio is expressed in higher terms.

Choose students to write for display the ratio 4 to 10 in the three forms. 4 to 10, 4:10, $\frac{4}{10}$

Writing ratios to describe comparisons

- Use a **discussion** to introduce types of ratios.

- Explain that ratios can describe different comparisons. The ratio $\frac{4}{10}$ compares part of the set (pennies) to the whole set (coins). Write "part-to-whole" for display.
Is the ratio of pennies to nickels a part-to-whole comparison? No; it compares one part of the set to another part of the set.
Write "part-to-part" for display.

What ratio compares the number of coins to the number of pennies? 10 to 4

Choose a student to write the ratio in the three forms. 10 to 4, 10:4, $\frac{10}{4}$
What does the ratio 10 to 4 mean? In the set of 10 coins, 4 are pennies.
What comparison does this ratio describe? the whole set to a part of the set
Write "whole-to-part" for display.

Finding equivalent ratios

- Use a **visual representation** to help the students write equivalent ratios.

- Display 4 dimes and 6 quarters.
What part-to-part ratio compares the number of dimes to the number of quarters? 4 to 6

LESSON 120

Determining & using the unit rate

- Use **problem solving** to introduce the term *unit rate*.

- Write "rate" for display. Explain that a *rate* is a special ratio that compares two quantities having different units.

  Write the following statement for display and choose a student to tell what units are being compared. Choose another student to write for display the rate in word form, ratio form, and fraction form.

  A car traveled 405 mi using 15 gal of gasoline. miles to gallons; 405 to 15, $405:15$, $\frac{405}{15}$

  If a car traveled 405 mi using 15 gal of gasoline, how can you find out how many miles it could travel using 1 gal of gasoline? I can find a ratio that is equivalent to $\frac{405 \text{ mi}}{15 \text{ gal}}$ with "1 gal" as the second term.

  What equivalent ratios could you write to find how many miles the car would travel using 1 gal of gasoline? $\frac{405 \text{ mi}}{15 \text{ gal}} = \frac{m}{1 \text{ gal}}$

- Write "$\frac{405 \text{ mi}}{15 \text{ gal}} = \frac{m}{1 \text{ gal}}$" for display. Point out that a unit rate tells how many of a quantity there are per 1 unit of another quantity. The second term of a unit rate is always 1.

  What relationship do you notice between these equivalent ratios? 15 gal is divided by 15 to get the unit rate of 1 gal.

  How could you find the first term in the unit rate? Since the second term in $\frac{405 \text{ mi}}{15 \text{ gal}}$ is divided by 15 to get 1 gal, the same operation must be performed on the first term; $405 \div 15 = m$.

  Point out that when students divide both terms by the same number, they are dividing by a name for 1 to find an equivalent ratio ($\frac{405}{15} \div \frac{15}{15} = \frac{m}{1}$).

- Direct the students to divide 405 mi by 15 gal to find m (the number of miles per gallon). $m = 27$ mi

  Complete the solution and write the final answer: $\frac{405 \text{ mi}}{15 \text{ gal}} = \frac{m}{1 \text{ gal}} = \frac{27 \text{ mi}}{1 \text{ gal}}$ or 27 mi/gal. Explain that it is not necessary to write the 1 in the second term of a unit rate.

- Follow a similar procedure for the following statements.

  Lisa earned \$42 in 7 hr. dollars to hours; 42 to 7, $42:7$, $\frac{42}{7}$; $\frac{\$42}{7 \text{ hr}} = \frac{d}{1 \text{ hr}} = \frac{6 \text{ dollars}}{1 \text{ hr}}$ or \$6/hr

  There are 141 calories in 10 potato chips. calories to potato chips; 141 to 10, $141:10$, $\frac{141}{10}$; $\frac{141 \text{ calories}}{10 \text{ chips}} = \frac{c}{1 \text{ chip}} = \frac{14.1 \text{ calories}}{1 \text{ chip}}$ or 14.1 calories/chip

- Follow a similar procedure to guide the students as they find an equivalent ratio using the unit rate in the following word problems. Remind students that both terms in the unit rate must be multiplied by the same number to find an equivalent ratio.

  Alex's go-cart can travel at a maximum speed of 15 mph. If he drives on a go-cart track at this speed, how many miles will he have driven in 0.5 hr? $\frac{15 \text{ mi}}{1 \text{ hr}} = \frac{m}{0.5 \text{ hr}} = \frac{7.5}{0.5}$; 7.5 mi

  Carrots are on sale for \$0.79 per pound. Anna is purchasing 3.75 lb of carrots. How much will she pay for the carrots? $\frac{\$0.79}{1 \text{ lb}} = \frac{d}{3.75 \text{ lb}} = \frac{\$2.96}{3.75}$; \$2.96

Recognizing evidence of design

- Guide the students in a **discussion** to help them recognize design in ratios.

- Write "$\frac{\$8}{1} = \frac{n}{5}$" for display.

Find the perimeter of the figure. *Equations may vary.*

1. 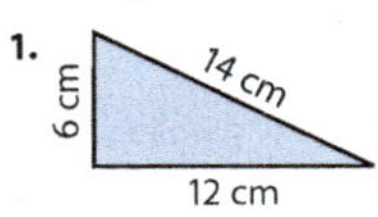

14 cm + 6 cm + 12 cm = 32 cm

2. 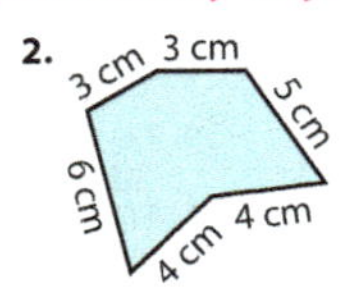

3 cm + 3 cm + 5 cm + 4 cm + 4 cm + 6 cm = 25 cm

3. 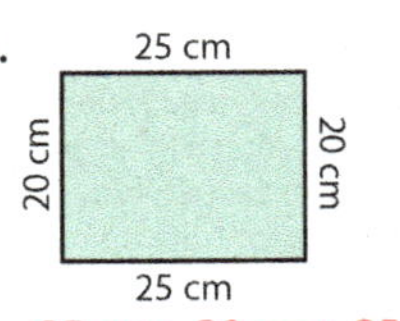

25 cm + 20 cm + 25 cm + 20 cm = 90 cm

Find the area of the figure.

4. 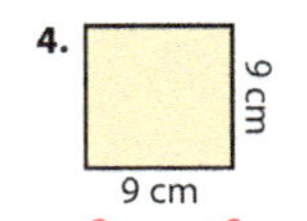

9 cm × 9 cm = 81 cm²

5. 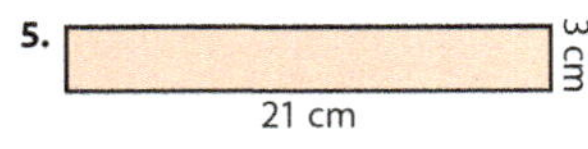

3 cm × 21 cm = 63 cm²

6. 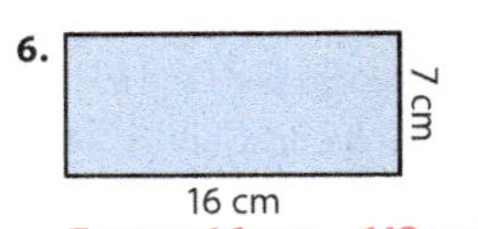

7 cm × 16 cm = 112 cm²

Find the volume of the figure.

7. 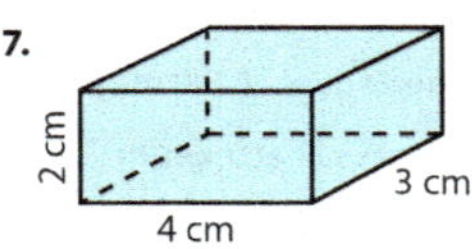

4 cm × 3 cm × 2 cm = 24 cm³

8. 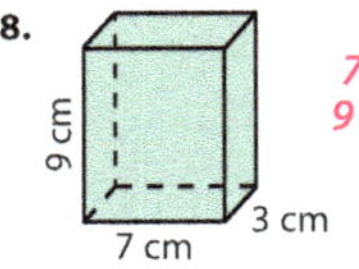

7 cm × 3 cm × 9 cm = 189 cm³

Solve.

9. Jerry made a square raised flower bed for his mother, using 20 ft boards. What is the area of the flower bed? 20 ft × 20 ft = 400 ft²

10. Amy built a rectangular birdhouse for bluebirds. It is 13 in. high, 5.5 in. wide, and 5 in. long. What is the volume of the birdhouse? 13 in. × 5.5 in. × 5 in. = 357.5 in.³

11. Sammy and Sally have a rectangular pool that is 6 ft long and 3 ft wide. What is its perimeter? (2 × 6 ft) + (2 × 3 ft) = 18 ft

480   Daily Review

- Point out that this ratio represents the amount of money Brendan earns per hour of work.

  If Brendan earns $8 for 1 hr of work, how much can he earn in 5 hr?

- Invite students to explain their reasoning for finding a solution (multiplying by a name for 1 to find higher terms).

  How much can Brendan earn in 5 hr? $40

  Could you find out how much Brendan could earn in 10 or 15 hr? Yes; I can multiply 8 by a fractional name for 1. I can multiply by $\frac{10}{10}$ to find that Brendan would earn $80 for 10 hr of work and multiply by $\frac{15}{15}$ to find that he would earn $120 for 15 hr of work.

  Why do these ratios work so well? They are consistent and dependable.

- Discuss the idea that consistency points to the orderly design planned by God at creation. He has given us His Word so that we can understand this design.

- Direct attention to the top of Student Edition page 266, and lead the students to answer the essential question, *"What evidence of design is found in ratios and rates?"* Rates show orderliness in that they are constant and dependable.

Apply

Student Edition pages 266–67

- Read and explain the directions for pages 266–67. Assist the students as they complete the pages independently.

Daily Review

- Students should complete Chapter 13, section *a*.

NOTES

LESSON 121

Student Edition pages 265, 268–69
Daily Review Chapter 13, section *b*

OBJECTIVES

- Complete a ratio table.
- Find equivalent ratios.
- Make a ratio table.
- Solve problems by using ratio tables.

TEACHER RESOURCES

- 72 *Pictured Ratios*
- 73 *Ratio Tables*

Engage

- **Read aloud** Student Edition page 265 to introduce and generate interest for this chapter's STEM activity.

Instruct

Completing a ratio table; finding equivalent ratios

- Use **visual analysis** to help the students find equivalent ratios.

- Display the *Pictured Ratios* page. Explain that each of the 6 pictures shows a relationship. A unit rate is shown when one of the pictured objects is being compared with a quantity of another object.

  What 2 objects are being compared in each picture? horse trailers to horses or horses to horse trailers

  Which of these pictures shows a unit rate? the picture showing 3 horses to 1 horse trailer

- Draw for display a ratio table with the first 2 ratio entries: $\frac{\text{horses}}{\text{trailers}}$ and $\frac{3}{1}$. Explain that a ratio table is a method for organizing information to show the comparison of 2 quantities or equivalent ratios. Unit rates are written at the beginning of each row, and the table is used to extend the pattern between the terms (a 3 to 1 ratio).

horses	3	6	9	12	15
trailers	1	2	3	4	5

Ratio Tables

How can a graph help me find my moon weight?

A **ratio table** is a method for organizing equivalent ratios. The table extends the pattern between the terms by multiplying or dividing both terms of the ratio by the same nonzero number (a name for 1).

Key Terms
- ratio table

		×2	×3	×4	×5
nickels	5	10	15	20	25
quarters	1	2	3	4	5

$$\frac{5}{1} = \frac{10}{2} = \frac{15}{3} = \frac{20}{4} = \frac{25}{5}$$

		÷2	÷4	÷8
pennies	80	40	20	10
dimes	8	4	2	1

$$\frac{80}{8} = \frac{40}{4} = \frac{20}{2} = \frac{10}{1}$$

Exercises

Complete the ratio table. Use the ratios to answer the question.

1. What is the unit rate of inches to feet? *12 : 1*
2. How many inches are in 3 ft? *36 in.*
3. How many inches are in 4 ft? *48 in.*
4. Sixty inches make up how many feet? *5 ft*

inches	12	24	*36*	*48*	60
feet	1	2	3	4	*5*

5. How many quarts can be made from 20 c? *5 qt*
6. How many cups are in 2.5 qt? *10 c*
7. What is the unit rate of cups to quarts? *4 : 1*

cups	40	20	*10*	*4*
quarts	10	*5*	2.5	1

Complete the ratio table.

8.

cm	2.54	*5.08*	10.16	20.32	*40.64*
in.	1	2	4	8	16

9.

km	1.61	*3.22*	4.83	*6.44*	*8.05*
mi	1	2	3	4	5

Use the unit rate 5,280 ft/mi to make a ratio table. Answer the question.

10. How many feet are in 2 mi?
    2 × 5,280 ft = 10,560 ft
11. How many feet are in 5 mi?
    5 × 5,280 ft = 26,400 ft
12. How many miles are 21,120 ft?
    21,120 ft ÷ 5,280 ft = 4 mi

State whether the ratio could be in a ratio table with $\frac{5}{7}$.

13. $\frac{25}{35}$ *yes*   14. $\frac{50}{60}$ *no*   15. $\frac{60}{84}$ *yes*

State whether the ratio could be in a ratio table with $\frac{3}{4}$.

16. $\frac{27}{32}$ *no*   17. $\frac{33}{44}$ *yes*   18. $\frac{42}{56}$ *yes*

Choose the ratio that could *not* be in a ratio table with the given ratio.

19. 5 to 10 $\frac{30}{60}$ $\frac{45}{90}$ $\boxed{\frac{25}{100}}$

20. 18 : 12   6 : 4 $\boxed{72 : 36}$ 36 : 24

268   Chapter 13

Which of the other pictures show the same 3 : 1 ratio of horses to horse trailers? the picture showing 6 horses to 2 horse trailers, and the picture showing 9 horses to 3 horse trailers

Write the next two ratios in the table: $\frac{6}{2}$ and $\frac{9}{3}$.

- Write "4" and "5" as the last two trailer entries. Choose students to complete the table and explain their reasoning: $\frac{12}{4}$; $\frac{15}{5}$. Point out that the Identity Property of Multiplication (multiplying by a name for 1) can be applied to the unit rate so that each term of a ratio is multiplied by the same number to find an equivalent ratio (e.g., $\frac{6}{2} \times \frac{2}{2} = \frac{12}{4}$ and $\frac{3}{1} \times \frac{5}{5} = \frac{15}{5}$).

What horse to trailer ratio can you write for picture number 2 on the *Pictured Ratios* page? $\frac{8 \text{ horses}}{2 \text{ trailers}}$

What is the unit rate for this ratio? $\frac{4 \text{ horses}}{1 \text{ trailer}}$

What is the unit rate of horses to trailers shown in picture number 5? $\frac{4 \text{ horses}}{1 \text{ trailer}}$, $\frac{6}{1.5} \div \frac{1.5}{1.5} = \frac{4}{1}$

What is the unit rate of horses to trailers shown in picture number 6? $\frac{4 \text{ horses}}{1 \text{ trailer}}$, $\frac{12}{3} \div \frac{3}{3} = \frac{4}{1}$

- Display only the first table on the *Ratio Tables* page. Guide students as they complete the table: $\frac{12}{3}$, $\frac{16}{4}$.

Make a ratio table to solve the problem. Extend the pattern by making equivalent ratios, or use combinations of the ratios.

A gallon of paint can cover a wall area of about 350 ft². How much area can 6 gal of paint cover? How much area can 15 gal cover?

$$\frac{350 \text{ ft}^2}{1 \text{ gal}} \times \frac{6}{6} = \frac{2,100 \text{ ft}^2}{6 \text{ gal}}$$

feet²	350	700	1,050	1,400	1,750	2,100
gallon	1	2	3	4	5	6

Use a combination of the ratios in the table to find the area for 15 gal.

15:n
15 gal:5,250 ft²
15 = 6 + 4 + 5
15:2,100 + 1,400 + 1,750

The sales tax on a purchase of $20 is $1.30. What will the tax be on a purchase of $70?

70 is not a multiple of 20 and would not extend this pattern. Finding the tax on a $10 purchase allows you to use a combination of the ratios in the table to find the tax on a $70 purchase.

70 = 60 + 10 or 70 = 80 − 10

tax	$1.30	$2.60	$3.90	$5.20	$0.65	$4.55
purchase	$20	$40	$60	$80	$10	$70

If $20:$1.30, then $10:$0.65.
$70:n 70 = 60 + 10
$70:$4.55 70:$3.90 + $0.65

Use the ratio table to answer the question. *Steps to solve may vary.*

A group of students is visiting the history museum. The table shows the price of admission that different groups will pay.

students	3	5	6	12
admission	$19.50	$32.50	$39.00	$78.00

21. How much will a group of 20 students pay?
$130.00
22. How much will a group of 24 students pay?
$156.00
23. How much will a group of 27 students pay?
$175.50

The gasoline tank in Rachel's car holds 15 gal. Her car can travel 420 mi on a tank of gas.

gallons	15	20	24	30
miles	420	560	672	840

24. How far can the car travel on 40 gal?
1,120 mi
25. How far can the car travel on 50 gal?
1,400 mi
26. How far can the car travel on 4 gal?
112 mi

There is less gravity on the moon than on the earth. A person weighing 120 lb on the earth would weigh approximately 20 lb on the moon.

earth weight (lb)	120	100	60	40
moon weight (lb)	20	17	10	7

27. About how much would a 200 lb person weigh on the moon?
34 lb
28. About how much would a 160 lb person weigh on the moon?
27 lb
29. Graph the ordered pairs from the earth and moon weights ratio table on a coordinate plane. Use the weights on the earth as the *x*-coordinates and the weights on the moon as the *y*-coordinates. Draw a line to connect the points. Locate your weight on the graph and find your approximate weight on the moon.

30. How did the graph in problem 29 help you find your moon weight?

30. *I found my earth weight (x-axis) and moved vertically until I reached the moon weight line. Then I moved left to find the corresponding moon weight (y-axis).*

Lesson 121 269

How many inches are in 3 yd? 108 in., $\frac{108}{3}$

How many inches are in 4 yd? 144 in., $\frac{144}{4}$

Solving problems using ratio tables

- Use a **graphic organizer** to help the students solve problems.

- Read the following word problem aloud.

  Patrick paid $0.90 sales tax for a $12 purchase. He wants to find out how much sales tax he would have to pay for a purchase of $42.00. $3.15

  What is being compared in the word problem? the amount of sales tax to the amount of a purchase; $0.90 to $12

- Display the $\frac{\text{Tax}}{\text{Purchase}}$ ratio table on the *Ratio Tables* page and write $\frac{\$0.90}{\$12}$ as the first entry. Explain that ratio tables can be used to find a specific ratio such as the amount of sales tax they need to pay when purchasing an item.

  What equivalent ratio can be made by multiplying each term of this first ratio by 2? $\frac{\$1.80}{\$24}$
  Write the ratios in the table.

- Follow a similar procedure for multiplying by 3 and 4. $\frac{\$2.70}{\$36}$ and $\frac{\$3.60}{\$48}$

  What property did you apply when renaming the ratio $\frac{\$0.90}{\$12}$ to higher terms? Identity Property of Multiplication; Each ratio was multiplied by a name for 1.

  Since the purchase price of $42 is $6 more than $36, how do you think you can find an equivalent ratio with $6 as its second term using the ratios in this table? I can divide the second term of any of the equivalent ratios by 6 to find the common divisor needed for dividing both the first and second term. (E.g., since I know that $12 ÷ 2 = $6, I can divide $\frac{\$0.90}{\$12}$ by $\frac{2}{2}$ to find an equivalent ratio of $\frac{\$0.45}{\$6}$.)

- Invite a student to write the ratio "$\frac{\$0.45}{\$6}$" in the table.

- Guide the students to use their knowledge of unit rates and equivalent ratios to complete the next two tables: $\frac{200}{2}$, $\frac{300}{3}$, $\frac{400}{4}$, $\frac{500}{5}$; and $\frac{32}{2}$, $\frac{64}{4}$, $\frac{128}{8}$, $\frac{256}{16}$.

  Would the ratio $\frac{600}{6}$ be in the same ratio table as $\frac{300}{3}$? Yes; when both are renamed to lowest terms, they share the same $\frac{100}{1}$ unit rate; therefore, they are equivalent ratios.

  Would the ratio $\frac{850}{8}$ go in the same ratio table as $\frac{300}{3}$? No; $\frac{850}{8}$ does not rename to a $\frac{100}{1}$ unit rate.

Making a ratio table

- Use a **discussion** to help the students make a ratio table to show the relationship of inches to yards.

- Direct the students to draw a ratio table as you draw one for display.

  What unit rate compares the number of inches to 1 yd? 36:1

  Instruct the students to write the unit rate as the first ratio in the table.

- Direct the students to write ratios in their tables as you ask the following questions.

  How many inches are in 2 yd? 72 in., $\frac{72}{2}$

LESSON 121

How can you use the ratio $\frac{\$0.45}{\$6}$ to help you find how much sales tax will be on a purchase of $42? Since $6 \times 7 = 42$, I can rename to higher terms by multiplying the ratio $\frac{\$0.45}{\$6}$ by $\frac{7}{7}$ and get an equivalent ratio with 42 as the second term.

Choose a student to write and solve the equation. Write "$\frac{\$3.15}{\$42}$" in the table.

- Display the $\frac{\text{Earth Weight}}{\text{Mars Weight}}$ ratio table. Explain that there is less gravity on Mars than on Earth. If an object weighs 100 lb on Earth, it would weigh approximately 40 lb on Mars.

The $\frac{\text{Earth Weight}}{\text{Mars Weight}}$ ratio of $\frac{1}{0.377}$ has been rounded to $\frac{1}{0.4}$ for this lesson.

- Guide the students in determining approximately how much people would weigh on Mars if they weighed the following numbers of pounds on Earth. For each weight, ask a student to tell how he would find how much the person would weigh on Mars and to explain his reasoning. Discuss each method as needed.

160 lb sample answers: I can multiply the ratio $\frac{10}{4}$ by $\frac{16}{16}$; since the first terms in the ratios $\frac{100}{40}$ and $\frac{60}{24}$ added together equal 160, I can add the second terms; 64 lb.

70 lb sample answers: I can multiply the ratio $\frac{10}{4}$ by $\frac{7}{7}$; since the first terms in the ratios $\frac{40}{16}$, $\frac{20}{8}$, and $\frac{10}{4}$ added together equal 70, I can add the second terms; 28 lb.

115 lb sample answers: I can divide both terms of $\frac{10}{4}$ by 2 to find that 5 lb on Earth would be about 2 lb on Mars ($\frac{5}{2}$). I can multiply $\frac{5}{2}$ by $\frac{23}{23}$ (since $\frac{115}{5} = 23$); since the first terms in the ratios $\frac{100}{40}$, $\frac{10}{4}$, and $\frac{5}{2}$ added together equal 115, I can add the second terms; 46 lb.

Write a comparison sentence by using >, <, or =.

1. 1.70 $\le$ 1.71   **2.** 0.8 $=$ 0.80   **3.** 8.465 $\le$ 8.645   **4.** 0.051 $\le$ 0.052

5. 1.60 $\ge$ 0.16   **6.** 0.653 $\le$ 0.66   **7.** 1.874 $\le$ 18.74   **8.** 3.09 $\ge$ 3.009

Solve. *Equations may vary.*

9. What is the cost of 6 lb of chicken if chicken is $2.89 per pound?
6 × $2.89 = $17.34

10. Kerri bought a two-cheeseburger meal including a drink and fries for $12.79. Cheeseburgers normally cost $3.99, and drinks are $2.39. Fries are $2.79. How much money did she save by buying the meal instead of buying the two burgers, the fries, and the drink separately?
(2 × $3.99) + $2.39 + $2.79 = $13.16; $13.16 − $12.79 = $0.37

Write an equation. Solve.

11. the amount Dad has left over after using $10.00 to purchase a drink that cost $2.89
$10.00 − $2.89 = $7.11

12. five tenths less than three and twenty-five hundredths *3.25 − 0.5 = 2.75*

13. thirteen hundredths more than thirteen thousandths *0.013 + 0.13 = 0.143*

14. the price of 1 can of beets when the price for five cans is $6.00 *$6.00 ÷ 5 = $1.20 each*

15. Estimate the product of 57 and 236. *60 × 200 = 12,000*

Daily Review   481

Apply

Student Edition pages 268–69

- Read and explain the directions for pages 268–69.

- Direct attention to problem 29 on page 269. Guide the students as they graph the ordered pairs from the ratio table and find their weight on the moon.

- Guide the students as they answer the essential question at the top of Student Edition page 268, "How can a graph help me find my moon weight?" I found my Earth weight (*x*-axis) and moved vertically until I reached the moon weight line. Then I moved left to find the corresponding moon weight (*y*-axis).

- Assist the students as they complete the pages independently.

Daily Review

- Students should complete Chapter 13, section *b*.

DIFFERENTIATED INSTRUCTION

Use the following to provide extra help for students who experience difficulty with the concepts taught in Chapter 13.

Write ratios.
Direct the students to read Matthew 14:17. Remind them that a ratio compares two quantities and can be written using the word *to* (word form), using a colon (:) to represent the word *to* (ratio form), and as a fraction (fraction form). Ask them to identify the objects that are mentioned in the verse. 5 loaves, 2 fish

Instruct the students to first write the ratio of loaves to fish in word form, and then to write it in ratio form and fraction form. 5 to 2, 5 : 2, $\frac{5}{2}$

Direct them to read aloud each form of the ratio. If necessary, allow the students to label the terms in each ratio form (e.g., 5 loaves : 2 fish) and to read aloud each ratio form.

Repeat the procedure using the following references: Luke 17:11–19 1 thankful leper to 10 healed lepers—1 to 10, 1 : 10, $\frac{1}{10}$ and Job 1:1–2 7 sons to 3 daughters—7 to 3, 7 : 3, $\frac{7}{3}$. You may further examine Job's substance before and after his trials as shown in Job 1:3 and Job 42:10–12.

Find equivalent ratios by multiplying and dividing.
Write "$\frac{7}{5} = \frac{n}{50}$" for display. Ask the students to identify the relationship between the second terms of the ratios. 50 is 10 times greater than 5 or 5 × 10 = 50.

Draw an arrow from 5 to 50 and write "× 10" below it. Remind the students that to find an equivalent ratio, the operation that is performed on one term must also be performed on the other term. Ask them to tell what operation needs to be performed to find the unknown term in the second ratio. multiplication

Direct the students to draw an arrow from the 7 to the *n* and write "× 10" above it. Then instruct them to erase the *n* and complete the ratio. 70

Follow a similar procedure for $\frac{20}{15} = \frac{4}{n}$.

4 is $\frac{1}{5}$ of 20 or 20 ÷ 5 = 4; division; 15 ÷ 5 = 3.

Continue the activity as needed using the following, or similar, problems.

$\frac{60}{6} = \frac{30}{3}$ $\frac{5}{2} = \frac{40}{16}$

$\frac{8}{100} = \frac{2}{25}$ $\frac{3}{20} = \frac{18}{120}$

$\frac{60}{25} = \frac{12}{5}$ $\frac{28}{4} = \frac{14}{2}$

NOTES

Student Edition pages 270–71
Daily Review Chapter 13, section c

OBJECTIVES

- Use counters to model a proportion.
- Determine whether two ratios are proportional.
- Solve problems by solving for a missing term in a proportion.
- Explain why proportions show reliable patterns. **BWS**

BIBLICAL WORLDVIEW SHAPING

- Design (Explain): Proportions are reliable because they are based on patterns observed in equivalent ratios that were designed by God.

ADDITIONAL MATERIALS

- black and red counters

Engage

- Direct attention to the top of Student Edition page 270 and direct the students to use **brainstorming** to prepare them to answer the essential question, "Why are proportions reliable?"

Instruct

Modeling a proportion

- **Model** proportions to help the students understand this skill.

- Place 4 black counters in one row, and then place 12 red counters in another row below the black counters.

  What part-to-part ratio compares the black counters to the red counters?
  4 to 12

  Write the ratio for display:
  "$\frac{4 \text{ black counters}}{12 \text{ red counters}} = \frac{4}{12}$."

- Invite a student to arrange the black and the red counters into 2 equal groups.

  What is the ratio of the black and red counters in each group? 2 to 6

Solving Proportions

Why are proportions reliable?

A **proportion** is an equation stating that two ratios are equivalent. Ratios are proportional when they are equivalent. The terms can be compared vertically, horizontally, or diagonally to test for equivalency.

Methods for Solving Proportions

Using Number Sense

$$\frac{1}{2} \quad\text{---}\quad \frac{3}{6}$$

This proportion can be read "1 is to 2 like 3 is to 6." Compare the relationship of the numerator to the denominator of each ratio.

1:2   1 is *one-half* of 2.
3:6   3 is *one-half* of 6.

$$\frac{1}{2} = \frac{3}{6}$$

$\frac{1}{2}$ is **proportional** to $\frac{3}{6}$.

Testing for Equivalent Ratios
Perform the same operation on both terms.

$$\overset{\times 3}{\underset{\times 3}{\frac{1}{2} = \frac{3}{6}}}$$

$\frac{1}{2}$ is **proportional** to $\frac{3}{6}$.

$$\overset{\times 4}{\underset{\times 4}{\frac{2}{3} \neq \frac{8}{10}}}$$

$\frac{2}{3}$ is **not proportional** to $\frac{8}{10}$.

Using Cross Multiplication
Cross multiply to compare the numerators of like fractions.

$$\frac{1}{2} \quad\text{---}\quad \frac{3}{6}$$

$6 \times 1 = 6 \quad \frac{1}{2} \times \frac{3}{6} \quad 2 \times 3 = 6$

$$\frac{1}{2} = \frac{3}{6}$$

$\frac{1}{2}$ is **proportional** to $\frac{3}{6}$.

$$\frac{2}{3} \quad\text{---}\quad \frac{8}{10}$$

$10 \times 2 = 20 \quad \frac{2}{3} \times \frac{8}{10} \quad 3 \times 8 = 24$

$$\frac{2}{3} \neq \frac{8}{10}$$

$\frac{2}{3}$ is **not proportional** to $\frac{8}{10}$.

Using Division

$$\frac{2}{3} \quad\text{---}\quad \frac{8}{10}$$

$$2 \div 3 \quad\underline{\quad}\quad 8 \div 10$$

$$0.\overline{6} \neq 0.8$$

$\frac{2}{3}$ is **not proportional** to $\frac{8}{10}$.

Comparing Lowest Terms

$$\frac{1}{2} = \frac{3}{6} \qquad \frac{3 \div 3}{6 \div 3} = \frac{1}{2}$$

$\frac{1}{2}$ is **proportional** to $\frac{3}{6}$.

$$\frac{2}{3} \neq \frac{8}{10} \qquad \frac{8 \div 2}{10 \div 2} = \frac{4}{5}$$

$\frac{2}{3}$ is **not proportional** to $\frac{8}{10}$.

Equivalent ratios will have the same lowest terms.

Exercises

Rename the ratios to lowest terms to compare. Write a fraction comparison, using = or ≠.

1. $\frac{12}{14}$ and $\frac{36}{42}$ $\frac{6}{7} = \frac{6}{7}$

2. $\frac{9}{27}$ and $\frac{7}{21}$ $\frac{1}{3} = \frac{1}{3}$

3. $\frac{22}{54}$ and $\frac{20}{26}$ $\frac{11}{27} \neq \frac{10}{13}$

4. $\frac{9}{15}$ and $\frac{6}{10}$ $\frac{3}{5} = \frac{3}{5}$

Write the operation that was performed on both terms.

5. $\frac{9}{14} = \frac{27}{42}$ $\times \frac{3}{3}$

6. $\frac{32}{80} = \frac{4}{10}$ $\div \frac{8}{8}$

7. $\frac{9}{11} = \frac{81}{99}$ $\times \frac{9}{9}$

8. $\frac{7}{12} = \frac{56}{96}$ $\times \frac{8}{8}$

Use cross multiplication to prove that the ratios are proportional.

9. $\frac{15}{20} = \frac{6}{8}$
   120 = 120

10. $\frac{2}{3} = \frac{12}{18}$
    36 = 36

11. $\frac{30}{12} = \frac{20}{8}$
    240 = 240

12. $\frac{5}{7} = \frac{40}{56}$
    280 = 280

Use division to find whether the ratios are proportional. Write a decimal comparison, using = or ≠.

13. $\frac{2}{4}$ and $\frac{15}{30}$
    0.5 = 0.5

14. $\frac{3}{8}$ and $\frac{4}{16}$
    0.375 ≠ 0.25

15. $\frac{2}{3}$ and $\frac{3}{5}$ *0.$\overline{6}$ ≠ 0.6*

16. $\frac{9}{72}$ and $\frac{6}{48}$
    0.125 = 0.125

Write the ratio for display:
"$\frac{2 \text{ black counters}}{6 \text{ red counters}} = \frac{2}{6}$."

- Repeat the procedure for 4 equal groups of red and black counters. Write the ratio: "$\frac{1 \text{ black counter}}{3 \text{ red counters}} = \frac{1}{3}$."

- Explain that the ratios $\frac{2}{6}$ and $\frac{1}{3}$ are equivalent to $\frac{4}{12}$. Write for display "$\frac{4}{12} = \frac{2}{6}$" and the term "proportion." Explain that a *proportion* is an equation stating that 2 ratios are equivalent. Ratios are proportional only if they are equivalent. This proportion is read "4 is to 12 like 2 is to 6." Read the proportion aloud together.

Determining proportional ratios

- Use **direct instruction** to help the students determine proportional relationships.

- Write "$\frac{4}{8} \text{---} \frac{1}{2}$" for display.

  Are these ratios equivalent? Yes; sample answer: relationships exist between the ratios: 4 is one-half of 8 and 1 is one-half of 2, and $4 \div 4 = 1$ and $8 \div 4 = 2$.

- Write "$\frac{a}{b} = \frac{c}{d}$." Explain that every proportion has vertical, horizontal, and diagonal relationships between its terms. Use the following activity to guide the students to the conclusion that if any one of these relationships is equivalent, then the ratios are proportional.

Solving Proportions to Find Equal Ratios

Comparing Unit Rates

Isabella earned $23 for 2 hr of babysitting. Felicia earned $31 for 3 hr of babysitting. Were the girls paid the same hourly rate?

$$\text{unit rate} = \frac{\text{pay}}{1 \text{ hr}}$$

Isabella

$$\frac{\$23}{2 \text{ hr}} = \frac{n}{1 \text{ hr}}$$

$n = \$23 \div 2$

$n = \$11.50$

Felicia

$$\frac{\$31}{3 \text{ hr}} = \frac{n}{1 \text{ hr}}$$

$n = \$31 \div 3$

$n = \$10.33$

Isabella earned more per hour than Felicia.

Finding the Equivalent Fraction

The coffee shop sells 4 doughnuts for $2. At this price, how much will 8 doughnuts cost?

$$\frac{\text{doughnuts}}{\text{cost}} \quad \frac{4}{2} = \frac{8}{c} \quad \begin{array}{l} 4 \times 2 = 8 \\ 2 \times 2 = c \\ c = 4 \end{array}$$

8 doughnuts will cost $4.

This strategy works well when the terms are related.

Cross Multiplying

The coffee shop sells 3 muffins for $6. At this price, how much will 8 muffins cost?

$$\frac{\text{muffins}}{\text{cost}}$$

$$\frac{3}{6} = \frac{8}{c}$$

$c \times 3 = 3c \quad \frac{3}{6} = \frac{8}{c} \quad 6 \times 8 = 48$

$3c = 48$

$$\frac{3c}{3} = \frac{48}{3}$$

$c = 16$

8 muffins will cost $16.

Solve the proportion. *Steps to solve may vary.*

17. $\frac{a}{24} = \frac{9}{36}$ *a = 6*

18. $\frac{14}{1} = \frac{b}{2}$ *b = 28*

19. $\frac{4}{16} = \frac{c}{12}$ *c = 3*

20. $\frac{33}{6} = \frac{d}{10}$ *d = 55*

21. $\frac{h}{2} = \frac{120}{80}$ *h = 3*

22. $\frac{6}{m} = \frac{18}{42}$ *m = 14*

Write a possible proportion for the situation. Write *yes* if the prices are equivalent. Write *no* if one price is a better buy.

23. 4 lb of apples for $6 or 10 lb of apples for $15 *yes*

24. a 12 oz can of corn for 55¢ or a 20 oz can of corn for 95¢ *no*

25. 12 eggs for $2 or 18 eggs for $3 *yes*

26. 1 qt of milk for $1 or 1 gal of milk for $4 *yes*

27. 5 bottles of soda for $7 or 6 bottles of soda for $8 *no*

28. a 5 oz candy bar for 50¢ or an 8 oz candy bar for 75¢ *no*

Solve the proportion to find an equivalent ratio.

29. Gabriel shoveled snow from 2 driveways in 3 hr. At this rate, how long will it take him to shovel 5 driveways? *7.5 hr*

30. During the election for class president, Colton received 3 votes for every vote cast for Bryce. Bryce received 5 votes. How many votes did Colton receive? *15 votes*

31. During the race, Driver 1 traveled 480 mi in 3 hr. At this rate, how far will Driver 1 travel in 5 hr? *800 mi*

32. Three servings of yogurt are 24 oz. How many ounces are in 4 servings? *32 oz*

33. Two pizzas cost $15. At this rate, how much would 7 pizzas cost? *$52.50*

34. A survey of middle-school students revealed that 36 out of 120 students have green eyes. How many students with green eyes would you expect to find out of 10 of these students? *3 students*

35. Factory workers can produce 25 items in 30 hr. How many hours will it take them to produce 60 items at this rate? *72 hr*

36. Why are proportions reliable? *Proportions are reliable because they are based on the patterns observed in equivalent ratios.*

Lesson 122   271

Remind the students that all three types of ratios (part-to-whole, part-to-part, and whole-to-part) can be written in fraction form, and that if ratios are equivalent, they are proportional.

- Explain that using strategies to solve proportions can be useful in real-world situations such as finding the best price when making a purchase.

Balloons come in different sized packages. A large package contains 15 balloons and costs $3. A smaller package contains 6 balloons and costs $2. Are the prices equivalent? no; $\frac{15 \text{ balloons}}{\$3} \neq \frac{6 \text{ balloons}}{\$2}$

What is being compared? the prices of balloons

Write "$\frac{\text{balloons}}{\text{price}}$" for display.

What ratio can you write for the larger package of balloons? 15 to $3

What ratio can you write for the smaller package? 6 to $2

Write "$\frac{15}{\$3}$ — $\frac{6}{\$2}$."

Are the costs of the packages the same? No; sample answers: the ratios renamed to lowest terms are different unit rates ($\frac{5 \text{ balloons}}{\$1} \neq \frac{3 \text{ balloons}}{\$1}$); the ratios renamed to common second terms are different ($\frac{30}{6} \neq \frac{18}{6}$); when I cross multiply 2×15 and 3×6, the products are different ($30 \neq 18$).

Write "$\neq$" to complete the problem.

Which package of balloons is the better deal? the larger package of balloons; I get 5 balloons for $1 in the larger package rather than just 3 balloons for $1 in the smaller package.

How else do you think you can compare the prices of the packages of balloons? I can find the price of 1 balloon in each package.

How do you think you can you find the price per balloon (the unit rate) for the packages? I can compare the ratio $\frac{\$3}{15 \text{ balloons}}$ to $\frac{\$2}{6 \text{ balloons}}$; I divide each first and second term by a common factor to rename the second term as 1 balloon, or divide the first term of the ratio by the second term.

Vertical: *a* is to *b* like *c* is to *d*; 4 is to 8 like 1 is to 2.

Ask the relationship between the first and second terms of each ratio to determine if the ratios are equivalent: 4 is one-half of 8, just as 1 is one-half of 2; $4 \div 8 = \frac{1}{2}$ and $1 \div 2 = \frac{1}{2}$.

Horizontal: *a* is to *c* like *b* is to *d*; 4 is to 1 like 8 is to 2.

Ask the relationship between the first terms of the two ratios and between the second terms of the two ratios to determine if the ratios are equivalent: 4 is four times greater than 1, just as 8 is four times greater than 2; $4 \div 4 = 1$ and $8 \div 4 = 2$.

Diagonal: (cross multiplication) $d \times a = b \times c$ or $da = bc$; $2 \times 4 = 8 \times 1$.

Guide the students in cross multiplying diagonal terms to write a possible equation to determine if the ratios are equivalent: 2×4 __ 8×1; $8 = 8$.

- Write "=" to complete the proportion: $\frac{4}{8} = \frac{1}{2}$.

- Follow a similar procedure for $\frac{6}{9}$ — $\frac{3}{4}$. $\frac{6}{9} \neq \frac{3}{4}$

Guide the students to the conclusion that if any one of the relationships (vertical, horizontal, or diagonal) is not equivalent, the ratios are not proportional.

To what can you relate the process of determining whether ratios are equivalent? It is similar to determining whether fractions are equivalent.

LESSON 122

- Write "$\frac{\$3}{15}$" and "$\frac{\$2}{6}$" for display. Direct the students to find the price per balloon in each package. $\frac{3}{15} \div \frac{15}{15} = \frac{\$0.20}{1 \text{ balloon}}$ and $\frac{2}{6} \div \frac{6}{6} = \frac{\$0.33}{1 \text{ balloon}}$

  Are the prices of the packages of balloons equivalent? No; the price per balloon in the larger package is less than the price per balloon in the smaller package.

  Write "$\frac{\$3}{15 \text{ balloons}} \neq \frac{\$2}{6 \text{ balloons}}$."

  Are the ratios $\frac{3}{15}$ and $\frac{2}{6}$ proportional? No; they are not equivalent.

- Follow a similar procedure for the following word problem.

  Last Saturday, Mark earned \$20 for working 4 hr. This Saturday, Mark earned \$30 for working 6 hr. Did Mark receive the same hourly rate on each Saturday?

  Yes; $\frac{\$20}{4 \text{ hr}} = \frac{\$30}{6 \text{ hr}}$; $\frac{\$5}{1 \text{ hr}} = \frac{\$5}{1 \text{ hr}}$; $\frac{20}{4}$ is proportional to $\frac{30}{6}$.

Solving proportion problems; explaining reliable patterns

- Use **problem solving** to help the students find a missing term.

- Read the following word problem aloud.

  A cookie recipe calls for 2 c of chocolate chips to make 48 cookies. How many cups of chocolate chips are needed to make 120 cookies? 5 c

  What is being compared? cups of chocolate chips to number of cookies

  Write "$\frac{\text{cups}}{\text{cookies}}$."

  What ratios can you write to compare the cups of chocolate chips to the number of cookies? 2 cups to 48 cookies and c to 120 cookies

  Write "$\frac{2 \text{ cups}}{48 \text{ cookies}} = \frac{c}{120 \text{ cookies}}$."

- Point out that when comparing ratios, the terms must show the same comparison; therefore, the order of terms is important.

- Direct the students to solve for the missing term in the proportion using the algorithm of their choice. c = 5

  Discuss the methods that were used to solve the problem.

Solve. Simplify if possible. *Answer is shown using cancellation.*

1. $4 \times \frac{4}{5} \; \frac{16}{5} = 3\frac{1}{5}$

2. $4 \times 2\frac{5}{6} \; \frac{4}{1} \times \frac{17}{6} = \frac{34}{3} = 11\frac{1}{3}$

3. $\frac{5}{7} \div \frac{1}{6} \; \frac{5}{7} \times \frac{6}{1} = \frac{30}{7} = 4\frac{2}{7}$

4. $6 \times \frac{2}{3} \; \frac{4}{1} = 4$

5. $3 \times 2\frac{1}{10} \; \frac{63}{10} = 6\frac{3}{10}$

6. $\frac{3}{4} \div \frac{3}{8} \; \frac{3}{4} \times \frac{8}{3} = 2$

7. $2 \times \frac{5}{12} \; \frac{5}{6}$

8. $7 \times 1\frac{3}{10} \; \frac{91}{10} = 9\frac{1}{10}$

9. $\frac{8}{12} \div \frac{2}{12} \; \frac{8}{12} \times \frac{12}{2} = 4$

10. $\frac{1}{4} \times \frac{2}{3} \; \frac{1}{6}$

11. $2 \div \frac{1}{6} \; \frac{2}{1} \times \frac{6}{1} = \frac{12}{1} = 12$

12. $\frac{3}{8} \div \frac{1}{2} \; \frac{3}{8} \times \frac{2}{1} = \frac{3}{4}$

13. $\frac{3}{5} \times \frac{1}{3} \; \frac{1}{5}$

14. $1 \div \frac{3}{12} \; \frac{1}{1} \times \frac{12}{3} = \frac{12}{3} = 4$

15. $\frac{1}{4} \div \frac{3}{5} \; \frac{1}{4} \times \frac{5}{3} = \frac{5}{12}$

16. $9 \times \frac{5}{7} \; \frac{45}{7} = 6\frac{3}{7}$

17. $4 \div \frac{2}{3} \; \frac{4}{1} \times \frac{3}{2} = 6$

18. $\frac{4}{6} \div 4 \; \frac{4}{6} \times \frac{1}{4} = \frac{1}{6}$

19. $\frac{4}{9} \times \frac{3}{8} \; \frac{1}{6}$

20. $3 \div \frac{1}{2} \; \frac{3}{1} \times \frac{2}{1} = \frac{6}{1} = 6$

21. $2 \div \frac{1}{2} \; \frac{2}{1} \times \frac{2}{1} = 4$

What does c = 5 represent? 5 c of chocolate chips are needed to make 120 cookies.

Erase the c in the proportion and write "5 c."

- Write the following table for display and read the following word problem aloud.

4			
10	30	40	50

In a survey, 4 out of every 10 people chose pizza as their favorite food. If 30 people were surveyed, how many people could you predict would choose pizza? 12 people

What is being compared? the number of people who chose pizza as their favorite food and the number of people surveyed

What ratios can you write to compare the number of people who chose pizza to the number of people surveyed? 4 to 10 and p to 30

Write "$\frac{4}{10} = \frac{p}{30}$."

- Direct the students to solve for the missing term in the proportion. p = 12

  Write each answer in the table as it is discussed.

  What does p = 12 represent? 12 people are likely to choose pizza if 30 people are surveyed.

How could you find the missing numbers for the other ratios? I could multiply by a fractional name for 1.

Guide the students as they complete the table. $\frac{4}{10} = \frac{16}{40}$; $\frac{4}{10} = \frac{20}{50}$.

What pattern do you see in the fractional names for 1 that you choose each time? $\frac{3}{3}, \frac{4}{4}, \frac{5}{5}$

Point out that patterns in equivalent ratios exist because God designed them. Proportions are reliable because they are based on these patterns.

- Follow a similar procedure for the following problems.

Labeling the terms in the ratios will help the students remember what the missing term represents.

Mom purchased 12 bottles of soft drink for $15. At this rate, how much will 14 bottles of soft drink cost? $\frac{bottles}{cost}$; $\frac{12}{15} = \frac{14}{c}$; $c = \$17.50$

During an election, Jared received 2 votes for every 3 votes that Kevin received. Kevin received 42 votes. How many votes did Jared receive? $\frac{votes\ for\ Jared}{votes\ for\ Kevin}$, $\frac{2}{3} = \frac{v}{42}$; $v = 28$ votes

If a farmer can raise 10 goats on 3 acres of pasture, how many acres of pasture would he need to raise 25 goats? $\frac{goats}{acres}$; $\frac{10}{3} = \frac{25}{g}$; $g = 7.5$ acres

Why are proportions reliable? Proportions are reliable because they are based on the patterns observed in equivalent ratios.

Apply

Student Edition pages 270–71
- Read and explain the directions for pages 270–71. Assist the students as they complete the pages independently.

Daily Review
- Students should complete Chapter 13, section c.

NOTES

Student Edition pages 272–74
Daily Review Chapter 13, section *d*

OBJECTIVES

- Determine whether two ratios are proportional.
- Find the unknown measure in similar figures using proportions.
- Find unknown measures in similar objects by using indirect measurement.
- Solve problems by using ratios to represent real-world situations.

TEACHER RESOURCES

- 11 *Graph Paper* (for the teacher and for each student)
- 74 *Missing Measurements*

ADDITIONAL MATERIALS

- a ruler or straight edge (for the teacher and for each student) (optional)

Engage

- Direct attention to the essential question at the top of Student Edition page 272, "How can I find the height of a tree without measuring the tree?" Direct the students to do a **Think-Pair-Share** to prepare the students to answer the essential question.

  Ideas could include the topic of shadows.
  1. How do the shadows cast by large and small trees differ?
  2. What other objects besides trees cast shadows?
  3. Could you use the height of an object and its shadow to help you find the height of a tree when its shadow is measured? How?

Instruct

Determining proportional ratios

- Use a **visual analysis** to help the students find proportional ratios.

Similar Figures

How can I find the height of a tree without measuring the tree?

Similar figures are geometric figures with the same shape but not necessarily the same size. Figures are similar when the corresponding angle measurements are equal *and* the ratios of corresponding side lengths are proportionate.

Key Terms
- similar figures
- indirect measurement

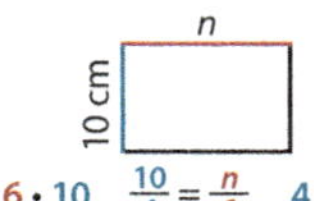

Finding the Unknown Measure

Solve a proportion to find an unknown measure in similar figures.

Ratios between Figures
Write a ratio for the corresponding sides of the two figures.

Ratios within a Figure
Write a ratio for the sides *within* the same figure.

Exercises

Write a proportion to find the unknown measure for the pair of similar figures. *Steps to solve may vary.*

1. 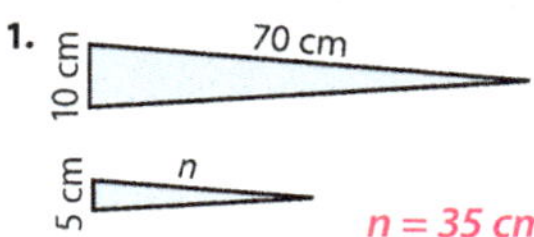 *n = 35 cm*

2. *n = 9 m*

3. 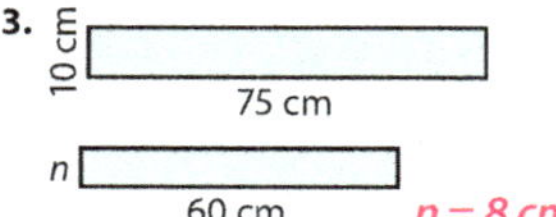 *n = 8 cm*

4. 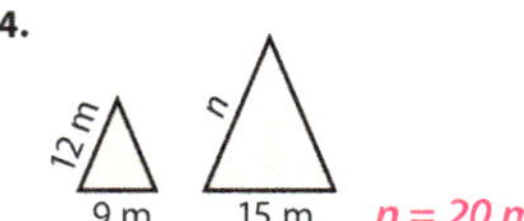 *n = 20 m*

5. 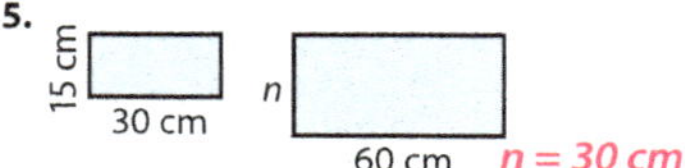 *n = 30 cm*

6. 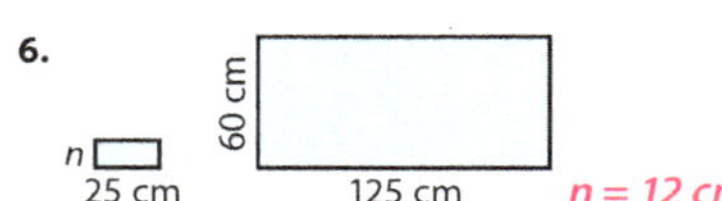 *n = 12 cm*

272   Chapter 13

- Distribute a *Graph Paper* page to each student and display your copy.

  What are similar figures? Similar figures are geometric figures with the same shape but not necessarily the same size.

- Draw on the displayed page a trapezoid with its dimensions as shown. Direct the students to draw a similar trapezoid on their page using the same number of units.

  Compare your trapezoid to a classmate's trapezoid. Are they identical? yes

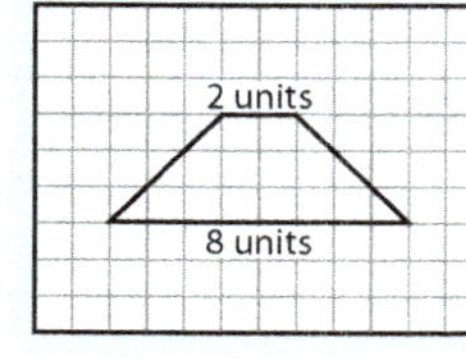

What word is used to describe 2 figures that are identical in shape and size? congruent

- Explain that a copier produces a congruent figure when it exactly copies a figure, and it produces a similar figure when it enlarges or reduces a figure. To produce a similar figure, the copier lengthens or shortens each side of the figure by the same amount without changing the angles of the figure.

- Direct the students to draw a trapezoid that is similar to their first trapezoid, doubling the length of each side, and then to write the dimensions. Demonstrate on the displayed page.

Solve. *Steps to solve may vary.*

7. Sydney enlarged a picture to be 3 times larger than the original. The original picture was 2 in. long by 4 in. wide. The enlarged picture has a length of 6 in. What is its width? $\frac{2}{4} = \frac{6}{n}$; *n = 12 in.*

9. The school photographs given in Ami's package are not similar in size. The larger photographs are 9 in. × 12 in. and 8 in. × 10 in. If the smaller of these photos remained 10 in. long, what would the width need to be so the photos would be similar? $\frac{9}{12} = \frac{n}{10}$; *n = 7.5 in.*

11. If 3 balloons cost 48¢, how much will 20 balloons cost? $\frac{\$0.48}{3 \text{ balloons}} = \frac{n}{20 \text{ balloons}}$; *n = $3.20*

8. A salmon swam 126 mi in 4.5 hr. At this rate, how far could it travel in 8 hr? $\frac{126 \text{ mi}}{4.5 \text{ hr}} = \frac{n}{8 \text{ hr}}$; *n = 224 mi*

10. The office has two sizes of envelopes. One size is 9 cm × 16.5 cm. The other size is 10.5 cm × 24 cm. Are these envelopes similar? *no;* $\frac{9}{16.5} \neq \frac{10.5}{24}$

12. A family wants to build a swimming pool that is similar in size to the standard Olympic-size pool. An Olympic-size pool is 50 m long by 25 m wide. If their pool will be 20 m long, how wide will it be? $\frac{50}{25} = \frac{20}{n}$; *n = 10 m*

Indirect measurement uses similar objects to find the measurement of an object difficult to measure. Solve a proportion to find the unknown measurement.

The sixth-grade class wanted to know the height of the flagpole in the schoolyard. The teacher taught them how to determine the height of the flagpole without measuring it by using the ratio of the length of the flagpole's shadow to the length of a meter stick's shadow. Let h = the height of the flagpole.

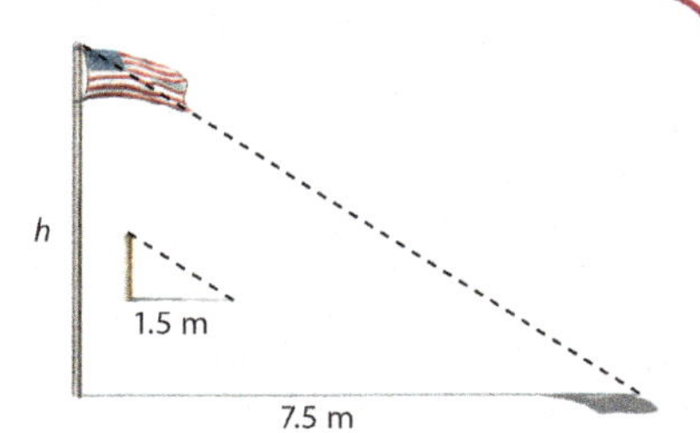

Ratios between Figures

$\dfrac{\text{flagpole height}}{\text{meter height}} = \dfrac{\text{flagpole shadow}}{\text{meter shadow}}$

$1.5 \cdot h \quad \dfrac{h}{1} = \dfrac{7.5}{1.5} \quad 1 \cdot 7.5$

$1.5h = 7.5$

$\dfrac{1.5h}{1.5} = \dfrac{7.5}{1.5}$

$h = 5 \text{ m}$

The height of the flagpole is 5 m.

Ratios within a Figure

$\dfrac{\text{flagpole height}}{\text{flagpole shadow}} = \dfrac{\text{meter height}}{\text{meter shadow}}$

$1.5 \cdot h \quad \dfrac{h}{7.5} = \dfrac{1}{1.5} \quad 7.5 \cdot 1$

$1.5h = 7.5$

$\dfrac{1.5h}{1.5} = \dfrac{7.5}{1.5}$

$h = 5 \text{ m}$

The height of the flagpole is 5 m.

What do you think is true about corresponding side lengths of similar figures? Corresponding side lengths are proportional.

- Explain that the proportion $\frac{2}{4} = \frac{8}{16}$ was written using ratios that compare measures *between* the corresponding sides of similar figures. A proportion can also be made using ratios that compare measures of sides *within* figures.

  What ratio describes the top : bottom relationship in the small trapezoid? 2 : 8

  What ratio describes the top : bottom relationship in the large trapezoid? 4 : 16

  Write "$\frac{2}{8}$ — $\frac{4}{16}$."

  Are these ratios proportional? Yes; sample answers: the second term in each ratio is 4 times greater than the first term; both terms of the second ratio are 2 times greater than the terms in the first ratio; when the terms of the ratios are cross multiplied, the products are the same.

  Write "=" to complete the proportion.

Using proportions to find the unknown measure

- Use an illustration to **demonstrate** the process of finding an unknown measure.

- Display the *Missing Measurements* page. Remind the students that if they know three terms in a proportion, they can find the value of the missing term. Since sides of similar figures are proportional, writing a proportion can help them find an unknown side measure.

- Write "$\frac{\text{length A}}{\text{length B}} = \frac{\text{width A}}{\text{width B}}$" for display.

  Point out that the ratios in the proportion compare corresponding sides *between* Figure A and Figure B.

  What ratios can be written between the corresponding sides of the figures? $\frac{7 \text{ m}}{14 \text{ m}}$ and $\frac{3 \text{ m}}{n}$

  Write "$\frac{7 \text{ m}}{14 \text{ m}} = \frac{3 \text{ m}}{n}$" for display.

  How can you find the value of n? sample answers: 7 is one half of 14, so $7 \times 2 = 14$ and $3 \times 2 = 6$; cross multiply: $7n = 14 \times 3$, $\frac{7n}{7} = \frac{42}{7}$, $n = 6$.

What ratio describes the top : top relationship between the small trapezoid and large trapezoid? 2 : 4

What ratio describes the bottom : bottom relationship? 8 : 16

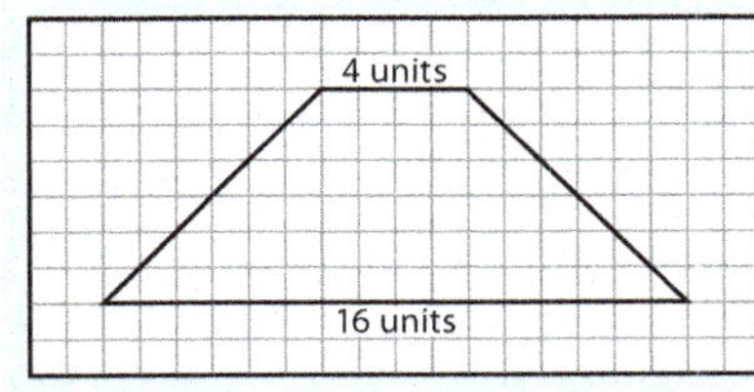

Write "$\frac{2}{4}$ — $\frac{8}{16}$" for display.

Are these ratios proportional? Yes; sample answers: the second term in each ratio is twice the first term; both terms of the second ratio are 4 times greater than the terms in the first ratio; when the terms of the ratios are cross multiplied, the products are the same.

Write "=" to complete the proportion.

Explain that corresponding angles in similar figures and congruent figures are congruent (have the same measure), and corresponding side lengths in congruent figures are congruent (equal).

Choose students to demonstrate solving the proportion and to explain the algorithms they used.

What does $n = 6$ represent? the width of Figure B, 6 m

- Write "$\frac{\text{length A}}{\text{width A}} = \frac{\text{length B}}{\text{width B}}$" for display.

Explain that a proportion can also be written using ratios for measures *within* each figure.

What ratios can be written within each figure? $\frac{7\,\text{m}}{3\,\text{m}}$ and $\frac{14\,\text{m}}{n}$

Write "$\frac{7\,\text{m}}{3\,\text{m}} = \frac{14\,\text{m}}{n}$."

How can you find the value of n using what you know about equivalent fractions? I can multiply $\frac{7}{3}$ by a name for 1 $(\frac{2}{2})$.

Choose a student to demonstrate solving the proportion. $\frac{7}{3} \times \frac{2}{2} = \frac{14}{6}$

What does $n = 6$ represent? the width of Figure B, 6 m

Does it matter which method you use to solve a proportion? no

What might make you choose one method over another method? sample answers: I might choose the method that seems easiest based on the ratios; it is easier to find equivalent fractions or ratios when the terms are related.

Allow students to use the terms *equivalent ratios* and *equivalent fractions* interchangeably.

Write and solve a proportion to find the unknown height. *Steps to solve may vary.*

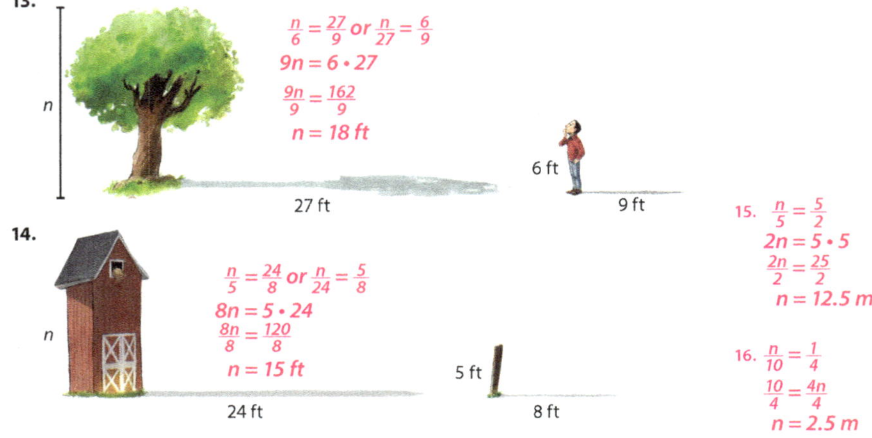

13. $\frac{n}{6} = \frac{27}{9}$ or $\frac{n}{27} = \frac{6}{9}$
$9n = 6 \cdot 27$
$\frac{9n}{9} = \frac{162}{9}$
$n = 18\text{ ft}$

27 ft   6 ft   9 ft

14. $\frac{n}{5} = \frac{24}{8}$ or $\frac{n}{24} = \frac{5}{8}$
$8n = 5 \cdot 24$
$\frac{8n}{8} = \frac{120}{8}$
$n = 15\text{ ft}$

24 ft   5 ft   8 ft

15. $\frac{n}{5} = \frac{5}{2}$
$2n = 5 \cdot 5$
$\frac{2n}{2} = \frac{25}{2}$
$n = 12.5\text{ m}$

16. $\frac{n}{10} = \frac{1}{4}$
$\frac{10}{4} = \frac{4n}{4}$
$n = 2.5\text{ m}$

15. A barn casts a shadow that is 5 m long. A house is 5 m high and casts a shadow that is 2 m long. What is the height of the barn?

16. A meter stick casts a shadow that is 4 m long. A tree casts a shadow that is 10 m long. How tall is the tree?

17. How can I find the height of a tree without measuring the tree? *I can measure the length of the shadow of the tree and the length of the shadow of a meter stick, then solve a proportion to find the height of the tree.*

MEET THE MATHEMATICIAN

The Greek mathematician **Thales**, who lived about 600 yr before Jesus was born, devised the system for finding the height of something that cannot be measured. He discovered the height of the Great Pyramid of Giza by using his height, the length of his shadow, and the length of the pyramid's shadow. Solve a proportion to find the height of the Great Pyramid.
$\frac{4}{10} = \frac{n}{822.5}$; *n = 329 cubits*

274   Chapter 13

- Point out that in the first proportion, $\frac{\text{length A}}{\text{length B}} = \frac{\text{width A}}{\text{width B}}$, the ratios each showed an $\frac{A}{B}$ comparison, and in the second proportion, $\frac{\text{length A}}{\text{width A}} = \frac{\text{length B}}{\text{width B}}$, the ratios each showed a $\frac{\text{length}}{\text{width}}$ comparison.

Write "$\frac{\text{length A}}{\text{width A}} \neq \frac{\text{width B}}{\text{length B}}$" for display.

Why are these ratios not proportional? The ratios are not showing the same comparison of figure to figure or length to width. The terms of the ratios within a proportion must make the same comparison.

Point out that the order of the terms of each ratio within a proportion must be the same to make the same comparison.

- Follow a similar procedure to find the unknown measurements in Figure D and Figure F. Proportions may vary.

Figure D: $\frac{8\,\text{m}}{10\,\text{m}} = \frac{12\,\text{m}}{n}$ and $\frac{8\,\text{m}}{12\,\text{m}} = \frac{10\,\text{m}}{n}$, $n = 15\text{ m}$

Figure F: $\frac{4\,\text{cm}}{10\,\text{cm}} = \frac{5\,\text{cm}}{n}$ and $\frac{5\,\text{cm}}{4\,\text{cm}} = \frac{n}{10\,\text{cm}}$, $n = 12.5\text{ cm}$

Using indirect measurement to find the unknown measure; solving problems using ratios

- Use **problem solving** to help the students find the unknown measure.

- Read the following word problem aloud.

Lee is 5 ft tall. He wondered how his height compared to the height of the tree in his front yard. Lee's father told him that they could find out the height of the tree without actually measuring it.

On a sunny day, Lee and his father went outside and measured the length of the tree's shadow and the length of Lee's shadow. The tree's shadow was 20 ft long, and Lee's shadow was 8 ft long. What is the height of the tree? 12.5 ft

- Explain that indirect measurement is the method of using similar figures and a proportion to find a measurement too difficult to measure directly, such as the height of a building or a tree. Remind the students that sometimes it is helpful to draw a picture of the situation. Draw for display a stick figure diagram of a

LESSON 123

Classify the triangle according to its angles. Write *acute*, *right*, or *obtuse*. Classify the triangle according to the length of its sides. Write *equilateral*, *isosceles*, or *scalene*.

1.  *right; scalene*

2.  *acute; equilateral*

3. 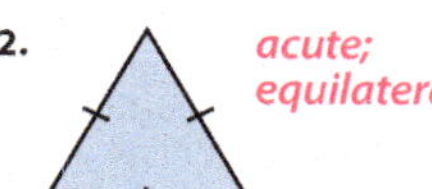 *acute; isosceles*

Find the unknown angle. *Equations may vary.*

4. 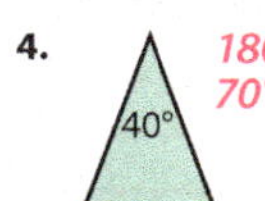 *$180° - (70° + 40°) = 70°$*

5. 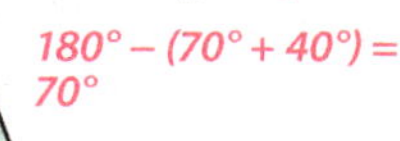 *$360° - (63° + 63° + 117°) = 117°$*

6. 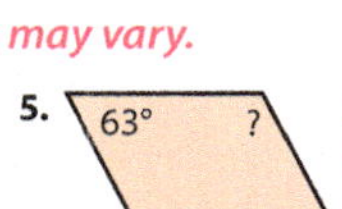 *$360° - (44° + 44° + 136°) = 136°$*

Find the measure of the complementary or supplementary angle. *Equations may vary.*

7. 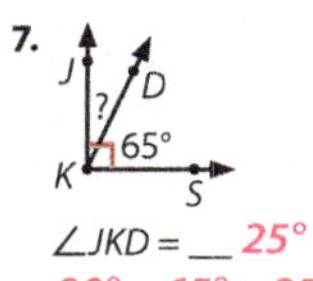 $\angle JKD = \underline{\ \ } \ 25°$
$90° - 65° = 25°$

8. 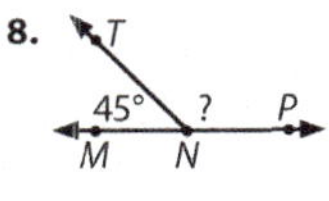 $\angle TNP = \underline{\ \ } \ 135°$
$180° - 45° = 135°$

9. 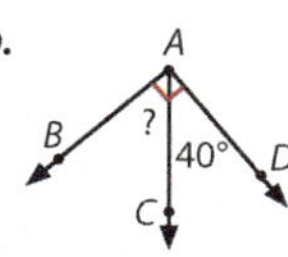 $\angle BAC = \underline{\ \ } \ 50°$
$90° - 40° = 50°$

Use the circle to answer the questions.

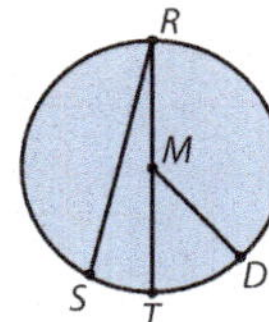

10. Name the circle. *circle M*
11. Name the diameter. *$\overline{RT}$*
12. Name a chord that is not a diameter. *$\overline{RS}$ or $\overline{SR}$*
13. Name a radius. *$\overline{MD}$, $\overline{MT}$, or $\overline{MR}$*

Daily Review 483

- Direct the students to write the proportion and solve it. $\frac{5\text{ ft}}{n} = \frac{8\text{ ft}}{20\text{ ft}}$; $n = 12.5$ ft

- Follow a similar procedure for the following word problems.

  A meter stick casts a shadow that is 3 m long. A house casts a shadow that is 24 m long. How tall is the house? 8 m

  A 6 ft tall man casts a shadow that is 15 ft long. A flagpole casts a shadow that is 35 ft long. How tall is the flagpole? 14 ft

- Guide the students to answer the essential question, "How can I find the height of a tree without measuring the tree?" I can measure the length of the shadow of the tree and the length of the shadow of a meter stick, then solve a proportion to find the height of the tree.

Students frequently make errors because they do not write proportions using corresponding ratios. You may write for display an incorrect proportion and direct the students to identify the error (e.g., $\frac{1}{3} = \frac{24}{n}$, $\frac{\text{object}}{\text{shadow}} = \frac{\text{shadow}}{\text{object}}$).

Apply

Student Edition pages 272–74

- Read and explain the directions for pages 272–74. Assist the students as they complete the pages independently.

Daily Review

- Students should complete Chapter 13, section *d*.

---

boy, a tree, and the shadows of each, similar to the ones pictured on Student Edition page 274.

What information is given? Lee is 5 ft tall, and his shadow is 8 ft long. The tree's shadow is 20 ft long.

Write the measurements.

How do you think you can solve the word problem? I can write ratios to describe the relationship of Lee's height and the length of his shadow and the relationship of the tree's height and the length of its shadow, and then write a proportion to compare the ratios within the figures.

- Write "$\frac{\text{Lee } h}{\text{Lee } s} = \frac{\text{tree } h}{\text{tree } s}$." Direct the students to write a proportion and solve it.

  $\frac{5\text{ ft}}{8\text{ ft}} = \frac{n}{20\text{ ft}}$; $n = 12.5$ ft

- Choose students to demonstrate solving the proportion and to explain the strategy they used. sample answer: cross multiply or find an equivalent ratio (E.g., 20 is $2\frac{1}{2}$ sets of 8, and $2\frac{1}{2}$ sets of 5 equals 12.5.)

  What ratios can you write to compare the measures *between* the figures? the ratios that compare Lee's height to the tree's height and Lee's shadow to the tree's shadow; $\frac{\text{Lee } h}{\text{tree } h} = \frac{\text{Lee } s}{\text{tree } s}$.

LESSON 124

Student Edition pages 275–77
Daily Review Chapter 13, section e

OBJECTIVES

- Find actual measurements using a scale and a scale drawing, map, or model.
- Determine the unknown measure on a scale drawing given the scale and the actual measurement.
- Solve word problems by using ratios.
- Describe how proportions are used in real-world applications.
  BWS

BIBLICAL WORLDVIEW SHAPING

- **Design (Explain):** Proportions are used to design and create a wide variety of useful objects.

ADDITIONAL MATERIALS

- modeling clay
- a map
- a ruler (for the teacher and for each student)
- a calculator (for each student)

ASSESSMENTS

- Chapter 13 Quiz 1

Allow the students to use calculators to solve proportion and percent problems throughout the remainder of this chapter.

Preparation

Form a piece of modeling clay into a ball that has a diameter of approximately 2 in. and a circumference of approximately $6\frac{1}{4}$ in.

Measure the length of a room that is familiar to the students (e.g., the classroom, the lunchroom, the library). Prepare for display a scale drawing of the room using a scale of $\frac{1\text{ in.}}{5\text{ ft}}$. (Do not write the actual room dimensions on the drawing.)

Scale

How are proportions used in life?

A **scale** is a ratio of measurements that compares the size of a drawing, a map, or a model with the size of the actual object. A **scale drawing** has dimensions proportionate to the size of the actual object. Maps and house plans are drawn smaller than the actual object. Illustrations of microscopic objects are drawn larger than the actual object. A **scale model** is a 3-dimensional model that is proportionate to the size of the actual object.

Key Terms
- scale
- scale drawing
- scale model

Finding the Actual Measurement

The distance between two cities on a map is 2.5 cm. Given a map scale of 1 cm : 100 km, what is the actual distance?

$$\frac{\text{map distance (cm)}}{\text{actual distance (km)}} = \frac{1}{100} \quad \frac{2.5}{n}$$

$$n = 250\text{ km}$$

Finding the Drawing or Model Measurement

The length of a car is 156 in. Find the length of a model car using the scale 1 in. : 52 in.

$$\frac{\text{model length}}{\text{actual length}} = \frac{1}{52} \quad \frac{n}{156}$$

$$156 = 52n$$

$$3\text{ in.} = n$$

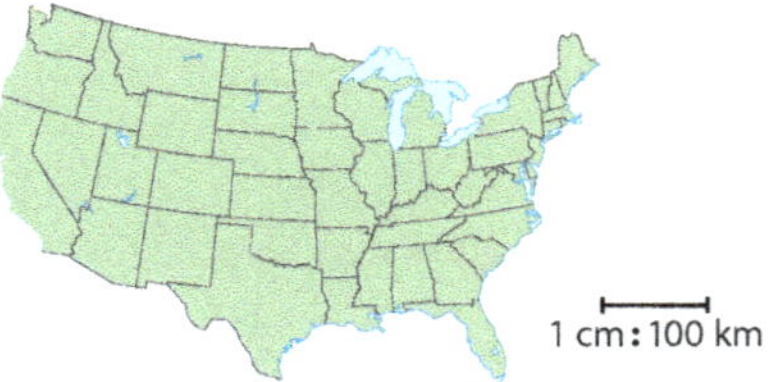

Exercises

Write a proportion using the scale 1 in. : 12 ft to find the answer.

1. If the length of the master bedroom is 1.2 in., what is the actual length? *14.4 ft*

2. If the outside wall from the kitchen to the master bedroom is 3.5 in., what is the actual length? *42 ft*

3. If the outside wall from the kitchen to the living room is 2.3 in., what is the actual length? *27.6 ft*

4. If the bathroom has a length of 0.5 in. and a width of 1 in., what are the actual dimensions? *6 ft long and 12 ft wide*

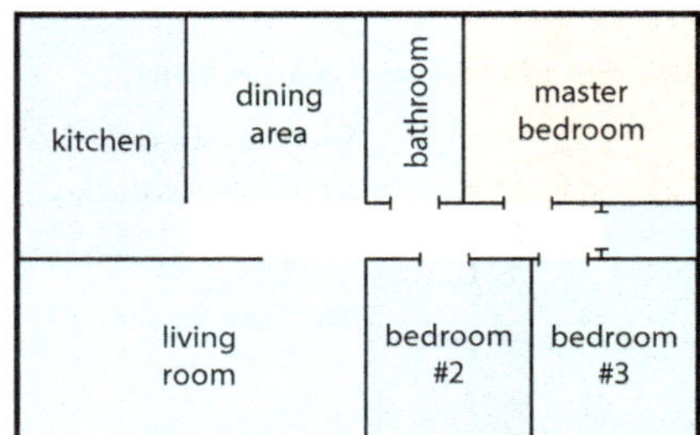

Lesson 124   275

Engage

- Direct attention to the top of Student Edition page 275, and direct the students to do a **Think-Pair-Share** to discuss the essential question, "How are proportions used in life?"

Instruct

Finding actual measurements using a scale & a model

- Use a **scale model** to help the students find a measurement.

- Display the ball of modeling clay and explain that it is a model representing the earth. The model has a diameter of 2 in., but the actual diameter of the earth is about 8,000 mi.

  What ratio compares the size of the model of the earth to the actual size of the earth? 2 in. to 8,000 mi

- Write "2 in. : 8,000 mi" and "$\frac{2\text{ in.}}{8{,}000\text{ mi}}$" for display. Explain that this ratio is a *scale,* a ratio of measurements that compares the size of a model, a drawing, or a map to the size of the actual object. Write "$\frac{\text{model measurement}}{\text{actual measurement}}$" for display.

- Explain that the diameter of the sun is about 865,000 mi.

  If you used the same scale, 2 in. to represent every 8,000 mi, to make a model of the sun, how could you find out how

Measure the *road distance* between cities to the nearest $\frac{1}{2}$ cm. Write a proportion using the scale 1 cm : 32 km to find the approximate distance.

5. Wray to Burlington $\frac{1}{32} = \frac{3}{n}$; $n = 96$ km
6. Deer Trail to Wild Horse $\frac{1}{32} = \frac{4.5}{n}$; $n = 144$ km
7. Kiowa to Burlington $\frac{1}{32} = \frac{7}{n}$; $n = 224$ km
8. Calhan to Stratton $\frac{1}{32} = \frac{5.5}{n}$; $n = 176$ km
9. Rush to Wild Horse $\frac{1}{32} = \frac{3.5}{n}$; $n = 112$ km
10. Deer Trail to Burlington $\frac{1}{32} = \frac{6}{n}$; $n = 192$ km
11. Calhan to Wild Horse $\frac{1}{32} = \frac{5}{n}$; $n = 160$ km

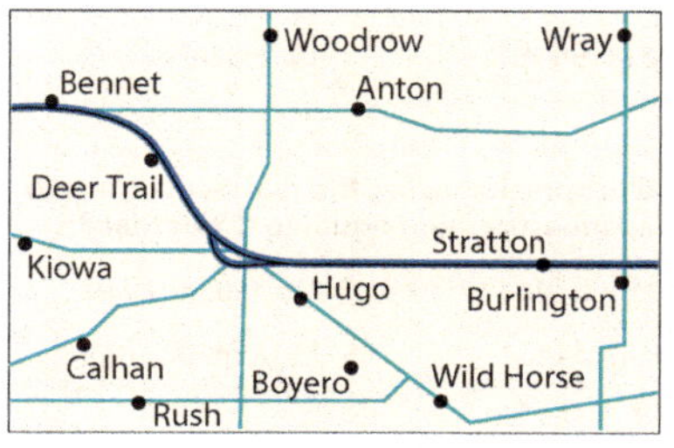

Write a proportion using the scale 1 cm : 35 cm to find the answer.

12. The height of a model Shetland pony is 2.6 cm. What is the actual height of this pony?
$\frac{1}{35} = \frac{2.6}{n}$; $n = 35 \cdot 2.6$; $n = 91$ cm

13. The height of a model Hackney pony is 3.6 cm. What is the actual height of this pony?
$\frac{1}{35} = \frac{3.6}{n}$; $n = 35 \cdot 3.6$; $n = 126$ cm

14. The height of a Palomino is 168 cm. What should the height of the model Palomino be?
$\frac{1}{35} = \frac{n}{168}$; $\frac{168}{35} = \frac{35n}{35}$; $n = 4.8$ cm

Palomino

Shetland pony

Hackney pony

276   Chapter 13

---

large to make your model? sample answers: write and solve a proportion; find an equivalent ratio

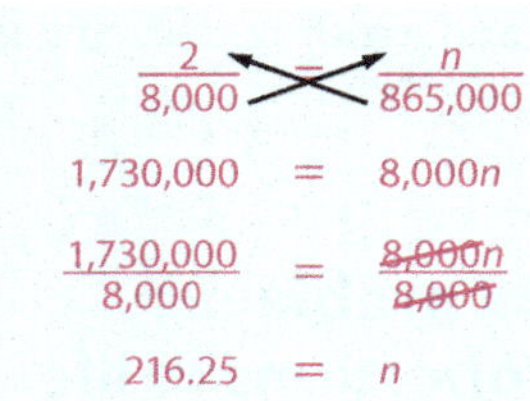

$$\frac{2}{8,000} \diagup \!\!\!\!\diagdown \frac{n}{865,000}$$

$$1,730,000 = 8,000n$$

$$\frac{1,730,000}{8,000} = \frac{8,000n}{8,000}$$

$$216.25 = n$$

- Write "$\frac{2 \text{ in.}}{8,000 \text{ mi}} = \frac{\text{sun model measurement}}{\text{sun actual measurement}}$" for display. Direct the students to write a proportion using ratios that compare the diameter measurement of the model with the actual diameter measurement

and then solve the proportion to find the diameter of the model of the sun. Discuss the answer.

What is the diameter of a model of the sun when a scale of 2 in. for every 8,000 mi is used? 216.25 in. or $216\frac{1}{4}$ in.

- Explain that a sphere with a diameter of $216\frac{1}{4}$ in. equals a diameter of approximately 18 ft. Discuss the size of the classroom compared to the size of the model of the sun.

Do you think the model of the sun would fit in this classroom? Answers will vary.

- Display the model of the earth and point out that the sun is much larger than the

earth; in fact, it would take over 1 million earths to equal the size of the sun.

Finding actual measurements using a scale & a scale drawing

- Use a **scale drawing** to help the students find a measurement.

- Explain that architects and engineers use scale drawings or blueprints when designing buildings.

- Write for display "drawing measurement : actual measurement = 1 in. : 5 ft" and display the prepared scale drawing of the selected room. Explain that 1 in. on the drawing represents 5 ft of the actual room measurement. Choose a student to measure the length of the room on the scale drawing and to write the measurement along the length.

Direct students to find the actual measurement of the room's length by solving a proportion that includes the scale and a ratio that compares the drawing length to the actual length. Answers will vary according to room dimensions.

- Repeat the procedure to find the actual width of the room.

Finding actual measurements using a scale & a map

- Use an **illustration** to help the students find a measurement.

- Display a map and explain that maps are a type of scale drawing. A map is created based on a scale that compares the map to the actual distance. A city map might have a scale of 1 in. to represent 3 mi, but a state map might have a scale of 2 in. to represent 60 mi. A map of a country might have a scale of 3 in. to represent 400 mi.

- Choose a student to locate the scale on the map and write it for display. Choose another student to measure the distance between two locations on the map and write it for display. Direct the students to find the actual distance between the locations by solving a proportion using the map scale. Answers will vary.

- Repeat the procedure using several other pairs of locations on the map.

Determining the unknown measure on a scale drawing

- Guide the students in a **practice exercise** to help them determine an unknown measure.

- Explain that house plans and maps are examples of scale drawings that are smaller than the actual object. In a science book, illustrations of microscopic cells and other objects are examples of scale drawings that are larger than the actual object.

  If you were to make a scale drawing, what information would you need to know? I need to know the actual measurements of the object and the scale for the drawing.

- Explain to the students that they will make scale drawings of their desktops using a scale of 1 in. for every 6 in. of the actual desk. Write "model measurement : actual measurement = 1 in. : 6 in." for display. Direct each student to measure the length and width of his desktop.

  What proportion could you write to find the length for your scale drawing? $\frac{1 \text{ in.}}{6 \text{ in.}} = \frac{n}{\text{actual length}}$

  What proportion could you write to find the width for your scale drawing? $\frac{1 \text{ in.}}{6 \text{ in.}} = \frac{n}{\text{actual width}}$

- Guide the students as they solve the proportions to find the length and width for the drawings. Answers will vary based on desk sizes. Then guide them as they make the scale drawings of their desktops, using their rulers to draw the correct length and width measurements. Point out that the scale tells them that their model drawing will be $\frac{1}{6}$ of the length and width of the actual desktop.

- How would the model drawing of your desktop change if the scale was 1 in. : 10 in.? The scale drawing would be smaller, $\frac{1}{10}$ of the length and width of the actual desktop rather than $\frac{1}{6}$ of the length and width.

- How would the model drawing of your desktop change if the scale was 1 in. : 2 in.? The scale drawing would be larger, $\frac{1}{2}$ of the length and width of the actual desktop rather than $\frac{1}{6}$ of the length and width.

  How would the model drawing of your desktop change if the scale was 2 in. : 1 in.? The length and width measurements of the scale drawing would be twice the length and width measurements of the actual desk; the scale drawing would be larger than the desk.

- Guide the students in determining the length and width of the scale drawing using the scale 2 in. : 1 in. and using the actual length and width measurements of their desks. $\frac{2 \text{ in.}}{1 \text{ in.}} = \frac{n}{\text{actual length}}$; $\frac{2 \text{ in.}}{1 \text{ in.}} = \frac{n}{\text{actual width}}$

Solving word problems; applying proportions in life

- Guide the students to use **problem solving** to make real-world applications.

- Guide the students in using proportions to solve the following word problems.

Write a proportion using the map scale 1 in. : 150 mi to find the actual distance represented by the measurement.

15. 4 in. *600 mi*
16. 9 in. *1,350 mi*
17. 3.6 in. *540 mi*
18. 0.6 in. *90 mi*
19. 5.8 in. *870 mi*
20. 13 in. *1,950 mi*

Write a proportion using the map scale 1 in. : 16 mi to find the map measurement equal to the actual distance.

21. 80 mi *5 in.*
22. 160 mi *10 in.*
23. 120 mi *7.5 in.*
24. 48 mi *3 in.*
25. 8 mi *0.5 in.*
26. 67.2 mi *4.2 in.*

Practice & Application

27. A playground is 72 ft wide. How wide would a scale drawing of the playground be in which 3 in. represents 12 ft?

28. A scale drawing of a plant cell is 2.4 cm in length. What is the actual length of the cell if the drawing uses a scale of 8 cm : 1 mm?

29. A model car is 12 in. long. How long is the actual car if the model was made with a scale of 1 in. : 16 in.? $\frac{1}{16} = \frac{12}{n}$; $n = 16 \cdot 12$; $n = 192$ in.

30. A dollhouse was built with a scale of 1 in. : 12 in. to model a real house. The dollhouse is 48 in. long. What is the actual length of the house that the dollhouse represents? $\frac{1}{12} = \frac{48}{n}$; $n = 12 \cdot 48$; $n = 576$ in.

31. The length of a highway is 225 mi long. How long would the highway be on a map with a scale of 2 in. : 75 mi?

32. A model train was made with a scale of 1 in. : 96 in. The model railroad car is 5.5 in. long. What is the actual length of the railroad car? $\frac{1}{96} = \frac{5.5}{n}$; $n = 96 \cdot 5.5$; $n = 528$ in.

33. A square microchip has an actual measure of 0.8 mm per side. Make two scale drawings of the microchip, using the given scales.

    scale = 1 cm : 0.1 mm *8 cm each side*
    scale = 5 mm : 0.2 mm *20 mm each side*

34. How are proportions used in life? *Proportions are used to design and create a wide variety of useful objects.*

27. $\frac{3}{12} = \frac{n}{72}$
    $72 \cdot 3 = 12n$
    $\frac{216}{12} = \frac{12n}{12}$
    $n = 18$ in.

28. $\frac{8}{1} = \frac{2.4}{n}$
    $\frac{8n}{8} = \frac{2.4}{8}$
    $n = 0.3$ mm

31. $\frac{2}{75} = \frac{n}{225}$
    $225 \cdot 2 = 75n$
    $\frac{450}{75} = \frac{75n}{75}$
    $n = 6$ in.

33. $\frac{1}{0.1} = \frac{n}{0.8}$
    $\frac{0.8}{0.1} = \frac{0.1n}{0.1}$
    $n = 8$ cm

    $\frac{5}{0.2} = \frac{n}{0.8}$
    $0.8 \cdot 5 = 0.2n$
    $\frac{4}{0.2} = \frac{0.2n}{0.2}$
    $n = 20$ mm

Rename the fraction, using a power of 10 as the denominator. Write the fraction as a decimal.

1. $\frac{3}{5}$ $\frac{6}{10}$; *0.6*
2. $\frac{1}{4}$ $\frac{25}{100}$; *0.25*
3. $\frac{1}{2}$ $\frac{5}{10}$; *0.5*
4. $\frac{12}{25}$ $\frac{48}{100}$; *0.48*

Solve. Use a bar to mark the repeating digits.

5. $3\overline{)2}$ *0.$\overline{6}$*
6. $6\overline{)139.50}$ *23.25*
7. $6\overline{)550}$ *91.$\overline{6}$*
8. $0.12\overline{)1.58}$ *13.1$\overline{6}$*

Solve. *Equations may vary.*

9. Karen's family vacationed at the beach. The first two days the motel charged them $89.95 each night. The rates went up to $107.55 on Friday and Saturday nights. How much did her family spend on the motel for four nights?
$(2 \times \$89.95) + (2 \times \$107.55) = \$395.00$

10. On Friday, Karen's family went to a fish fry on the beach. Her dad and mom bought 2 adult plates for $7.95 each and 3 child plates for $4.95 each. How much did her family spend on that meal?
$(2 \times \$7.95) + (3 \times \$4.95) = \$30.75$

484   Daily Review

LESSON 124

Apply

Student Edition pages 275–77

- Read and explain the directions for pages 275–77. Assist the students as they complete the pages independently.

Daily Review

- Students should complete Chapter 13, section *e.*

Assess

Quiz 1

- Use the **summative assessment** to evaluate the students' progress at this point in the chapter.

The length of a car is 258 in. How long would a model of this car be if it were built using a scale of 1 in. : 43 in.? $\frac{1 \text{ in.}}{43 \text{ in.}} = \frac{n}{258 \text{ in.}}$; $n = 6$ in.

The floor plan of a house has a scale of 1 in. : 6 ft. If the actual living room is 24 ft long and 18 ft wide, what are the length and width dimensions on the floor plan? $\frac{1 \text{ in.}}{6 \text{ ft}} = \frac{l}{24 \text{ ft}}$, $l = 4$ in.; $\frac{1 \text{ in.}}{6 \text{ ft}} = \frac{w}{18 \text{ ft}}$, $w = 3$ in.

The distance between 2 cities is 75 mi. If the map scale is 1 in. : 15 mi, what is the map measurement? $\frac{1 \text{ in.}}{15 \text{ mi}} = \frac{n}{75 \text{ mi}}$; $n = 5$ in.

A plant cell is 0.5 mm in length. How long is the drawing of the cell if the scale is 5 cm : 1 mm? $\frac{5 \text{ cm}}{1 \text{ mm}} = \frac{l}{0.5 \text{ mm}}$; $l = 2.5$ cm

- Remind the students that proportions are used to design many practical objects.
Can you summarize some of the real-world uses for proportions that we have discussed in this lesson? sample answer: They are used in designing houses, cars, buildings, and city or state maps.

- Lead the students to answer the essential question, "How are proportions used in life?" Proportions are used to design and create a wide variety of useful objects.

LESSON 125

Student Edition pages 278–79
Daily Review Chapter 13, section ƒ

OBJECTIVES

- Express percents as ratios, decimals, and fractions in lowest terms.
- Express decimals and fractions as percents.
- Compare percents to decimals and fractions.
- Solve percent word problems.

TEACHER RESOURCES
- 11 *Graph Paper* (extra copies as needed)
- 75 *Percent*

ADDITIONAL MATERIALS
- a calculator (for each student)

Engage

- Direct attention to the top of Student Edition page 278. Assign the students a **bell ringer** to initially answer the essential question, "How can I find my grade on a test?"

Instruct

Expressing percents as ratios, decimals & fractions

- Use **illustrations** to help the students write percents in equivalent forms.

- Write "percent" and a percent sign (%) for display. Explain that percent is a part-to-whole ratio in which a part is compared to a whole that is made up of 100 equal parts. The term *percent* and the percent sign (%) mean "out of 100" or "per 100."

- Draw a large square (10 × 10) for display on the *Graph Paper* page and explain that it represents 1 whole or 100 hundredths. Shade the entire square.

  What fraction tells the part of this whole that is shaded? $\frac{100}{100}$

Percent

How can I find my grade on a test?

Percent is a ratio in which a quantity is compared to 100. The symbol for percent is **%**. Percent means "out of 100," "per 100," or "÷100."

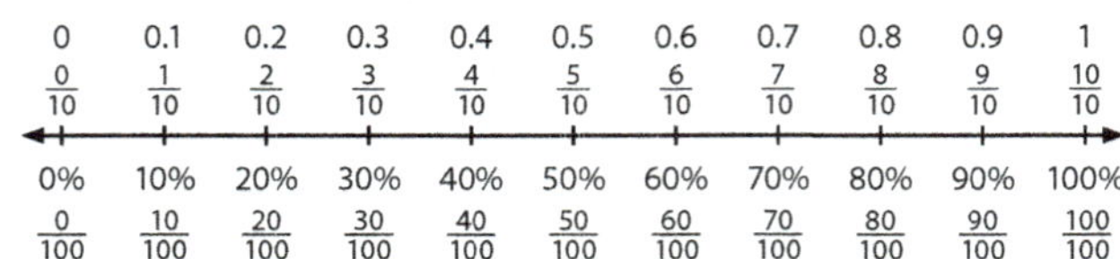

1 Whole	32 Out of 100	50 Divided by 100
$100\% = \frac{100}{100} = 1.0$	$\frac{32}{100}$ 0.32 32%	$50\% = \frac{50}{100} = 50 \div 100 = 0.5$
	fraction decimal percent	50

This number line shows equivalent fractions, decimals, and percents.

0	0.1	0.2	0.3	0.4	0.5	0.6	0.7	0.8	0.9	1
$\frac{0}{10}$	$\frac{1}{10}$	$\frac{2}{10}$	$\frac{3}{10}$	$\frac{4}{10}$	$\frac{5}{10}$	$\frac{6}{10}$	$\frac{7}{10}$	$\frac{8}{10}$	$\frac{9}{10}$	$\frac{10}{10}$
0%	10%	20%	30%	40%	50%	60%	70%	80%	90%	100%
$\frac{0}{100}$	$\frac{10}{100}$	$\frac{20}{100}$	$\frac{30}{100}$	$\frac{40}{100}$	$\frac{50}{100}$	$\frac{60}{100}$	$\frac{70}{100}$	$\frac{80}{100}$	$\frac{90}{100}$	$\frac{100}{100}$

Exercises

Draw a number line. Divide the number line into 10 equal parts. Graph a point on the number line to represent the percent.

1. 40%   **2.** 70%   **3.** 85%

Write the decimal for the shaded part of the grid. Write the decimal as a percent.

4. 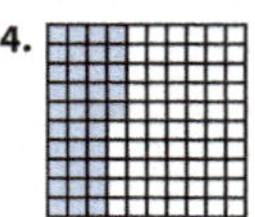 *0.35 = 35%*   **5.** 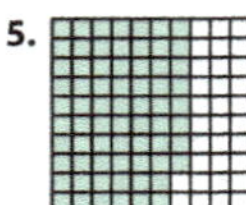 *0.68 = 68%*   **6.** 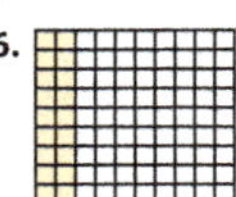 *0.2 = 20%*

Write the fraction for the shaded part of the circle. Write the fraction as a percent.

7.  *$\frac{1}{10}$ = 10%*   **8.** 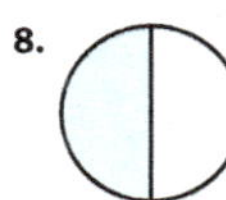 *$\frac{1}{2}$ = 50%*   **9.**  *$\frac{1}{4}$ = 25%*

Estimate what percent of the rectangle is shaded.

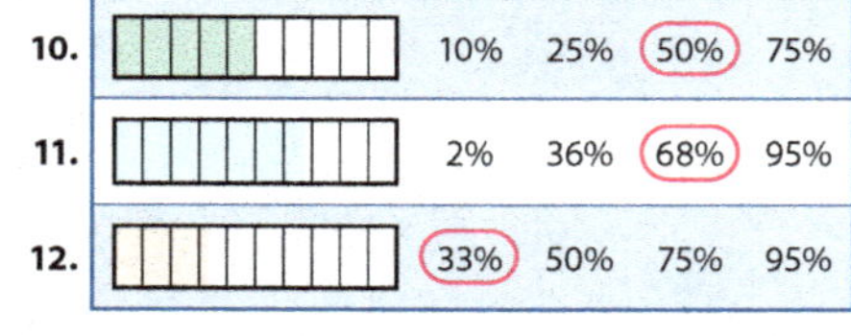

10. 10%   25%   (50%)   75%

11. 2%   36%   (68%)   95%

12. (33%)   50%   75%   95%

Draw a rectangle. Shade the rectangle to represent an approximation of the percent.

13. 25%

14. 48%

15. 90%

How can you write the value of 1 whole or $\frac{100}{100}$ in ratio form? 1 : 1 or 100 : 100

How can you write the value of 1 whole in decimal form? 1 or 1.00

Write "$1 = \frac{100}{100} = 100 : 100 = 1.00$" for display.

Since percent means "out of 100" or "per 100," how can you write the value in percent form? 100%

Write "= 100%" after the other equivalencies. Point out that anything greater than 100% indicates more than one whole.

Is it possible to model 110% of the displayed square? Yes; 110% = 1 whole square and $\frac{1}{10}$ of a second square.

Invite a student to extend the figure to illustrate 110%.

What would 200% of the square look like? 2 whole squares

Invite a student to draw 2 large squares (10 × 10) for display.

Draw another large square partitioned into 10 rows and 10 columns and explain that it represents 100 hundredths. Shade 10 hundredths (e.g., the first column).

How many of the 100 "squares" are shaded? 10

Write "10 per 100" for display.

How can you write a value of 10 per 100 in ratio form? 10 : 100

How can you write a value of 10 per 100 in percent form? 10%

Changing a Decimal to a Percent

Use mental math: move the decimal point two places to the right when multiplying by 100.

$$0.71 \times 100 = 71\%$$
$$0.2 \times 100 = 20\%$$

Changing a Percent to a Decimal

Use mental math: move the decimal point two places to the left when dividing by 100.

$$60\% = \frac{60}{100} = 60 \div 100 = 0.6$$
$$4\% = \frac{4}{100} = 4 \div 100 = 0.04$$

Changing a Fraction to a Percent

1. Divide. Round the answer to the nearest hundredth.
2. Change the decimal to a percent.

$$\frac{3}{5} = 3 \div 5 = 0.6 = 60\%$$

Changing a Percent to a Fraction

1. Rename the percent as a fraction with 100 as the denominator.
2. Simplify the fraction by renaming to lowest terms.

$$40\% = \frac{40}{100} = \frac{2}{5}$$

Write the percent in decimal form. Write the decimal in percent form.

16. 52% *0.52* **17.** 89% *0.89*

18. 2% *0.02* **19.** 0.07 *7%*

20. 0.54 *54%* **21.** 23% *0.23*

Write the percent as a fraction in lowest terms. Write the fraction as a percent. Round to the nearest percent if needed.

22. 30% $\frac{30}{100} = \frac{3}{10}$

23. 80% $\frac{80}{100} = \frac{4}{5}$

24. 15% $\frac{15}{100} = \frac{3}{20}$

25. $\frac{2}{5}$ *2 ÷ 5 = 0.4 = 40%*

26. $\frac{3}{4}$ *3 ÷ 4 = 0.75 = 75%*

27. $\frac{3}{8}$ *3 ÷ 8 = 0.375 ≈ 38%*

Write a comparison sentence by using >, <, or =.

28. $\frac{5}{6}$ *≥* 60% **29.** 0.73 *=* 73% *Steps to solve may vary.*

30. $\frac{3}{10}$ *≥* 3% **31.** 0.4 *≥* 4%

32. 0.48 *≥* 5% **33.** $\frac{2}{3}$ *≤* 85%

34. 0.7 *≤* 75% **35.** 0.8 *=* 80%

Use the circle graph to find the answer.

36. What fraction of those surveyed chose cats as their favorite pet? $\frac{1}{4}$

37. What fraction chose an animal other than cats as their favorite pet? $\frac{3}{4}$

38. What percent chose dogs as their favorite pet? *65%*

39. What percent chose a pet other than dogs? *35%*

40. What percent represents the whole graph? *100%*

Favorite Pet

cats 25%

dogs 65%

birds 5%

ferrets 2%

other 3%

Solve.

41. The art teacher took 50 students on a field trip to the museum. Thirty of the students bought a post card while there. What percent of the students bought a post card? $\frac{30}{50} = \frac{60}{100} = 60\%$

42. Dillon spelled 24 words correctly out of 25 on a spelling test. What percent of the words did he spell correctly? $\frac{24}{25} = \frac{96}{100} = 96\%$

43. How can I find my grade on a test?
I can divide the number of correct answers by the number of total answers to find the percent of correct answers.

How can you write a value of 10 per 100 in decimal form? 0.10

How can you write a value of 10 per 100 in fraction form? $\frac{10}{100}$

Write the equivalencies beside 10 per 100: $= 10:100 = 10\% = 0.10 = \frac{10}{100}$.

What is $\frac{10}{100}$ written in lowest terms? $\frac{1}{10}$

Guide the students to the conclusion that the other forms can also be written in lowest terms: $10:100 = 1:10$, $0.10 = 0.1$.

- Repeat the procedure with or without the drawings. 40 per 100 = 40:100 = 40% = 0.40 = $\frac{40}{100}$; lowest terms: $\frac{40}{100} = \frac{2}{5}$; 40:100 = 2:5; 0.40 = 0.4

Point out that although the decimal 0.4 has the same value as $\frac{2}{5}$ (2 ÷ 5 = 0.4), a decimal cannot be written as fifths because decimals are a base 10 system.

- Follow a similar procedure for the following values.

25 per 100 = 25:100; 25%; 0.25; $\frac{25}{100}$
lowest terms: $\frac{25}{100} = \frac{1}{4}$; 25:100 = 1:4
4 per 100 = 4:100 = 4% = 0.04 = $\frac{4}{100}$
lowest terms: $\frac{4}{100} = \frac{1}{25}$; 4:100 = 1:25

- Display the *Percent* page and direct attention to the number line.

Which fractional parts does this number line show? The number line shows tenths and hundredths; the longest marks partition the length of the 1 whole into 10 equal parts, and the shortest lines partition it into 100 equal parts.

How can you write the value of point *A* as a fraction? $\frac{1}{10}$ or $\frac{10}{100}$

How can you write the value of point *A* as a decimal? 0.1 or 0.10

How can you write the value of point *A* as a percent? 10%; Accept all reasonable explanations.

- Follow a similar procedure for the other points. *B* $\frac{25}{100}$; 0.25; 25% *C* $\frac{4}{10}$ or $\frac{40}{100}$; 0.4 or 0.40; 40% *D* $\frac{58}{100}$; 0.58; 58% *E* $\frac{8}{10}$ or $\frac{80}{100}$; 0.8 or 0.80; 80%

What percent does the 0 represent? 0%
What percent does the 1 represent? 100%

- Write for display "*A* = 40%, *B* = 90%, *C* = 35%." Instruct the students to draw a number line, partition it into tenths, and graph the three points.

- Direct attention to the first rectangular model on the displayed page.

What percent of the whole rectangle does each part represent? 10%; Since there are 10 parts in the rectangle, each part is $\frac{1}{10}$ or 10% of the rectangle.

How many parts of the rectangle are shaded? $2\frac{1}{2}$

What percent of the whole rectangle do the $2\frac{1}{2}$ shaded parts represent? 25%; Each of the 2 completely shaded parts is 10% of the rectangle, and the half-shaded part is 5%; 10% + 10% + 5% = 25%.

Since the shading is halfway to 50%, what percent is half of 50%? 25%

- Choose students to shade the last rectangle to show 26%, 49%, 50%, 51%, and 98%.

Expressing decimals & fractions as percents

- Use **models** of decimal and fraction forms to help the students write them as percents.

LESSON 125

- Draw a circle for display and shade one-half of it.

  What fraction represents the shaded part of the circle? $\frac{1}{2}$

  Write "$\frac{1}{2}$" for display.

  How can you write the fraction $\frac{1}{2}$ as a percent? 50%; Since percent is a "per 100" ratio, I can rename $\frac{1}{2}$ as an equivalent fraction with 100 as its second term ($\frac{50}{100}$).

  Point out that the terms of the fraction can also be divided to find the decimal equivalent, which can be renamed as a percent: $\frac{1}{2} = 1 \div 2 = 0.5 = 0.50 = 50\%$. Point out that 5 tenths must be renamed as 50 hundredths to determine the percent. Choose a student to write "= 50%" after the $\frac{1}{2}$.

- Follow a similar procedure for $\frac{3}{4}$ of a circle: $\frac{3}{4} \times \frac{25}{25} = \frac{75}{100} = 75\%$ or $\frac{3}{4} = 3 \div 4 = 0.75 = 75\%$.

- Follow a similar procedure to guide the students in finding the percent for $\frac{1}{3}$ and $\frac{2}{3}$ of a circle by dividing the terms of each fraction. $1 \div 3 = 0.\overline{3} \approx 33\%$; $2 \div 3 = 0.\overline{6} \approx 67\%$

  Point out that when the second term of a ratio is related to 100, as with halves and fourths, it is easy to rename the ratio as an equivalent fraction. However, whether the terms are related or not, the first term of a ratio can always be divided by its second term to find a decimal (e.g., $\frac{1}{3} = 1 \div 3 = 0.\overline{3} \approx 33\%$). If the terms do not divide equally, the ratio and the percent are approximately equal.

Comparing percents to decimals & fractions

- Guide the students in a **practice exercise** to compare fractions and percents.

- Write "0.6," "0.14," and "0.08" for display. Direct the students to write each decimal as a percent. Choose students to share their answers and explain their reasoning. 60%, 14%, 8%

  What do you notice about the decimal point when a decimal is renamed as a percent? The decimal point moves 2 places to the right.

Explain that a percent can be found by multiplying a decimal by 100 to move the decimal point two places to the right and then writing a percent sign.

Choose students to demonstrate multiplying 0.6, 0.14, and 0.08 by 100 to find the percents. $0.6 \times 100 = 60\%$, $0.14 \times 100 = 14\%$, $0.08 \times 100 = 8\%$

- Follow a similar procedure to guide the students in writing 43%, 30%, and 2% as decimals. 0.43, 0.3, 0.02; The decimal point moves 2 places to the left. Conclude with the students that a decimal value can be found by dividing a percent by 100 to move the decimal point two places to the left. $43\% = \frac{43}{100} = 43 \div 100 = 0.43$; $30\% = \frac{30}{100} = 30 \div 100 = 0.3$; $2\% = \frac{2}{100} = 2 \div 100 = 0.02$

- Write "$\frac{3}{5}$," "$\frac{7}{10}$," and "$\frac{2}{7}$" for display. Direct the students to write each fraction as a decimal and as a percent. Choose students to write their answers for display and to explain their reasoning. $\frac{3}{5} = 0.6 = 60\%$; $\frac{7}{10} = 0.7 = 70\%$; $\frac{2}{7} \approx 0.29 \approx 29\%$

- Write for display "$\frac{7}{10}$ — 7%." Instruct each student to decide whether >, <, or = completes the statement. Choose a student to complete the comparison and explain his answer. $\frac{7}{10} > 7\%$

Complete the table, using the given values to evaluate the expression.

1.

b	$5b + 8$
6	38
12	68
29	153
45	233

2.

x	$\frac{x}{4} - 2$
8	0
24	4
48	10
64	14

3.

n	$6 + n^2$
4	22
12	150
16	262
20	406

Evaluate the expression. Let $m = 4$.

4. $5m - 7$ *13*

5. $(6 + 2m) - 3$ *11*

6. $3 + m + 6$ *13*

7. $\frac{m}{2} + 7$ *9*

8. $m + 7 - 6$ *5*

9. $105 - 12m$ *57*

Simplify the expression.

10. $4(8x)$ *32x*

11. $9 + (6 + 2x)$ *15 + 2x*

12. $8x + (2 + 4x)$ *12x + 2*

13. $6(n + 2)$ *6n + 12*

14. $5(4x + 3.1)$ *20x + 15.5*

15. $9b + 3b + 12b$ *24b*

Daily Review 485

- Follow a similar procedure for the following comparisons. Review finding equivalent fractions, decimals, and percents as needed.

$\frac{2}{5} < 50\%$ $0.75 = \frac{6}{8}$ $\frac{1}{3} > 25\%$

Solving percent word problems

- Use **collaboration** to help the students solve percent word problems.

- Pair the students and guide them in writing proportions to solve word problems. Explain that to find a percent using a proportion, the second term in the ratio with the unknown value must be 100 because the second term in the known ratio is equivalent to 100%. (E.g., the 15 points scored by Carmen's team is equal to 100%.)

Some students may realize that the word problems can also be solved by writing a ratio to compare the information given in the word problem and dividing the terms. This will rename the ratio as a decimal that can be renamed as a percent.

- Read the following word problems aloud.

Carmen scored 3 of her team's 15 points during a volleyball game. What percent of her team's points did she score? $\frac{3}{15} = \frac{n}{100}$; 20%

Samuel paid the sale price of $35 for a $50 game. What percent of the original cost did Samuel pay? $\frac{35}{50} = \frac{n}{100}$; 70%

Dad left a tip of $5 for the server after purchasing a meal for $25. What percent of the cost of the meal did he leave for a tip? $\frac{5}{25} = \frac{n}{100}$; 20%

On a math test, Tony answered 40 out of 45 questions correctly. What percent of the test questions did he answer correctly? $\frac{40}{45} = \frac{n}{100}$; approximately 89%

- Guide the students as they answer the essential question, "How can I find my grade on a test?" I can divide the number of correct answers by the number of total answers to find the percent of correct answers.

Apply

Student Edition pages 278–79
- Read and explain the directions for pages 278–79. Assist the students as they complete the pages independently.

Daily Review
- Students should complete Chapter 13, section *f*.

NOTES

Student Edition pages 280–81
Daily Review Chapter 13, section *g*

OBJECTIVES

- Find the percent of a number by using an equation, a model, and a proportion.
- Solve percent word problems.

TEACHER RESOURCES

- 76 *Percent Models: Finding the Part*

ADDITIONAL MATERIALS

- a calculator (for each student)

Engage

- Direct attention to the top of Student Edition page 280 and direct the students to **turn and talk** to initiate discussion of the essential question, "How can I find the amount of tip I want to give a server?"

Instruct

Finding the percent of a number by using an equation

- Guide a **discussion** to help the students use an equation to find the percent of a number.

- Read the following word problem aloud.

If 50% of the 20 students in Miss Gardner's class are boys, how many students are boys? 10 boys

What does "percent" mean? out of 100 or per 100

What do you think 50% of 20 is? 10; sample answers: 50 is half of 100, and 10 is half of 20; since 50% of a number is equal to $\frac{1}{2}$ of that number, 50% of 20 = 10.

Point out that their reasoning indicated the statement "50 is to 100 as __ is to 20." Guide the students to conclude that the word *of* in the word problem means "multiply" and the word *are* means "equals." Write "50% × 20 = __" for display.

How can you write 50% as a numerical value without the percent sign? $\frac{50}{100}$ or 0.50

- Write "n% of a number = $\frac{n}{100}$ × the number" and write "$\frac{50}{100}$ × 20 = __" below the formula. Explain that students can find the part that is the percent of a number by writing the percent as a fraction and multiplying it by the whole amount.

- Choose a student to solve the equation. Point out that he can use cancellation or write the percent as a fraction in lowest terms before multiplying. $\frac{50}{100}$ × 20 = $\frac{50}{5}$ = 10 or $\frac{1}{2}$ × 20 = 10
How many students are boys? 10

- Follow a similar procedure to find the percent of these numbers.
30% of 40 $\frac{30}{100}$ × 40 = 12
70% of 50 $\frac{70}{100}$ × 50 = 35
25% of 60 $\frac{25}{100}$ × 60 = 15 or $\frac{1}{4}$ × 60 = 15

- Read the following word problem aloud.

Piper has saved 36% of the money she needs for a new computer that costs $500. How much money has she saved for the computer? $180

How much money is Piper trying to save? $500

What part of the money has Piper saved? 36%

Finding the Percent of a Number

How can I find the amount of tip I want to give a server?

Percents are used to find the discount of a sale item, the sales tax on a purchase, or an appropriate tip for the cost of a meal at a restaurant. 100% represents all of a given number. Less than 100% of a number represents part of that number. Use the formula to find the percent of a number.

$$n\text{\% of a number} = \frac{n}{100} \times \text{the number}$$

Rename the percent as a fraction or a decimal to solve an equation.

Finding the Percent of a Number

Elise wanted a sweater that cost $50. She saved $27 toward the purchase and waited for a sale. The first sale was 25% off the original price. Does Elise have enough money to buy the sweater with the discount of 25%? What is 25% of $50?

$\frac{25}{100} \times 50 =$ __

$\frac{25}{2\,100} \times \frac{50^{1}}{1} = \frac{25}{2} = 12\frac{1}{2}$ discount = $12.50

sale price: $50.00 − $12.50 = $37.50

Elise does *not* have enough money to buy the sweater with a discount of 25%.

Near the end of the season, the sale increased to 60% off the original price. Does Elise have enough money to buy the sweater with the discount of 60%? What is 60% of $50?

0.60 × 50 = __
0.60 × 50 = 30.00   discount = $30.00

sale price: $50.00 − $30.00 = $20.00

Elise has enough money to buy the sweater with a discount of 60%.

Making a Model

Garrett wanted basketball shoes that cost $125. The first sale price was 30% off. Then the shoes went on sale for 60% off. What was the discount of the shoes during each sale?

0   30% $125

10%	10%	10%	10%	10%	10%	10%	10%	10%	10%

Each section represents 10% of 125.
10% of 125 = $12.50

30%: 3 × $12.50 = $37.50
30% discount = $37.50

60%: 6 × $12.50 = $75.00
60% discount = $75.00

Exercises

Find the percent of the number. Solve by renaming the percent as a fraction. *Cancellation steps may vary.*

1. 50% of 78 *39*   2. 30% of 80 *24*   3. 40% of 200 *80*   4. 25% of 48 *12*   5. 60% of 25 *15*
6. 75% of 52 *39*   7. 20% of 85 *17*   8. 33% of 100 *33*   9. 10% of 250 *25*   10. 70% of 15 *10$\frac{1}{2}$*

Find the percent of the number. Rename the percent as a decimal if needed.

11. 15% of 80 *12*   12. 35% of 120 *42*   13. 24% of 400 *96*   14. 5% of 64 *3.2*   15. 100% of 25 *25*
16. 18% of 65 *11.7*   17. 52% of 65 *33.8*   18. 39% of 200 *78*   19. 99% of 50 *49.5*   20. 45% of 20 *9*

Make a model to find the percent of the number.

21.	10% of 70	20% of 70	80% of 70	*7; 14; 56*
22.	10% of 50	40% of 50	90% of 50	*5; 20; 45*
23.	10% of 85	30% of 85	70% of 85	*8.5; 25.5; 59.5*

280   Chapter 13

Problems involving percents can be solved by setting up a proportion. An unknown in a proportion can be found by cross multiplying or by finding the equivalent ratio.

$$\frac{\text{part}}{100} = \frac{\text{part}}{\text{whole}}$$

Using Proportions to Find the Percent

A survey showed that 6 of 23 sixth-grade students preferred Gooey Cluster candy bars over Nutty Crunch. What percent of students preferred Gooey Cluster?

$$\frac{n}{100} = \frac{6}{23} \text{ (part liking Gooey Cluster)} \text{ (whole class)}$$

Cross Multiplying

1. Write the proportion.
2. Multiply to find the cross products.
3. Solve the equation.

$$\frac{n}{100} \times \frac{6}{23}$$

$$23 \cdot n = 100 \cdot 6$$
$$23n = 600$$
$$\frac{23n}{23} = \frac{600}{23}$$
$$n = 26.09$$

About 26% of the sixth-grade students preferred Gooey Cluster bars.

The survey was given to 25 fifth-graders. Eight students preferred Gooey Cluster. What percent of the class preferred Gooey Cluster?

$$\frac{n}{100} = \frac{8}{25} \text{ (part liking Gooey Cluster)} \text{ (whole class)}$$

Finding Equivalent Ratios

1. Write the proportion.
2. Multiply or divide by a form of 1.
3. Solve the equation.

$$\frac{n}{100} \overset{\div 4}{\underset{\div 4}{=}} \frac{8}{25}$$

$$n \div 4 = 8$$
$$\frac{n}{4} \cdot 4 = 8 \cdot 4$$
$$n = 32$$

32% of the fifth-grade class preferred Gooey Cluster bars.

Write a proportion to find the percent of the number. *Method used to solve may vary.*

24. 55% of 80 $\frac{55}{100} = \frac{n}{80}$; $n = 44$ **25.** 48% of 20 $\frac{48}{100} = \frac{n}{20}$; $n = 9.6$ **26.** 16% of 45 $\frac{16}{100} = \frac{n}{45}$; $n = 7.2$

27. 60% of 95 $\frac{60}{100} = \frac{n}{95}$; $n = 57$ **28.** 2% of 100 $\frac{2}{100} = \frac{n}{100}$; $n = 2$ **29.** 25% of 25 $\frac{25}{100} = \frac{n}{25}$; $n = 6.25$

30. 80% of 60 $\frac{80}{100} = \frac{n}{60}$; $n = 48$ **31.** 35% of 50 $\frac{35}{100} = \frac{n}{50}$; $n = 17.5$ **32.** 10% of 300 $\frac{10}{100} = \frac{n}{300}$; $n = 30$

Practice & Application

33. Miles spends 5% of his day practicing basketball. How many hours does he practice each day *1.2 hr*

34. Miles made 35% of his 40 free-throw shots during the basketball season. How many free-throw shots did he make? *14 shots*

35. Mr. Callahan bought a cordless screwdriver during a $\frac{1}{4}$ off sale. The original price was $64. What was the discount? What was the sale price? *$16; $48*

36. If sales tax is 8%, how much tax would Evelyn pay on a purchase of $37.50? *$3.00*

37. The cost of dinner at a restaurant was $60. If Charles gave a 20% tip to the server, how much was the tip? *$12*

38. Manuel is 80% of the height of his father. If his father is 6 ft tall, how tall is Manuel? *4.8 ft*

39. Maggie baby-sat and made $80 during the week. She plans to give 10% to the church on Sunday and to save 40%. How much will she give to the church and how much will she save? *$8; $32*

40. Alicia deposited $150 in a simple savings account that earns 2% each year. If she does not deposit or withdraw any money, how much interest will it earn in one year? *$3*

41. On a survey, 65% of the respondents said they prefer dogs over cats as pets. If 200 people completed the survey, how many people prefer dogs? *130 people*

42. How can I find the amount of tip I want to give a server? *I can multiply the total cost of my meal by the percent of tip I want to give. Ex. 20% x $15.00 bill = $3.00 tip*

Lesson 126   281

---

by mentally moving the decimal point one place to the left.

- Guide the students in using a **model** to find the percent of a number.

- Display the first word problem and model on the *Percent Models: Finding the Part* page. Read aloud the word problem and explain that this type of model is helpful for picturing how to find the percent of a number.

What percent does the entire model represent? 100%

What whole amount is 100% in this problem? $60

How much money does each part of the model represent? $6.00; Each part of the model is 10% of the whole and $6 is 10% of $60.

What part of the money did Brad give to the church? 10% of $60, $\frac{1}{10}$ of $60, $6

Shade 10% of the model and write $6 above the 10% line.

Since 10% of 60 is 6, how can you find the amount of money Brad saved? He saved 40%, and 40% is 4 times more than 10%, so $4 \times \$6 = \$24$.

Shade to the 40% line of the model and write "$24" above the 40% line. Point out that if students know 10% or $\frac{1}{10}$ of a price, they can find 20%, 30%, 40%, and so on.

How can you find the total amount that Brad gave to the church and saved? sample answers: 10% + 40% ($6 + $24) or 50% = 5 × 10% (5 × $6)

How much money did Brad give to the church and save? $30

- Direct attention to the next word problem.

What is the whole amount or 100% in this word problem? $75

Write "$75" above the 100% line of the model.

What part of the $75 did Shannon save? 60%

Write "60%" below the model and shade the 60%.

What is 10% of $75? $7.50; Since 10% is $\frac{1}{10}$ of the whole, I can mentally divide $75 by 10, moving the decimal point in $75 one place to the left.

How can you use the formula "n% of a number $= \frac{n}{100} \times$ the number" to solve this word problem? I can substitute 36% for the percent and 500 for the whole, then multiply to find the number that is equal to 36% of $500.

- Write for display "36% of $500 is __."
What is the decimal form for 36%? 0.36

Choose a student to use the formula and the decimal form of 36% to solve the problem. $0.36 \times 500 = \$180$

- Follow a similar procedure for these problems.
40% of 75 $0.40 \times 75 = 30$
8% of 35 $0.08 \times 35 = 2.8$

- Guide the students to answer the essential question, "How can I find the amount of tip I want to give a server?" I can multiply the total cost of my meal by the percent of tip I want to give (e.g., 20% × $15.00 bill = $3.00 tip)

Finding the percent of a number by using a model

- Use a **discussion** to review the process of finding a percent of a number.

What is 10% of 30? 3

What is 10% of 70? 7

What is 10% of 25? 2.5

How can you mentally find 10% of a number? I can divide the number by 10

LESSON 126

Write "$7.50" above the 10% line.

How can you find 60% of $75, using mental math? $6 \times 10\% = 6 \times \$7.50 = \$45$ [sample mental solution: $(6 \times \$7) + (6 \times \$0.50) = \$45$]

Write "$45" above the 60% section.

- Direct the students to draw a model to solve the third word problem. Choose a student to complete the model on the displayed page. Select another student to write the equations he used to solve the problem and to explain his solutions.

Discuss the solutions.

discount solution: 10% of $180 = $18; 30% = 3 × 10% = 3 × $18 = $54

sale price solution: $180 − $54 = $126

Point out that this model is best used when a multiple of 10 is the percent. However, the model could be partitioned into more parts to solve problems in which the percent is not a multiple of 10: 20 sections of 5% or 100 sections of 1%.

Finding the percent of a number by using a proportion; solving percent word problems

- Guide a **discussion** to help the students use a proportion to find the percent of a number.

- Read the following word problem aloud.

During the flu season, 26% of the 250 students were absent from school. How many students were absent? 65 students

- Explain that another method of finding the percent of a number is using a proportion. Write: "$\frac{\text{part}}{100} = \frac{\text{part}}{\text{whole}}$" for display.

What is the fraction form for 26%? $\frac{26}{100}$

What is the whole or 100% of the students? 250

- Choose a student to write a proportion using the information. $\frac{26}{100} = \frac{s}{250}$

Direct the students to solve the proportion to find the number of students that were absent. $6,500 = 100s$; $\frac{6,500}{100} = \frac{100s}{100}$; $s = 65$

How many students were absent? 65 students

Write the time.

1. 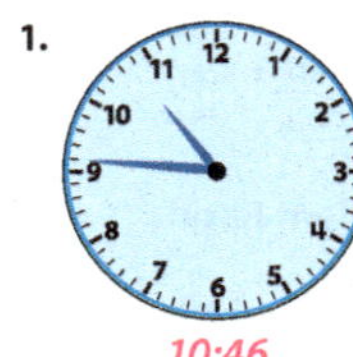

10:46

2. 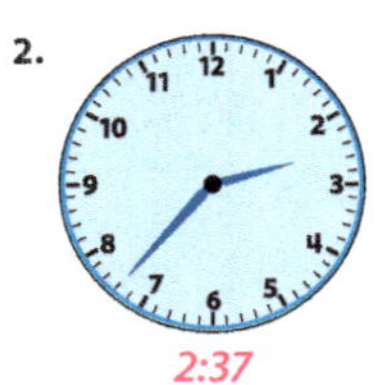

2:37

3. 

6:09

Write the temperature in °F.

4. 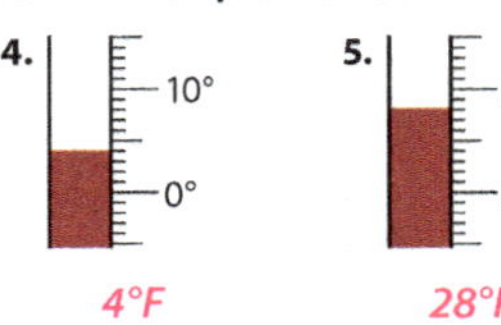

4°F

5.

28°F

6. freezing point of water 32°F

7. normal body temperature 98.6°F

8. boiling point of water 212°F

Complete the table.

9.

pounds	ounces
1	16
4	64
7	112
10	160

10.

inches	feet
12	1
48	4
72	6
108	9

11.

tons	pounds
1	2,000
2	4,000
6	12,000
8	16,000

12.

yards	inches
1	36
4	144
6	216
7	252

- Follow a similar procedure to guide the students in solving 75% of 96 and 52% of 25.

$\frac{75}{100} = \frac{n}{96}$; $7,200 = 100n$; $\frac{7,200}{100} = \frac{100n}{100}$; $n = 72$

$\frac{52}{100} = \frac{n}{25}$; $1,300 = 100n$; $\frac{1,300}{100} = \frac{100n}{100}$; $n = 13$

- Explain that students can also find 52% of 25 by renaming $\frac{52}{100}$ as an equivalent ratio since the second terms, 100 and 25, are related. Guide the students in using this method to solve the problem.

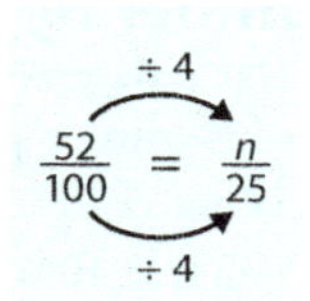

- Direct the students to solve the following word problems; give guidance as needed. Encourage them to use the method of their choice: writing an equation, using a model, or using a proportion. Discuss the answers and the methods that were used.

Adi purchased a book during a 25% off sale. The original price was $13.50. What was the amount of discount? What was the sale price? $3.38; $10.12

If the sales tax is 7%, how much tax would Clay pay on his purchase of $43.25? $3.03

Lyndsi is 85% of the height of her older sister. If her older sister is 60 in. tall, how tall is Lyndsi? 51 in. tall

---

Apply

Student Edition pages 280–81
- Read and explain the directions for pages 280–81. Assist the students as they complete the pages independently.

Daily Review
- Students should complete Chapter 13, section *g*.

NOTES

LESSON 127

Student Edition pages 282–84
Daily Review Chapter 13, section *h*

OBJECTIVES

- Find the unknown whole in a percent problem by using a model, an equation, and a proportion.
- Solve percent word problems.

TEACHER RESOURCES

- *77 Percent Models: Finding the Whole*

ADDITIONAL MATERIALS

- a calculator (for each student)

ASSESSMENTS

- Chapter 13 Quiz 2

Engage

- Direct attention to the top of Student Edition page 282, and guide the students in a **review** to answer the essential question, "How can I find the discount price of a pair of shoes?" I can multiply the original price by the percent of the discount and subtract that amount from the original price.

- Point out that students have learned to use percents to find the part of the whole, and now they will learn how to use percents to find the whole when given the part.

Instruct

Finding the unknown whole by using a model

- Guide the students as they use a **model** to find the unknown whole.

- Display the *Percent Models: Finding the Whole* page. Point out that a percent model can be made using any figure that can easily be partitioned into 10 equal parts.

  What does each of the 10 equal sections in a percent model represent? $\frac{1}{10}$ or 10% of the 100% in the whole model

  Read aloud the first word problem.

Finding the Unknown Whole

How can I find the discount price of a pair of shoes?

Simon spent $24 in the sporting goods store. This is 30% of his birthday money. How much birthday money did Simon have to begin with?

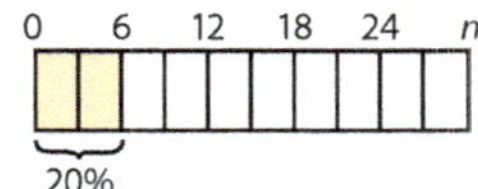

Solving with an Equation	**Solving with a Proportion**

Solving with an Equation

Substitute known information into the formula to find an unknown.

percent × whole = part **whole = part ÷ percent**

$$30\% \times n = \$24 \qquad n = \$24 \div 30\%$$
$$0.3n = \$24 \qquad n = \$24 \div 0.3$$
$$\frac{0.3n}{0.3} = \$24 \div 0.3 \qquad n = \$80$$
$$n = \$80$$

The original amount was $80.

Solving with a Proportion

Find the unknown whole by solving a proportion. Cross multiply or find the equivalent ratio to solve.

$$\frac{part}{100} = \frac{part}{whole}$$

$$\frac{30}{100} = \frac{\$24}{n}$$
$$30n = \$2{,}400$$
$$\frac{30n}{30} = \frac{\$2{,}400}{30}$$
$$n = \$80$$

The original amount was $80.

Making a Model

Allison scored 20% of the Blue Jays' points during the basketball game. If she scored 6 points, what was her team's score?

0   6   12   18   24   *n*

20%

Each section represents 3 points.
20% of the Blue Jays' score = 6 points
100% = 5 × 20%
5 × 6 points = 30 points
100% of the Blue Jays' score = 30 points

Exercises

Write an equation to find the unknown whole.

1. 15% of what number is 12? *n* = 80
2. 75% of what number is 9? *n* = 12
3. 20% of what number is 50? *n* = 250
4. 16 is 25% of what number? *n* = 64
5. 60% of what number is 15? *n* = 25
6. 14 is 35% of what number? *n* = 40

Write a proportion to find the unknown whole. *Proportions may vary.*

7. 35% of what number is 42? *n* = 120
8. 3% of what number is 6? *n* = 200
9. 60% of what number is 24? *n* = 40
10. 7 is 14% of what number? *n* = 50
11. 52% of what number is 78? *n* = 150
12. 36 is 45% of what number? *n* = 80

282   Chapter 13

What is the question asking you to find? the whole amount or the total amount Dori earned during the week

What information is given? $120 is 60% of the whole amount that Dori earned.

Point out 60% and $120 on the model.

What do you think the *n* on the model represents? the whole amount that Dori earned during the week

- Shade 60% of the model. Remind the students that the shaded sections are the part that equals the percent of the whole.

How many equal parts make up 60%? 6

How can you find the value of 10%? I can divide 120 by 6; dividing 120 by 6 will give the value of each 10% section of the model.

What is the value of 10%? $20; $120 ÷ 6 = $20

Write "10%" below its line in the model and $20 above the line.

Since 10% of the unknown whole is $20, what equation can you write to find 100%? 10 × $20 = __; 10 sections of 10% make up 100%, so I multiply the value of 10% ($20) by 10 to find the whole.

What does 10 × $20 equal? $200

Write "$200" above the 100% line of the model.

What whole amount did Dori earn? $200

- Direct attention to the second word problem and read it aloud.

Make a model to find the unknown whole.

13. 40% of what number is 8? *n = 20*

14. 70% of what number is 84? *n = 120*

15. 80% of what number is 56? *n = 70*

16. 117 is 90% of what number? *n = 130*

17. 50% of what number is 75? *n = 150*

18. 15 is 25% of what number? *n = 60*

Find the percent of the number. *Method used to solve may vary.*

19. 10% of 75 *7.5*

20. 5% of 60 *3*

21. 40% of 120 *48*

22. 20% of 75 *15*

23. 40% of 9 *3.6*

24. 63% of 200 *126*

25. 40% of 75 *30*

26. 18% of 50 *9*

27. 96% of 40 *38.4*

Write the fraction as a percent.

28. $\frac{3}{4}$ *75%*

29. $\frac{3}{50}$ *6%*

30. $\frac{7}{10}$ *70%*

31. $\frac{1}{2}$ *50%*

32. $\frac{2}{5}$ *40%*

33. $\frac{4}{25}$ *16%*

Practice & Application

34. Caden mowed 4 lawns in 6 hr. At this rate, how long will it take him to mow 5 lawns? *7.5 hr*

35. During the race, Car 14 traveled 240 mi in 2 hr. At this rate, what distance will the car travel in 5 hr? *600 mi*

36. Aubrey bought a skirt during a $\frac{1}{3}$ off sale. The original price was $45. What was the discount? What was the sale price? *$15; $30*

37. If sales tax is 8%, how much tax would be charged on a purchase of $6.50? *$0.52*

38. Catherine spends 35% of her day at work. How many hours does she spend at work each day? *8.4 hr*

39. All items in the sports department were on sale for 30% off. Weston received a discount of $12 on the basketball he purchased. What was the original price of the basketball? *$40*

40. Nicholas scored 20% of his team's goals during a soccer game. He scored 2 goals. How many goals did his team score? *10 goals*

41. During the election for class representative, Stella received 2 votes for every vote cast for her opponent. Her opponent received 8 votes. How many votes did Stella receive? *16 votes*

42. A house casts a shadow 28 ft long. A 6 ft man casts a shadow that is 8 ft long. How high is the house? *21 ft*

43. A meter stick casts a shadow that is 3 m long. A tree casts a shadow that is 12 m long. How tall is the tree? *4 m*

44. Brad got 29 answers correct out of 33 problems on a math test. About what percent of the problems did he get right? *88%*

45. Caiti wanted a pair of shoes that cost $65. She watched for them to go on sale. At first, the sale price was 20% off the original price. Caiti bought the shoes when they went on sale for 45% off. What was the discount of the shoes during each sale? What did she pay for the shoes? *$13.00; $29.25; $35.75*

46. How can I find the discount price of a pair of shoes? *I can multiply the original price by the percent of the discount and subtract that amount from the original price.*

What information is given in this word problem? $14 is 40% of the whole (the original price of the game).

Write 40% below its line and $14 above the line. Shade 40% of the model.

How many equal parts make up 40%? 4

What is the value of 10%? $3.50; $14 (40%) divided by 4 tells me the value of each 10% section; $14 ÷ 4 = $3.50.

Write "10%" below its line and $3.50 above the line.

Since 10% of the unknown whole is $3.50, how can you find 100%? I can multiply $3.50 by 10.

Remind students that when they multiply a number by 10, the decimal point moves one place to the right.

What was the original price of Eric's game? $35.00

Write "$35" above the 100% line of the model.

- Direct attention to the third word problem and read it aloud. Direct the students to draw a model to solve the problem. Invite a student to complete the model on the displayed page. Choose another student to write and explain the equations he used to solve the problem. Discuss the solution. $72 ÷ 3 = $24; 10 × $24 = $240

Remind the students that this model is best used when a multiple of 10 is the percent. However, the model could be partitioned into more parts: 20 sections of 5% or 100 sections of 1%.

How could you use the model to find 85% of $240? sample answers: 8.5 × $24 = $204; partition the whole model into twice as many parts so that each section is $\frac{1}{20}$ or 5% of the whole and multiply the value of 5% by 17 (17 × $12 = $204)

Finding the unknown whole by using an equation

- Guide a **discussion** to help the students use an equation to find the unknown whole.

- Write "*percent × whole = part*" for display and explain that they can use this formula to find the whole or 100%.

- Read the following word problem aloud.

Brian counted 8 trucks in the parking lot. If 20% of the vehicles in the parking lot are trucks, how many vehicles are in the parking lot? 40 vehicles

What information are you given in the word problem? 8 trucks is 20% of the total number of vehicles.

What equation can you write in words to solve the problem? 20% of *v* is 8; 20% of the unknown number of vehicles are trucks (8).

Write for display "20% of *v* is 8."

How can you write 20% as a numerical value without the percent sign? $\frac{20}{100}$ or 0.20

- Direct the students to rewrite an equation using mathematical symbols and an equivalent numerical value for 20%. 0.20 × *v* = 8 or $\frac{20}{100}$ × *v* = 8 Choose a student to write both equations for display.

How can you find the value of *v*? I can divide each side of the equation by 20 hundredths to isolate the *v*.

Which equation do you think is easier to solve? sample answer: The equation that has the percent expressed as a decimal is easier; when dividing by a fraction, I multiply by the reciprocal to solve the problem.

- Instruct the students to find the total number of vehicles by solving one of the equations. Discuss the answer and then choose a student to complete both of the equations written for display. $\frac{0.20v}{0.20} = \frac{8}{0.20}$; *v* = 40 vehicles

- Follow a similar procedure for the following problems.

  70% of what number is 42? 60

  45% of what number is 18? 40

  21 is 14% of what number? 150

Finding the unknown whole by using a proportion

- Use a **guided discovery** to help the students find the unknown whole.

- Write "$\frac{\text{part}}{100} = \frac{\text{part}}{\text{whole}}$" for display. Explain that a proportion can also be used to find the unknown whole in a percent problem. Since a percent is a part-to-whole comparison with 100 as the whole, it can be compared to another part-to-whole comparison.

- Read the following word problem aloud.

  A church gave $240 to a missionary family. This amount is 30% of the church's weekly missions budget. What is the weekly missions budget of the church? $800

  What does the word *this* refer to in the word problem? It refers to the amount that was given to the missionary family; $240.

  What is the fraction form for 30%? $\frac{30}{100}$

  What part-to-whole relationship will help you find the weekly missions budget amount? $240 is a part of the weekly missions budget.

- Choose a student to write the part-to-whole ratios to make a proportion.

  $\frac{30}{100} = \frac{\$240}{n}$

  How can you solve for the missing term in the proportion? I can find an equivalent ratio or cross multiply.

  Direct the students to solve the proportion. $800

  Discuss the answer and the method students used to solve the proportion. Conclude that $800 represents the whole or the church's weekly missions budget.

- Follow a similar procedure for these problems.

  40% of what number is 20? 50

  15% of what number is 24? 160

  45 is 90% of what number? 50

John Napier (1550–1617) was a Scot who fervently defended the great Protestant reformer John Knox. As a Christian, he was deeply involved in political and religious struggles and turned to mathematics for relaxation. Although he hoped to be remembered for a commentary he wrote on the book of Revelation, he is chiefly remembered as the man who invented "Napier's bones," a multiplication table that was made from bone or ivory. This was one of the earliest forms of calculators used to do multiplication and division! John Napier also introduced the idea of a point to separate the whole number part of a number from its fraction part. This point is the decimal point.

Solving percent word problems

- Guide the students in a **practice exercise** to help them solve percent word problems.

- For each problem, ask whether students need to find the part or the whole and encourage them to use the method of their choice: using an equation, a model, or a proportion. Discuss the answers and the methods that the students used.

  Daniel scored 30% of his team's points during a basketball game. Daniel scored 6 points. What was the team's score? 20 points

  At a Christian school, there are 15 boys in the 6th grade. If 60% of the students in the 6th grade are boys, how many students are in the 6th grade? 25 students

  A $350 recliner is on sale for 45% off the original price. What is the discount? $157.50 What is the sale price? $192.50

  If sales tax is 8%, how much tax would be charged for a purchase of $21.95? $1.76

Write the ordered pair for the point.

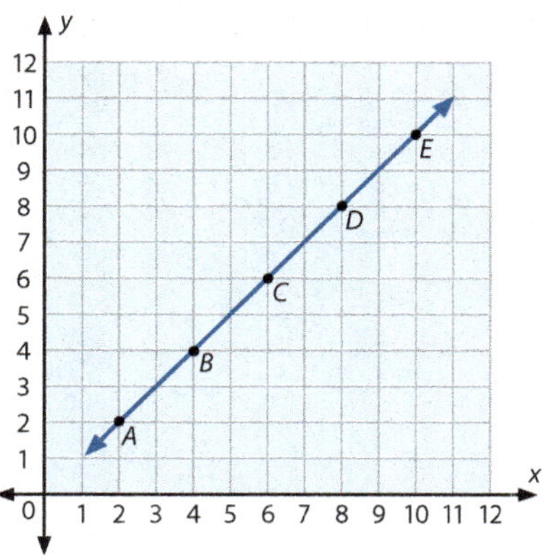

1. A *(2, 2)*
2. B *(4, 4)*
3. C *(6, 6)*
4. D *(8, 8)*
5. E *(10, 10)*

Name the point on the graph represented by the ordered pair.

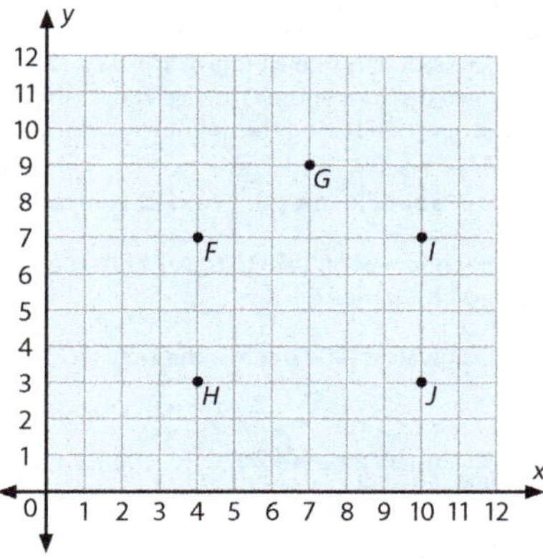

6. (4, 7) *F*
7. (7, 9) *G*
8. (4, 3) *H*
9. (10, 7) *I*
10. (10, 3) *J*

Apply

Student Edition pages 282–84
- Read and explain the directions for pages 282–83. Assist the students as they complete the pages independently.

Daily Review
- Students should complete Chapter 13, section *h*.

Assess

Quiz 2
- Use the **summative assessment** to evaluate the students' progress at this point in the chapter.

LESSON 128

Student Edition pages 285–87
Daily Review Chapter 13, section *i*

OBJECTIVES

- Calculate the distance, the rate of speed, and the time.
- Find an equivalent rate by using a proportion.
- Evaluate the use of proportions in solving real-world problems. **BWS**
- Rename units to calculate distance, rate of speed, or time.

BIBLICAL WORLDVIEW SHAPING

- Design (Evaluate): A non-biblical worldview has no adequate explanation for the dependability of proportions.

ADDITIONAL MATERIALS

- a calculator (for each student)

Engage

- Direct attention to the question presented on Student Edition page 264, "If I can make 8 baskets in 40 seconds, how many can I make in 120 seconds?" Guide the students to do a **Think-Pair-Write** to answer the question. $\frac{8}{40} = \frac{n}{120}$; Since $40 \times 3 = 120$, $8 \times 3 = 24$. I predict that I can make 24 baskets in 120 seconds.

- You may point out the fact that because proportions are reliable, they can be used to solve many different problems.

Instruct

Calculating distance, rate of speed, or time; finding equivalent rates

- Use **direct instruction** to help the students apply their knowledge of proportions to the formula $r = \frac{d}{t}$.
- Write r (*rate of speed*) $= \frac{d \ (distance)}{t \ (time)}$ for display. A rate of speed is expressed as a distance measurement divided by a time measurement and is written using the word "per" such as "miles *per* hour."

- Guide the students to conclude that since speed is a rate, a proportion can be used to find equivalent rates of speed when they know the distance or the time traveled.

- Assist the students in realizing that their knowing the formula $r = \frac{d}{t}$ helps them to write a proportion for finding the unit rate of speed for the following word problems. Instruct them to solve for the rate in the following problems by finding an equivalent ratio or by cross multiplying. (See Lesson 122.)

Alexa walked 5 mi in 2 hr. What was her average speed per hour? $\frac{n \ \text{mi}}{1 \ \text{hr}} = \frac{5 \ \text{mi}}{2 \ \text{hr}}$; $n = 2.5$; $r = 2.5$ mph

The ranger traveled 78.6 mi through the park in 3 hr on his 4-wheeler. What was his average speed per hour? $\frac{n \ \text{mi}}{1 \ \text{hr}} = \frac{78.6 \ \text{mi}}{3 \ \text{hr}}$; $n = 26.2$; $r = 26.2$ mph

- Read aloud the following word problem.

In 1852, Henri Giffard built a steam airship that flew at a speed of 5 mph. The airship traveled 17 mi. About how many hours did it take to travel that distance? $t = 3.4$ hr

Could you use a proportion to find the amount of time it took the airship to travel 17 mi? Yes; since I know 3 of the 4

Speed, Distance & Time

Why might some unbelievers struggle to explain the dependability of proportions in real-world problem solving?

Key Terms
- speed
- rate
- distance
- time

Speed is a **rate**, a ratio which compares distance to time and is usually written as a unit: miles per hour, kilometers per hour, or meters per second. A known rate of speed can be used to find an unknown distance or time traveled at the same rate of speed. Make a proportion to find the unknown distance or time.

r (**rate of speed**): how *fast*   *d* (**distance**): how *far*   *t* (**time**): how *long*

How many hours will it take to travel a distance of 1,500 mi at a speed of 100 mph?

Find the equivalent ratio or cross multiply to solve.

$$\frac{\text{distance}}{\text{time}} = \frac{100 \text{ mi}}{1 \text{ hr}} = \frac{1{,}500 \text{ mi}}{n \text{ hr}} \qquad \begin{aligned} 100n &= 1{,}500 \\ \frac{100n}{100} &= \frac{1{,}500}{100} \\ n &= 15 \text{ hr} \end{aligned}$$

It will take 15 hr to travel 1,500 mi.

Exercises   *Steps to solve may vary.*

Find the distance traveled.

1. How many miles can you walk in 2 hr if your average speed is 4 mi/hr? *8 mi*

2. How many km can you drive in 5 hr if your car's average speed is 52 km/hr? *260 km*

3. How many feet can you run in 12 sec at a speed of 7 ft/sec? *84 ft*

Find the time traveled.

4. How many hours would it take to walk 5 mi at 3 mi/hr? *1.67 hr*

5. How many hours would it take to drive 200 mi at 50 mi/hr? *4 hr*

6. How many hours would it take to fly 165 mi at 330 mi/hr? *0.5 hr*

Lesson 128   285

Find the average speed (unit rate).

7. What is the average speed of a car that traveled 224 km in 3.5 hr? *64 km/hr*

8. What is the average speed of a car that traveled 140 mi in 4 hr? *35 mi/hr*

9. What is the average speed of a train that traveled 270 mi in 3 hr? *90 mi/hr*

Solve.

10. Anderson's airplane trip was $2\frac{1}{4}$ hr long. If the plane averaged 350 mi/hr, how far did Anderson travel? *787.5 mi*

11. Justin is flying 1,190 mi in a cross-country flight. If the plane averages 340 mi/hr, how long will the flight take? *3.5 hr*

12. Sean and his friend Tristan biked 24 mi at an average speed of 12 mi/hr. How many hours did they bike? *2 hr*

13. The average speed of a truck was 50 mi/hr, and the average speed of a car was 60 mi/hr. How many fewer hours did it take the car than the truck to travel 600 mi? *2 hr*

14. During a race, Lorie's horse averaged 30 mi/hr and Sadie's horse averaged 25 mi/hr. About how many fewer minutes did it take Lorie's horse than Sadie's horse to go 5 mi? *2 min*

numbers in the proportion (*speed* = 5 mi per 1 hr and *distance* = 17 mi), I can solve for the unknown time.

Using the formula $r = \frac{d}{t}$, what proportion can you write to find the unknown time? $\frac{5\text{ mi}}{1\text{ hr}} = \frac{17\text{ mi}}{t}$

Direct the students to use cross multiplication to solve for the unknown time. $5t = 17; t = 3.4$ hr

• Guide the students as they use a proportion to find the unknown time by using the following information.

$r = 65$ mi/hr; $d = 195$ mi $\frac{65\text{ mi}}{1\text{ hr}} = \frac{195\text{ mi}}{t}$, $65t = 195; t = 3$ hr

Mrs. Collins walked at a speed of 4 mph. If she maintained the same speed, how far did she walk in 3 hr? $d = 12$ mi

What information is given in this word problem? The rate of speed is 4 mph, and the time is 3 hr.

What information is not given? the distance; The question is asking how far she walked.

• Direct the students to write a proportion to find the unknown distance. Choose a student to give the answer and explain the solution. $\frac{4\text{ mi}}{1\text{ hr}} = \frac{d}{3\text{ hr}}$; $d = 12$; Accept explanations of finding an equivalent ratio or cross multiplying.

What measuring unit is used to find the distance in this problem? miles; The rate of speed was measured in miles per hour.

Write the answer: distance = 12 mi.

• Follow a similar procedure to guide the students as they solve the following problems.

Ashley and her friend biked 15 mi at an average speed of 6 mph. About how long did it take to travel that distance? $t = 2.5$ hr

Robert drove his car an average speed of 40 mph. Jason drove his car an average speed of 50 mph. How many more hours did it take Robert to travel 200 mi? Robert: $t = 5$ hr; Jason: $t = 4$ hr; $5 - 4 = 1$ hr

Juanita traveled 528 mi by train in 6 hr. What was the average speed of the train? $r = 88$ mph

Evaluating the use of proportions

• Guide a **discussion** to help the students evaluate a biblical worldview truth.

• Direct attention to the previous problem about the train.

How did you begin to solve this word problem? by writing a proportion $\frac{528}{6} = \frac{n}{1}$

How did you solve the proportion? by dividing by 1 in the form of $\frac{6}{6}$

Do these procedures work every time? sample answer: Yes; they work for distance, speed, time, and other kinds of problems. They are consistent and reliable.

• Discuss the idea that an unbeliever thinks the world happened by chance or natural processes. The Bible teaches that God designed the world. Certain mathematical principles hold true because His design is dependable.

Why might some unbelievers struggle to explain the dependability of proportions in real-world problem solving? Some unbelievers do not believe the world was created by a designer. Without a designer, proportions would not be dependable.

Renaming to calculate distance, rate, or time

- Use **critical thinking questions** to help the students rename units.

- Read the following word problem aloud.

  A white-tailed deer ran 30 mph for 2 min. How far did the deer run? *d* = 1 mi

  What information is given in this word problem? r = 30 mph; t = 2 min

  What measuring units are used to measure the rate of speed in the problem? miles per hour

  What measuring unit is used to measure time? minutes

- Explain that a proportion compares *like* units. When the units of measure in a proportion are not alike, students need to rename a unit of measure so that they can compare like units before they begin to solve the problem.

  What two units are related? How are they related? minutes and hours; There are 60 min in 1 hr.

  Write "$\frac{30\text{ mi}}{1\text{ hr}} = \frac{30\text{ mi}}{60\text{ min}}$" for display.

- Direct the students to use the renamed rate of speed and the formula $r = \frac{d}{t}$ to find the unknown distance. Choose a student to tell the answer and explain his solution. Solutions will vary; $\frac{30}{60} = \frac{d}{2}$; d = 1.

  What measuring unit is used to measure distance in this problem? miles; The rate of speed was measured in miles per minute.

  Write the answer: d = 1 mi.

- Follow a similar procedure for the following word problem. Guide the students to conclude that they need to find the rate or speed of a train, which is typically measured in miles per hour, and that 1 day can be renamed as 24 hr.

  The Fuller family went on a cross-country trip by train. They traveled 2,040 mi in 1 day. How fast did the train travel? $r = \frac{2{,}040}{24\text{ hr}}$; r = 85 mph

---

Renaming Units to Make a Proportion

A proportion compares *like* units. Use equivalent times and distances to rename when units are *not* alike.

Caitlyn's grandmother lives 10 mi from her house. If Caitlyn travels at an average speed of 60 mi/hr, how many minutes will it take her to travel to Grandmother's house?

$$r = 60\text{ mi/hr}$$
$$d = 10\text{ mi}$$
$$t = \underline{}$$

Minutes are needed to measure the time it takes to travel to Grandmother's house, but the known rate of speed is per hour.

$$1\text{ hr} = 60\text{ min}$$
$$\frac{60\text{ mi}}{1\text{ hr}} = \frac{60\text{ mi}}{60\text{ min}}$$

Write a proportion using the renamed units of speed.

$$\frac{\text{distance}}{\text{time}} = \frac{\text{mi}}{\text{min}} \qquad \frac{60}{60} = \frac{10}{n}$$
$$n = 10\text{ min}$$

If Caitlyn travels at a speed (rate) of 60 mi/hr (or 60 mi/60 min), she will travel 10 mi in 10 min.

The packaging from Jeffrey's rocket boasts an average rocket speed of 195 ft/sec. If his rocket flies upward for 8.2 sec, how many yards high will it fly?

$$r = 195\text{ ft/sec}$$
$$d = \underline{}$$
$$t = 8.2\text{ sec}$$

A distance in yards is needed to measure the flight, but the known rate of speed is feet per second.

$$1\text{ yd} = 3\text{ ft}$$
$$\frac{195\text{ ft}}{1\text{ sec}} = \frac{65\text{ yd}}{1\text{ sec}}$$

Write a proportion using the renamed units of speed.

$$\frac{\text{distance}}{\text{time}} = \frac{\text{yd}}{\text{sec}} \qquad \frac{65}{1} = \frac{n}{8.2}$$
$$n = 533\text{ yd}$$

If the rocket travels at a speed (rate) of 195 ft/sec (or 65 yd/sec), it will travel 533 yd in 8.2 sec.

Find the rate (r), distance (d), or time (t).

15. r = 50 mi/hr
d = ___ *50 mi*
t = 60 min

16. r = 5 ft/min
d = 20 yd
t = ___ *12 min*

17. $r = \frac{1}{2}$ mi/hr
d = ___ *$2\frac{1}{2}$ mi*
t = 5 hr

18. r = ___ mi/min *0.1 mi/min*
d = 0.5 mi
t = 5 min

19. r = 250 m/min
d = 3 km
t = ___ *12 min*

20. r = ___ mi/hr *45 mi/hr*
d = 540 mi
$t = \frac{1}{2}$ day

21. r = 21 ft/sec
d = ___ *315 ft*
t = 15 sec

22. r = ___ mi/hr *25 mi/hr*
d = 600 mi
t = 1 day

23. r = 40 ft/sec
d = 1 mi
t = ___ *132 sec*

24. r = 70 km/hr
d = ___ *35 km*
t = 30 min

25. Robert's remote-control car can travel at a rate of 66 ft per minute. How far could the car travel in 15 min? At this rate, how long would it take the car to travel 1 mi?

26. Why might some unbelievers struggle to explain the dependability of proportions in real-world problem solving? *Some unbelievers do not believe the world was created by a designer. Without a designer, proportions would not be dependable.*

25. $\frac{66\text{ ft}}{1\text{ min}} = \frac{d}{15\text{ min}}$; *d* = 990 ft; $\frac{66\text{ ft}}{1\text{ min}} = \frac{5{,}280\text{ ft}}{t}$; *t* = 80 min

- Direct the students to find the unknown measurement for each of the following groups of information. Guide the students as they rename units of measure as needed.

For the following problems and the problems on Student Edition page 287, allow the students to refer to the equivalency charts in the Measurement section of the student Handbook. For your convenience the student Handbook is available on TeacherToolsOnline.com.

r = 12 km/min	r = 20 ft/min
t = 5 min	t = 90 sec
d = 60 km	d = 30 ft
d = 208 km	d = 60 ft
t = 3.2 hr	t = 90 sec
r = 65 km/hr	r = 40 ft/min
r = 6 km/hr	r = 15 ft/min
d = 3 km	d = 20 yd
t = 0.5 hr	t = 4 min

Solve.

1. 971 + 136 + 538 + 818 + 881   *3,344*   **2.** 766 + 245 + 952 + 446 + 312   *2,721*

3. 228 + 347 + 474 + 146 + 359   *1,554*   **4.** 873 + 721 + 979 + 619 + 648   *3,840*

5. 95,939
  − 59,962
    35,977

6. 62,884
  − 10,611
    52,273

7. 91,315
  − 87,795
    3,520

8. 47,386
  − 25,668
    21,718

9. 358
  × 711
 254,538

10. 471
  × 512
 241,152

11. 948
  × 343
 325,164

12. 324
  × 460
 149,040

Solve. Round the quotient to the nearest tenth.

13. $69\overline{)854}$ *12.37 ≈ 12.4*   **14.** $21\overline{)389}$ *18.52 ≈ 18.5*   **15.** $25\overline{)514}$ *20.56 ≈ 20.6*   **16.** $13\overline{)624}$ *48*

Apply

Student Edition pages 285–87
- Read and explain the directions for pages 285–87. Assist the students as they complete the pages independently.

Daily Review
- Students should complete Chapter 13, section i.

Student Edition pages 288–89

CHAPTER REVIEW

OBJECTIVES

- Find equivalent ratios.
- Determine the unit rate.
- Determine whether two ratios are proportional.
- Find the unknown measure in similar figures using proportions.
- Use an indirect measurement to find the unknown measure in similar objects.
- Find actual measurements using a scale.
- Determine the unknown measure on a scale drawing given the scale and the actual measurement.
- Express percents as ratios, decimals, and fractions.
- Express fractions and ratios as percents.
- Find a percent of a number or the unknown whole.

TEACHER RESOURCES

- 78 *Circle Graph: Elements in the Earth's Crust*

ADDITIONAL MATERIALS

- a calculator (for each student)

CHAPTER REVIEW

Complete the ratio table.

1.

centimeters	2.54	5.08	*7.62*	10.16	*12.7*
inches	1	2	3	4	5

State whether the ratio could be in a ratio table with $\frac{4}{7}$.

2. $\frac{16}{28}$ *yes*

3. $\frac{48}{80}$ *no*

4. $\frac{56}{98}$ *yes*

Find the unit rate.

5. Reese used 17 gal of gas to drive 510 mi.
30 mi/gal
6. Jonah earned $56.00 in 8 hr.
$7.00/hr
7. Allie read 30 pages in 20 min.
1.5 pg/min
8. Sarah bought 5 lb of apples for $4.25.
$0.85/lb

Find the distance traveled.

9. 4 min at 15 yd/min *60 yd*

10. 3.5 hr at 50 mi/hr *175 mi*

11. 18 sec at 20 ft/sec *360 ft*

12. 6 days at 21 mi/day *126 mi*

Write a possible proportion for the situation. Write "yes" if the prices are equivalent. Write "no" if one price is a better buy.
Steps to solve may vary.
13. 8 ears of corn for $1 or 15 ears of corn for $2
$\frac{\$1}{8} \neq \frac{\$2}{15}$; *no*
14. 4 pencils for $2 or 12 pencils for $6 $\frac{\$2}{4} = \frac{\$6}{12}$; *yes*

15. 22 oz drink for $1.50 or 28 oz drink for $1.75
$\frac{\$1.50}{22} \neq \frac{\$1.75}{28}$; *no*
16. 12 eggs for $2 or 18 eggs for $3 $\frac{\$2}{12} = \frac{\$3}{18}$; *yes*

Proportions may vary.
Write a proportion to find the unknown measure for the pair of similar figures.

17.
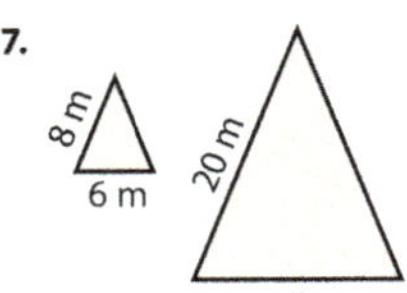

$\frac{8}{6} = \frac{20}{n}$
$\frac{8n}{8} = \frac{120}{8}$
$n = 15\ m$

18.
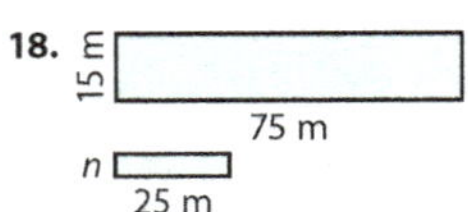

$\frac{15}{n} = \frac{75}{25}$
$\frac{375}{75} = \frac{75n}{75}$
$n = 5\ m$

Use a map scale of 1 in.:75 mi to find the actual distance represented by the measurement.

19. 4 in. $\frac{1\,in.}{75\,mi} = \frac{4\,in.}{d}$; $d = 4 \cdot 75\ mi$; $d = 300\ mi$

20. 2.6 in. $\frac{1\,in.}{75\,mi} = \frac{2.6\,in.}{d}$; $d = 2.6 \cdot 75\ mi$; $d = 195\ mi$

Write a proportion to find the unknown height.

21.
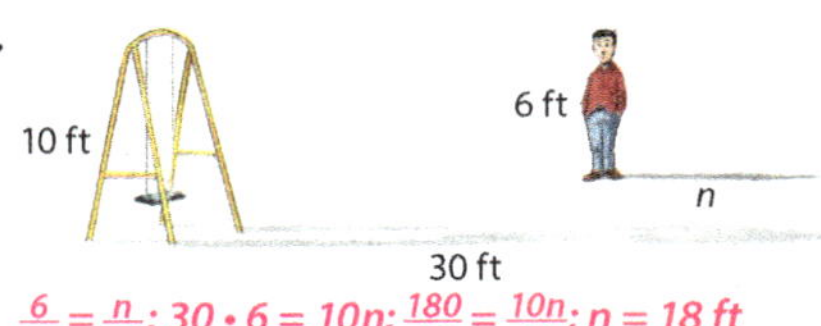

$\frac{6}{10} = \frac{n}{30}$; $30 \cdot 6 = 10n$; $\frac{180}{10} = \frac{10n}{10}$; $n = 18\ ft$

22.

$\frac{5}{9} = \frac{n}{36}$; $36 \cdot 5 = 9n$; $\frac{180}{9} = \frac{9n}{9}$; $n = 20\ ft$

The Chapter Review offers an opportunity for students to discuss the concepts they have learned in the chapter. They may work collaboratively or independently as you review concepts. Circulate among the students, giving individual help as needed. Students who demonstrate proficiency with the discussion, the modeling, and the Student Edition pages are ready for the Chapter Test. Students who encounter difficulties with the review concepts would benefit from additional coaching and practice before testing.

Finding equivalent ratios; determining the unit rate

What is a ratio? a comparison of two quantities

- Inform the students that there are 240 mL in 1 c. Choose a student to write the ratio that compares the number of milliliters to 1 c. 240 mL : 1 c

- Draw a ratio table similar to the ones used in Lesson 121. Write "milliliters" and the first two entries, 240 and 480, in the top row. Write "cups" and the entries 1, 2, 3, 4, and 6 in the bottom row. Remind the students that all the ratios in a ratio table are equivalent or proportional.

Method used to solve may vary.

Write the percent in decimal form. Write the decimal in percent form.

23. 64% *0.64* **24.** 4% *0.04*

25. 0.09 *9%* **26.** 0.83 *83%*

Write the percent as a fraction in lowest terms. Write the fraction as a percent.

27. 40% $\frac{2}{5}$ **28.** 10% $\frac{1}{10}$

29. $\frac{3}{5}$ *60%* **30.** $\frac{5}{10}$ *50%*

Find the percent of the number.

31. 30% of 70
0.30 • 70 = 21

32. 42% of 75
0.42 • 75 = 31.5

33. 25% of 64
0.25 • 64 = 16

34. 60% of 30
0.60 • 30 = 18

Find the unknown whole.

35. 5% of what number is 3? *n = 60*

36. 40% of what number is 32? *n = 80*

37. 25% of what number is 5? *n = 20*

Solve. *Steps to solve may vary.*

38. A survey revealed that 84 out of 124 students have brown eyes. About how many students with brown eyes would you expect to find out of 10 of these students? *7 students*

39. Factory workers can produce 18 items in 15 hr. How many hours will it take them to produce 12 items at this rate? *10 hr*

40. A playground is 144 ft wide. How wide would a scale drawing of the playground be in which 2 in. represents 12 ft? *24 in.*

41. Jenna made $200 during the week. She plans to give 10% to the church and to put 40% in her savings account. How much will she give to the church and how much will she save? *$20; $80*

42. Alicia deposited $500 in a simple savings account that earns 3% each year. If she does not deposit or withdraw any money, how much interest will she earn in one year? *$15*

43. On a survey, 85% of the respondents said there are 2 vehicles in their households. If 200 people completed the survey, how many people own 2 vehicles? *170 people*

44. Sienna enlarged a picture to be 5 times larger than the original. The original picture was 2 in. long by 4 in. wide. The enlarged picture has a length of 10 in. What is the width? *20 in.*

45. Zane answered 42 out of 50 questions correctly on a science test. What percent of the questions did he answer correctly? *84%*

46. Erin bought a purse during a 25% off sale. The original price was $45. What was the discount? What was the sale price? *$11.25; $33.75*

47. If sales tax is 7%, how much tax would be charged on a purchase of $11? *$0.77*

48. All items in the store were marked 30% off. Tucker received a discount of $21 on a soccer ball he bought. What was the original price of the ball? *$70*

49. Elliot scored 60% of his free-throw attempts. If he made 3 free throws during the game, how many attempts did he make? *5 attempts*

Mom used 20 gal of gas to travel 500 mi. How many miles per gallon did her vehicle get? 25 mi/gal

- Follow a similar procedure for the following information.

  14 gal of gas costs $47.60 $3.40/gal

  3 lb of grapes costs $7.05 $2.35/lb

- Follow a similar procedure for the following problems. Conclude with the students that both terms of the unit rate can be multiplied by a name for 1 to find an equivalent ratio. Remind the students that speed is a rate, a special ratio that compares distance to time.

  3 min at 12 yd/min 36 yd

  5 days at 17 mi/day 85 mi

  2.7 lb at $1.25/lb $3.38

Determining whether two ratios are proportional

When are two ratios proportional? when they are equivalent

How can you determine whether two ratios are proportional? Various strategies can be used to compare the terms of the ratios vertically, horizontally, or diagonally.

Review the strategies as needed. (See Lesson 122.) Direct the students to write ratios for the information in each of the following word problems and to determine whether the ratios are proportional. Choose students to answer the questions and to write a mathematical statement that supports their answer. Instruct each student to explain the strategy he used to determine whether the statement is a proportion.

- Read the following word problem aloud.

  Alyssa earned $15 for working 3 hr, and Mackenzie earned $24 for working 5 hr. Did they receive the same hourly rate?
  no; $\frac{15}{3} \neq \frac{24}{5}$

  Aiden bought a 16 oz drink for $1.92. Dylan bought a 12 oz drink for $1.44. Were the prices per ounce equivalent?
  yes; $\frac{16}{\$1.92} = \frac{12}{\$1.44}$

How can you complete this table? I can multiply the number of cups by 240 or write and solve a proportion.

Choose students to write the number of milliliters in 3 c 720 and 4 c 960.

What other ways could you find the number of milliliters in 6 c? sample answers: I could double the number of milliliters in 3 c or add the number of milliliters in 2 c to the number of milliliters in 4 c.

Complete the table. 1,440

- Write the ratios "$\frac{25}{36}$" and "$\frac{60}{72}$" for display.

How can you find out if these ratios could be in a ratio table with $\frac{5}{6}$? sample answer: I can determine if each ratio is equivalent to $\frac{5}{6}$ by cross multiplying and comparing the products.

Choose students to write possible proportions and cross multiply to determine if the ratios are equivalent to $\frac{5}{6}$. $\frac{5}{6} \neq \frac{25}{36}$ and $\frac{5}{6} = \frac{60}{72}$

- Instruct the students to solve the following word problem. Guide the students to conclude that the second term in a unit rate is 1. The unit rate can be found by dividing both terms of a ratio by a name for 1. Discuss the solution as needed. (See Lesson 120.)

- Read the following word problem aloud.

<table>
<tr><td>

Finding the unknown measure in similar figures; using indirect measurement

- Draw for display a small rectangle and write "40 cm" and "60 cm" along its sides. Also draw a similar rectangle and write "100 cm" along its shorter side and "*n*" along its longer side. Remind the students that the rectangles are similar figures; they are the same shape, but are not the same size (See Lesson 123).

  What proportion can you write to find the unknown measurement using ratios *between* these similar figures? $\frac{40\text{ cm}}{100\text{ cm}} = \frac{60\text{ cm}}{n}$

  What proportion can you write to find the unknown measurement using ratios *within* the figures? $\frac{40\text{ cm}}{60\text{ cm}} = \frac{100\text{ cm}}{n}$

- Write both proportions for display and instruct the students to solve them. Compare the answers and allow students to explain the method they used to solve the proportions. $n = 150$ cm

- Guide the students as they solve the following word problem. Discuss the idea that the method of indirect measurement uses similar objects and a proportion to find the unknown measure of an object that is difficult to measure.

  A building casts a shadow that is 3.6 m long. A sign that is 2 m tall casts a shadow that is 1.2 m long. What is the height of the building? 6 m; $\frac{h}{2\text{ m}} = \frac{3.6\text{ m}}{1.2\text{ m}}$, $h = 6$ m; or $\frac{2\text{ m}}{1.2\text{ m}} = \frac{h}{3.6\text{ m}}$, $h = 6$ m

Finding actual measurements using a scale; determining the unknown measure on a scale drawing

  What is a scale? a ratio of measurements that compares the size of a drawing or model to the size of the actual object

  What are examples of scale drawings and models used in everyday life? sample answers: maps, floor plans, toy horses, dollhouses, cars, trains, airplanes

  Guide the students as they solve the following word problems. Review the solutions as needed. (See Lesson 124.)

</td><td>

A map has a scale of 1 in. : 25 mi. If the distance between 2 cities is 4.5 in., what is the actual distance? $\frac{1\text{ in.}}{25\text{ mi}} = \frac{4.5\text{ in.}}{n}$; $n = 112.5$ mi

The distance between 2 cities is 90 mi. If the map scale is 1 in. : 15 mi, what is the map measurement? $\frac{1\text{ in.}}{15\text{ mi}} = \frac{n}{90\text{ mi}}$; $n = 6$ in.

Expressing percents as ratios, fractions & decimals

  What is the definition of a percent? a ratio in which a quantity (part) is compared to 100 (whole)

- Display the *Circle Graph* page. Explain that the table and graph show the major elements that are found in the earth's crust. Complete the table as students give the ratio, fraction, and decimal for each percent (e.g., 28%: $\frac{28}{100}$; $\frac{7}{25}$; 0.28).

- Review the process for writing a ratio as a percent. (See Lesson 125.) Instruct the students to solve the following word problems.

  On a math test, Anthony answered 46 out of 50 questions correctly. What percent of the questions did he answer correctly? $\frac{46}{50} = 0.92$ (or $\frac{92}{100}$) $= 92\%$

  During the summer, 4 out of 12 students attend summer school. What percent of the students attend summer school? $\frac{4}{12} \approx 0.33 \approx 33\%$

Finding a percent of a number or the unknown whole

- Read the following word problem aloud.

  During basketball season Jacob made 25% of his 32 three-point shots. How many three-point shots did he make? 8 shots

  How can you solve this word problem? I can find 25% of 32 (*n*% of a number = $\frac{n}{100} \times$ the number); the answer represents a part or a percent of the total (32) three-point shots that Jacob attempted.

  Point out that the percent can be renamed as a fraction or a decimal.

- Direct the students to solve the problem. Discuss the solutions. (See Lesson 126.) Equations will vary.

</td><td>

- Follow a similar procedure for the following word problem. Explain that since the answer represents the whole (100% of the shots that John attempted), the formula *percent × whole = part* can be used to write an equation or a proportion. (See Lesson 127.) Equations will vary; $\frac{30}{100} = \frac{6}{s}$, $s = 20$.

  John made 30% of his three-point shots during the basketball season. If John made 6 three-point shots that season, what was the total number of three-point shots attempted by John? 20 shots

- Direct the students to solve these percent problems using the method of their choice.
  2% of 500 = 10
  18% of 25 = 4.5
  70% of *n* is 35 $n = 50$
  36 is 45% of *n* $n = 80$

Student Edition pages 288–89

- Read and explain the directions for pages 288–89. Assist the students as they complete the pages independently.

MATH TALK

Follow the General Procedure for Math Talks as outlined in Lesson 5 on Teacher Edition page 13b.

- Represent $8\frac{1}{2}\%$ in as many different ways as you can.

- Compare: 59% 0.60 $\frac{2}{3}$

</td></tr>
</table>

CHAPTER 13 TEST

CUMULATIVE REVIEW

CONCEPT REVIEW

- Reading and interpreting a circle graph (Lesson 61)
- Solving word problems (Chapters 1–13)
- Simplifying square roots, exponents, and expressions (Chapter 2)
- Solving for a missing term in a proportion (Lesson 122)
- Calculating the area of a complex figure (Lessons 106–7)
- Calculating the circumference of a circle (Lesson 105)
- Calculating the volume of a cylinder (Lesson 116)
- Identifying the equation for the volume of a rectangular prism (Lesson 114)
- Identifying the equation for the area of a triangle (Lesson 107)

- To prepare the students for the format of achievement tests, instruct them to work on a separate sheet of paper, if necessary, and to mark their answers on the *Cumulative Review Answer Sheet.*

Student Edition pages 290–92

The Cumulative Review provides additional practice of previously learned concepts. These pages may be completed during this lesson or anytime after this lesson, since they require limited or no teaching.

Use the circle graph to find the answer.

The circle graph represents examples of 3-dimensional figures found at home.

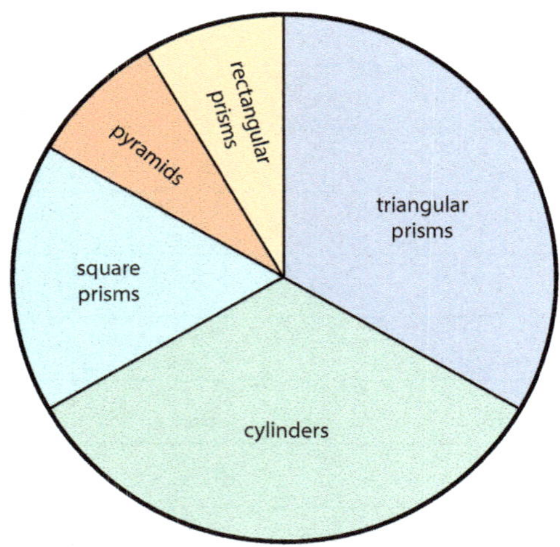

1. What categories are least represented?
   A. pyramids and triangular prisms
   B. cylinders and rectangular prisms
   C. square prisms and pyramids
   D. pyramids and rectangular prisms

2. Cylinders represent what part of the graph?
   A. $\frac{1}{3}$ C. $\frac{1}{6}$
   B. $\frac{1}{2}$ D. $\frac{1}{9}$

3. Square prisms represent what part of the graph?
   A. $\frac{1}{3}$ C. $\frac{1}{6}$
   B. $\frac{1}{2}$ D. $\frac{1}{9}$

4. What part of the graph is made up of cylinders and square prisms?
   A. $\frac{1}{2}$ C. $\frac{3}{4}$
   B. $\frac{2}{3}$ D. $\frac{5}{6}$

5. What type of figure is as equally represented as cylinders?
   A. pyramids
   B. triangular prisms
   C. square prisms
   D. rectangular prisms

Choose the answer.

6. A roll of quarters has a value of $10.00. Ana has $16\frac{3}{4}$ rolls of quarters. How much money does she have?

- **A.** $165.75
- **B.** $167.50
- **C.** $170.25
- **D.** $175

7. Dad bought grass seed for the lawn. Each bag covers 1,000 ft². How many bags did he buy if the yard is 120 ft × 60 ft and the house takes up about $\frac{1}{2}$ of the area?

- **A.** 3 bags
- **B.** 4 bags
- **C.** 6 bags
- **D.** 8 bags

8. Mariah collected 1 dozen eggs on Monday and twice as many on Tuesday. How many eggs did she collect in all?

- **A.** 2 dozen
- **B.** 30 eggs
- **C.** 36 eggs
- **D.** $1\frac{1}{2}$ dozen

9. Jude learned 15 Bible verses for the Bible quiz team. Dylan learned 3 times as many verses as Jude. How many verses did the two boys learn altogether?

- **A.** $15 + 15 + 15 = 45$ verses
- **B.** $15 + (15 + 3) = 33$ verses
- **C.** $15 + (3 \cdot 15) = 60$ verses
- **D.** none of the above

10. Working together, it takes Jace and Spencer $2\frac{1}{2}$ hr to mow and trim the Allens' lawn. It takes them $3\frac{3}{4}$ hr to mow and trim the Reas' lawn. How many hours will it take them on Saturday to mow and trim both lawns?

- **A.** 5 hr
- **B.** $5\frac{1}{4}$ hr
- **C.** $5\frac{1}{2}$ hr
- **D.** $6\frac{1}{4}$ hr

11. $\sqrt{121}$

- **A.** 10
- **B.** 11
- **C.** 12
- **D.** 13

12. 5^4

- **A.** 125
- **B.** 200
- **C.** 500
- **D.** 625

13. $3 + 5 \times 8 - 2$

- **A.** 23
- **B.** 32
- **C.** 41
- **D.** 43

14. $^-2 + 6$

- **A.** $^-4$
- **B.** 0
- **C.** 4
- **D.** 8

15. 31×15

- **A.** 375
- **B.** 405
- **C.** 455
- **D.** 465

16. $590 \div 14$

- **A.** $42.\overline{1}$
- **B.** 42.04
- **C.** 42.14
- **D.** 42.4

17. $600 \div 0.25$

- **A.** 2.4
- **B.** 24
- **C.** 240
- **D.** 2,400

18. $\frac{3}{11} = \frac{n}{44}$

- **A.** $n = 12$
- **B.** $n = 15$
- **C.** $n = 18$
- **D.** $n = 33$

Cumulative Review   291

Choose the answer.

19.

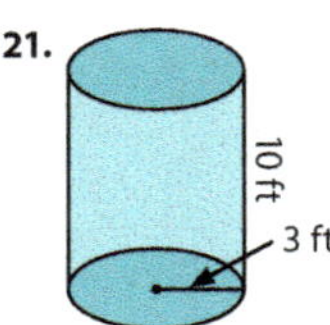

Area is found using the formula $A = l \cdot w$. What is the area of the figure?

A. 267 cm²

B. 303 cm²

C. 258 cm²

D. 315 cm²

20.

The circumference of a circle is found using the formula $C = \pi d$. What is the circumference of circle H?

A. 63.59 cm

B. 14.13 cm

C. 28.26 cm

D. 7.07 cm

21.

The formula for the volume of a cylinder is $V = (\pi r^2) \times h$. What is the volume of this cylinder?

A. 28.26 ft³

B. 94.2 ft³

C. 124.2 ft³

D. 282.6 ft³

22.

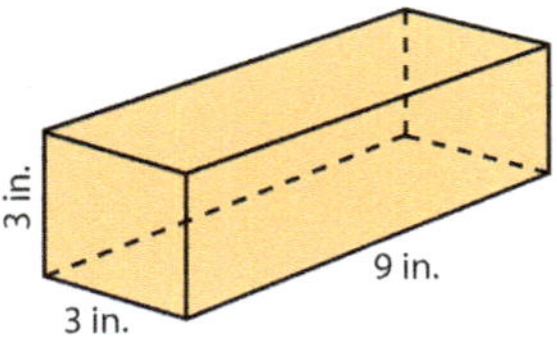

Which equation shows the volume of the rectangular prism?

A. $3 \times 9 = 27$ in.²

B. $(3 \cdot 9) + (3 \cdot 3) = 36$ in.²

C. $2(3 \cdot 3) + 2(3 \cdot 9) = 72$ in.²

D. $3 \times 3 \times 9 = 81$ in.³

23.

Which equation can be used to find the area of the shaded part?

A. $4 \text{ ft} \times 6 \text{ ft} = 24 \text{ ft}^2$

B. $\frac{1}{2}(6 \text{ ft} \times 4 \text{ ft}) = 12 \text{ ft}^2$

C. $(2 \cdot 4 \text{ ft}) + (2 \cdot 6 \text{ ft}) = 20 \text{ ft}^2$

D. $\frac{1}{3}(4 \text{ ft} + 6 \text{ ft}) = 3.3 \text{ ft}^2$

24.

Which equation can be used to find the area of the triangle?

A. $\frac{1}{2}(2.5 \times 1.8) = 2.25 \text{ ft}^2$

B. $2.5 \times 1.8 = 4.5 \text{ ft}^2$

C. $2(2.5 \times 1.8) = 9 \text{ ft}^2$

D. $(2 \cdot 2.5) + (2 \cdot 1.8) = 8.6 \text{ ft}^2$

Building a Bridge
Why can I build a sturdy bridge?

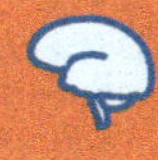

Student Edition pages 265, 293–94

STEM

OBJECTIVES

- Identify the problem.
- Research the problem.
- Suggest bridge design solutions.
- Collaboratively design a bridge prototype.

TEACHER RESOURCES

- 11 *Graph Paper* (2 for the teacher and 2 for each student; extra copies as needed)
- 25 *Engineering Design Process*
- 79 *Bridge & Truss Designs* (for the teacher and for each student)

ADDITIONAL MATERIALS

- pictures of bridges
- transparent tape
- wooden craft sticks ($4\frac{1}{2}$" × $\frac{3}{8}$" × $\frac{1}{16}$"; 200 for each group)

Ask a question to identify the problem.

1. What problem do I want to solve? *designing and building the most efficient bridge from the specified materials that will span a 12 in. opening*

2. Give the stated requirements for your structure. *Answers may vary.*
   a. distance bridge must span *12 in.*
   b. number of sticks that may be used *100*
   c. maximum length bridge may measure *15 in.*
   d. other requirements *sample answer: top suitable for bearing a load for testing*

Imagine possible solutions.

3. Give the ratio equation for measuring a bridge's efficiency. $\text{bridge efficiency} = \dfrac{\text{mass of load}}{\text{mass of bridge}}$

4. How could I solve the stated problem? *sample answer: I could design my bridge with structures that would make it as strong as possible using the fewest sticks and draw a plan for its construction.*

Plan to solve the problem.

5. Explain your group's plan to build an efficient bridge, using words and/or drawings and including measurements. Give the structures you plan to use to make your bridge sturdy. *Answers will vary.*

Lessons 131–33   293

Preparation

Visit TeacherToolsOnline.com for additional STEM resources on bridge design.

Locate pictures of bridges of varying design, including historic, unusual (very long, innovatively designed), and artistically beautiful bridges, to display during the lesson. Include pictures of the bridges mentioned but not pictured on Student Edition page 265, if available. A list of suggested bridges to display is provided at the end of this lesson.

To save time during the lesson, prepare the *Graph Paper* pages by folding under one of the pages along the top of the grid. Tape it to the bottom of the grid on the other page to provide a $17\frac{3}{4}$ in. continuous grid that the students can use for drawing their part of a bridge plan.

Display the *Engineering Design Process* page and refer to it as you proceed through the steps of the STEM project.

Ask

Bridge design

- Guide a **visual analysis** of bridges to help the students connect bridge design with bridge efficiency in the bridge they will build.

- Direct the students to read the career link about Mr. Rosales at the bottom of Student Edition page 265 silently.

  What are Mr. Rosales's bridge designs known for? *their beauty and how they fit into their surroundings*

  Besides beauty, what other qualities of a bridge do you think are important to a bridge designer? *sample answers: reliability (strength and safety), cost of the materials, length required, how well it will function in the setting in which it is needed, availability of materials, durability (how long it will last)*

- Distribute a copy of the *Bridge & Truss Designs* page to each student and display your copy. Discuss the drawing and description of each type of bridge. Point out the 3 common truss designs.

  Point out that the Liberty Bridge designed by Mr. Rosales, shown in the inset picture on Student Edition page 265, is an innovatively designed cable-stayed bridge.

- Display the pictures of bridges that you located, guiding the students as they identify the structural design of each bridge by comparing it with the *Bridge & Truss Designs* page.

- Point out that because of the different ways different bridge designs handle the forces or load put on them, certain designs are better suited to certain settings.

  Why do you think a bridge designer might choose a suspension bridge over a beam bridge for a longer span? A beam bridge must rest on pillars beneath its deck, which might not be practical or even possible for a long distance. A suspension bridge is supported by cables above the deck connected to towers which help bear the load. A suspension bridge can typically stretch a longer distance than a beam bridge.

  Point out that many bridges combine different structures in their design. For example, suspension bridges sometimes also use trusses under the deck for support.

The problem

- Use **direct instruction** to help the students identify the problem they need to solve.

- Write the word "efficiency" for display. Explain that a bridge's efficiency is a measurement of how well the bridge performs. Point out that the bridges the students build for the STEM project will be tested for efficiency.

  The suggested requirements for the project may be adjusted as desired.

---

6. Collaborate to divide the construction process among your group members. Record the responsibilities of each person. *Answers will vary.*

7. Using graph paper, draw a full-size model of the section of the bridge that you are responsible to build. *Answers will vary.*

Create to implement your solution.

8. Build a bridge according to the specifications of your project.

Test and improve your solution.

9. Weigh the bridge and record its weight in grams. Round to the nearest tenth. *Answers will vary.*

10. Test the bridge and calculate its efficiency.
    a. Record the weight of the load the bridge held until failure. Round to the nearest whole number. *Answers will vary.*
    b. Convert the weight of the load from pounds to kilograms. (2.2 lb ≈ 1 kg) Round to the nearest whole number. Rename kilograms to grams for the final answer. (1 kg = 1,000 g)
    c. Calculate the bridge's efficiency, using a ratio. Round to the nearest whole number.

11. How could I improve my design?

Reflect.

12. Why can I build a sturdy bridge? *The reliability of ratios in the design of God's world makes building sturdy bridges possible (Hebrews 1:3).*

- Explain to the students that they will work with their group to design and build a bridge made of only 100 or fewer popsicle sticks and all-purpose glue. The bridge must span a 12 in. opening but be no longer than 15 in. The bridge must be able to support a load placed on top of it so that it can be tested to failure (until it breaks and no longer supports the load) to determine its efficiency.

  What problem do you need to solve? how to design and build the most efficient bridge from the specified materials that will span a 12 in. opening

  Direct the students to answer problems 1–2 on Student Edition page 293.

---

10. b. *Divide the weight of the load (lb) by 2.2 to convert to kilograms (round to the nearest whole number); multiply the number of kilograms by 1,000 to rename to grams.*

   c. *Answers will vary.*
   bridge efficiency = $\frac{\text{mass of load}}{\text{mass of bridge}}$

11. *sample answers: I could change my truss design, add more cross bracing, strengthen my beams, or use fewer sticks for greater efficiency.*

Imagine

Bridge efficiency

- Guide a **discussion** to help the students identify an efficient design for their bridge. Emphasize the importance of good planning in bridge design. Refer to Student Edition page 265.

  What type of planning does Mr. Rosales do when designing a bridge? He makes a lot of 3D drawings and models and computer drawings.

- Explain that building the strongest bridge possible with the least amount of materials makes it efficient. A bridge's efficiency can be calculated mathematically by comparing the mass of the load the bridge can bear to the bridge's mass.

- Display the following equation.

$$\text{bridge efficiency} = \frac{\text{mass of load}}{\text{mass of bridge}}$$

  How could you solve the stated problem? I could design my bridge with structures that would make it the strongest possible using the fewest sticks and draw a plan for its construction.

- Direct attention to the *Bridge & Truss Designs* page once again.

  Which bridge design do you think would work best for the project? sample answer: a truss bridge; Trusses are a strong structure that can be made from popsicle sticks. Suspension and cable-stayed bridges rely on cables which are not materials specified for my project. An arch would be difficult to construct from rigid sticks. Strong pillars for a beam bridge would be difficult to construct from sticks, and the deck of a popsicle-stick beam bridge might not be strong enough to support a heavy load across an opening.

- Direct the students to answer problems 3–4 on Student Edition page 293.

Plan

Designing a bridge

- Use **direct instruction** to provide some guiding principles for popsicle-stick bridge design.

  1. Use the fewest sticks needed for a strong structure to increase the efficiency of the bridge.
  2. Triangles are a strong shape, so trusses are a good idea.
  3. Breaking or cutting sticks weakens them.
  4. Gluing two sticks together provides more strength.
  5. Symmetry is important.
  6. Draw a plan using sticks and graph paper to picture the elements of the structure (e.g., trusses, lower beams, upper beams, cross bracing [a *v*- or an *x*-structure across an opening]).

Although cutting sticks is discouraged, students may feel that their design requires it. Encourage them to glue cut sticks to another stick to provide stability.

- Guide the students in a **collaborative activity** to design a popsicle-stick bridge prototype.

- Direct the students' attention to problems 5–7 on Student Edition pages 293–94. Instruct them to complete the problems during their group planning time. Assist the students as they complete the pages independently.

- Distribute 2 *Graph Paper* pages to each student. If the pages were not connected before class, guide the students as they prepare a longer grid with 2 pages.

  Instruct the students to take their copy of the Student Edition, paper for note taking and for completing the Student Edition pages, and the *Bridge & Truss Designs* and *Graph Paper* pages with them for the group planning time.

- Assemble the students in groups and distribute 200 popsicle sticks to each group to use for planning.

  Direct the groups to discuss their plan and record notes about their strategy for building a sturdy, efficient bridge.

Since the bridge will be loaded from the top for testing, a solid deck is not required.

- When each group has a basic plan in mind, instruct the group to divide the design of each surface of the bridge among the group members. For example, the design might require a different planning page for each side (or one page for identical sides) and the top, bottom, and ends of the structure.

  Instruct the students to use the sticks to plan and draw a full-size model of their section of the group's design on the *Graph Paper* page. Remind them to include the measurements of their design.

- When the groups' designs are completed, direct the students to check their design against the specifications for the project. Instruct them to make any needed adjustments to their design.

- Direct the students to label the *Graph Paper* page model with their name. Collect and store the models and sticks for the next lesson.

Student Edition pages 293–94

- Problems 1–7 were assigned during the lesson.

Resources

Following are suggested bridges to display during the visual analysis activity.

- Arch bridges:
  - Roman stone arch bridge over the river Meles in Turkey
  - New River Gorge Bridge in WV
  - Natchez Trace Parkway Bridge in TN

- Beam bridges:
  - Lake Pontchartrain Causeway Bridge in LA
  - The Lego Bridge in Germany
  - Chengyang Bridge in China

- Cable-stayed bridges:
  - Margaret Hunt Hill Bridge in TX
  - Arthur Ravenel Jr. Bridge in SC
  - Langkawi Sky Bridge in Malaysia

- Suspension bridges:
  - Inca rope bridge at Q'eswachaka, Peru
  - Akashi Kaikyo Bridge in Japan
  - Brooklyn Bridge (hybrid cable-stayed/suspension bridge) in NY

- Truss bridges:
  - Helix "DNA" Bridge (curved tubular truss bridge) in Singapore
  - The Rolling Bridge (movable curling truss bridge) in England
  - Don N. Holt Bridge in SC

STEM

OBJECTIVE
- Build a bridge prototype according to specifications.

TEACHER RESOURCES
- 80 *STEM Rubric: Building a Bridge*

ADDITIONAL MATERIALS
- paper to cover work area
- prepared models of the *Graph Paper* page (from Lesson 131)
- wooden craft sticks ($4\frac{1}{2}$" × $\frac{3}{8}$" × $\frac{1}{16}$"; 200 for each group; from Lesson 131)
- all-purpose glue (for each student)
- cotton swabs (for each student)
- scissors or shears suitable for cutting wooden sticks (optional)
- transparent tape
- binder clips (0.75", 1.25", and 2"; 50+ for each group)

Provide an assortment of binder clips for holding the sticks together on the assembled bridge sections while the glue dries. Smaller clips are easier to open but may not stretch wide enough to provide enough pressure. Larger ones are more secure but may be more difficult to open. One clip for each joint on the bridge is helpful.

Preparation
Cover the work areas with paper and assemble the supplies for each group before class.

Create

Building a bridge
- Guide a **collaborative activity** in which the students implement their designs by constructing a bridge.

- Explain to the students that they will follow their design to construct their bridge during today's lesson. After their bridge is assembled, they will allow the glue to dry overnight and then test the bridge for efficiency in the next lesson.

It may be helpful to divide today's lesson into 2 periods to allow time for the glue in some structures to dry before attaching the remaining structures for a completed bridge.

Encourage the students to work carefully but to view the project as an opportunity to learn what makes a bridge efficient.

- Mention some helpful procedures for constructing a bridge:
  1. Examine sticks for any obvious signs of weakness or deformity and use only strong sticks in the structure.
  2. It may be helpful to lay out the structure on the *Graph Paper* page model to align the sticks as they are glued together, being careful to glue only the sticks together.
  3. Use only as much glue as is needed by using a cotton swab to spread it thinly on both surfaces being glued together. Excessive glue will add extra weight to the bridge and reduce its efficiency.
  4. Depending on the design, it may be easier to construct the sides of the bridge first and let them dry for a while before joining them to the top, bottom, and ends.
  5. Clamp wherever 2 sticks are glued together while they dry to help strengthen the structure. Handle assembled structures carefully.
  6. Some structures, such as cross bracing, may not be possible to clamp for drying. Transparent tape may be carefully applied to hold those sticks in place to dry. Remove the tape carefully.
  7. Allow the completed bridge to dry overnight before testing.

- Assemble the students in their collaborative groups and distribute the supplies to each group. Guide the students as they complete problem 8 on Student Edition page 294.

Student Edition page 294
- Problem 8 was assigned during the lesson.

Assess

Rubric
- Use the prepared *STEM Rubric: Building a Bridge* page or design a **rubric** to include your chosen criteria. Evaluate the student's progress in the Model category to evaluate the bridge design before testing the bridge to failure in the next lesson.

- Retain the rubric to evaluate the student's progress in the other categories when the project is completed.

Preparation

If you plan to use books for loading the bridge to test its efficiency, it may be helpful to record their weight ahead of class to save time. A bathroom scale may be helpful for this.

Arrange to have a helper photograph or video the testing process to free you to guide the students.

Test & Improve

Testing design efficiency

- Guide a **collaborative activity** for the students to test their bridges and suggest improvements.

- Assemble the students in their groups with their bridges. Instruct each group to carefully remove any tape or clamps remaining on the bridge.

  Before testing begins, take pictures of each bridge from different angles, including a photo of the bridge with the group members who constructed it.

- Direct the students' attention to problems 9–10 on Student Edition page 294. Discuss the procedure they will follow to test their bridge.

- Guide the students as they weigh each bridge by using the digital gram scale or metric balance scale to find its weight in grams.

  If needed, balance the bridge on a clear plastic ruler or sheet of plastic across the scale in such a way that the weight can be read. Weigh the ruler first and subtract its weight from the total weight to find the weight of the bridge alone, rounded to the nearest tenth.

  Direct the students to record the weight of the bridge in grams for problem 9.

- Set up the testing site by arranging 2 chairs or other supports of equal height at a 12 in. distance from one another.

  Balance the ends of the bridge on the supports.

- It may be helpful to lay a flat plywood or metal baking sheet across the top of the bridge to help support the load evenly. Be sure to add the weight of the sheet to the total load.

Direct each observer to put on his safety goggles or glasses before you begin testing. Caution the person loading the weight to keep his feet away from beneath the bridge.

- Direct each group to select a member to record the weight as it is loaded on the bridge. Use a scale, as needed, to weigh the items in pounds or ounces before they are added to the load.

- Count to 10 after each addition of weight before adding more weight to the load.

- When the bridge fails, total the weight the bridge held before failure and record the weight of the load rounded to the nearest whole number (nearest pound) for problem 10a.

  Collect the pieces from the bridge and return them to the group for analysis.

Calculating efficiency

- Guide a **discussion** to help the students analyze the design of their bridge by calculating its efficiency using a ratio.

- After all the bridges have been tested, direct the students to take their Student Edition and paper on which their answers are recorded and to assemble with their group.

- Guide the students to use a calculator to complete problems 10b–c.

  What does problem 10b prompt you to do? convert the weight of the load my bridge held in pounds (rounded to the nearest whole number) to kilograms and then to grams

  How will you convert from pounds to kilograms? Since 2.2 lb ≈ 1 kg, I can divide the weight in pounds by 2.2 to get kilograms.

  Note that the problem directs them to round their answer to the nearest whole number.

  How do you convert kilograms to grams? Since 1 kg = 1,000 g, I can multiply my answer by 1,000 to get grams.

  Display the efficiency score equation, "bridge efficiency $= \frac{\text{mass of load}}{\text{mass of bridge}}$." Explain that this number represents their bridge's efficiency score. Note that problem 10c directs them to round their answer to the nearest whole number.

  Direct the groups to calculate and record their bridge's efficiency for problem 10c.

STEM

OBJECTIVES

- Test design.
- Analyze design by using math.
- Solve a real-world problem by using a ratio.
- Suggest improvements.
- Apply learned design and engineering practices to appreciate God's design in creation. **BWS**

BIBLICAL WORLDVIEW SHAPING

- Design (Apply): Using design and engineering practices to build fosters appreciation for the reliability of mathematical design in creation.

ADDITIONAL MATERIALS

- completed bridges
- camera
- metric scale, measuring grams
- rigid clear plastic ruler or sheet of plastic
- 2 chairs or other supports
- bathroom scale, measuring pounds
- plywood sheet or metal baking sheet (optional)
- safety goggles (for each observer)
- weights, heavy books, or bucket and sand
- rug or pad (optional)
- calculator
- *STEM Rubric: Building a Bridge* pages (partially completed copies from Lesson 132)

- Instruct the groups to examine the remains of their bridge and to take note of which parts held and which parts of it failed under stress. Direct them to discuss as a group any improvements they would make to their design and to record their answer to problem 11.

Appreciating God's design

- Guide a **discussion** to help the students apply what they have learned about designing an efficient bridge to appreciate God's design in creation.
- Read aloud the following account of the 1940 Tacoma Narrows Bridge collapse.

On July 1, 1940, the first Tacoma Narrows Bridge, a suspension bridge crossing the Tacoma Narrows strait of Puget Sound in the state of Washington, opened to traffic. On November 7, 1940, a short 5 months later, the bridge collapsed into the water below it. What happened?

Even during its construction, bridge workers observed that the bridge would move during winds. They nicknamed the bridge Galloping Gertie. Although the engineers who designed it believed that the bridge was solid enough to withstand the windy conditions, they had failed to consider certain scientific principles that governed the stability of the bridge. Winds caused the bridge to twist unexpectedly and fail.

Thankfully, no human lives were lost in the collapse, and Galloping Gertie serves as an example of the unchanging principles of physics and design which cannot be ignored without disastrous consequences.

- Choose a student to read aloud Hebrews 1:2–3.

Who made the world and upholds all things by the word of His power? the Son of God, Jesus Christ

- Point out that although bridges are often considered marvels of engineering, it is God who designed the world to reflect reliable ratios, such as the one they used to calculate the efficiency of their bridge. Men can use those reliable ratios to design reliable structures.

Why are you able to build a sturdy bridge? A bridge is sturdy when its design respects reliable ratios like the one I used to determine the efficiency of my bridge. This is a helpful value that I am able to calculate because of the reliability of the design in God's world.

Student Edition page 294

- Problems 9–11 were assigned during the lesson. Assist the students as they complete problem 12 independently.

Assess

Rubric

- Use the prepared *STEM Rubric: Building a Bridge* page or design a **rubric** to include your chosen criteria.
- If using the prepared rubric, evaluate the student's progress in the categories other than the Model category when the project is completed. The Data Record category may be used to evaluate the student's completion of the Student Edition pages for the STEM project.

Enrichment topics

Following are ideas for further study related to the STEM project.

- Research the ways forces (loads) act on bridges and how the different structural parts deal with load.
- Research the design and construction of the Brooklyn Bridge in New York City, NY, the longest suspension bridge in the world at the time of its construction. Investigate the problems encountered and the innovations employed.
- Watch a video of the Tacoma Narrows Bridge collapse of 1940. Why did the bridge fail, and how did its collapse affect the engineering of future bridges?
- What is the oldest bridge still in existence? What is the oldest bridge still in use? Where is the longest bridge in the world located?
- Research a bridge designer, either historic or contemporary. How has his work influenced bridge design? Contact a living designer to let him know of your interest in his work.
- Visit local bridges and photograph them. Publish your images to a photoblog.

Lesson Plan Overview

Day	Lesson	Student Edition Pages	Teacher Edition Pages	Teacher Resources & Additional Materials	Topics, Skills & Biblical Worldview Shaping
136	134	295–97	295–97b	**Teacher Resources** • 11 *Graph Paper* • 81 *Measurement Flashcards* **Additional Materials** • rulers • a yardstick • a measuring tape (customary) • colored pencils	**Topic:** Linear Measurement **Skills:** estimating, measuring, and converting linear measurements; finding a fraction of a measurement unit; adding and subtracting linear measurements **BWS:** Knowledge
137	135	298–99	298–99b	**Teacher Resources** • 82 *Customary Measurement Craze* **Additional Materials** • a spring scale • a 1 lb loaf of bread • objects to weigh • customary measurement containers • waterproof containers • water (2 gal) • food coloring (optional)	**Topic:** Weight & Capacity **Skills:** estimating, weighing, measuring, and converting customary weight and capacity measurements; adding and subtracting customary weight and capacity measurements
138	136	300–301	300–301b	**Teacher Resources** • 83 *Metric Measurement* • 84 *Metric Measure Mania* **Additional Materials** • a meter stick • a yardstick • measuring tapes (metric) • paper clips	**Topic:** Metric Linear Measurement **Skills:** estimating, measuring, and converting metric linear measurements; determining the appropriate linear measurement; comparing measurements by using > or <; adding and subtracting metric linear measurements
139	137	302–3	302–3b	**Teacher Resources** • 83 *Metric Measurement* **Additional Materials** • 1 L bottles of water • a 1 L beaker • a 5 mL plastic syringe • water for measuring • food coloring (optional) • waterproof containers • additional containers for measuring • a balance or metric scale **Assessments** • Chapter 14 Quiz 1	**Topic:** Metric Capacity & Mass **Skills:** estimating, measuring, and converting metric capacity and mass measurements **BWS:** Knowledge

Day	Lesson	Student Edition Pages	Teacher Edition Pages	Teacher Resources & Additional Materials	Topics, Skills & Biblical Worldview Shaping
140	138	304–5	304–5b	Teacher Resources • 81 *Measurement Flashcards* • 85 *Customary Measurement Word Problems* • 86 *Metric Measurement Word Problems*	**Topic:** Customary & Metric Systems **Skills:** solving measurement word problems
141	139	306–7	306–7b	Teacher Resources • 87 *Thermometer* • 88 *Temperature Hunt* • 89 *Double-Scale Thermometer* Additional Materials • waterproof containers • water • waterproof thermometers • additional thermometers • rulers	**Topic:** Fahrenheit & Celsius **Skills:** identifying standard Celsius and Fahrenheit temperatures, determining a reasonable temperature, reading a thermometer, determining the difference between temperatures, measuring temperatures, converting temperatures between systems
142	140	308–9	308–9b	Teacher Resources • 81 *Measurement Flashcards* • 90 *Customary & Metric Conversions* Additional Materials • a larger set of measurement flashcards • measurable objects • rulers	**Topic:** Relating Customary & Metric Units **Skills:** identifying approximate equivalencies between customary and metric units of measurement, comparing customary and metric measurements, estimating conversions between customary and metric measurements **BWS:** Knowledge
143	141	310–11	310–11b	Teacher Resources • 91 *Time Measurement* Additional Materials • a demonstration clock	**Topic:** Telling & Renaming Time **Skills:** differentiating between a.m. and p.m., converting units of time, finding a fraction of a unit of time, adding and subtracting time
144	142	312–13	312–13b	Teacher Resources • 91 *Time Measurement* • 92 *Time Zones of the World* • 93 *Time Zones* Additional Materials • demonstration clocks Assessments • Chapter 14 Quiz 2	**Topic:** Elapsed Time & Time Zones **Skills:** using a table to determine world time zones, determining elapsed time **BWS:** Knowledge
145	143	314–15	314–15b	Teacher Resources • 92 *Time Zones of the World* • 94 *Map Key* • 95 *Blank Thermometer* Additional Materials • labeled measurable objects from Lesson 140 • measuring tools • rulers	**Topic:** Renaming Units of Measure **Skills:** estimating customary and metric measurements, calculating measurements, solving measurement word problems, determining mileage by using a map scale

Day	Lesson	Student Edition Pages	Teacher Edition Pages	Teacher Resources & Additional Materials	Topics, Skills & Biblical Worldview Shaping
146	144	316–17	316–17b		**Topic:** Unit Multipliers **Skills:** writing an equivalency as a unit multiplier, converting measurements by using a unit multiplier
147	145	318–19	318–19b	Teacher Resources • 95 *Blank Thermometer* Additional Materials • a larger set of measurement flashcards from Lesson 140 • demonstration clocks	**Topic:** Chapter Review
148	146	320–23	320–23	Additional Materials • demonstration clocks (optional)	**Topic:** Test & Cumulative Review

CHAPTER OBJECTIVES

- Estimate and measure using customary and metric linear, weight (mass), and capacity measurements.
- Convert measurements to smaller or larger units.
- Solve problems with time and measurement.
- Solve problems with Fahrenheit and Celsius temperatures.
- Explain how the act of measuring is rooted in Scripture.

To determine whether your students are prepared for the concepts taught in this chapter, you may wish to consult the preassessment checklist found on TeacherToolsOnline.com.

A *Parent Letter* page informing parents of the items needed for this chapter was provided in Chapter 13. Check the teacher resources and additional materials lists for each lesson ahead of teaching the lesson.

In this edition we have abbreviated customary measurements without a final period (e.g., ft, yd, mi), except for inches (in.), which otherwise might be confused with the word *in*.

Make fact practice, both oral and written, part of your daily math routine to help the students with mastery.

Visit AfterSchoolHelp.com for math practice resources, or visit TeacherToolsOnline.com for additional resources to enhance the lessons.

Student Edition pages 295–97
Daily Review Chapter 14, section *a*

OBJECTIVES

- Explain when measurement began. **BWS**
- Estimate linear measurements by using benchmarks.
- Measure to the nearest inch, half inch, fourth inch, eighth inch, and sixteenth inch.
- Convert linear measurements to smaller or larger units.
- Find a fraction of a measurement unit.
- Add and subtract linear measurements.

BIBLICAL WORLDVIEW SHAPING

- **Knowledge (Explain):** Measurement began when God created an orderly, measurable world.

TEACHER RESOURCES

- 11 *Graph Paper* (for the teacher and for each student; printed 100% size)
- 81 *Measurement Flashcards* (for each student)

ADDITIONAL MATERIALS

- a ruler (measuring to $\frac{1}{16}$ in.; for the teacher and for each student)
- a yardstick
- a measuring tape (customary)
- a colored pencil (for each student)

Preparation

Determine a familiar location 1 mi from your school.

Prepare a set of flashcards for each student from the *Measurement Flashcards* page.

Throughout this chapter, use the measurement flashcards to daily review equivalencies and abbreviations in preparation for the Chapter 14 Test. Allow the students to refer to the equivalencies in the Measurement section of the student Handbook until mastery is achieved. For your convenience the student Handbook is available on TeacherToolsOnline.com.

Linear Measurement

When did the act of measuring begin?

Knowing linear equivalents allows you to rename any measurement as larger or smaller units.

Key Terms
- linear equivalents
- renaming measurements

Linear Equivalents

12 inches (in.) = 1 foot (ft)	36 in. = 1 yard (yd)	5,280 ft = 1 mile (mi)
	3 ft = 1 yd	1,760 yd = 1 mi

Renaming Measurements

Rename larger units as smaller units. Determine the equivalency and then *multiply*.

$$3 \text{ yd} = \underline{\ \ } \text{ in.}$$
$$1 \text{ yd} = 36 \text{ in.}$$
$$3 \times 36 = 108$$
$$3 \text{ yd} = 108 \text{ in.}$$

$$\begin{array}{r} 36 \\ \times\ 3 \\ \hline 108 \end{array}$$

Rename smaller units as larger units. Determine the equivalency and then *divide*.

$$50 \text{ in.} = \underline{\ \ } \text{ ft}$$
$$12 \text{ in.} = 1 \text{ ft}$$
$$50 \div 12 = 4 \text{ r2}$$
$$50 \text{ in.} = 4\tfrac{2}{12} \text{ ft} = 4\tfrac{1}{6} \text{ ft}$$
$$\text{or}$$
$$50 \text{ in.} = 4 \text{ ft } 2 \text{ in.}$$

$$12\overline{)50} \quad 4\tfrac{2}{12}$$
$$\underline{-48}$$
$$2$$

50 is *not* divisible by 12.

Determine the equivalency and solve.

$$\tfrac{3}{4} \text{ of a mile} = \underline{\ \ } \text{ ft}$$
$$1 \text{ mi} = 5,280 \text{ ft}$$
$$\tfrac{3}{4} \times \cancel{5,280}^{\,1,320} = 3,960$$
$$\tfrac{3}{4} \text{ mi} = 3,960 \text{ ft}$$

$$\begin{array}{r} 1,320 \\ \times\ \ \ \ 3 \\ \hline 3,960 \end{array}$$

Exercises

Measure the coin. Use the most precise measurement.

1.  *1 in.*

2.  $\frac{3}{4}$ *in. (or* $\frac{6}{8}$ *in.)*

3. $\frac{7}{8}$ *in. (or* $\frac{14}{16}$ *in.)*

Write the best unit of measurement: in., ft, yd, or mi.

4. height of a cell phone tower *ft or yd*
5. distance from New York to Florida *mi*
6. length of a football field *yd*
7. length of a robin *in.*

Rename.

8. 5 yd = *180* in.
9. 117 in. = ___ yd *3 $\frac{1}{4}$*
10. $\frac{3}{4}$ of a yard = *27* in.
11. $\frac{1}{9}$ of a yard = *4* in.

12. 12 ft = *4* yd
13. 30 in. = ___ ft *2 $\frac{1}{2}$*
14. $\frac{2}{3}$ of a foot = *8* in.
15. $\frac{1}{2}$ of a yard = *18* in.

16. 2 mi = ___ ft *10,560*
17. 7 ft = *2* yd *1* ft
18. $\frac{1}{4}$ of a mile = ___ ft *1,320*
19. $\frac{1}{2}$ of a mile = ___ ft *2,640*

20. 108 in. = *3* yd
21. 6 ft 6 in. = *78* in.
22. $\frac{3}{4}$ of a foot = *9* in.
23. 2,640 yd = ___ mi *1 $\frac{1}{2}$*

24. 1$\frac{2}{3}$ yd = *5* ft
25. 5,290 ft = *1* mi *10* ft
26. $\frac{2}{3}$ of a mile = ___ ft *3,520*
27. 82 in. = ___ ft *6 $\frac{5}{6}$*

Engage

- Assign a **Ticket in the Door** to help the students explore the chapter essential question on Student Edition page 295, "How does the act of measuring point to Scripture?"

Direct the students to write down a Bible example of the act of measuring, such as a linear, weight, or capacity measurement; telling time; or ordering and counting things. Discuss the students' answers.

sample answers: linear measurement—the dimensions of Noah's ark (Genesis 6:15); weight—Mary's pound of spikenard ointment (John 12:3); capacity—the omer and ephah, units of dry volume (Exodus 16:33, 36); telling time—references to noon (1 Kings 18:26) or the hour of the day (Luke 23:44); ordering and counting things—the ordering of the days of creation (Genesis 1) and the counting of the days and the animals during the Flood (Genesis 7)

Explain that Bible examples of measurement teach us when the act of measuring began, why standardized measurements are important, and how measuring should be used.

Adding & Subtracting Measurements

Adding		Subtracting
3 yd 22 in. + 2 yd 16 in. **5 yd 38 in. =** **6 yd 2 in.** 1 yd = 36 in.	1. Add or subtract the smaller units. When subtracting, rename a larger unit as a smaller unit if needed. 2. Add or subtract the larger units. 3. Simplify the answer by renaming smaller units as larger units if possible.	$\overset{3}{\cancel{4}}$ ft $\overset{18}{\cancel{6}}$ in. − 1 ft 8 in. **2 ft 10 in.** 1 ft = 12 in.

Add or subtract. Simplify the answer when possible.

28. 3 ft 6 in.
  + 2 ft 3 in.
  5 ft 9 in.

29. 5 yd 2 ft
  + 1 yd 1 ft
  6 yd 3 ft =
  7 yd

30. 47 ft 8 in.
  + 21 ft 9 in.
  68 ft 17 in. =
  69 ft 5 in.

31. 8 yd 2 ft
  − 3 yd 1 ft
  5 yd 1 ft

32. 11 ft 7 in.
  − 5 ft 9 in.
  5 ft 10 in.

Complete the table.

33.

feet	3	5	8	10
inches	*36*	*60*	*96*	*120*

34.

feet	*6*	*7*	*8*	*9*
yards	2	$2\frac{1}{3}$	$2\frac{2}{3}$	3

35.

inches	*36*	*54*	*72*	*90*
yards	1	$1\frac{1}{2}$	2	$2\frac{1}{2}$

36.

yards	880	1,760	2,640	3,520
miles	$\frac{1}{2}$	1	$1\frac{1}{2}$	2

46. $\overline{ST} = 2\frac{1}{8}$ in., $\overline{TU} = 2\frac{1}{8}$ in., $\overline{US} = \frac{3}{8}$ in.
$P = 2\frac{1}{8}$ in. $+ 2\frac{1}{8}$ in. $+ \frac{3}{8}$ in. $= 4\frac{5}{8}$ in.

Practice & Application

Process used to solve may vary.

37. Christian jogged $\frac{2}{3}$ of a mile. How many feet did he jog? *$\frac{2}{3} \cdot 5{,}280 = 3{,}520$ ft*

38. Syrie used $2\frac{3}{4}$ yd of material to make a skirt. How many inches of material did she use?
$2\frac{3}{4} \cdot 36 = \frac{11}{4} \cdot 36 = 99$ in.

39. Terra ran $\frac{3}{4}$ of a mile. Sawyer ran 900 yd. Who ran farther?
Terra; $\frac{3}{4} \cdot 1{,}760 = 1{,}320$ yd; 1,320 yd > 900 yd

40. The height of the wall from the ceiling to the light switch is 2 yd. The height from the floor to the light switch is 36 in. What is the measurement of the wall from the floor to the ceiling in feet?
2 yd + 36 in. = (2 · 3) + (36 ÷ 12) = 6 + 3 = 9 ft

41. Write a comparison statement using > or < to compare the measurements of a 3 ft wide desk and a $1\frac{1}{2}$ yd wide table.
$1\frac{1}{2} \cdot 3 = \frac{3}{2} \cdot 3 = \frac{9}{2} = 4\frac{1}{2}$ ft; 3 ft < $4\frac{1}{2}$ ft

42. Draw a line segment $2\frac{3}{4}$ in. long. Write equivalent measurements using eighths and sixteenths.
$2\frac{3}{4}$ in. $= 2\frac{6}{8}$ in. $= 2\frac{12}{16}$ in.

43. Write a comparison statement using > or < to compare the measurements of a $10\frac{1}{2}$ in. wide book and a $10\frac{3}{16}$ in. wide book. *$10\frac{1}{2} = 10\frac{8}{16}$; $10\frac{1}{2}$ in. > $10\frac{3}{16}$ in.*

44. Write a comparison statement using > or < to compare the distances of 1,500 yd and 1 mi.
1,500 yd < 1,760 yd

45. Write a comparison statement using > or < to compare the wingspans of 2 butterflies. One measures $1\frac{1}{2}$ in. and the other measures $\frac{7}{4}$ of an inch. *$\frac{7}{4} = 1\frac{3}{4}$; $1\frac{1}{2}$ in. < $\frac{7}{4}$ in.*

46. Find the measurements of $\overline{ST}$, $\overline{TU}$, and $\overline{US}$. Find the perimeter of $\triangle STU$.

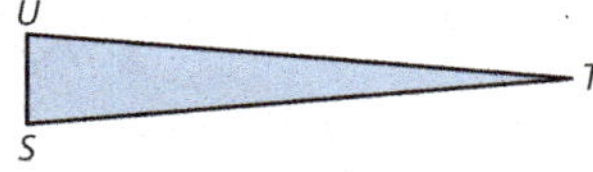

47. When did the act of measuring begin?
Measuring began when God created an orderly, measurable world. (Genesis 1)

Lesson 134   297

- Choose 3 students to place their rulers, one at a time, on top of the yardstick to show the following fractional comparisons: $\frac{1}{3}$ ($\frac{12}{36}$), $\frac{2}{3}$ ($\frac{24}{36}$), and $\frac{3}{3}$ ($\frac{36}{36}$) of the whole.

  How far is a mile? sample answers: the distance from __ (a location with which the student is familiar) to __ (another location); 5,280 ft; A mile is a long unit of measurement that is used for measuring distance.

  What is used to measure a mile? sample answers: a car's odometer or a pedometer

- Display the following equivalencies for 1 mi: 1 mi = 5,280 ft = 1,760 yd. Explain that the average person can walk 1 mi in about 20 min. Point out that __ (a predetermined location) is 1 mi from their school.

- Explain that when an exact measurement is not needed, they can estimate.

  What part of your hand could you use to estimate 1 in.? sample answer: the midsection of my index finger

  What could you use to estimate 1 ft? sample answers: the distance from my elbow to my knuckles; the width of my body

  What could you use to estimate 1 yd? sample answer: the distance from my nose to the end of my arm stretched out sideways

- Choose students to use a benchmark to estimate the length of the room in yards.

  Choose 2 students to collaboratively use a yardstick along the base of the wall to measure its length.

  Does the wall measure to an exact yard? Answers will vary.

  Point out that unless the wall measures to an exact yard, the measurement is not precise.

Instruct

When measurement began

- **Discuss** the lesson essential question at the top of Student Edition page 296, "When did the act of measuring begin?" to connect mankind's act of measuring with the origin of measuring as related in Scripture.

  Remind the students that God the Creator exists outside the limits of time (He is eternal) or space (He is infinite). He cannot be measured.

  When did the act of measuring begin? It began when God created an orderly, measurable world. (Genesis 1)

- Direct the students' attention back to Psalm 90:12 on Student Edition page 295. Point out that time is one example of an act of measuring that began at creation.

  In what way are you responsible for the gift of time? God wants me to recognize the shortness of my life so that I will use my earthly days wisely.

Estimating linear measurements

- Guide an **interactive activity** to help the students estimate measurements.

  Display the ruler, yardstick, and measuring tape.

LESSON 134

How could you get a more precise measurement? I could measure using smaller units, such as feet or inches.

- Repeat the procedure using 1 ft rulers to measure the wall in feet.

  Is this a precise measurement? It is more precise than yards, but unless the wall measures to an exact foot, it is still not precise.

- Follow a similar procedure to measure the wall in inches using the measuring tape.

  What customary unit of measurement is more precise than an inch? Although the inch is the smallest linear customary unit, a fractional part of an inch would be a more precise measurement.

- Ask students to choose the most precise unit of measure for the following:
  length of a pencil inches
  height of a student inches or feet
  length of carpet feet or yards
  distance between cities miles

Measuring to a fractional part of an inch

- Use **guided practice** to help the students measure precisely.

- Distribute a *Graph Paper* page to each student. Direct the students to use a colored pencil to lightly outline 2 figures on the *Graph Paper* page, a 3×3 square that contains 9 units^2 and a 1×25 rectangle that contains 25 units^2.

- Guide the students as they use their rulers to measure one side of the outlined square to the nearest $\frac{1}{2}$ in., $\frac{1}{4}$ in., and $\frac{1}{8}$ in.

  What did the side of the square measure? sample answers: $\frac{1}{2}$ in.; $\frac{3}{4}$ in.; $\frac{5}{8}$ in. or $\frac{6}{8}$ in.

  Point out that the actual length is between eighth marks on the ruler.

- Point out the sixteenth marks on your ruler.

  What does each sixteenth mark on your ruler measure? $\frac{1}{16}$ of an inch; The marks divide each inch into 16 equal parts.

Write the ratio as a fraction in lowest terms.

1. 10 peppermints to 6 lemon drops $\frac{10}{6} = \frac{5}{3}$
2. 2 c sugar to 10 c water $\frac{2}{10} = \frac{1}{5}$
3. 8 elephants to 7 giraffes $\frac{8}{7}$
4. 54 cookies to 6 students $\frac{54}{6} = \frac{9}{1}$

Use the table to write the ratio. *Ratio form may vary.*

5. cats to dogs *6:4*
6. lizards to birds $\frac{3}{12}$
7. fish to total animals *50:93*
8. dogs to hamsters *4 to 7*
9. animals with fur to animals without fur $\frac{20}{73}$
10. reptiles to fish *11:50*

Andrew's Pet Store			
cats	6	fish	50
dogs	4	hamsters	7
lizards	3	gerbils	3
turtles	8	birds	12

Complete the ratio table.

11.

cars	10	*20*	40	*80*
trucks	6	12	*24*	48

12.

students	19	*57*	95	*171*
girls	10	30	*50*	90

Daily Review   489

Repeat the procedure for students to measure the square to the nearest $\frac{1}{16}$ in. $\frac{11}{16}$ in.

Which measurement is the most precise? $\frac{11}{16}$ in.; $\frac{1}{16}$ in. is a smaller unit of measurement than $\frac{1}{2}$ in., $\frac{1}{4}$ in., or $\frac{1}{8}$ in.; therefore, sixteenths can be used to more precisely measure shorter lengths to get a closer estimate.

- Direct the students to estimate to the nearest inch the length of the long side of the 1×25 rectangle. about 6 in. Then instruct them to measure the length as precisely as possible. $5\frac{12}{16}$ in. or $5\frac{3}{4}$ in.

Converting linear measurements

- Use **discussion** to guide the students as they use an equivalency to convert (rename) linear measurements.

- Write "7 ft = __ in." for display.

  When you rename a larger unit to a smaller unit, will you have more or fewer units? more; It takes more of the smaller units to measure the same length as 1 larger unit.

  When applied to whole numbers, does multiplication or division give you more units? multiplication

  What ratio or equivalency do you know for feet to inches? 1 ft : 12 in. or 1 ft = 12 in.

Write "1 ft = 12 in." for display.

Point out that writing the units of the standard equivalency in the same order that they appear in the problem can help them as they determine whether to multiply or divide.

How many inches are in 7 ft? 84 in.; 7 ft is 7 sets of 12 in.; 7 × 12 in. = 84 in.

Write "7 × 12 in. = 84 in." for display.

Explain that to rename larger units (ft) to smaller units (in.), they determine the equivalency and multiply. Complete the initial equivalency statement.

- Write "5 yd 2 ft = ___ ft" for display.

How can you determine the number of feet in 5 yd 2 ft? I can identify the equivalency (1 yd = 3 ft), multiply to find the number of feet in 5 yd (5 × 3 ft = 15 ft), and then add 2 ft to the product.

What multistep equation can you write to find the number of feet in 5 yd 2 ft? (5 × 3 ft) + 2 ft = ___ ft

Choose a student to write the equation for display and solve it. 17 ft Complete the equivalency statement.

- Repeat the procedure for 5 yd 2 ft = ___ in. 1 yd = 36 in. and 1 ft = 12 in.; (5 × 36 in.) + (2 × 12 in.) = 204 in.

- Follow a similar procedure for 3' 11" = ___" 47, 4 yd 5 in. = ___ in. 149, and 2 mi 350 ft = ___ ft 10,910. Explain that the symbol (') represents foot or feet and the symbol (") represents inch or inches.

Finding a fraction of a unit

- Use **guided practice** to help the students rename a fractional part of a unit.
- Read aloud the following word problem.

Kalee needs $\frac{1}{3}$ of a yard of ribbon for a picture frame. How many inches of ribbon does she need? 12 in.

Explain that "a yard" implies 1 yd, and "of " means the part of 1 yd that they are dealing with. Therefore, $\frac{1}{3}$ of a yard means the same as $\frac{1}{3}$ × 1 yd. To solve the word problem, they rename the unit (yd) as an equivalent number of the desired unit (in.) and then multiply by the fraction; $\frac{1}{3}$ yd = ___ in.; 1 yd = 36 in.; $\frac{1}{3}$ × 36 in. = ___ in.

- Instruct the students to write and solve an equation to find the number of inches in $\frac{1}{3}$ of a yard. $\frac{1}{3}$ × 36 in. = $\frac{36 \text{ in.}}{3}$ = 12 in.; $\frac{1}{3}$ of a yard = 12 in.

- Follow a similar procedure for $2\frac{1}{3}$ yd of material. $2\frac{1}{3}$ yd = ___ in.; $\frac{7}{3}$ yd = ___ in.; 1 yd = 36 in.; $\frac{7}{3}$ × 36 in. = 84 in.; $2\frac{1}{3}$ yd = 84 in.

What generalization can you make when converting larger units to smaller units? I find the equivalency and then multiply.

- Follow a similar procedure to convert (rename) smaller units to larger units using division. Point out that remainders represent a part of the next set of the larger unit.

80 in. = ___ yd 36 in. = 1 yd; 80 in. ÷ 36 in. = $2\frac{8}{36}$ yd or $2\frac{2}{9}$ yd

42 in. = ___ yd 36 in. = 1 yd; 42 in. ÷ 36 in. = $1\frac{6}{36}$ yd or $1\frac{1}{6}$ yd

58" = ___' 12 in. = 1 ft; 58" ÷ 12" = $4\frac{10}{12}$' or $4\frac{5}{6}$'

Adding & subtracting linear measurements

- Use **guided practice** to help the students add and subtract linear measurements.
- Write the following problems for display and guide the students as they add and subtract like units, beginning with the smaller units. Direct them to rename the sum to larger units when possible and to rename in the minuend as needed.

6 ft 3 in.	7 yd 2 ft
+4 ft 9 in.	+9 yd 2 ft
10 ft 12 in.	16 yd 4 ft
11 ft	17 yd 1 ft

7 yd 5 in.	12 ft 5 in.
−2 yd 8 in.	− 7 ft 8 in.
4 yd 33 in.	4 ft 9 in.

- Guide the students as they rename the answers as only the larger unit and then as only the smaller unit. 11 ft 132 in.; 17 yd 1 ft $17\frac{1}{3}$ yd or 52 ft; 4 yd 33 in. $4\frac{11}{12}$ yd or 177 in.; 4 ft 9 in. $4\frac{3}{4}$ ft or 57 in.

Apply

Student Edition pages 296–97

- Read and explain the directions for pages 296–97. Assist the students as they complete the pages independently.

Daily Review

- Students should complete Chapter 14, section *a*.

Student Edition pages 298–99
Daily Review Chapter 14, section *b*

OBJECTIVES

- Estimate and weigh items to the nearest pound or ounce.
- Convert weight and capacity measurements to smaller or larger units.
- Find a fraction of a measurement unit.
- Measure with units of capacity.
- Add and subtract weight and capacity measurements.

TEACHER RESOURCES

- 82 *Customary Measurement Craze*

ADDITIONAL MATERIALS

- a spring scale (measuring pounds and ounces)
- a 1 lb loaf of bread with 16 1-oz slices (or 1 lb of cheese with 16 1-oz slices)
- objects to weigh (less than, equal to, and more than 1 lb)
- customary measurement containers (measuring cup and/or pitcher)
- waterproof containers: four 1 c, four 1 pt, four 1 qt, two 1 gal
- water (2 gal)
- food coloring (optional)

The objects to weigh and the waterproof containers used in this lesson are items that have been brought to class by the students in response to the parent letter that was sent home at the beginning of Chapter 13.

Add food coloring to the water for demonstration purposes if desired.

Weight & Capacity

Why would I weigh less on the moon than on the earth?

Knowing **weight** and **capacity** equivalents allows you to rename measurements as larger or smaller units.

Key Terms
- weight
- capacity

Weight Equivalents

1 pound (lb) = 16 ounces (oz)
1 ton (tn) = 2,000 lb

Capacity Equivalents

1 cup (c) = 8 fluid ounces (fl oz)   1 quart (qt) = 2 pt
1 pint (pt) = 2 c   1 gallon (gal) = 4 qt

Renaming Measurements

Rename larger units as smaller units. Determine the equivalency and then *multiply*.

30 lb = ___ oz
1 lb = 16 oz
30 × 16 = 480
30 lb = 480 oz

$$\begin{array}{r} 30 \\ \times\,16 \\ \hline 480 \end{array}$$

Patrick's pickup truck has a curb weight of $2\frac{1}{4}$ tn. How many pounds does his truck weigh?

$2\frac{1}{4}$ tn = ___ lb
1 tn = 2,000 lb

$$\frac{9}{4} \times \overset{500}{\cancel{2{,}000}} = 4{,}500$$

$2\frac{1}{4}$ tn = 4,500 lb

Rename smaller units as larger units. Determine the equivalency and then *divide*.

19 qt = ___ gal
4 qt = 1 gal
19 ÷ 4 = 4 r3
19 qt = $4\frac{3}{4}$ gal

$$4\overline{)19}\;\,^{4\frac{3}{4}}$$
$$-\underline{16}$$
$$3$$

or

19 qt = 4 gal 3 qt

Claire drank $\frac{1}{4}$ of a gallon of milk. How many quarts did she drink?

$\frac{1}{4}$ gal = ___ qt
1 gal = 4 qt

$$\frac{1}{\underset{1}{\cancel{4}}} \times \overset{1}{\cancel{4}} = 1$$

$\frac{1}{4}$ gal = 1 qt

Exercises

Write the best unit of measurement: oz, lb, tn, fl oz, c, pt, qt, or gal.

1. glass of milk *fl oz or c*   **2.** pitcher of punch *qt or gal*   **3.** yogurt for family *pt or qt*

4. bag of potatoes *lb*   **5.** truckload of bricks *lb or tn*   **6.** water in bathtub *gal*

Complete the table.

7.

pounds	1	2	$2\frac{1}{2}$	3
ounces	16	32	40	48

8.

cups	2	4	5	6
pints	1	2	$2\frac{1}{2}$	3

Rename.

9. 6 lb = 96 oz   **10.** 6 lb 9 oz = 105 oz   **11.** 2 tn 25 lb = 4,025 lb   **12.** 3 tn = 6,000 lb

13. 12 lb 9 oz = 201 oz   **14.** 9 c = 4 pt 1 c   **15.** 4 lb 2 oz = 66 oz   **16.** 10 qt = 2 gal 2 qt

Engage

- Guide a **discussion** of the essential question at the top of Student Edition page 298, "Why would I weigh less on the moon than on the earth?"
- Explain that weight is the measure of the pull of gravity on the mass of their body. The student's body mass is the same everywhere, but more gravity means a larger weight measurement. Larger astronomical bodies, such as the earth, exert more pull of gravity than smaller ones do.

Why would you weigh less on the moon than on the earth? Though the mass of my body would not change, the pull of gravity on the moon is less than the pull of gravity on the earth.

Instruct

Reviewing equivalencies

- Use **guided practice** to help the students review customary measurement equivalencies. Write the following equivalencies for display and choose students to complete them. The students may consult the Measurement section in the student Handbook. Leave the completed equivalencies displayed.

1 pt = ___ c 2   1 qt = ___ pt 2

1 gal = ___ qt 4   1 lb = ___ oz 16

1 tn = ___ lb 2,000

Rename.

17. $\frac{1}{2}$ of a ton = __ lb **1,000**
18. $\frac{5}{8}$ of a pound = **10** oz
19. $\frac{3}{4}$ of a gallon = **3** qt
20. $\frac{3}{8}$ of a cup = **3** fl oz

21. $\frac{1}{2}$ of a pound = **8** oz
22. $\frac{1}{4}$ of a ton = __ lb **500**
23. $\frac{3}{4}$ of a cup = **6** fl oz
24. $\frac{1}{2}$ of a gallon = **2** qt

25. $3\frac{1}{2}$ qt = **7** pt
26. $1\frac{1}{2}$ lb = **24** oz
27. $10\frac{1}{4}$ tn = __ lb **20,500**
28. $3\frac{1}{2}$ gal = **14** qt

Add or subtract. Simplify the answer when possible.

29.   2 tn 345 lb
+ 4 tn 536 lb
6 tn 881 lb

30.   3 gal 2 qt
+ 4 gal 3 qt
7 gal 5 qt =
8 gal 1 qt

31.   2 qt 1 pt
+ 1 qt 1 pt
3 qt 2 pt =
4 qt = 1 gal

32.   2 lb 2 oz
− 1 lb 4 oz
14 oz

33.   3 qt 2 pt
− 1 qt 1 pt
2 qt 1 pt

Practice & Application

34. What fraction of a gallon is a quart? $\frac{1}{4}$

35. Which has the greater capacity: 6 fl oz or $\frac{1}{2}$ of a cup? **6 fl oz**

36. Which amount of hamburger weighs more: $\frac{1}{4}$ of a pound or 8 oz? **8 oz**

37. What fraction of a quart is a pint? $\frac{1}{2}$

38. Which is larger: 2 pt or 3 c? **2 pt**

39. Ben measured the game area by walking heel-to-toe 9 steps. Did he measure 9 in., 9 ft, or 9 yd? **9 ft**

40. Anna measured a ribbon that could stretch from her nose to her fingertips twice. Did Anna measure 2 in., 2 ft, or 2 yd? **2 yd**

41. Coach Willis had the team walk around the track for 20 min. Did the team walk 1 mi, 2 mi, or 3 mi? **1 mi**

42. Kaylee poured two 12 fl oz cans of tomato sauce into a bowl. Did she use a bowl that held 1 c, 1 pt, or 1 qt? **1 qt**

43. Jamile knows that 1 lb equals 16 oz. Which ratio shows the relationship of pounds to ounces: 16:16, 1:16, or 16:1? **1:16**

44. Mother needed $\frac{3}{4}$ of a pound of hamburger for her meat pie. Did she use 4 oz, 8 oz, or 12 oz? **12 oz**

45. Mr. Pennington had a board $2\frac{1}{4}$ ft long. He used 21 in. of the board to build a shelf. How much of the board did he have left: $\frac{1}{4}$ ft, $\frac{1}{2}$ ft, or $\frac{3}{4}$ ft? **6 in. = $\frac{1}{2}$ ft**

46. Why would I weigh less on the moon than on the earth?

DID YOU KNOW?

Weight is a measure of the pull of gravity on an object. *Gravity* is the force of one object pulling on another object. Lunar gravity is $\frac{1}{6}$ of the earth's gravity, so the average astronaut in a lunar suit weighs only 60 lb on the moon.

Mass is the measurement of the amount of matter an object has. Although the astronaut's weight is different on the moon from what it is on the earth, the astronaut's mass—the amount of matter making up his or her body—does not change.

Even though some people use the terms *weight* and *mass* interchangeably, scientists are careful to make a distinction between the words. Scientists use the terms *gram* and *kilogram* as units of mass, but for weight they use the term *newton*.

Lesson 135   299

LESSON 135

Guide the students as they estimate the weight of each item by using the slice of bread (1 oz) and the loaf (1 lb) as benchmarks. List the item and its estimated weight as each estimate is given. Choose students to then weigh each item and to record the weight.

Which objects weigh less than a pound? Which weigh more than a pound? Which weigh exactly a pound? Answers will vary.

- Point out the displayed ton equivalency. Note that a ton is a standard customary unit that is used to weigh very heavy objects.

  What objects might be weighed in tons? sample answers: cars, ships, trains, aircraft, a concrete beam bridge, whales

  What is the ratio or equivalency of tons to pounds? 1:2,000 or 1 tn = 2,000 lb

Converting weight measurements; finding a fraction of a unit

- Use **guided practice** to help the students convert (rename) weight measurements.

- Guide the students as they identify the equivalency and then multiply to convert the following units of weight to smaller units. Follow a procedure similar to the one used to convert units of linear measurement in Lesson 134.

  Remind the students that it takes more of the smaller units to measure the same weight as the larger units.

  5 lb = __ oz

  1 lb = 16 oz

  5 × 16 oz = 80 oz

  3 tn 1,500 lb = __ lb

  1 tn = 2,000 lb

  (3 × 2,000 lb) + 1,500 lb = 7,500 lb

- Follow a similar procedure to convert the following weights to larger units.

  Remind the students that the remainders represent a part of the next set of the larger unit.

  4,350 lb = __ tn __ lb

  4,350 lb ÷ 2,000 lb = $2\frac{350}{2,000}$ tn = 2 tn 350 lb

  95 oz = __ lb __ oz

  95 oz ÷ 16 oz = $5\frac{15}{16}$ lb = 5 lb 15 oz

Estimating & measuring weight

- Guide an **interactive activity** to help the students estimate and weigh items to the nearest pound or ounce.

  Display the spring scale. Explain that a spring scale is used to find the weight of an object in pounds and/or ounces. Weight is used to determine the price for mailing a package, the better unit price, or a dose of medicine.

  Where have you seen a scale? sample answers: at home (bathroom or food scale), a doctor's office, a grocery store, a post office, a recycling plant

How can you determine the weight of a loaf of bread? I can weigh the loaf on a scale or check the packaging label.

- Guide the students as they weigh a loaf of bread and a slice of the bread. Then allow a student to count the slices in the loaf. 16 Write "1 lb of bread = 16 1-oz slices" for display.

  How many ounces does a 1 lb loaf of bread weigh? 16

  What fractional part of a pound does 1 slice of bread weigh? $\frac{1}{16}$ lb

  Note that ounce and pound are standard customary units for measuring weight.

- Display the *Customary Measurement Craze* page and the objects to weigh.

- Read aloud the following word problem.

Faith's cookie recipe calls for 12 oz of chocolate chips. She buys chocolate chips in 1 lb bags. If Faith makes 4 batches of cookies, how many bags of chocolate chips does she need?

How can you solve this problem? I can find the number of ounces of chocolate chips needed for the 4 batches and then find the number of bags needed.

How many ounces of chocolate chips does Faith need? 48 oz; 4 batches of 12 oz is 48 oz; 4 × 12 oz = 48 oz.

Will Faith need more than 4 bags of chocolate chips? No; 12 oz < 16 oz (1 bag), so 4 bags would be more than enough for 4 batches.

How could you find the exact number of bags Faith needs? I can divide the total number of ounces needed by the 16 oz in each bag; 48 oz = ___ 1 lb bags; 16 oz = 1 lb; 48 oz ÷ 16 oz = 3 (1 lb) bags of chocolate chips.

- Follow a similar procedure for the following word problem. Remind the students that to find a fractional part of a unit, they must first rename the unit to the desired smaller units and then multiply by the fraction.

Dad ordered $\frac{3}{4}$ of a ton of gravel for a landscape project. How many pounds of gravel did he order? $\frac{3}{4}$ tn = ___ lb; 1 tn = 2,000 lb; $\frac{3}{4}$ × 2,000 lb = 1,500 lb

Measuring capacity; converting capacity measurements

- Guide an **interactive activity** to help the students measure the capacity of containers.

Explain that capacity is the amount a container can hold. Point out that fluid ounce, cup, pint, quart, and gallon are standard customary units for measuring capacity.

- Display the various waterproof containers. Remind the students that containers that have the same capacity (volume) may vary in shape. (See Lesson 117.) Guide students as they measure the units of water with the measuring cups or pitchers to fill the containers and determine their capacity.

Write a comparison sentence by using = or ≠.

1. $\frac{3}{5}$ ≠ $\frac{1}{3}$
2. $\frac{4}{5}$ = $\frac{16}{20}$
3. $\frac{40}{80}$ ≠ $\frac{1}{4}$
4. $\frac{12}{27}$ ≠ $\frac{4}{7}$

Find the unit rate.

5. 15 gal of gas to drive 450 mi *30 mi/gal*
6. $84 earned in 7 hr *$12/hr*
7. 135 pages read in 45 min *3 pg/min*
8. 5 cans of peas for $2.00 *$0.40/can*
9. 4 lb meat for $8.76 *$2.19/lb*
10. 12 pencils for $6.00 *$0.50/pencil*

Write the missing term that completes the equivalent ratio.

11. $\frac{1}{7} = \frac{n}{49}$ *n = 7*
12. $\frac{2}{7} = \frac{10}{n}$ *n = 35*
13. $\frac{36}{42} = \frac{6}{n}$ *n = 7*
14. $\frac{5}{9} = \frac{n}{36}$ *n = 20*
15. $\frac{3}{4} = \frac{18}{n}$ *n = 24*
16. $\frac{30}{16} = \frac{n}{8}$ *n = 15*

490   Daily Review

- Write the following ratios and/or equivalencies for display. Choose students to show the equivalency by pouring water from a full larger container into smaller containers. Choose students to record the numerical equivalency for display.

gallons to quarts 1 gal : 4 qt or 1 gal = 4 qt

quarts to pints 1 qt : 2 pt or 1 qt = 2 pt

pints to cups 1 pt : 2 c or 1 pt = 2 c

cups to fluid ounces 1 c : 8 fl oz or 1 c = 8 fl oz

What is the difference between a fluid ounce and an ounce? A fluid ounce is used to measure capacity, and an ounce is used to measure weight.

How could you find the number of fluid ounces in 1 qt? I can count (or add) the number of fluid ounces as I pour measured fluid ounces of water into a quart container, or I can solve the equivalency statement 1 qt = ___ fl oz; sample solutions: 1 c = 8 fl oz; 1 pt = 2 c or 16 fl oz (2 × 8 fl oz = 16 fl oz); 1 qt = 2 pt or 32 fl oz (2 × 16 fl oz = 32 fl oz).

How could you find the number of fluid ounces in 1 gal? I can solve the equivalency statement 1 gal = ___ fl oz; sample solution: since 1 qt = 32 fl oz and 4 qt = 1 gal, I can multiply 4 × 32 fl oz; 128 fl oz.

- Guide the students as they solve the following equivalencies.

  How will you know whether to multiply or divide? If I am renaming from larger to smaller units, I will multiply. If I am renaming from smaller to larger units, I will divide.

  8 qt = 16 pt   10 pt = 5 qt
  62 fl oz = 7 c 6 fl oz   6 gal = 24 qt
  21 c = 5 qt 1 c   3 gal 2 qt = 14 qt

- Read aloud the following word problem.

  Jorgen is to provide 20 fl oz of sports drink for each of the 12 players on his team. The powdered drink mix gives instructions for mixing 1 gal. Will 1 gal be enough for the team? How much drink is needed? Does Jorgen need to fill the 3 gal drink cooler? No; he needs 240 fl oz or $1\frac{7}{8}$ gal; no.

  How many fluid ounces are in 1 gal? 1 gal = 128 fl oz

  How many fluid ounces of sports drink are needed? 12×20 fl oz = 240 fl oz

  Will 1 gal be enough? no; 128 fl oz < 240 fl oz

  How can you find out if Jorgen needs to fill the 3 gal drink cooler? sample answers: 240 fl oz needed − 128 fl oz (1 gal) = 112 fl oz more than 1 gal needed; 240 fl oz ÷ 128 fl oz = $1\frac{7}{8}$ gal; 2 gal × 128 fl oz per gal = 256 fl oz per 2 gal; make an equivalency (ratio) table for gal/fl oz ($\frac{1}{128} = \frac{2}{256} = \frac{3}{384}$)

- Direct the students to find out if Jorgen needs to fill the 3 gal drink cooler. Choose students to explain their answer. No; the 240 fl oz needed is equivalent to $1\frac{7}{8}$ gal, which is less than 2 gal (256 fl oz).

Adding & subtracting weight & capacity measurements

- Use **guided practice** to help the students add and subtract weight and capacity measurements. Follow procedures similar to those used in Lesson 134 to guide the students as they solve the following problems and then rename the answers as only the larger unit and only the smaller unit.

$$3 \text{ gal } 2 \text{ qt}$$
$$+\ 7 \text{ gal } 4 \text{ qt}$$
10 gal 6 qt
11 gal 2 qt
$11\frac{1}{2}$ gal
46 qt

$$7 \text{ tn } 200 \text{ lb}$$
$$+\ 12 \text{ tn } 300 \text{ lb}$$
19 tn 500 lb
$19\frac{1}{4}$ tn
38,500 lb

$$6 \text{ tn } 250 \text{ lb}$$
$$-\ 1 \text{ tn } 750 \text{ lb}$$
4 tn 1,500 lb
$4\frac{3}{4}$ tn
9,500 lb

Apply

Student Edition pages 298–99

- Read and explain the directions for pages 298–99. Assist the students as they complete the pages independently.

Daily Review

- Students should complete Chapter 14, section *b*.

NOTES

Student Edition pages 300–301
Daily Review Chapter 14, section c

OBJECTIVES

- Estimate and measure to the nearest meter, centimeter, and millimeter.
- Determine the appropriate linear unit.
- Convert metric linear measurements to smaller or larger units.
- Find a fraction of a measurement unit.
- Compare metric linear measurements by using > or <.
- Add and subtract metric linear measurements.

TEACHER RESOURCES

- 83 *Metric Measurement*
- 84 *Metric Measure Mania* (for each group)

ADDITIONAL MATERIALS

- a meter stick
- a yardstick
- a metric measuring tape (for each group)
- a paper clip (for each group)

Engage

- Direct the students to do a **Think-Pair-Share** to explore the essential question at the top of Student Edition page 300, "How can completing a table help me rename metric measurements?"

Instruct

Reviewing the metric system

- Use **direct instruction** to review the structure of the metric system.

Display the *Metric Measurement* page. Explain that the metric measurement system (like the decimal system) is a base 10 system, unlike the customary measurement system.

The metric system of measurement is used in the United States and in most other countries. It includes linear, capacity, and mass units. Point out that the meter, liter, and gram are metric units that are used to measure linear distance (length), capacity, and mass.

- Explain that since the metric system is based on the number 10, each larger unit is 10 times greater than the next smaller unit to its right in the table, and each smaller unit is $\frac{1}{10}$ of the next larger unit to its left.

- Point out that each prefix is combined with a metric unit to identify larger and smaller measuring units and to tell the value of each unit. Guide the students as they read each prefix at the top of the table. Choose students to give the meaning of each prefix (the number written below it).

Estimating & measuring with metric linear units

- Guide an **interactive activity** to help the students become familiar with metric measurements. Display the meter stick. Explain that the meter is the basic unit for metric linear measurement (length or distance).

- Write the following metric units and their meter equivalents for display.

Explain that they are the most commonly used metric units of length.

Point out that the prefix before the unit *meter* indicates that the unit is either a

---

Metric Linear Measurement

How can completing a table help me rename metric measurements?

Knowing metric equivalents allows you to rename measurements as larger or smaller units. The **meter** (m) is the basic unit of length in the metric system.

Key Terms
- millimeter
- centimeter
- meter
- kilometer

Metric Linear Equivalents

1 kilometer (km) = 1,000 m
1 m = 100 centimeters (cm)
1 m = 1,000 millimeters (mm)

Decimal Metric Linear Equivalents

1 kilometer (km) = 1,000 m
1 centimeter (cm) = 0.01 m
1 millimeter (mm) = 0.001 m

Renaming Metric Measurements

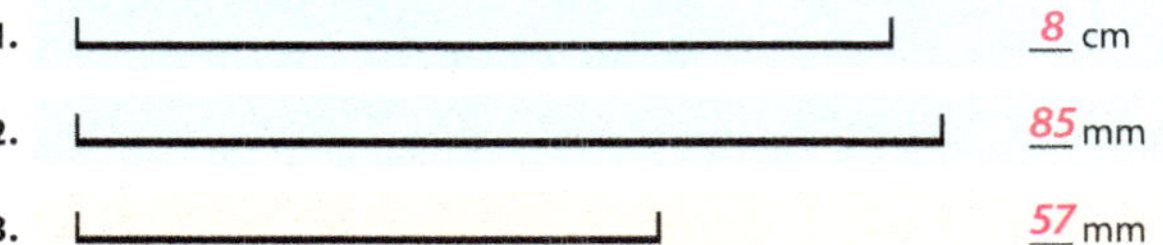

Rename larger units as smaller units. Determine the equivalency and then *multiply*.

$$4 \text{ km} = \underline{\quad} \text{ m}$$
$$1 \text{ km} = 1{,}000 \text{ m}$$
$$4 \times 1{,}000 = 4{,}000$$
$$4 \text{ km} = 4{,}000 \text{ m}$$

$$\begin{array}{r} 1{,}000 \\ \times \quad 4 \\ \hline 4{,}000 \end{array}$$

Rename smaller units as larger units. Determine the equivalency and then *divide*.

$$150 \text{ cm} = \underline{\quad} \text{ m}$$
$$100 \text{ cm} = 1 \text{ m}$$
$$150 \div 100 = 1.5$$
$$150 \text{ cm} = 1.5 \text{ m}$$

$$\begin{array}{r} 1.5 \\ 100\overline{)150.0} \\ -100 \\ \hline 500 \\ -500 \\ \hline 0 \end{array}$$

Determine the equivalency and solve.

$$\frac{3}{4} \text{ km} = \underline{\quad} \text{ m}$$
$$1 \text{ km} = 1{,}000 \text{ m}$$
$$\frac{3}{\overset{1}{\cancel{4}}} \times \overset{250}{\cancel{1{,}000}} = 750$$
$$\frac{3}{4} \text{ km} = 750 \text{ m}$$

$$\begin{array}{r} 250 \\ \times \quad 3 \\ \hline 750 \end{array}$$

Exercises

Write the measurement of the line segment, using the given unit.

1. ________________ __8__ cm

2. ________________ __85__ mm

3. ________________ __57__ mm

Write the best unit of measurement: km, m, cm, or mm.

4. distance to the ocean *km*
5. length of an umbrella *cm or m*
6. length of an ant *mm*
7. width of a wedding ring *mm*
8. height of a door *cm or m*
9. distance of a bike-a-thon *km*

Write a comparison sentence by using > or <.

10. 99 cm $\leq$ 1 m
11. 2 km $\leq$ 2,410 m
12. 89 m $\geq$ 9 cm
13. 56 cm $\geq$ 75 mm
14. 3 m $\leq$ 420 cm
15. 100 cm $\leq$ 1 km

300   Chapter 14

Rename.

16. 2 m = __200__ cm
17. 780 cm = __7.8__ m
18. 5 m = __5,000__ mm
19. 2,100 mm = __2.1__ m
20. 423 cm = __4.23__ m
21. 3,000 m = __3__ km
22. 6,500 m = __6.5__ km
23. 3.76 m = __376__ cm
24. $\frac{1}{2}$ km = __500__ m
25. $\frac{1}{4}$ m = __25__ cm
26. $\frac{2}{5}$ cm = __4__ mm
27. $\frac{3}{4}$ m = __75__ cm

Add or subtract.

28. 25 m
+ 8 m
33 m

29. 4.7 m
− 3.2 m
1.5 m

30. 10.5 km
+ 2.1 km
12.6 km

31. 981 mm
− 245 mm
736 mm

32. 795 m
+ 349 m
1,144 m

Complete the table.

33.
m	1,000	1,500	2,000
km	1	1.5	2

34.
cm	300	350	400
m	3	3.5	4

35.
m	1	1.5	2
mm	1,000	1,500	2,000

Place a decimal point in the answer to make a reasonable measurement.

36. height of a doghouse = 082 m **0.82 m**
37. width of a bracelet = 12 cm **1.2 cm**

Find the length in millimeters and centimeters.

38.  __53__ mm or __5.3__ cm

39.  __80__ mm or __8__ cm

Practice & Application

40. Is the length of a cell phone closer to 10 mm or 10 cm? **10 cm**

41. Which trail is shorter: River Trail at 1.2 km or Mountain Trail at 958 m? **Mountain Trail**

42. Would the length of a minivan be closer to 2 m or 5 m? **5 m**

43. Is the wingspan of a butterfly about 55 mm or 55 cm? **55 mm**

44. Mrs. Poole planted a square flower bed 100 cm long. Is each side of the flower bed greater than, less than, or equal to a meter? **equal to a meter**

45. Preston scribbled the number 25 when he measured the length of his computer. What metric unit of measurement did he use to measure the computer? **cm**

46. The Haas family visited Canada on vacation. What metric unit rate would follow the number 60 on a speed limit sign? **km/h**

47. Jade went to the eye doctor to get contacts. The doctor measured her pupil at 4. What metric unit of measure did he use to measure her pupil? **mm**

48. Megan trained for a 5K race. How many meters are in 5 km? **5,000 m**

49. How can completing a table help me rename metric measurements? **A table helps me rename smaller units to larger units by showing the pattern of increase.**

---

multiple or a fraction of 1 m. Underline each prefix as you discuss the unit.

kilometer = 1,000 m

meter = 1 m

centimeter = $\frac{1}{100}$ m

millimeter = $\frac{1}{1,000}$ m

- Display the yardstick beside the meter stick.

  How does 1 m compare to 1 yd? 1 m is slightly longer than 1 yd.

  Inform the students that 1 m is about $39\frac{1}{4}$ in.

  Arrange the students in groups and distribute the metric measuring tapes and paper clips. Direct each student to use the tape to find a personal benchmark for 1 m, 1 cm, and 1 mm. Explain that they can use the paper clip to find a benchmark for a millimeter. sample benchmarks: arms outstretched or height of a doorknob from the floor (1 m), width of the tip of a finger (1 cm), width of the wire in a paper clip (1 mm)

- Distribute a *Metric Measure Mania* page to each group, and instruct each group to choose a record keeper to record the group's answers as they are given.

  Direct each group to estimate the height of a door to the nearest meter and then to choose a student to use the measuring tape to measure the height of the door to the nearest meter.

Is your measured height of the door a precise or exact measurement? The measurement is precise if it is to an exact meter, but if it is longer or shorter than an exact meter, it is not precise.

What smaller units on the measuring tape could you use to measure more precisely than a meter? centimeter or millimeter

What is the ratio or equivalency of centimeters to meters? 100 cm : 1 m or 100 cm = 1 m

- Direct attention to the shortest marks on the meter tape.

  What unit is $\frac{1}{10}$ of a centimeter? a millimeter

  What is the ratio or equivalency of millimeters to centimeters? 10 mm : 1 cm or 10 mm = 1 cm

  What is the ratio or equivalency of millimeters to meters? 1,000 mm : 1 m or 1,000 mm = 1 m

- Instruct each group to estimate the height of the same door to the nearest centimeter and the nearest millimeter and then to choose a student to measure the height to the nearest centimeter and the nearest millimeter.

- Follow a similar procedure, allowing each group to estimate and measure other objects in the room. Guide the students as they compare the estimate and measurement of each object.

- Explain that a kilometer is used for measuring long distances. Remind the students of the location that is a mile from their school (Lesson 134), and tell them that 1 mi ≈ 1.6 km.

  What is the ratio or equivalency of kilometers to meters? 1 km : 1,000 m or 1 km = 1,000 m

  How could you measure a kilometer using meter sticks? I could place 1,000 meter sticks end to end.

  What are some distances that could be measured in kilometers? sample answers: the distance from home to school, the length of a bridge, the distance of a race (5K)

LESSON 136

Determining the appropriate linear unit

- Use **guided practice** to help the students determine the appropriate linear unit. Write the abbreviations "km," "m," "cm," and "mm" for display. Choose students to tell the unit that would likely be used to measure the following items. Discuss why they might choose one unit instead of another.

  height of a ceiling m or cm

  length of a race m or km

  width of a button mm or cm

  width of a door m or cm

  distance between cities km

  thickness of a dime mm

- Guide the students as they determine the better estimate.

  distance across a state—500 m or 500 km? 500 km

  length of a book cover—25 cm, 25 mm, or 25 m? 25 cm

  face of a quarter—25 m, 25 cm, or 25 mm? 25 mm

  width of a garage door—5 cm, 5 m, or 5 km? 5 m

Converting metric linear measurements; finding a fraction of a unit

- Guide a **discussion** to help the students convert (rename) metric linear measurements.

  Read aloud the following word problem.

  Marco and his dad ran a 5K race. Since 5K stands for 5 km, how many meters did Marco run? 5,000 m

  What equation can you write to find the answer? 5 km = __ m

  What equivalency can you use to solve the problem? 1 km = 1,000 m

  What operation is used to rename larger units (km) to smaller units (m)? multiplication; It takes more of the smaller units to measure the same distance as 1 larger unit.

  How can you solve the problem? 5 × 1,000 m = 5,000 m; 5 km is equal to 5 sets of 1,000 m.

  Complete the equivalency.

- Follow a similar procedure for the following word problem. Remind the students that to find a fraction of a unit, they must first rename the unit as the desired smaller unit.

  Marco's younger sister ran half of a kilometer. How many meters did she run? 500 m; $\frac{1}{2}$ km = __ m; 1 km = 1,000 m; $\frac{1}{2}$ × 1,000 m = 500 m

- Use the following questions to guide the students as they solve another word problem.

  Marco's brother ran 7,500 m. How many kilometers did he run? 7.5 km

  What equation can you write to find the answer? 7,500 m = __ km

  What equivalency can you use to solve the problem? 1,000 m = 1 km

  What operation is used to rename smaller units (m) to larger units (km)? division; 1 larger unit can measure the same distance as a number of smaller units.

  How can you solve the problem? 7,500 m ÷ 1,000 m = 7.5 km

Write the percent in decimal form.

1. 52% *0.52*   2. 17% *0.17*   3. 19% *0.19*   4. 2% *0.02*   5. 75% *0.75*

Write the decimal in percent form.

6. 0.58 *58%*   7. 0.8 *80%*   8. 0.09 *9%*   9. 0.27 *27%*   10. 0.93 *93%*

Write the percent in fraction form in lowest terms.

11. 60% $\frac{3}{5}$ 12. 20% $\frac{1}{5}$ 13. 50% $\frac{1}{2}$ 14. 25% $\frac{1}{4}$ 15. 75% $\frac{3}{4}$

Find the percent of the number.

16. 20% of 100 *20*   17. 50% of 8 *4*   18. 50% of 90 *45*   19. 10% of 30 *3*   20. 25% of 100 *25*

Daily Review   491

- Guide the students as they identify the needed equivalency and use multiplication or division to find the number of renamed units.

Students will be taught to use mental math to multiply or divide by a power of 10 when converting metric units of measurement in the next lesson.

Draw a table for display and fill in the row and column headings from Student Edition page 301, problem 33.

Direct attention to the essential question at the top of student page 300.

How can completing a table help you rename metric measurements? A table helps me rename metric units by showing the pattern of increase or decrease.

Choose students to convert the measurements to complete the table.

- Guide the students as they convert the following measurements.

265 cm = __ m 2.65

3.4 m = __ cm 340

2,471 mm = __ m 2.471

3,000 mm = __ m 3

Comparing measurements

- Guide a **Turn and Talk** to help the students complete the following comparison statements. Point out that either unit can be converted so that like units are being compared.

952 m < 8 km	1 m > 100 mm
675 cm > 6 m	5 cm < 54 mm
4 km > 49.82 cm	25.3 mm < 3 cm

Adding & subtracting metric linear measurements

- Use **guided practice** to help the students solve the following problems.

$$5.2 \text{ km} + 3.0 \text{ km} = 8.2 \text{ km}$$

$$763 \text{ m} + 819 \text{ m} = 1{,}582 \text{ m}$$

$$465 \text{ mm} - 243 \text{ mm} = 222 \text{ mm}$$

$$21.7 \text{ cm} - 4.9 \text{ cm} = 16.8 \text{ cm}$$

Apply

Student Edition pages 300–301

- Read and explain the directions for pages 300–301. Assist the students as they complete the pages independently.

Daily Review

- Students should complete Chapter 14, section *c*.

DIFFERENTIATED INSTRUCTION

Use the following to provide extra help for students who experience difficulty with the concepts taught in Chapter 14.

Convert units.

Knowing whether to multiply or divide when converting units may be difficult for some students. Write "3 lb = __ oz" for display and write "__ lb = __ oz" below it. Ask the students to identify the equivalent relationship between pounds and ounces. 1 lb = 16 oz Write "1" and "16" in the appropriate blanks. Ask the students whether multiplication or division is needed to convert 1 lb to more of the smaller unit (ounces). multiplication Ask them what they should multiply 3 lb by to find the equivalent number of ounces and to explain their answer. 3 lb × 16 oz; Since there are 16 oz in 1 lb, I multiply 3 × 16 oz to find the number of ounces in 3 lb. Direct the students to complete the original problem. 3 × 16 oz = 48 oz Repeat the procedure for 3.98 cm = __ m using division. 100 cm = 1 m; 3.98 cm ÷ 100 = 0.0398 m

NOTES

Student Edition pages 302–3
Daily Review Chapter 14, section *d*

OBJECTIVES

- Estimate and measure with metric units of capacity.
- Convert metric capacity and mass measurements to smaller or larger units.
- Estimate and measure with metric units of mass.
- Explain how the activity of standardizing measurements is justified by Scripture. **BWS**

BIBLICAL WORLDVIEW SHAPING

- **Knowledge (Explain):** Scripture requires accurate and honest measurement based on a standard.

TEACHER RESOURCES

- 83 *Metric Measurement*

ADDITIONAL MATERIALS

- a 1 L bottle of water (for the teacher and for each student)
- a 1 L beaker with measurements or metric measuring container
- a 5 mL plastic syringe with measurements
- water for measuring
- food coloring (optional)
- waterproof containers (from Lesson 135)
- additional containers for measuring capacity (e.g., a utility bucket, nail polish bottle, soup can, pitcher, drink cooler, teacup)
- a balance or metric scale

ASSESSMENTS

- Chapter 14 Quiz 1

Metric Capacity & Mass

What does the Bible say about standardized measurements?

Capacity is the amount of liquid a container will hold. The **liter** (L) is the basic unit of capacity in the metric system.

1 L = 1,000 milliliters (mL)
1 mL = 0.001 L

Mass is the amount of matter an object has. The **gram** (g) is a basic unit of mass.

1,000 g = 1 kilogram (kg)
1 g = 0.001 kg
1 g = 1,000 milligrams (mg)

Key Terms
- capacity
- liter
- milliliter
- mass
- gram
- milligram

Renaming Metric Units

Rename larger units as smaller units. Determine the equivalency and then *multiply*.

3.25 L = ___ mL
1 L = **1,000** mL
3.25 × **1,000** = 3,250
3.25 L = **3,250** mL

Rename smaller units as larger units. Determine the equivalency and then *divide*.

2,500 g = ___ kg
1,000 g = 1 kg
2,500 ÷ **1,000** = 2.5
2,500 g = **2.5** kg

1 L = 1,000 mL	1,000 g = 1 kg
×1,000	÷1,000
3.250 L = 3,250 mL	2,500 g = 2.5 kg

Exercises

Write the best unit of mass: mg, g, or kg.

1. box of books *kg* 2. mouse *g*
3. grain of salt *mg* 4. snowflake *mg*
5. child *kg* 6. bag of potatoes *kg*

Write the better unit of capacity: mL or L.

7. fish aquarium *L* 8. spoonful of medicine *mL*
9. pitcher of punch *L* 10. mug of hot cocoa *mL*
11. glass of milk *mL* 12. cooler of water *L*

Use the table to find the answer.

13. Which owl is the smallest? *elf owl*
14. Which owl has a mass of 1.7 kg? *great horned owl*
15. Which owl is the largest? *snow owl*
16. How much larger is the spotted owl than the barn owl? *125 g*
17. What is the size difference between the smallest owl and the largest owl? *2,257.5 g*

Sizes of Owls	
elf owl	42.5 g
barn owl	470 g
spotted owl	595 g
great horned owl	1,700 g
snow owl	2.3 kg

Rename.

18. 3 kg = ___ g   *3,000*
19. 5,700 g = ___ kg   *5.7*
20. 4.5 L = ___ mL   *4,500*
21. 2,000 mg = ___ g   *2*
22. 7.1 kg = ___ g   *7,100*
23. 2,430 mg = ___ g   *2.43*
24. 8,000 mL = ___ L   *8*
25. 9,500 mL = ___ L   *9.5*
26. 2.75 L = ___ mL   *2,750*

The bottles of water and the waterproof containers used in this lesson are items that have been brought to class by the students.

Add food coloring to the water for demonstration purposes if desired.

of a meter, $\frac{500}{100}$ m or 5 m; 30 **mm** = 30 **thousandths** of a meter, 0.030 m or 0.03 m).

2 km = ___ m *2,000*   500 cm = ___ m *5*
30 mm = ___ cm *3*

Engage

- **Review** renaming metric linear measurements. Display the *Metric Measurement* page and choose students to solve the following problems. Point out that the meaning of the prefix can help them to rename (e.g., 2 **km** = 2 **thousand** meters or 2,000 m; 500 **cm** = 500 **hundredths**

Instruct

Estimating & measuring with metric units of capacity

- Guide an **interactive activity** to help the students become familiar with metric units of capacity.

Display the *Metric Measurement* page. Point out the units of capacity and their abbreviations. Distribute the 1 L bottles of water.

Use the table to find the answer.

27. Which has the greater mass: a cock or a hen? *cock*
28. What is the difference between the mass of a cock and the mass of a cockerel? *0.45 kg*
29. What is the pullet's mass in kilograms? *2.7 kg*
30. A Cornish cockerel has a mass of 3.86 kg. Which is larger: the Plymouth Rock cockerel or the Cornish cockerel? *Cornish cockerel*
31. A farmer ordered 15 pullets. Give the mass of the shipment in kilograms.
    15 × 2,700 = 40,500 g = 40.5 kg

Sizes of Plymouth Rock Chickens	
pullet	2,700 g
hen	3.4 kg
cockerel	3.65 kg
cock	4.1 kg

Add or subtract.

32. 2.5 kg + 1.4 kg
    3.9 kg
33. 4,271 g − 1 kg
    3,271 g or 3.271 kg
34. 3 L − 2,500 mL
    500 mL or 0.5 L

Complete the table.

35.
m	cm
9	*900*
10	1,000
11	*1,100*

36.
mL	L
2,000	2
3,500	*3.5*
4,200	4.2

37.
g	kg
7,000	*7*
9,500	*9.5*
1,500	1.5

38.
m	km
4,000	*4*
6,500	6.5
1,500	*1.5*

DID YOU KNOW?

Capacity is the amount of liquid that a container can hold. The capacity of this container is 1 L.

Mass is the amount of matter an object has. The mass of this container filled with water is 1 kg.

Volume is the number of cubic units within a 3-dimensional figure. The volume of this container is 1,000 cm³.

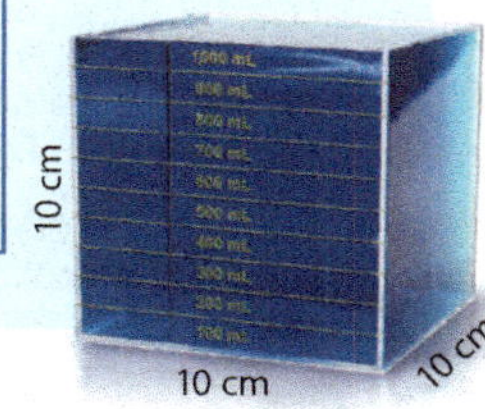

Practice & Application

39. Would a math book have a mass of 15 kg or 1.5 kg? *1.5 kg*
40. Would Alexander drink 1 L or 100 mL during a hike? *1 L*
41. Would a man walk 25 m or 1 km in 20 min? *1 km*
42. Would a basketball player be 150 cm tall or 2 m tall? *2 m*
43. Would the most precise measurement of the width of a penny be in centimeters or millimeters? *millimeters*
44. Would an aquarium hold 20 mL or 20 L of water? *20 L*
45. Would a bag of 1,000 small candies have a mass of 1 g or 1 kg? *1 kg*
46. Would the average sixth-grader weigh 40 g or 40 kg? *40 kg*
47. What does the Bible say about standardized measurements? *God's Word requires that my measurements be accurate and honest according to a standard.*

Lesson 137   303

- Display the additional containers, such as the examples listed below. Direct the students to choose the better unit (milliliters or liters) for measuring the contents of each container and to write an estimate of each container's capacity to the nearest 100 mL or the nearest 1 L: utility bucket L, nail polish bottle mL, soup can mL, pitcher L, drink cooler L, and teacup mL.

Demonstrate filling each container to find its capacity. Guide the students as they determine which estimates are the closest to the actual capacities. Continue to display the filled containers for the mass estimating activity later in the lesson.

Converting metric capacity measurements

- Guide a **discussion** to help the students convert metric capacity measurements. Write "3.254 L = __ mL" for display.

Should you rename larger units as smaller units or smaller units as larger units to solve this problem? larger units as smaller units (liters as milliliters)

Should you multiply or divide? multiply; It takes more of the smaller units to measure the same capacity as 1 larger unit.

What equivalency can you use to rename liters as milliliters? 1 L = 1,000 mL

Remind the students that writing the equivalency in the same order that the units appear in the problem will help them to determine which operation to use when renaming units of measurement.

How can you estimate the number of milliliters in 3.254 L? Since 1 L = 1,000 mL, 3 L = 3 × 1,000 mL or 3,000 mL.

What should you multiply 3.254 L by to rename it as milliliters? 1,000; Each liter is equal to 1,000 times as many milliliters.

- Write "1,000 × 3.254" vertically. Demonstrate solving the problem by annexing 3 zeros in the product and writing the decimal point in the product. 3,254.000 mL or 3,254 mL

What is capacity? the amount a container can hold

What is the standard metric unit of capacity? liter

What abbreviation is used for liter? L

What abbreviation is used for milliliter? mL

What items are measured in liters? sample answers: water, soft drinks

What items are measured in milliliters? sample answers: medicine, lotion, liquid food items

Can you drink 1 L of water? Answers will vary.

- Display the 1 L beaker or container. Point out the measurement markings on the beaker: 100 mL, 200 mL, . . . 1,000 mL or 1 L.

Display the syringe and explain that it has a 5 mL capacity. Fill the syringe to the 1 mL mark and display it.

How many 1 mL measures of water are needed to fill the 1 L beaker? 1,000; 1,000 mL equals 1 L.

What is the ratio or equivalency of liters to milliliters? 1 L : 1,000 mL or 1 L = 1,000 mL

- Fill the 1 L beaker with water. Allow students to use the beaker, the syringe, and the 1 L bottles of water to measure the capacity of the waterproof containers in liters and milliliters for display. Provide additional water if needed.

LESSON 137

What do you notice about the decimal point in a number when it is multiplied by a power of 10? The decimal point moves one place to the right for each power of 10.

How many places did the decimal move when multiplying by 1,000? 3 places; $1,000 = 10^3$

Is 3,254 a reasonable answer? Yes; 3.254 L = 3 L (or 3,000 mL) + $\frac{254}{1,000}$ L (or 254 mL). Choose a student to complete the problem.

- Write "8,417 mL = __ L" for display.

Should you rename larger units as smaller units or smaller units as larger units to solve this problem? smaller units as larger units (milliliters as liters)

Should you multiply or divide? divide; It takes 1 larger unit to measure the same capacity as a number of smaller units.

What equivalency can you use to rename milliliters as liters? 1,000 mL = 1 L

How many liters do you estimate 8,417 mL to be? 8 L; The amount is a little more than 8,000 mL, and since 1,000 mL = 1 L, 8,000 mL ÷ 1,000 mL = 8 L.

Some students may find it easier to estimate by writing a proportion: $\frac{1,000 \text{ mL}}{1 \text{ L}} = \frac{8,000 \text{ mL}}{8 \text{ L}}$.

What could you divide 8,417 mL by to rename it as liters? 1,000; Each mL is equal to $\frac{1}{1,000}$ of 1 L.

How could you use mental math to solve the problem? Since the decimal moves one place to the left for each power of 10 in the divisor, I can divide 8,417 mL by 1,000 mL by moving the decimal point 3 places to the left; 8,417 mL ÷ 1,000 mL = 8.417 L. The quotient will be a smaller number of liters (larger units).

Choose a student to demonstrate the moving of the decimal from the right of the ones place in 8,417 to 8.417.

Is 8.417 a reasonable answer? Yes; 8,417 mL = 8,000 mL (or 8 L) + 417 mL (or $\frac{417}{1,000}$ L).

- Guide the students as they solve problems similar to the ones below. Point out that the unlike units must be renamed as either unit so that like units are being added or subtracted.

Find the volume of a prism with the given dimensions. *Equations may vary.*

1. rectangular prism: l = 3 cm, w = 2 cm, h = 6 cm *3 cm • 2 cm • 6 cm = 36 cm³*

2. square prism: s = 7 m *(7 m)³ = 343 m³*

3. rectangular prism: l = 7 m, w = 8 m, h = 6 m *7 m • 8 m • 6 m = 336 m³*

Find the volume of a cylinder with the given dimensions.

4. r = 2 m, h = 7 m *3.14 • (2 m)² • 7 m = 87.92 m³*

5. r = 4 m, h = 9 m *3.14 • (4 m)² • 9 m = 452.16 m³*

6. r = 5 m, h = 10 m *3.14 • (5 m)² • 10 m = 785 m³*

Solve.

7. Jason filled a rectangular planter with potting soil. His planter is 4 ft long, 2 ft wide, and 0.5 ft high. How much potting soil did it take to fill his planter? *4 ft • 2 ft • 0.5 ft = 4 ft³*

8. Sarah made a vanilla cake in a pan that is 13 in. by 9 in. by 2 in. What is the volume of half of her pan? *13 ft • 9 ft • 2 ft = 234 in.³; $\frac{234 \text{ in.}^3}{2}$ = 117 in.³*

9. The fish tank in Dr. Goforth's office is cube shaped with equal dimensions of 3.3 ft. What is the volume of his fish tank? *(3.3 ft)³ = 35.937 ft³*

492 Daily Review

---

3 L = 3 L	3 L = 3,000 mL
+ 26 mL = 0.026 L	+ 26 mL = 26 mL
3.026 L	3,026 mL

Estimating & measuring with metric units of mass

- Guide a **discussion** to help the students become familiar with metric units of mass. Point out the metric measurement units of mass and their abbreviations on the displayed *Metric Measurement* page. Display the balance (or metric scale). Explain that the balance is a tool for measuring mass, the amount of matter an object has.

- Display the 5 mL syringe and the 1 L beaker. Measure the mass of the empty 5 mL syringe and record it. Fill the syringe to the 1 mL mark with water and weigh it again. Subtract to find the mass of the water in the syringe. 1 mL of water weighs 1 g.

What are some other items that would have a mass of about 1 gram? sample answers: a raisin, a paper clip

Repeat the procedure with the 1 L beaker. 1 L of water weighs 1 kg.

What are some other items that would have a mass of about 1 kilogram? sample answers: 1 L bottle of water, a book

Point out that a 1 L bottle of water can be used as a benchmark for 1 kg.

- Direct attention to each of the filled containers used earlier in the lesson. Follow

---

303a • Lesson 137   Math 6

a procedure similar to the one used in the capacity activity to guide the students as they determine the better unit of measure (kg or g), the estimated mass to the nearest 100 g or the nearest 1 kg, and the actual mass of each filled container: utility bucket kg, nail polish bottle g, soup can g, pitcher kg, drink cooler kg, and teacup g.

What metric unit for mass is smaller than a gram and a kilogram? milligram

- Explain that milligrams are used to measure the mass of very light objects. The mass of one crystal of salt is about 1 mg.

How many milligrams are equal to 1 g? 1,000

What unit is 1 thousandth (0.001) of a gram? milligram

What are some other items that would be measured in milligrams? sample answers: medicine, a tiny insect

Converting metric mass measurements

- Guide a **discussion** to help the students use mental math to convert metric mass measurements. Write "2.476 kg = __ g" for display.

How can you use mental math to solve the problem? Since I need to find more of a smaller unit, I can think of the equivalency 1 kg = 1,000 g and multiply 2.476 times 1,000 g, moving the decimal point 3 places to the right; 2.476 × 1,000 g = 2,476 g.

Is 2,476 g a reasonable answer? yes; 2,476 g = 2,000 g + 476 g = 2 kg + $\frac{476}{1,000}$ kg = 2.476 kg

- Follow a similar procedure for 265 g = __ kg. Point out that they can mentally divide by 1,000 g to find the number of larger units. 1,000 g = 1 kg; 265 g ÷ 1,000 g = 0.265 kg

Standardizing measurements

- **Discuss** the essential question on Student Edition page 302, "What does the Bible say about standardized measurements?" to help the students explain how the act of standardizing measurements is justified by Scripture.

What does it mean to standardize measurements? It means to make them the same so that they always measure the same amount.

Read aloud Deuteronomy 25:13. Explain that in that culture a person would have a set of weights to put on one side of the balance to measure against what he was buying or selling on the other side of the balance.

What would be wrong with a person using a set of weights that was lighter than they appeared for selling or a heavier set for purchasing? It would be dishonest. He could sell less product for the money with lighter weights and purchase more for the money with heavier weights.

Read aloud Proverbs 16:11.

What does the Bible say about standardized measurements? God's Word requires that my measurements be accurate and honest according to a standard.

Apply

Student Edition pages 302–3

- Read and explain the directions for pages 302–3. Assist the students as they complete the pages independently.

Daily Review

- Students should complete Chapter 14, section d.

Assess

Quiz 1

- Use the **summative assessment** to evaluate the students' progress at this point in the chapter.

NOTES

LESSON 138

Student Edition pages 304–5
Daily Review Chapter 14, section e

OBJECTIVES

- Add, subtract, multiply, and divide measurements.
- Solve measurement word problems.

TEACHER RESOURCES

- 85 *Customary Measurement Word Problems*
- 86 *Metric Measurement Word Problems*

ADDITIONAL MATERIALS

- *Measurement Flashcards* (from Lesson 134)

Engage

- **Review** measurement equivalencies for mastery by allowing pairs of students to quiz one another using their measurement flashcards.

- Review renaming measurements to help the students master the concept. Write the following problems for display and choose students to solve them.

3 L = __ mL 3,000   12 ft = __ yd 4

3 gal = __ qt 12   200 cm = __ m 2

3 c = __ fl oz 24   32 oz = __ lb 2

4 tn = __ lb 8,000   6,000 mg = __ g 6

Instruct

Solving measurement word problems with math operations

If you guide your students through solving the problems on the *Metric Measurement Word Problems* page rather than the problems on the *Customary Measurement Word Problems* page, follow procedures similar to the ones provided for solving the customary word problems and refer to the metric solutions at the end of this lesson.

Customary & Metric Systems

Why do you think scientists use the metric system?

	Addition	Subtraction	Multiplication	Division
Customary	11 in. +1 ft 7 in. 1 ft 18 in. = 2 ft 6 in. Simplify the answer when 2 units of measure are used, if possible.	1 lb = 16 oz 34 24 35 lb 8 oz − 1 lb 13 oz 33 lb 11 oz 16 +8 24	1 lb 4 oz × 7 7 lb 28 oz = 8 lb 12 oz	12 ft 4 in. ÷ 4 = __ 3 ft 1 in. 4)12 ft 4 in. −12 0 ft 4 in. − 4 0
Metric	1.8 kg + 1,500 g = __ 1.8 kg 1,800 g +1.5 kg or +1,500 g 3.3 kg 3,300 g	7,000 mg − 3 g = __ 7 g 7,000 mg −3 g or −3,000 mg 4 g 4,000 mg	3.61 L × 7 25.27 L	72.48 m ÷ 8 = __ 9.06 m 8)72.48 m −72 048 − 48 0

Exercises

Solve. Simplify the answer when possible.

1. 10 ft 4 in.
 + 5 ft 6 in.
 15 ft 10 in.

2. 3 gal 2 qt
 +7 gal 3 qt
 10 gal 5 qt = 11 gal 1 qt

3. 6 yd 2 ft
 +8 yd 2 ft
 14 yd 4 ft = 15 yd 1 ft

4. 9 ft 9 in.
 +8 ft 7 in.
 17 ft 16 in. = 18 ft 4 in.

5. 12 lb 4 oz
 − 7 lb 9 oz
 4 lb 11 oz

6. 23 ft 7 in.
 − 8 ft 10 in.
 14 ft 9 in.

7. 6,000 mL
 − 1,500 mL
 4,500 mL

8. 3,906 L
 − 2,879 L
 1,027 L

9. 8 L + 14 L + 72 mL
 22,072 mL or 22.072 L

10. 703 g + 4 kg + 65 g
 4,768 g or 4.768 kg

11. 6 qt − 5 pt
 3 qt 1 pt

12. 4 yd − 5 ft
 2 yd 1 ft

13. 1.47 g
 × 6
 8.82 g

14. 4 pt 3 c
 × 3
 12 pt 9 c = 16 pt 1 c

15. 6 ft 6 in.
 × 4
 24 ft 24 in. = 26 ft

16. 16 gal 2 qt
 × 8
 128 gal 16 qt = 132 gal

17. 48.6 m ÷ 3 *16.2 m*

18. 4)16 ft 8 in.
 4 ft 2 in.

19. 2)6 tn 210 lb
 3 tn 105 lb

20. 3)18 gal 9 pt
 6 gal 3 pt

21. Mrs. DeYoung used a long roll of paper to cover 9 tables for the neighborhood picnic. Each table cover was 5 ft 3 in. long. What was the total length of paper she used?
 9 • 5 ft 3 in. = 45 ft 27 in. = 47 ft 3 in.

22. The church shipped a garden tiller and a garden seeder to a missionary in South America. The tiller weighed 75 lb, and the seeder weighed 7 lb 11 oz. How much more did the tiller weigh than the seeder? *75 lb − 7 lb 11 oz = 67 lb 5 oz*

- Use **guided practice** to help the students solve measurement word problems.

  Display the *Customary Measurement Word Problems* page and direct the students to silently read the first word problem.

  The Poole family raises broiler chickens. They feed the baby chicks starter feed. How much starter feed is needed for 1 chick for the first 3 weeks? *1 lb 14 oz*

  Week 1   4 oz of starter feed per chick

  Week 2   9 oz of starter feed per chick

  Week 3   1 lb 1 oz of starter feed per chick

  How could you solve the word problem? I could add the amounts of starter feed needed for each of the first 3 weeks.

- Write the problem vertically for display and choose a student to solve it. *4 oz + 9 oz + 1 lb 1 oz = 1 lb 14 oz*

  What fraction of a pound does 14 oz represent? $\frac{14}{16}$ lb; Since 16 oz equals 1 lb, 14 oz represents 14 of the 16 equal parts of a pound and can be written $\frac{14}{16}$ lb or $\frac{7}{8}$ lb.

  How can you find the decimal equivalent for 14 oz? Since 14 oz is $\frac{7}{8}$ lb, I can divide the numerator 7 by the denominator 8.

  Direct the students to find the decimal equivalent for $\frac{7}{8}$ lb. *0.875 lb*

  How many pounds is 1 lb 14 oz? $1\frac{7}{8}$ lb or 1.875 lb

23. The average dairy cow produces 7 gal of milk per day. How many 8 oz glasses are in 7 gal of milk?

24. Dairy cows need to drink 200 L of water to produce 30 L of milk. How many liters of water would a cow need to drink to produce 300 L of milk? *$\frac{200}{30} = \frac{x}{300}$; x = 2,000 L of water*

25. The average Holstein cow needs 22 kg of feed to produce 30 L of milk. If a cow received 10 kg of hay, how many kilograms of corn would she need to produce 30 L of milk? *22 − 10 = 12 kg of corn*

26. The average Holstein cow produces 9 gal of milk per day. How many cows would it take to fill a 1,000 gal milk tank in one day? *1,000 ÷ 9 ≈ 112 cows*

27. Mr. Well's prized Holstein weighs 1,598 lb. Mr. Bradstreet's top Jersey weighs 987 lb. What is the difference in the weight of the two cows? *1,598 − 987 = 611 lb*

28. The Hayner Dairy Farm houses 1,500 cows in a barn that is 200 ft wide and 750 ft long. What is the perimeter and the area of the barn?

29. If a sixth-grader drank 8 oz of milk at each meal, how many ounces would the student consume in a day? *3 • 8 = 24 oz*

23. *8 oz = 1 c*
2 c = 1 pt
2 pt = 1 qt
4 qt = 1 gal
16 c = 1 gal
7 • 16 = 112 eight-ounce glasses

30. Everett drinks 3 qt of milk each week. Wesley drinks 13 c of milk each week. Who drinks more milk? *4 c = 1 qt; 3 qt • 4 c = 12 c; Wesley; 13 c > 12 c*

31. Adriana bought 36 oz of cheese. Did she purchase 4 eight-ounce bags or 3 twelve-ounce bags? *3 • 12 = 36; 3 twelve-ounce bags*

32. The menu offered 1 L of chocolate milk and 750 mL of chocolate milk. Which choice offered more milk? *1 L; 1L = 1,000 mL*

33. The total amount of milk produced in one day on the Belmont Farm is 2,250 L. There are 75 cows on the farm. What is the average amount of milk each cow produces per day? *2,250 ÷ 75 = 30 L of milk*

34. A cow needs 0.92 m of space while eating. How many meters of space are needed for 50 cows? *50 • 0.92 = 46 m*

35. The Carson family drinks 2 gal of milk each week. How many quarts of milk do they drink? Use the number of quarts to find an equivalent number of pints and cups. *2 gal • 4 qt = 8 qt; 8 qt • 2 pt = 16 pt; 16 pt • 2 c = 32 c*

36. Why do you think scientists use the metric system? *The metric system is easier to use for converting measurements because it is a base 10 system.*

28. *P = (2 • 200 ft) + (2 • 750 ft)*
= 400 + 1,500 = 1,900 ft
A = 200 ft • 750 ft = 150,000 ft²

How many ounces is 1 lb 14 oz? 30 oz; 1 lb = 16 oz and 16 oz + 14 oz = 30 oz

- Repeat the procedure for the second word problem, using the following questions.

One of the Poole's broiler chickens weighs 5 lb 4 oz after seven weeks of intense feeding. One of their laying hens that is the same age weighs 4 lb 8 oz. What is the difference in the weights of the two chickens? 5 lb 4 oz − 4 lb 8 oz = 12 oz; $\frac{3}{4}$ lb or 0.75 lb

How could you solve this word problem? I can compare the weights of the chickens by subtracting the weight of the laying hen from the weight of the broiler chicken.

How can you subtract 8 oz from 4 oz? I must rename 1 of the pounds in 5 lb 4 oz as 16 oz and add it to the 4 oz.

What is the difference in weight of the two chickens? 12 oz

- Follow a similar procedure for the third word problem, using the information from problem number 1. Point out that since they already know that 30 oz of starter feed is needed to feed 1 chick for the first three weeks, they can multiply the 30 oz by 12 to find the amount of starter feed that is needed for the 12 chicks.

The Pooles had 12 baby chicks to raise. How many pounds of starter feed were needed to feed the 12 chicks for the first

3 weeks? 12×30 oz = 360 oz; 360 oz ÷ 16 oz = $22\frac{1}{2}$ lb or 22.5 lb

- Read aloud word problem number 4.

The baby chicks were kept in a small square area that is easy to keep warm. The length of the area is 5 ft 8 in. What is its perimeter? 20 ft 32 in.; 22 ft 8 in.; or $22\frac{2}{3}$ ft

How can you find the perimeter of the square area? Since a square has 4 congruent sides, I can add the length 4 times or multiply the length by 4.

- Write "4 × 5 ft 8 in." vertically for display. Explain that when multiplying a measurement, they multiply each unit by the multiplier.

- Choose a student to demonstrate the multiplication. 20 ft 32 in. Point out that the Distributive Property of Multiplication was used to solve the problem: 4(5 ft + 8 in.) = (4 × 5 ft) + (4 × 8 in.) = 20 ft + 32 in.

What do you notice about the product? The number of inches in the product can be renamed as feet.

How can you find the number of feet that is equal to 32 in.? Since the equivalency of feet to inches is 1 ft = 12 in. (a ratio of 1 : 12), I can divide the 32 in. by 12 in.; 32 in. ÷ 12 in. = $2\frac{8}{12}$ ft = $2\frac{2}{3}$ ft or 2 ft 8 in.

What is the perimeter of the chick's area? 22 ft 8 in. or $22\frac{2}{3}$ ft; 2 ft 8 in. (the renamed 32 in.) added to the 20 ft equals 22 ft 8 in. or $22\frac{2}{3}$ ft.

- Follow a similar procedure for word problem number 5. Explain that the average weight is found by dividing the total weight by the number of chickens. When dividing a measurement, the students divide each unit by the divisor and rename the quotient if needed.

At the end of seven weeks, the twelve broiler chickens had a total weight of 60 lb 12 oz. What was the average weight of one broiler chicken? 60 lb 12 oz ÷ 12 = (60 lb ÷ 12) + (12 oz ÷ 12) = 5 lb 1 oz

- Direct the students to solve the last word problem. Choose students to write their solutions for display. Discuss the solutions as needed.

LESSON 138

The chicks ate half of a 5 lb bag of starter feed. How many pounds of starter feed were eaten? $2\frac{1}{2}$ lb How many ounces of starter feed were eaten? 40 oz; sample solutions: $\frac{1}{2} \times 5$ lb $= 2\frac{1}{2}$ lb, $2\frac{1}{2}$ lb $\times 16$ oz $= 40$ oz

- Follow a similar procedure to guide the students as they solve the word problems on the *Metric Measurement Word Problems* page.
  1. 12×5.74 kg $= 68.88$ kg
  2. 24×0.5 L $= 12$ L
  3. 2.4 kg $- 1.9$ kg $= 0.5$ kg or 500 g
  4. 37.2 kg $\div 12 = 3.1$ kg
  5. $1 - \frac{2}{3} = \frac{1}{3}$; $\frac{1}{3} \times 3.27$ kg $= 1.09$ kg; 1,090 g

If you taught the *Metric Measurement Word Problems* page during the lesson, you may guide the students as they solve the word problems on the *Customary Measurement Word Problems* page now.

- Guide a **discussion** of the essential question at the top of Student Edition page 304, "Why do you think scientists use the metric system?" to emphasize a benefit of using the metric system.

Which measurement conversion is easier for you to perform mentally, renaming 3 mi as feet or 3 km as meters? 3 km $= 3 \times 1,000$ m $= 3,000$ m is easier to perform mentally than 3 mi $= 3 \times 5,280$ ft $= 15,840$ ft.

What makes it easier to convert metric measurements? The metric system, like the decimal system, is a base 10 system. Multiplying and dividing by tens is easier than with other numbers.

Why do you think scientists use the metric system? The metric system is easier to use for converting measurements because it is a base 10 system.

Explain that scientists' use of the metric system allows scientists all over the world to understand one another's work, even when they speak different languages.

Find the ratio. Answer the question. *Ratio form may vary.*

1. Write the ratio of blue balls to total balls. *4 : 18*
2. Write the ratio of red balls to total balls. *5 : 18*
3. Which color ball is most likely to be chosen from the bag? *green*

4. Write the ratio that tells the probability that the spinner will land on blue. *3 : 8*
5. Write the ratio that tells the probability that the spinner will land on green. *4 : 8*
6. Which color has the lowest probability that the spinner will land on it? *red*

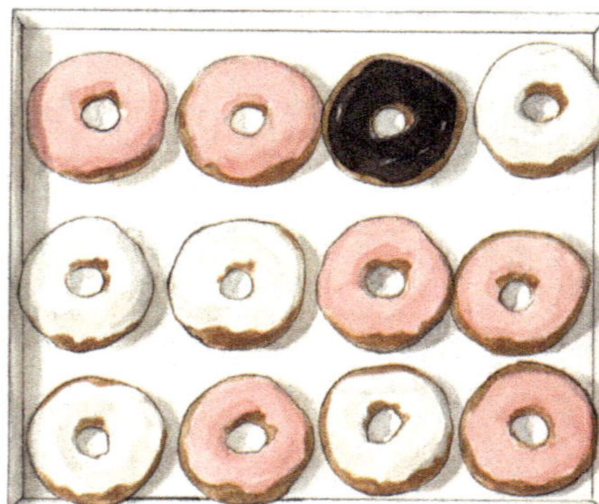

7. Write the ratio in fraction form to show the number of white-frosted doughnuts to total doughnuts. $\frac{5}{12}$
8. Write the ratio in word form to show the number of white-frosted doughnuts to pink-frosted doughnuts. *5 to 6*
9. Write the ratio to show the number of chocolate-frosted doughnuts to white-frosted and pink-frosted doughnuts. *1 : 11*
10. If someone takes one doughnut without looking, what type of doughnut will be the least likely taken? *chocolate-frosted*

Apply

Student Edition pages 304–5

- Read and explain the directions for pages 304–5. Assist the students as they complete the pages independently.

Daily Review

- Students should complete Chapter 14, section *e*.

LESSON 139

Student Edition pages 306–7
Daily Review Chapter 14, section *f*

OBJECTIVES

- Identify standard Celsius and Fahrenheit temperatures.
- Determine the more reasonable temperature based on standard temperatures.
- Read a Celsius and a Fahrenheit thermometer.
- Determine the amount of increase or decrease between two temperatures.
- Measure temperature by using a thermometer.
- Convert temperatures: Celsius to Fahrenheit and Fahrenheit to Celsius.

TEACHER RESOURCES

- 87 *Thermometer* (for the teacher and for each student)
- 88 *Temperature Hunt* (for each group)
- 89 *Double-Scale Thermometer* (for the teacher and for each student)

ADDITIONAL MATERIALS

- 3 waterproof containers
- water (3 different temperatures to fill 3 containers)
- 3 waterproof thermometers
- 3 additional thermometers
- a ruler (for each student)

Preparation

During the lesson you will fill 3 containers with water having the following approximate temperatures: 145°F (hot), 85°F (lukewarm), and 33°F (ice water). Place a thermometer in each container. Place the 3 additional thermometers at 3 different locations in the room so that the temperatures will vary.

An adult may volunteer to be a "location" and have his temperature taken by the students.

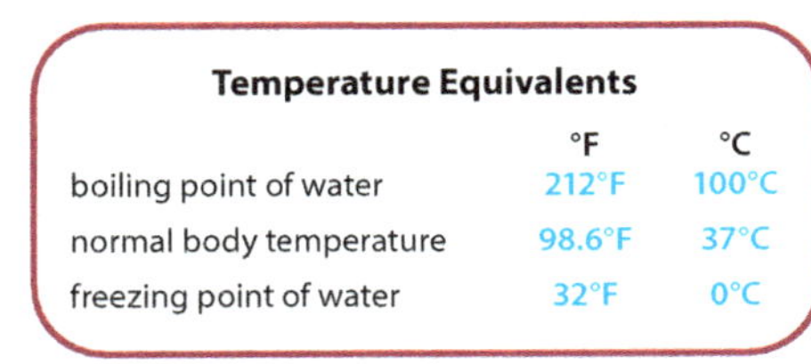

Fahrenheit & Celsius

How can I determine the Celsius temperature if I know the Fahrenheit temperature?

Use the conversion facts for temperature and the formulas to convert between **Fahrenheit** and **Celsius**.

Key Terms
- Fahrenheit
- Celsius

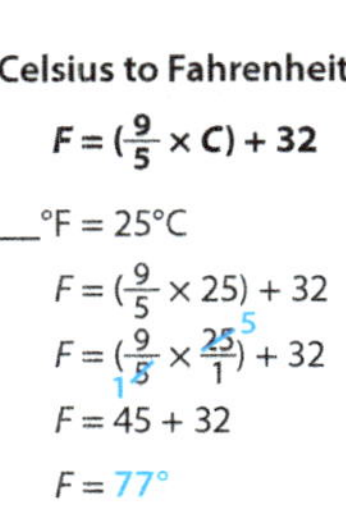

Temperature Equivalents

	°F	°C
boiling point of water	212°F	100°C
normal body temperature	98.6°F	37°C
freezing point of water	32°F	0°C

Converting Temperature

Celsius to Fahrenheit

$$F = \left(\frac{9}{5} \times C\right) + 32$$

___°F = 25°C

$$F = \left(\frac{9}{5} \times 25\right) + 32$$
$$F = \left(\frac{9}{5} \times \frac{25}{1}\right) + 32$$
$$F = 45 + 32$$
$$F = 77°$$

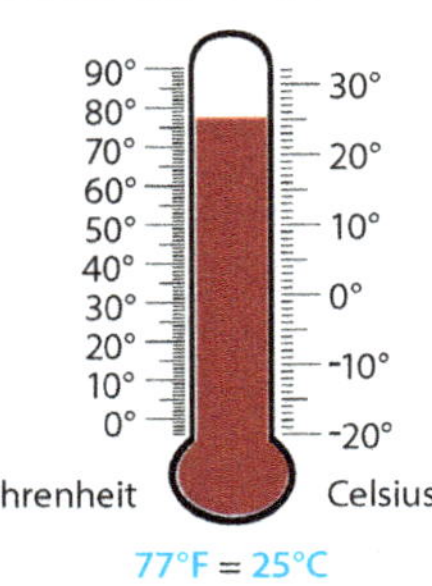

Fahrenheit to Celsius

$$C = \frac{5}{9} \times (F - 32)$$

___°C = 77°F

$$C = \frac{5}{9} \times (77 - 32)$$
$$C = \frac{5}{9} \times 45$$
$$C = \frac{5}{9} \times \frac{45}{1}$$
$$C = 25°$$

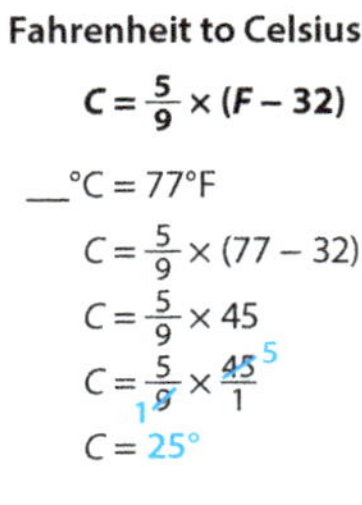

77°F = 25°C

Exercises

Choose the more reasonable temperature.

1. bowl of ice cream — (32°F) — 32°C
2. day at the beach — 50°F — (38°C)
3. room temperature — 20°F — (20°C)
4. water for boiling eggs — (212°F) — 50°C
5. sick child — 98.6°F — (103°F)
6. snowy day — (−3°C) — 30°C
7. fall day — (65°F) — 25°F
8. normal body temperature — 30°C — (37°C)
9. mowing the yard — (25°C) — 80°C
10. ice skating — 40°F — (30°F)
11. boiling water — (100°C) — 200°C

Determine the scale of the thermometer. Write the temperature.

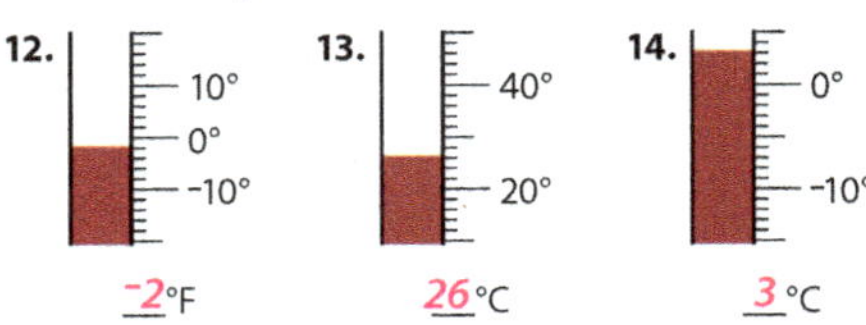

12. −2°F 13. 26°C 14. 3°C

Use the appropriate formula to convert the temperature. *Answers based on cancellation.*

15. 59°F = ___°C *15*   16. 10°C = ___°F *50*

17. 41°F = ___°C *5*   18. 20°C = ___°F *68*

19. 95°F = ___°C *35*   20. 50°C = ___°F *122*

306   Chapter 14

Engage

- Direct the students to do a **Think-Pair-Share** to explore the essential question at the top of Student Edition page 306, "How can I determine the Celsius temperature if I know the Fahrenheit temperature?"

Instruct

Identifying standard temperatures; determining the more reasonable temperature

- Guide a **discussion** to help the students choose a reasonable temperature based on a standard.

What unit is used to measure temperature? a degree

- Distribute a *Thermometer* page to each student and display your copy. Direct attention to the Celsius scale and the Fahrenheit scale, and note the degree sign (°) followed by a C and the degree sign (°) followed by an F at the bottom of the thermometer. Explain that in the metric system temperature is measured in degrees Celsius (°C), and in the customary system it is measured in degrees Fahrenheit (°F).

What does each line on both sides of the thermometer represent? 1 degree in that system

Steps to solve may vary.
Use the journal entries to find the answer.

21. What was the temperature in Fahrenheit on Day 1? *15°C = 59°F*

22. How many grams of food did Aaron take on his hike? *8 × 1,000 = 8,000; 8 kg = 8,000 g*

23. How many kilometers is Camp 1 above sea level? *2,438 ÷ 1,000 = 2.438; 2,438 m = 2.438 km*

24. What is a mild temperature for this part of Alaska? *14°C or 15°C*

25. What is the total mass of the sled and the backpack that Aaron took from Camp 2 to Camp 3? *30 + 25 = 55 kg*

26. What is the difference in elevation from Camp 2 to Camp 3? *11,500 − 7,770 = 3,730 ft*

27. If the ratio of melted snow to water is 10 : 1, how many milliliters of water did Aaron prepare for drinking on Day 6? $\frac{10}{1} = \frac{2.5\,L}{n}$; *n = 0.25 L = 250 mL*

28. How many degrees warmer was it on Day 6 than it was on Day 9? *10 + 14 = 24°*

29. On Day 9 what was the temperature inside the tent? *Answers may vary, but should be above 0°C.*

30. How much elevation did they gain on the hike on Day 12? *17,200 − 14,400 = 2,800 ft*

31. On Day 12 the team encountered a long ice slope. Was the ice slope a kilometer long? Explain your answer. *no; 700 m = 0.7 km*

32. On Day 16 Aaron looked up from the trail and saw the summit. About how many meters was the summit above him at that point? *330 ft ÷ 3 ≈ 100 yd; approximately 100 m*

33. How many meters higher is the summit than Camp 1? *6,194 − 2,438 = 3,756 m*

34. How can I determine the Celsius temperature if I know the Fahrenheit temperature? *I can use a thermometer with Celsius and Fahrenheit number lines. I can also use a formula to calculate the Celsius temperature.*

28. 

Reading a thermometer

- Use **guided practice** to help the students identify corresponding Fahrenheit and Celsius temperatures on a thermometer.

  Instruct the students to place their ruler across the thermometer on the *Thermometer* page at 10°F.

  What is the corresponding Celsius temperature of 10°F? *⁻13°C–⁻14°C*

- Repeat the procedure for 32°F *0°C*, 79°F *26°C–27°C*, and ⁻5°F *⁻22°C–⁻23°C*.

  As you move down the thermometer, does it show colder or hotter temperatures? *colder temperatures*

  How much colder is ⁻1°C than 0°C? *1 degree colder*

  How much colder is ⁻2°C than 0°C? *2 degrees colder*

- Direct the students to locate the following temperatures on their thermometers and to tell how many degrees warmer or colder than zero each temperature is.

  ⁻7°C *7°C colder than zero*

  8°F *8°F warmer than zero*

  13°C *13°C warmer than zero*

  ⁻5°F *5°F colder than zero*

Determining amount of increase or decrease between two temperatures; measuring temperature by using a thermometer

- Guide a **collaborative activity** for the students to compile temperature data and note the changes in temperature as the activity proceeds.

  Point to 18° Celsius on your displayed thermometer, and direct the students to locate it on the thermometer on their *Thermometer* page.

  If the temperature at noon was 18°C and later that same day it was 6°C, how much did the temperature change? *12°; I can count the number of degrees between the two temperatures, or I can subtract to find the difference between the two temperatures.*

What is the freezing point of water in degrees Celsius? *0°C*

What is the freezing point of water in degrees Fahrenheit? *32°F*

What is the boiling point of water in degrees Celsius? *100°C*

What is the boiling point of water in degrees Fahrenheit? *212°F*

What is the normal body temperature in degrees Celsius? *37°C*

What is the normal body temperature in degrees Fahrenheit? *98.6°F*

- Explain that these standard temperatures can be used as a reference when considering other Celsius and Fahrenheit temperatures.

- Choose students to select the more reasonable Celsius and Fahrenheit temperatures for each of the following items, using the standard temperatures as reference.

  frozen yogurt: 4°C or ⁻4°C *⁻4°C*; 25°F or 35°F *25°F*

  room temperature: 23°C or 3°C *23°C*; 70°F or 100°F *70°F*

  a child with a fever: 38.7°C or 50°C *38.7°C*; 96°F or 101.7°F *101.7°F*

  a fall day: 15°C or 80°C *15°C*; 10°F or 60°F *60°F*

  lava: 95°C or 700°C *700°C*; 199°F or 1,300°F *1,300°F*

LESSON 139

- Explain that thermometers can use intervals other than 1 degree between the lines (marks) on their scales, so it is important to determine the scale used before attempting to read a thermometer.

- Arrange the students in groups and distribute the *Temperature Hunt* page to each group. Instruct each group of students to read the temperatures of the prepared liquids and of the predetermined locations and to record the data on the page.

  Following the activity, allow each group to share and compare its data with the class. Discuss the changes in temperature that occurred at each station throughout the activity.

Converting temperatures

- Direct a **guided activity** to help the students convert temperatures between two scales.

  Distribute the *Double-Scale Thermometer* page and display your copy. Explain that the double-scale thermometer compares Celsius to Fahrenheit degrees. Point out that although water freezes at the same temperature and boils at the same temperature, the temperatures are expressed differently on each scale (32°F = 0°C and 212°F = 100°C). On the Fahrenheit scale, the degree difference between freezing and boiling (180°F) is divided into 20 equal parts of 9°F. On the Celsius scale the difference between freezing and boiling (100°C) is divided into 20 equal parts of 5°C.

- Guide the students as they complete the lower section of both scales, adding 9° to find the next warmer Fahrenheit temperatures 50°, 59°, 68° and adding 5° to find the next warmer Celsius temperatures 10°, 15°, 20°. Compare the Fahrenheit and Celsius temperatures. Point out that the Celsius freezing point is a good reference point because it is 0°C.

  What do you think is the Fahrenheit equivalent of 105°C? 221°F; 105°C is 5° more than boiling, which is the same as 9° more than the Fahrenheit boiling temperature; 100°C + 5°C = 212°F + 9°F.

  Complete both scales.

Write the numbers in order from least to greatest.

1. 4 -4 0 7
   $^-4$ 0 4 7

2. 8 -7 0 -8
   $^-8$ $^-7$ 0 8

3. -7 -10 -2 0
   $^-10$ $^-7$ $^-2$ 0

4. 0 1 -2 -5
   $^-5$ $^-2$ 0 1

Write a comparison sentence by using > or <.

5. 2 $>$ -2

6. -3 $\le$ -2

7. -50 $>$ -75

8. -8 $\le$ 4

9. -6 $>$ -9

10. -12 $>$ -16

11. 5 $>$ 4

12. 6 $\le$ 7

Use the number line to solve.

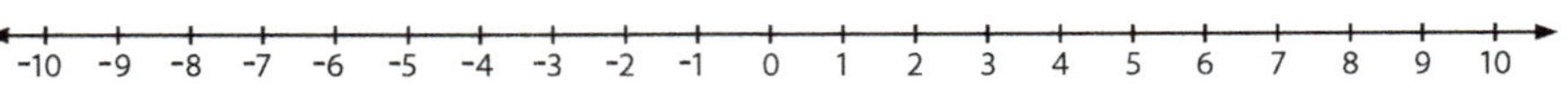

13. $^-4 + {}^-3$ *-7*

14. $5 + 1$ *6*

15. $^-3 + {}^-6$ *-9*

16. $^-5 + 5$ *0*

17. $4 + {}^-9$ *-5*

18. $8 + {}^-4$ *4*

494   Daily Review

What ratio represents the relationship of Fahrenheit to Celsius? 9 to 5; For every 9°F that the temperature increases, the equivalent Celsius temperature increases by 5°C.

- Write the formulas "$F = (\frac{9}{5} \times C) + 32$" and "$C = \frac{5}{9} \times (F - 32)$" for display.

  Explain that these are the formulas used to convert a known Celsius temperature to an equivalent Fahrenheit temperature and a known Fahrenheit temperature to an equivalent Celsius temperature.

  Why do you think 32° is added when converting from Celsius to Fahrenheit? The difference in the freezing points of Celsius and Fahrenheit is 32°.

Is the difference between the other equivalent temperatures 32°? No; the Fahrenheit scale increases by 9° for every 5° the Celsius scale increases.

Point out that the 9 : 5 ratio is the reason for multiplying by $\frac{9}{5}$ or $\frac{5}{9}$ when converting from Fahrenheit to Celsius or from Celsius to Fahrenheit.

- Direct attention to the essential question at the top of Student Edition page 306 to help the students summarize the strategies available to them for converting temperatures between systems.

How can you determine the Celsius temperature when you know the Fahrenheit temperature? I can use a thermometer with Celsius and Fahrenheit number lines. I can also use a formula to calculate the Celsius temperature.

- Demonstrate using the formula to find the Fahrenheit temperature that is equivalent to 105°C. Point out that the answer is the same as when they counted on the double-scale thermometer.

$F = (\frac{9}{5} \times C) + 32°$

$F = (\frac{9}{5} \times 105°) + 32°$

$F = (\frac{9}{5} \times \frac{105°}{1}) + 32°$

$F = (9 \times 21°) + 32°$

$F = 189° + 32°$

$F = 221°$

- Guide the students as they use the formulas to solve the following word problems.

On Monday morning, Reagan checked the temperature to see if she would need to wear a heavy coat to school. It was 20°C. Did she need a heavy coat? What was the Fahrenheit temperature? No coat was needed; $F = (\frac{9}{5} \times 20°) + 32° = 68°F$.

During a science experiment, Damon read a Fahrenheit thermometer as 59°F. He needed to record the temperature in Celsius. What was the Celsius temperature? $C = \frac{5}{9} \times (59° - 32°) = 15°C$

Apply

Student Edition pages 306–7
- Read and explain the directions for pages 306–7. Assist the students as they complete the pages independently.

Daily Review
- Students should complete Chapter 14, section f.

LESSON 139

NOTES

Student Edition pages 308–9
Daily Review Chapter 14, section g

OBJECTIVES

- Identify approximate equivalencies between customary and metric units of measurement.
- Compare customary and metric measurements.
- Estimate conversions between customary and metric measurements.
- Estimate a conversion between customary and metric units to help solve a real-world problem. **BWS**

BIBLICAL WORLDVIEW SHAPING

- **Knowledge (Apply):** Estimating conversions helps us understand our world as we solve real-world problems.

TEACHER RESOURCES

- 81 *Measurement Flashcards*
- 90 *Customary & Metric Conversions*

ADDITIONAL MATERIALS

- a larger set of measurement flashcards
- measurable objects (length)
- measurable objects (weight/mass and capacity; from those used in Lessons 135 and 137, optional)
- a ruler (dual metric & customary; for each student)

Preparation

Prepare a larger flashcard of each equivalency on the *Measurement Flashcards* page to use for the review activity.

Write the following equivalencies for display so that they can be revealed one at a time during the review game.

3 mi = __ ft

$\frac{2}{3}$ yd = __ ft

4 kg = __ g

2.5 L = __ mL

$\frac{7}{10}$ m = __ cm

30 in. = __ ft

Engage

- Use **games** to review standard measurement equivalencies and converting measurements.

Prepare several measurable objects (from those used in Lessons 135 and 137 or others you choose) by labeling them on the back or underneath with both customary and metric measurements of length, weight/mass, or capacity. Measurements provided on the labels of manufactured items may be used. Prepare at least 2 objects in each measurement category (length, weight/mass, and capacity) so they can be compared during the activity.

- Use the prepared measurement flashcards to conduct a review game. Assemble the class into 2 teams. Direct 1 student at a time from each team to compete as you show an incomplete equivalency and they try to complete it first. Keep score.

- Reveal one equivalency at a time from those prepared for display. Allow the teams to choose a spokesperson to give their answer as they collaborate in solving the problem for points.

3 mi = __ ft 15,840

$\frac{2}{3}$ yd = __ ft 2

4 kg = __ g 4,000

2.5 L = __ mL 2,500

$\frac{7}{10}$ m = __ cm 70

30 in. = __ ft $2\frac{1}{2}$

Relating Customary & Metric Units

What real-world problem can a unit conversion help me solve?

Approximate Customary and Metric Conversions

Length	Capacity	Weight or Mass
1 in. ≈ 2.5 cm	1 fl oz ≈ 30 mL	1 oz ≈ 30 g
1 ft ≈ 30 cm	1 L ≈ 1 qt	1 kg ≈ 2.2 lb
1 mi ≈ 1.6 km		
1 m ≈ 39 in.		

Estimating Approximate Conversions between Customary and Metric Units

Rename larger units as smaller units. Determine the approximate equivalency and then *multiply*.	Rename smaller units as larger units. Determine the approximate equivalency and then *divide*.
3 ft ≈ __ cm	11 lb ≈ __ kg
1 ft ≈ 30 cm	1 kg ≈ 2.2 lb
3 × 30 = 90	11 ÷ 2.2 = 5
3 ft ≈ 90 cm	11 lb ≈ 5 kg

Exercises

Write the approximate measurement.

1. Vivian's umbrella is 1 yd long. About how many meters long is it? *1 m*

2. Dan took 1 fl oz of antibiotic medicine. How many milliliters of medicine did he take? *30 mL*

3. Laura drank 1 qt of water. About how many liters of water did she drink? *1 L*

4. Lee ran 1.6 km. About how many miles did he run? *1 mi*

5. Mom bought 2 lb of chocolates. About how many kilograms of chocolates did she purchase? *1 kg*

6. A worm is 2.5 cm long. How many inches long is the worm? *1 in.*

Write a comparison sentence by using >, <, or ≈.

7. 4 in. ≥ 4 cm

8. 3 kg ≥ 3 lb

9. 1 qt ≈ 1 L

10. 2 m ≈ 2 yd

11. 2 ft ≥ 30 cm

12. 30 mL ≤ 5 fl oz

13. 2 lb ≤ 4 kg

14. 3 m ≤ 3 km

Use the conversion facts above to find an approximate equivalency.

15. 3 in. ≈ __ cm *3 × 2.5 = 7.5*

16. 2 fl oz ≈ __ mL *2 × 30 = 60*

17. 6 ft ≈ __ cm *6 × 30 = 180*

18. 4 m ≈ __ in. *4 × 39 = 156*

19. 7 L ≈ __ qt *7 × 1 = 7*

20. 60 mL ≈ __ fl oz *60 ÷ 30 = 2*

21. 78 in. ≈ __ m *78 ÷ 39 = 2*

22. 120 cm ≈ __ ft *120 ÷ 30 = 4*

23. 8.8 lb ≈ __ kg *8.8 ÷ 2.2 = 4*

Write the answer. *Steps to solve may vary.*

24. When you convert kilograms to pounds, do you get a higher or lower number of pounds? *higher*

25. Sara's kitten weighed 3 lb. Mira's kitten was 2 kg. Who had the smaller kitten? *2 • 2.2 lb = 4.4 lb; 3 lb < 4.4 lb; Sara*

26. The Colemans traveled 160 km. How many miles did they travel? *160 km ÷ 1.6 = 100 mi*

27. Brady's rope measured 2 m. Rory's rope measured 80 in. Whose rope was longer? *2 • 39 in. = 78 in.; 78 in. < 80 in.; Rory's*

Rename.

28. 1 ft = _12_ in.

1 yd = _3_ ft

1 mi = _1,760_ yd

29. 1 c = _8_ fl oz

1 pt = _2_ c

1 gal = _4_ qt

30. 1 lb = _16_ oz

1 tn = _2,000_ lb

31. 1 L = _1,000_ mL

1 m = _100_ cm

1 kg = _1,000_ g

32. $\frac{1}{4}$ of a kilometer = _250_ m   **33.** $\frac{3}{4}$ of a foot = _9_ in.   **34.** $\frac{1}{2}$ of a meter = _50_ cm   **35.** $\frac{1}{3}$ of a yard = _1_ ft

Complete the table.

36.

ounces	16	48	80
pounds	1	3	5

37.

grams	2,000	7,000	10,000
kilograms	2	7	10

Add or subtract. Simplify the answer when possible.

38. 3 ft 8 in.
  + 1 ft 9 in.
  4 ft 17 in. =
    5 ft 5 in.

39. 6 gal 3 qt
  + 2 gal 2 qt
  8 gal 5 qt =
    9 gal 1 qt

40. 9 lb 8 oz
  − 3 lb 10 oz
  5 lb 14 oz

41. 3 kg 760 g
  − 1 kg 820 g
  1 kg 940 g

Practice & Application

42. Andre needed $\frac{2}{3}$ of a yard of rope for his project. How many feet did he use?
$\frac{2}{3} \cdot 3$ ft = 2 ft

43. Marco ate $\frac{1}{4}$ of a gallon of ice cream. How many quarts did he eat? *$\frac{1}{4} \cdot 4$ qt = 1 qt*

44. Dad grilled 6 quarter-pound hamburgers. How many ounces of hamburger did he grill? *$6 \cdot (\frac{1}{4} \cdot 16) = 6 \cdot 4$ oz = 24 oz*

45. Kendra measured the width of her bracelet as 13. What would be a reasonable metric unit? *mm*

46. Mom needed half of a foot of ribbon for a craft project. How many inches did she need? *$\frac{1}{2} \cdot 12$ in. = 6 in.*

47. Kristen drank half of a liter of water. How many milliliters did she drink? *$\frac{1}{2} \cdot 1,000$ mL = 500 mL*

48. Mr. Lehman biked $1\frac{3}{4}$ km. How many meters did he bike?
$1\frac{3}{4} \cdot 1,000$ m = $\frac{7}{4} \cdot 1,000$ m = 1,750 m

49. Miss Cherie used 6 ft 8 in. of border for a bulletin board. How many inches did she use?
(6 • 12 in.) + 8 in. = 72 in. + 8 in. = 80 in.

50. Nathaniel measured his foot as 28. What would be a reasonable metric unit? *cm*

51. A 1 kg box holds 1,000 paper clips. What is the mass of 5 paper clips?
$5 \times (1,000$ g ÷ 1,000) = 5 g

52. The printing company ordered $2\frac{1}{4}$ tn of paper. How many pounds did it order?
$2\frac{1}{4} \cdot 2,000$ lb = $\frac{9}{4} \cdot 2,000$ = 4,500 lb

53. Devon compared his body mass to 40, the body mass of an average sixth-grader. What would be a reasonable metric unit? *kg*

54. Josie prepared a quart of punch. How many fluid ounces did she prepare?

55. What real-world problem can a unit conversion help you solve? Estimate the approximate conversion. Show your work.
Answers will vary.

54. *1 qt = 2 pt;*
2 pt = 4 c;
4 c = 32 fl oz

---

1 m (height of a doorknob from the floor) ≈ 39 in.

1 fl oz (2 tablespoons of liquid) ≈ 30 mL

1 L (a 1 L bottle of water) ≈ 1 qt

1 oz (a slice of bread) ≈ 30 g

1 kg (a 1 L bottle of water) ≈ 2.2 lb

Which is longer, an inch or a centimeter? an inch

Which is longer, a mile or a kilometer? a mile

Which is longer, a yard or a meter? a meter

Which has the greater capacity, a quart or a liter? The capacity of a quart and a liter are approximately the same, but the capacity of the liter is slightly greater than the quart; 1 qt = 32 oz and 1 L ≈ 33.81 oz.

Which has the greater mass, a pound or a kilogram? a kilogram

What do you notice about the number of customary units and the number of metric units in each approximation? Answers will vary; a greater number of smaller units are needed to measure approximately the same amount as 1 larger unit.

Comparing customary & metric measurements

- Use **guided practice** to help the students compare measurements between systems.

Write the following comparisons for display. Choose students to write > or < to complete the comparisons. Allow them to refer to the displayed page.

4 in. > 4 cm		2 kg > 2 lb
5 mi > 5 km		2 ft < 100 cm
3 m < 4 yd		4 L > 2 qt
2 m > 40 in.		5 fl oz > 5 mL
1 oz < 50 g		

- Display 2 or 3 of the measurable objects labeled for length by turning them so that their labels cannot be seen by the class. Read the length of one of the objects in metric units. Allow the students to choose the correct object by its measurement. Once the object has been correctly identified, guide the students as they estimate its approximate length in customary units. Compare the estimate with the actual measurement.

Instruct

Identifying approximate equivalencies

- Guide an **interactive activity** for the students to identify approximate equivalencies between 2 measurement systems.

- Review the personal benchmarks that were established in Lessons 134 and 136 for 1 in., 1 ft, 1 mi, and 1 m.

Example benchmarks for the following equivalencies are given in parentheses.

- Display the *Customary & Metric Conversions* page with the equivalencies covered.

Direct each student to use his metric ruler to measure the length of the midsection of his index finger (a benchmark for 1 in.) to find a close metric equivalent. $2\frac{1}{2}$ or 2.5 cm

Uncover the equivalency "1 in. ≈ 2.5 cm" on the displayed page.

- Follow a similar procedure to guide the students as they identify the other equivalencies on the displayed page. Allow the students to refer to the page throughout the remainder of the lesson.

1 ft (distance from a student's elbow to his knuckles) ≈ 30 cm

1 mi (the predetermined distance used in Lesson 134) ≈ 1.6 km

LESSON 140

Repeat the procedure with the objects labeled for weight/mass and capacity.

Retain the labeled items to be used again in Lesson 143.

Estimating conversions between customary & metric measurements

- Use **guided practice** to help the students solve word problems to estimate measurement conversions between the customary and metric systems.

- Follow procedures similar to the ones used in Lessons 134–37 to rename the units of measurement in the following word problems, using the approximate equivalencies on the displayed *Customary & Metric Conversions* page. Choose students to tell whether multiplication or division should be performed to rename the units.

Seth ran a 5 mi race. Approximately how many kilometers did he run? *8 km; 5 × 1.6 km ≈ 8 km*

After the race, Seth drank 960 mL of sports drink. Approximately how many fluid ounces did he drink? *32 fl oz; 960 mL ÷ 30 mL ≈ 32 fl oz*

While training for the race, Seth lost 3 kg. Approximately how many pounds did he lose? *6.6 lb; 3 × 2.2 lb ≈ 6.6 lb*

Kira jumped 510 cm in the long jump competition. Approximately how many feet did she jump? *17 ft; 510 cm ÷ 30 cm ≈ 17 ft*

Using measurement conversions

- Guide the students in a **Think-Pair-Share** to answer the essential question at the top of Student Edition page 308, "What real-world problem can a unit conversion help me solve?"

Remind the students that solving conversion problems helps them to obey God's command to exercise dominion in His world (Genesis 1:26) and prepares them to fulfill God's plan for their lives.

Allow the students a few minutes to think about their answer to the question, then instruct them to work with a partner to write a problem that can be solved

by using an estimated unit conversion between the customary and metric systems. Direct the students to solve their problem. Continue to display the approximate equivalencies on the *Customary & Metric Conversions* page for the students to reference.

Allow each pair to share their problem for the class to solve together. *sample problems:*

You are training for a 5K race. Your dad is helping you estimate the equivalent distance in miles using his car odometer. Approximately how many miles would you need to run to train for the race? 5 km ≈ __ mi; 1 mi ≈ 1.6 km; 5 ÷ 1.6 ≈ 3.1 mi; I would need to run a little over 3 miles to train.

Solve.

1.

Year	Campers
2014	200
2015	350
2016	400
2017	425
2018	450
2019	473

Camp Silver records the number of campers that attend each year. What is the average attendance of campers for the years shown on the table? *(200 + 350 + 400 + 425 + 450 + 473) ÷ 6 = 2,298 ÷ 6 = 383 campers*

2. Find the average grade for each student. Round the average to the nearest whole number.

	Test 1	Test 2	Test 3	Average
Kara	75	85	90	*83*
Jason	92	100	85	*92*
Abigail	85	95	90	*90*
Robert	100	100	97	*99*

3. Calculate Jim's average bowling score for Saturday's four games.

1	2	3	4
156	128	134	150

(156 + 128 + 134 + 150) ÷ 4 = 568 ÷ 4 = 142

4. Jessica saw 5 birds on Monday, 6 on Tuesday, 3 on Wednesday, 4 on Thursday, and 2 on Friday. What is the average number of birds she saw each day? *(5 + 6 + 3 + 4 + 2) ÷ 5 = 20 ÷ 5 = 4 birds*

5. In 2007 the Chicago Cubs won 85 baseball games. They won 97 games in 2008, 83 in 2009, and 75 in 2010. What is their average number of games won? *(85 + 97 + 83 + 75) ÷ 4 = 340 ÷ 4 = 85 games*

Daily Review   495

Your insulated water bottle holds 26 fl oz. You are filling it from a 1 L container of bottled water. Will your water bottle hold the whole liter of water? *1 L ≈ __ fl oz; 1 L ≈ 1 qt; 1 qt = 4 c = 4 × 8 fl oz = 32 fl oz; 1 L ≈ 32 fl oz. My bottle will not hold the whole liter of water.*

Your friend is a missionary to a foreign field. His mother orders five 1 L plastic bags of milk to be delivered to her home each week. Your mother purchases 2 gal of milk at the store each week for your family. Whose mother purchases more milk in a week? *5 L __ 2 gal; 1 L ≈ 1 qt, 5 L ≈ 5 qt; 1 gal = 4 qt, 2 gal = 2 × 4 qt = 8 qt; 5 L < 2 gal; My mother orders more milk than my missionary friend's mother does.*

Your pen pal shares her favorite metric chocolate chip cookie recipe. The recipe calls for two 168 g packages of chocolate chips. How many whole ounces of chocolate chips do you need to make the recipe? 2×168 g $= 336$ g; 1 oz ≈ 30 g; 336 g $\div 30 = 11\frac{1}{5}$ oz; I would need 12 oz of chocolate chips.

Apply

Student Edition pages 308–9

- Read and explain the directions for pages 308–9. Assist the students as they complete the pages independently.

Daily Review

- Students should complete Chapter 14, section *g*.

NOTES

LESSON 141

Student Edition pages 310–11
Daily Review Chapter 14, section *h*

OBJECTIVES

- Differentiate between a.m. and p.m.
- Convert units of time to smaller or larger units.
- Find a fraction of a unit of time.
- Add and subtract time.

TEACHER RESOURCES

- 91 *Time Measurement*

ADDITIONAL MATERIALS

- a demonstration clock with movable hands

Engage

- Direct the students to **brainstorm** to answer the essential question at the top of Student Edition page 310, "Where does time begin?" Time begins at the prime meridian (0° longitude) in Greenwich, England.

  Discuss the students' answers. Direct their attention to the "Did You Know" section at the bottom of Student Edition page 311 to discuss the prime meridian (0° longitude) in Greenwich, England, where each new day begins on the earth.

Instruct

Reviewing time

- Use **guided practice** to review time measurement concepts.

  Write the following equivalencies for display and choose students to complete them. Use the *Time Measurement* page to review the equivalencies as needed.

  1 minute = __ seconds 60
  1 hour = __ minutes 60
  1 day = __ hours 24
  1 week = __ days 7
  1 month = __ days 28–31
  1 year = __ months 12
  1 year = __ weeks 52
  1 year = __ days 365

  1 leap year = __ days 366
  1 decade = __ years 10
  1 century = __ years 100
  1 millennium = __ years 1,000

  Students may refer to the *Time Measurement* page or the time equivalents in the Measurement section of the student Handbook throughout the lesson.

- Choose students to tell whether hours, minutes, or seconds is the best unit for estimating the amount of time needed for each of these activities.

  a sneeze seconds
  an educational video hours or minutes
  a trip hours or minutes
  repeating a math fact seconds
  a song minutes
  a lunch break minutes or hours

- Review telling time to the minute. Set the demonstration clock for 10:34. What time is shown? 10:34

  How do you count the minutes to reach 34 min after the hour? I begin at 10:00, count by 5s until I reach the number 6 on the clock, and then I count each additional minute by 1s.

- Set the clock for 2:30. What ways can you read this time? 2:30; half past 2

Telling & Renaming Time

Where does time begin?

Key Terms
- second
- minute
- hour
- day

Time Equivalents

1 minute (min) = 60 seconds (sec) 1 hour (hr) = 60 min 1 day (d) = 24 hr

Renaming Time

Rename larger units as smaller units. Determine the equivalency and then *multiply*.

7 hr = __ min
1 hr = 60 min
7 × 60 = 420
7 hr = 420 min

Rename smaller units as larger units. Determine the equivalency and then *divide*.

72 hr = __ d
24 hr = 1 d
72 ÷ 24 = 3
72 hr = 3 d

Adding & Subtracting Time

Hailey spent 3 hr 25 min working at her job before her break. After her break she worked 4 hr 50 min. How long did she work?

  3 hr 25 min   Add the minutes.
+ 4 hr 50 min   Add the hours.
  7 hr 75 min = Rename the
  8 hr 15 min   answer if possible.

Hailey's work hours totaled 16 hr 40 min on Tuesday. On Wednesday her work hours totaled 23 hr 30 min. How many hours did she work on Wednesday?

  $\overset{22}{2}\overset{90}{3}$ hr 30 min   Subtract the minutes.
− 16 hr 40 min   Rename if needed.
   6 hr 50 min   Subtract the hours.

A Fraction of Time

Hailey spent $\frac{1}{4}$ of an hour helping her dad in the garden. How many minutes did she spend helping her dad?

$\frac{1}{4}$ of an hour = __ min
1 hr = 60 min
$\frac{1}{4} \times 60 = 15$ min

Exercises

Write the time.

1.  6:15
2.  10:43
3. 4:17
4.  11:30

Rename the time.

5. 3 d = __ hr *3 • 24 = 72*
6. 480 min = __ hr *480 ÷ 60 = 8*
7. 7 hr = __ min *7 • 60 = 420*
8. 10 d = __ hr *10 • 24 = 240*
9. 300 sec = __ min *300 ÷ 60 = 5*
10. 120 hr = __ d *120 ÷ 24 = 5*
11. 9 min = __ sec *9 • 60 = 540*
12. 264 hr = __ d *264 ÷ 24 = 11*
13. 720 min = __ hr *720 ÷ 60 = 12*

Write the equivalent unit of time.

14. 1 d = *24* hr
15. 1 hr = *60* min
16. 1 min = *60* sec

Rename the time.

17. $\frac{1}{2}$ of an hour = *30* min
19. $\frac{3}{4}$ of an hour = *45* min
21. $\frac{1}{10}$ of a minute = *6* sec

18. $\frac{1}{4}$ of a day = *6* hr
20. $\frac{1}{8}$ of a day = *3* hr
22. $\frac{1}{4}$ of a minute = *15* sec

17. $\frac{1}{2} • 60 = 30$
18. $\frac{1}{4} • 24 = 6$
19. $\frac{3}{4} • 60 = 45$
20. $\frac{1}{8} • 24 = 3$
21. $\frac{1}{10} • 60 = 6$
22. $\frac{1}{4} • 60 = 15$

310 Chapter 14

Add or subtract. Simplify the answer when possible.

23. 5 hr 15 min
+ 5 hr 26 min
10 hr 41 min

24. 7 hr 45 min 20 sec
+ 6 hr 39 min 15 sec
13 hr 84 min 35 sec =
14 hr 24 min 35 sec

25. 10 hr 35 min 15 sec
+ 6 hr 29 min 45 sec
16 hr 64 min 60 sec =
17 hr 5 min

26. 3 min 15 sec
+ 4 min 49 sec
7 min 64 sec =
8 min 4 sec

27. 15 hr 43 min
− 8 hr 50 min
6 hr 53 min

28. 18 hr 17 min 45 sec
− 9 hr 46 min 21 sec
8 hr 31 min 24 sec

29. 23 hr 32 min 10 sec
− 20 hr 40 min 30 sec
2 hr 51 min 40 sec

30. 15 min 10 sec
− 8 min 21 sec
6 min 49 sec

Write the best unit of measurement: hours, minutes, or seconds.

31. to use speed dial *seconds*

32. to take a trip *hours*

33. to bake cookies *minutes*

34. to finish math homework *minutes*

35. to run a mile *minutes*

36. to open a door *seconds*

37. to read a chapter book *hours*

38. to ride to the store *minutes*

Practice & Application

39. The temperature inside the room was 71°___. *F*

40. The pair of shoes had a mass of 1 ___. *kg*

41. The ballplayer drank 1 ___ of a sports drink. *L or qt*

42. The width of the wedding ring was 4 ___. *mm*

43. The boiling water was 212°___. *F*

44. The football game lasted about 2 ___. *hr*

45. The backpack weighed about 4 ___. *lb or kg*

46. Hayden ate 1 ___ of ice cream. *c or pt*

47. Eleanor walked 1 ___ in 20 min. *mi*

48. Mom worked in her garden for $\frac{1}{4}$ of an hour; she worked *15* min.

49. Paul needed $\frac{3}{4}$ of a foot of rope; he needed *9* in.

50. Dad cut $\frac{2}{3}$ of a yard from a board; he cut *24* in.

51. Use the information below about the 24 hr clock to convert 6:00 p.m., 10:00 p.m., and 9:00 p.m. to 24 hr time. *18:00; 22:00; 21:00*

52. Where does time begin?
Time begins at the prime meridian (0° longitude) in Greenwich, England.

DID YOU KNOW?

Time begins at the **prime meridian** (0° longitude) that passes through Greenwich, England. At the Royal Greenwich Observatory, a 24-hour clock runs from midnight to midnight. The last hour of the day is 24 o'clock, or is that the first hour? It can be both times!

On a 24 hr clock, the first 12 hr of the day are numbered from 1 to 12. Then 1 in the afternoon is numbered 13, 2 is numbered 14, and so on. To convert an afternoon time from 12 hr time to 24 hr time, add 12 to the afternoon time. If it is 3 p.m., add 12. On a 24 hr clock, 3 p.m. is 15 o'clock in the afternoon. The 24 hr clock is used in many countries. In the United States, the military uses the 24 hr clock, and it is commonly called "military time."

Shepherd Gate Clock, Greenwich, England

Lesson 141   311

- Repeat the procedure using the following times.
  6:20 *6:20; 20 min after six*
  9:45 *9:45; 15 min before 10; quarter to 10*
  12:15 *12:15; 15 min after 12; quarter after 12*
  8:55 *8:55; 5 min before 9*

Differentiating between a.m. & p.m.

- Use **direct instruction** to explain the use of a.m. and p.m. and the 24-hour clock.
  Display 12:00 on the clock. Point out that when it is 12:00 during the day, we say it is 12 noon; when it is 12:00 at night, we say it is 12 midnight.

- Write "a.m." and "p.m." for display. Explain that the abbreviation a.m. is for *ante meridiem*, which is Latin for *before noon*. The abbreviation p.m. is for *post meridiem*, which is Latin for *after noon*.

- Explain that a day begins at midnight and continues until the next midnight; therefore, 12:00 at night is 12 a.m. because the minutes following it begin the morning hours, and 12:00 during the middle of the day is 12 p.m. because the minutes following it begin the afternoon hours.

  Explain that in military time, which uses a 24-hour clock, midnight is referred to as 0000 (zero) hours, 1 a.m. as 0100 (zero one hundred) hours, noon is 1200 (twelve hundred) hours, 1 p.m. is 1300 (thirteen

hundred) hours, and 11 p.m. is 2300 (twenty-three hundred) hours.

- Choose students to tell whether the following activities would occur during the a.m. hours or during the p.m. hours.
  7:00 breakfast a.m.
  12:00 lunch p.m.
  3:00 school dismissal p.m.
  9:00 bedtime p.m.
  7:00 basketball game p.m.
  12:00 celebration of a new year a.m.

- Explain the 24-hour clock using the information at the bottom of Student Edition page 311. Point out that midnight simultaneously ends one day (2400 hours) and begins the next day (0000 hours). Time on a 24-hour clock can also be read using the word *o'clock* (e.g., 13 o'clock [1 p.m.] and 23 o'clock [11 p.m.]).

Converting units of time

- Use **discussion** to guide the students as they convert units of time.
  Write "120 min = ___ hr" for display.
  How do you rename 120 min to hours?
  To rename smaller units (minutes) to larger units (hours), I determine the equivalency (60 min = 1 hr) and then divide by 60 min.

- Choose a student to demonstrate renaming 120 min as hours and to complete the expression. 120 min ÷ 60 min = 2 hr

- Repeat the procedure for 3 d = ___ hr. To rename larger units (days) to smaller units (hours), they determine the equivalency (1 d = 24 hr) and then multiply by 24 hr. 3 × 24 hr = 72 hr

- Write the following equivalencies for display. Choose students to solve the equivalencies for display while the other students solve them independently. Instruct the students to explain the renaming process in their displayed solutions.
  420 min = ___ hr 7
  96 hr = ___ d 4
  5 min = ___ sec 300
  36 hr = ___ d $1\frac{1}{2}$

LESSON 141

14-H

Finding a fraction of a unit of time

- Use **guided practice** to help the students solve time word problems involving a fraction of a unit of time.

  Read aloud the following word problem.

  Faith spent $\frac{1}{3}$ of an hour practicing her piano lesson. How much time did she spend practicing the piano? 20 min

  What smaller unit of time can be used to show part of an hour? minute

  What ratio or equivalency do you know for hours to minutes? 1 hr : 60 min or 1 hr = 60 min

  How can you find $\frac{1}{3}$ of an hour? Since I need to find part of an hour or minutes, and "of" means times or multiply, I can substitute 60 min for 1 hr and multiply $\frac{1}{3} \times 60$ min.

- Choose a student to write the equation and solve it. $\frac{1}{3} \times 60$ min = 20 min

- Follow a similar procedure for the following word problem. Point out that the smaller unit *hour* is needed to find a part ($\frac{1}{4}$) of a day. 1 d : 24 hr or 1 d = 24 hr; substitute 24 hr for 1 d

  Eli spends $\frac{1}{4}$ of his day at school. How much time does he spend at school? $\frac{1}{4} \times 24$ hr = 6 hr

Adding & subtracting time

- Use **guided practice** to help the students solve word problems that use addition and subtraction of time.

  Read aloud the following word problem.

  Maria was training to run a half marathon. She recorded her running times for three training sessions: 2 hr 25 min 22 sec, 2 hr 24 min 14 sec, and 2 hr 20 min 41 sec. How much time did Maria run during these training sessions? 7 hr 10 min 17 sec

  How can you find the amount of time Maria spent running during the three training sessions? I can add the recorded times and rename in the sum if needed.

Solve. Rename to lowest terms. *Answer is shown using cancellation.*

1. $\frac{1}{3} \div \frac{1}{5}$ $\frac{1}{3} \times \frac{5}{1} = \frac{5}{3} = 1\frac{2}{3}$

2. $\frac{3}{4} \div \frac{1}{2}$ $\frac{3}{4} \times \frac{2}{1} = \frac{3}{2} = 1\frac{1}{2}$

3. $2\frac{4}{7} \div \frac{3}{4}$ $\frac{18}{7} \times \frac{4}{3} = \frac{24}{7} = 3\frac{3}{7}$

4. $\frac{3}{5} \div \frac{2}{3}$ $\frac{3}{5} \times \frac{3}{2} = \frac{9}{10}$

5. $\frac{9}{18} \div \frac{3}{6}$ $\frac{9}{18} \times \frac{6}{3} = \frac{3}{3} = 1$

6. $3\frac{3}{8} \div \frac{4}{8}$ $\frac{27}{8} \times \frac{8}{4} = \frac{27}{4} = 6\frac{3}{4}$

7. $\frac{4}{8} \div \frac{1}{4}$ $\frac{4}{8} \times \frac{4}{1} = \frac{4}{2} = 2$

8. $5\frac{1}{3} \div 2\frac{1}{6}$ $\frac{16}{3} \times \frac{6}{13} = \frac{32}{13} = 2\frac{6}{13}$

9. $5\frac{6}{7} \div \frac{1}{3}$ $\frac{41}{7} \times \frac{3}{1} = \frac{123}{7} = 17\frac{4}{7}$

10. $\frac{9}{12} \div \frac{1}{6}$ $\frac{9}{12} \times \frac{6}{1} = \frac{9}{2} = 4\frac{1}{2}$

11. $9\frac{2}{4} \div 3\frac{1}{6}$ $\frac{38}{4} \times \frac{6}{19} = \frac{6}{2} = 3$

12. $4\frac{3}{8} \div 1\frac{2}{6}$ $\frac{35}{8} \times \frac{6}{8} = \frac{105}{32} = 3\frac{9}{32}$

Solve.

13. Miss Snow teaches ice skating to beginners. Each lesson is $\frac{1}{2}$ of an hour long. How many lessons can she give in 3 hr? *$3 \div \frac{1}{2} = \frac{3}{1} \times \frac{2}{1} = \frac{6}{1} = 6$ lessons*

14. David is planning to grill burgers for a cookout. He uses 1 lb of hamburger to make 4 burgers. How many burgers can he make with $4\frac{1}{2}$ lb of meat? *$4\frac{1}{2} \times 4 = \frac{9}{2} \times \frac{4}{1} = \frac{18}{1} = 18$ burgers*

496   Daily Review

- Write the problem for display and choose a student to solve it. Point out that the like units are added first, beginning with the smallest unit. Then the seconds and minutes in the sum are renamed to larger units because each of their totals is greater than 60; 60 sec = 1 min and 60 min = 1 hr.

  2 hr 25 min 22 sec
  2 hr 24 min 14 sec
  + 2 hr 20 min 41 sec
  6 hr 69 min 77 sec = 7 hr 10 min 17 sec

The Pilgar family ran a marathon. Mr. Pilgar ran it in 3 hr 59 min 53 sec. Mrs. Pilgar ran the race in 4 hr 34 min 14 sec. Who ran the race faster? How much faster? 34 min 21 sec

Who ran the race faster? Mr. Pilgar; His time is less than Mrs. Pilgar's time.

How can you find how much faster he ran the race? I can subtract Mr. Pilgar's time from Mrs. Pilgar's time.

- Write the problem for display and choose a student to solve it. Point out that in order to subtract 53 sec they must rename 1 min as 60 sec so that there are 74 sec in the minuend. Then, in order to subtract 59 min, they must rename 1 hr as 60 min so that there are 93 min in the minuend.

   4 hr 34 min 14 sec
− 3 hr 59 min 53 sec
       34 min 21 sec

- Write the following problems for display and direct the students to solve them. Then choose students to demonstrate solving the problems, explaining each step as they solve the problem.

   3 hr 25 min   55 min 28 sec
+ 4 hr 50 min   + 40 min 59 sec
   7 hr 75 min   95 min 87 sec
   8 hr 15 min   1 hr 36 min 27 sec

   5 hr   9 min 35 sec
− 2 hr 25 min 45 sec
   2 hr 43 min 50 sec

Student Edition pages 310–11

- Read and explain the directions for pages 310–11. Assist the students as they complete the pages independently.

Daily Review

- Students should complete Chapter 14, section *h*.

NOTES

Student Edition pages 312–13
Daily Review Chapter 14, section _i_

OBJECTIVES

- Explain how the act of measuring is rooted in Scripture. **BWS**
- Use a table to determine world time zones.
- Determine the elapsed time.
- Add and subtract time.

BIBLICAL WORLDVIEW SHAPING

- **Knowledge (Formulate):** Measuring is rooted in Scripture because it tells us when it began, why it is possible, and how it should be used.

TEACHER RESOURCES

- 91 _Time Measurement_
- 92 _Time Zones of the World_ (for the teacher and for each student)
- 93 _Time Zones_

ADDITIONAL MATERIALS

- a demonstration clock with movable hands (for the teacher and for each student)

ASSESSMENTS

- Chapter 14 Quiz 2

Elapsed Time & Time Zones

How does the act of measuring point to Scripture?

Elapsed time is the amount of time that passes between two points in time.

How much time passes between 8:20 a.m. and 1:55 p.m.? Count the hours first and then count the minutes.

Count the hours.   8:20 a.m. to 1:20 p.m. = 5 hr
Count the minutes. 1:20 a.m. to 1:55 p.m. = 35 min
elapsed time: 5 hr 35 min

Key Terms
- elapsed time
- time zones

8:20 a.m.   1:55 p.m.

Exercises

Write the elapsed time.

1. 9:15 a.m. to 9:45 a.m.
   30 min
2. 3:20 a.m. to 5:20 p.m.
   _14 hr_
3. 7:30 a.m. to 4:45 p.m.
   9 hr 15 min
4. 8:20 a.m. to 9:00 a.m.
   40 min
5. 11:45 p.m. to 12:30 a.m.
   45 min
6. 8:10 a.m. to 12:35 p.m.
   4 hr 25 min
7. 4:17 p.m. to 4:59 p.m.
   42 min
8. 2:18 a.m. to 5:42 a.m.
   3 hr 24 min
9. 1:50 p.m. to 12:10 a.m.
   10 hr 20 min
10. 6:25 p.m. to 6:51 p.m.
    26 min
11. 10:11 a.m. to 1:31 p.m.
    3 hr 20 min
12. 5:30 p.m. to 7:45 p.m.
    2 hr 15 min

Write the time.

13. 2 hr 15 min after 3:00
    _5:15_
14. 3 hr 20 min after 12:00
    _3:20_
15. 25 min after 2:45
    _3:10_
16. 1 hr 50 min before 5:00
    _3:10_
17. 45 min before 3:15
    _2:30_
18. 3 hr before 7:00
    _4:00_

Complete the elapsed time on the schedule. Use the table to find the answer.

19. Which bus trip is the longest? _#1_

20. Which bus trip is the shortest? _#2_

21. Which bus trip begins in the afternoon and ends before midnight? _#4_

22. The four bus trips depart from San Diego and arrive at San Francisco. What accounts for the difference in elapsed times? _Answers may vary. Possible answers include heavy traffic times, different routes, and extra stops._

Bus Schedule			
Bus	Departure: San Diego	Arrival: San Francisco	Elapsed Time
#1	1:30 a.m.	2:40 p.m.	_13 hr 10 min_
#2	8:15 p.m.	7:15 a.m.	_11 hr_
#3	10:15 p.m.	10:50 a.m.	_12 hr 35 min_
#4	12:15 p.m.	11:35 p.m.	_11 hr 20 min_

Add or subtract. Simplify the answer when possible.

23.   6 hr 21 min 50 sec
    + 1 hr 45 min 20 sec
    7 hr 66 min 70 sec =
    8 hr 7 min 10 sec

24.   8 hr 29 min
    + 3 hr 50 min
    11 hr 79 min =
    12 hr 19 min

25.   4 hr
    − 2 hr 17 min
    1 hr 43 min

26.   7 hr 0 min 50 sec
    − 2 hr 30 min 15 sec
    4 hr 30 min 35 sec

312   Chapter 14

The students will use the _Time Zones of the World_ page again in Lesson 143.

If you do not have a demonstration clock available for each student, you may make clocks from paper plates using card stock hands attached with paper fasteners. Alternately, group students to use the clocks you have available or emphasize mental calculation in this lesson and future lessons.

Engage

- Repeat the **guided practice** activity used near the beginning of Lesson 141

to review equivalent units of time. Use the _Time Measurement_ page as needed.

Instruct

The act of measuring in Scripture

- Guide a **discussion** of the essential question at the top of Student Edition page 312 to help the students answer the question, "How does the act of measuring point to Scripture?" Review the discussions from Lessons 134, 137, and 140 that provide a scriptural basis for viewing measurement. (Genesis 1, Deuteronomy 25:13, Proverbs 16:11, Genesis 1:26)

When did the act of measuring begin? Measurement began when God created an orderly, measurable world.

How does the beginning of measurement at creation point to Scripture? I would not know where measurement began without the Scripture record.

What does the Bible say about standardized measurements? God's Word requires that my measurements be accurate and honest according to a standard.

How do standardized measurements point to Scripture? God's image in me makes it possible for me to understand the importance of standardized measurements.

Time zones help the world identify the same time of day. They were created so that noon is the middle of the day in each time zone. The zones make a number line starting with 0 at Greenwich, England. To the east, or right, of Greenwich (0), the numbers *increase* by one up to 12. To the west, or left, of Greenwich (0), the numbers *decrease* by one down to -12. There are 24 time zones around the world; 6 of these are in the United States. When it is 3:00 p.m. in Florida, it is 2:00 p.m. in Texas.

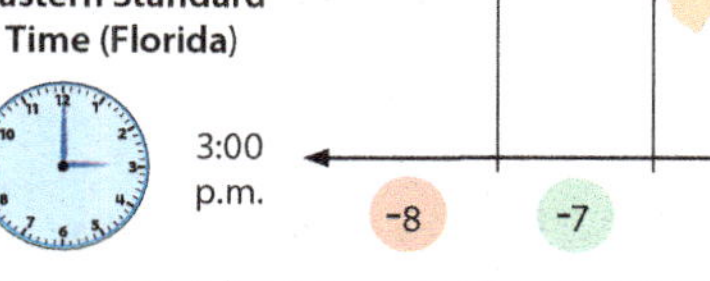

Use a time zone map to find the answer.

27. What time zone is represented by 0° longitude, the prime meridian? *0 or Greenwich mean time*

28. What is the time in the middle of the day in each time zone? *12:00 p.m. or noon*

29. What time zone is 1 hr earlier than eastern standard time? *central standard time*

30. Why is the Pacific standard time zone represented by -8? *because it is 8 time zones west of 0, or Greenwich mean time*

31. The eastern standard time zone is -5 on the number line. How many time zones is the eastern standard time zone from Greenwich (0°)? *5 time zones*

32. If Greenwich mean time is 12:00, what time is it in eastern standard time? mountain standard time? *5:00; 7:00*

33. If it is 5:00 Pacific standard time, then it is 6:00 mountain standard time. What time would it be in central standard time? *7:00*

34. If it is 6:00 in eastern standard time, what time is it in the following places?

New York City, NY	*6:00*
Chicago, IL	*5:00*
Denver, CO	*4:00*
Portland, OR	*3:00*

35. How does the act of measuring point to Scripture?

Practice & Application

36. Mrs. Alier placed a pie in the oven at 1:45 p.m. The pie needed to bake 50 min. What time will the pie be finished? *2:35 p.m.*

37. Dominic left school at 10:20 a.m. for a dentist appointment. The drive to the dentist took 10 min. The appointment lasted 1 hr 15 min. The drive back to school took 15 min. What time did Dominic return to school? *12:00 p.m.*

38. Hudson starts school at 8:00 a.m. His first class is reading, which lasts 45 min. Next he has a 50 min math class. Then he has a break. What time is Hudson's break? *9:35 a.m.*

39. Macy practiced her flute for 20 min on Monday. She practiced from 3:30 p.m. to 4:15 p.m. on Wednesday. She got in an extra hour of practice on Friday. What was Macy's total practice time? *2 hr 5 min*

35. *Scripture records God's creation of an orderly, measurable world. It tells me that God requires my measurements to be accurate and honest according to a standard. I fulfill the creation mandate by solving real-world measurement problems.*

Lesson 142   313

longitude. The clocks show the time relationship when it is 12 noon in Greenwich. Note that time increases by 1 hr for each time zone east of Greenwich (0 on the number line) and decreases by 1 hr for each time zone west of Greenwich.

How many time zones are there in the world? 24

- Guide the students as they count the zones. Point out that +12 and -12 are the same zone and are both written as -12+. The International Date Line, an imaginary line running through the Pacific Ocean at about 180° longitude, is located in this time zone. A person traveling west across the date line gains a day and a person traveling east across the date line loses a day.

- Display the *Time Zones* page and write "12:00 noon" in the Greenwich Mean Time column. Guide the students as they complete the table by determining the time in the designated locations when it is 12 noon in Greenwich, England. Allow the students to use the number line on their *Time Zones of the World* page to count the time zones.

Time Zones of Major Cities			
Greenwich Mean Time	City	Zone	Time
12:00 noon	London	0	12:00 p.m.
	Moscow	+3	3:00 p.m.
	Tokyo	+9	9:00 p.m.
	New York	-5	7:00 a.m.
	Los Angeles	-8	4:00 a.m.

- Repeat the procedure to complete the table using 3:00 p.m. Greenwich mean time.

Time Zones of Major Cities			
Greenwich Mean Time	City	Zone	Time
3:00 p.m.	London	0	3:00 p.m.
	Moscow	+3	6:00 p.m.
	Tokyo	+9	12:00 a.m.
	New York	-5	10:00 a.m.
	Los Angeles	-8	7:00 a.m.

What real-world problem did a unit conversion help you solve? Answers will vary.

How does solving a measurement problem point to Scripture? The Scripture tells me to exercise dominion in God's world, and solving measurement problems is one way I can do that.

Determining world time zones

- Use **direct instruction** to help the students determine the time in different locations around the world.

Distribute a copy of the *Time Zones of the World* page to each student and display your copy. Explain that the time zones begin at the prime meridian (0° longitude) in Greenwich, England. The time in this time zone is referred to as Greenwich mean time. All other time zones are ahead of or behind Greenwich mean time (0° longitude) and were created so that noon is the middle of the day in each time zone.

Point out the different time zones on the map. Explain that each time zone is approximately a 15-degree longitudinal segment. Locations in each time zone share the same time.

- Direct attention to the number line at the bottom of the map. Point out that the number line shows each time zone and its relationship to Greenwich, England, or 0°

LESSON 142

Determining elapsed time

- Use **guided practice** to help the students determine elapsed time.

  Write "1:15 p.m. to 5:31 p.m." for display.

  How can you determine the elapsed time or the amount of time that passed from 1:15 p.m. to 5:31 p.m.? I can first count the full hours and then count the remaining minutes that passed between the beginning time (1:15 p.m.) and the ending time (5:31 p.m.).

- Set the demonstration clock at 1:15. Guide the students as they count the elapsed time as you advance the time to 2:15, 3:15, 4:15, 5:15, 5:16, 5:17, . . . 5:31: 1 hr, 2 hr, 3 hr, 4 hr, 4 hr 1 min, 4 hr 2 min, . . . 4 hr 16 min.

  How many hours elapsed between 1:15 p.m. and 5:15 p.m.? 4 hr

  How many minutes elapsed between 5:15 p.m. and 5:31 p.m.? 16 min

  What is the total amount of elapsed time between 1:15 p.m. and 5:31 p.m.? 4 hr 16 min

- Write "10:45 a.m. to 2:14 p.m." Remind the students that whenever time elapses over the 12:00 hour, the label of a.m. or p.m. changes. Repeat the previous procedure as you guide the students to count the elapsed time.

  How much time elapsed between 10:45 a.m. and 2:14 p.m.? 3 hr 29 min

- Distribute the student clocks and write the following times for display. Direct each student to use his or her clock to find the elapsed times.

  5:17 p.m. to 7:19 p.m. 2 hr 2 min

  12:15 a.m. to 4:30 a.m. 4 hr 15 min

  10:20 a.m. to 5:10 p.m. 6 hr 50 min

  1:40 p.m. to 2:15 a.m. 12 hr 35 min

- Instruct the students to use their clocks to determine the elapsed time in each of the following word problems.

  The Pattons left on a trip at 1:50 p.m. and arrived at their destination at 7:30 p.m. How long did their trip take them? 5 hr 40 min

  Daniel played basketball from 10:10 a.m. until 10:42 a.m. How long did he play? 32 min

Solve.

1. $9.4\overline{)0.1316}$ 0.014
2. $5.4\overline{)3.186}$ 0.59
3. $67\overline{)20.77}$ 0.31
4. $1\overline{)0.05}$ 0.05

5. $8.9\overline{)436.1}$ 49
6. $7.5\overline{)0.375}$ 0.05
7. $1.3\overline{)0.429}$ 0.33
8. $27\overline{)7.83}$ 0.29

Rename the fraction, using a power of 10 as the denominator. Write the fraction as a decimal.

9. $\frac{2}{5}$ $\frac{4}{10}$; 0.4
10. $\frac{5}{25}$ $\frac{20}{100}$; 0.20
11. $\frac{3}{4}$ $\frac{75}{100}$; 0.75
12. $\frac{1}{2}$ $\frac{5}{10}$; 0.5

Write the fraction as a decimal. Use a bar to mark the repeating digits.

13. $\frac{3}{4}$ 0.75
14. $\frac{8}{9}$ $0.\overline{8}$
15. $\frac{2}{3}$ $0.\overline{6}$
16. $\frac{1}{4}$ 0.25

Daily Review   497

- Guide the students to collaborate with a partner as they find the following times. Allow them to use their clocks to count on or to count back the hours and minutes.

  2 hr 12 min after 3:35 a.m. 5:47 a.m.

  40 min after 11:50 a.m. 12:30 p.m.

  3 hr 30 min before 7:00 p.m. 3:30 p.m.

Adding & subtracting time

- Use **guided practice** to help the students solve the following word problems. Follow a procedure similar to the one used in Lesson 141.

  Morgan raked leaves for his elderly neighbor. He raked for 2 hr and 35 min on Saturday and again on Monday afternoon for 1 hr and 50 min. How much time did he spend raking leaves? 2 hr 35 min + 1 hr 50 min = 4 hr 25 min

  Sophie's trip to the museum took 1 hr 50 min. The trip home took 2 hr 35 min because her family stopped at a restaurant. How much time did they spend at the restaurant? 2 hr 35 min − 1 hr 50 min = 45 min

- Write 4 hr 9 min − 1 hr 15 min vertically for display and instruct the students to solve. 2 hr 54 min

Apply

Student Edition pages 312–13
- Read and explain the directions for pages 312–13. Assist the students as they complete the pages independently.

Daily Review
- Students should complete Chapter 14, section *i*.

Allow the students to use their clocks while solving the elapsed time problems on both pages. Also, allow them to refer to the *Time Zones of the World* page to answer questions 27–34 on Student Edition page 313.

Assess

Quiz 2
- Use the **summative assessment** to evaluate the students' progress at this point in the chapter.

NOTES

LESSON 143

Student Edition pages 314–15
Daily Review Chapter 14, section j

OBJECTIVES

- Estimate customary and metric measurements of objects.
- Add, subtract, multiply, and divide measurements.
- Solve measurement word problems.
- Determine mileage by using a map scale.
- Convert temperatures: Celsius to Fahrenheit and Fahrenheit to Celsius.
- Determine the time in various time zones.

TEACHER RESOURCES

- 92 *Time Zones of the World* (for the teacher and for each student)
- 94 *Map Key*
- 95 *Blank Thermometer* (for the teacher and for each student)

ADDITIONAL MATERIALS

- labeled measurable objects from Lesson 140
- measuring tools
- a ruler (for each student)

This lesson reinforces concepts taught previously in this chapter and further develops the skill of using a map scale to determine mileage. Refer to Lesson 124 in Chapter 13 and Lessons 134–42 in this chapter as needed.

For the estimating activity in this lesson, provide the measuring tools needed to determine the length, capacity, and weight/mass of the objects that you choose to use.

Engage

- Use **guided practice** to help the students review measurement conversions.

Renaming Units of Measure

How can a number line help me determine time in various time zones?

Write the equivalent unit of measurement.

1.

Linear
1 cm = *10* mm
1 mi = *1,760* yd
1 m = *100* cm
1 yd = *3* ft

2.

Capacity
1 c = *8* fl oz
1 L = *1,000* mL
1 gal = *4* qt
1 pt = *2* c

3.

Weight/Mass
1 lb = *16* oz
1 kg = *1,000* g
1 tn = *2,000* lb
1 g = *0.001* kg

4.

Part of a Unit
$\frac{2}{3}$ of a yard = *2* ft
$\frac{1}{2}$ of a kilometer = *500* m
$\frac{1}{4}$ of a meter = *25* cm
$\frac{1}{3}$ of a foot = *4* in.

Choose the best unit of measurement.

5.	a baseball bat	1 ft	(1 yd)	1 km
6.	the mass of a paper clip	1 lb	1 kg	(1 g)
7.	a large jug of milk	(1 gal)	1 fl oz	1 L
8.	a 20 min walk	1 ft	1 m	(1 km)

Solve. Simplify the answer when possible.

9. 4 hr 23 min
+ 2 hr 59 min
6 hr 82 min =
7 hr 22 min

10. 6 yd 2 ft 8 in.
− 1 yd 2 ft 9 in.
4 yd 2 ft 11 in.

11. 3.7 L
× 4
14.8 L

12. 3 gal 1 qt 1 pt
+ 4 gal 2 qt 1 pt
7 gal 3 qt 2 pt =
8 gal

13. 9 m
− 3 m 87 cm
5 m 13 cm

14. 1.760 yd
× 2
3.52 yd

15. 3 hr 25 min 40 sec
+ 2 hr 45 min 35 sec
5 hr 70 min 75 sec =
6 hr 11 min 15 sec

16. 7 hr 30 min 5 sec
− 2 hr 45 min 10 sec
4 hr 44 min 55 sec

17. 3 ft 8 in.
× 5
15 ft 40 in. =
18 ft 4 in.

Use a centimeter ruler and the map to find the answer.

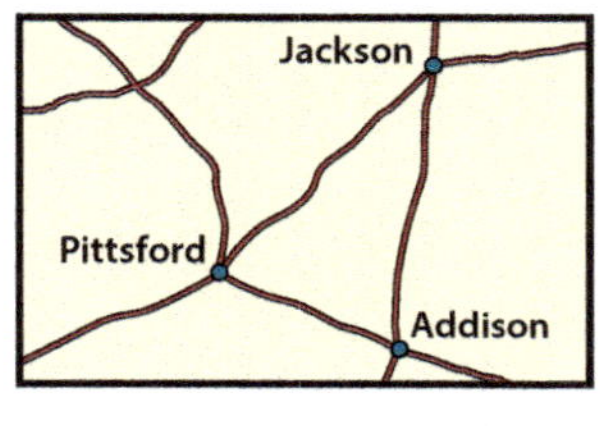

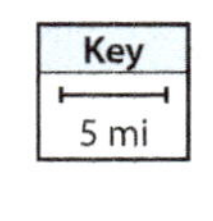

18. What is the mileage from Jackson to Addison?
15 mi

19. The Brickmans live in Jackson. Last week they traveled to Addison for piano lessons and then went shopping in Pittsford. They returned home by the shortest route. How many miles did they travel? *40 mi*

Write the following problems for display and direct the students to solve them. Remind them that thinking of the standard equivalency in the order that the units appear in the problem will help them determine whether to multiply or divide.

180 sec = __ min 3

5 m = __ cm 500

5 gal = __ qt 20

2 gal 3 qt = __ qt 11

1,600 g = __ kg 1.6

3 lb 7 oz = __ oz 55

2 c = __ fl oz 16

3 km = __ m 3,000

48 in. = __ ft 4

21 in. = __ ft __ in. 1; 9

16 ft = __ yd __ ft 5; 1

1 mi 500 ft = __ ft 5,780

Instruct

Estimating customary & metric measurements of objects

- Guide the students in a **Think-Pair-Share** to estimate measurements of objects.

Display the labeled objects with the labels turned away from the students. Hold up one of the objects and direct each student to write a customary or metric estimate of the length, capacity, or weight/mass for the item. Allow the

Use the time zone map to find the answer.

20. At noon Pacific standard time, Sebastian called his sister Tory in South Carolina. What time was it at Tory's house? *3:00 p.m.*

21. Callie lives in Virginia. She made a video call to her family in Colorado at 4:00 p.m. eastern standard time. At what time of day did her family receive the video call? *2:00 p.m.*

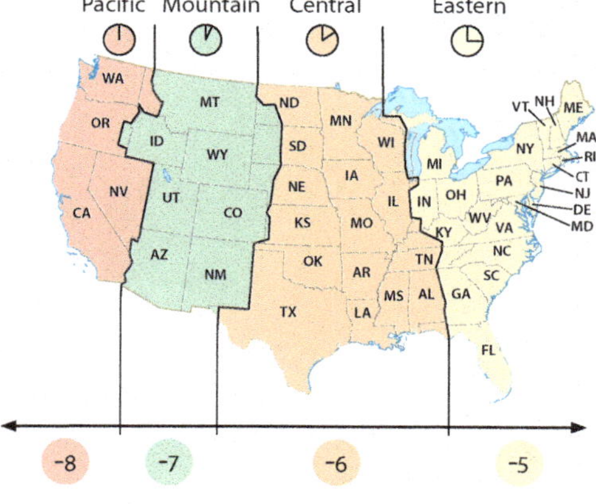

Rename.

22. 6 yd = __ in. *6 • 36 = 216*

23. 8 ft = _2_ yd _2_ ft *8 ÷ 3 = 2 r2*

24. 26 in. = _2_ ft _2_ in. *26 ÷ 12 = 2 r2*

25. 48 in. = __ yd *1 1/3*

26. 6 qt = _1_ gal _2_ qt *6 ÷ 4 = 1 r2*

27. 2 m = __ cm *2 • 100 = 200*

28. $2\frac{1}{2}$ gal = __ qt *5/2 • 4 = 10*

29. 3.51 m = __ cm *351*

30. 3,500 mL = __ L *3.5*

31. 4.1 kg = __ g *4,100*

32. 68°F = __°C *20*

33. 95°C = __°F *203*

34. 213 cm ÷ 3 *71 cm*

35. 8 gal 4 qt ÷ 4 *2 gal 1 qt*

36. 12 ft 6 in. ÷ 2 *6 ft 3 in.*

25. *$48 ÷ 36 = 1\frac{12}{36} = 1\frac{1}{3}$*   **29.** *3.51 × 100 = 351*

30. *3,500 ÷ 1,000 = 3.5*   **31.** *4.1 × 1,000 = 4,100*

Practice & Application

37. Mrs. Lynch needs a quart of milk. She could only find milk in liter containers. Would she have enough milk if she purchased a liter? *yes; 1 qt ≈ 1 L*

38. The nurse took Autumn's temperature and found that she had a fever. The thermometer read 39°. What scale was the thermometer? *Celsius*

39. Dad bought 1 kg of hamburger from the store. Would that be enough meat to feed a family of 5? *yes; 1 kg ≈ 2.2 lb*

40. Travis had a yardstick that was 36 in. He measured his dad's desk at $1\frac{1}{2}$ yd. Is Travis's measurement of $1\frac{1}{2}$ yd precise? What measurement is more precise? *no; 54 in. or 4 ft 6 in.*

41. How can a number line help me determine time in various time zones? *I can place the time zones along a number line to help me find the time at different points on the earth.*

32. *$C = \frac{5}{9} × (68° − 32°)$*
$C = \frac{5}{9} × 36°$
$C = \frac{5}{9} • \frac{36°}{1}$
$C = 20°$

33. *$F = (\frac{9}{5} × 95°) + 32°$*
$F = (\frac{9}{5} • \frac{95°}{1}) + 32°$
$F = 171° + 32°$
$F = 203°$

Lesson 143   315

---

large bottle of soft drink 2 L; Most soft drinks are in 2 L bottles.

Calculating measurements

- Use **guided practice** to help the students solve measurement problems.

Write the following problems for display and direct the students to solve them. Direct the students to rename each answer in terms of its larger unit and as its smaller unit.

5 hr 35 min
+ 3 hr 45 min
9 hr 20 min
$9\frac{1}{3}$ hr; 560 min

7 km 500 m
− 3 km 450 m
4 km 50 m
4.050 km; 4,050 m

5 kg
− 1 kg 750 g
3 kg 250 g
3.250 kg; 3,250 g

8 lb   4 oz
− 2 lb 12 oz
5 lb 8 oz
$5\frac{1}{2}$ lb; 88 oz

2 ft 8 in.
×   2
5 ft 4 in.
$5\frac{1}{3}$ ft; 64 in.

3 gal 2 qt
×   3
10 gal 2 qt
$10\frac{1}{2}$ gal; 42 qt

Solving measurement word problems

- Use **guided practice** to help the students solve the following word problems. Review the standard equivalencies as needed and discuss the solutions.

John and his dad walk a quarter of a mile from their home to the golf range each Saturday. How many feet is it from their home to the golf range? $\frac{1}{4} × 5,280$ ft = 1,320 ft

How many yards have John and his dad walked upon returning home from their $\frac{1}{4}$ mi walk to the golf range? 880 yd; sample solutions: $\frac{1}{2}(1,760$ yd$) = \frac{1,760}{2}$ yd = 880 yd; $2(\frac{1}{4} × 1,760$ yd$) = 2(440$ yd$) = $ 880 yd; $2(1,320$ ft $÷ 3$ ft$) = 2(440$ yd$) = $ 880 yd

If John and his dad left home at 8:30 a.m., played a round of golf, and returned home at 12:15 p.m., how long were they gone from home? 3 hr 45 min or $3\frac{3}{4}$ hr; I count the hours and then the minutes from the beginning time to the ending time to get the total; 12 hr 15 min − 8 hr 30 min = 3 hr 45 min or $3\frac{3}{4}$ hr

---

students to consult with a partner to compare their estimates and to determine the best estimate. Choose a student to measure the item or refer to its label to find the exact measurement (e.g., mass [g] of a pudding cup estimates will vary, 100 g; customary capacity [fl oz] of a pudding cup estimates will vary, 3.5 fl oz; metric length [cm] of a standard index card estimates will vary, 12.5 cm).

- Repeat the procedure using the other displayed objects.

- Allow the students to continue to work with a partner to suggest a reasonable estimate for measuring each of the following typical items. Choose a student to explain how he determined his esti-

mate. Encourage the students to refer to benchmarks and equivalencies when giving their explanations. sample estimates and explanations:

5 paper clips 5 g; 1 paper clip is 1 g.

distance from the floor to the ceiling 9 ft, 3 yd, or 3 m; The door is 6 ft tall, so the ceiling is 9 ft from the floor.

tall glass of water 2 c, 1 pt, or 16 fl oz; Larger amounts are not the typical capacity of a tall glass.

loaf of bread 16 oz or 1 lb; Some bread loaves weigh 1 lb.

walk around a park 1 mi or 1 km; Mile and kilometer are the best units for measuring distance.

LESSON 143

Dad and John ate lunch at the Biggie Burger Restaurant. They each ordered the Gigantic Burger. The hamburger in each Gigantic Burger weighed $\frac{3}{4}$ of a pound. How many ounces of meat were in each burger? $\frac{3}{4} \times 16$ oz = 12 oz

Last week, Mom and Katie walked 2 km in 30 min on Monday, 4,900 m in 1 hr 15 min on Wednesday, and 3,850 m in $\frac{3}{4}$ of an hour on Friday. How many meters did they walk? 2,000 m + 4,900 m + 3,850 m = 10,750 m

How many kilometers did they walk? 2 km + 4.9 km + 3.85 km = 10.75 km; 10,750 m = 10.75 km

How much time did they walk? sample solutions: 30 min + 1 hr 15 min + 45 min = 1 hr 90 min or $2\frac{1}{2}$ hr; 30 min + 75 min + 45 min = 150 min or $2\frac{1}{2}$ hr

What is the average time per day that Mom and Katie walked last week? 150 min ÷ 3 = 50 min

Determining mileage by using a map scale

- Guide an **interactive activity** to help the students use a map scale to determine mileage.

  Display the *Map Key* page. Point out that the key on the page shows the scale that is needed to determine the actual distance or mileage between locations shown on the map.

- Choose a student to measure the distance along the route from Peaceful Valley to Stonefield using a ruler. $2\frac{1}{2}$ in. or 2.5 in. What is the mileage (or the number of miles) from Peaceful Valley to Stonefield? 25 mi

  Choose a student to complete the line in the table for the measured distance.

- Repeat the procedure using other pairs of locations. Complete the table for each distance.

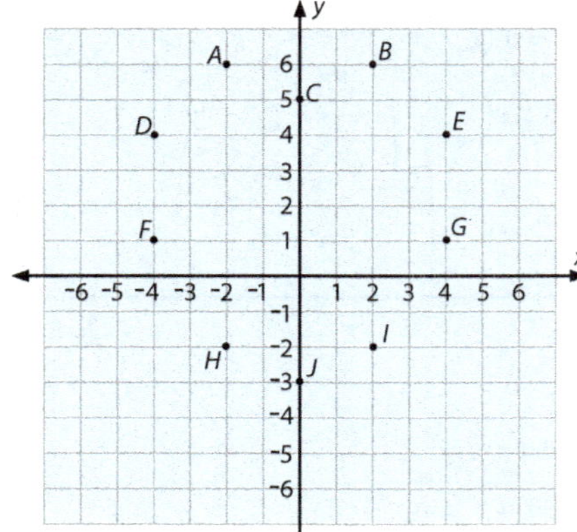

Name the point represented by the coordinates.

1. (2, 6) *B*
2. (-4, 1) *F*
3. (-2, -2) *H*
4. (4, 4) *E*
5. (0, 5) *C*

Write the coordinates for the point.

6. A *(-2, 6)*
7. D *(-4, 4)*
8. G *(4, 1)*
9. I *(2, -2)*
10. J *(0, -3)*

498   Daily Review

Converting temperatures

- Use **guided practice** to help the students identify and convert temperatures.

  Distribute the *Blank Thermometer* page to each student and display your page. Guide the students as they label the standard Fahrenheit and Celsius equivalents for the following standard temperatures on the displayed page.

  freezing point of water 32°F; 0°C

  boiling point of water 212°F; 100°C

  normal body temperature 98.6°F; 37°C

- Instruct the students to locate the following temperatures on their *Blank Thermometer* page. Choose students to read the approximate Celsius temperature for each Fahrenheit temperature and the approximate Fahrenheit temperature for each Celsius temperature.

  38°F 3°C   -16°C 6°F

  75°F 24°C   -5°C 24°F

- Write the formulas "$F = (\frac{9}{5} \times C) + 32°$" and "$C = \frac{5}{9} \times (F - 32°)$" for display. Remind the students that the ratio of Fahrenheit temperatures to Celsius temperatures is 9 : 5. Remind them that 32° is added or subtracted because 32° is the difference in the freezing points on the Fahrenheit and Celsius scales.

  Guide the students as they use the formulas to convert the Celsius temperatures to equivalent Fahrenheit temperatures and the Fahrenheit temperatures to equivalent Celsius temperatures.

50°C $F = (\frac{9}{5} \times 50°) + 32°; F = 122°$

15°C $F = (\frac{9}{5} \times 15°) + 32°; F = 59°$

104°F $C = \frac{5}{9} \times (104° - 32°); C = 40°$

68°F $C = \frac{5}{9} \times (68° - 32°); C = 20°$

Determining the time in various time zones

- Distribute a copy of the *Time Zones of the World* page to each student and display your copy. Point out the number line at the bottom of the map, and guide a **discussion** of the essential question at the top of Student Edition page 314, "How can a number line help me determine time in various time zones?" I can place the time zones along a number line to help me find the time at different points on the earth.

  Point out that the number line shows each time zone and its relationship to Greenwich, England, or 0° longitude.

- Use **guided practice** to help the students review determining the time in different time zones.

  When it is 5:00 a.m. in New York, what time is it in California? 2:00 a.m.

  When it is 8:00 p.m. in Denver, what time is it in Rio de Janeiro? 12:00 a.m. (midnight)

  When it is 2:00 p.m. in Halifax, what time is it in London? 6:00 p.m.

Student Edition pages 314–15

- Read and explain the directions for pages 314–15. Assist the students as they complete the pages independently.

Daily Review

- Students should complete Chapter 14, section *j*.

LESSON **143**

NOTES

Student Edition pages 316–17
Daily Review Chapter 14, section *k*

OBJECTIVES

- Write an equivalency as a unit multiplier.
- Convert measurements by using a unit multiplier.

Engage

- Direct the students to **brainstorm** to explore the essential question at the top of Student Edition page 316, "How can a unit multiplier help me when I am converting one unit to another unit?"

  Instruct them to listen carefully during the lesson to learn what a unit multiplier is and how it can help them.

Instruct

Writing an equivalency as a unit multiplier

- Guide a **discussion** to introduce the concept of a unit multiplier.

  Write "$\frac{3}{3}$," "$\frac{5}{5}$," "$\frac{24}{24}$," and "$\frac{1,000}{1,000}$" for display.

  What do these ratios or fractions have in common? Each has a value of 1.

  What does the Identity Property of Multiplication state? When a number is multiplied by 1, the product is always the other factor.

- Remind the students that a fraction can be renamed as an equivalent fraction in higher terms by multiplying the fraction by a fraction form of 1 (e.g., $\frac{2}{3} \times \frac{4}{4} = \frac{8}{12}$). The terms of a fraction change when multiplied by 1, but its value does not change ($\frac{2}{3} = \frac{8}{12}$).

  What is $\frac{8}{12}$ renamed to lowest terms? $\frac{2}{3}$

- Read aloud the following word problem.

  Mrs. Roberts distributed to her students lengths of string for them to measure. Some students said their strings measured 1 ft. Other students said their strings measured 12 in. What do you know about the lengths of string? The lengths of string were all the same length

because the measurements are equivalent; 1 ft = 12 in. and 12 in. = 1 ft.

- Choose students to write the equivalencies for display. 1 ft = 12 in.; 12 in. = 1 ft

  How can you write these equivalencies as ratios in fraction form? $\frac{1\,ft}{12\,in.}$ and $\frac{12\,in.}{1\,ft}$

  Write "$\frac{1\,ft}{12\,in.}$" and "$\frac{12\,in.}{1\,ft}$" for display. Remind the students that a ratio is a comparison of two quantities or amounts.

  What do you notice about these equivalencies when written in fraction form? They have a value of 1.

- Explain that it is important to keep the unit labels in equivalencies that are written with two different units. Point out

that if the ratio is written without unit labels, the terms are no longer equal. Write "1 ≠ 12" for display. With the unit labels, the ratio written in fraction form is still an equivalency and has a value of 1.

- Explain that equivalencies written in fraction form are called *unit multipliers*. A unit multiplier is a fraction, written with 2 different units, that equals 1. Unit multipliers are used to rename or convert the units of a number.

A unit multiplier may also be called a *conversion factor* because it is a factor in a multiplication equation that is used to convert units.

Unit Multipliers

How can a unit multiplier help me when I am converting one unit to another unit?

We sometimes need to rename a unit of measurement as another one to find the equivalent value. This is called a **unit conversion**. We can convert inches to feet or grams to kilograms without changing the original value.

12 in. = 1 ft and 1 ft = 12 in.
The value is the same; the units are different.

Equivalencies written in fraction form are called unit multipliers. A **unit multiplier** is a fraction, written with two different units, that equals 1.

1 ft = 12 in.; $\frac{1\,ft}{12\,in.}$ or $\frac{12\,in.}{1\,ft}$

12 in. and 1 ft are equivalent.

Key Terms
- unit conversion
- unit multiplier

7. $\frac{1\,yd}{3\,ft}$; $\frac{3\,ft}{1\,yd}$

8. $\frac{1\,kg}{1,000\,g}$; $\frac{1,000\,g}{1\,kg}$

9. $\frac{1\,mi}{5,280\,ft}$; $\frac{5,280\,ft}{1\,mi}$

10. $\frac{1\,lb}{16\,oz}$; $\frac{16\,oz}{1\,lb}$

11. $\frac{1\,yd}{36\,in.}$; $\frac{36\,in.}{1\,yd}$

Exercises

Complete the unit multiplier. Write the equal unit multiplier.

1. $\frac{8\,fl\,oz}{1\,c}$ or $\frac{1\,c}{8\,fl\,oz}$
2. $\frac{1,000\,mL}{1\,L}$ or $\frac{1\,L}{1,000\,mL}$
3. $\frac{36\,in.}{1\,yd}$ or $\frac{1\,yd}{36\,in.}$
4. $\frac{1\,mi}{1,760\,yd}$ or $\frac{1,760\,yd}{1\,mi}$
5. $\frac{2\,c}{1\,pt}$ or $\frac{1\,pt}{2\,c}$
6. $\frac{60\,sec}{1\,min}$ or $\frac{1\,min}{60\,sec}$

Write the unit multiplier 2 ways in fraction form.

7. 1 yd = 3 ft
8. 1 kg = 1,000 g
9. 1 mi = 5,280 ft
10. 1 lb = 16 oz
11. 1 yd = 36 in.
12. 1 hr = 60 min
13. 1 gal = 4 qt
14. 1 m = 100 cm
15. 1 g = 1,000 mg
16. 1 tn = 2,000 lb

12. $\frac{1\,hr}{60\,min}$; $\frac{60\,min}{1\,hr}$
13. $\frac{1\,gal}{4\,qt}$; $\frac{4\,qt}{1\,gal}$
14. $\frac{1\,m}{100\,cm}$; $\frac{100\,cm}{1\,m}$
15. $\frac{1\,g}{1,000\,mg}$; $\frac{1,000\,mg}{1\,g}$
16. $\frac{1\,tn}{2,000\,lb}$; $\frac{2,000\,lb}{1\,tn}$

316   Chapter 14

The unit multiplier is used to convert one unit to another unit.

Convert 4.5 ft to inches.   4.5 ft = ___ in.

1. Write 4.5 ft as a fraction. $\frac{4.5\ ft}{1}$

2. Find the unit multiplier.
   1 ft = 12 in. $\frac{1\ ft}{12\ in.}$ or $\frac{12\ in.}{1\ ft}$

3. Choose the unit multiplier that shows the original unit in the denominator.
   $\frac{12\ in.}{1\ ft}$

4. Write a multiplication equation. Solve. Use cancellation where possible.

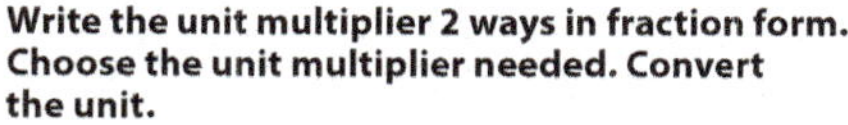

$$\frac{4.5\ ft}{1} \times \frac{12\ in.}{1\ ft} = \frac{54.0\ in.}{1}$$

In this problem only the labels cancel.

4.5 ft = **54 in.**

Write the unit multiplier 2 ways in fraction form. Choose the unit multiplier needed. Convert the unit.

17. 7 yd = ___ in. *252*   **18.** 64 oz = ___ lb *4*

19. 8 gal = ___ qt *32*   **20.** 3 hr = ___ min *180*

21. 3 kg = ___ g *3,000*   **22.** 2,500 mL = ___ L *2.5*

23. 450 cm = ___ m *4.5*   **24.** 3 c = ___ fl oz *24*

Use a unit multiplier to convert the unit.

25. 5 tn = ___ lb *10,000*   **26.** 13 gal = ___ qt *52*   **27.** 20 c = ___ pt *10*   **28.** 8 mi = ___ yd *14,080*

29. 108 in. = ___ ft *9*   **30.** 120 fl oz = ___ c *15*   **31.** 4 yd = ___ in. *144*   **32.** 128 oz = ___ lb *8*

33. Use cancellation to convert.

2 mi = ___ in.

$$\frac{2\ mi}{1} \times \frac{5{,}280\ ft}{1\ mi} \times \frac{12\ in.}{1\ ft} = \text{___ in.}\ \textit{126,720}$$

12 gal = ___ c

$$\frac{12\ gal}{1} \times \frac{4\ qt}{1\ gal} \times \frac{2\ pt}{1\ qt} \times \frac{2\ c}{1\ pt} = \frac{192\ c}{1} = \text{___ c}\ \textit{192}$$

34. How can a unit multiplier help me when I am converting one unit to another unit? *A unit multiplier reduces the number of equations I have to use to convert units.*

Lesson 144   317

LESSON 144

What measurement do you need to convert? 108 in.

What unit do you need to convert 108 in. to? yards

What equivalency do you know for inches to yards? 36 in. = 1 yd

What unit multipliers can you write? $\frac{36\ in.}{1\ yd}$, $\frac{1\ yd}{36\ in.}$

Write "$\frac{36\ in.}{1\ yd}$" and "$\frac{1\ yd}{36\ in.}$" for display.

- Write "$\frac{108\ in.}{1} \times \frac{1\ yd}{36\ in.}$" for display.

- Remind the students that a number can be written as a fraction with 1 in the denominator without changing its value.

  Point out that since unit multipliers are used to rename or convert the units of a number, the students must choose the unit multiplier with the same unit label in the denominator (in.) so that the two labels (in.) cancel out one another, leaving the desired label (yd) in the answer.

- Demonstrate the multiplication, showing the cancellation. Point out that the two labels (in.) cancel out one another so that the label in the product is "yd."

$$\frac{\overset{3}{\cancel{108\ in.}}}{1} \times \frac{1\ yd}{\underset{1}{\cancel{36\ in.}}} = \frac{3\ yd}{1} = 3\ yd$$

- Write "12.5 gal = ___ qt" for display.

  What equivalency can you think of to solve this problem? 1 gal = 4 qt or 4 qt = 1 gal

  What unit multiplier is needed? $\frac{4\ qt}{1\ gal}$; Since the unit *gallons* needs to be converted to *quarts*, *gallons* needs to be in the denominator of the unit multiplier.

  What multiplication equation can you write to convert 12.5 gal to an equivalent number of quarts? $\frac{12.5\ gal}{1} \times \frac{4\ qt}{1\ gal} = \text{___ qt}$

- Direct the students to solve the problem. Then choose a student to write the equation for display and to demonstrate solving it.

$$\frac{12.5\ \cancel{gal}}{1} \times \frac{4\ qt}{1\ \cancel{gal}} = \frac{50\ qt}{1} = 50\ qt$$

- Use **guided practice** to help the students write unit multipliers. Guide them as they write 2 unit multipliers for each of the following equivalencies.

1 yd = 3 ft $\frac{1\ yd}{3\ ft}$, $\frac{3\ ft}{1\ yd}$

1 mi = 5,280 ft $\frac{1\ mi}{5{,}280\ ft}$, $\frac{5{,}280\ ft}{1\ mi}$

1 m = 100 cm $\frac{1\ m}{100\ cm}$, $\frac{100\ cm}{1\ m}$

1 lb = 16 oz $\frac{1\ lb}{16\ oz}$, $\frac{16\ oz}{1\ lb}$

1 hr = 60 min $\frac{1\ hr}{60\ min}$, $\frac{60\ min}{1\ hr}$

What is the value of each unit multiplier that was written? 1; The terms in a unit multiplier are equivalent.

- Write the following problems for display. Guide the students as they identify the number of units needed to complete the unit multiplier.

$\frac{\text{___ kg}}{1{,}000\ g}$ 1 $\frac{2{,}000\ lb}{\text{___ tn}}$ 1 $\frac{1\ min}{\text{___ sec}}$ 60

$\frac{1\ mi}{\text{___ yd}}$ 1,760 $\frac{2\ c}{\text{___ pt}}$ 1 $\frac{1\ gal}{\text{___ qt}}$ = 4

Using a unit multiplier to convert measurements

- Use **guided practice** to help the students convert measurements by using a unit multiplier.

- Write "108 in. = ___ yd" for display.

LESSON 144

- Follow a similar procedure for the following problems.

$$540 \text{ mL} = \_ \text{ L} \quad \frac{540 \text{ mL}}{1} \times \frac{1 \text{ L}}{1,000 \text{ mL}} = \frac{540}{1,000} \text{ L} = 0.54 \text{ L}$$

$$32 \text{ yd} = \_ \text{ ft} \quad \frac{32 \text{ yd}}{1} \times \frac{3 \text{ ft}}{1 \text{ yd}} = 96 \text{ ft}$$

$$68 \text{ oz} = \_ \text{ lb} \quad \frac{68 \text{ oz}}{1} \times \frac{1 \text{ lb}}{16 \text{ oz}} = \frac{68}{16} \text{ lb} = 4.25 \text{ lb}$$

- Write "4 hr = __ sec" for display. Arrange the students in groups and instruct the students in each group to work together to solve the problem. Allow students from each group to share how they arrived at their answer.

$$\frac{4 \text{ hr}}{1} \times \frac{60 \text{ min}}{1 \text{ hr}} = \frac{240 \text{ min}}{1} = 240 \text{ min};$$

$$\frac{240 \text{ min}}{1} \times \frac{60 \text{ sec}}{1 \text{ min}} = \frac{14,400 \text{ sec}}{1} = 14,400 \text{ sec}$$

> Most students will solve this problem using the multistep solution provided above. You may show that the problem can also be solved using 1 equation with 2 unit multipliers: $\frac{4 \text{ hr}}{1} \times \frac{60 \text{ min}}{1 \text{ hr}} \times \frac{60 \text{ sec}}{1 \text{ min}} = 14,400 \text{ sec.}$

- Use **discussion** to help the students answer the essential question.

Remind the students of the Chapter 13 STEM bridge-building project.

What units did you use to measure the weight of your popsicle-stick bridge? grams

What units did you use to measure the weight of the load your bridge held? pounds

Point out that before the students could calculate the efficiency of their bridge, they had to convert the mass of the load from pounds to grams. Remind them that they did this by solving more than one equation to rename pounds to grams. (See problem 10b on Student Edition page 294.)

- Direct attention to the essential question at the top of Student Edition page 316. Rephrase the question to fit the context of the bridge problem.

How can a unit multiplier help you when you are converting pounds to grams? A unit multiplier reduces the number of equations I have to use to convert units.

Solve.

1. 643,564
   + 246,203
   889,767

2. 391,715
   − 96,639
   295,076

3. 493
   × 321
   158,253

4. 14)994 → **71**

5. 228,258
   + 552,220
   780,478

6. 793,151
   − 150,895
   642,256

7. 141
   × 998
   140,718

8. 18)108 → **6**

9. 734,280
   + 154,745
   889,025

10. 26,956
    − 25,666
    1,290

11. 860
    × 775
    666,500

12. 21)126 → **6**

13. 571,900
    + 648,843
    1,220,743

14. 472,320
    − 205,663
    266,657

15. 106
    × 215
    22,790

16. 16)368 → **23**

17. 826,520
    + 862,498
    1,689,018

18. 453,388
    − 436,850
    16,538

19. 124
    × 842
    104,408

20. 23)920 → **40**

Write the equation "bridge efficiency = $\frac{\text{mass of load}}{\text{mass of bridge}}$" for display.

- Use **guided practice** to help the students use fewer equations to convert units by using unit multipliers.

What is the efficiency score of a bridge that weighs 60 g and holds a load of 66 lb before it fails?

$$66 \text{ lb} = \_ \text{ g}; \quad \frac{66 \text{ lb}}{1} \times \frac{1 \text{ kg}}{2.2 \text{ lb}} \times \frac{1,000 \text{ g}}{1 \text{ kg}} = \frac{66,000}{2.2} \text{ g} = 30,000 \text{ g};$$

$$\text{bridge efficiency} = \frac{30,000 \text{ g}}{60 \text{ g}} = 500$$

Guide the students to first convert the weight of the load from pounds to grams by using unit multipliers and then to use the renamed units to calculate the bridge's efficiency.

What unit multipliers do you know that could help you convert pounds to grams? $\frac{1 \text{ kg}}{2.2 \text{ lb}}$ or $\frac{2.2 \text{ lb}}{1 \text{ kg}}$ and $\frac{1,000 \text{ g}}{1 \text{ kg}}$ or $\frac{1 \text{ kg}}{1,000 \text{ g}}$

Write "66 lb = __ g" for display. Guide the students as they choose the first unit multiplier to use. $\frac{1 \text{ kg}}{2.2 \text{ lb}}$

Point out that they can use 2 unit multipliers in a row to convert pounds to the unit they need (grams) by using the equivalencies they know. Complete the equation to rename the units and then to calculate the bridge's efficiency.

Apply

Student Edition pages 316–17

- Read and explain the directions for pages 316–17. Assist the students as they complete the pages independently.

Daily Review

- Students should complete Chapter 14, section *k*.

NOTES

Student Edition pages 318–19

CHAPTER REVIEW

OBJECTIVES

- Determine the appropriate unit of measurement.
- Convert units of measurement within the same system.
- Read a Fahrenheit and a Celsius thermometer.
- Identify standard Fahrenheit and Celsius temperatures.
- Add, subtract, multiply, and divide measurements.
- Solve measurement word problems.
- Tell time to the minute.
- Determine the elapsed time.

TEACHER RESOURCES

- 95 *Blank Thermometer*

ADDITIONAL MATERIALS

- a larger set of measurement flashcards from Lesson 140
- a demonstration clock (for the teacher and for each student)

CHAPTER REVIEW

Write the equivalent unit of measurement.

1. 1 yd = _3_ ft
   1 mi = _1,760_ yd
   1 m = _100_ cm

2. 1 gal = _4_ qt
   1 c = _8_ fl oz
   1 L = _1,000_ mL

3. 1 lb = _16_ oz
   1 tn = _2,000_ lb
   1 kg = _1,000_ g

4. 1 hr = _60_ min
   1 d = _24_ hr
   1 min = _60_ sec

5. $\frac{1}{2}$ of a mile = _2,640_ ft
   $\frac{1}{5}$ of a kilometer = _200_ m
   $\frac{1}{4}$ of a meter = _25_ cm

6. $\frac{1}{6}$ of an hour = _10_ min
   $\frac{1}{3}$ of a yard = _1_ ft
   $\frac{1}{8}$ of a pound = _2_ oz

Choose the best unit of measurement.

7. length of a pencil — liter · meter · (centimeter) · kilometer
8. amount of liquid in a glass — meter · (milliliter) · gram · liter
9. mass of a bunch of bananas — liter · centimeter · meter · (kilogram)
10. distance to town — inches · gallons · (miles) · feet
11. weight of a baby — ounces · feet · cups · (pounds)
12. amount of sports drink for a team — cups · (gallons) · yards · tons

Rename.

13. 120 min = _2_ hr   *120 ÷ 60 = 2*

14. $2\frac{1}{2}$ hr = _150_ min

15. 1 hr 45 min = _105_ min

16. 2 mi = _10,560_ ft

17. $1\frac{1}{2}$ yd = _4_ ft _6_ in.

18. 30 in. = _2_ ft _6_ in.

19. 15 qt = _3_ gal _3_ qt

20. 10 gal = _40_ qt   *10 · 4 = 40*

21. 32 oz = _2_ lb   *32 ÷ 16 = 2*

22. 3,000 lb = _1_ tn _1,000_ lb

23. $1\frac{1}{4}$ lb = _20_ oz $\frac{5}{4} \cdot 16 = 20$

24. 10 pt = _1_ gal _1_ qt

25. 250 cm = _2.5_ m

26. 3.5 kg = _3,500_ g

27. 4,000 mL = _4_ L

14. $\frac{5}{2}$ *hr · 60 = 150 min*

15. *60 min + 45 min = 105 min*

16. *2 × 5,280 ft = 10,560 ft*

17. $\frac{3}{2}$ *yd · 3 = $\frac{9}{2}$ ft = $4\frac{1}{2}$ ft*

18. *30 in. ÷ 12 = 2 ft r6 in.*

19. *15 qt ÷ 4 = 3 gal r3 qt*

22. *3,000 lb ÷ 2,000 = 1 tn r1,000 lb*

24. *10 pt ÷ 2 = 5 qt; 5 qt ÷ 4 = 1 gal r1 qt*

25. *250 cm ÷ 100 = 2.5 m*

26. *3.5 kg · 1,000 = 3,500 g*

27. *4,000 mL ÷ 1,000 = 4 L*

318   Chapter 14

The Chapter Review offers an opportunity for students to discuss the concepts they have learned in the chapter. They may work collaboratively or independently as you review concepts. Circulate among the students, giving individual help as needed. Students who demonstrate proficiency with the discussion, the modeling, and the Student Edition pages are ready for the Chapter Test. Students who encounter difficulties with the review concepts would benefit from additional coaching and practice before testing.

Determining the appropriate unit of measurement

- Review the customary and metric units of measurement, the standard equivalents, and the abbreviations by using the larger measurement flashcards.
- Choose students to tell the customary unit that would likely be used to measure the following items.

length of a pencil inches
height of a student inches or feet
length of a rug feet or yards
distance between cities miles
capacity of a can of soda fluid ounces
capacity of a jug of milk quarts or gallons
weight of a piece of candy ounces
weight of a loaf of bread ounces or pounds
weight of a whale tons

- Choose students to tell the metric unit that would likely be used to measure the following items.

height of a door meters or centimeters
length of a race meters or kilometers
width of a penny millimeters or centimeters
distance across a state kilometers

Choose the more reasonable temperature.

28. a child with a fever — (100°F) · 100°C

29. a cup of boiling soup — 95°F · (100°C)

30. a glass of ice water — 50°F · (0°C)

31. a comfortable room — (70°F) · 40°C

Write the time. Write what time it is after the given elapsed time.

32. 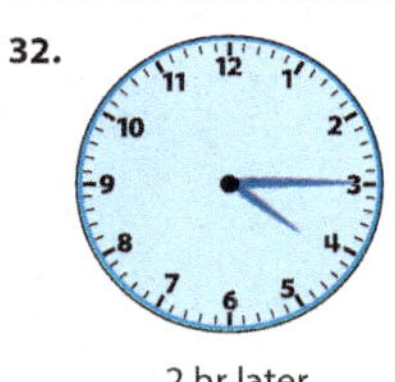
2 hr later
4:15; 6:15

33. 
45 min later
6:25; 7:10

34. 
$3\frac{1}{2}$ hr later
11:50; 3:20

35. 
1 hr 20 min later
9:40; 11:00

Solve. Simplify the answer when possible.

36.
$$\begin{array}{r} 10\text{ hr }25\text{ min }20\text{ sec} \\ +\ 4\text{ hr }35\text{ min }52\text{ sec} \\ \hline \end{array}$$
14 hr 60 min 72 sec =
15 hr 1 min 12 sec

37.
$$\begin{array}{r} 8\text{ hr} \\ -\ 3\text{ hr }25\text{ min} \\ \hline \end{array}$$
4 hr 35 min

38.
$$\begin{array}{r} 23.5\text{ L} \\ \times\quad 4 \\ \hline \end{array}$$
94.0 L

39.
$$\begin{array}{r} 12\text{ ft }10\text{ in.} \\ +\ 3\text{ ft }\ 8\text{ in.} \\ \hline \end{array}$$
15 ft 18 in. =
16 ft 6 in.

40.
$$\begin{array}{r} 15\text{ gal }2\text{ qt} \\ -\ 4\text{ gal }3\text{ qt} \\ \hline \end{array}$$
10 gal 3 qt

41.
$$\begin{array}{r} 3,457\text{ kg} \\ -\ 2,149\text{ kg} \\ \hline \end{array}$$
1,308 kg

42.
$$\begin{array}{r} 8\text{ m }20\text{ cm} \\ +3\text{ m }45\text{ cm} \\ \hline \end{array}$$
11 m 65 cm

43.
$$\begin{array}{r} 9\text{ lb }\ 7\text{ oz} \\ +3\text{ lb }10\text{ oz} \\ \hline \end{array}$$
12 lb 17 oz =
13 lb 1 oz

44.
$$\begin{array}{r} 4\text{ lb} \\ -2\text{ lb }10\text{ oz} \\ \hline \end{array}$$
1 lb 6 oz

Rename.

45. 54 in. = __ ft *4$\frac{1}{2}$*
54 ÷ 12 = 4$\frac{1}{2}$

46. 1$\frac{1}{2}$ gal = __ qt *6*
$\frac{3}{2}$ • 4 = 6

47. 3.5 m = __ cm *350*
3.5 • 100 = 350

Point out that writing the standard equivalency in the same order that the units appear in the problem helps the students to see whether multiplication or division is needed to find the equivalent.

Remind the students that the meaning of the metric prefix can help them determine metric equivalencies (e.g., 263 cm = __ m; 263 hundredths of a meter $[\frac{263}{100}]$ = 2.63 m).

- Direct the students to find the equivalents.

5 gal = 20 qt	263 cm = 2.63 m
180 sec = 3 min	5 kg = 5,000 g
4,265 m = 4.265 km	2 c = 16 fl oz
3 yd = 9 ft	4,138 g = 4.138 kg
2 L = 2,000 mL	84 in. = 7 ft
240 oz = 15 lb	4 mi = 7,040 yd

Reading a Fahrenheit and Celsius thermometer; identifying standard temperatures

- Display the *Blank Thermometer* page. Choose a student to identify ⁻2°C on the displayed page.

  What is the corresponding Fahrenheit temperature? 29°F

- Choose a student to locate 37°C on the displayed page.

  What is the corresponding Fahrenheit temperature? 98.6°F

  What do you know about 98.6°F? It is the normal body temperature.

- Choose a student to locate 0°C on the displayed page.

  What is the corresponding Fahrenheit temperature? 32°F

  What happens to water at this temperature? Water freezes.

  At what Celsius and Fahrenheit temperatures does water boil? 100°C and 212°F

Calculating measurements

- Write "3 m 32 cm + 14 m 89 cm" vertically for display and instruct the students to solve it. Choose a student to demonstrate solving the problem. 17 m 121 cm = 18 m 21 cm

---

thickness of a dime millimeters

capacity of a bottle of water milliliters or liters

capacity of a medicine bottle milliliters

mass of a candy bar grams

mass of a student kilograms

Converting units of measurement

How do you know whether to multiply or divide when converting from larger units to smaller units? When I rename larger units as smaller units, it takes more of the smaller units to measure the same amount as 1 larger unit, so I multiply.

How do you know whether to multiply or divide when converting from smaller units to larger units? When I rename smaller units to larger units, it takes fewer larger units to measure the same amount as smaller units, so I divide.

How do you know what to multiply or divide by? I can think of the relationship in the standard equivalency for the units that are being converted and then multiply or divide accordingly.

- Write the following problems for display. Guide the students as they determine how they will convert the units.

- Repeat the procedure for the following problems.

  66 kg 245 g
  − 10 kg 763 g
  55 kg 482 g

  14 ft 8 in.
  ×   5
  70 ft 40 in.
  73 ft 4 in.

Solving measurement word problems

- Read aloud the following word problem.

  Carson played soccer for $\frac{1}{2}$ of an hour, ran for 45 min, and practiced with his team for an hour. How many hours was Carson involved in sports activities? 2 hr 15 min or $2\frac{1}{4}$ hr

  How can you solve this problem? I add the different times and rename to find the total number of hours.

- Read the word problem again and instruct the students to solve it.

- Choose students to write their equation and solution for display and to explain their problem-solving strategy. sample solutions: 30 min + 45 min + 60 min = 135 min = 2 hr 15 min; $\frac{1}{2}$ hr + $\frac{3}{4}$ hr + 1 hr = $2\frac{1}{4}$ hr or 2 hr 15 min

- Follow a similar procedure for the following word problem.

  Travis needs 3 pieces of wire for a science project. Each piece of wire must be 2 m long. He has a roll of wire that contains 1,000 cm. Does the roll contain enough wire for his science project? Will Travis have wire left over? 3 × 2 m = 6 m, 6 m = 600 cm; 1,000 cm − 600 cm = 400 cm; The roll contains more than enough wire; there will be 400 cm of wire left over.

Telling time to the minute

- Set the demonstration clock for the following times and choose students to tell the times: 1:34, 5:17, and 12:39.

- Set the clock for 12:00.

  When the clock shows 12:00 and it is lunch time, is it 12 noon or 12 midnight? 12 noon

  Is 12 noon 12 a.m. or 12 p.m.? 12 p.m.; The minutes following 12 noon begin the afternoon hours.

  When the clock shows 12:00 at night, is it 12 noon or 12 midnight? 12 midnight

  Is 12 midnight 12 a.m. or 12 p.m.? 12 a.m.; The minutes following 12 midnight begin the morning hours.

  What abbreviation refers to the hours from 12 midnight to just before 12 noon? a.m.

  What abbreviation refers to the hours from 12 noon to just before 12 midnight? p.m.

Determining the elapsed time

- Distribute the demonstration clocks. Allow each student to use his clock to determine the elapsed time for the following problems.

  Jodi went to the park from 11:20 a.m. to 1:37 p.m. How much time did she spend at the park? 2 hr 17 min

  Mr. Jones took his suit to the dry cleaners at 9:10 a.m. The clerk said that his suit would be ready to pick up in 6 hr. What time can Mr. Jones pick up his suit? 3:10 p.m.

Student Edition pages 318–19

- Read and explain the directions for pages 318–19. Assist the students as they complete the pages independently.

You may allow the students to use their clocks to find the elapsed time for problems 32–35.

LESSON 146

Student Edition pages 320–23

CHAPTER 14 TEST

CUMULATIVE REVIEW

CONCEPT REVIEW

- Reading and interpreting a line graph (Lesson 101)
- Rounding numbers to a given place (Lessons 1, 4)
- Estimating a sum or a quotient (Lessons 2, 22–24)
- Evaluating an expression by using substitution (Lesson 94)
- Simplifying an expression (Lesson 96)
- Determining the value of a variable in an expression or an equation (Lessons 97–99)
- Identifying common factors and common multiples of two numbers (Lessons 31–32)
- Naming parts of a circle using symbols: diameter, radius, chord (Lesson 61)
- Identifying a 3-dimensional figure (Lesson 62)
- Choosing an inequality statement for a number line (Lesson 99)

ADDITIONAL MATERIALS

- a demonstration clock (for each student) (optional)

Use the line graph to find the answer.

The line graph shows the cost of a first-class postage stamp in January of the given year.

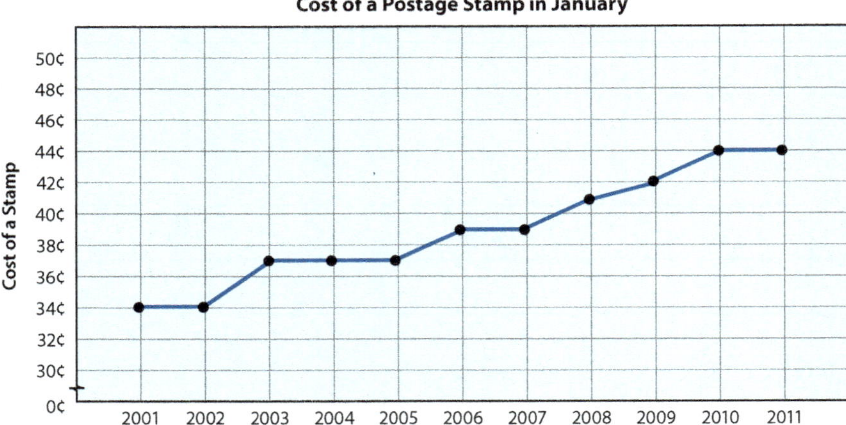

1. What is the difference in the cost of the 2001 and the 2011 stamps?
   - **A.** 10 cents
   - **B.** 15 cents
   - **C.** 20 cents
   - **D.** 25 cents

2. How many times is a price increase shown between 2001 and 2011?
   - **A.** 0 times
   - **B.** 2 times
   - **C.** 5 times
   - **D.** 10 times

3. How many times is a price decrease shown between 2001 and 2011?
   - **A.** 0 times
   - **B.** 2 times
   - **C.** 5 times
   - **D.** 10 times

4. The greatest increase in the cost of a stamp was between which two years?
   - **A.** 2002 to 2003
   - **B.** 2005 to 2006
   - **C.** 2008 to 2009
   - **D.** 2010 to 2011

5. In which three years was the price of stamps the same?
   - **A.** 2001, 2002, 2003
   - **B.** 2003, 2004, 2005
   - **C.** 2006, 2007, 2008
   - **D.** 2009, 2010, 2011

6. What does this graph indicate about the cost of a postage stamp?
   - **A.** The cost stayed the same.
   - **B.** The cost steadily decreased.
   - **C.** The cost steadily increased.
   - **D.** There is not enough data represented.

320   Chapter 14

You may allow the students to use their clocks to solve the elapsed time problems on the Chapter 14 Test.

- To prepare the students for the format of achievement tests, instruct them to work on a separate sheet of paper, if necessary, and to mark their answers on the *Cumulative Review Answer Sheet*.

Student Edition pages 320–23

The Cumulative Review provides additional practice of previously learned concepts. These pages may be completed during this lesson or anytime after this lesson, since they require limited or no teaching.

MAIL SYSTEM

YOUR LETTER
BJU PRESS
1430 WADE HAMPTON BLVD
GREENVILLE, SC 29609
MAIL COLLECTION BOX
PROCESSING PLANT
CULLING
BAR CODES LINKED WITH SPECIFIC ZIP CODES ARE SPRAYED ON EVERY LETTER
POSTMARK CANCELLATION STAMP
LETTERS THIS WAY
METRIC
METRIC
MAGAZINES + CATALOGUES THIS WAY
PACKAGES THIS WAY
LETTERS SORTED BY RANGE OF ZIP CODES
TRANSPORT
SORTED BY SPECIFIC ZIP CODE AND PREPARED FOR DELIVERY
BJU PRESS
BJU PRESS
1430 WADE HAMPTON BLVD
GREENVILLE, SC 29609
DELIVERED!

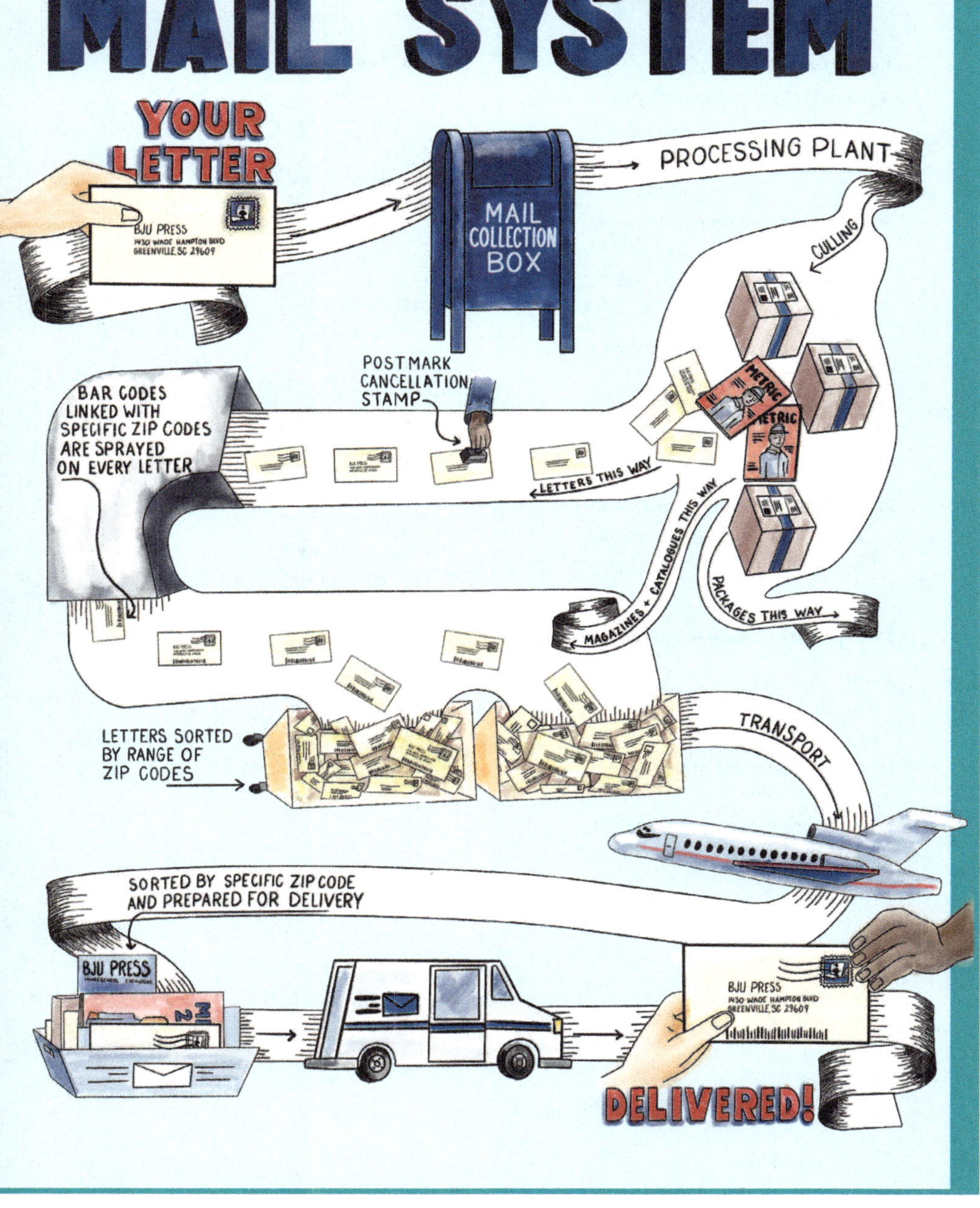

Choose the answer.

7. Round 13.43 to the nearest whole number.

- **A.** 10
- **B.** 13
- **C.** 14
- **D.** 15

8. Estimate the quotient for 3,625 ÷ 43.

- **A.** 80
- **B.** 85
- **C.** 90
- **D.** 900

9. Use front-end estimation to find the sum.

283,498 + 690,785

- **A.** 500,000
- **B.** 600,000
- **C.** 873,000
- **D.** 970,000

10. Round 987,642 to the nearest one thousand.

- **A.** 987,600
- **B.** 988,000
- **C.** 990,000
- **D.** 1,000,000

11. If $\frac{1}{4}$ of 60 is 15, what is $\frac{3}{4}$ of 60?

- **A.** 20
- **B.** 30
- **C.** 45
- **D.** 90

12. What is the price of 10 lb of grapes if they cost $1.79 a pound?

- **A.** $17.90
- **B.** $28.40
- **C.** $34.00
- **D.** $179

13. $3n + 8$ if $n = 1.6$

- **A.** 5.6
- **B.** 12.8
- **C.** 32.4
- **D.** 40

14. $9.83 \div n$ if $n = 10$

- **A.** 0.983
- **B.** 98.3
- **C.** 983
- **D.** 9,830

15. $n - 56 = 49$

- **A.** $n = 7$
- **B.** $n = 13$
- **C.** $n = 99$
- **D.** $n = 105$

16. $3y = 108$

- **A.** $y = 25$
- **B.** $y = 30$
- **C.** $y = 36$
- **D.** $y = 105$

17. $y + y + y + y$

- **A.** $y + 4$
- **B.** $4y$
- **C.** $4 - y$
- **D.** $y \div 4$

18. $4.99 + m = 5$

- **A.** $m = 0.1$
- **B.** $m = 0.01$
- **C.** $m = 0.001$
- **D.** $m = 1$

Math 6

Choose the answer.

19. What factors are common to 18 and 24?

   A. 3, 6, 9   **C.** 3, 4, 6

   B. 2, 3, 6   **D.** 2, 4, 6

20. What is the LCM of 6 and 4?

   A. 6   **C.** 18

   B. 12   **D.** 24

21.

n		
3.8	6.7	0.4

   A. $n = 1.8$ **C.** $n = 9.9$

   B. $n = 7.2$ **D.** $n = 10.9$

22. Use the factor tree to choose the prime factorization.

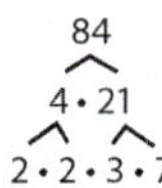

$$84$$
$$4 \cdot 21$$
$$2 \cdot 2 \cdot 3 \cdot 7$$

   A. $2 \cdot 3 \cdot 7$

   B. $4 \cdot 7$

   C. $2^2 \cdot 3 \cdot 7$

   D. $2 \cdot 3^2 \cdot 7$

23. Choose the true statement about circle M.

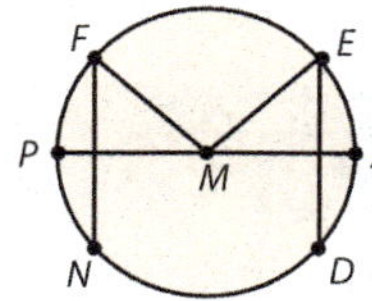

   A. The diameter is $\overline{PA}$.

   B. $\overline{MA} = \overline{MF}$

   C. $\overline{FN}$ and $\overline{ED}$ are chords.

   D. all of the above

24. Name the figure.

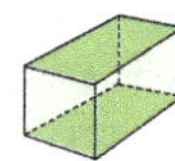

   A. rectangular pyramid

   B. square prism

   C. rectangular prism

   D. none of the above

25. Choose the inequality statement for the number line.

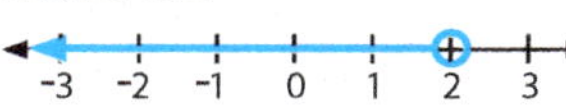

   A. $x > 2$

   B. $x < 2$

   C. $x = 2$

   D. $x > 3$

Lesson Plan Overview

				Chapter 15: Statistics	
Day	Lesson	Student Edition Pages	Teacher Edition Pages	Teacher Resources & Additional Materials	Topics, Skills & Biblical Worldview Shaping
149	147	324, 326–27	324, 326–27b	Teacher Resources • 96 *Frequency Tables* Additional Materials • calculators	**Topic:** Statistics **Skills:** completing a frequency table; determining the range, median, and mode for a set of data; calculating the mean **BWS:** Service
150	148	325, 328–29	325, 328–29b	Teacher Resources • 97 *Double Bar Graph* • 98 *Double Line Graph* • 99 *Types of Graphs*	**Topic:** Double Bar & Double Line Graphs **Skills:** interpreting a double bar and a double line graph
151	149	330–31	330–31b	Teacher Resources • 99 *Types of Graphs* • 100 *Stem-and-Leaf Plot*	**Topic:** Stem-and-Leaf Plots **Skills:** interpreting and completing a stem-and-leaf plot
152	150	332–33	332–33b	Teacher Resources • 99 *Types of Graphs* • 101 *Line Plot* Additional Materials • 3 × 5 cards Assessments • Chapter 15 Quiz 1	**Topic:** Line Plots **Skills:** interpreting and recording data on a line plot, determining the effects of an outlier
153	151	334–35	334–35b	Teacher Resources • 99 *Types of Graphs* • 102 *Restaurant Statistics* • 103 *Histogram: Piano Practice Statistics* • 104 *Histogram* Additional Materials • rulers	**Topic:** Histograms **Skills:** interpreting and constructing a histogram **BWS:** Service
154	152	336–37	336–37b	Teacher Resources • 99 *Types of Graphs* • 105 *Box-and-Whisker Plot: Points Scored* • 106 *Box-and-Whisker Plot* Additional Materials • rulers	**Topic:** Box-and-Whisker Plots **Skills:** interpreting a box-and-whisker plot; determining the lower, middle, and upper quartiles of a set of data; constructing a box-and-whisker plot
155	153	338–39	338–39b	Teacher Resources • 107 *Graph: Double Bar Graph* • 108 *Graph: Double Line Graph* • 109 *Stem-and-Leaf Plot & Line Plot* • 110 *Graph: Histogram* • 111 *Graph: Box-and-Whisker Plot* Assessments • Chapter 15 Quiz 2	**Topic:** Graph Review **Skills:** interpreting graphs **BWS:** Service

Day	Lesson	Student Edition Pages	Teacher Edition Pages	Teacher Resources & Additional Materials	Topics, Skills & Biblical Worldview Shaping
156	154	340–41	340–41b	Teacher Resources • 112 *Data* Additional Materials • prepared terms and definitions from *Types of Graphs*, pages 1–4 • 3 × 5 cards	**Topic:** Comparing Graphs **Skills:** choosing a graph to display a set of data
157	155	342–43	342–43b	Teacher Resources • 97 *Double Bar Graph* • 98 *Double Line Graph* • 103 *Histogram: Piano Practice Statistics* • 105 *Box-and-Whisker Plot: Points Scored* Additional Materials • prepared terms and definitions from *Types of Graphs*, pages 1–4 • Bible	**Topic:** Chapter Review **BWS:** Service
158	156	344–46	344–46		**Topic:** Test & Cumulative Review
159	157	325, 347–48	347–48a	Teacher Resources • 25 *Engineering Design Process* • 113 *Business Idea Organizer* Additional Materials • examples of successful young entrepreneurs (optional)	**Topic:** STEM: Entrepreneurship **Skills:** identifying and researching a problem, recording possible solutions, identifying a workable solution
160	158	347–48	348b–c	Teacher Resources • 114 *Rover's Romps Business Plan* • 115 *Business Plan* Additional Materials • completed *Business Idea Organizer* pages	**Topic:** STEM: Entrepreneurship **Skills:** assembling business data, writing a business plan **BWS:** Service
161	159	347–48	348d–e	Teacher Resources • 116 *Rover's Romps Profit Projection* • 117 *Business Profit Projection* • 118 *STEM Rubric: Entrepreneurship* Additional Materials • completed *Business Plan* pages	**Topic:** STEM: Entrepreneurship **Skills:** testing and improving a plan

CHAPTER 15

CHAPTER OBJECTIVES

- Interpret a double bar graph, double line graph, stem-and-leaf plot, line plot, histogram, and box-and-whisker plot.
- Complete a frequency table and a stem-and-leaf plot using given data.
- Determine the range, median, mean, and mode for a set of data.
- Determine the lower, middle, and upper quartiles of a set of data.
- Construct a histogram and a box-and-whisker plot using given data.
- Propose ways that statistics make people's lives better.
- STEM: Help others by using statistics to solve a business problem.

To determine whether your students are prepared for the concepts taught in this chapter, you may wish to consult the preassessment checklist found on TeacherToolsOnline.com.

Throughout this chapter, select problems from the list of mental math problems provided on the 134–36 *Mental Math Strings* pages.

Prepare the *Types of Graphs* pages (4) by cutting along the borders of each term and description to use as each term is introduced in the chapter, beginning in Lesson 148.

Make fact practice, both oral and written, part of your daily math routine to help the students with mastery.

Visit AfterSchoolHelp.com for math practice resources, or visit TeacherToolsOnline.com for additional resources to enhance the lessons.

"

© Abby Ivy Studio

ENTREPRENEURSHIP

How can I help others by using statistics to solve a business problem?

Have you ever had a really great idea? Not a just-okay idea, but a really GREAT one? And you thought, "Somebody ought to invent that/produce that/provide that." Why not you?

You might be surprised to learn that young entrepreneurs are changing the world—and making money—with everything from wildly popular mobile game apps to bee-friendly lemonade to an ingenuous plastic cup that won't spill even when shaken.

Not every savvy kid business owner invents something. Some simply use their God-given gifts and abilities to meet needs or solve problems faithfully and consistently.

School break will be here before you know it! What are your plans? You might be thinking, "But I don't know where to start."

Have we got a great idea for you!

Amanda Brown
entrepreneur, creative
director & editor

Amanda is the creative director and head editor of print magazine and devotional-style blog *Oh Beloved One* for Christian girls. Amanda started the magazine in her tween years because of the lack of spiritually edifying reading materials for her age group. Originally a print newsletter filled with articles written by her middle-school-age friends, Amanda's venture has matured with her, stretching her abilities. Social media polls, quizzes, and website analytics provide her with data for creating better content for her readers. In a personal interview, Amanda said, "It's time to stop thinking of ourselves as 'artsy' or 'technology-minded' or 'number-smart.' Life takes a combination of these skills: creatively choosing a solution, organizing chaos, and using math to tie everything together."

LESSON 147

Student Edition pages 324, 326–27
Daily Review Chapter 15, section *a*

OBJECTIVES

- Complete a frequency table using given data.
- Determine the range, median, and mode for a set of data.
- Calculate the mean for a set of data.
- Explain how statistics help with managing data. **BWS**

BIBLICAL WORLDVIEW SHAPING

- Service (Explain): Statistics make data easier to understand and interpret.

TEACHER RESOURCES

- 96 *Frequency Tables* (for the teacher and for each student)

ADDITIONAL MATERIALS

- a calculator (for each student)

Throughout this chapter, allow the students to use a calculator as needed for managing data during the lesson and for problems on the Student Edition pages.

Statistics

How are statistics helpful in managing data?

Statistics is the branch of mathematics that deals with the collection, organization, analysis, and interpretation of data. The data is more easily interpreted when displayed in a table, a chart, or a graph.

Brooke organized her math test scores from least to greatest in a **frequency table**. From this table she can find the range, the mean, the median, and the mode of test scores.

test scores: 90, 84, 78, 82, 90, 87, 82, 93, 90

Math Test Scores		
Score	Tally	Frequency
78	I	1
82	II	2
84	I	1
87	I	1
90	III	3
93	I	1

range: 93 − 78 = 15
mean: 78 + 82 + 82 + 84 + 87 + 90 + 90 + 90 + 93 = 776; 776 ÷ 9 = 86.2
median: ~~78 82 82 84~~ (87) ~~90 90 90 93~~ 87
mode: 90

Key Terms

- statistics
- frequency table
- data
- range
- mean
- median
- mode

- The **data** is collected information or facts.
- The **range** is the difference between the greatest value (number) and the least value.
- The **mean** (average) is the sum of the data divided by the number of addends.
- The **median** is the middle value (or an average of the two middle values) of a set of data when ordered from least to greatest.
- The **mode** is the value that occurs most often, or has the greatest frequency. Some sets may have more than one mode, and some sets may not have a mode.

Exercises

Use the frequency table to find the answer.

1. How many campers are represented on the frequency table?
   13 campers
2. What is the range in ages of the campers? *14 − 10 = 4*
3. How many of the campers are 11? *4 campers*
4. What age groups have the same frequency? *10, 13, and 14*
5. What age has the greatest frequency (mode)? *11*

Campers		
Age	Tally	Frequency
10	II	2
11	IIII	4
12	III	3
13	II	2
14	II	2

6. Find the mean and the median for this data. Round the mean to the nearest tenth.
   mean: 11.8; median: 12

Make a frequency table showing the high temperatures during spring vacation. Use the data to find the answer.

7. What is the range in temperature? *88° − 80° = 8°*

High Temperatures during Spring Vacation
83°   81°   86°   88°   80°   82°   85°

8. What is the mean? Round to the nearest tenth. *83.6°*
9. What is the median? Use < to write a mathematical sentence comparing the median to the mean.
   median: 83°; 83° < 83.6°
10. Why does this data not have a mode? *because each number appears only once*

Engage

- Share a **surprising fact** to introduce the topic of data and statistics. Explain that in 2013 the Norwegian research organization SINTEF was quoted as saying that "90% of all the data in the world has been generated over the last two years." (SINTEF. "Big Data, for better or worse: 90% of world's data generated over last two years." ScienceDaily. ScienceDaily, 22 May 2013. www.sciencedaily.com/releases/2013/05/130522085217.htm)

  What is data? Accept any reasonable answer; a set of collected information or facts.

  What kind of data do you produce? sample answers: test scores, game scores, texts, videos, purchase records, online activity, health information (doctor's visits)

  How do you think the amount of data in the world has changed in the years since 2013 when that statistic was shared? sample answer: It has continued to grow very quickly.

- Point out that because of the great amount of data being generated and recorded in the digital age in which we live, data must be managed to be usable. Explain that the branch of mathematics that deals with collecting, organizing, analyzing, and interpreting data is called statistics.

- Direct attention to the chapter essential question on Student Edition page 324, "How can statistics make people's lives better?" Encourage the students to be alert during the lesson to 1 of 3 ways that they can use statistics to help make people's lives better.

Instruct

Completing a frequency table

- Guide a **discussion** to help the students organize data by completing a frequency table. Distribute a copy of the *Frequency Tables* page to each student, and display your copy.

  What data is listed at the top of the *Frequency Tables* page? spelling test scores

  What is the lowest recorded score? 82

Find the range and the mode(s) for the set of data.

11. 3 8 2 6 5 2 4 2
range: 8 − 2 = 6; mode: 2

12. 33 35 31 30 33 33 35
range: 35 − 30 = 5; mode: 33

13. 18 17 14 16 17 15 13 18
range: 18 − 13 = 5;
mode: 17 and 18

Find the mean and the median for the set of data.

14. 82 85 79 81 85 87
mean: 83.2; median: 83.5

15. 48 52 41 51 40
mean: 46.4; median: 48

16. 17 12 15 10 12 18 19 12
mean: 14.4; median: 13.5

Practice & Application

The Franseens have started a new church. The frequency table shows the attendance for the first Sunday. Use the data to find the answers for problems 17–21.

First Sunday Attendance

Age Group	Tally	Frequency
0–5	卌 ‖	7
6–12	卌 卌 ‖‖	14
13–17	卌 卌 ‖	11
18–25	卌 卌	10
26+	卌 卌 卌 卌 ‖‖	24

17. What is the least number of chairs that could have been set up for ages 6 and up? *59 chairs*

18. The church uses a ratio of 1 teacher to 4 students in the 0–5 age group. How many teachers were needed? *2 teachers*

19. How many of the people that attended the new church were 26 or older? *24 people*

20. What ratio can be written comparing people 18 and older to those under 18? *34 : 32*

21. Write the frequency data for the table from least to greatest. *7, 10, 11, 14, 24*

Use the rectangular prism to find the answer.

22. How many faces does a rectangular prism have? *6*

23. How many faces have an area of 9 ft²? *2*

24. How many faces have an area of 6 ft × 3 ft? *4*

25. What is the total surface area of the rectangular prism? *90 ft²*

26. What is the volume of the rectangular prism? *54 ft³*

27. How are statistics helpful in managing data?
Statistics make data easier to interpret.

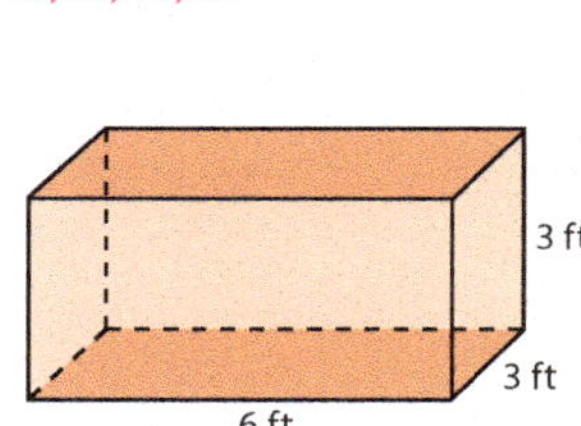

Lesson 147   327

How many test scores of 82 were recorded? 1

Draw 1 tally in the Tally column in the same row as the score 82.

- Explain that the tallies in a frequency table represent the number of times that an event occurred. The number of occurrences (tallies) in a category is called the frequency.

What is the number of times that 82 was scored on a spelling test? 1

Write "1" in the Frequency column.

- Repeat the procedure to record the remaining data on the frequency table: 86 1, 89 2, 90 1, 93 1, 95 3, 98 1, 100 1. Explain that if they had more than 4 tallies for a score, they would draw a diagonal line through 4 tallies to indicate 5 tallies.

Why are the scores 83, 84, 87, 88, and 91 not recorded on the frequency table?
Those scores are not listed in the data.

Determining the range, median & mode for a set of data; calculating the mean

- Use a **guided activity** to help the students analyze a set of data. Direct attention to the word *range* below the Spelling Test Scores frequency table. Explain that the range is the difference between the greatest value and the least value. It is a measure of how much the data varies (its variability) or is spread out (its span).

Which spelling test score has the greatest value? 100

Which score has the least value? 82

What is the difference between the greatest value and the least value in this set of data? 18; 100 − 82 = 18

What is the range of the data? 18

Direct the students to write the range "18" as you write it on the displayed page.

- Direct attention to the word *mean*. Explain that the word *mean* is a statistical term for the average.

How can you find the average test score? I can add all of the scores and divide the sum by the number of scores.

What is the highest recorded score? 100
Would it be easy to accurately know how well you are doing in spelling by simply looking at the unorganized list of scores? no

- Remind the students that statistics help them analyze data.

What statistic do teachers prepare by using your test scores to tell you how well you are doing in a subject? sample answer: the average of my test scores in that subject

- Explain that to analyze a set of data they first need to organize it to make it more manageable. A frequency table is one way to organize data.

What do you think would be a good way to organize the spelling test scores? sample answer: I could order the test scores from least to greatest.

Choose a student to write the data in least to greatest order for display. 82, 86, 89, 89, 90, 93, 95, 95, 95, 98, 100

- Use the following procedure to guide the students as they complete the first frequency table on the page. Direct them to write each entry in the table as you write it on the displayed page.

What is the lowest test score? 82

Write "82" as the first test score in the Score column.

LESSON 147

- Instruct the students to use their calculators to find the average test score. 1,012 ÷ 11 = 92

What is the average test score? 92

Direct the students to write "92" as the mean as you write it for display.

How could you use the data in the frequency table to write 1 equation to find the mean? I could write multiplication expressions inside parentheses to show the test scores that occurred more than 1 time and write brackets around the expression with all of the test scores to show that the scores must be added before I divide by the number of scores.

- Write the equation for display and guide the students as they follow the order of operations to solve it: "[82 + 86 + (2 × 89) + 90 + 93 + (3 × 95) + 98 + 100] ÷ 11 = 92." Point out that although this method requires more steps to solve the equation and may appear to require more time, solving 1 equation is more efficient when they need to find the mean (average) of a large amount of data.

- Direct attention to the word *median*. Explain that the median is the middle value in an ordered set of data.

How do you think you could determine the median in this ordered set of data? sample answer: I could repeatedly cross out the least value and the greatest value in the ordered list of test scores until only the middle value remains.

Demonstrate the process on the displayed page to guide the students as they cross out pairs of values on their ordered lists of test scores. Point out that this method of finding the median can be used whether values are written in ascending order or descending order.

82, 86, 89, 89, 90, 93, 95, 95, 95, 98, 100

How many numbers are on each side of the median? 5

What is the median? 93

Direct the students to write "93" as the median as you write it for display.

- Direct attention to the word *mode*. Explain that the mode is the number that occurs most frequently in a set of data. A set of data may have more than

1 mode. If none of the numbers appears more frequently than the other numbers, a set of data has no mode.

Which test score occurs most frequently in the data listed in this frequency table? 95 occurs 3 times, which is more frequently than any of the other scores.

Direct the students to write "95" as the mode as you write it for display.

Point out that the range, mean, median, and mode for a set of data are also referred to as statistics.

- Direct attention to the data at the bottom of the page.

What data or collected information is listed here? daily low temperatures

- Guide a **collaborative activity** to help the students analyze a set of data. Direct the students to collaborate with a partner to organize the temperature data at the bottom of the page in ascending order 69°, 69°, 71°, 72°, 73°, 74°, 74°, 78° and then to complete the frequency table: 69 2, 71 1, 72 1, 73 1, 74 2, 78 1. Choose students to write their answers for display.

- Direct the students to continue to collaborate to find the range and the mean of the data. Choose students to demonstrate finding the range 78° − 69° = 9° and the mean 580° ÷ 8 = 72.5° for display.

Write the ratio in word form, ratio form, and fraction form.

1. 1 computer for every 3 students *1 to 3, 1 : 3, $\frac{1}{3}$*

2. 2 workers for every 15 children *2 to 15, 2 : 15, $\frac{2}{15}$*

3. 4 tables for every 32 people *4 to 32, 4 : 32, $\frac{4}{32}$*

4. 6 servings for every pie *6 to 1, 6 : 1, $\frac{6}{1}$*

5. 6 cookies for every 3 lunches *6 to 3, 6 : 3, $\frac{6}{3}$*

Write the ratio as a fraction in lowest terms.

6. 2 to 8 $\frac{1}{4}$ 7. 4 to 12 $\frac{1}{3}$ 8. 5 to 10 $\frac{1}{2}$

9. 8 to 20 $\frac{2}{5}$ 10. 10 to 100 $\frac{1}{10}$

Use equivalent ratios to find the missing term.

11. $\frac{4}{8} = \frac{n}{16}$ *n = 8*   12. $\frac{1}{4} = \frac{n}{100}$ *n = 25*

13. $\frac{2}{3} = \frac{4}{n}$ *n = 6*   14. $\frac{1}{5} = \frac{n}{100}$ *n = 20*

Solve.

15. 249.71
   + 84.09
   333.80

16. $3.75
   × 5
   $18.75

17. $20.00
   − $12.75
   $7.25

18. 1,287 ÷ 3 *429*

19. 1.2 + 39.764 *40.964*

500   Daily Review

- Instruct each student to repeatedly cross out the lowest temperature and the highest temperature in his ordered list until only the middle temperature remains.

  What do you notice about finding the median with this set of data? There is no middle number because there is an even number of temperatures.

- Explain that when there is an even number of data, the mean or average of the 2 middle values is the median.

  What are the 2 middle values in this set of data? 72° and 73°

  Draw a circle around the 2 middle temperatures for display: 72° and 73°.

  Choose a student to demonstrate finding the mean (average) of the middle temperatures using 1 equation. $(72° + 73°) \div 2 = 72.5°$

- Remind the students that a data set may have more than 1 mode.

  What are the modes in this set of data? 69° and 74°; Both temperatures occurred twice, which is more frequently than any of the other temperatures.

Helping with statistics

- Guide a **discussion** of the essential question at the top of Student Edition page 326, "How are statistics helpful in managing data?" Remind the students of the spelling test score data that they analyzed at the beginning of the lesson.

  How does a teacher help you when he uses statistics to manage your test score data? sample answer: The statistics that he finds help me interpret the data more easily so that I can see whether I need to make changes to learn the material better to improve my scores.

Explaining the Gospel

- Remind the students that to be pleasing to God, our service to others must be based upon our personal relationship to Jesus Christ. Use Explaining the Gospel (see the Appendix) to present our need to repent and trust Christ for salvation. Point out that only those who repent of their sin and place their faith in Jesus Christ as Savior are found acceptable to God. Share your salvation testimony or invite a student to share his testimony.

Apply

Student Edition pages 326–27

- Read and explain the directions for pages 326–27. Assist the students as they complete the pages independently.

Daily Review

- Students should complete Chapter 15, section *a*.

DIFFERENTIATED INSTRUCTION

Use the following to provide extra help for students who experience difficulty with the concepts taught in Chapter 15.

Find the mean for a set of data.
Remind the students that the mean is the average of a set of data; it is the sum of the data divided by the number of addends. Provide the students with 24 self-adhesive notes. Explain that each note represents $1. Instruct them to illustrate that John has $4 by placing 1 note above the other, making a column similar to a bar on a vertical bar graph. Direct the students to make other columns to illustrate that Paul has $5, Steven has $8, and Kevin has $7. Explain that the mean can be found by moving 1 or more dollars from one column and putting them in another column until all columns have an equal number of dollars. Direct the students to move only the dollars that are necessary to move to make the columns equal. After they have completed the task, point out that they redistributed some of the dollars from the taller columns to the shorter columns to make all of the columns level or equal. Ask them what the mean of $4, $5, $8, and $7 is. $6 Continue the activity as needed using other sets of data.

Student Edition pages 325, 328–29
Daily Review Chapter 15, section *b*

OBJECTIVES

- Interpret a double bar graph and a double line graph.
- Determine the range, median, and mode for a set of data.
- Calculate the mean for a set of data.

TEACHER RESOURCES

- 97 *Double Bar Graph*
- 98 *Double Line Graph*
- 99 *Types of Graphs*, pages 1–2

Throughout the chapter as each graph is introduced, build a word bank for display by using the name of the graph and its description from the *Types of Graphs* pages.

Engage

- **Read aloud** Student Edition page 325 to introduce and generate interest for this chapter's STEM activity.

- Direct the students to **turn and talk** to explore the essential question at the top of Student Edition page 328, "Which type of graph should I use to track my friend's and my training progress for a 5K run?" Take a poll to see how many students choose each graph. Challenge them to see if what they learn in today's lesson confirms or changes their vote.

Instruct

Interpreting a double bar graph

- Guide a **discussion** to help the students interpret a double bar graph.

  Display the *Double Bar Graph* page. Explain that a bar graph summarizes data in pictorial form by using bars that can be drawn horizontally or vertically. All bar graphs have 2 axes. Each axis is labeled with categories to show the data that is being graphed. The labeling of the axes determines the direction of the bars on the graph.

- Point out that the categories listed on the horizontal axis of this double bar graph are first grade through sixth grade. The categories listed on the vertical axis represent the frequency (number of occurrences), the number of students per grade.

  According to the title, what information is recorded or pictured on this double bar graph? Grace Christian School attendance

  For what grades does the graph show attendance? first, second, third, fourth, fifth, and sixth

- Point out that the scale on the left side of the graph has a range of 16 (ranges from 0 to 16). Explain that the interval of a scale is the amount between the numbers of the scale. The interval is often determined using the range of the data and the space that is available for making the graph.

  What is the interval of the scale for this graph? 2

- Remind the students that bar graphs are used to compare data. Explain that a double bar graph compares 2 sets of related or similar data on the same graph.

Double Bar & Double Line Graphs

Which type of graph should I use to track my friend's and my training progress for a 5K run?

Graphs are pictorial forms used to summarize data. Bar graphs and line graphs are used to compare or to observe changes in data. The horizontal and vertical axes are reference lines that are labeled with units in equal **intervals**.

> A **double bar graph** compares two sets of related data. The bars can be drawn either horizontally or vertically.

> A **double line graph** shows changes over time for two sets of related data.

A **key** shows the data that each bar or line represents. The range, the mean, the median, and the mode can be found using the graph data.

Key Terms
- graphs
- intervals
- double bar graph
- double line graph
- key

Exercises

Use the bar graph to find the answer.

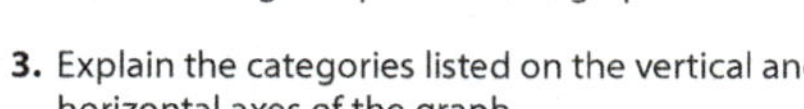

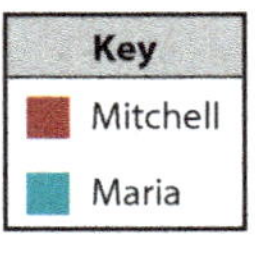

1. *the amount of time that Mitchell and Maria each spend doing homework*

3. *vertical: days of the week homework is done; horizontal: minutes spent on homework*

1. What is being compared in this graph?
2. What interval is used for tracking homework time? *15 min*
3. Explain the categories listed on the vertical and horizontal axes of the graph.
4. What is the average (mean) amount of time spent on homework for Maria? for Mitchell? *51 min for both*
5. On which day did Maria and Mitchell spend the most amount of time doing homework? *Thursday*
6. Is there a mode of the data for either Mitchell or Maria? If so, what is it? *yes; Mitchell: 60*
7. Find the range of homework time for Maria. *90 − 15 = 75*
8. Find the median time spent on homework for Mitchell and for Maria. *Mitchell: 60; Maria: 45*
9. Compare the range of homework time for Maria and Mitchell. *The range is the same for both: 90 − 15 = 75.*
10. What days of the week are not shown on the graph? *Saturday and Sunday*

328   Chapter 15

Use the line graph to find the answer.

Colin and Elisha began preparing for a 5K run. They recorded their progress and placed the number of kilometers they ran each week on a line graph. The double line graph shows the change in the number of kilometers they ran in a 6 week period.

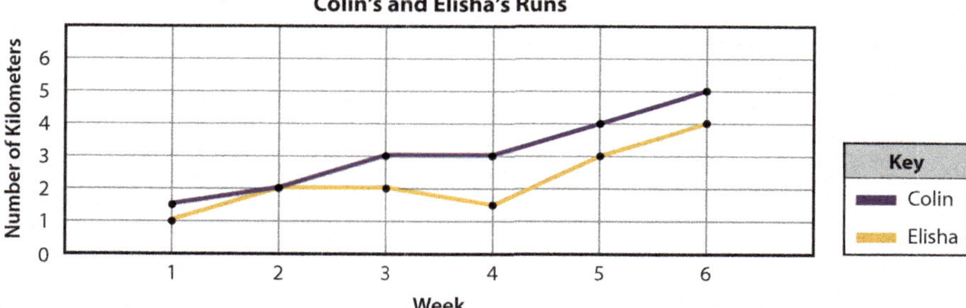

11. What information is recorded on this graph?

12. Over how much time was the data collected? *6 weeks*

13. Which week shows the greatest difference in kilometers run? What is the difference? *week 4; 1.5 km*

14. How many weeks show only a 1 km difference between the two runners? *3 weeks*

15. Where is the decrease shown in the number of kilometers run by Elisha? *from week 3 to week 4*

16. Find the range, the mean, the median, and the mode for the kilometers run by Colin. Round the mean to the nearest tenth.

17. During which week did the two boys run the same number of kilometers? *week 2*

18. Find the range, the mean, the median, and the mode for the kilometers run by Elisha. Round the mean to the nearest tenth.

11. *the kilometers run each week by Colin and Elisha*

Practice & Application

Church Attendance		
Date	Morning	Evening
March 1	212	184
March 8	203	180
March 15	225	190
March 22	240	196
March 29	240	183

16. *range: 3.5; mean: 3.1; median: 3; mode: 3*

18. *range: 3; mean: 2.3; median: 2; mode: 2*

24. *The bars compare the morning service to the evening service.*

19. Make a double line graph showing church attendance for the month of March. Begin the numerical data with 175 and use intervals of 10. Use the line graph to find the answers for problems 20–23.

20. What is the range for the evening attendance? *196 − 180 = 16*

21. What is the mean for the morning attendance? *224*

22. Write the data for the morning attendance from least to greatest. What is the median? *203, 212, 225, 240, 240; median: 225*

23. What is the mode for the morning attendance? *240*

24. Make a double bar graph using the same church attendance data. Explain the comparison being made between each set of bars.

25. Which type of graph should I use to track my friend's and my training progress for a 5K run? *I should create a double line graph to track our training progress over time.*

Lesson 148   329

---

of boys in the first and the fifth grades is the same; however, the lighter bars show that there are 4 fewer girls in the first grade.

What equation can you write to find the number of students in grades 1 through 6? (12 + 8) + (9 + 13) + (11 + 15) + (14 + 9) + (12 + 12) + (12 + 13) = __

- Write the equation for display and direct the students to find the number of students in grades 1 through 6. 20 + 22 + 26 + 23 + 24 + 25 = 140 students

Which grade has the most students? third grade

How many students are in the third grade? 26

Which grade has the fewest students? first grade

How many students are in the first grade? 20

What equation can you write to find the average number of students in grades 1 through 6? 140 students ÷ 6 = __

- Direct the students to find the average number of students. Guide them to round the quotient to the nearest whole number. 23 students per grade; 140 ÷ 6 ≈ 23.3

Determining the range, median & mode; calculating the mean

- Guide a **discussion** to help the students analyze the data from the *Double Bar Graph* page. Guide the students as they order the number of girls in each grade from least to greatest. 8, 9, 12, 13, 13, 15

What is the range for the number of girls? 7; The range is the difference between the greatest value and the least value; 15 − 8 = 7.

What is the median of a set of data? The median is the middle value in an ordered list that has an odd number of values or the average of the 2 middle values in an ordered list that has an even number of values.

What is the median for the number of girls in grades 1 through 6 at Grace Christian School? 12.5; The average of the 2 middle values, 12 and 13, is 12.5.

What data do the bars on this graph compare? the number of students in each grade of Grace Christian School

What 2 sets of related data do the bars show? The darker bars show the number of boys in each grade, and the lighter bars show the number of girls in each grade; the key shows what the darker and lighter bars represent.

Which grade has the same number of boys and girls? fifth grade

Which grade has the most boys? fourth grade

Which has the most girls? third grade

Which grades have more girls than boys? second, third, and sixth grades

Which grade has the greatest difference in the number of boys and girls? fourth grade; The difference in the height of the bars is the greatest (more than 2 intervals) at the fourth grade level.

Are there more boys or girls in the fourth grade class? boys

How many more boys are there? 5 more boys; 14 boys − 9 girls = 5 more boys

Which grade has a total of 20 students? first grade; 12 boys + 8 girls = 20 students

How many more students are in the fifth grade than the first grade? 4 students; sample answer: When I compare the darker bars, I can see that the number

LESSON 148

What is the mode of a set of data? the value or number in a list of data that occurs most often or has the greatest frequency

What is the mode for the number of girls? 13

Which 2 grades each have 13 girls? second grade and sixth grade

How can you find the mean (average) of the set of girls in grades 1 through 6? I can add the number of girls in all 6 grades and divide the sum by 6.

- Direct the students to find the mean for the set of girls. $(8 + 9 + 12 + 13 + 13 + 15) \div 6 \approx 11.7$; 12 girls

- Follow a similar procedure to guide the students as they order the number of boys in each grade and then find the following statistics for the set of data. Point out that since the 2 middle values are the same (12), it is not necessary to calculate the average to find the median.

ordered data: 9, 11, 12, 12, 12, 14

range: $14 - 9 = 5$

median: 12

mode: 12

mean: $(9 + 11 + 12 + 12 + 12 + 14) \div 6 \approx 11.7$; 12 boys

Interpreting a double line graph

- Guide a **discussion** to help the students interpret a double line graph. Display the *Double Line Graph* page. Explain that a double line graph compares changes that occur over a period of time for 2 sets of related data.

What information is recorded on this double line graph? the high and low temperatures of Peaceful Valley for 1 week

What information is listed along the horizontal axis? the days of the week, Sunday through Saturday

What information is listed along the vertical axis? the Fahrenheit temperatures

What is the interval of the scale of this graph? 2; There are 2 degrees between the temperatures on the scale.

Write a proportion to find the unknown measure of the similar figure.

1. 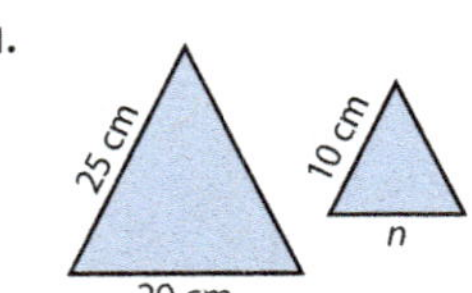
$\frac{25}{20} = \frac{10}{n}$; $n = 8$ cm

2. 12 cm — n — 9 cm — 12 cm
$\frac{12}{9} = \frac{n}{12}$; $n = 16$ cm

3. 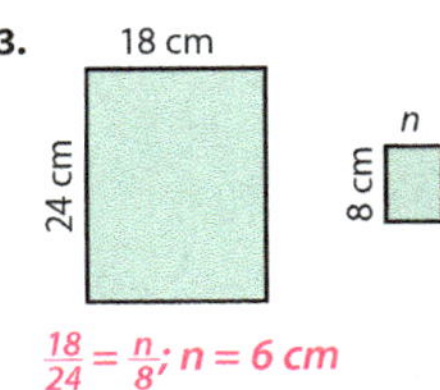
$\frac{18}{24} = \frac{n}{8}$; $n = 6$ cm

4. 20 cm — 100 cm — 6 cm — n
$\frac{20}{100} = \frac{6}{n}$; $n = 30$ cm

5. 1.5 cm — 6 cm — 2.5 cm — n
$\frac{1.5}{6} = \frac{2.5}{n}$; $n = 10$ cm

6. 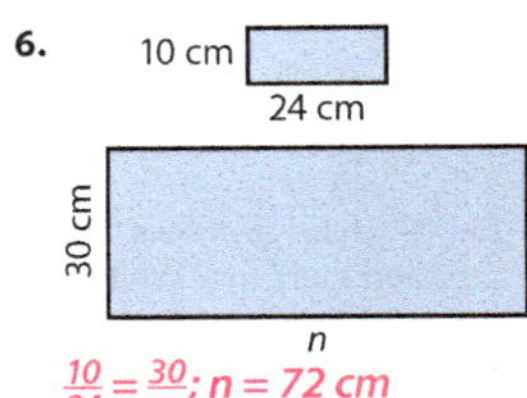
$\frac{10}{24} = \frac{30}{n}$; $n = 72$ cm

Write a proportion to solve.

7. A parking meter that is 1.5 m tall casts a shadow of 3 m. A light pole in the parking lot casts a shadow of 12 m. How tall is the light pole?
$\frac{1.5}{3} = \frac{n}{12}$; $n = 6$ m

8. A tree casts a shadow of 1.2 m. A meter stick casts a shadow of 0.4 m. What is the height of the tree?
$\frac{1}{0.4} = \frac{n}{1.2}$; $n = 3$ m

- Explain that since there is no data for 0° to 59°, the broken scale on the vertical axis between 0° and 60° indicates a larger interval than the rest of the scale.

What does the key tell you? The points of the high temperatures are connected by a solid line, and the points of the low temperatures are connected by a broken or dashed line.

What does the line graph show you about the changes in the high temperatures during the week? sample answers: Only once during the week was the daily high temperature lower than the previous day's high temperature; the range of the high temperatures was 5° (74° − 69° = 5°).

What does the line graph show you about the changes in the low temperatures during the week? sample answers: Only once during the week was the daily low temperature significantly lower than the previous day's low temperature; the range of the low temperatures was 10° (70° − 60° = 10°).

On which day was the warmest temperature recorded? Saturday

On which day was the coolest temperature recorded? Friday

What was the range of temperatures for Peaceful Valley during the week? 14°; The difference between the warmest high temperature and the coolest low temperature was 14° (74° − 60° = 14°).

Determining the range, median & mode; calculating the mean

- Use **guided practice** to help the students analyze the data from the *Double Line Graph* page. Discuss the steadiness of the high and low temperatures. Guide the students as they compare the ranges of the high and low temperatures and conclude that the high temperatures were quite steady (range of 5°) compared to the low temperatures (range of 10°), which showed a greater fluctuation.

  What was the range of the temperatures on Sunday? 8°

  Monday? 4°

  Tuesday? 3°

  Wednesday? 6°

  Thursday? 2°

  Friday? 12°

  Saturday? 5°

- Guide the students as they find the median, mode, and mean for the high temperatures and the low temperatures. Direct them to round each mean to the nearest hundredth. Choose students to explain how they found the answers.

  high temperatures:

  median 72; mode 72; mean 71.29; 499 ÷ 7 ≈ 71.285

  low temperatures:

  median 66; mode no mode; mean 65.57; 459 ÷ 7 ≈ 65.571

- **Discuss** how double bar graphs and double line graphs summarize data to help the students answer the essential question. Display the prepared bar graph, line graph, double bar graph, and double line graph terms and descriptions from the *Types of Graphs* pages. Point out that a double bar graph is used to compare 2 sets of data but that a double line graph compares changes that occur over a period of time for 2 sets of data.

  How would you answer the essential question now? I can create a double line graph to track our training progress over time.

  Did you learn something about the graphs that changed your answer? Answers will vary.

- Continue to display the terms throughout the chapter as more are added.

Apply

Student Edition pages 328–29

- Read and explain the directions for pages 328–29. Assist the students as they complete the pages independently. You may guide them as they make the double line graph for problem 19 or allow them to work in groups.

Daily Review

- Students should complete Chapter 15, section *b*.

NOTES

Student Edition pages 330–31
Daily Review Chapter 15, section c

OBJECTIVES

- Interpret a stem-and-leaf plot.
- Complete a stem-and-leaf plot.

TEACHER RESOURCES

- 99 *Types of Graphs*, page 3
- 100 *Stem-and-Leaf Plot* (for the teacher and for each student)

Engage

- Direct the students to **brainstorm** to explore the essential question at the top of Student Edition page 330, "How can I organize my math test scores?"

Use questions such as the following to stimulate the students' thinking.

What ways of organizing data have been discussed so far during Chapter 15? ascending or descending order; frequency table; double bar graph; double line graph

Which method would you choose? Answers will vary; the double bar and double line graphs are used to compare 2 sets of data rather than the data in just 1 set.

Are there other methods that you could use? Answers will vary.

Encourage the students to be looking for another way of organizing test scores in today's lesson.

Instruct

Interpreting a stem-and-leaf plot

- Use a **guided activity** to help the students interpret a stem-and-leaf plot. Distribute a copy of the *Stem-and-Leaf Plot* page to each student, and display your page.

- Read aloud the following word problem.

The Christian school Carter attends has classes for K-4 through 6th grade. Carter conducted a survey to find the number of students enrolled in each of the 8 grades. He collected the following data: 20, 17, 21, 19, 22, 18, 20, and 9. He later recorded his data on a stem-and-leaf plot.

Stem-and-Leaf Plots

How can I organize my math test scores?

Another way to display data is on a **stem-and-leaf plot**. In a stem-and-leaf plot, each piece of data is displayed using the tens digit of the value as a stem and the ones digit of the value as a leaf. The leaves indicate the number of items in the data set. A title and a key are provided to give guidance for reading and understanding the graph.

Juan recorded his math test scores to evaluate his progress: 89, 92, 87, 95, 79, 98, 94, 92, 89, and 92. He ordered the scores from least to greatest and put them in a stem-and-leaf plot. Then he found the range, the mean, the median, and the mode.

79, 87, 89, 89, 92, 92, 92, 94, 95, 98

Range
98 − 79 = 19

Mean
907 ÷ 10 = 90.7

Median
92

Mode
92

Math Test Scores

Stem	Leaf
7	9
8	7 9 9
9	2 2 2 4 5 8

| Key | 8\|7 = 87 |

There are 10 leaves, or 10 pieces of data.

Key Terms

- stem-and-leaf plot
  - stem: tens
  - leaf: ones

Exercises

Make a stem-and-leaf plot showing the number of books sold. Use the plot to find the answer.

1. Write the title and the key for the stem-and-leaf plot.
   Answers may vary.
2. How many days are represented? *12*

3. Write the number of books sold from least to greatest.
   9, 10, 12, 14, 15, 18, 19, 20, 21, 21, 23, 30
4. What is the range of the data?
   30 − 9 = 21
5. Find the mean of the data. Round to the nearest tenth.
   The sum of the data is 212; 212 ÷ 12 ≈ 17.7
6. What is the median? *18.5*

7. What is the mode? *21*

The store manager recorded the number of books sold in his Christian bookstore for a time period of two weeks. The store is closed on Sundays.

books sold: 12, 21, 30, 14, 10, 9, 15, 18, 23, 20, 21, 19

Make a stem-and-leaf plot showing the ages of missionaries. Use the plot to find the answer.

8. Write the title and the key for the stem-and-leaf plot.
   Answers may vary.
9. Write the ages of the missionaries from least to greatest.

10. What is the range of the data? *61 − 28 = 33*

11. Find the mean. Round to the nearest tenth.
    mean: sum of data = 777; 777 ÷ 19 ≈ 40.9
12. What is the median? *39*

13. What is the mode? *30*

14. Which stem has the most leaves? *3*

15. What ages are represented by stem 4? *40, 41, 43, and 44*

The Community Bible Chapel compiled a list of the ages of the missionaries it supports. Each piece of data represents the oldest person in each missionary family.

ages of missionaries: 44, 53, 41, 37, 39, 44, 30, 40, 60, 32, 43, 28, 30, 61, 35, 57, 35, 30, 38

9. *28, 30, 30, 30, 32, 35, 35, 37, 38, 39, 40, 41, 43, 44, 44, 53, 57, 60, 61*

330   Chapter 15

What is the title of this survey? Enrollment at Carter's Christian School

What data did Carter collect? the number of students in each of the 8 grades

- Write the data for display to the left of the graph: 20, 17, 21, 19, 22, 18, 20, 9.

- Add the prepared stem-and-leaf-plot term and its description from the *Types of Graphs* page to the displayed word bank from the previous lesson. Explain that a stem-and-leaf plot is a way to display the frequency of data by using the actual digits from the data to record the results. Each piece of data is displayed using the tens digit as the stem and the ones digit as a leaf.

What tens digits from each piece of the data are used as the stem values in the graph? 0, 1, 2; The digit in the Stem column indicates the number of tens in each number in the data; 0 in the Stem column shows that the tens place has no value for the number 9.

Why are there no numbers greater than 2 in the Stem column? The data that Carter collected had no values containing more than 2 tens.

- Choose a student to count the number of leaves or pieces of data recorded in the Leaf column. 8

Why do you think there are 8 digits in the Leaf column? There is 1 leaf or digit for each piece of data (each number) that Carter recorded on the graph.

Use the stem-and-leaf plot to find the answer.

Many missionaries go on deputation, which is a time of visiting churches to raise prayer and financial support. This stem-and-leaf plot shows the number of months some missionaries have been on deputation.

Months of Deputation	
Stem	Leaf
1	3 5 6 7 7
2	4 5 5 5 7 9 9
3	0 6 7
4	4
5	5 7
6	3

Key $1|3 = 13$

16. What is the greatest amount of time spent on deputation? *63 mo*

17. What is the least amount of time spent on deputation? *13 mo*

18. What is the range of the data? *63 − 13 = 50*

19. A total of 584 mo are represented on the stem-and-leaf plot. Find the mean for this data. Round to the nearest tenth. *584 ÷ 19 ≈ 30.7*

20. What is the median? *27*

21. What is the mode? *25*

Practice & Application

Emergency Service Calls		
Year	Fox Hills Squad	Cool Springs Squad
2016	515	589
2017	522	513
2018	533	506
2019	516	533
2020	559	509

22. Make a double line graph showing the number of service calls over time for each emergency squad. Begin the numerical data with 500 and use intervals of 10. Use the line graph to find the answers for problems 23–24.

23. What is the range, the mean, and the median for the Fox Hills Squad data? *range: 559 − 515 = 44; mean: 529; median: 522*

24. What is the range, the mean, and the median for the Cool Springs Squad data? *range: 589 − 506 = 83; mean: 530; median: 513*

25. 14.86 × 27 *401.22*

26. 3.79 × 5.2 *19.708*

27. Find the quotient of 495 ÷ 0.34. Round to the nearest hundredth. *1,455.88*

28. Find the quotient of 733 ÷ 6.4. Round to the nearest thousandth. *114.531*

29. $1\frac{3}{4} \times \frac{5}{8}$ $\frac{7}{4} \times \frac{5}{8} = \frac{35}{32} = 1\frac{3}{32}$ *Answers are shown using cancellation.*

30. $\frac{2}{3} \times 3\frac{1}{2}$ $\frac{2}{3} \times \frac{7}{2} = \frac{7}{3} = 2\frac{1}{3}$

31. $\frac{5}{8} \div \frac{1}{2}$ $\frac{5}{8} \times \frac{2}{1} = \frac{5}{4} = 1\frac{1}{4}$

32. $5\frac{1}{6} \div \frac{5}{12}$ $\frac{31}{6} \times \frac{12}{5} = \frac{62}{5} = 12\frac{2}{5}$

33. Write the definitions for the terms *mean*, *median*, and *mode*. List your math test grades for this quarter. Find the mean, the median, and the mode of your test grades. *Answers will vary.*

34. How can I organize my math test scores? *I can list my math scores on a stem-and-leaf plot.*

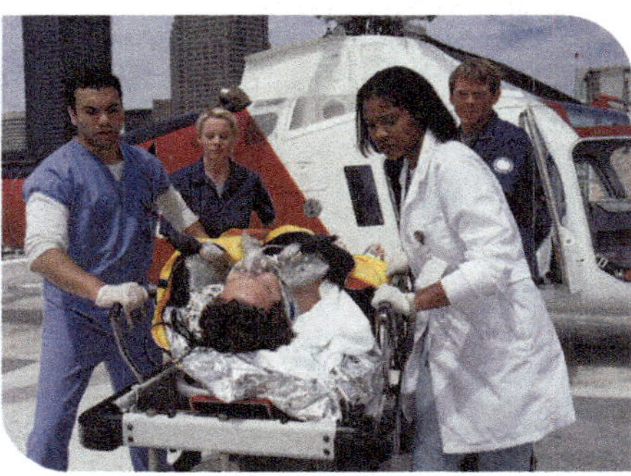

Lesson 149   331

---

find the middle leaf. If the number of leaves is even, I can count from the top leaf down and from the bottom leaf up to find the middle 2 leaves and then average the numbers represented by the 2 middle leaves (digits).

How many leaves are in the stem-and-leaf plot? 8

Since there is an even number of leaves (digits), what 2 numbers must you average? 19 and 20; 9 and 0 (the fourth and fifth leaves) are part of the numbers 19 and 20; there are 3 leaves (digits) representing values that are less than the 2 middle leaves and 3 leaves (digits) representing values that are greater than the 2 middle leaves.

- Point out that since 19 and 20 are consecutive numbers, they can determine that the median for the set of data is 19.5 without calculating the average of 19 and 20. Write "19.5" as the median.

What is another word for the average of a set of data? mean

How do you find the mean? I add all of the data (the numbers of students in all of the grades) and divide the sum by 8 (the number of grades or addends).

- Direct the students to find the mean. $(9 + 17 + 18 + 19 + 20 + 20 + 21 + 22) ÷ 8 = 146 ÷ 8 = 18.25$

What is the mean for this set of data? 18.25

What does a mean or average of 18.25 students per class mean? $18\frac{1}{4}$ students are enrolled per class; although there is not $\frac{1}{4}$ of a student in each class, the total enrollment (146) was not equally distributed among the 8 classes; 146 is not divisible by 8.

- Remind the students that the range, mean, median, and mode can be referred to as statistics for the set of data.

Completing a stem-and-leaf plot

- Guide the students to **collaborate** to compile statistics for a set of data by using a stem-and-leaf plot. Read aloud the following word problem.

- Choose a student to explain the key. sample explanation: "1|7 = 17 students" shows that 1|7 indicates 1 ten and 7 ones or 10 + 7 or 17 students.

What is the least number and the greatest number of students in a grade at Carter's Christian school? 9 and 22

What is the term given for the difference between the greatest value and the least value in a set of data? range

What is the range of the number of students in a grade? 13; 22 − 9 = 13

- Remind the students that the range (13) measures the variability or the span of the data (how spread out it is). Direct them to write "13" as the range on their page as you write it for display. Instruct

them to write the mode, median, and mean as you write them for display during the following discussion.

Which number of students in each grade occurred most often? 20; There were 20 students in 2 different grades.

What term is used to refer to the value that occurs most often or has the greatest frequency in a set of data? mode

Write "20" as the mode.

What is the middle value in a set of data? the median

How do you think you can find the median for the digits in a stem-and-leaf plot? If the data has an odd number of leaves, I can count from the top leaf down and from the bottom leaf up to

LESSON 149

Lennae asked each of her friends to record the ages of their siblings. Then she compiled the following data: 3, 8, 14, 2, 5, 10, 14, 16, 4, 10, 8, 7, 4, 15, 24, 20, 10, 9.

- Write the given data to the left of the blank graph at the bottom of your page for display. Direct the students to write the data on their copy of the page.

  What could you title this stem-and-leaf plot? sample answer: Ages of Friends' Siblings

  Direct the students to write the title above the stem-and-leaf plot as you write it for display. Instruct them to complete each of the following steps on their own page with a partner or group. Give guidance as needed.

- Direct the students to organize the data from least to greatest. Choose a student to write his organized list for display. 2, 3, 4, 4, 5, 7, 8, 8, 9, 10, 10, 10, 14, 14, 15, 16, 20, 24

  What digits should you write in the Stem column to represent the tens values in the data? 0, 1, and 2; There are data that have no tens, 1 ten, and 2 tens.

- Direct the students to write "0," "1," and "2" in the Stem column as you write the stem values for display.

The digits representing the stem values may be written in ascending or descending order.

How many leaves will be on this stem-and-leaf plot? 18; There will be 1 leaf or digit for each piece of data (each number) in the list of data.

- Instruct the students to write the leaves on the graph in order from least to greatest for each stem. Choose students to enter the data on your page for display.

- Direct the students to make a key similar to the one at the top of the page. Remind them that the key is an example of how to read the stems and leaves. Answers will vary.

  Choose several students to write their key for display and explain it. Write one of the keys for display below the graph.

Write the percent as a decimal and as a fraction in lowest terms.

1. 53% *0.53;* $\frac{53}{100}$ **2.** 8% *0.08;* $\frac{2}{25}$ **3.** 70% *0.70 or 0.7;* $\frac{7}{10}$

Write the ratio as a percent.

4. $\frac{8}{100}$ *8%*   **5.** 20 : 100 *20%*   **6.** 5 per 100 *5%*

Write the decimal as a percent. Annex 0s as needed.

7. 0.01 *1%*   **8.** 0.1 *10%*   **9.** 0.69 *69%*

Write the percent as a fraction with a denominator of 100. Rename the fraction in lowest terms.

10. 50% $\frac{50}{100};$ $\frac{1}{2}$ **11.** 6% $\frac{6}{100};$ $\frac{3}{50}$ **12.** 10% $\frac{10}{100};$ $\frac{1}{10}$

Solve.

As part of a class project, Daniel surveyed 40 people to find out whether they preferred basketball or baseball.

Sport	Tally	Frequency
baseball	卌 卌 II	12
basketball	卌 卌 卌 卌 卌 III	28

13. What percent of the people preferred baseball? *30%*

Ages of Friends' Siblings

Stem	Leaf
0	2 3 4 4 5 7 8 8 9
1	0 0 0 4 4 5 6
2	0 4

Key   1|0 = 10 years old

What is the age of the youngest sibling? 2
What is the age of the oldest sibling? 24
Point out that these numbers measure the variability of the data.

What is the difference between the greatest value and the least value in a set of data called? the range
What is the range of the siblings' ages? 22; 24 − 2 = 22

Direct the students to write "22" as the range as you write it for display.

- Instruct the students to continue to collaborate to use the stem-and-leaf plot to find the mode, median, and mean (rounded to the nearest hundredth) and to record the answers below the graph on their copy of the page. Choose students to write the answer on your copy for display.
  mode 10, median 9.5, mean 10.17; 183 ÷ 18 ≈ 10.17

- Remind the students of the essential question that they brainstormed at the beginning of the lesson.

  What is another way you can organize your math test scores? I can list my math scores on a stem-and-leaf plot.

Apply

Student Edition pages 330–31

- Read and explain the directions for pages 330–31. Assist the students as they complete the pages independently.

Daily Review

- Students should complete Chapter 15, section *c*.

NOTES

Student Edition pages 332–33
Daily Review Chapter 15, section *d*

OBJECTIVES

- Interpret a line plot.
- Record data on a line plot.
- Determine the effects of an outlier.

TEACHER RESOURCES

- 99 *Types of Graphs*, page 3
- 101 *Line Plot* (for the teacher and for each student)

ADDITIONAL MATERIALS

- a 3 × 5 card (for each student)

ASSESSMENT

- Chapter 15 Quiz 1

Engage

- Assign a **Ticket in the Door** to assess the students' prior knowledge of the term *outlier* with reference to a line plot. Distribute a 3 × 5 card to each student. Direct their attention to the essential question at the top of Student Edition page 332, "What is an outlier?" Instruct the students to answer the question on their 3 × 5 card ticket. Collect the tickets and read several answers. Explain that they will learn what an outlier is in today's lesson.

Instruct

Interpreting a line plot

- Use a **guided activity** to help the students organize data using a line plot. Distribute a copy of the *Line Plot* page to each student, and display your copy. Explain that a line plot is another type of graph that is used to record data. An *X* is placed above a number on a number line each time that number, which indicates a piece of data, appears in the set of data. A line plot for a set of data helps the user to quickly identify the frequency of the data. Add the prepared line plot term and its description from the *Types of Graphs* page to the displayed word bank from the previous lessons.

Line Plots

What is an outlier?

A **line plot** uses the range of data as its **scale**. Each piece of data is indicated by an X above the number it represents.

A **cluster** is a tight grouping of data on the line plot.

There is a cluster from 1 mi to 5 mi.

A **gap** is an empty space with no data on the line plot.

There is a gap from 6 mi to 7 mi.

An **outlier** is a piece of data that is much greater or much less than the other data.

The outlier is 8 mi jogged.

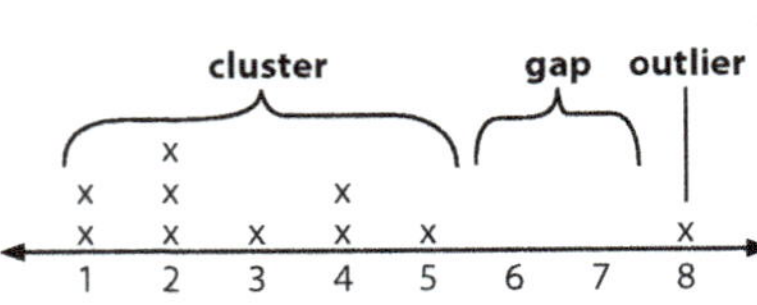

Key Terms
- line plot
- scale
- cluster
- gap
- outlier

Exercises

Use the line plot to find the answer.

Each track team member kept a record of how many cups of water he drank daily. Jeremy collected the data and made this line plot.

1. How many members are on the team? *19 members*
2. How many members drank 6 c of water per day? *3 members*
3. Name and explain the outlier. *1; 1 is much less than the other data.*
4. How do you determine the range? *Find the difference between the greatest value and the least value.*
5. What is the range? *9 − 1 = 8*
6. What is the mode? *5*
7. What is the median? *5*
8. Find the mean. Round to the nearest tenth. *The sum of the data is 103; 103 ÷ 19 ≈ 5.4*

Use the line plot to find the statistic, excluding the outlier. Round to the nearest tenth if necessary.

9. range *9 − 4 = 5*   10. mode *5*   11. median *5*   12. mean *The sum of the data is 102; 102 ÷ 18 ≈ 5.7*

13. Analyze how the range, the mode, the median, and the mean were affected by excluding the outlier. *Without the outlier, the median and the mode remained the same, the range was less, and the mean was greater.*

332   Chapter 15

- Read aloud the following scenario for the graph at the top of the *Line Plot* page.

The Sunday school teacher recorded the number of verses that each of her students memorized for the memory verse contest.

How many students are in the Sunday school class? 21; Each *X* on the line plot represents 1 student.

What is the least number of verses that any student memorized? 1

What is the greatest number of verses memorized by any student? 15

How can you find the range of the number of verses that were memorized? I can find the difference between the greatest number and the least number of verses that were memorized; 15 − 1.

What is the range? 14

Guide the students as they write the range in the blank just below the line plot. Remind them that the range shows the span or the variability of the data.

Throughout this lesson, direct each student to write the answers on his page as you write them for display.

- Point out the scale of the line plot (1–15). Explain that the scale for a line plot must include the range of the data being recorded.

What is the mode for this set of data? 7; Since the mode is the number that occurs most frequently in a set of data,

---

Make a line plot showing the number of museum visitors. Use the line plot to find the answer.

> **Museum Visitors**
> 16, 17, 19, 21, 20, 25, 20, 21, 20, 18, 20, 21, 21, 19, 18, 21, 16, 18, 19

14. How many days were there 21 visitors? *5 days*

15. What is the mode? *21*

16. Where is the cluster on the line plot? *16 to 21*

17. Find the mean. Round to the nearest tenth. *19.5*

18. Which number is the outlier? *25*

19. What is the median? *20*

20. Where is the gap on the line plot? *22 to 24*

21. What is the range of the data? *25 − 16 = 9*

22. Find the range, the mean, the median, and the mode of the data, excluding the outlier. Round the mean and the median to the nearest tenth.

22. range: 21 − 16 = 5;
mean: sum of data = 345; 345 ÷ 18 ≈ 19.2;
median: 19.5; mode: 21

Practice & Application

Complete the table. Use the table to find the answer.

> Sports activities, such as baseball, soccer, swimming, and tennis, all use math in some way. Information such as scores, time, or distance is recorded. These statistics can be analyzed to evaluate individual or team performance.

Soccer Team Records			
Season	Games Won	Games Lost	Percent Won
2016	10	5	*67%*
2017	8	6	*57%*
2018	12	5	*71%*
2019	11	6	*65%*
2020	10	8	*56%*

23. In which season did the soccer team have its best record? *2018*

24. In which season did the team have its worst record? *2020*

27. Answers may vary; touchdown + field goal; touchdown + extra point + safety; field goal + field goal + field goal

Solve.

25. Write a ratio that shows that Jasper made 9 of 15 attempted field goals. What percentage of his attempted field goals did Jasper make? $\frac{9}{15} = 9 ÷ 15 = 0.6; 0.6 \times 100 = 60\%$

26. C. J. made 12 of his 19 field goal attempts. Who has a better success percentage, Jasper or C. J.? (Use the solution from problem 25.) $\frac{12}{19} = 63\%; 63\% > 60\%; C. J.$

27. Sherman High scored 9 points in the first half of the football game. Use the table to list all the possible ways the team could score 9 points. An extra point can be scored only immediately after a touchdown.

Football Points			
touchdown	6	field goal	3
extra point	1	safety	2

28. Make a line plot and a line graph to show the consecutive science test scores for Rico. Explain which graph allows you to better visualize how his grade changes with each test.

> **test scores:** 100, 98, 96, 92, 95, 89, 90, 95, 80, 93, 91

The line graph shows change over time.

29. What is an outlier?
An outlier is a piece of data that is much greater or much less than the other data.

- Point out that in this set of data the mode (7), median (7), and mean (6.67) are all close in value.

  Where do you notice a grouping of the majority of the data? The majority of the Xs are grouped above the numbers 4 to 9 on the line plot.

- Explain that a large group of data that are close together on a line plot is called a *cluster*.

  What does the cluster on this line plot indicate? The grouping of 2 or more Xs above each number from 4 to 9 indicates that 2 or more students memorized anywhere from 4 to 9 verses per student.

  Where do you notice an empty space on the line plot? 12–14

  What do you think this empty space indicates? No data was recorded for 12–14; no students memorized 12, 13, or 14 verses.

- Explain that a space in which no data has been recorded on a line plot is called a *gap*.

- Point out the X above 15. Explain that this piece of data is an *outlier*, a value that is much greater than or much less than the values of the other data and is separated from the other data by a gap.

  Remind the students of the essential question for this lesson.

  Is an outlier what you thought it would be? Answers will vary.

  How can the parts of the word *outlier* help you to remember its meaning? An outlier is a piece of data that lies outside of the area in which the majority of the related data in a set lies.

Recording data on a line plot; determining the effects of an outlier

- Use a **guided activity** to help the students determine the effects of an outlier on a set of data.

  Do you think that an outlier in a set of data will affect the range, mode, median, and mean of the data? Answers will vary.

- Guide the students as they find the range, mode, median, and mean for the number of verses memorized without the outlier. Allow them to use a calculator if needed.

---

the number above which the most Xs are recorded (7) indicates the greatest frequency (i.e., the greatest number of students memorized 7 verses); 4 students memorized 7 verses.

Write "7" as the mode.

- Review the meaning of the median, the middle value of a set of data that has been ordered from least to greatest.

  What number is the median? 7; Xs represent the number of verses memorized by each of 21 students; there are 10 Xs above the numbers that are less than 7 (1–6) and 10 Xs above the numbers that are equal to or greater than 7; the middle X is above 7.

Write "7" as the median.

- Review the meaning of the mean, the average for a set of data.

  How can you find the mean? I add all of the data (the numbers of verses that were memorized) and divide the sum by 21 (the number of students or addends).

Direct the students to find the mean for the set of data. Instruct them to round the answer to the nearest hundredth. $140 ÷ 21 ≈ 6.666; 6.67$

Write "6.67" as the mean on the displayed page.

What does the mean of 6.67 indicate for this set of data? If the verses memorized were equally distributed among the 21 students, each student would have memorized $6\frac{67}{100}$ verses.

LESSON 150

What is the range without the outlier?
10; 11 − 1 = 10

Write "10" as the range without the outlier.

Did omitting the outlier affect the mode?
No; the mode is still 7; the outlier does not affect the number with the greatest frequency.

Write "7" as the mode.

What is the median without the outlier?
6.5

What is the mean? 6.25; 125 ÷ 20 = 6.25

Write "6.5" as the median and "6.25" as the mean.

What do you notice about the statistics when they were calculated without the outlier? The mode remained the same, but the range, median, and mean decreased.

- Point out that in the set of verses-memorized data, the outlier has a greater value than the other data. The range decreased when the outlier was omitted because omitting it reduced the span of the data; and the median and the mean decreased when they were determined without the outlier. The mode was not affected by the outlier because the outlier does not occur most frequently.

  Point out that omitting outliers allows the statistics to better represent the data.

- Guide the students as they list the science test scores in the following situation. Read aloud the following scenario.

The teacher recorded the following scores for a science test: 90, 93, 95, 90, 81, 90, 95, 91, 90, 92, 93, 94.

What is the range of the test scores? 14; The difference between the highest score (95) and the lowest score (81) is 14; 95 − 81 = 14.

Write "14" as the range.

- Direct attention to the blank line plot at the bottom of the *Line Plot* page. Remind the students that the scale for a line plot should include the range of the data.

What do you think the scale for this line plot should be? Accept any correct scale; the scale must include the range of the data (the science test scores).

Write the equivalent measurement.

1. 1 ft = __ in.   **2.** 1 mi = __ ft   **3.** 1 gal = __ qt   **4.** 1 tn = __ lb   **5.** 1 pt = __ c   **6.** 1 lb = __ oz
    12   5,280   4   2,000   2   16

Rename the units.

7. 18 in. = __ ft   **8.** 12 ft = __ yd *4*   **9.** 2 tn 1,280 lb = __ lb   **10.** 24 fl oz = __ c *3*
   $1\frac{1}{2}$ *or 1.5*   *5,280*

Solve.

11.   1 ft 11 in.   **12.**   3 lb 12 oz   **13.**   1,760 yd   **14.**   3 gal 1 qt
    + 2 ft 16 in.   −   20 oz   +   845 yd   − 1 gal 2 qt
    3 ft 27 in. =   *2 lb 8 oz*   *2,605 yd*   *1 gal 3 qt*
    5 ft 3 in.

15. yards in $\frac{1}{2}$ of a mile *880 yd*   **16.** feet in $\frac{2}{3}$ of a yard *2 ft*   **17.** inches in $\frac{1}{4}$ of a foot *3 in.*

18. Mother used $2\frac{1}{2}$ lb of ground beef to make meatloaf. How many ounces were left from the 3 lb package? *$3 lb − 2\frac{1}{2} lb = \frac{1}{2} lb; \frac{1}{2} \times 16 oz = 8 oz$*

19. Claire placed six 18 in. pieces of ribbon across her bulletin board. How many yards of ribbon did she use? *6 × 18 in. = 108 in.; 108 in. ÷ 36 = 3 yd of ribbon*

20. Jordan cut an 8 ft board into 3 equal pieces. How many inches long were the pieces?
8 × 12 in. = 96 in.; 96 in. ÷ 3 = 32 in.

- Guide the students as they write the scale 81–95 for the line plot.

- Direct the students to record each test score as you record it on the displayed page. Point out that the Xs need to be uniform in size to easily compare the data.

How many students took the science test? 12; Each X represents 1 student.

Where do you see a cluster in the line plot? 90–95

Is there a gap in the line plot? Where? yes; from 82 to 89

What is the outlier? 81

What is the mode? 90; 90 is the score above which the most Xs were recorded; 4 students earned a score of 90 on their test.

Write "90" as the mode.

- Instruct the students to find the median and the mean for the set of data, rounding the mean to the nearest hundredth.

What is the median? 91.5; The 2 middle scores are 91 and 92; 5 scores (Xs) are less than 91 and 5 scores are greater than 92. The average of 91 and 92 is 91.5; therefore, the median is 91.5.

Write "91.5" as the median.

What is the mean? 91.17; The sum of all the test scores is 1,094; 1,094 divided by 12 (the number of scores) is 91.17.

Write "91.17" as the mean.

Since the outlier is a lesser value than the other data, how do you predict omitting it will affect the statistics for this set of data? sample answer: The mode will remain the same because its frequency is not affected by the lesser outlier; the range will be greatly decreased, because the outlier is so far from the cluster of the remaining data; and the median and mean will increase when they are determined without the lesser outlier.

- Guide the students as they find and record the range, mode, median, and mean without the outlier (81).

What is the range? 5

What is the mode? 90

What is the median? 92

What is the mean? 92.09

Were your predictions about calculating the statistics without the outlier correct? Yes; as predicted, the mode remained the same, the range decreased, and both the median and the mean increased.

Which statistic showed the greatest change? the range

What change in the range was shown? The range changed from 14 with the outlier to 5 without the outlier.

Apply

Student Edition pages 332–33
- Read and explain the directions for pages 332–33. Assist the students as they complete the pages independently.

Daily Review
- Students should complete Chapter 15, section *d*.

Assess

Quiz 1
- Use the **summative assessment** to evaluate the students' progress at this point in the chapter.

NOTES

When there is a greater amount of data to graph, **histograms** are used instead of the typical bar graph. In a histogram, the data is separated into equal **intervals** on the horizontal axis. The vertical axis is used to show the frequency. A histogram always uses vertical bars with no spaces between them. A frequency table helps to organize the data before the histogram is constructed.

Key Terms
- histograms
- interval

Since 1983, May has been declared as Fitness Month. To encourage physical fitness, Mr. Hawkins assigned the students in his physical education class to keep track of the number of miles they ran during the first week of May. He posted each student's results at the end of the week.

miles run: 8, 7, 12, 7, 1, 5, 10, 2, 3, 10, 12, 9, 4, 6, 9, 4, 3, 7, 8, 5, 7, 4, 8

P.E. Class's First Week of Miles Run

Interval	Tally	Frequency
1–2	||	2
3–4	⊮	5
5–6	|||	3
7–8	⊮ ||	7
9–10	||||	4
11–12	||	2

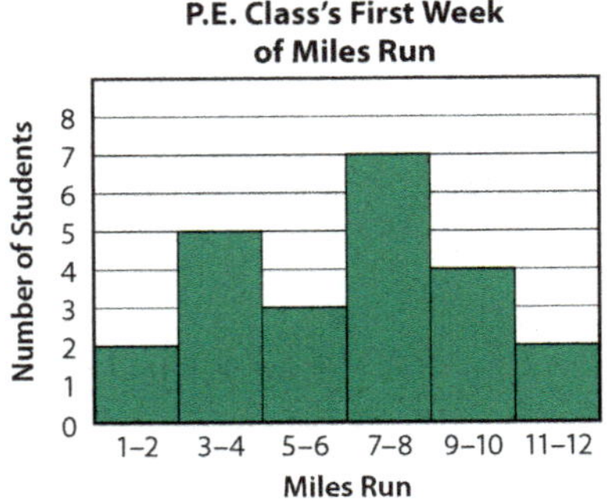

Exercises

Use the frequency table and the histogram to find the answer.

1. What data is represented by the frequency table and the histogram? *running totals for one week*

2. How many students ran 1 to 6 mi? *10 students*

3. How many students are in the physical education class? *23 students*

4. The largest group of runners is represented in which interval? *7–8*

5. What is the range of miles run? *12 − 1 = 11*

6. How many students ran 9 mi or more? *6 students*

7. How many students ran 3–4 mi in the first week of May? *5 students*

8. What interval is the mode for this set of data? *7–8*

9. How many more students ran in the 7–8 mi interval than in the 1–2 mi interval? *5 more students*

10. Brodie ran $9\frac{1}{4}$ mi. Which interval is he in? *9–10*

OBJECTIVES

- Explain how statistics help people make predictions. **BWS**
- Interpret a histogram.
- Complete a frequency table using given data.
- Construct a histogram using given data.

BIBLICAL WORLDVIEW SHAPING

- **Service (Explain):** Statistics show patterns and trends that help people make predictions.

TEACHER RESOURCES

- 99 *Types of Graphs*, page 4
- 102 *Restaurant Statistics*
- 103 *Histogram: Piano Practice Statistics*
- 104 *Histogram* (for the teacher and for each student)

ADDITIONAL MATERIALS

- a ruler (for the teacher and for each student)

Print the *Histogram* page at 100% scale so the histogram will be the size specified during the instruction.

Engage

- Direct attention to the essential question at the top of Student Edition page 334, "How do statistics help people make predictions?" Read aloud the following scenario and direct the students to **turn and talk** to explore the answer.

Suppose that it is 5:00 p.m. and your family is really hungry! You are wondering how long you might have to wait to get a table at your favorite restaurant.

How can statistics help you? Accept any reasonable answers.

Allow the students to share their ideas before continuing with the following activity.

Instruct

Using statistics to help make predictions

- Guide a **visual analysis** of a graph to prepare the students to answer the essential question. Display the *Restaurant Statistics* page.

Instead of displaying the *Restaurant Statistics* page, you may display a popular-times graph from a digital map app to show the wait times from a local restaurant and adjust the following discussion. If the frequency of the data (wait time in minutes) is not given, provide estimates.

What data does this graph present? statistics about the busiest times at a popular restaurant

Why would a restaurant want you to know when their restaurant is more busy or less busy? sample answer: It helps the restaurant to serve more customers during slow periods or to attract customers who want to arrive at a convenient time to avoid a long wait. The graph also conveys the popularity of the restaurant and can be used as an advertisement for it.

Use the set of data to complete the frequency table. Use the table to make a histogram. Use the frequency table and the histogram to find the answer.

History Test Scores
79, 82, 86, 91, 97, 80, 87, 93, 91, 90, 89, 81, 83, 95, 92, 88, 85, 73, 68, 81, 76, 70, 85, 73, 83, 79

History Test Scores		
Interval	Tally	Frequency
60–69			*1*
70–79			*6*
80–89			*12*
90–99			*7*

11. How many students took the history exam? *26 students*

12. Ashton made a 90.3 on the test. In which interval is his test grade? *90–99*

13. What is the range of test scores? *97 − 68 = 29*

14. Which interval has the highest frequency? *80–89*

15. How many intervals are on the histogram? *4*

16. What interval is the mode? *80–89*

17. How many students scored 80 or above? *19 students*

18. Mrs. Clater allowed any student who made lower than a 70 to retake the test. How many students took the test again? *1 student*

19. How many students scored 79 or less? *7 students*

Practice & Application

Use the frequency table to find the percent for the set of data.

Tanner surveyed his class to find which after-school activity his classmates preferred most.

Class Preferences for After-School Activities		
Activity	Tally	Frequency
bike riding			6
skateboarding			8
playing sports			11

20. percentage of students that ride bikes $\frac{6}{25} = \frac{n}{100}$; *n = 24; 24%*

21. percentage of students that skateboard $\frac{8}{25} = \frac{n}{100}$; *n = 32; 32%*

22. percentage of students that play sports $\frac{11}{25} = \frac{n}{100}$; *n = 44; 44%*

Use the table to find the answer.

23. Which country has the largest urban population? *Iceland*

24. Which country has the largest rural population? *South Africa*

25. Write the percentages of rural population from least to greatest. What is the median? *7, 14, 21, 25, 48; median: 21*

26. Find the mean of the rural population. *The sum of the data is 115; 115 ÷ 5 = 23*

27. Is there a mode for this data? *no*

% of Population Distribution		
Country	Urban	Rural
United States	79	21
Iceland	93	7
South Africa	52	48
Colombia	75	25
Japan	86	14

28. Create a double bar graph for the table. Use intervals of 10.

29. Explain how a frequency table can help you construct a histogram. *The frequency table helps determine the intervals and organize the data.*

30. How do statistics help people make predictions? *Statistics show patterns and trends in data that help people make predictions.*

Lesson 151　335

---

between the bars. Each bar represents an interval. Add the histogram term and description from the *Types of Graphs* page to the displayed word bank from the previous lessons.

What is the title of this histogram? Piano Students' Weekly Practice

What are the intervals on this histogram? 0–1, 2–3, 4–5, 6–7, 8–9, and 10–11

What do you notice about the intervals? The intervals are of equal size, they are continuous, and they do not overlap.

- Explain that unlike bar graphs that can have horizontal or vertical bars, a histogram always has vertical bars. The horizontal axis shows the data separated into equal intervals, and the vertical axis shows the frequency of the data.

What do the intervals along the horizontal axis of this histogram represent? the number of practice hours

What do the numbers along the vertical axis represent? the number of students practicing the piano

- Direct attention to the bars on the histogram.

What does the first bar illustrate? 2 students practiced 1 hr or less during the week.

How many students practiced 2–3 hr during the week? 5

How many students practiced 4–5 hr? 8

How many students practiced 6–7 hr? 4

How many students practiced 8–9 hr? 2

How many students practiced 10–11 hr? 1

How many piano students does the teacher have? 22; Adding the number of students that were recorded for each interval of practice time gives me the total number of students; 2 + 5 + 8 + 4 + 2 + 1 = 22 students.

How do you think you can find the mode for this set of data? The tallest bar in the histogram indicates the interval of time that the most students practiced (the interval with the greatest frequency).

What interval is the mode for this set of data? 4–5 hr

How many hours did the least number of students practice? 10–11 hr; The 10–11 interval has the shortest bar, showing that 1 student practiced 10–11 hr.

- Direct attention to the horizontal and vertical axis labels and the bars representing the data.

When would you expect to have the least wait time? 4–5 p.m.

How long might you have to wait to be seated? 15 min

When would you expect to have the longest wait time? 7–8 p.m.

How long might your wait be? 60 min (1 hr)

Do you think these times are always precisely true? No; they are only a prediction based on previous wait times.

How could this graph help your family? It could help us predict the best time to arrive to avoid a long wait or predict whether we have time to wait during the restaurant's busy time.

How do statistics help people make predictions? Statistics show patterns and trends in data that help people make predictions.

Explain that the graph they have been examining is called a histogram.

Interpreting a histogram

- Guide a **discussion** to help the students interpret a histogram. Display the *Histogram: Piano Practice Statistics* page. Explain that although it is similar to a bar graph, a histogram, rather than a typical bar graph, is used to graph large amounts of data. The data is separated into equal intervals so there is no space

What is the range of the practice hours in this set of data? **11; The intervals begin with 0 and end with 11; 11 − 0 = 11.**

Completing a frequency table; constructing a histogram

- Use a **guided activity** to help the students construct a histogram. Distribute a copy of the *Histogram* page to each student and display your copy.

  What data is given on the page? **the ages of campers that attended the first week of camp**

- Remind the students that making a frequency table can be used to organize data so that the data is more manageable. A histogram can be constructed using the organized data in a frequency table.

- Direct attention to the frequency table on the page. Explain that since the data represents a continuous range of numbers, they can list the ages in intervals. Knowing the range of a set of data is useful in determining the scale of a histogram and the scale's intervals for graphing the data.

  What is the range of the ages of the campers that attended the first week of camp? **7 years; The difference in the ages of the youngest camper (8) and the oldest camper (15) is 7; 15 − 8 = 7.**

  What do you think the scale of the histogram should be? **The scale should be 8–15; the scale must include the least number to the greatest number in the data.**

- Explain that intervals of a histogram do not need to begin with 0 or 1. The range (7) indicates the greatest number of intervals possible (1 for each year), but 4 intervals of 2 years each can also be used.

  What intervals of 2 years each will include the range of 7 years from 8 to 15? **8–9, 10–11, 12–13, and 14–15**

- Direct the students to write the age intervals in the Age column of the frequency table. Then instruct them to draw a tally beside the appropriate age interval as you cross out each piece of data (each age) listed on the page.

Write the equivalent measurement.

1. 1 m = ___ cm *100*
2. 1 L = ___ mL *1,000*
3. 1 kg = ___ g *1,000*
4. 1 km = ___ m *1,000*

Rename the units.

5. 3 m = ___ cm *300*
6. 7,250 m = ___ km *7.250*
7. 5,000 g = ___ kg *5*
8. 2 L = ___ mL *2,000*

Solve.

9. $\frac{1}{2}$ of a kilometer *500 m*
10. $\frac{1}{4}$ of a meter *25 cm*
11. $\frac{3}{4}$ of a liter *750 mL*

12. 2,500 mL + 1,500 mL = *4,000 mL*
13. 3,417 kg − 2,750 kg = *667 kg*
14. 3 L − 2,750 mL *250 mL*
15. 8,341 g + 978 g *9,319 g*

16. The punch recipe calls for 1 L of orange juice, 2 L of lemon-lime soda, 300 mL of lemonade concentrate, and 1.5 L of water. How much punch does the recipe make? *4.8 L or 4,800 mL of punch*

17. The nurse said Carissa's temperature was normal. What was her temperature in Celsius? *37°C*

You may guide the students to order the data from least to greatest before you instruct them to draw the tallies on the frequency table.

- Instruct the students to count the tallies for each interval and write the total in the Frequency column: ages 8–9 7, ages 10–11 6, ages 12–13 9, ages 14–15 3.

- Use the following procedure to guide the students as they construct a histogram using the data recorded in the frequency table. Demonstrate each step.

  What part of the histogram is used to illustrate the intervals? **the horizontal axis**

How many bars are needed in the histogram? **4; One bar is drawn for each of the 4 age intervals.**

- Guide the students as they use a ruler to mark four 1 in. wide intervals on the horizontal axis (the bottom line of the histogram). Instruct them to write the age intervals below the horizontal axis.

  What label could you write below the histogram to show what the intervals represent? **sample answer: Campers' Ages**

- Direct the students to write the label "Campers' Ages" on the blank below the intervals for the histogram.

What part of the histogram illustrates the frequency? the vertical axis

What label could you write for the vertical axis? sample answer: Number of Campers

- Direct the students to write the label "Number of Campers" on the blank to the left of the histogram.

What could you title this histogram? sample answers: Ages of Campers: Week 1; First Week of Camp Attendance

- Guide the students as they select a title from the answers that were given and direct them to write the title above the histogram.

- Guide the students as they draw bars on the histogram to illustrate the data recorded in the frequency table. Remind them that there are no spaces between the bars of a histogram. Direct the students to shade the bars to complete the histogram.

What is the total number of campers that attended the first week of camp? 25; When the numbers of campers that were recorded for each age interval are added, the total number of campers is 25; 7 + 6 + 9 + 3 = 25 campers.

Which age group had the greatest number of campers during the first week of camp? the campers that were 12–13 years old

What age group had the fewest number of campers during that week? the campers that were 14–15 years old

- Choose a student to tell whether the following statement is a true statement and to explain his answer: More than half of the campers that attended the first week of camp were 8–11 years old. Yes, the statement is true; 13 of the campers were 8–11 years old, and 13 is more than half of the total number of campers (25); $\frac{1}{2}$ of 25 = 12.5.

Apply

Student Edition pages 334–35

- Read and explain the directions for pages 334–35. Assist the students as they complete the pages independently.

Daily Review

- Students should complete Chapter 15, section *e*.

NOTES

LESSON 152

Student Edition pages 336–37
Daily Review Chapter 15, section ƒ

OBJECTIVES

- Interpret a box-and-whisker plot.
- Determine the lower, middle, and upper quartiles of a set of data.
- Construct a box-and-whisker plot using given data.

TEACHER RESOURCES

- 99 *Types of Graphs*, page 4
- 105 *Box-and-Whisker Plot: Points Scored*
- 106 *Box-and-Whisker Plot* (for the teacher and for each student)

ADDITIONAL MATERIALS

- a ruler (for each pair of students)

Engage

- Direct the students to do a **Think-Pair-Share** to explore the essential question at the top of Student Edition page 336, "What is a whisker?" Take a poll to determine which answer most of the students agree with, then record the answer for display. Revisit the answer at the end of the lesson and take another poll to see if the students still agree with the answer or if they want to amend it.

Instruct

Interpreting a box-and-whisker plot

- Guide a **discussion** to help the students interpret a box-and-whisker plot. Choose a student to list the points that Joel scored (2–16) for display as you slowly read aloud the values in the following situation.

- Read aloud the following scenario.

  Joel's basketball team played in 7 games of a basketball tournament. The basketball coach recorded the number of points that Joel scored in each of the 7 games. The points that the coach recorded for Joel were 2, 4, 7, 10, 11, 12, and 16.

Box-and-Whisker Plots

What is a whisker?

A **box-and-whisker plot** summarizes data on a number line, using a list of data that is organized from least to greatest.

Use the least value of the data as the beginning point of the number line and the greatest value of the data as the ending point.

Find the median of the whole set of data and plot that point on the number line. This point is the **middle quartile**. The median separates the data into two sets.

Find and plot the median for the lower half of the data. This point is the **lower quartile**.

Find and plot the median for the upper half of the data. This point is the **upper quartile**.

Draw a box from the lower quartile to the upper quartile. Draw a line from the lower quartile to the least value and a line from the upper quartile to the greatest value. These lines are the *whiskers* of the plot.

Key Terms
- box-and-whisker plot
- middle quartile
- lower quartile
- upper quartile

> **number of points scored by the boys' basketball team:** 72, 85, 91, 87, 79, 78, 93

Plot the middle, upper, and lower quartiles.

middle quartile (median of data):
72, 78, 79, ⟨85⟩, 87, 91, 93   85
lower quartile: 72, ⟨78⟩, 79   78
upper quartile: 87, ⟨91⟩, 93   91

Remember to average the two middle values to find the median.

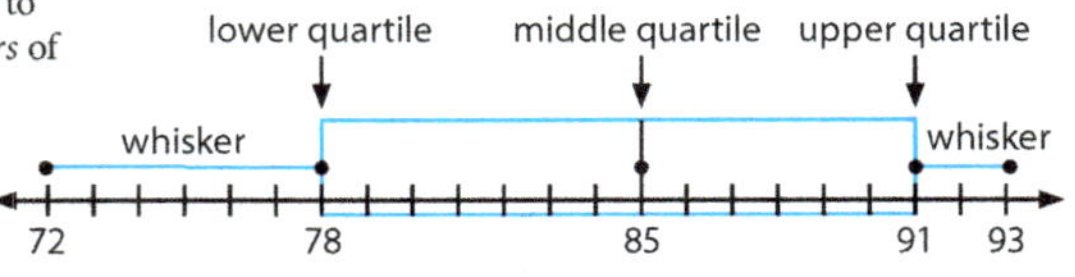

Exercises

Make a box-and-whisker plot. Use the plot to find the answer.

> **ages of the children who attended Jesse's birthday party:** 3, 7, 7, 9, 11, 11, 12, 12, 12, 13, 15

1. In what order is the data? *least to greatest*
2. What is the median of the data? *11*
3. What is the lower quartile? *7*
4. What is the upper quartile? *12*
5. Which side has the longer whisker? *lower whisker*
6. Which box shows the smaller amount of data? *upper quartile*
7. What is the range of the data? *15 − 3 = 12*

> **math test scores listed by Mrs. Peyton on the board:** 83, 91, 95, 80, 100, 88, 92, 98, 100, 85, 93, 90, 75, 92, 85

8. Write the data from least to greatest. *75, 80, 83, 85, 85, 88, 90, 91, 92, 92, 93, 95, 98, 100, 100*
9. What is the median of the data? *91*
10. What is the lower quartile? *85*
11. What is the upper quartile? *95*
12. What is the range of the data? *100 − 75 = 25*
13. Describe the length of the whiskers. *lower: 75–85; upper: 95–100*
14. Is the median in the center of the box? *no*
15. What part of the plot shows the smaller amount of data? *upper quartile*

336   Chapter 15

What is the range of the points that Joel scored? 14; 16 − 2 = 14
Write the range for display.

What is the mode for this set of data? There is no mode; none of the values occurs more frequently than the other values.

How can you find the mean of the points that were scored by Joel? Since the mean is the average of all the points scored, I can divide the sum of the points scored by the number of games that Joel played in (the number of addends).

- Direct the students to find the mean. Instruct them to round the answer to the nearest whole number. 62 ÷ 7 ≈ 8.8; 9 points Write the mean for display.

- Remind the students that when a set of data has been ordered from least to greatest, the median is the middle value or an average of the 2 middle values.

What do you notice about the displayed list of the points that Joel scored? The points are listed in order from least to greatest.

What is the median of the set of data? 10; 10 is the middle value of the set of data; there are 3 numbers of lesser value and 3 numbers of greater value.

Write the median for display.

- Display the *Box-and-Whisker Plot: Points Scored* page. Explain that a box-and-whisker plot is used to summarize data on a number line by using a list of data

Use the box-and-whisker plot to find the answer.

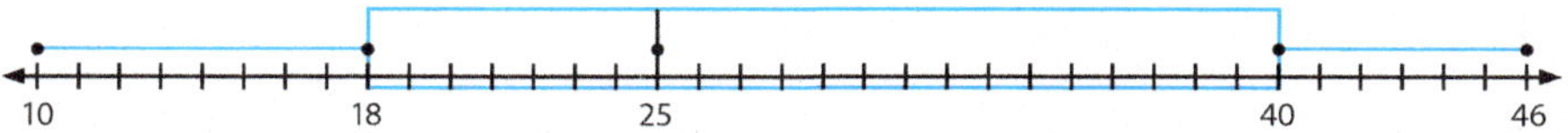

10   18   25   40   46

16. What is the greatest number of miles traveled?
46 mi
17. What are the upper and lower quartiles?
upper: 40; lower: 18
18. What is the least number of miles traveled?
10 mi
19. What is the range of the data?
46 − 10 = 36
20. Find the median of the data. *25*
21. What is true about the length of the whiskers?
The lower whisker is longer than the upper whisker.

Use the data to find the middle, lower, and upper quartiles. Round decimal answers to the nearest tenth. Make a box-and-whisker plot to display the set of data.

22. 5, 8, 19, 21, 23, 27, 30, 31, 33
*middle quartile: 23
lower quartile: (8 + 19) ÷ 2 = 13.5
upper quartile: (30 + 31) ÷ 2 = 30.5*

23.   71, 72, 75, 79, 80, 81, 82
*middle quartile: 79
lower quartile: 72
upper quartile: 81*

Practice & Application

24. Make a stem-and-leaf plot to show the number of pages read by the students. Use the data and the stem-and-leaf plot to find the answers for problems 25–27.

  52, 45, 60, 65, 30, 48, 41, 37, 62, 43

25. Find the mean. Round to the nearest tenth.
The sum of the data is 483; 483 ÷ 10 = 48.3
26. Use the stem-and-leaf plot to find the median.
(45 + 48) ÷ 2 = 46.5
27. What is the range? *65 − 30 = 35*

28. Use intervals of 5 to make a frequency table and a histogram to show the number of miles traveled by families to church. Use the data from the table or the histogram to find the answers for problems 29–31.

  17, 5, 3, 10, 24, 2, 14, 7, 15, 7, 30,
  8, 16, 20, 10, 20, 33, 11

29. What mile interval shows the mode? *6–10*
30. The Chapin family walks $\frac{1}{4}$ of a mile to church. In which interval are the Chapins? *0–5*
31. How many families drive more than 10 mi?
10 families
32. What is a whisker? *The whisker is the line in a box-and-whisker plot that extends from the box to the least or greatest value of the data.*

24.

Pages Read	
Stem	**Leaf**
3	0 7
4	1 3 5 8
5	2
6	0 2 5

Key
3|0 = 30

28.

Miles Traveled to Church		
Miles	**Tally**	**Frequency**
0–5	III	3
6–10	|||||	5
11–15	III	3
16–20	IIII	4
21–25	I	1
26–30	I	1
31–35	I	1

---

that has been organized from least to greatest. A box is drawn around the middle half of the data and whiskers are drawn from the box to the least value of the data and from the box to the greatest value. Add the box-and-whisker term and description from the *Types of Graphs* page to the displayed word bank from the previous lessons.

What is the range of the data listed on the page? 14

Point out that the range of the data helps them to determine the scale and the interval when drawing a number line for a box-and-whisker plot. Note that every mark (tick) on the number line does not need to have a number written below it.

Determining the lower, middle & upper quartiles

- Guide a **visual analysis** of a box-and-whisker plot to help the students identify the lower, middle, and upper quartiles of a set of data.

Explain that besides the least value and the greatest value of the data, 3 other statistics are needed to construct a box-and-whisker plot: the lower quartile, the middle quartile, and the upper quartile. The middle quartile is determined first. It is the median of the whole set of data and separates the data into 2 equal sets.

What is the median of the ordered set of data on the page? 10

Point out the middle quartile (10) in the box-and-whisker plot.

- Explain that the upper and lower quartiles are the medians of the upper and lower halves of the ordered set of data.

What numbers in the ordered set of data are to the left of the middle quartile? 2, 4, and 7

What is the median of 2, 4, and 7? 4

What is the lower quartile? 4; 4 is the median of the lower half of the ordered set of data.

Point out the lower quartile (4) in the box-and-whisker plot.

What numbers in the ordered set of data are to the right of the middle quartile? 11, 12, and 16

What is the median of 11, 12, and 16? 12

What is the upper quartile? 12; 12 is the median of the upper half of the ordered set of data.

Point out the upper quartile (12) in the box-and-whisker plot.

- Explain that the points indicating the middle, lower, and upper quartiles have been drawn above the number line. A vertical line has been drawn at the location of each quartile and horizontal lines have been drawn to form a box with the lower quartile and the upper quartile as its sides.

Point out that the whiskers of the plot were formed by drawing a point above the least value of the set of data (2) and another point above the greatest value of the set of data (16). Then a horizontal line was drawn from the lower quartile to the least value and another horizontal line was drawn from the upper quartile to the greatest value. The value of the points at the end of each whisker can be used to determine the range of the data pictured on a box-and-whisker plot.

What do you notice about the whiskers in this box-and-whisker plot? The upper whisker is longer (12–16) than the lower whisker (2–4).

LESSON 152

Constructing a box-and-whisker plot

- Guide a **collaborative activity** to help the students construct a box-and-whisker plot. Distribute a copy of the *Box-and-Whisker Plot* page and display your copy. Instruct the students to collaborate with a partner as they construct a plot. Direct them to have their ruler available for the activity.

- Use the following procedure to guide the students as they construct a box-and-whisker plot for the given data.
  What data is given on the page? test scores

- Direct the students to find the median of the test scores. 87
  How did you find the median of this set of data? To find the median, I needed to find the average of the 2 middle numbers in the ordered list of data; $(85 + 89) \div 2 = 87$.
  What is the median of the whole set of data called on a box-and-whisker plot? the middle quartile

- Instruct the students to write "87" in the blank as the middle quartile as you write it on the displayed page.

- Direct attention to the term *lower quartile* on the page. Direct the students to find the lower quartile of the scores.
  How do you find the lower quartile? I find the median of the lower half of the data.
  Since 87 is the middle quartile, which test scores are in the lower half of the data? 76, 80, 82, and 85

- Direct the students to find the median of the test scores in the lower half of the data. the average of the 2 middle numbers in the lower half of the data: $(80 + 82) \div 2 = 81$
  What is the lower quartile? 81

- Instruct the students to write "81" as the lower quartile as you write it for display.

- Repeat the procedure to guide the students as they find the upper quartile. $(91 + 95) \div 2 = 93$

- Guide the students as they draw points above the number line for the middle quartile (the median), the lower quartile, and the upper quartile. Then guide them as they use a ruler to draw the vertical lines through each quartile, beginning with the middle quartile, and the horizontal lines to form the box.

Direct the students as they draw the points to show the least value and the greatest value and then draw the whiskers from the box to the values.

What is the range of the test scores? 22; The difference between the greatest value and the least value on the box-and-whisker plot is 22; $98 - 76 = 22$.

What do you notice about the whiskers in this box-and-whisker plot? The whiskers are the same length.

Describe the length of the whiskers. lower whisker: 76–81; upper whisker: 93–98

Is the median in the center of the box? yes

Is the median (the middle quartile) included in the ordered list of data? no

Is the lower quartile included in the ordered list of data? no

Is the upper quartile included in the ordered list of data? no

---

Use the circle graph to find the answer.

The sixth-grade class surveyed 100 students to find their favorite subjects.

1. What percent of students surveyed liked heritage the best? *50%*

2. Of the 100 students surveyed, how many chose math? *25 students*

3. What percent of the students surveyed chose science? *25%*

Favorite Subjects

Mrs. Hancock made a circle graph to show the percentages of the different kinds of flowers in her garden.

4. List the kinds of flowers in order from the largest percentage to the smallest percentage. *rose, daisy/tulip, lily, violet*

Garden Flowers

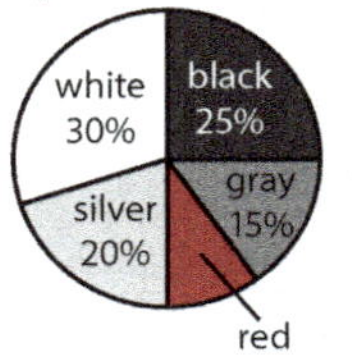

The car dealership made a circle graph of the most popular car colors. The dealership used the information to order new cars.

5. Based on the graph, what color car would the dealership order the most of? *white*

6. If they ordered 100 cars, how many cars would they order in black? *25 cars*

7. Does this graph show how many red vans to order? *no*

8. List the colors in order from the greatest percentage to the smallest percentage. *white, black, silver, gray, red*

Popular Car Colors

Math 6

- Review the list of data at the top of the page.

  If you were examining the box-and-whisker plot and the lists of data were not given, what do you think are the only 2 pieces of data that you can be certain are included in the list of test scores? I could know only that the least value (76) and the greatest value (98) would be included in the list of test scores (the data), because they were the only actual test scores for which points were drawn. The middle, lower, and upper quartiles were all determined by averaging 2 test scores.

- Remind the students of the essential question and of the answer that most of them voted for.

  Is your answer still the same? Answers will vary.

  If not, how would you answer the question now? In statistics, a whisker is the line in a box-and-whisker plot that extends from the box to the least or greatest value of the data.

Apply

Student Edition pages 336–37

- Read and explain the directions for pages 336–37. Assist the students as they complete the pages independently.

Daily Review

- Students should complete Chapter 15, section *f*.

NOTES

Student Edition pages 338–39
Daily Review Chapter 15, section *g*

OBJECTIVES

- Interpret a double bar graph, double line graph, stem-and-leaf plot, line plot, histogram, and box-and-whisker plot.
- Explain how statistics help people make decisions. **BWS**

BIBLICAL WORLDVIEW SHAPING

- **Service (Explain):** Statistics present data so that the information can be understood and can inform decisions.

TEACHER RESOURCES

- 107 *Graph: Double Bar Graph*
- 108 *Graph: Double Line Graph*
- 109 *Stem-and-Leaf Plot & Line Plot*
- 110 *Graph: Histogram*
- 111 *Graph: Box-and-Whisker Plot*

ASSESSMENT

- Chapter 15 Quiz 2

Engage

- Guide the students in a **Think-Pair-Share** to prepare them to answer the essential question at the top of Student Edition page 338, "How can statistics help people make decisions?"

Direct them to choose 1 of the types of graphs they have studied during the chapter and to think of how statistics presented by that type of graph might help someone make a decision.

Instruct

Interpreting a double bar graph

- Guide a **visual analysis** to review interpreting a double bar graph. Display the *Graph: Double Bar Graph* page. Remind the students that a bar graph is used to compare data. A double bar graph compares 2 sets of related or similar data.

Graph Review

How can statistics help people make decisions?

Exercises

Use the histogram to find the answer.

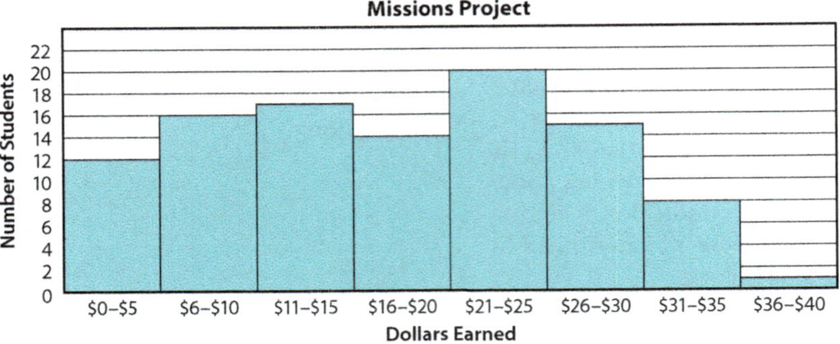

1. How many students helped to earn money for the missions project? *103 students*

2. How many students earned $26–$30? *15 students*

3. What is the mode for the amount earned? *$21–$25*

4. In which interval was the greatest amount of money earned by a student for the missions project? *$36–$40*

5. Explain the data shown on the graph. *The data shows the amount of money earned by 103 students for a missions project.*

Use the double line graph to find the answer.

6. In which game did Luke score his highest number of points? *game 4*
7. In which game did Cal *not* score? *game 1*

8. What is the range of Luke's scores? *9 − 2 = 7*

9. What is the range of Cal's scores? *6 − 0 = 6*

10. In which game did Luke score only one more point than Cal? *game 5*
11. What was the mean of Cal's scores? Round to the nearest hundredth. *The sum of the data is 22; 22 ÷ 6 ≈ 3.67*

12. What was the mean of Luke's scores? Round to the nearest hundredth. *The sum of the data is 37; 37 ÷ 6 ≈ 6.17*

13. What trend do you see in Luke's and Cal's games? *The number of points scored improved as more games were played.*

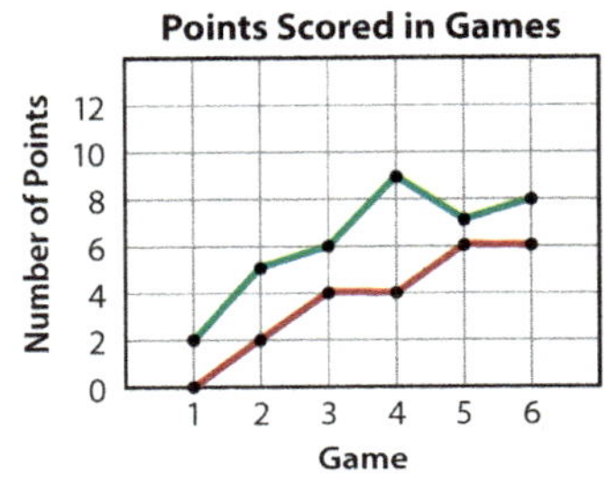

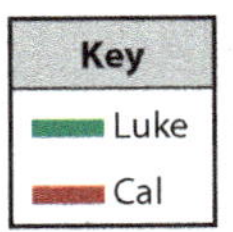

What data is being compared in this double bar graph? The urban and rural populations of countries A, B, C, and D are being compared; sample answers: the key shows that the darker bars indicate urban populations and the lighter bars indicate rural populations; the categories below the bars indicate 4 different countries.

What type of community does the word *urban* refer to? a city

What type of community does the word *rural* refer to? a town or a village in the country

What comparison is shown for country A? 75% of the population lives in an urban area (a city), and 25% of the population lives in a rural area (the country).

What comparison is shown for country B? 90% of the population lives in a city, and 10% of the population lives in the country.

Which 2 countries have the same distribution of urban and rural populations? countries A and D; Both countries have a population that is 75% urban and 25% rural.

Which country has the greatest percentage of rural population? country C

What percentage of country C's population lives in a rural area? 40%

What is the estimated mean of the rural populations of all 4 countries? approximately 25%; Accept any reasonable explanation.

Use the stem-and-leaf plot to find the answer.

14. How many children attended VBS? *21 children*

15. What is the range in ages of the children at VBS?
12 − 6 = 6

16. What is the mode of the data? *11*

17. What is the median of the data? *10*

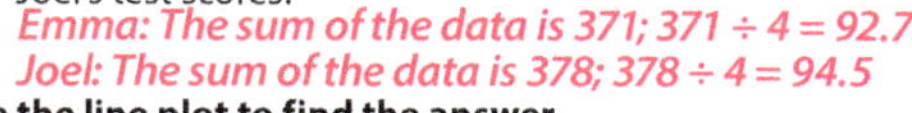

Ages of Children at VBS		
Stem	**Leaf**	
0	6 6 6 7 7 8 8 8 9 9	
1	0 0 0 0 1 1 1 1 1 2 2	

Key 1|0 = 10

Use the double bar graph to find the answer.

18. What is being compared in the graph? *Joel's and Emma's math test scores*

19. On which test did Joel and Emma receive the same score? *test 3*

20. What is the range of the test scores for Joel? for Emma? *Joel: 98 − 88 = 10; Emma: 96 − 90 = 6*

21. Which student had the highest test score? *Joel; he had a score of 98.*

22. What is the mean of Emma's test scores? of Joel's test scores? *Emma: The sum of the data is 371; 371 ÷ 4 = 92.75; Joel: The sum of the data is 378; 378 ÷ 4 = 94.5*

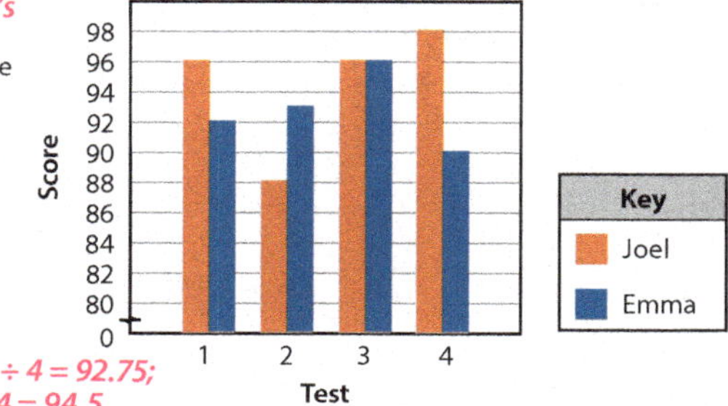

Use the line plot to find the answer.

23. What is the range in the miles walked? *9 − 1 = 8*

24. What is the mode of the data? *1*

25. How many people walked 3 or fewer miles? *9 people*

26. What is the median of the data? *3*

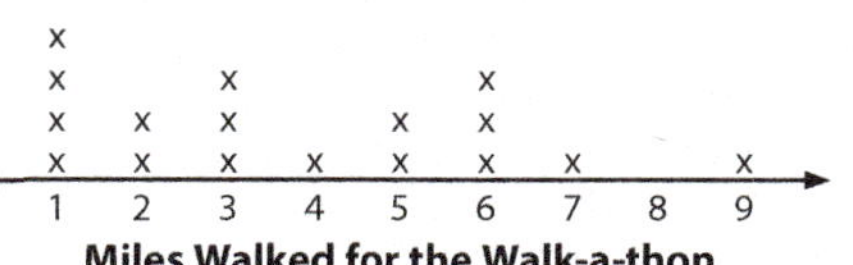

Use the box-and-whisker plot to find the answer.

Mrs. Alier used history test data for her box-and-whisker plot.

27. What is the lowest test score? *78*

28. What is the highest test score? *98*

29. What is the median test score? *86*

30. What is the lower quartile? *81*

31. What is the upper quartile? *93*

32. How can statistics help people make decisions? *Statistics make data understandable so that people can make informed decisions.*

Lesson 153 339

What do you notice about the populations in each of the 4 countries? There is a greater percentage of urban population than rural population in each country.

What decision might someone from country B make after viewing the statistics for their country? sample answers: They might decide to move to the country from the city because of the dense population in their urban area, or they might choose to start a business in a city because of the larger customer base there.

Interpreting a double line graph

- Guide a **visual analysis** to review interpreting a double line graph.

Display the *Graph: Double Line Graph* page. Remind the students that a line graph is used to show changes in data over a period of time. A double line graph compares the changes in 2 sets of data over the same period of time.

What information is recorded on the graph? the monthly precipitation in inches for city Y and city Z from January through August; sample answers: The key shows that the solid line is used to illustrate the changes in the amount of monthly precipitation that city Y received and the dashed line is used to illustrate the changes in the amount of monthly precipitation that city Z

received; the scale indicates that the amounts of precipitation were recorded in inches.

What interval was used for the precipitation scale? Half-inch intervals were used.

During which months did city Y receive the least amount of precipitation? July and August

What amount was recorded? 2 in.

What was the greatest amount of precipitation recorded? 4 in.

When was it received? in March and April

Where was it received? city Y

When was the greatest amount of precipitation received in city Z? March

What amount was recorded? 2 in.

During what month did city Z receive the lowest amount of precipitation? August

What amount of precipitation was recorded? 0.25 in. or a quarter of an inch

What is the mean of the amounts of precipitation city Y received? 3 in.

What is the mean of the amounts of precipitation city Z received? approximately 1.03 in.

Considering this double line graph, what prediction could you make about the amount of precipitation that will be received in these cities in September? sample answers: City Y will receive more precipitation than city Z; both city Y and city Z will show an increase in the amount of rain received because the amount of rain that they received during the previous 2 or 3 mo was below their averages for the past 8 mo.

- Read aloud the following scenario.

Dakota's family is planning a sight-seeing trip to either city Y or city Z.

Which city might they decide to visit based on the statistics? Why? sample answer: city Z, because lower rainfall would be better for sight-seeing

Which month do you think they might decide to visit the city? Why? sample answer: August; The chance of rain is lowest in August.

LESSON 153

Interpreting a stem-and-leaf plot & a line plot

- Guide a **visual analysis** to review interpreting a stem-and-leaf plot and a line plot. Display the *Stem-and-Leaf Plot & Line Plot* page. Direct attention to the stem-and-leaf plot on the displayed page. Remind the students that a stem-and-leaf plot is a way to display the frequency of data by using the actual digits in each piece of data to record the results.

What do the digits in the Stem column represent? the number of tens in each number of the data

What do the digits in the Leaf column represent? the digits in the ones place of each piece of data

What is the range of the employees' ages? 43 yr; The least value on the stem-and-leaf plot is 19 and the greatest value is 62; 62 − 19 = 43.

What age bracket shows the greatest frequency? the 20s; The 8 leaves for the 2 stem indicate that there are 8 employees in their 20s.

What is the median? 34

- Instruct the students to find the mean age of the employees to the nearest year. Guide the students as they recall that the mean is the average age. 773 ÷ 22 ≈ 35.1; The average age is 35.

- Direct attention to the line plot on the page. Remind the students that a line plot is also used to identify the frequency of a set of data.

What is the range of the number of miles that were hiked? 8 mi; The line plot shows that the greatest number of miles hiked was 9 and the least number of miles hiked was 1; 9 − 1 = 8.

What is the mode for the number of miles hiked? 5; The four *X*s above the number 5 show that 5 mi was the distance that was most frequently hiked.

What is the median? 4

What is the outlier on this line plot? 9

Does the outlier affect the mode? No; the mode is still 5.

Does the outlier affect the median? No; the median is still 4.

What kind of decisions might be made based on the data in the stem-and-leaf plot or the line plot? sample answers: An employer might decide what type of activities to include in an employee picnic based on the frequency of the ages represented. An event planner might avoid a hike of greater than 6 mi in the activities schedule because the cluster of the data on the line plot indicates that most of the people represented have not hiked in that range of miles.

Interpreting a histogram

- Guide a **visual analysis** to review interpreting a histogram. Display the *Graph: Histogram* page. Remind the students that histograms are useful when there is a large amount of data to graph. Unlike a bar graph, there is no space in between the bars of a histogram. Each bar represents an equal interval of the data. Point out the consecutive intervals.

What does this histogram illustrate? the number of credits earned by college juniors for the first semester

What were the least credits earned? 1–2

What were the most credits earned? 23–24

What interval is the mode for this histogram? 15–16; The tallest bar indicates the interval of credits that were earned by the most students in the junior class (the interval with greatest frequency).

How many students in the junior class earned 9–10 credits? 16

How many earned 17–18 credits? 15

Write the improper fraction as a mixed number or a whole number.

1. $\frac{4}{3}$ $1\frac{1}{3}$ 2. $\frac{7}{2}$ $3\frac{1}{2}$ 3. $\frac{12}{4}$ 3 4. $\frac{6}{6}$ 1 5. $\frac{9}{4}$ $2\frac{1}{4}$

Solve. Write the answer in lowest terms. *Answer is shown using cancellation.*

6. $\frac{2}{3} + \frac{1}{3} = \frac{3}{3} = 1$

7. $\frac{4}{5} = \frac{8}{10}$; $+\frac{2}{10} = \frac{2}{10}$; $\frac{10}{10} = 1$

8. $6\frac{1}{2} = 6\frac{2}{4}$; $-4\frac{1}{4} = 4\frac{1}{4}$; $2\frac{1}{4}$

9. $5 = 4\frac{3}{3}$; $-2\frac{2}{3} = 2\frac{2}{3}$; $2\frac{1}{3}$

10. $\frac{8}{10} = \frac{4}{5}$; $-\frac{3}{15} = \frac{1}{5}$; $\frac{3}{5}$

11. $3 \times \frac{4}{5}$ $\frac{12}{5} = 2\frac{2}{5}$

12. $1\frac{1}{2} \times 2\frac{3}{6}$ $\frac{3}{2} \times \frac{15}{6} = \frac{15}{4} = 3\frac{3}{4}$

13. $4\frac{2}{8} \times 3\frac{1}{5}$ $\frac{34}{8} \times \frac{16}{5} = \frac{68}{5} = 13\frac{3}{5}$

14. $3 \div \frac{1}{2}$ $\frac{3}{1} \times \frac{2}{1} = 6$

15. $4\frac{1}{5} \div 1\frac{1}{4}$ $\frac{21}{5} \times \frac{4}{5} = \frac{84}{25} = 3\frac{9}{25}$

16. $\frac{6}{8} \div \frac{1}{4}$ $\frac{6}{8} \times \frac{4}{1} = \frac{6}{2} = 3$

17. Jackson filled bags with candy to give to his classmates. He filled each bag with $\frac{1}{4}$ of a pound of candy. He had 3 lb of candy. Would he have enough bags to give to 20 students? $3 \div \frac{1}{4} = \frac{3}{1} \times \frac{4}{1} = 12$; no

18. Missy placed $\frac{3}{4}$ of a yard of ribbon around a bouquet of flowers. She had $5\frac{1}{2}$ yd of ribbon. How many bouquets could she put ribbon around? $5\frac{1}{2} \div \frac{3}{4} = \frac{11}{2} \times \frac{4}{3} = \frac{22}{3} = 7\frac{1}{3}$; 7 bouquets

How many earned 19–20 credits? 12

What does an overall view of the histogram show you? Answers will vary; most of the junior class earned 9–20 credits.

Do you think it is reasonable for a college to decide to require its students to complete at least 13 credits per semester? Why? sample answer: Yes; the average number of credits taken is 15–16, and 13 is fewer than that.

Interpreting a box-and-whisker plot; explaining how statistics help people make decisions

• Guide a **visual analysis** to review interpreting a box-and-whisker plot. Display the *Graph: Box-and-Whisker Plot* page. Remind the students that a box-and-whisker plot summarizes data on a number line. A box is drawn around the middle half of the data and whiskers are drawn from the box to the least value of the data and from the box to the greatest value. Each whisker contains an equal amount of the other half of the data.

What pieces of data are used to construct a box-and-whisker plot? The data used are those having the least and greatest values, the middle quartile (the median of all the ordered data), the lower quartile (the median of the ordered data that are less than the middle quartile), and the upper quartile (the median of the ordered data that are greater than the middle quartile).

What is the lowest recorded score for Science Test 6? 80

What is the highest score? 97

What is the median test score? 90

What is the lower quartile? 81.5

What is the upper quartile? 94

You may point out that the box of this box-and-whisker plot shows that half of the test scores are between 81.5 and 94 and that the whiskers show that the lower one-fourth of the scores are between 80 and 81.5 and the upper one-fourth of the scores are between 94 and 97.

What are the only 2 pieces of data on this box-and-whisker plot that you can be certain are included in the list of recorded test scores? 80 and 97; They are the high and low test scores. The other numbers shown on the box-and-whisker plot are medians. They may be the middle value listed in the lower and upper halves of the data, or they may be the average of the 2 middle values listed in the lower and upper halves of the data.

• Write for display the following list of Science Test 6 scores: 80, 80, 80, 83, 86, 89, 91, 92, 93, 95, 96, 97. Guide the students as they find the middle quartile 90 (the average of 89 and 91), the lower quartile 81.5 (the average of 80 and 83), and the upper quartile 94 (the average of 93 and 95).

Guide the students to conclude that none of the quartiles are included in the list of recorded test scores.

• Guide the students as they use the displayed list of Science Test 6 scores to verify the answers given to the previous questions about the box-and-whisker plot.

What decision might these statistics help the science teacher make? sample answer: He might decide that his test was a fair test for his students.

How can statistics help people make decisions? Statistics make data understandable so that people can make informed decisions.

Apply

Student Edition pages 338–39

• Read and explain the directions for pages 338–39. Assist the students as they complete the pages independently.

Daily Review

• Students should complete Chapter 15, section *g*.

Assess

Quiz 2

• Use the **summative assessment** to evaluate the students' progress at this point in the chapter.

NOTES

Student Edition pages 340–41
Daily Review Chapter 15, section *h*

OBJECTIVES

- Calculate the mean for a set of data.
- Determine the range, median, and mode for a set of data.
- Record data in a frequency table.
- Choose a graph to display a set of data.

TEACHER RESOURCES

- 112 *Data* (for the teacher and for each student)

ADDITIONAL MATERIALS

- prepared terms and definitions from *Types of Graphs*, pages 1–4
- a 3 × 5 card (for each student)

Preparation

Remove the terms and descriptions from the *Types of Graphs* pages from display. Arrange them in random order for distribution later in the lesson.

Engage

- Distribute a 3 × 5 card to each student. Assign a **Ticket in the Door** to introduce the essential question at the top of Student Edition page 340, "Which type of graph should I use to track my group's training for a 5K run?" Scan their responses and use them to activate the students' prior knowledge and assess their readiness for the lesson.

Instruct

Calculating the mean for a set of data; determining the range, median & mode

- Guide an **interactive activity** to help the students analyze a set of data. Distribute a copy of the *Data* page to each student and display your copy. Point out the chart at the top of the page.

Comparing Graphs

Which type of graph should I use to track my group's training for a 5K run?

Exercises

Use the table to find the answer. *Answers may vary.*

Chase and Chloe recorded the amount of savings they have on the first day of each month.

	Jan.	Feb.	Mar.	April	May	June
Chase	$100	$140	$115	$125	$135	$145
Chloe	$100	$110	$120	$130	$140	$150

1. Which type of graph would you use to compare how Chase's and Chloe's savings totals change over a 6 mo span? *double line graph*

2. Which type of graph would you use to compare their monthly totals? *double bar graph*

3. Make the graph you chose for problem 1 or 2. *Answers will vary.*

Use the data to find the answer.

ages of the first 15 presidents of the United States on the day of their inauguration: 57, 61, 57, 57, 58, 57, 61, 54, 68, 51, 49, 64, 50, 48, 65

Cameron pitched 7 innings in Saturday's game. He faced the following number of batters.
3, 5, 4, 7, 4, 3, 3

4. Which type of graph would you use to show the number of ages that repeat? *stem-and-leaf plot*
5. How is the data listed in the type of graph you chose for problem 4? *by digits (tens, ones) in numerical order*
6. Make the graph you chose for problem 4. *Answers will vary.*

7. Which type of graph would you use to show the number of batters in each inning? *a line plot or a bar graph*
8. Make the graph you chose for problem 7.
9. What is the mode of the data? Does your graph for problem 8 easily depict the mode? *3; yes*

340   Chapter 15

- Read aloud the following scenario.

Mr. Boles hired 4 students to work at his shop after school and on Saturdays.

How many weeks did each of the 4 students work? 4 weeks

Which student worked the most hours during the 4 weeks? Logan

How many hours did he work? 58 hr

- Direct the students to find the mean of the number of hours that each student worked. Guide the students to recall that the mean is the average of a set of data; it can be found by dividing the sum of the data by the number of addends.

What is the mean of the number of hours that each student worked?

Ethan, 7.5 hr
Logan, 14.5 hr
Samuel, 7.75 hr
Zachary, 11 hr

Guide the students to conclude that each mean represents the average number of hours that each student worked per week because the hours that each student worked were recorded on a weekly basis.

- Instruct the students to find the mean of the number of hours that all 4 of the students worked during each of the 4 weeks.

Use the set of data to find the middle, lower, and upper quartiles. Make a box-and-whisker plot to display the data.

10. 4, 6, 6, 7, 8, 12, 13
middle quartile: 7
lower quartile: 6
upper quartile: 12

11. 16, 18, 19, 23, 24, 25
middle quartile: 21
lower quartile: 18
upper quartile: 24

Find the range and the mean for the set of data. Round decimal answers to the nearest tenth.

12. 16, 12, 9, 8, 2
range: 16 − 2 = 14;
mean: sum of data = 47;
47 ÷ 5 = 9.4

13. 24, 20, 16, 18, 23, 25, 21
range: 25 − 16 = 9;
mean: sum of data = 147; 147 ÷ 7 = 21

14. 56, 42, 50, 48, 42, 40
range: 56 − 40 = 16;
mean: sum of data = 278; 278 ÷ 6 ≈ 46.3

15. 8, 3, 9, 5, 7, 1, 8, 4
range: 9 − 1 = 8;
mean: sum of data = 45; 45 ÷ 8 ≈ 5.6

Use the frequency table to make a histogram.

16.

Ages Enrolled in Summer Swimming Lessons		
Interval	Tally	Frequency
3–5	|||	3
6–9	|||| |||| |||| ||	17
10–12	|||| |||| ||	12
13–15	|||| |||	8

The frequency table shows the ages of children taking summer swimming lessons.

Use the line graph to find the answer.

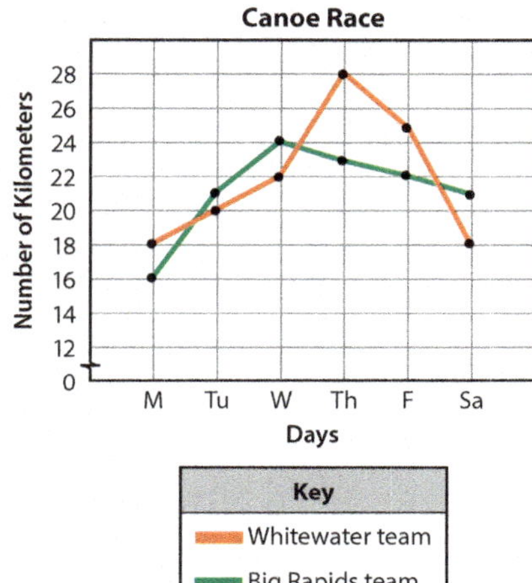

17. On which days did the Big Rapids team paddle the greater distance? *Tuesday, Wednesday, and Saturday*

18. How far did the Whitewater team paddle on Monday? *18 km*

19. On which 2 days did the Whitewater team paddle the same distance? *Monday and Saturday*

20. About how many kilometers did the Big Rapids team paddle for the week? *16 + 21 + 24 + 23 + 22 + 21 = 127 km*

21. Which team paddled the greater distance?

22. Which type of graph should I use to track my group's training for a 5K run? *I should use a multiple line graph.*

21. *the Whitewater team; 18 + 20 + 22 + 28 + 25 + 18 = 131 km; 131 > 127*

Guide the students to conclude that 10.19 represents the average number of hours worked per student per week.

What is the range for a set of data? The range is the difference between the greatest and the least numbers in a set of data.

What is the range of the data on the chart? 11 hr; The greatest number of hours worked is 16 and the least number of hours worked is 5; 16 − 5 = 11 hr.

- Direct each student to list the data (hours worked) from least to greatest. 5, 6, 6, 6, 8, 8, 10, 10, 10, 10, 10, 12, 14, 16, 16, 16

What is the median of this ordered set of data? 10 hr; The median is the middle value or the average of the 2 middle values in a set of data. The average of the 2 middle values (10 and 10) in the ordered set of data is 10.

What is the mode of this set of data? 10; The mode is the value in a set of data that has the greatest frequency (occurs most often); 10 occurred the most often in the set of data; its frequency is 5.

Recording data in a frequency table

- Use **direct instruction** to guide the students as they complete a frequency table. Direct the students to list the hours worked from greatest to least in the frequency table on their page. Then instruct them to complete the frequency table. Give guidance and demonstrate on the displayed page as needed.
Frequency: 16 3; 14 1; 12 1; 10 5; 8 2; 6 3; 5 1

Choosing a graph to display a set of data

- Use an **interactive review** to prepare the students to choose the best graph to display a set of data. Explain that different types of graphs are used to display data in various ways so that the data can be quickly and easily interpreted or understood. A particular type of graph is chosen based on how clearly the information can be displayed on the graph. Different types of graphs organize data differently for statistical reports.

What is the mean of the number of hours that all 4 students worked during each of the weeks?

Week 1, 12 hr
Week 2, 10.5 hr
Week 3, 9.25 hr
Week 4, 9 hr

Guide the students to conclude that each mean represents the average number of hours worked per student for that week because the total number of hours that the students worked that week was divided by the number of students.

- Direct the students to find the mean of the number of hours that all 4 of the students worked during all 4 weeks. Instruct them to round their answer to the nearest hundredth. Choose students to write their solution for display and explain it. sample answers: I can divide the total number of hours worked by 16 (the number of pieces of data): (8 + 6 + 10 + 6 + 16 + 16 + 10 + 16 + 10 + 10 + 5 + 6 + 14 + 10 + 12 + 8) ÷ 16 ≈ 10.19 or [5 + (3 × 6) + (2 × 8) + (5 × 10) + 12 + 14 + (3 × 16)] ÷ 16 ≈ 10.19; or I can add the average number of hours worked per week for all 4 students and divide the sum by 4 (the number of weeks): (12 + 10.5 + 9.25 + 9) ÷ 4 ≈ 10.19; or I can add the average number of hours worked per student for all 4 weeks and divide the sum by 4 (the number of students): (7.5 + 14.5 + 7.75 + 11) ÷ 4 ≈ 10.19.

LESSON 154

- Arrange the students in groups and instruct them to take their *Data* sheets with them. Distribute the terms and descriptions from the *Types of Graphs* pages among the groups. As each type of graph is discussed, instruct the groups who have the term or the description to send students to display the term next to the description to reassemble the word bank that was previously displayed.

  What kind of graph shows changes over a period of time for a set of data? a line graph

  What kind of graph allows you to easily compare a set of related or similar data? a bar graph

  What type of graph is used to arrange data by using the actual digits of the data to clearly display the frequency of the data? a stem-and-leaf plot

  What type of graph could you use to clearly display the frequency of data by creating columns of *X*s above numbers in the scale as you record each piece of data? a line plot

  What kind of graph shows changes over a period of time for 2 sets of related data? a double line graph

  What kind of graph compares 2 sets of related data on the same graph? a double bar graph

  What type of graph allows you to display a large amount of data by separating the data into equal intervals? a histogram

  What type of graph can you use to summarize data by finding the median of a set of data that has been ordered from least to greatest and then finding the median of each half of the set of data? a box-and-whisker plot

- Direct the students' attention to the essential question.

  Which graph would you choose? Why? sample answer: I could use a multiple line graph because it would allow me to show each person's progress as we train for the 5K run.

- Guide the students as they analyze the data recorded on the *Data* page to decide which graph could be used to best display the following.

  a) Compare the hours that Ethan and Logan worked each week. double bar graph

  b) Compare how the number of hours that Ethan and Zachary worked changed during the 4 weeks. double line graph

  c) Show the frequency of the number of hours worked. stem-and-leaf plot or line plot

  d) Display the quartiles of the data. box-and-whisker plot

- Guide a discussion about how to best display the data on each of the graphs. Then direct each student to choose 1 of the graphs and construct it. Allow the students to collaborate with their groups as they construct their graph.

Apply

Student Edition pages 340–41

- Read and explain the directions for pages 340–41. Assist the students as they complete the pages independently.

Daily Review

- Students should complete Chapter 15, section *h*.

Solve.

1. $1,285.79 + $2,391.82 = **$3,677.61**

2. 32.105 − 15.019 = **17.086**

3. 50.12 × 3 = **150.36**

4. $150.00 − $79.35 = **$70.65**

5. 4 × 2.175 **8.7**

6. $\frac{3}{4} \times \frac{5}{6}$ **$\frac{5}{8}$**

7. 1,518 ÷ 6 **253**

8. $\frac{6}{9} \div \frac{1}{3}$ **$\frac{6}{9} \times \frac{3}{1} = \frac{6}{3} = 2$**

9. 3)4,560 **1,520**

10. 25)8,175 **327**

11. 47)16.215 **0.345**

12. 19)116.28 **6.12**

Use the number line to solve.

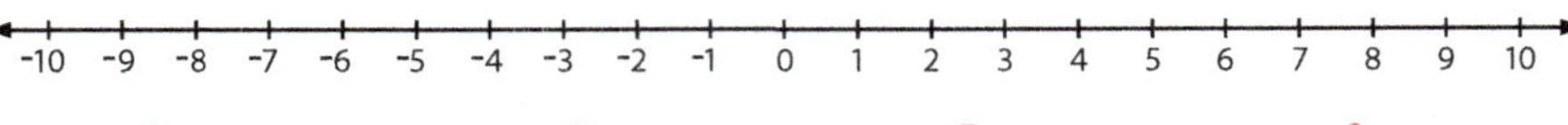

13. 3 + ⁻1 **2**

14. ⁻4 + ⁻5 **⁻9**

15. ⁻6 + 1 **⁻5**

16. ⁻4 + 4 **0**

Solve.

17. $n + 8 = 12$ **$n = 4$**

18. $\frac{n}{4} = \frac{25}{100}$ **$n = 1$**

19. $3n = 18$ **$n = 6$**

20. $36 \div 9 = n$ **$n = 4$**

Daily Review 507

Student Edition pages 342–43

CHAPTER REVIEW

OBJECTIVES

- Propose ways that statistics make people's lives better. **BWS**
- Complete a frequency table using given data.
- Determine the range, median, and mode for a set of data.
- Calculate the mean for a set of data.
- Interpret graphs: double bar graph, double line graph, stem-and-leaf plot, line plot, histogram, box-and-whisker plot.

BIBLICAL WORLDVIEW SHAPING

- **Service (Formulate):** Statistics can help people avoid bad decisions and help them make wise purchases to save money and improve quality.

TEACHER RESOURCES

- 97 *Double Bar Graph*
- 98 *Double Line Graph*
- 103 *Histogram: Piano Practice Statistics*
- 105 *Box-and-Whisker Plot: Points Scored*

ADDITIONAL MATERIALS

- prepared terms and definitions from *Types of Graphs*, pages 1–4
- Bible (for the teacher and for each student)

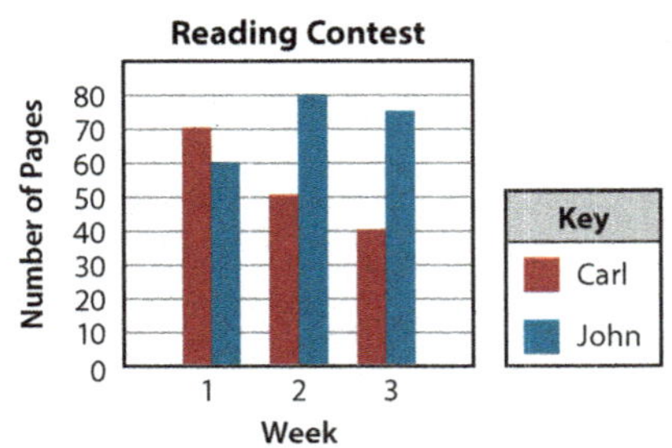

CHAPTER REVIEW

Use the set of data to find the answer.

basketball game scores:
63, 72, 59, 52, 69, 62, 59, 58, 60

1. Make a frequency table.

2. Find the mean game score. Round to the nearest tenth. *The sum of the data is 554; 554 ÷ 9 ≈ 61.6*

3. Find the median of the scores. *60*

4. What is the range of the game scores? *72 − 52 = 20*

5. What is the mode? *59*

Use the double bar graph to find the answer.

Reading Contest

Number of Pages / Week

Key: Carl, John

6. Which week shows the greatest difference in the number of pages read? *week 3*

7. What is the mean number of pages read by Carl? Round to the nearest tenth. *The sum of the data is 160; 160 ÷ 3 ≈ 53.3*

8. Who read the most pages? *John*

9. How many pages did John read? *60 + 80 + 75 = 215 pages*

1.

Basketball Game Scores		
Score	Tally	Frequency
52	I	1
58	I	1
59	II	2
60	I	1
62	I	1
63	I	1
69	I	1
72	I	1

342   Chapter 15

Use the double line graph to find the answer.

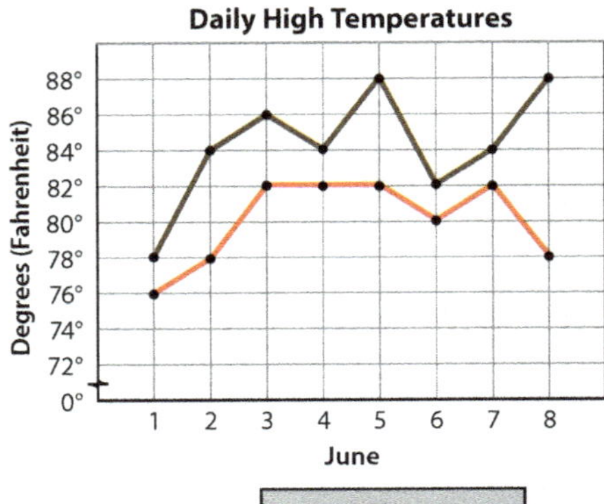

Daily High Temperatures

Degrees (Fahrenheit) / June

Key: Sunnyville, Peaceful Valley

10. On which days were the temperature differences between the cities 2°? *June 1, 4, 6, and 7*

11. What day shows the greatest temperature difference between the two cities? *June 8*

12. What is the range in temperatures for Sunnyville? *88 − 78 = 10*

13. What is the mode for Peaceful Valley? *82*

14. What is the mean temperature for Sunnyville? Round to the nearest tenth. *The sum of the data is 674; 674 ÷ 8 ≈ 84.3*

Use the set of data to find the answer.

pages of term paper: 8, 4, 6, 3, 4, 3, 5, 5, 4, 5, 4, 2, 4

15. Make a line plot.

16. What is the mode for the number of pages? *4*

17. What is the range? *8 − 2 = 6*

18. What is the median? *4*

15. 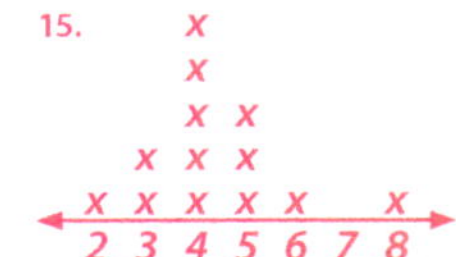

The Chapter Review offers an opportunity for students to discuss the concepts they have learned in the chapter. They may work collaboratively or independently as you review concepts. Circulate among the students, giving individual help as needed. Students who demonstrate proficiency with the discussion, the modeling, and the Student Edition pages are ready for the Chapter Test. Students who encounter difficulties with the review concepts would benefit from additional coaching and practice before testing.

Preparation

Display the terms and descriptions from the *Types of Graphs* page as needed as a reference for the students during the lesson.

Proposing ways that statistics make people's lives better

- Direct attention to the picture on Student Edition page 324 and to the chapter essential question, "How can statistics make people's lives better?"

  What data do you think the student is viewing? sample answers: data about running shoes, possibly including things such as style, color, price, and reviews

Use the set of data to find the answer.

hours worked: 8, 15, 20, 20, 20, 25, 30, 30, 35, 35, 35, 40, 40, 42

19. Make a stem-and-leaf plot.

20. What is the range of hours worked? *42 − 8 = 34*

21. Find the mean. Round to the nearest tenth.
The sum of the data is 395; 395 ÷ 14 ≈ 28.2

22. What is the median? *30*

19. 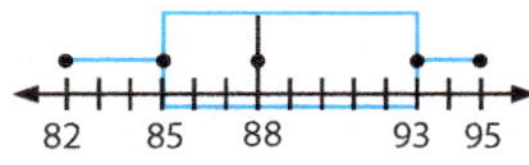

Hours Worked	
Stem	Leaf
0	8
1	5
2	0 0 0 5
3	0 0 5 5 5
4	0 0 2

Key	1|5 = 15

Use the box-and-whisker plot to find the answer.

Mr. Darsey used the math test scores as data for his box-and-whisker plot.

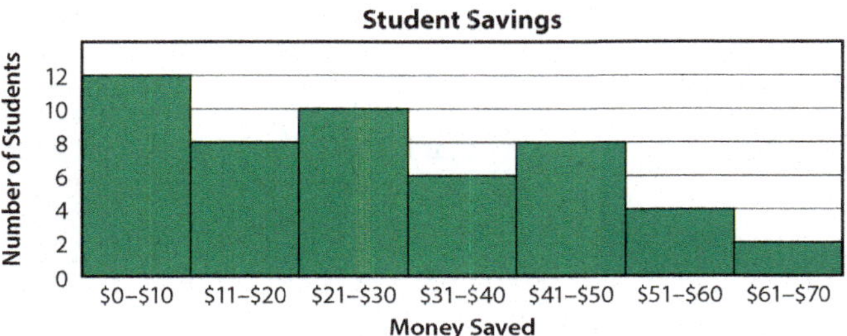

82 85 88 93 95

23. What is the highest score? *95*

24. What is the median score? *88*

25. What is the lower quartile? *85*

Use the histogram to find the answer.

Student Savings

(Number of Students vs. Money Saved)
- $0–$10: 12
- $11–$20: 8
- $21–$30: 10
- $31–$40: 6
- $41–$50: 8
- $51–$60: 4
- $61–$70: 2

26. What is the range of the data? *70 − 0 = 70*

27. State whether this statement is true or false: More than half of the students have $30 or less in their savings. Explain your answer. *True; of the 50 students represented in the graph, 30 students are in the ranges $0 to $30.*

28. What is the mode? *$0–$10*

29. Describe situations in which statistics makes someone's life better. *sample answer: Statistics could help someone get the best buy on an item or choose a better quality product for the money.*

Why would she view the data? to help her make a cost-effective and wise purchase for the type of shoe she wants

Have you ever viewed data about a purchase you planned to make? Answers will vary.

- Choose a student to read aloud Proverbs 18:13.

What practice does the writer of Proverbs say is foolish and shameful? to give an answer (or state an opinion) before I have enough information

Point out Proverbs 18:15.

What does Proverbs 18:15 say are marks of intelligence and wisdom? seeking and acquiring knowledge

- Remind the students that statistics deals with the collection, organization, analysis, and interpretation of data. Reliable statistics make information easier to understand and interpret.

How can statistics help people to avoid folly and shame and to grow in intelligence and wisdom? It can help them interpret data more easily so that they can be informed to make sound predictions and wise decisions based on reliable data.

Point out to the students that when they use their skill in math to provide accurate statistics so that people have access to reliable data, they are serving others and making their lives better.

Completing a frequency table using given data; determining the range, median & mode; calculating the mean

- Write for display, "Ages of young people in the youth orchestra: 10, 15, 12, 9, 10, 11, 11, 12, 16, 14, 12, 13, 11, 13, 14, 12, 15."

What is a good way to organize data to make it more manageable? sample answers: order the data from least to greatest; record the data in a frequency table; display the data on a graph

- Direct the students to order the data from least to greatest. Then instruct them to record the data on a frequency table: 9 1; 10 2; 11 3; 12 4; 13 2; 14 2; 15 2; 16 1.

What is the range of ages in this set of data? 7; The range is the difference between the greatest value and the least value in a set of data; 16 − 9 = 7.

What is the mean of a set of data? the average of the data

How can you find the mean of the ages of the students in the youth orchestra? I can add all the ages and divide the sum by the number of students in the data.

- Direct the students to find the mean. Instruct them to round their answer to the nearest whole number if needed. 210 ÷ 17 ≈ 12.3; The average age is 12.

What is the median of the ages? 12; The median is the middle value or the average of the 2 middle values in an ordered list of data. Since there is an odd number of ages, the median is the middle number (12).

What number is the mode in this set of data? 12; The mode is the number that occurs most frequently in a set of data; 12 is recorded 4 times, more frequently than any of the other ages.

Interpreting a double bar graph

- Display the *Double Bar Graph* page. Remind the students that double bar graphs can be drawn horizontally or vertically. Guide the students as they recall how double bar graphs are used. to compare 2 sets of related or similar information

---

What information is recorded on this graph? the number of boys and girls in grades 1–6 at Grace Christian School

What information (category) is on the horizontal axis? the grades in the school

What is listed on the vertical axis? The number of students is listed in intervals of 2.

Which grade has the greatest difference between the number of boys and the number of girls? fourth grade

- Direct the students to find the mean of the number of girls in grades 1–6 and the mean of the number of boys in grades 1–6. Instruct them to round their answer to the nearest tenth if needed. $70 \div 6 \approx$ 11.7 girls per grade; $70 \div 6 \approx 11.7$ boys per grade

How does the average number of boys compare with the average number of girls? The average number of boys is the same as the average number of girls, 11.7.

What is the mode for the girls? 13

What is the mode for the boys? 12

What is the range for the girls? 7

What is the range for the boys? 5

Interpreting a double line graph

- Display the *Double Line Graph* page. Guide the students as they recall how a double line graph is used. to show changes over a period of time for 2 related sets of data

Choose a student to explain the changes in the temperatures in Peaceful Valley during 1 week.

On what day did the least change in temperature occur? Thursday

What was the range of temperatures on Thursday? 2°

On what day did the greatest change in temperature occur? Friday

What was the range of temperatures on Friday? 12°

What was the median high temperature? 72°

What was the median low temperature? 66°

Interpreting a stem-and-leaf plot & a line plot

- Write for display "Years taught: 5, 6, 8, 9, 10, 11, 11, 12, 12, 12, 20." Explain that this data was gathered during a survey that was taken to determine the statistics of the faculty's teaching experience.

How is data displayed on a stem-and-leaf plot? Each piece of data is displayed using the tens digit of the piece of data as its stem and the ones digit of the piece of data as its leaf. Since the leaves in a set of data indicate the number of items in the data, the ones digit of each piece of data must be written in the Leaf column.

- Choose a student to make a stem-and-leaf plot for display for the years taught by the faculty while the other students construct a plot at their desk.

What number has the greatest frequency? 12

Guide the students to conclude that 12 is the mode of the set of data; 12 is the value that occurs most often.

What does a mode of 12 tell you about the faculty's years of teaching experience? More teachers had 12 years of teaching experience than the other numbers of years listed.

What other type of graph is used to display the frequency of data? a line plot

How do you display data on a line plot? An *X* is drawn above a number on a number line each time that number appears in the data.

How do you determine the scale of the number line used to make a line plot? The scale for the line plot must include the range of the data.

What is the range of this set of data? 15; The data ranges from 5 to 20; 20 − 5 = 15.

- Choose a student to make a line plot for display for the years taught by the faculty while the other students construct a plot at their desk.

Where is a cluster located on this line plot? 8–12; There is a tight grouping of data from 8 to 12.

Where is a gap located on this line plot? at 7 and from 13 to 19; There is an empty space with no data at those locations on the line plot.

What number is an outlier on the line plot? 20; It has a value that is much greater than the values of the other data.

- Remind the students that an outlier can affect the mean, range, and median; but the mode is not affected.

Interpreting a histogram

- Display the *Histogram: Piano Practice Statistics* page.

When would you graph data on a histogram? when there is a large amount of data to graph

Describe a histogram. Although a histogram is similar to a bar graph, a histogram always uses vertical bars, and there are no spaces between the bars. The data is separated into equal intervals on the horizontal axis, and each bar represents an interval.

How many piano students practiced 4–5 hr during the week? 8

How many practiced 8–9 hr? 2

What interval is the mode of the set of data that is displayed on this histogram? How do you know? 4–5; The bar for that interval indicates the greatest frequency.

What does an overall view of the histogram show you? sample answer: Most of the piano students practiced 2–7 hr during the week.

Interpreting a box-and-whisker plot

- Display the *Box-and-Whisker Plot: Points Scored* page. Guide the students as they recall how a box-and-whisker plot is used. to summarize data on a number line using a list of data that has been ordered (organized) from least to greatest

What do the quartiles of a box-and-whisker plot represent? The middle quartile is the median of the set of data, the lower quartile is the median of the lower half of the set of data, and the upper quartile is the median of the upper half of the set of data.

- Review the lists of data on the displayed page.

What is the least number of points that Joel scored during a basketball game? 2

What is the greatest number of points Joel scored? 16

What is the range of the data? How do you know? 14; 16 − 2 = 14

What is the median? How do you know?
10; It is the middle quartile, the middle value in the ordered set of data.

If the lists of data were not given, what are the only 2 pieces of data that you can be certain would be included in the list of data? Why? **I can know only that the least value (2) and the greatest value (16) would be included in the list of data. By examining just the box-and-whisker plot I cannot know whether the middle, lower, and upper quartiles are included in the list of data or if they were determined by averaging 2 pieces of data.**

Student Edition pages 342–43

- Read and explain the directions for pages 342–43. Assist the students as they complete the pages independently. Allow the students to work with a partner or group to answer problem 29.

NOTES

LESSON 156

Student Edition pages 344–46

CHAPTER 15 TEST

CUMULATIVE REVIEW

CONCEPT REVIEW

- Adding, subtracting, multiplying, and dividing whole numbers, decimals, and fractions (Chapters 1–3, 5, 7–9)
- Finding the value of a variable in an equation (Lessons 97–99)
- Adding customary measurements (Lesson 138)
- Converting customary measurements (Lesson 135)
- Expressing a percent as a fraction in lowest terms (Lesson 125)
- Finding the decimal equivalent of a fraction (Lesson 85)
- Identifying the value of a digit in a decimal (Lesson 4)
- Estimating the difference by rounding (Lessons 3, 5)
- Expressing a fraction as a percent (Lesson 125)
- Measuring a line to the nearest sixteenth inch (Lesson 134)
- Identifying a chord and a diameter in a circle (Lesson 61)
- Finding the unknown measure of an angle in a pair of supplementary angles and in a triangle (Lessons 55, 57)
- Calculating the volume of a cylinder (Lesson 116)

Choose the answer.

1. $\begin{aligned} 58.509 \\ -49.932 \end{aligned}$

   A. 7.437   C. 9.437

   B. 8.577   D. 18.437

2. $\begin{aligned} 4.6897 \\ +68.3974 \end{aligned}$

   A. 74.3071   C. 73.0121

   B. 73.0871   D. 72.0971

3. $\begin{aligned} \$37.24 \\ \times \quad 342 \end{aligned}$

   A. $12,736.08   C. $12,068.80

   B. $12,638.08   D. $11,763.08

4. $8\overline{)783}$

   A. $95\frac{7}{8}$ C. $97\frac{7}{8}$

   B. $96\frac{5}{8}$ D. $98\frac{1}{8}$

5. $76.3 \div 100$

   A. 0.0763   C. 7.63

   B. 0.763   D. none of the above

6. $\frac{5}{8} - \frac{7}{32}$

   A. $\frac{2}{32}$ C. $\frac{13}{32}$

   B. $\frac{7}{32}$ D. $\frac{15}{32}$

7. $6\frac{1}{10} - 3\frac{3}{5}$

   A. $3\frac{4}{5}$ C. $2\frac{7}{10}$

   B. $3\frac{2}{5}$ D. $2\frac{1}{2}$

8. $\frac{2}{7} + \frac{1}{4}$

   A. $\frac{5}{7}$ C. $\frac{11}{14}$

   B. $\frac{3}{28}$ D. $\frac{15}{28}$

9. $\frac{5}{8} \div \frac{2}{3}$

   A. $\frac{5}{6}$ C. $\frac{5}{16}$

   B. $\frac{5}{8}$ D. $\frac{15}{16}$

10. $\frac{5}{8} \times 6$

   A. $3\frac{3}{4}$ C. 3

   B. $3\frac{1}{2}$ D. $2\frac{7}{8}$

344   Chapter 15

- To prepare the students for the format of achievement tests, instruct them to work on a separate sheet of paper, if necessary, and to mark their answers on the *Cumulative Review Answer Sheet.*

Student Edition pages 344–46

The Cumulative Review provides additional practice of previously learned concepts. These pages may be completed during this lesson or anytime after this lesson, since they require limited or no teaching.

Choose the answer.

11. $5{,}177 = 31 \times n$

- **A.** $n = 166$
- **B.** $n = 167$
- **C.** $n = 169$
- **D.** $n = 170$

12. $n \div 251 = 36$

- **A.** $n = 9{,}036$
- **B.** $n = 8{,}306$
- **C.** $n = 8{,}936$
- **D.** $n = 1{,}036$

13.
$$\begin{array}{r} 16 \text{ lb } 10 \text{ oz} \\ +\ 18 \text{ lb } \ \ 6 \text{ oz} \\ \hline \end{array}$$

- **A.** 35 lb
- **B.** 35 lb 4 oz
- **C.** 36 lb 2 oz
- **D.** 37 lb

14. Twenty quarts can make how many gallons?

- **A.** 4 gal
- **B.** 6 gal
- **C.** 5 gal
- **D.** 7 gal

15. What is 15% as a fraction in lowest terms?

- **A.** $\frac{1}{4}$
- **B.** $\frac{1}{5}$
- **C.** $\frac{3}{20}$
- **D.** $\frac{5}{17}$

16. What is the value of $\frac{1}{7}$ in decimal form? Round to the nearest thousandth.

- **A.** 0.143
- **B.** 1.429
- **C.** 1.144
- **D.** 7.1

17. What is the value of 4 in 649,782,163.51?

- **A.** 40 thousand
- **B.** 4 million
- **C.** 40 million
- **D.** 4 billion

18. Round to the greatest place value to estimate the answer.

$7.014 - 4.38$

- **A.** 3
- **B.** 2
- **C.** 4
- **D.** 1

19. What is $\frac{4}{5}$ as a percent?

- **A.** 70%
- **B.** 90%
- **C.** 80%
- **D.** 100%

20. Measure the line segment to the nearest sixteenth inch.

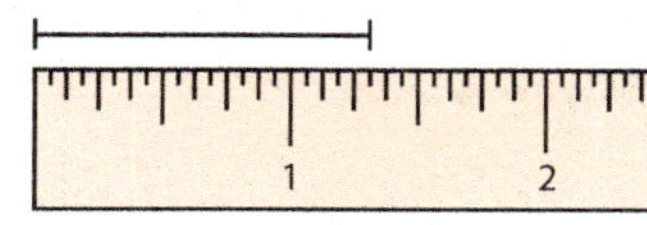

- **A.** $1\frac{5}{16}$
- **B.** $1\frac{7}{16}$
- **C.** $1\frac{9}{16}$
- **D.** $1\frac{15}{16}$

Use ⊙A to find the answer.

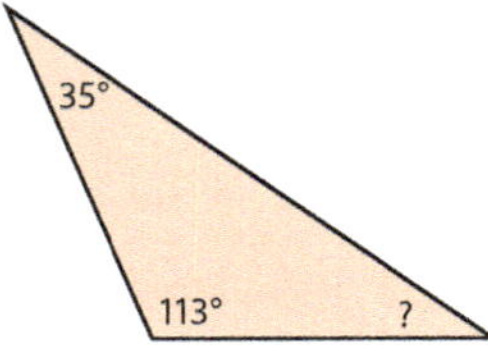

21. What is the diameter of circle *A*?

A. $\overline{AD}$
B. $\overline{CB}$
C. $\overline{EF}$
D. none of the above

22. If ∠*CAD* is 73°, what is the measure of ∠*BAD*?

A. 75°
B. 90°
C. 107°
D. none of the above

23. Name a chord that is *not* the diameter.

A. $\overline{EF}$
B. $\overline{AD}$
C. $\overline{AC}$
D. none of the above

Choose the answer.

24. What is the measure of the unknown angle?

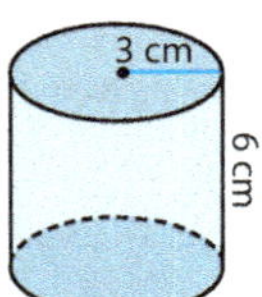

A. 42°
B. 32°
C. 22°
D. none of the above

25. What is the volume of the figure?

A. 148.45 cm³
B. 169.56 cm³
C. 16.966 cm³
D. none of the above

Math 6

Entrepreneurship

How can I help others by using statistics to solve a business problem?

Answers to #3–6 and 8–12 will vary.

Write the answer.

1. Define entrepreneurship. *setting up a business with the goal of making a profit*

Ask a question to identify the problem.

2. What problem would you like to solve? *identifying a good business I could start*

Imagine possible solutions.

3. What do you enjoy doing? List as many God-given gifts, abilities, and interests as possible to help you identify possible business ideas.

4. What kind of business could help you solve a problem?

Plan to solve the problem.

5. Identify and record your best business idea.

Create to implement your solution.

6. Write a business plan.

Reflect.

7. Tell how you could help others by using statistics to solve a problem in your business. Give a statistics strategy you would use, the business problem you are addressing, and how solving the problem would help someone.

7. *Answers will vary. Examples may include using a survey or poll to gather the data I need to identify a needed product or service that makes people's lives better, presenting my projected expenses and potential income to my parents to increase their confidence in my business plan, or communicating statistics about the need for my product or service to my potential clients so they can see ways my business can serve them.*

Lessons 157–59   347

Preparation

Read the Amanda Brown interview, located at the end of this lesson, and choose excerpts to share during the opening discussion.

Visit TeacherToolsOnline.com for more resources on youthful entrepreneurship.

Display the *Engineering Design Process* page and refer to it as you proceed through the steps of the STEM project.

Ask

Identifying the problem

• Guide a **discussion** to introduce the topic of entrepreneurship from Student Edition page 325. Direct the students' attention to the career link about Amanda Brown on the page. Point out that Amanda was about their age when she began publishing a magazine that later became a business.

What problem did Amanda want to solve? the lack of spiritually edifying reading materials for girls her age

---

LESSON 157

Student Edition pages 325, 347–48

STEM

OBJECTIVES

• Identify the problem.
• Research the problem.
• Record possible solutions.
• Identify a workable solution.

TEACHER RESOURCES

• 25 *Engineering Design Process*
• 113 *Business Idea Organizer* (for the teacher and for each student)

ADDITIONAL MATERIALS

• examples of successful young entrepreneurs (optional)

Read aloud and discuss excerpts from Amanda's story that you have chosen from the Amanda Brown interview.

Why do you think starting her own magazine was an attractive idea to Amanda? She loved writing, reading, and art, activities which are involved in publishing a magazine.

When did Amanda's magazine become a business? in 2018 when she decided to print and sell copies of her magazine

Have you ever thought of starting a business? Answers will vary.

• Write the term "entrepreneurship" for display. Explain that entrepreneurship is the activity of setting up a business with the goal of making a profit. Point out that a person who does this is called an entrepreneur.

Explain to the students that the goal of the STEM project is for them to choose a business idea with their group and to develop a plan for making their business successful.

What problem would you like to solve? identifying a good business that I could start

Imagine

Researching the problem

- Use **brainstorming** to help the students explore business ideas. Guide their thinking with the following points about starting a business.

  1. **Find a problem to solve.** A successful business will provide a service or a product that meets a need so that people are willing to pay for it. The students' business must solve a problem for its customer.

  Discuss the following example. Encourage the students to think about the problem from the perspective of different groups of people and not just from their own perspective.

  What problem might a food business solve for a customer? sample answers: satisfying his hunger; providing food for someone who is not skilled at cooking or baking, someone who is busy and does not have time to cook, or someone who is too tired to cook; satisfying his desire for interesting food choices to give him a varied diet; providing a unique home-made gift for him to give; helping him celebrate a special occasion; providing special foods that might be difficult for him to find elsewhere or to prepare himself

  2. **Identify your gifts.** Point out to the students that choosing an activity that they love doing which uses their God-given gifts and abilities will help them stick with their business even when it is not easy. Explain that besides their natural abilities, their God-given gifts also include their circumstances, such as where they live, the things they own, and the family and friend network of which they are a part.

  What are some examples of gifts, abilities, or strong interests that might guide someone in his choice of a business to start? sample answers: A person lives on a farm, in the city, or in a neighborhood; he or she enjoys helping people, is artistic, enjoys working with his or her hands, loves reading and writing, is good with math or science, enjoys being outside, loves animals, is a good time-manager, is a diligent worker, loves baking, enjoys taking things apart and repairing them, or likes working with children.

- As desired, share examples of successful young entrepreneurs to stimulate your students' thinking.

Recording possible solutions

- Display your copy of the *Business Idea Organizer* page and distribute a copy to each student to help him **brainstorm** promising business ideas.

  Direct the students' attention to the questions in problems 3 and 4 on Student Edition page 347. Help the students answer the questions by guiding them as they share ideas to record on their *Business Idea Organizer* page.

Test and improve your solution.

8. Research to determine a realistic price for your product or service.

9. Calculate your projected income.

10. Research to estimate your total start-up expense.

11. Prepare a 4-week profit projection. Express your weekly summary data in a graph.

12. Make changes to improve your business plan as needed.

Instruct the students to write on their own page any interesting ideas that are suggested by other students, as well as any ideas they themselves have, as you record for display the ideas the students share. Encourage the students to write down every idea that comes to mind, even if it seems unimportant or unlikely.

Plan

Identifying a workable solution

- Guide the students as they **collaborate** with a group to identify a business idea. Emphasize the importance of having a team of people with varying abilities working together to run the different parts of a business.

More than one group may develop the same business idea. Direct the students to take their completed organizers with them to their group.

Encourage the students to collaborate with their group to identify the best business idea that works well for the whole team. Direct them to answer problem 5 on Student Edition page 347 by recording their group's best idea in the blank at the bottom of the organizer page.

- Explain that in the next lesson, they will develop a plan for starting their business. Retain the organizers from each group for the next lesson.

Student Edition page 347

- Problems 3–5 were assigned during the lesson.

- Assist the students as they complete problems 1–2 independently.

Amanda Brown interview

- **How/why did you get started?**

I started *Oh Beloved One*, a blog and magazine, back in 2012 simply out of my love for reading and writing. American Girl and Nancy Drew were "too babyish" for me as I reached the tween years, but I discovered that literature options for my age were extremely worldly. So I decided to start my own magazine, combining my loves of writing, reading, and art. Back then, *Oh Beloved One* looked like a bimonthly newsletter made in Microsoft Office Word, full of articles written by my middle-school-age friends. But, after a journey through depression and anxiety, this venture took on a new face and purpose. We're now an in-print magazine and devotional-style blog. During my journey, I realized just how amazing it is that we are beloved by the most powerful being in the cosmos—that, because of this love, we can cast out all fear (1 John 4:18).

- **What is your mission statement?**

Our mission is to create visually stunning, practical resources that aid young women in their spiritual walks.

- **What insights do you have about the nitty-gritties of starting a business?**

Sometimes people write a five-year-plan, get a loan, and wake up wearing a suit and tie. For me, starting a business sneaked up on me. In 2018, I decided to start printing and selling the magazines and that's when the business aspect began. I am not a numbers person, and because we started small, I assumed that I didn't need to keep track of numbers. This, of course, was a terrible idea. Now I use an online software to track all my income and expenses.

Pricing products is also very interesting for someone who's never been formally taught how to do that. In the beginning, I would constantly underestimate or forget about shipping, which would reduce my profit.

Lastly, finding "manufacturers" has been very educational. Printing magazines isn't cheap, and I stumbled my way through the process in the beginning; who knew magazines had to have a page number that is a multiple of four? To reduce costs, I changed my printer and found a local shop. I take the 30-minute drive at least once every month to pick up my order so I don't have to pay for shipping. When you're starting out, every penny counts!

- **How have statistics impacted your business?**

Statistics are very important to me on many levels! Regarding the magazine, I always keep in mind statistics regarding the print industry. Many reports warn that fewer people are buying magazines. With this information, as a marketer and graphic designer, I find ways to creatively convince people that purchasing print magazines will make them happier than reading a digital version. I also use statistics in some devotionals I've published under the *Oh Beloved One* name; statistics from reputable sources that support your content give you credibility. Statistics on social media also verify that I am reaching my target audience: the majority of my followers are between the ages of 18–24 and 91% are women. Life is about figuring what works and what doesn't, readjusting, and then getting ready to repeat the process.

- **Do you use analytics, or do you do any surveys to gauge interest?**

Surveys are changing their look and functionality. Three years ago, people would take weeks to formulate the perfect survey and then share the link with customers, hoping they would not only click the link but also take the survey. Now, however, we can be sneaky: I use Instagram and mix in fun and informative polls, quizzes, and open questions on the stories to get the information I need while giving the user a fun experience. For example, right now I'm working on developing a prayer journal. So, on Instagram, I asked girls to choose between bullet or lined paper, spiral notebooks or binders, etc. People are lazy, and we have to work with people's mindsets. People would rather interact with a poll on Instagram than bother clicking on a link to a seven-question survey that requires deep thinking.

Analytics are also very important for me! I monitor site data such as what terms people are using to find our website, what social media platforms are driving traffic to our website, which blog posts and pages are receiving the most views, and how many people are visiting our website. It's very fun to play around with naming blog posts and see how that affects site traffic. For example, a post entitled "Jesus Isn't Your Friend" (which has a slight shock factor) will probably receive more views than a post called "About Friends." I can also use analytics to correlate marketing tactics with website hits. Whenever I post a blog post, I advertise it on my social media to drive traffic. Finally, analytics are very important for my mailing list. Mailing lists give you direct access into people's inboxes. I use analytics to track who is subscribing and unsubscribing, which emails are opened the most, which links are clicked, and even which emails led to a sale. Using these numbers helps me to hone my emails, creating better content for the customer.

    – Amanda Brown. E-mail interview. 10 August 2020.

STEM

OBJECTIVES

- Assemble business data.
- Write a business plan.
- Help others by using statistics to solve a business problem. **BWS**

BIBLICAL WORLDVIEW SHAPING

- Service (Apply): Statistics can help identify needs of customers or show the value of a service to customers to make people's lives better.

TEACHER RESOURCES

- 114 *Rover's Romps Business Plan* (for the teacher and for each student)
- 115 *Business Plan* (for the teacher and for each student)

ADDITIONAL MATERIALS

- *Business Idea Organizer* pages (completed student copies)

Plan

Assembling business data

- Guide a **discussion** to help the students assemble the data they need to prepare a business plan. Explain that today they will write a business plan that will help them think through their idea and give their business a better chance of success.

  What questions do you have about starting a business? sample answers: How will I get the supplies needed to start my business? Who will I sell to and how will I reach them with my product or service? Will I have enough customers? How will I advertise my business? Do I have any competitors? When will I do the work? What part of the business will I be responsible for, and what parts will my team be responsible for?

Explain that a business plan helps an entrepreneur evaluate his business idea to identify the risks and benefits of the business and see if the business makes sense. Once his plan is skillfully prepared, an entrepreneur then presents it to those whose support he needs, such as family members or others who might help him get his business started.

- Display the *Rover's Romps Business Plan* page to give the students an example to follow. Use prompts such as the following to help you discuss the topic of each box and the content about Rover's Romps in the data boxes on the displayed page.

  Business Name: What will you name your business? Choose a descriptive or catchy name that explains your product or service. Make sure your business name is not being used by another business, perhaps by conducting an internet search.

  What we will do: What is the purpose of your business?

  Problem: What problem does your business address?

  Solution/wow factor: What solution to a problem will your business offer? What sets your business apart from similar businesses?

  Customers: Who will buy your products or service?

  Competition: Are there others who offer a similar solution that would compete for the same customers?

  Where/how sold: How or where will your customers purchase your product or service?

  Marketing activities: How will you reach your potential customers?

  Expenses: What do you have to buy to start and continue your business?

  Team and responsibilities: Who will operate the business and what is each person's job?

  Goals for growth: How would you like to see your business grow?

- Guide the students as they consider who will see their business plan.

  Whose approval and support do you need to help you start a business? I need the approval and support of my parents (or guardians) and of other friends or family who might help me.

  What questions might your parents ask you about your business idea? sample answers: Who will be involved? When will you do it? How much time will it require? What supplies will you need? How will you pay for the supplies? Will you need transportation? Can you make a profit?

  What preparation and attitude might help you get your parents' approval and support? My business plan should be prepared with the data needed to answer the questions that I expect them to ask me, and I should be humbly open to their counsel.

  Encourage the students to keep these things in mind as they prepare their business plan.

Create

Writing a business plan

- Guide the students as they use a **graphic organizer** to write a business plan.

  Assemble the students into their groups. Return the completed copies of the *Business Idea Organizer* (from the previous lesson) to each group.

  Distribute a copy of the *Rover's Romps Business Plan* page and of the *Business Plan* page to each student.

  Give the groups several minutes to collaborate to complete Student Edition page 347, problem 6, by filling out a first draft of their plan on the *Business Plan* page. Encourage them to consult the *Rover's Romps Business Plan* page as a guide to help them fill out as much as possible about their business.

Helping others by using statistics to solve a business problem

- At the end of the allotted time, guide a **discussion** of the essential question on Student Edition page 347, "How can I help others by using statistics to solve a business problem?"

  Read aloud Proverbs 18:15. Remind the students that the Lord commends those who seek knowledge for being prudent or intelligent. Point out that statistics helps the entrepreneur, his supporters, and his future customers to solve problems by having access to data that can help them make informed, intelligent choices.

- Direct the students' attention to the *Rover's Romps Business Plan* page.

  What are some ways that statistics might have helped the Rover's Romps entrepreneurs with their business plan? The entrepreneurs might have identified the problem by reading statistics about the need. They might have determined that their business would have prospective customers by taking a poll or survey of their neighbors. They likely researched statistics to learn the number of competing businesses in the area.

  How might entrepreneurs help others by using statistics? Entrepreneurs might consult statistics to help them identify a product or service that will make their customers' lives better. The entrepreneurs might provide statistics to their prospective customers to show them how the entrepreneurs' business can help them. The entrepreneurs might help their business team by researching statistics to learn how to make their business successful. The entrepreneurs might help their supporters by providing them with statistics that answer their questions and increase their confidence in the entrepreneur's business.

- Explain that they will use statistics to solve a business problem in the next lesson. Retain the students' *Business Plan* pages for the next lesson.

Student Edition page 347

- Problem 6 was assigned during the lesson.

- Assist the students as they complete problem 7 independently.

NOTES

STEM

OBJECTIVES

- Test a plan.
- Improve a plan.

TEACHER RESOURCES

- 116 *Rover's Romps Profit Projection*, pages 1–3 (for the teacher and for each student)
- 117 *Business Profit Projection* (for the teacher and for each student; additional copies as needed)
- 118 *STEM Rubric: Entrepreneurship*

ADDITIONAL MATERIALS

- *Business Plan* pages (completed student copies; clean copy for each group and additional copies as needed)

Test

Testing a plan

- Use **direct instruction** to help the students prepare a profit projection to include with their business plan.

  What is the main goal of starting a business? to make a profit

  What are some important financial questions that your parents or other supporters might ask about your business plan? sample answers: How much will your expenses cost? How do you plan to pay those expenses? Can your business make a profit?

- Explain to the students that one way for an entrepreneur to test whether his business can make a profit is to prepare a profit projection. Explain that a profit projection is a presentation of data that forecasts the financial performance of a business. It is a way of helping others by using statistics to solve a business problem.

Point out that a profit projection helps the entrepreneur both to plan wisely (Luke 14:28) and to demonstrate to his supporters that his business idea is a worthy one.

Can you be sure that your business will perform as you forecast it to? No; only God knows the future.

How do you think you could estimate how much profit you plan to make from your business? I could estimate my income and expenses and then subtract my expenses from my income to figure my estimated profit.

- Distribute a copy of *Rover's Romps Profit Projection*, pages 1–3, to each student; and display a copy of each page as you discuss it. Use the following discussion points as needed.

- Note the profit formula (*Income – Expenses = Profit*) on page 1.

  What does Table 1 help the Rover's Romps team estimate or project? their income

  What does Table 2 help them estimate or project? their expenses for the 1st week

- Choose a student to read aloud the paragraph at the top of the page that explains how the team decided what to charge for their service and how it determines their projected income.

  How did Rover's Romps determine what they should charge for their service? They researched the competitors' prices.

  How would knowing the competitors' prices help them? They would know what a reasonable price would be and what people were used to paying for a service like theirs. It would help them to set their price competitively to attract customers but also to provide them with profit.

  How many dogs will each person walk each day? 4 dogs; 2 hr/d × 2 dogs/hr = 4 dogs/d per person

  If the walks last only 20 min each, why do you think Rover's Romps planned only 2 walks each hour? 2 walks would take 40 min; 60 min – 40 min = 20 min; that would give them 20 min to return the dogs to their owners and to get to the next location.

  Do you think 2 walks/hr is reasonable? Answers will vary.

- Direct the students' attention to the first income projection based on 12 walks/d.

  Do you think it is realistic for the team to expect that they will get to walk 12 dogs each day for 5 d? Answers will vary. It would require a lot of work to line up 60 customers each week.

  Point out the adjusted income projection for 3 walks/d or 15 walks/wk in Table 1.

  Why do you think the team also estimated their income for only 3 walks/d (15 walks/wk)? to see what their income would be if they had fewer customers; to get a more realistic income projection

- Point out that their income will depend in large part on their marketing activities. Explain that marketing is critically important for building a business.

  What marketing activity does Rover's Romps have planned? passing out fliers

  Can you think of other ways they could advertise their business? telling friends and neighbors about it in person; making phone calls, emailing, or texting; posting fliers in public places

  Can marketing alone build your business? No; my business needs God's blessing to succeed (Matthew 6:33).

- Point out that to get their business started, the Rover's Romps team will have expenses. Direct the students' attention to the section titled, "Estimated expense for 1st week @ 60 walks" in Table 2.

  Explain that the team researched each item on their list and estimated their costs for the first week. They added the expected sales tax to the cost of each item and then rounded up the cost to the next dollar.

  Why do you think they rounded up to the next dollar? to be sure that they had enough money to pay their expenses

  Are you surprised at their total expenses? Answers may vary.

  Do you think their expenses will be the same every week? Why or why not? No; they will need to purchase supplies to get their business started that they will not have to purchase every week.

- Display Table 3 on page 2. Direct the students to compare the estimated expenses for 15 walks in Table 3 with the expenses for 60 walks in Table 2. Note that their expenses did not change.

Why did their expenses stay the same? They need the same things to get their business started whether they are walking 60 dogs/wk or 15 dogs/wk.

What could they do if they do not have enough money for those expenses? sample answers: They could wait until they have the cash on hand to start their business, they could look for ways to lower their expenses, or they could adjust their business idea to one that has fewer expenses. They could ask their parents for a loan until they start earning income.

- Point out the entry in Table 4 that explains how the team cut their expenses. Discuss the adjusted expenses to see what they cut.
  How much could they save by purchasing some used items and by using what they already have? $118 − $53 = $65

- Display Table 5 on page 3.
  Which income projection did the team decide to use, 12 walks/d or 3 walks/d? 3 walks/d

  Why do you think they decided to use the lower level of projected income rather than the higher one? They probably wanted to be safe and have a more realistic idea of their possible income.

  Are these figures the actual amounts that the team earned? No; this is a projection of what they estimate they will earn.

- Point out that positive amounts (income or profit) in the table are added to the total, but that negative amounts (expenses; shown in parentheses) are subtracted from it.
  Note that the team projects that Week 1 will end with a profit.
  Will they start Day 1 with a profit? No; they will have a balance of ⁻$53 at the beginning of the day because of their start-up expenses.

  When might they expect to be out of debt? If they earn at least the $30/d that they project, they will be out of debt by the end of Day 2; 2 × $30 = $60; $60 − 53 = $7 profit.

- Point out the recurring advertising expense each week.
  Will they have to advertise every week? Answers may vary. Advertising is important to building their business and making it more profitable, so it is a good idea to continue to advertise.

Besides the advertising expense each week, what other recurring expenses will they have and how often? Dog treats last 10 wk; 5 days × (3 dogs/d × 1 treat/dog) = 15 treats/wk; 150 treats ÷ 15 treats/wk = 10 wk. Disposable bags last 5 wk, with 5 bags left over; 5 days × (3 walks/d × 1 bag/walk) = 15 bags/wk; 80 bags ÷ 15 bags/wk = 5 wk, r5 bags)

Do you think it will be easy to find 15 customers each week? Answers may vary. The quality of their service and reasonableness of their price will make their service more attractive to new customers and win them repeat business from existing customers. The weather, sickness, and other variables will also affect their business.

What will happen if the team does not find enough customers to live up to their income projection? Their income will be lower. They might not be able to pay their expenses.

- Explain that there is always a risk in starting a business. If their business starts with high expenses and the business does not perform as they anticipate, they might lose money in the venture.
  How could you minimize the risk of losing money in your business? sample answers: I could pray for wisdom before starting a business, research to identify a promising business idea, start a business that requires fewer or lower expenses, avoid borrowing money to start, and work hard to make a profit.

- Point out the line graph showing the team's profit projection for their first month.
  Why do you think they chose a line graph to display their data? A line graph shows change over time.

  Why is a graph an effective way to present statistics related to your business plan? A graph presents the data in a visual way that makes it easy to understand.

- Distribute a copy of the *Business Profit Projection* page to each student, and display your copy. Explain that their profit projection will be a valuable addition to their business plan.

  Point out that the lines on the *Business Profit Projection* page correspond to those in Table 5, which is an example of how their completed projection should look.

Improve

Improving a plan

- Guide the students as they **collaborate** with their group to improve their business plan.

  Assemble the students in their business teams. Direct them to take their *Business Profit Projection* page with them to their group. Distribute the groups' completed *Business Plan* pages. Direct the students to complete Student Edition page 348, problems 8–12.

  Encourage the groups to divide the research and financial preparation responsibilities among the members of the team. As time permits, schedule each team to present its business plan, including the profit projection, to the class. Encourage them to present their business plan and profit projection, including the summary graph, to their parents.

Student Edition page 348

- Problems 8–12 were assigned during the lesson.

Assess

Rubric

- Use the prepared *STEM Rubric: Entrepreneurship* page or design a **rubric** to include your chosen criteria.

Lesson Plan Overview

Chapter 16: Probability

Day	Lesson	Student Edition Pages	Teacher Edition Pages	Teacher Resources & Additional Materials	Topics, Skills & Biblical Worldview Shaping
162	160	349–51	349–51b	**Teacher Resources** • 119 *Probability* • 120 *Theoretical Probability* **Additional Materials** • colored markers: red and blue • colored pencils or crayons: red and blue • calculators (optional)	**Topic:** Theoretical Probability **Skills:** deciding whether a given statement is impossible, unlikely, likely, or certain; finding the theoretical probability of an event **BWS:** Design
163	161	352–53	352–53b	**Teacher Resources** • 121 *Sample Spaces* • 122 *Tree Diagram* • 123 *Multiplication Counting Principle* **Additional Materials** • calculators (optional)	**Topic:** Sample Spaces **Skills:** finding the sample space for and the probability of an event, making a tree diagram, determining the number of possible outcomes using the Multiplication Counting Principle
164	162	354–55	354–55b	**Teacher Resources** • 124 *Experimental Probability* • 125 *Spinning Penny Experiment* **Additional Materials** • paper clips • calculators (optional) • pennies (optional)	**Topic:** Experimental Probability **Skills:** finding the theoretical probability of an event, conducting a probability experiment, finding the experimental probability of an event **BWS:** Design
165	163	356–57	356–57b	**Teacher Resources** • 126 *Fair or Unfair Games* **Additional Materials** • 6 marbles: 4 orange, 1 brown, 1 yellow • an opaque bag • two 1–6 number cubes **Assessments** • Chapter 16 Quiz 1	**Topic:** Fair or Unfair? **Skills:** using probability to determine whether a game is fair or unfair, conducting a probability experiment, finding the experimental probability of an event, making predictions using probability
166–67	164	358–59	358–59b	**Additional Materials** • 6 marbles: 4 orange, 1 brown, 1 yellow • an opaque bag • a 1–6 number cube • calculators	**Topic:** Independent & Dependent Events **Skills:** differentiating between independent and dependent compound events, finding the probability of compound events using a formula **BWS:** Design
168	165	360–61	360–61b	**Teacher Resources** • 127 *Probability Spinner* **Additional Materials** • paper clips • calculators (optional)	**Topic:** Chapter Review
169	166	362–64	362–64		**Topic:** Test & Cumulative Review

CHAPTER 16

CHAPTER OBJECTIVES

- Determine the number of possible outcomes for a game.
- Find the experimental probability and the theoretical probability of an event.
- Conduct a probability experiment.
- Find the sample space for and the probability of an event and its complement.
- Make predictions using probability.
- Defend the claim that Christians should study probability.

16 PROBABILITY

Why are we studying probability?

To determine whether your students are prepared for the concepts taught in this chapter, you may wish to consult the preassessment checklist found on TeacherToolsOnline.com.

Throughout this chapter, select problems from the list of mental math problems provided on the 134–36 *Mental Math Strings* pages.

Make fact practice, both oral and written, part of your daily math routine to help the students with mastery.

Visit AfterSchoolHelp.com for math practice resources, or visit TeacherToolsOnline.com for additional resources to enhance the lessons.

Student Edition pages 349–51
Daily Review Chapter 16, section *a*

OBJECTIVES

- Decide whether a given statement is impossible, unlikely, likely, or certain.
- Recall that the design evident in our world allows us to make useful predictions. **BWS**
- Find the theoretical probability of an event and its complement.

BIBLICAL WORLDVIEW SHAPING

- **Design (Recall):** We can make useful predictions because the world is designed.

TEACHER RESOURCES

- 119 *Probability* (for the teacher and for each student)
- 120 *Theoretical Probability*

ADDITIONAL MATERIALS

- colored markers: red and blue (for the teacher)
- colored pencils or crayons: red and blue (for each student)
- a calculator (for each student) (optional)

Permitting the students to use their calculators, as needed, throughout this chapter will allow them to focus on the concepts that are being taught.

Engage

- Direct attention to the chapter opener and essential question on Student Edition page 349 and guide a **visual analysis** of the illustration.

  Use the following prompts:
  1. How could you describe this scene? Students are looking at a picture of a storm or weather alert.
  2. What is pictured on the screen the students are viewing? choppy waves and wind-blown trees, an approaching tropical storm or hurricane
  3. What questions are posed by the students? "How should we prepare for the storm?" "Should we evacuate?"
  4. If a hurricane is several hundred miles away and is heading in your direction, do you think it is probable, or likely, that it will hit your city? sample answer: The direction that a hurricane heads is based on many different factors, but it would definitely be worth watching.
  5. Based on the picture clues, what is one way you could answer the chapter essential question, "Why are we studying probability?" to be able to prepare for predicted events such as storms

---

Theoretical Probability

Why are we able to make useful predictions?

Probability is the likelihood that an event will occur. **Theoretical probability** is found when the total possible outcomes of an event are known and all outcomes are equally likely to occur. Probability is written as a ratio or a percent.

Key Terms
- probability
- theoretical probability
- complementary events

What is the probability of drawing a blue marble from each bag?

$$P(\text{event}) = \frac{\text{number of favorable outcomes}}{\text{number of possible outcomes}}$$

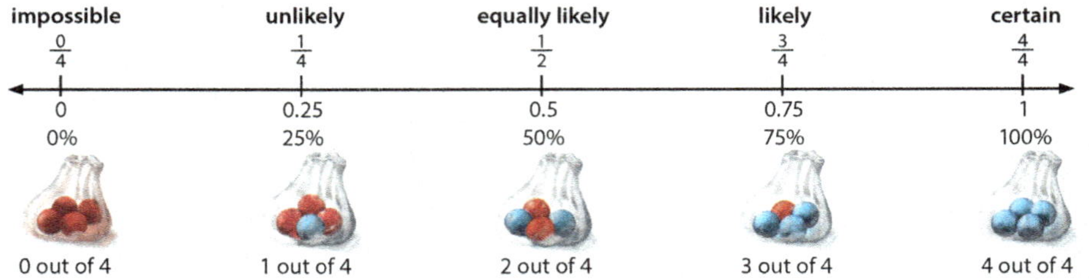

impossible	unlikely	equally likely	likely	certain
$\frac{0}{4}$	$\frac{1}{4}$	$\frac{1}{2}$	$\frac{3}{4}$	$\frac{4}{4}$
0	0.25	0.5	0.75	1
0%	25%	50%	75%	100%
0 out of 4	1 out of 4	2 out of 4	3 out of 4	4 out of 4

Complementary events are two events that could happen but not at the same time. The sum of the two events must equal 1 or 100%.

If there is a 25% chance that it will snow, what is the chance that it will *not* snow? Write the probability of the event as a percent. The complement of it snowing today is *not* snowing today.

25% + P(*not* snowing) = 100%   100% − 25% = 75%

When there is a 25% chance that it will snow, there is a 75% chance that it will *not* snow.

Exercises

A marble is drawn from the pictured bag. Write the probability of the event as a fraction and as a percent.

1. P(green) $\frac{1}{10}$; *10%*
2. P(blue) $\frac{3}{10}$; *30%*
3. P(red) $\frac{4}{10}$; *40%*
4. P(yellow) $\frac{2}{10}$; *20%*
5. P(either red or blue) $\frac{7}{10}$; *70%*
6. P(not red) $\frac{6}{10}$; *60%*
7. P(purple) $\frac{0}{10}$; *0%*
8. P(not purple) $\frac{10}{10}$; *100%*

Use the spinner to find the probability of the event. Write it as a fraction in lowest terms.

9. P(blue) $\frac{4}{6} = \frac{2}{3}$
10. P(green) $\frac{2}{6} = \frac{1}{3}$
11. P(even) $\frac{3}{6} = \frac{1}{2}$
12. P(odd) $\frac{3}{6} = \frac{1}{2}$
13. P(green and odd) $\frac{1}{6}$
14. P(green and even) $\frac{1}{6}$
15. P(blue and even) $\frac{2}{6} = \frac{1}{3}$
16. P(blue and odd) $\frac{2}{6} = \frac{1}{3}$

350   Chapter 16

Instruct

Determining the likelihood of a given statement

- Guide the students as they color a **circle graph** to match a probability scale.
- Write for display: "It will rain today."

  Is it possible or impossible that it will rain today? possible; It has rained other days.

  Since it is possible, is it unlikely or likely it will rain today? sample answers: likely since it rains often or unlikely since it seldom rains
- Repeat the procedure using these statements.

Use the spinner on page 350 to find the probability of the event and its complement. Write both as percents. Round to the nearest percent.

17. P(blue) and P(not blue) *67%; 33%*

18. P(green) and P(not green) *33%; 67%*

19. P(odd) and P(not odd) *50%; 50%*

20. P(5) and P(not 5) *17%; 83%*

A 1–6 number cube is rolled once. Write the probability of the event as a fraction in lowest terms and as a percent. Round to the nearest percent.

21. P(3) $\frac{1}{6}$; *17%*

22. P(multiple of 2) $\frac{1}{2}$; *50%*

23. P(2 or 5) $\frac{1}{3}$; *33%*

24. P(7) *0; 0%*

25. P(greater than 1) $\frac{5}{6}$; *83%*

26. P(less than 5) $\frac{2}{3}$; *67%*

27. P(4) $\frac{1}{6}$; *17%*

28. P(not 6) $\frac{5}{6}$; *83%*

29. P(odd) $\frac{1}{2}$; *50%*

30. P(composite) $\frac{1}{3}$; *33%*

The probability of event A is given. Find the complement, P(not A).

31. P(A) = 30% *P(not A) = 70%*

32. P(A) = $\frac{1}{4}$ *P(not A) = $\frac{3}{4}$*

33. P(A) = 40% *P(not A) = 60%*

34. P(A) = 45% *P(not A) = 55%*

35. P(A) = $\frac{2}{5}$ *P(not A) = $\frac{3}{5}$*

36. P(A) = 29% *P(not A) = 71%*

Identify the event as certain, equally likely, or impossible. Each choice will be used only once.

37. If a 1–6 number cube is rolled one time, what word or phrase best describes the probability of the event?
a. rolling an even number *equally likely*
b. rolling a number 1–6 *certain*
c. rolling a number greater than 6 *impossible*

Write the probability as a fraction and as a percent. Round to the nearest percent.

38. Eighty people attended the Sunday morning service at Regency Bible Church. Twenty people sang in the choir. What is the probability that a person attending is a choir member?
$\frac{20}{80} = \frac{1}{4}$; $\frac{1}{4}$ = 25%

39. You are given 5 choices for a multiple-choice test question. If you do not know the answer, what is the probability of guessing the correct answer? *$\frac{1}{5}$ = 20%*

40. The Heritage Christian School soccer team has a record of 8 wins and 3 losses. Victory Christian School has a record of 6 wins and 2 losses. Which team is more likely to win its next game?

Soccer Records		
	HCS	**VCS**
wins	8	6
losses	3	2
games played	11	8

Heritage: $\frac{8}{11} \approx 73\%$
Victory: $\frac{6}{8} = 75\%$
Victory Christian School is more likely to win.

41. Why are we able to make useful predictions?
We can make useful predictions because the world is designed.

---

will occur. A probability of 0, or 0%, means it is impossible for the event to occur. A probability of $\frac{1}{2}$ means an event will occur 1 out of 2 times or approximately 50% of the time.

- Explain that the circles below the probability scale represent spinners. Ask the following questions as you guide the students as they use their red and blue colored pencils (or crayons) to color the first two spinners so that the probability of landing on blue matches the probability above it on the scale. Demonstrate on the displayed page.

  If the probability of landing on blue is 0, what part of the spinner will be blue? none or 0%

  Since there are only two options, red or blue, what part of the spinner will you color red? all 4 sections ($\frac{4}{4}$) or 100%

  If the probability of landing on blue is 25%, what part of the spinner will be blue? $\frac{1}{4}$; 25% = $\frac{25}{100}$ or $\frac{1}{4}$

  What part of the spinner will be red? $\frac{3}{4}$; Whatever is not blue must be red (the only other color).

- Follow a similar procedure for the 3 remaining spinners. $\frac{1}{2}$ blue and $\frac{1}{2}$ red; $\frac{3}{4}$ blue and $\frac{1}{4}$ red; $\frac{4}{4}$ blue

Evidence of design

- Guide the students in a **discussion** to help them recall that a designed world makes predictions possible.

- Besides predicting spinner results and forecasting weather, in what other situations would it be helpful to make predictions? sample answers: chances of winning certain games (board games, video games); in sports—batting average in baseball or field goal success rate in football

- You may also mention that car insurance companies use predictions to determine rates. (What is the likelihood of a teen driver having a car accident? What is the probability of hitting a deer on the road?)

- Read aloud Isaiah 45:18.

  What do we learn from this verse? God created the heavens and the earth.

"It will snow today." possible; Likelihood is dependent on the time of year and climate.

"The chair is alive." impossible; It is certain that chairs are not alive; they do not eat or breathe.

"Most people in this room are left-handed." possible; It is unlikely because most people are right-handed.

Where might it be likely that most people are left-handed? in a setting where there are many left-handed people

"Most people in this room have brown eyes." possible; Likelihood is dependent on the physical characteristics of the people in the room.

"God rules supreme." certain; It is impossible for God not to rule supreme because of who He is.

- Distribute a copy of the *Probability* page to each student and display your copy. Explain that probability is the likelihood (or uncertainty) that an event will occur.

- Direct attention to the probability scale as represented on the number line on the page. Point out that probability can be expressed as a fraction or as a decimal between 0 and 1. It can also be expressed as a percent between 0% and 100%. The closer a probability is to 1, the more likely the event will happen. A probability of 1, or 100%, means it is certain that the event

LESSON 160

- Remind the students that the Bible teaches that God designed the world in a certain way. As we figure out His design, we can make accurate predictions about the world.

- Direct attention to the top of Student Edition page 350, and guide the students as they answer the essential question, "Why are we able to make useful predictions?" We can make useful predictions because the world is designed.

Finding theoretical probability

- Use **direct instruction** to help the students understand and find the theoretical probability of an event.

- Display the *Theoretical Probability* page. Explain that theoretical probability is calculated when all of the possible outcomes are known and are equally likely to occur.

  What is the ratio of 1-tiles to all the tiles in the bag? 3 to 6

  What fraction describes the probability of drawing a 1-tile from the bag? $\frac{3}{6}$

  What is the probability in lowest terms? $\frac{1}{2}$

  Point out that $P(1)$ on the table represents the probability of drawing a 1-tile from the bag. Write "$\frac{1}{2}$" in the fraction column. Explain that the probability can be read "one out of two."

  What decimal is equivalent to $\frac{1}{2}$? 0.5

  What percent is equivalent to $\frac{1}{2}$? 50%

  Write "0.5" and "50%" to complete the first row of the table.

- Explain that since theoretical probability assumes that all outcomes are equally likely to occur, $P(event)$ can be expressed as a part-to-whole ratio: the number of favorable (desired) outcomes divided by the number of possible outcomes.

Make a stem-and-leaf plot with the data. Use the data to answer the questions.

> Mr. Arnold recorded the number of emergency calls that were placed over a 10-day period in March.

calls	70	82	74	70	69	76	75	80	78	73
day	1	2	3	4	5	6	7	8	9	10

1. What is the range of the calls? *82 – 69 = 13*

2. What is the mean? *74.7 ≈ 75*

3. What is the mode? *70*

4. What is the median? *74.5*

Emergency Calls Recorded	
Stem	Leaf
6	9
7	0, 0, 3, 4, 5, 6, 8
8	0, 2

Key 6|9 = 69

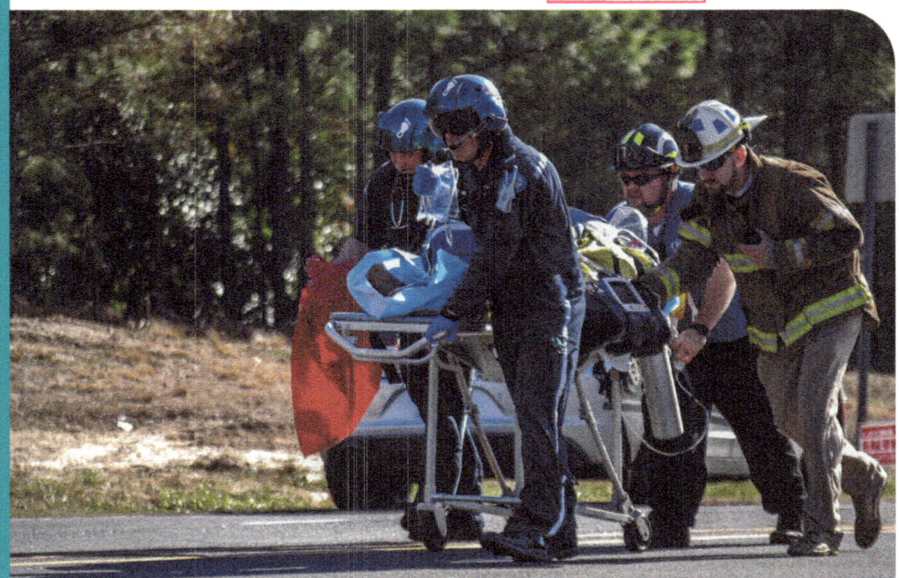

- Write the following statements for display: "It will rain today" and "It will not rain today." Explain that these events are complementary events; only one or the other can occur. Point out that $P(1)$ and $P(not\ 1)$ are complementary events; they cannot happen at the same time.

  What is the ratio of non-1-tiles to all the tiles in the bag? 3 to 6

  What fraction describes the probability of drawing a non-1-tile from the bag? $\frac{3}{6}$

  What is the probability in lowest terms? $\frac{1}{2}$

  What is the decimal equivalent to $\frac{1}{2}$? 0.5

  What is the percent equivalent to $\frac{1}{2}$? 50%

Write all three forms of the probability in the table.

- Direct attention to the equation: $P(event) + P(not\ event) = 1$ or 100%. Explain that the sum of the probability of complementary events will always equal 1 or 100%. When they know the probability of an event, they can subtract the known probability from 1 or 100% to find the probability of its complement.

  Guide the students as they determine the related equations specific to the probability of drawing the 1-tile from the bag. sample answers: $P(1) + P(not\ 1) = 1$; $1 - P(not\ 1) = P(1)$; $P(not\ 1) + P(1) = 100\%$; $100\% - P(1) = P(not\ 1)$

351a • Lesson 160

Math 6

- Follow a similar procedure to complete the table. Assist the students in concluding that since probabilities can be written as a percent, the decimal should be rounded to the nearest hundredth when dividing the number of favorable outcomes by the number of possible outcomes. The percent (probability) is an approximation.

  $P(2)$ $\frac{1}{3}$, 0.33, 33%
  $P(not\ 2)$ $\frac{2}{3}$, 0.67, 67%

  $P(3)$ $\frac{1}{6}$, 0.17, 17%
  $P(not\ 3)$ $\frac{5}{6}$, 0.83, 83%

  $P(odd\ number)$ $\frac{2}{3}$, 0.67, 67%
  $P(not\ odd\ number)$ $\frac{1}{3}$, 0.33, 33%

- Draw for display a bag containing the following number squares: two 1-tiles, three 2-tiles, and four 3-tiles. Repeat the activity.

  $P(1)$ $\frac{2}{9}$, 0.22, 22%
  $P(not\ 1)$ $\frac{7}{9}$, 0.78, 78%

  $P(2)$ $\frac{1}{3}$, 0.33, 33%
  $P(not\ 2)$ $\frac{2}{3}$, 0.67, 67%

  $P(3)$ $\frac{4}{9}$, 0.44, 44%
  $P(not\ 3)$ $\frac{5}{9}$, 0.56, 56%

- Display the *Probability* page again. Guide the students to conclude the following complement [$P(not$ blue)] for each spinner.

  $P(blue) = 0$; $P(not$ blue) = 1 or 100%
  Point out that when it is impossible to spin blue, it is certain that they will spin red (*not* blue).

  $P(blue) = \frac{1}{4}$; $P(not$ blue) = $\frac{3}{4}$ or 75%
  Guide the students to conclude that when it is unlikely that they will spin blue, it is likely that they will spin red (*not* blue).

  $P(blue) = \frac{1}{2}$; $P(not$ blue) = $\frac{1}{2}$ or 50%
  Guide the students to reason that it is equally likely that they will spin blue or red (*not* blue).

  $P(blue) = \frac{3}{4}$; $P(not$ blue) = $\frac{1}{4}$ or 25%
  Guide the students to conclude that when it is likely that they will spin blue, it is unlikely that they will spin red (*not* blue).

  $P(blue) = 1$; $P(not$ blue) = 0 or 0%
  Guide the students to reason that when it is certain that they will spin blue, it is impossible that they will spin red (*not* blue).

Apply

Student Edition pages 350–51
- Read and explain the directions for pages 350–51. Assist the students as they complete the pages independently.

Daily Review
- Students should complete Chapter 16, section *a*.

MATH TALK

Follow the General Procedure for Math Talks as outlined in Lesson 5 on Teacher Edition page 13b.

How many dots are shown?

NOTES

Student Edition pages 352–53
Daily Review Chapter 16, section *b*

OBJECTIVES

- Find the sample space for and the probability of an event.
- Make a tree diagram to list the sample space for an event.
- Determine the number of possible outcomes using the Multiplication Counting Principle.

TEACHER RESOURCES

- 121 *Sample Spaces*
- 122 *Tree Diagram* (for the teacher and for each student)
- 123 *Multiplication Counting Principle*

ADDITIONAL MATERIALS

- a calculator (for each student) (optional)

Engage

- Direct attention to the top of Student Edition page 352 and lead the students in a **brainstorming activity** to explore the essential question, "How can a tree help me pack for vacation?"

- Allow several students to share their ideas with the class. Write down some of the responses to refer to at the end of the lesson.

Instruct

Finding the sample space for & the probability of an event

- Direct the students to **use a table** to help them to determine the probability of an event.

- Display the *Sample Spaces* page.

  If you were to make 1 spin using this spinner, what are the possible outcomes? A, B, or C

- Explain that the sample space for an event is the set of all possible outcomes for the event. The sample space can be listed

---

Sample Spaces

How can a tree help me pack for vacation?

Knowing the number of possible outcomes is necessary when calculating probability. The **sample space** for an event is the set of all possible outcomes. A **tree diagram** is an organized way to show all the possible outcomes.

$$P(\text{event}) = \frac{\text{number of favorable outcomes}}{\text{number of possible outcomes}}$$

The sample spaces show the outcomes of flipping 1, 2, and 3 coins. Each coin has a head (**H**) and a tail (**T**).

flipping 1 coin: 2 possible outcomes {**H, T**}

flipping 2 coins: 4 possible outcomes {**HH, HT, TH, TT**}

flipping 3 coins: 8 possible outcomes {**HHH, HHT, HTH, HTT, THH, THT, TTH, TTT**}

Key Terms
- sample space
- tree diagram
- Multiplication Counting Principle

Tree Diagram					
1st Coin	2nd Coin	3rd Coin	Outcome for Coins		
			1	**2**	**3**
		H	H	HH	HHH
	H	T	T	HT	HHT
H		H		TH	HTH
	T	T		TT	HTT
		H			THH
	H	T			THT
T		H			TTH
	T	T			TTT

Exercises

Use the sample spaces above to write the probability of the event as a fraction and as a percent. *Students may or may not use simplest fraction form.*

1. 1 coin, P(heads) $\frac{1}{2}$; *50%*
2. 3 coins, P(at least one tail) $\frac{7}{8}$; *88%*
3. 2 coins, P(at least one head) $\frac{3}{4}$; *75%*
4. 3 coins, P(at least 2 tails) $\frac{4}{8} = \frac{1}{2}$; *50%*
5. 2 coins, P(*not* tails) $\frac{1}{4}$; *25%*
6. 3 coins, P(heads and tails) $\frac{6}{8} = \frac{3}{4}$; *75%*

Use the spinner to find the answer.

7. List the sample space for spinning the spinner two times. *{rr, rb, rg, br, bb, bg, gr, gb, gg}*
8. Use the sample space from problem 7 to find the probability of landing on the same color both times. $\frac{3}{9} = \frac{1}{3}$ *or 33%*
9. What is the probability of landing on a different color both times? $\frac{6}{9} = \frac{2}{3}$ *or 67%*

Use the spinner and/or a 1–6 number cube to find the answer.

10. List the sample space for one spin of the spinner. *{A, B, C}*
11. List the sample space for one roll of the number cube. *{1, 2, 3, 4, 5, 6}*
12. List the sample space for one spin of the spinner *or* one roll of the number cube. *{A, B, C, 1, 2, 3, 4, 5, 6}*
13. List the sample space for one spin of the spinner *and* one roll of the number cube. *{A1, A2, A3, A4, A5, A6, B1, B2, B3, B4, B5, B6, C1, C2, C3, C4, C5, C6}*
14. Write the probability (as a fraction and as a percent) for rolling a 6 after any letter for one spin on the spinner *and* one roll of the number cube. $\frac{3}{18} = \frac{1}{6}$; *17%*

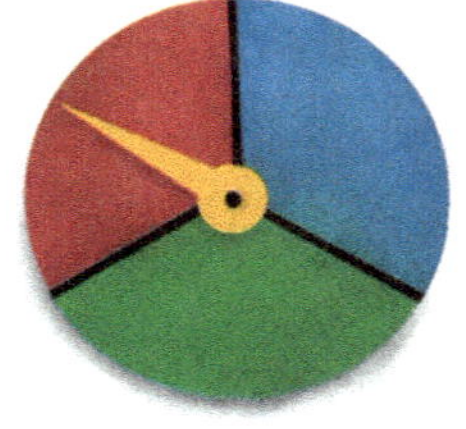
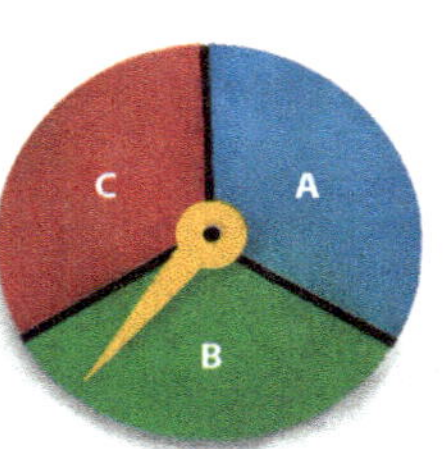

---

using brackets (set notation). Point out that the sample space {A, B, C} in the table lists the 3 possible outcomes for 1 spin of this spinner.

If you desired to spin an A with 1 spin of this spinner, what fraction represents the probability of spinning an A? $\frac{1}{3}$, 0.33, or 33%

Point out that the ratio of favorable (desired) outcomes (A) to possible outcomes (A, B, C) is 1 to 3.

Write "P(A) = $\frac{1}{3}$" for display. Explain that the probability can be read "1 out of 3."

How can you write P(A) as a decimal? 0.33

How can you write P(A) as a percent? 33%

- Follow a similar procedure for P(B) $\frac{1}{3}$, 0.33, 33%; P(C) $\frac{1}{3}$, 0.33, 33%; and P(*not* A) $\frac{2}{3}$, 0.67, 67%.

- Explain that if they were to spin the spinner 2 times, they could land on A for spin 1 and land on A again for spin 2, so one possible outcome for 2 spins would be AA. Guide the students as they find the sample space for 2 spins and list it on the displayed page. {AA, AB, AC, BA, BB, BC, CA, CB, CC}

How many possible outcomes are there for 2 spins? 9

Write "9" in the table.

Use the **Multiplication Counting Principle** to find the number of outcomes when choices are given.

Multiplication Counting Principle: Given event A and event B, the number of possible outcomes of both events is A × B.

Jaden has 5 different-colored shirts and 3 different-colored pants. How many different outfits can he make?

5 shirt choices × 3 pant choices
5 × 3 = **15 outfit choices**

Clark's Catering offers 3 meats, 4 vegetables, and 2 desserts. How many different meal combinations can be made?

3 meat choices × 4 vegetable choices × 2 dessert choices
3 × 4 × 2 = **24 meal combinations**

Use the Multiplication Counting Principle to find the number of possible outcomes.

15. Lorenzo is purchasing a new shirt. The store has short-sleeve and long-sleeve shirts in 8 different colors. How many different shirts does Lorenzo have to choose from? *16 shirt choices*

16. Julian bought a combination lock with 4 dials. Each dial contains the numbers 1 to 9. How many possible combinations can be made for this lock? *6,561 combinations*

17. Morgan is ordering an egg sandwich for breakfast. Her choices are biscuit or croissant; sausage, bacon, or ham; and with cheese or without cheese. How many sandwich choices does Morgan have? *12 sandwich choices*

18. Sasha is redecorating her bedroom. From the store's showroom, she can choose one of 4 beds, one of 2 nightstands, and one of 3 desks. How many possible bedroom sets can Sasha choose from? *24 bedroom sets*

Find the number of possible outcomes and list the sample space. Write the probability as a fraction in lowest terms and as a percent.

19. Options on a new car are a standard or an automatic transmission in a 2- or 4-door model. Find P(automatic transmission, 4-door).
19. $2 \times 2 = 4$ combinations; $\frac{1}{4}$; 25%

20. White and black cars come with a red, tan, or black interior. Find P(white or black car, black interior). *$2 \times 3 = 6$ combinations; $\frac{1}{3}$; 33%*

21. You may choose one kind of ice-cream cone (sugar or regular), one scoop of ice cream (chocolate or vanilla), and one topping (sprinkles, peanuts, or chocolate chips). Find P(cone with chocolate ice cream). *$2 \times 2 \times 3 = 12$ combinations; $\frac{1}{2}$; 50%*

22. Carson packed 3 shirts (red, orange, green), 2 pants (blue, khaki), and 2 sweatshirts (solid, print). If Carson wears a shirt, a pair of pants, and a sweatshirt, how many different combinations can he make? Find P(red shirt, blue pants, solid sweatshirt). *$3 \times 2 \times 2 = 12$ combinations; $\frac{1}{12}$; 8%*

Practice & Application

Katelyn has enough money to buy a small pizza with one topping. Thick and thin crusts cost the same price.

23. Make a tree diagram of Katelyn's choices.

24. List the sample space of her choices.

25. Find P(small pizza, thick crust).

26. Find P(small pizza, thick crust, pepperoni).

27. How can a tree help me pack for vacation?
A tree diagram can help me organize all possible combinations of clothes for my trip.

Bochi's Pizzeria		
Sizes	**Crusts**	**Toppings**
small medium large extra large	thick thin	mushrooms olives spinach pepperoni ham extra cheese

Lesson 161   353

- Guide the students as they conclude the following probabilities for 2 spins on the spinner. Choose students to calculate the decimal form to the nearest hundredth. P(AA) $\frac{1}{9}$, 0.11, 11%; P(AC) $\frac{1}{9}$, 0.11, 11%; P(at least one A) $\frac{5}{9}$, 0.56, 56%

Making a tree diagram to list the sample space for an event

- Direct the students to use a **graphic organizer** to help them list the sample space.

- Point out that AAA on the *Sample Spaces* page is one possible outcome if they were to spin the spinner 3 times. Assist the students as they suggest several other outcomes.

- Distribute a copy of the *Tree Diagram* page to each student and display your copy. Explain that a tree diagram can be used to organize all the possible outcomes and to list the sample space of an event. The first branch lists the possible outcomes for the first event or 1 spin. Point out that the number of possible outcomes for 1 spin (3) is recorded below the tree.

If you spin A on the first spin, what are the possible outcomes for the second spin? A, B, or C

- Trace each dotted line to draw a branch from A to A, A to B, and A to C.

Throughout the activity, instruct the students to draw and label the branches after you draw and label them for display.

If you spin B on the first spin, what are the possible outcomes for the second spin? A, B, or C

Draw 3 similar branches from B. Label the branches A, B, C.

If you spin C on the first spin, what are the possible outcomes for the second spin? A, B, or C

Draw 3 branches from C and label them A, B, C. Model reading each branch of the tree to find each possible outcome in the sample space for 2 spins and list them for display. {AA, AB, AC, BA, BB, BC, CA, CB, CC}

How many possible outcomes are there for 2 spins? 9

Write "9" in the table below the second spin.

- Follow a similar procedure to extend the tree diagram to show the third spin. Guide the students as they read aloud each branch of the tree diagram and guide them as they list all the possible outcomes to create the sample space for 3 spins. {AAA, AAB, AAC, ABA, ABB, ABC, ACA, ACB, ACC, BAA, BAB, BAC, BBA, BBB, BBC, BCA, BCB, BCC, CAA, CAB, CAC, CBA, CBB, CBC, CCA, CCB, CCC}

How many possible outcomes are there for 3 spins? 27

Write "27" in the table.

- Guide the students as they find the following probabilities for 3 spins: P(AAA) $\frac{1}{27}$, 0.04, or 4%; P(CCC) $\frac{1}{27}$, 0.04, or 4%; P(at least 2 Bs) $\frac{7}{27}$, 0.26, or 26%.

- Direct the students to draw a circular spinner on the back of their page and to shade half of the spinner. Then guide them as they draw a tree diagram and identify the sample spaces (black or white) for 3 spins of the spinner.

1 spin {B, W}
2 spins {BB, BW, WB, WW}
3 spins {BBB, BBW, BWB, BWW, WBB, WBW, WWB, WWW}

LESSON 161

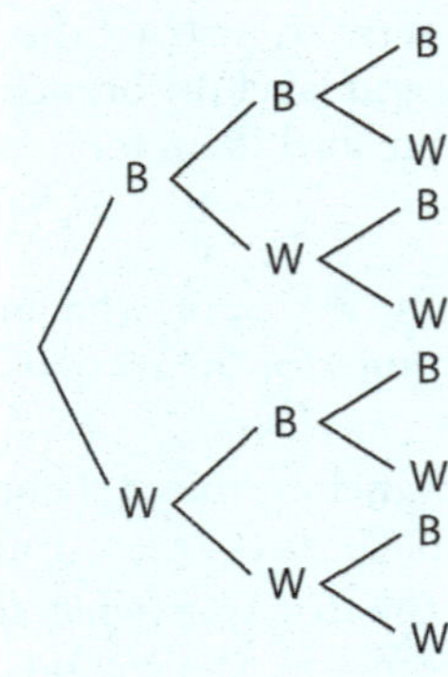

- Refer to the essential question and responses given at the beginning of the lesson and guide the students as they answer the question, "How can a tree help me pack for vacation?" sample answer: The organizational chart I have used is shaped like a tree. A tree diagram can help me organize all the possible combinations of clothes for my trip.

Determining the number of possible outcomes

- Use a **guided practice** to help the students find the number of possible outcomes.

- Display the first word problem on the *Multiplication Counting Principle* page with the tree diagram covered. Instruct the students to think of the different combinations of peanut butter sandwiches if one other spread and one type of bread were chosen. Assist the students as they list several possible combinations.

  What can help you list the sample space for this word problem? a tree diagram

- Uncover the tree diagram. Point out the 1 choice of peanut butter, the 3 choices of spread (G, S, H), and the 2 choices of bread (W, M).

  How many possible combinations of peanut butter (P) and a spread are there? 3

  Guide the students as they read each tree branch to list the possible combinations of peanut butter and a spread: PG, PS, PH.

- Follow a similar procedure to list the possible combinations of peanut butter, a spread, and a type of bread. 6 possible combinations {PGW, PGM, PSW, PSM, PHW, PHM}

Use the picture to answer the question.

1. What is the ratio of vegetables to tuna? *3 : 4*

2. What is the ratio of animal crackers to chips? *2 : 1*

3. What is the ratio of rice mix to animal crackers? *2 : 2*

4. What is the ratio of canned food to total food items? *7 : 12*

Write each ratio as a fraction in lowest terms.

5. 6 boys to 8 girls $\frac{6}{8} = \frac{3}{4}$

6. 1 c brown sugar to 2 c orange juice $\frac{1}{2}$

7. 2 c gelatin to 5 c strawberries $\frac{2}{5}$

8. 3 adults to 18 children $\frac{3}{18} = \frac{1}{6}$

9. 15 elephants to 25 mice $\frac{15}{25} = \frac{3}{5}$

10. 3 piano players to 21 brass players $\frac{3}{21} = \frac{1}{7}$

Daily Review   509

What is the probability of having a peanut butter sandwich with jelly? $\frac{4}{6} = \frac{2}{3}$, 0.67, or 67%; Assist the students as they conclude that there are 4 sandwiches that contain jelly out of 6 possible combinations (outcomes).

Point out that 2 of the 3 spreads are jellies, and 4 of the 6 possible sandwiches have jelly. Guide the students as they find the complement $P(not\ \text{jelly}) = \frac{2}{6}$ or $\frac{1}{3}$.

- Guide the students as they compare the number of branches from each choice to the number of possible outcomes. Assist the students as they reason that multiplying the number of choices and the

number of branches from those choices gives the number of possible outcomes (e.g., 1 peanut butter choice × 3 spread choices = 3 possible combinations of peanut butter and a spread; 1 peanut butter choice × 3 spread choices × 2 bread choices = 6 possible combinations of peanut butter, a spread, and a type of bread).

- Explain that the process of multiplying the number of choices and the number of branches from those choices to find the total number of possible outcomes is called the Multiplication Counting Principle.

- Read aloud the second word problem. What multiplication equation can you write to show the total number of fruit basket combinations Mrs. Parker can make? 2 × 2 × 3 = 12; 2 basket sizes × 2 cellophane choices × 3 bow choices = 12 combinations.

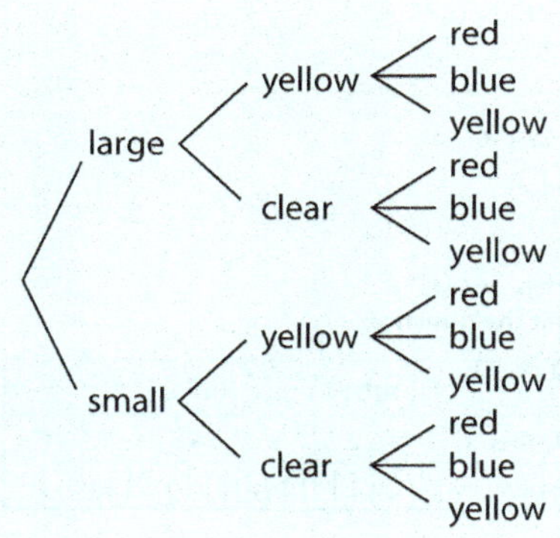

- Guide the students as they make a tree diagram to list the sample space as you demonstrate.

  What is the probability of Mrs. Parker making a fruit basket with a red bow? sample answer: 2 basket sizes × 2 cellophane choices × 1 bow choice = 4 combinations (outcomes) with a red bow; $\frac{4}{12} = \frac{1}{3}$, 0.33, or 33%

- Display the *Sample Spaces* page again. Guide students as they apply the Multiplication Counting Principle to write a multiplication equation for finding the number of possible outcomes for 2 spins. 3 possible outcomes on the first spin (A, B, C), and 3 possible outcomes on the second spin (A, B, C); 3 × 3 = 9

  Compare the product to the number of possible outcomes listed in the sample space for 2 spins.

  Repeat the procedure for 3 spins. 3 × 3 × 3 = 27

- Guide the students as they use the Multiplication Counting Principle to answer the following questions.

  How many different outfits can you make from 7 shirts and 3 pairs of pants? 7 × 3 = 21 outfits

  How many different lunch combinations of a drink, a burger, and fries can you make from 5 types of drinks, 4 types of burgers, and 2 types of fries? 5 × 4 × 2 = 40 lunch combinations

  How many possible outcomes are there if you roll a 1–6 number cube two times? 6 × 6 = 36 possible outcomes

Apply

Student Edition pages 352–53

- Read and explain the directions for pages 352–53. Assist the students as they complete the pages independently.

Daily Review

- Students should complete Chapter 16, section *b*.

DIFFERENTIATED INSTRUCTION

Use the following to provide extra help for students who experience difficulty with the concepts taught in Chapter 16.

Predict the results for a large sample of data using probability.
Students who are having difficulty using probability to predict the results for a large sample of data may find it easier to write and solve a missing term proportion to predict the results. Draw for display the "Favorite Number from 0 to 5" chart and assist the students as they conclude that the tallies indicate that a total of 20 people were surveyed. Guide the students as they determine the probability of each number being chosen as a favorite.

Favorite Number from 0 to 5		
0	||	$P(0) = \frac{2}{20} = \frac{1}{10}$
1	⫴⫴ |||	$P(1) = \frac{8}{20} = \frac{2}{5}$
2	|||	$P(2) = \frac{3}{20}$
3		$P(3) = \frac{0}{20} = 0$
4	||	$P(4) = \frac{2}{20} = \frac{1}{10}$
5	⫴⫴	$P(5) = \frac{5}{20} = \frac{1}{4}$

Remind the students that the probability of the sample of 20 people surveyed can be used to predict the results if a larger sample of 100 people were surveyed (e.g., $P(0) = \frac{1}{10}$; therefore, $\frac{1}{10} \times 100$ = the number of people out of 100 that are expected to choose 0 as their favorite number). Direct the students to write and solve a proportion that has a missing term to find the expected number of people out of 100 that would choose 0 as their favorite number. $\frac{1}{10} = \frac{n}{100} = 10$ Guide the students as they conclude

that since 1 out of 10 people chose 0 as their favorite number when 20 people were surveyed, they could predict that 10 people would choose 0 as their favorite number if 100 people were surveyed. Follow a similar procedure for each of the other numbers.

Student Edition pages 354–55
Daily Review Chapter 16, section c

OBJECTIVES

- Find the theoretical probability of an event.
- Predict the results of an experiment using the theoretical probability of an event.
- Conduct a probability experiment.
- Find the experimental probability of an event.
- Explain why we cannot predict outcomes exactly. **BWS**
- Create a line plot for the results of a probability experiment.

BIBLICAL WORLDVIEW SHAPING

- **Design** (Explain): God's world is so complex that we often cannot predict exactly what will happen.

TEACHER RESOURCES

- 124 *Experimental Probability* (for the teacher and for each student)
- 125 *Spinning Penny Experiment* (for each pair of students, optional)

ADDITIONAL MATERIALS

- a paper clip (for the teacher and for each student)
- a calculator (for each student) (optional)
- a penny for every two students (optional)

A paper clip can be used as the pointer for a spinner. Place the paper clip so that one end is in the center of the spinner. In order to spin the paper clip, place a pencil point inside the end of the paper clip so that the pencil point is touching the center of the spinner, holding the paper clip in place.

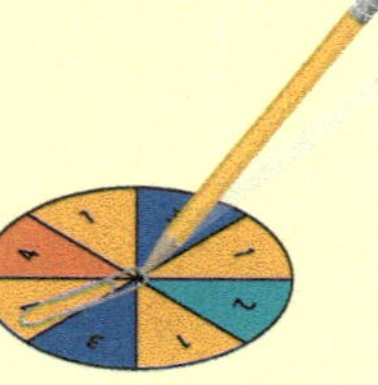

Experimental Probability

Why can't we predict outcomes exactly?

Experimental probability is found using data collected from an experiment or a survey. The experimental probability of an event is the number of observed occurrences of an event in relation to the total number of trials (or people surveyed).

$$P(\text{event}) = \frac{\text{number of favorable outcomes}}{\text{total number of trials}}$$

Theoretical probability tells the outcome of an experiment if all the given outcomes are equally likely to occur. The actual experimental results may not be the same as the expected (or predicted) results.

Key Terms
- experimental probability

Exercises

Write the theoretical probability of the event when rolling a 1–6 number cube.

1. $P(5)$ $\frac{1}{6}$ *or 17%*
2. $P(3)$ $\frac{1}{6}$ *or 17%*
3. $P(6)$ $\frac{1}{6}$ *or 17%*

Use the tally chart to find the experimental probability of the event. Write it as a fraction in lowest terms and as a percent. Answer the question.

4. $P(1)$ $\frac{2}{20} = \frac{1}{10}$; *10%*
5. $P(4)$ $\frac{4}{20} = \frac{1}{5}$; *20%*
6. $P(2)$ $\frac{3}{20}$; *15%*
7. $P(5)$ $\frac{5}{20} = \frac{1}{4}$; *25%*
8. $P(3)$ $\frac{5}{20} = \frac{1}{4}$; *25%*
9. $P(6)$ $\frac{1}{20}$; *5%*

Number Cube Rolls						
Number	1	2	3	4	5	6
Results	II	III	ʬII	IIII	ʬII	I

10. How would variables such as an unbalanced cube, a toss from a different angle, or the placement of each number on the cube before the toss cause a difference in the theoretical and experimental probability for the same experiment?

Any variable would most likely change the outcome of experimental probability. The theoretical probability would most likely stay the same.

Use the spinner to find the theoretical probability of the event. Write it as a fraction in lowest terms and as a percent. Round to the nearest percent.

11. $P(\text{red})$ $\frac{3}{6} = \frac{1}{2}$; *50%*
12. $P(\text{blue})$ $\frac{2}{6} = \frac{1}{3}$; *33%*
13. $P(\text{yellow})$ $\frac{1}{6}$; *17%*
14. $P(not \text{ blue})$ $\frac{4}{6} = \frac{2}{3}$; *67%*
15. $P(\text{red and blue})$ *0%; The spinner cannot land on 2 colors.*
16. $P(not \text{ red})$ $\frac{3}{6} = \frac{1}{2}$; *50%*

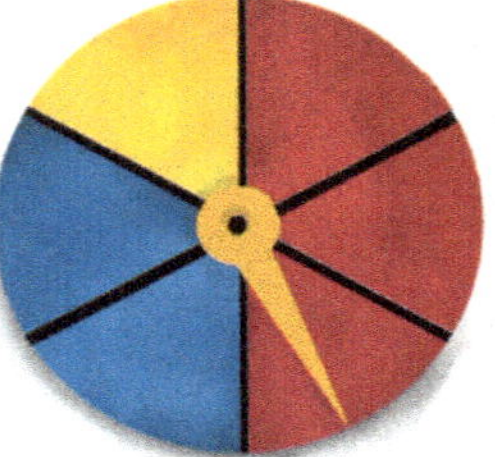

Use your findings from problems 11–16 to find the answer.

17. Which line plot would you expect to best represent the results of an experiment that consisted of spinning the spinner 6 times? Explain your answer.

B; $P(red) = \frac{3}{6} = \frac{1}{2}$; $P(blue) = \frac{2}{6} = \frac{1}{3}$; $P(yellow) = \frac{1}{6}$

A	B	C
x x x x x x red blue yellow	x x x x x x red blue yellow	x x x x x red blue yellow

354   Chapter 16

Engage

- Direct attention to the top of Student Edition page 354 and direct the students to do a **Think-Pair-Share** to explore the essential question, "Why can't we predict outcomes exactly?"

Instruct

Finding theoretical probability & predicting results

- Guide the students as they **complete a table** in order to determine theoretical probability.

What is probability? It is the likelihood that an event will occur.

- Distribute an *Experimental Probability* page and a paper clip to each student and display your copy of the page. Direct attention to the spinner and remind students that the theoretical probability of an event is a ratio of the number of favorable (desired) outcomes divided by the number of possible outcomes that are equally likely to occur.

If you were to make 1 spin using this spinner, what are the possible outcomes? 1, 2, 3, or 4

What fraction represents the probability of spinning 1? $\frac{4}{8}$ or $\frac{1}{2}$; The ratios express the number of favorable outcomes (1) to the number of possible outcomes.

Create a spinner similar to the model given. Write the theoretical probability for the expected results. Conduct the experiment and complete the table.

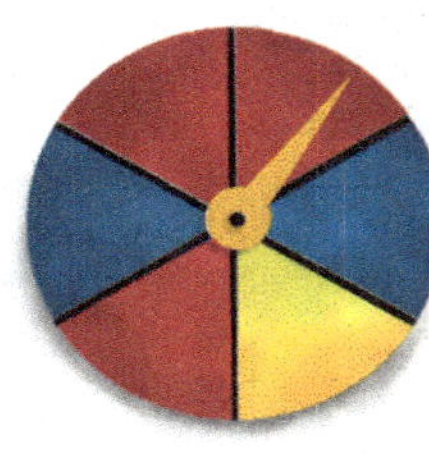

18. Spinner Experiment: Spin a paper clip on the spinner 12 times.

Theoretical Probability	Number of Trials	Expected Results	Actual Results	Experimental Probability
$P(\text{red}) = \frac{1}{2}$	12	$\frac{1}{2} = \frac{6}{12}$		
$P(\text{blue}) = \frac{1}{3}$	12	$\frac{1}{3} = \frac{4}{12}$		
$P(\text{yellow}) = \frac{1}{6}$	12	$\frac{1}{6} = \frac{2}{12}$		

Use the survey results to find the probability of the event. Write it as a fraction in lowest terms and as a percent. Round to the nearest percent.

Notebook Survey	
Color	**Results**
blue	6
red	3
green	4
mixed	6
other	1

19. $P(\text{red notebook})$ $\frac{3}{20}; \textit{15\%}$

20. $P(\text{green notebook})$ $\frac{4}{20} = \frac{1}{5}; \textit{20\%}$

21. $P(\text{red or blue notebook})$ $\frac{9}{20}; \textit{45\%}$

22. $P(\text{mixed colors})$ $\frac{6}{20} = \frac{3}{10}; \textit{30\%}$

Practice & Application

23. Five crayons (red, blue, green, yellow, and orange) were placed in a bag. Find $P(\text{orange})$. $\frac{1}{5}$ *or 20%*

24. Make a list or a tree diagram to determine the sample space for flipping the same coin 3 times. Find $P(\text{at least 2 heads})$. *{HHH, HHT, HTH, HTT, THH, THT, TTH, TTT}; $\frac{4}{8} = \frac{1}{2}$ or 50%*

25. Suppose you roll a 1–6 number cube 2 times. Use the Multiplication Counting Principle to find the number of possible outcomes. Find $P(\text{both rolls result in the same number})$. *$6 \times 6 = 36$ combinations; $\frac{6}{36}$ or 17%*

26. Mrs. Larson surveyed her class and found that 13 out of 25 students have brown eyes. Find $P(\text{not brown eyes})$. *$\frac{12}{25}$ or 48%*

27. Mr. Hernandez surveyed his students to find their favorite type of book. He found that 37% of the class prefers historical fiction. What is $P(\text{complement})$? *$P(\text{not prefer historical fiction}) = 63\%$*

28. Meteorologists provide weather forecasts expressed in percents. What does the meteorologist mean by a 50% chance of rain on Monday? *There is a 1 in 2 chance that it will rain on Monday.*

29. What percentage would a meteorologist say if $P(\text{snowstorm on Tuesday}) = \frac{3}{5}$? *There is a 60% chance of a snowstorm on Tuesday.*

30. Write the value of n if $\frac{10}{7} = \frac{n}{21}$. *$n = 30$*

31. Find the product of $15 \times 16 \times 20$. *4,800*

32. How long will it take to travel 330 mi at a speed of 60 mph? *5.5 hr, $5\frac{1}{2}$ hr, or 5 hr 30 min*

33. Roll a 1–6 number cube 18 times. Draw a line plot showing the frequencies of rolling a 1, 2, 3, 4, 5, and 6 during the experiment. *Answers will vary.*

34. Why can't we predict outcomes exactly? *God's world is so complex that we often cannot predict exactly what will happen.*

33.
1 2 3 4 5 6   *number showing on number cube*

Write "$P(1) = 8$" in the Expected Results column.

What do you notice about the probability of landing on 1 for 1 spin ($\frac{1}{2}$), 8 spins ($\frac{4}{8}$), and 16 spins ($\frac{8}{16}$)? They are equivalent fractions or ratios.

- Direct the students to predict the expected results (probability) of each event if the spinner is spun 16 times and to write their predictions on the page. $P(2) = \frac{1}{8} \times 16 = 2$, $P(3) = \frac{1}{4} \times 16 = 4$, $P(4) = \frac{1}{8} \times 16 = 2$

Conducting a probability experiment; finding experimental probability

- Guide the students as they **conduct an experiment** to find the experimental probability.

- Explain that experimental probability is the probability found during an experiment. Demonstrate spinning the paper clip on the spinner and recording the result by drawing a tally in the Actual Results column of the table for the appropriate number. Allow the students to make several practice spins using their paper clips.

- Instruct each student to spin his paper clip 1 time and to record the result on the table. Direct the students to continue the experiment for a total of 16 spins.

While the students are conducting the experiment, you may complete the experiment and record the results which will be used to demonstrate recording the data on a line plot later in the lesson.

Are the actual results the same as your expected results? Answers will vary.

- Point out that there are a number of variables (conditions) that affect the actual results of an experiment so that the actual results are not exactly the same as the expected results. In this experiment, students may not have spun their paper clip the same way, held their pencil at the same angle, begun with the pointer on the same space, and so on.

- Direct the students to complete the first column of the table by writing the theoretical probability for each event as a fraction in lowest terms: $P(1) = \frac{1}{2}$, $P(2) = \frac{1}{8}$, $P(3) = \frac{1}{4}$, $P(4) = \frac{1}{8}$.

If you were to make 8 spins using this spinner, what do you predict is the number of times you would land on 1? 4; Since the probability of landing on 1 is $\frac{1}{2}$, it is likely that 1 out of every 2 spins (one-half of the spins) would result in 1; 4 is one-half of 8 spins.

What mathematical expression can you write for one-half of 8? $\frac{1}{2} \times 8$

Choose a student to write the expression for display and solve it while the other students solve it. $\frac{1}{2} \times 8 = \frac{8}{2} = 4$

Since $P(1)$ is $\frac{1}{2}$, what proportion could you write to find the expected number out of 12 spins that would result in 1? $\frac{1}{2} = \frac{n}{12}$; $n = 6$ spins

- Explain that probability can be used to predict the expected results of an experiment.

What equation can you write to predict the number of times you would land on 1 if you made 16 spins of the spinner? $\frac{1}{2} \times 16 = 8$ or $\frac{1}{2} = \frac{n}{16}$; $n = 8$ spins

---

LESSON 162

- Explain that experimental probability can be found by taking the actual number of favorable outcomes of the event (the frequency) and dividing it by the total number of trials (16). Demonstrate calculating the experimental probability for 1.

- Direct each student to use his results to calculate the experimental probability for each number and then to record it on his table.

- Guide each student as he compares the theoretical probability to his experimental probability. Guide a discussion about the differences and similarities. Explain that if accurate spinners were used and more trials were performed, the experimental probability would be closer to the theoretical probability.

Predicting exact outcomes

- Guide the students in a **discussion** to answer the essential question.

- Remind the students that the design we see in creation helps us to make useful predictions.

  Although Christians should study creation to learn about God's design, can people ever know the world as God does? no

- Read aloud Romans 11:33–34.

  What do these verses say about God's knowledge, wisdom, and judgments? They are deep and unsearchable. We cannot know them.

- Discuss the ideas that God's thoughts are still high above ours even when we detect evidence of His orderly design. We use probability to predict what is likely to happen, not to guarantee what will happen.

  Why can't we predict outcomes exactly? sample answer: God's world is so complex that we often cannot predict exactly what will happen.

Write a comparison sentence by using = or ≠.

1. $\frac{1}{2} = \frac{2}{4}$

2. $\frac{1}{3} \neq \frac{3}{7}$

3. $\frac{81}{72} \neq \frac{17}{26}$

4. $\frac{9}{12} \neq \frac{3}{5}$

Find the missing measurement.

5.
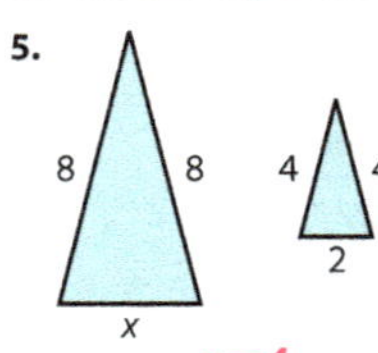
$x = 4$

6.
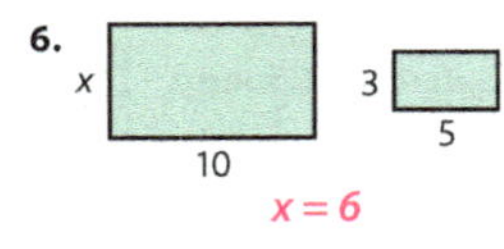
$x = 6$

7.
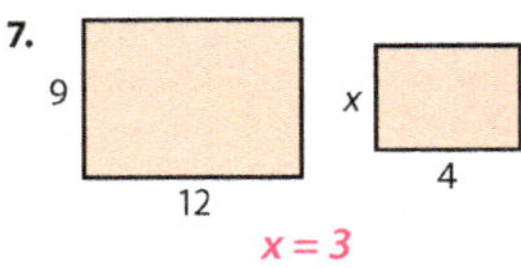
$x = 3$

Find the missing term that completes the equivalent ratio.

8. $\frac{3}{4} = \frac{q}{100}$ $q = 75$

9. $\frac{2}{q} = \frac{4}{16}$ $q = 8$

10. $\frac{2}{3} = \frac{6}{q}$ $q = 9$

11. $\frac{65}{85} = \frac{13}{q}$ $q = 17$

12. $\frac{84}{108} = \frac{q}{9}$ $q = 7$

13. $\frac{q}{56} = \frac{6}{8}$ $q = 42$

510   Daily Review

Creating a line plot for the results of an experiment

- Use **guided practice** to help the students complete a line plot.

- Direct attention to the line at the bottom of the *Experimental Probability* page and remind the students that a line plot is used to record data. Demonstrate counting the tallies for 1 to find the frequency for 1, and drawing the appropriate number of *X*s above the number 1 to correspond with the frequency for 1.

  How many *X*s are needed to complete the line plot? 16; There were 16 trials.

- Direct the students to complete the line plot using their results. Then guide a discussion about the varying line plots.

  Which numbers have a similar frequency? Why do you think this is true? sample answer: The numbers that will most likely have a similar frequency are 2 and 4; each has the same theoretical probability, a 1 out of 8 probability of being landed on.

  Which number has the highest frequency? sample answer: It will most likely be 1; it has the greatest theoretical probability, a 4 out of 8 probability of being landed on.

- Arrange the students in pairs and distribute a *Spinning Penny Experiment* page to each pair. Direct each pair of students to conduct the experiment.

  What variables exist when conducting the spinning penny experiment? sample answers: We may not have spun the pennies at the same speed or on the same type of surface, held our fingers at the same place, and so on.

- Allow students to tell their experimental probabilities for the spinning penny experiment. Guide the students to the conclusion that an expected result or probability for an experiment is similar to a scientific hypothesis for a science experiment or an estimate for a math problem. The expected results are educated guesses.

Apply

Student Edition pages 354–55

- Read and explain the directions for pages 354–55. Assist the students as they complete the pages independently.

Daily Review

- Students should complete Chapter 16, section *c.*

NOTES

LESSON 163

Student Edition pages 356–57
Daily Review Chapter 16, section *d*

OBJECTIVES

- Determine whether all players have an equal chance of winning a game.
- Conduct a probability experiment.
- Find the experimental probability and the theoretical probability of an event.
- Make predictions using probability.

TEACHER RESOURCES

- 126 *Fair or Unfair Games* (for the teacher) (for each pair or group of students, optional)

ADDITIONAL MATERIALS

- 6 marbles (4 orange, 1 brown, 1 yellow) in an opaque bag
- two 1–6 number cubes (for the teacher) (for each pair or group of students, optional)

ASSESSMENTS

- Chapter 16 Quiz 1

Engage

Determining whether all players have an equal chance of winning

- Direct the students to play a **game** to help them determine whether a game is fair.

- Display the bag of marbles. Explain that the bag contains orange, brown, and yellow marbles. Choose 3 students (A, B, and C) to play "Draw a Marble." Explain that if an orange marble is drawn, Student A receives 1 point; if a brown marble is drawn, Student B receives 1 point; and if a yellow marble is drawn, Student C receives 1 point. The first person to receive 5 points wins the game. Shake the bag, allow Student A to draw a marble, and direct the appropriate student to record his point by writing a tally. Return the marble to the bag and shake the bag. Allow Student B to draw a marble and

direct the appropriate student to record his point. Repeat the procedure, allowing Student C to draw a marble. Continue the activity until a winner has been declared. Show the students the 6 marbles.

Does each student have an equal opportunity of winning? In other words, is this game fair or unfair? This game is unfair; Student A had more opportunities to receive points since there were more orange marbles than brown and yellow marbles.

- Direct the students to give the theoretical probabilities of drawing an orange marble, a brown marble, or a yellow marble from the bag. $P(\text{orange}) = \frac{4}{6} = \frac{2}{3}$, $P(\text{brown}) = \frac{1}{6}$, $P(\text{yellow}) = \frac{1}{6}$

What would you do to make the game fair? I would place the same number of each color marble in the bag.

- Explain that in a fair game all the players are equally likely to win; no player has an advantage. Remove 3 of the orange marbles and return the other 3 marbles to the bag.

- Guide the students as they find the theoretical probabilities of drawing an orange marble, a brown marble, or a yellow marble in a fair game. $P(\text{orange}) = \frac{1}{3}$, $P(\text{brown}) = \frac{1}{3}$, $P(\text{yellow}) = \frac{1}{3}$

- Direct attention to the top of Student Edition page 356 and guide the students as they answer the essential question,

---

Fair or Unfair?

How can I use math to determine whether a game is fair?

In a **fair game**, all players are *equally likely* to win. No player has an unfair advantage. Compare the probabilities of winning for each player to determine whether a game is fair.

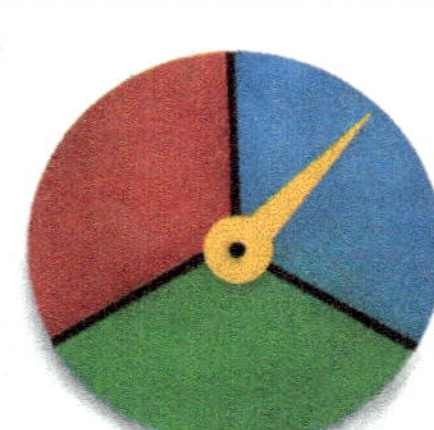

Number Cube Roll

1. Roll a 1–6 number cube.
2. If the number is even, Player A wins. If the number is odd, Player B wins.

sample space: {1, 2, 3, 4, 5, 6}

$P(\text{even}) = \frac{3}{6} = \frac{1}{2}$ or 50%

$P(\text{odd}) = \frac{3}{6} = \frac{1}{2}$ or 50%

Since $P(\text{even}) = P(\text{odd})$, this game is fair for the two players.

Three Coin Toss

1. Toss 3 coins.
2. If the coins all match, Player A wins. If the coins do not all match, Player B wins.

sample space: {HHH, HHT, HTH, HTT, THH, THT, TTH, TTT}

$P(\text{all match}) = \frac{2}{8} = \frac{1}{4}$ or 25%

$P(\text{do not all match}) = \frac{6}{8} = \frac{3}{4}$ or 75%

Since $P(\text{all match}) < P(\text{do not all match})$, this game is unfair. The players are not equally likely to win. Player B has an unfair advantage.

Exercises

3. explanation: The game is fair because each player has a 50% chance of winning.

Use the spinner to find the answer. Write the probability as a fraction in lowest terms.

1. Three players choose a color on the spinner. If the spinner lands on red, Player A wins. If it lands on blue, Player B wins. If it lands on green, Player C wins. Find $P(\text{red})$, $P(\text{blue})$, and $P(\text{green})$. Is this game fair or unfair? *$P(\text{blue}) = \frac{1}{3}$; $P(\text{red}) = \frac{1}{3}$; $P(\text{green}) = \frac{1}{3}$; fair*

2. The three players change colors. This time Player A chooses blue, and Player B chooses red. Player C does not want green again, so he chooses *not* green. Find $P(\text{blue})$, $P(\text{red})$, and $P(\text{not green})$. Is this game fair or unfair? *$P(\text{blue}) = \frac{1}{3}$; $P(\text{red}) = \frac{1}{3}$; $P(\text{not green}) = \frac{2}{3}$; unfair*

5. *sample space: {1, 2, 3, 4, 5, 6}; $P(\text{less than 4}) = \frac{3}{6} = \frac{1}{2}$; $P(\text{greater than 4}) = \frac{2}{6} = \frac{1}{3}$; $P(\text{less than 4}) > P(\text{greater than 4})$; unfair*

List the sample space to find the answer. Write the probability as a fraction in lowest terms.

3. A bag contains 1 red marble and 1 blue marble. One marble is drawn from the bag. If the marble is red, Player A wins. If the marble is blue, Player B wins. Find $P(\text{drawing a red marble})$ and $P(\text{drawing a blue marble})$. Is this game fair or unfair? Explain your answer. *sample space: {r, b}; $P(\text{red}) = \frac{1}{2}$; $P(\text{blue}) = \frac{1}{2}$; $P(\text{red}) = P(\text{blue})$; fair*

5. A 1–6 number cube is rolled once. Player A wins if the number is less than 4. Player B wins if the number is greater than 4. Find $P(\text{less than 4})$ and $P(\text{greater than 4})$. Is this game fair or unfair?

7. How can I use math to determine whether a game is fair? *I can compare the possibilities of winning for each player to determine whether each player has an equal opportunity of winning.*

4. Cory and Edward have to decide who will empty the dishwasher. Edward suggests that they toss 2 coins. If no heads come up, Edward will empty the dishwasher. If at least 1 head comes up, Cory will empty the dishwasher. Find $P(\text{no heads})$ and $P(\text{at least 1 head})$ for tossing 2 coins. Is Edward's proposal fair or unfair?

6. Cory suggests that they toss 3 coins. If at least 2 heads come up, Edward will empty the dishwasher. If at least 2 tails come up, Cory will empty the dishwasher. Find $P(\text{at least 2 heads})$ and $P(\text{at least 2 tails})$ for tossing 3 coins. Is Cory's proposal fair or unfair?

4. *sample space: {HH, HT, TH, TT}; $P(\text{no heads}) = \frac{1}{4}$; $P(\text{at least 1 head}) = \frac{3}{4}$; $P(\text{no heads}) < P(\text{at least 1 head})$; unfair*

356   Chapter 16

The probability of event A is given. Find the complement, P(not A).

8. $P(A) = \frac{6}{10}$ **9.** $P(A) = 35\%$

10. $P(A) = \frac{2}{3}$ **11.** $P(A) = 85\%$

12. $P(A) = \frac{3}{8}$ **13.** $P(A) = 37\%$

Solve.

20. At Peggy's Pancake House, you can order regular or whole-wheat pancakes with a choice of blueberry, strawberry, or maple syrup. How many different choices of pancakes with syrup do you have? *2 pancakes × 3 syrups = 6 choices*

21. The combination lock on the family's storage building has 3 dials. Each dial contains the numbers 1 to 9. How many possible combinations are there for this lock? *9 digits × 9 digits × 9 digits = 729 combinations*

22. Tessa attempted 15 free throws and made 9 of them. What is the probability that she will be successful and make the next throw? $\frac{9}{15} = \frac{3}{5}$ *or 60%*

23. You are given 4 choices for a multiple-choice question. If you do not know the answer, what is the probability of guessing the correct answer? $\frac{1}{4}$ *or 25%*

24. You are answering a true or false question. If you do not know the answer, what is the probability of guessing the correct answer? $\frac{1}{2}$ *or 50%*

Use the survey results to find the probability of the event. Write the probability as a decimal and as a percent.

> Scientists use the probability of an event to make predictions. A survey using a sample of 100 sixth-graders can be used to predict the genetic traits of other sixth-graders.

Genetic Survey Results				
Trait	Dimples	Straight Hair	Attached Earlobes	Widow's Peak
yes	40	50	30	50
no	60	50	70	50

8. *P(not A) = $\frac{4}{10}$*   **9.** *P(not A) = 65%*

10. *P(not A) = $\frac{1}{3}$*   **11.** *P(not A) = 15%*

12. *P(not A) = $\frac{5}{8}$*   **13.** *P(not A) = 63%*

A 1–6 number cube is rolled once. Write the probability of the event as a fraction in lowest terms and as a percent. Round to the nearest percent.

14. P(6) $\frac{1}{6}$; *17%*   **15.** P(a number other than 4)

16. P(9) *0; 0%*   **17.** P(a multiple of 3) $\frac{1}{3}$; *33%*

18. P(odd) $\frac{1}{2}$; *50%*   **19.** P(a prime number) $\frac{2}{3}$; *67%*

15. $\frac{5}{6}$; *83%*

25. P(dimples) *0.4; 40%*

26. P(straight hair) *0.5; 50%*

27. P(attached earlobes) *0.3; 30%*

28. P(widow's peak) *0.5; 50%*

29. Use the information given for the sample of 100 sixth-graders to predict the number of students that will and will *not* have a genetic trait. Make a chart to show your predictions for a sample of 200 students.

Lesson 163   357

"How can I use math to determine whether a game is fair?" *I can compare the possibilities of winning for each player to determine whether each player has an equal opportunity of winning.*

Conducting a probability experiment; finding the experimental probability of an event

- Direct the students to play a **game** to help them determine the probability of rolling a certain sum.

- Display the *Fair or Unfair Games* page and read aloud the procedure for the "Two-Cube Sum" game. Instruct each student to choose the number from 2 to 12 that he thinks will be the winning number, write the number on a small piece of paper, and exchange the paper with a classmate.

Why do you think there is not a column for 1 or 13 in the table? *sample answer: The smallest number on a cube is 1, and the largest number is 6. The smallest sum will be 2 (1 + 1 = 2), and the largest sum will be 12 (6 + 6 = 12). The sample space for this event is {2, 3, 4, 5, 6, 7, 8, 9, 10, 11, 12}.*

LESSON 163

- Allow the students to take turns rolling the cubes and giving the sum of the two numbers. Record each result by drawing a tally in the table on the displayed page. Continue the activity until a number has a frequency of 10 tallies.

  You may allow pairs or small groups of students to play the game. Provide a *Fair or Unfair Games* page and 2 number cubes for each group.

  What was the winning number? *Answers will vary.*

- Direct the students to find the experimental probability of the number with the highest frequency and the number with the lowest frequency. Instruct them to write the probability as a fraction in lowest terms, a decimal, and a percent.

> If pairs or groups of students played the game, each group should calculate the experimental probability using their data.

How can you determine whether this game is fair or unfair? *I can determine whether all the sums are equally likely to occur by listing the possible pairs of addend combinations for each sum.*

How many possible outcomes are there for 1 cube? *6*

If you apply the Multiplication Counting Principle, what multiplication equation can you write to find the number of possible outcomes for 2 cubes? *6 × 6 = 36*

- Invite students to give the possible addend combinations for each sum and list them for display.

  2 = 1 + 1
  3 = 1 + 2, 2 + 1
  4 = 1 + 3, 3 + 1, 2 + 2
  5 = 1 + 4, 4 + 1, 2 + 3, 3 + 2
  6 = 1 + 5, 5 + 1, 2 + 4, 4 + 2, 3 + 3
  7 = 1 + 6, 6 + 1, 2 + 5, 5 + 2, 3 + 4, 4 + 3
  8 = 2 + 6, 6 + 2, 3 + 5, 5 + 3, 4 + 4
  9 = 3 + 6, 6 + 3, 4 + 5, 5 + 4
  10 = 4 + 6, 6 + 4, 5 + 5
  11 = 5 + 6, 6 + 5
  12 = 6 + 6

Is this game fair or unfair? *unfair; All of the sums are not equally likely to occur.*

Which number has the most possible combinations? *7 (6 combinations)*

LESSON 163

Which numbers have the least possible combinations? 2 and 12 (1 combination each)

- Direct students to determine the theoretical probability of 2 rolled number cubes having a sum of each number from 2 to 12. $P(2 \text{ or } 12) = \frac{1}{36}$, 0.03, 3%; $P(3 \text{ or } 11) = \frac{2}{36} = \frac{1}{18}$, 0.06, 6%; $P(4 \text{ or } 10) = \frac{3}{36} = \frac{1}{12}$, 0.08, 8%; $P(5 \text{ or } 9) = \frac{4}{36} = \frac{1}{9}$, 0.11, 11%; $P(6 \text{ or } 8) = \frac{5}{36}$, 0.14, 14%; $P(7) = \frac{6}{36} = \frac{1}{6}$, 0.17, 17%

- Instruct the students to use the list of possible combinations to find $P(odd$ sum) and $P(not$ odd), expressing the probability as a fraction in lowest terms. Remind them that $P(not$ odd) is the complement of $P(odd$ sum). $P(\text{odd sum}) = \frac{18}{36} = \frac{1}{2}$, $P(\text{not odd}) = \frac{18}{36} = \frac{1}{2}$

  Would a game of rolling an odd sum or a *not* odd sum be fair or unfair? fair; The probability of rolling an odd sum is equal to the probability of rolling a *not* odd sum.

  What is the probability of rolling an even sum using 2 cubes? $\frac{18}{36}$ or $\frac{1}{2}$; $P(\text{even})$ is the same as $P(not$ odd).

- Guide the students as they play "Two-Cube Even or Odd Sum" to determine the experimental probability of even and odd sums.

Making predictions using probability

- Assist the students in conducting a **survey** to help them make predictions using probability.

- Explain that meteorologists look at past conditions and use probability to forecast weather; insurance companies use probabilities to decide the cost of insuring a car, a house, or a person's life; and testing companies look at the probability of students guessing correct answers when writing standardized tests.

Write the fraction as a percent.

1. $\frac{1}{4}$ *25%*   2. $\frac{1}{2}$ *50%*   3. $\frac{3}{4}$ *75%*   4. $\frac{1}{5}$ *20%*

Find the percent of the number.

5. 50% of 80 *40*   6. 25% of $4.00 *$1.00*   7. 10% of $8.00 *$0.80*   8. 75% of 40 *30*

Estimate the percent shaded for the rectangle.

9. 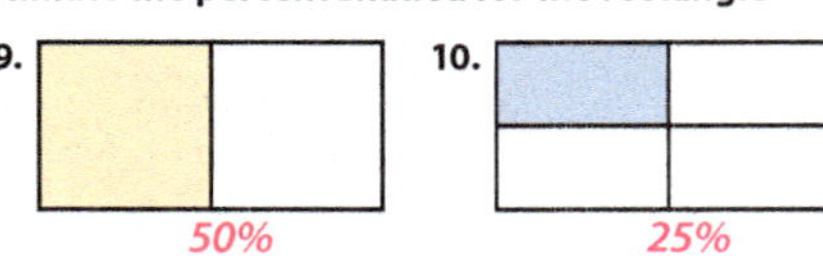 *50%*   10. *25%*   11. 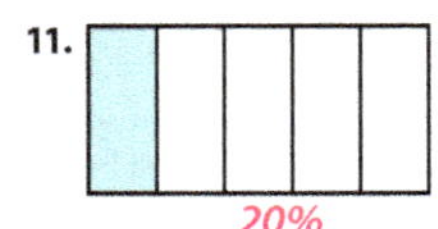 *20%*   12. 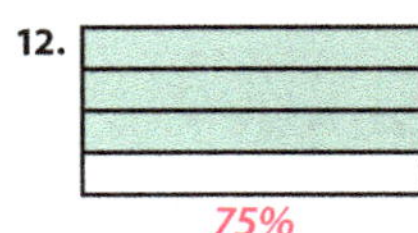 *75%*

Write the number as a percent.

13. $\frac{50}{100}$ *50%*   14. 0.64 *64%*   15. $\frac{15}{100}$ *15%*   16. 0.09 *9%*

Solve.

17. John got 85% of his test correct. What percent did he miss? *15%*

18. Five out of 25 children play soccer. What percent of children play soccer? *20%*

19. Kyle earned $16.00. He wants to put 10% of it in the offering. How much money will he put in the offering? *$1.60*

20. Annie scored 25% of the game points. The total number of points was 40. How many points did she score? *10 points*

- Read aloud the following scenario.

  Jack likes to eat Cinnamon Rings cereal for breakfast every morning. His mother wants him to have more variety, so he eats it every 3 days. Jack made a spinner and divided it into 3 equal sections hoping that his mother would let him use the spinner to decide his breakfast. He labeled the sections Cinnamon Rings, Oatmeal, and Eggs. Jack hopes that he designed the spinner so that it lands on Cinnamon Rings the most often.

  Will this spinner provide Jack with more chances to eat Cinnamon Rings than the other breakfast items? No; all the outcomes are equally likely to occur; $P(\text{CR}) = \frac{1}{3}$, $P(\text{O}) = \frac{1}{3}$, and $P(\text{E}) = \frac{1}{3}$.

- Instruct the students to draw a spinner that will give Jack twice the opportunity to eat Cinnamon Rings as each of the other breakfast items. sample drawing: a spinner divided into fourths; 2 fourths labeled Cinnamon Rings, 1 fourth labeled Oatmeal, and 1 fourth labeled Eggs.

- Survey 5 sixth-grade students to determine their favorite pets. Guide all of your students as they record the responses in a frequency table and identify the sample space [e.g., cat (2), dog (2), rabbit (1)]. Direct the students to calculate the probability of each type of pet being chosen as a favorite (e.g., $P(\text{cat}) = \frac{2}{5}$, $P(\text{dog}) = \frac{2}{5}$, $P(\text{rabbit}) = \frac{1}{5}$).

- Explain that the probability calculated using the data gathered from a small sample of people can be used to make predictions about a larger population of the same people. Scientists use probability when studying genetic traits. Pollsters use probability when predicting the winner in an election.

- Direct the students to use the survey data to predict the expected results for 100 students. Remind them that this prediction is theoretical (an educated guess). Then guide the students as they conduct an actual survey of 100 students' favorite pets using the same animal choices. Guide them to the conclusion that as the number of students surveyed increases, the actual results (experimental probability) will be closer to the predicted results (theoretical probability).

Apply

Student Edition pages 356–57

- Read and explain the directions for pages 356–57. Assist the students as they complete the pages independently.

Daily Review

- Students should complete Chapter 16, section *d*.

Assess

Quiz 1

- Use the **summative assessment** to evaluate the students' progress at this point in the chapter.

NOTES

Student Edition pages 358–59
Daily Review Chapter 16, section *e*

OBJECTIVES

- Defend the claim that Christians should study probability. **BWS**
- Define independent and dependent compound events.
- Differentiate between independent and dependent compound events.
- Find the probability of compound events using a formula.

BIBLICAL WORLDVIEW SHAPING

- **Design (Formulate):** Christians study probability to make predictions that glorify God for His wisdom.

ADDITIONAL MATERIALS

- 6 marbles (4 orange, 1 brown, 1 yellow) in an opaque bag (from Lesson 163)
- a 1–6 number cube (from Lesson 163)
- a calculator (for each student)

Lesson 164 is designed to be a 2-day lesson. Student Edition page 358 problems 1–6 may be assigned on Day 1.

Independent & Dependent Events

Key Terms
- compound event
- independent event
- dependent event

Why are we studying probability?

A **compound event** involves two or more simple events that can be independent or dependent. An **independent event** occurs when the sample space of one event remains the same regardless of the outcome of a previous event. A **dependent event** occurs when the outcome of one event affects the sample space of a later event. You can find the probability of a compound event by multiplying the individual theoretical probabilities.

Independent Event

Alberto spins the spinner. He lands on a red section. He spins a second time and lands on blue.
sample space for each spin: {r, b, g}
$P(\text{red}) = \frac{1}{3}$ $P(\text{blue}) = \frac{1}{3}$
Find the probability of landing on red on the first spin and blue on the second spin.

$P(A, B) = P(A) \times P(B)$
sample space for 2 spins: {rr, rb, rg, bb, br, bg, gg, gr, gb}
$P(\text{red, blue}) = \frac{1}{3} \times \frac{1}{3} = \frac{1}{9}$

Dependent Event

Rosalie draws a marble from the bag. She keeps the red marble and passes the bag to Jolene to draw a marble. Jolene chooses a blue marble.
sample space for the first draw: {r, b, g}
$P(\text{red}) = \frac{1}{3}$

sample space for the second draw: {b, g}
$P(\text{blue after 1 red marble drawn}) = \frac{1}{2}$

Find the probability of drawing a red marble followed by a blue marble.

$P(A, B) = P(A) \times P(B \text{ after } A)$
$P(\text{red, blue}) = \frac{1}{3} \times \frac{1}{2} = \frac{1}{6}$

Exercises

State whether the compound events are independent or dependent.

1. rolling a 5 and then rolling a 6 on a 1–6 number cube *independent*
2. selecting a marble from a bag and keeping it; then selecting a second marble *dependent*
3. choosing a soft drink and then choosing a sandwich from a menu *independent*
4. drawing names from a container to form soccer teams *dependent*
5. selecting a marble from a bag and returning it; then selecting a second marble *independent*
6. flipping a coin 3 times *independent*

Write an equation to find the probability of the event. Write the probability as a fraction in lowest terms.

> A marble is drawn from the bag and then replaced before the next selection.

7. What is the sample space of each draw from the bag? *{r, r, b, b, y, g}*
8. $P(\text{red, yellow})$ $\frac{2}{6} \times \frac{1}{6} = \frac{2}{36} = \frac{1}{18}$
9. $P(\text{yellow, green})$ $\frac{1}{6} \times \frac{1}{6} = \frac{1}{36}$
10. $P(\text{red, not yellow})$ $\frac{2}{6} \times \frac{5}{6} = \frac{10}{36} = \frac{5}{18}$
11. $P(\text{red, blue})$ $\frac{2}{6} \times \frac{2}{6} = \frac{4}{36} = \frac{1}{9}$
12. $P(\text{yellow, blue})$ $\frac{1}{6} \times \frac{2}{6} = \frac{2}{36} = \frac{1}{18}$
13. $P(\text{green, not blue})$ $\frac{1}{6} \times \frac{4}{6} = \frac{4}{36} = \frac{1}{9}$

358   Chapter 16

Engage

Defending a claim

- Direct attention to Student Edition page 358 and guide a **discussion** to answer the chapter essential question, "Why are we studying probability?"
- Review the following ideas previously presented in this chapter:
  1. We can make useful predictions because the world is designed. (Lesson 160)
  2. God's world is so complex that we often cannot predict exactly what will happen. (Lesson 162)

- Read aloud Psalm 40:5.

  What does verse 5*a* say God does? wonders or wonderful works

  How should we respond when we learn of God's wonderful works or deeds? We should declare or proclaim them, even though there are too many to count.

- Use the following ideas to guide your students to answer the chapter essential question.
  1. As we study probability, we recognize God's design and our weaknesses.
  2. We are able to make reasonable predictions, but our reasoning is also limited.

  3. When we acknowledge His wisdom and complex design, we glorify our Creator.

  Why are we studying probability? sample answer: We study probability to make reasonable predictions in order to glorify our Creator by acknowledging His wisdom.

Instruct

Defining & differentiating between independent & dependent compound events

- Use **direct instruction** to help the students distinguish types of compound events.

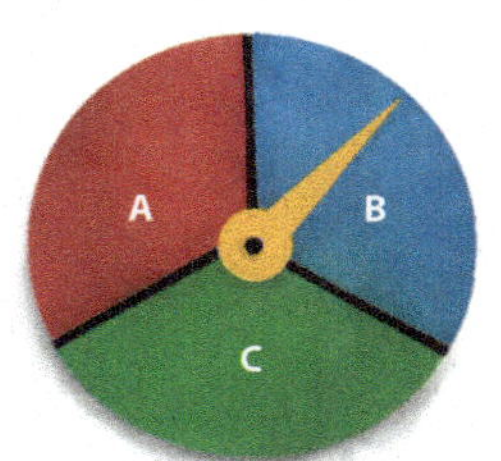

14. P(red, yellow) $\frac{1}{15}$ **15.** P(yellow, green) $\frac{1}{30}$

16. P(red, *not* yellow) $\frac{4}{15}$ **17.** P(red, blue) $\frac{2}{15}$

18. P(yellow, blue) $\frac{1}{15}$ **19.** P(green, *not* blue) $\frac{1}{10}$

Spin the spinner and/or roll a 1–6 number cube for the event. Write an equation to find the probability of the event as a fraction.

20. P(A, 2) $\frac{1}{3} \times \frac{1}{6} = \frac{1}{18}$ **21.** P(A, B, C) $\frac{1}{3} \times \frac{1}{3} \times \frac{1}{3} = \frac{1}{27}$

22. P(B, 3, 4) $\frac{1}{3} \times \frac{1}{6} \times \frac{1}{6} = \frac{1}{108}$ **23.** P(B, C, 5) $\frac{1}{3} \times \frac{1}{3} \times \frac{1}{6} = \frac{1}{54}$

24. Are the compound events above dependent or independent?
independent

Write an equation to find the probability of the following events. Write the probability as a fraction in lowest terms. *Students may use calculator to find product.*

The 20 students in class were each assigned a number, 1–20. The numbers were written on slips of paper and placed in a container. The teacher drew numbers to determine which students would be grouped together for a science project. She did *not* return any numbers to the container.

26. *The sample space will decrease by one name at the end of each individual event.*

25. Is each drawn number dependent or independent?
dependent
27. P(1, 4) $\frac{1}{20} \times \frac{1}{19} = \frac{1}{380}$

26. What will happen to the number of outcomes in the sample space for each drawing?

28. P(3, 17) $\frac{1}{18} \times \frac{1}{17} = \frac{1}{306}$

29. P(2, 6, 10) $\frac{1}{16} \times \frac{1}{15} \times \frac{1}{14} = \frac{1}{3,360}$

30. P(16, 18, 20) $\frac{1}{13} \times \frac{1}{12} \times \frac{1}{11} = \frac{1}{1,716}$

31. P(11, 12, 13, 14) $\frac{1}{10} \times \frac{1}{9} \times \frac{1}{8} \times \frac{1}{7} = \frac{1}{5,040}$

32. P(5, 7, 8, 9) $\frac{1}{6} \times \frac{1}{5} \times \frac{1}{4} \times \frac{1}{3} = \frac{1}{360}$

33. What numbers are left in the container? *15, 19*

Students may use calculator to find product.
If you select 2 socks at random, find the probability of picking each pair. Write the probability as a fraction in lowest terms.

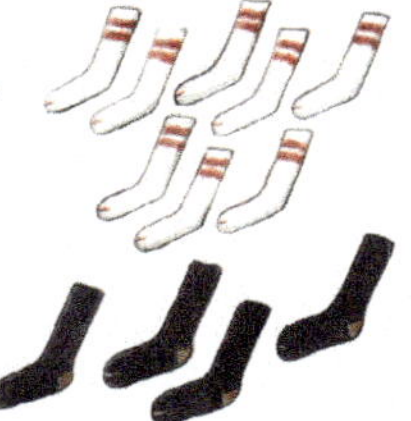

34. P(white, white) $\frac{8}{12} \times \frac{7}{11} = \frac{56}{132} = \frac{14}{33}$ **35.** P(white, black) $\frac{8}{12} \times \frac{4}{11} = \frac{8}{33}$

36. P(black, black) $\frac{4}{12} \times \frac{3}{11} = \frac{4}{44} = \frac{1}{11}$ **37.** P(black, white) $\frac{4}{12} \times \frac{8}{11} = \frac{8}{33}$

During the last 2 years in June, it rained 12 out of 60 days. Find the probability of the independent event(s). Write it as a fraction in lowest terms.

38. P(1 rain day) $\frac{12}{60} = \frac{1}{5}$ **39.** P(1 rain day, 1 rain day) $\frac{1}{5} \times \frac{1}{5} = \frac{1}{25}$ **40.** P(1 rain day, 1 no-rain day) $\frac{1}{5} \times \frac{4}{5} = \frac{4}{25}$

41. P(1 no-rain day) $\frac{48}{60} = \frac{4}{5}$ **42.** P(1 no-rain day, 1 no-rain day) $\frac{4}{5} \times \frac{4}{5} = \frac{16}{25}$ **43.** P(1 no-rain day, 1 rain day) $\frac{4}{5} \times \frac{1}{5} = \frac{4}{25}$

44. Why are we studying probability? *We study probability to make reasonable predictions in order to glorify our Creator by acknowledging His wisdom.*

Lesson 164   359

What is the probability of the same color marble being drawn again? $\frac{0}{2}$ or 0%; Since the marble was not returned, there are no more marbles of that color.

What is the probability of choosing one or the other of the colored marbles? $\frac{1}{2}$ or 50%; Since there are only 2 marbles remaining in the bag, the probability has changed.

- Follow a similar procedure for drawing 1 of the 2 remaining marbles from the bag. Point out that these events are dependent because the sample space (possible outcomes) was affected by what was drawn during the first event.

- Return the marbles to the bag. Guide the students to conclude that the probability of drawing any of the colors is back to $\frac{1}{3}$ or 33%. Repeat the experiment.

- Return the marbles to the bag again. Choose a student to draw 1 marble from the bag, show the marble to the other students, and then return the marble to the bag.

What is the probability of choosing each color marble on a second draw? P(orange) $= \frac{1}{3}$ or 33%, P(brown) $= \frac{1}{3}$ or 33%, P(yellow) $= \frac{1}{3}$ or 33%

Choose another student to draw 1 marble, show the marble to the other students, and then return the marble to the bag.

Why did the probability remain the same for both drawings? The bag still contained 3 marbles, 1 marble of each color. The events are independent because the outcome of the first draw did not affect the outcome of the second draw.

- Direct the students to determine whether the following descriptions of compound events are independent or dependent. If the events are dependent, instruct a student to explain what changed.

Roll a number cube 3 times. independent

Draw names from a container to form different math teams. dependent; I would not return the name to possibly be drawn again.

Flip a coin 2 times. independent

Predict whether it will rain or not rain during the next 2 days. independent

---

- Explain that a compound event involves 2 or more simple events. Spinning a spinner more than once, drawing more than one item from a bag, tossing 2 coins, or tossing 1 coin 3 times are examples of compound events. A compound event can be an *independent event* or a *dependent event*.

  1. In an independent event, the outcome of one event does not affect the outcome of the other(s).

  For example, if a coin lands on heads for the first toss, it does not affect the outcome of how the coin will land on a second or third toss. The color on which a spinner lands on spin 1 does not affect where the spinner will land on spin 2.

  2. In a dependent event, the outcome of the first event affects the outcome of a later event.

- Display 3 of the marbles (1 orange, 1 brown, and 1 yellow) and then place them in the bag.

What is the probability of drawing the orange marble from the bag? $\frac{1}{3}$ or 33%

What is the probability of drawing the brown marble? $\frac{1}{3}$ or 33%

What is the probability of drawing the yellow marble? $\frac{1}{3}$ or 33%

Invite a student to draw 1 marble from the bag without returning it to the bag.

How many marbles remain in the bag? 2

LESSON 164

Choose 3 tiles from a bag without returning the tile after each draw. dependent; There is 1 fewer tile to choose from each time.

You may end the lesson here and teach the remainder of the lesson and use the remainder of Student Edition pages 358–59 on Day 2.

Finding the probability of compound events

- Use a **demonstration** to help the students understand compound events.

- Display the 1–6 number cube.

  What is the probability of rolling 5? $\frac{1}{6}$; 1 out of 6 sides will produce the favorable outcome.

  What is the probability of rolling 3? $\frac{1}{6}$

  How can you find the probability of rolling a 5 on the first roll and a 3 on the second roll? I can list the sample space (all possible outcomes) or make a tree diagram.

- Demonstrate making a tree diagram to find $P(5 \text{ and } 3) = \frac{1}{36}$.

  The word "and" can be substituted for the comma in compound event notation.

- Explain that the probability of compound events can be found by multiplying the probability of the individual events. Write "$P(A \text{ and } B) = P(A) \times P(B)$" for display.

  What equation can you use to find $P(5 \text{ and } 3)$? $P(5) \times P(3) = \underline{}$

  Invite a student to write the equation for display and solve it. $\frac{1}{6} \times \frac{1}{6} = \frac{1}{36}$ Write "$P(5 \text{ and } 3) = \frac{1}{36}$." Continue to display the equations.

  Are these events independent or dependent? How do you know? independent; The outcome of the first event does not affect the outcome of the second event.

- Follow a similar procedure for the following independent events.

Write the ordered pair for the point.

1. A *(-4, 2)*

2. B *(2, 1)*

3. C *(-1, -3)*

4. D *(3, -3)*

Name the quadrant in which the point is located.

5. A *quadrant II*

6. B *quadrant I*

7. C *quadrant III*

8. D *quadrant IV*

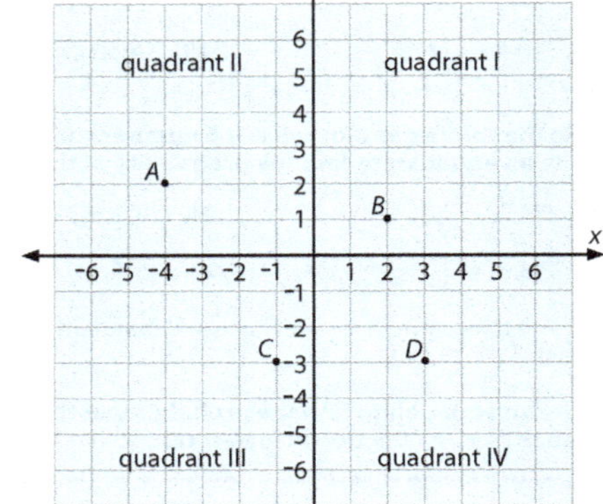

512 Daily Review

---

toss a coin 2 times: $P(\text{heads, heads}) = \frac{1}{2} \times \frac{1}{2} = \frac{1}{4}$

toss a coin 3 times: $P(\text{tails, heads, tails}) = \frac{1}{2} \times \frac{1}{2} \times \frac{1}{2} = \frac{1}{8}$

roll a 1–6 number cube 3 times: $P(1, 2, 3) = \frac{1}{6} \times \frac{1}{6} \times \frac{1}{6} = \frac{1}{216}$

- Display 3 of the marbles (1 orange, 1 brown, and 1 yellow) and then place them in the bag.

  What is the probability of drawing the brown marble from the bag? $\frac{1}{3}$

  What is the probability of drawing the orange marble? $\frac{1}{3}$

What is the probability of drawing the yellow marble? $\frac{1}{3}$

If you were to choose the brown marble on the first draw and not return it to the bag, what is the probability of drawing the yellow marble? $\frac{1}{2}$

- Write "$P(A, B) = P(A) \times P(B \text{ after } A)$." Direct attention to the previously displayed equations and point out that the comma (A, B) represents the word *and*. Explain that when a marble is not returned to the bag, the choices in the bag change, which also changes the probability of future (subsequent) draws.

Are these events independent or dependent? How do you know? **dependent; The outcome of the first event affects the possible outcomes of the second event.**

What equation can you use to find $P(\text{brown, yellow})$ if you are *not* returning the marble to the bag after each draw? **$P(\text{brown, yellow}) = P(\text{brown}) \times P(\text{yellow after brown})$; $\frac{1}{3} \times \frac{1}{2} = \underline{\quad}$**

- Choose a student to write the equation and solve it. $\frac{1}{6}$
  Write "$P(\text{brown, yellow}) = \frac{1}{3} \times \frac{1}{2} = \frac{1}{6}$."

- Follow a similar procedure using all 6 of the marbles.

  What is the probability of drawing an orange marble from the bag? **$\frac{4}{6}$ or $\frac{2}{3}$**

  If you were to choose an orange marble on the first draw and not return the marble to the bag, what is the probability of drawing a yellow marble? **$\frac{1}{5}$**

  What equation can you use to find $P(\text{orange, yellow})$ when not returning the drawn marble to the bag? **$\frac{2}{3} \times \frac{1}{5} = \underline{\quad}$**

  Invite a student to write the equation and solve it. $\frac{2}{15}$

- Repeat the procedure to find $P(\text{yellow, orange, brown})$. $\frac{1}{6} \times \frac{4}{5} \times \frac{1}{4} = \frac{4}{120} = \frac{1}{30}$

- Direct each student to write his name on a slip of paper. Remove the marbles from the bag and place the slips of paper in the bag.

  What is the probability of drawing a specific student's name from the bag? **$\frac{1}{n}$ (n = number of students in the class)**

- Write $P(\text{student 1, student 2})$ for display using 2 students' names.

  What equation do you think you could use to find the probability of drawing two specific students to be partners on a field trip? **$\frac{1}{n} \times \frac{1}{n-1}$ (where n = number of students in the class)**

  Choose a student to write the equation for display and solve it using a calculator, while the other students solve it independently.

- Follow a similar procedure to find the probability of drawing the names of 4 specific students to be on the same team. $\frac{1}{n} \times \frac{1}{n-1} \times \frac{1}{n-2} \times \frac{1}{n-3}$

Apply

Student Edition pages 358–59

- Read and explain the directions for pages 358–59. Assist the students as they complete the pages independently, in pairs, or in small groups.

Daily Review

- Students should complete Chapter 16, section *e*.

NOTES

CHAPTER REVIEW

OBJECTIVES

- Predict the results of an experiment using the theoretical probability of an event.
- Determine whether a game is fair or unfair using probability.
- Find the complement of the probability of an event.
- Make a tree diagram to list the sample space for an event.
- Find the number of possible outcomes using the Multiplication Counting Principle.
- Make predictions using probability.

TEACHER RESOURCES

- 127 *Probability Spinner* (for the teacher and for each student)

ADDITIONAL MATERIALS

- a paper clip (for the teacher and for each student)
- a calculator (for each student) (optional)

CHAPTER REVIEW

A marble is drawn from the pictured bag. Write the probability of the event as a fraction and as a percent.

1. P(green) $\frac{1}{10}$; *10%*
2. P(red) $\frac{4}{10}$; *40%*
3. P(red or blue) $\frac{7}{10}$; *70%*
4. P(blue) $\frac{3}{10}$; *30%*
5. P(yellow) $\frac{2}{10}$; *20%*
6. P(not green) $\frac{9}{10}$; *90%*

Use the spinner to find the probability of the event. Write it as a fraction in lowest terms and as a percent.

7. P(blue) $\frac{1}{3}$; *33%*
8. P(green) $\frac{1}{6}$; *17%*
9. P(red and odd) $\frac{1}{3}$; *33%*
10. P(not blue) $\frac{2}{3}$; *67%*
11. P(red) $\frac{1}{2}$; *50%*
12. P(odd) $\frac{1}{2}$; *50%*
13. P(red or blue) $\frac{5}{6}$; *83%*
14. P(multiple of 2) $\frac{1}{2}$; *50%*

The probability of event A is given. Find the complement, P(not A).

15. P(A) = $\frac{1}{4}$ $\frac{3}{4}$
16. P(A) = $\frac{1}{5}$ $\frac{4}{5}$
17. P(A) = 20% *80%*

22. *P(yellow and yellow) = $\frac{1}{4}$*
   P(at least 1 green) = $\frac{3}{4}$; unfair

Make a list or a tree diagram to find the possible outcomes. Write the probability as a fraction in lowest terms.

18. Moriah has a spinner with 2 equal sections labeled A and B and another spinner with 3 equal sections labeled 1, 2, and 3. If she spins both spinners, what is the probability of the outcome being B and an odd number? *$\frac{1}{3}$*

19. Reuben has a spinner with 3 equal sections labeled A, B, and C and another spinner with 3 equal sections labeled 1, 2, and 3. If he spins both spinners, what is the probability of the outcome being A and an odd number? *$\frac{2}{9}$*

20. Luis has a spinner with 2 equal sections labeled A and B and a number cube labeled 1–6. If he rolls the number cube and spins the spinner, what is the probability of the outcome being A and an even number? *$\frac{1}{4}$*

21. A spinner has 2 equal sections of red and blue. If you spin the spinner 2 times, what is the probability of landing on blue exactly 2 times? *$\frac{1}{4}$*

Make a list or a tree diagram to find the sample space for the game. Write the probability as a fraction in lowest terms. Determine whether the game is fair or unfair.

22. A spinner has 2 equal sections of yellow and green. Spin the spinner 2 times. Player A wins if the spinner lands on yellow 2 times. Player B wins if it lands on green at least 1 time. What is the probability of both spins being yellow? What is the probability of at least one spin being green? Is this game fair or unfair?

23. A 1–6 number cube is rolled once. Player A wins if the number is less than 4. Player B wins if the number is greater than 3. What is the probability of rolling a number less than 4? What is the probability of rolling a number greater than 3? Is this game fair or unfair?
   P(less than 4) = $\frac{1}{2}$
   P(greater than 3) = $\frac{1}{2}$; fair

The Chapter Review offers an opportunity for students to discuss the concepts they have learned in the chapter. They may work collaboratively or independently as you review concepts. Circulate among the students, giving individual help as needed. Students who demonstrate proficiency with the discussion, the modeling, and the Student Edition pages are ready for the Chapter Test. Students who encounter difficulties with the review concepts would benefit from additional coaching and practice before testing.

Predicting & determining the experimental probability of an event; determining whether a game is fair or unfair

- Distribute a *Probability Spinner* page and a paper clip to each student and display your copy of the page. Point out that the spinner on the page will be used for both activities: "Spin 1, 2, or 3" and "Even or Odd Spin."

   What is the set of all possible outcomes for an event called? the sample space
   What is the sample space for 1 spin of this spinner? {1, 2, 3}

- Instruct the students to determine the theoretical probability of spinning each number and to write the probability as a fraction in lowest terms to complete the first column of the table: $P(1) = \frac{1}{6}$, $P(2) = \frac{1}{2}$, $P(3) = \frac{1}{3}$.

   Then direct them to use the theoretical probability to predict the expected results of each event for 12 spins. $P(1) = \frac{1}{6} \times 12 = 2$, $P(2) = \frac{1}{2} \times 12 = 6$, $P(3) = \frac{1}{3} \times 12 = 4$ or $\frac{1}{6} = \frac{n}{12}$, $\frac{1}{2} = \frac{n}{12}$, $\frac{1}{3} = \frac{n}{12}$

- Instruct the students to spin the paper clip 12 times and to record each result by drawing a tally in the Actual Results column of the table.

Use the Multiplication Counting Principle to find the number of possible outcomes.

24. Damien is buying an ice cream sundae. He has a choice of ice cream (vanilla, chocolate, or strawberry), a choice of syrup (chocolate, caramel, or strawberry), and a choice of one topping (nuts, chocolate chips, gummy bears, or whipped cream). How many different combinations does Damien have to choose from?

25. Elisa is buying a milkshake. She has a choice of ice cream (vanilla or chocolate) and a choice of add-ins (candies, cookie crumbles, mint pieces, chocolate chips, or caramel). How many different combinations does Elisa have to choose from?
2 ice creams × 5 add-ins = 10 combinations

Use the menu to find the answer.

Pasta Palace		
Shape	**Type**	**Sauce**
spaghetti	egg	classic Italian
linguine	wheat	Alfredo
fettuccine	spinach	pesto

26. At the Pasta Palace, you can choose the shape and type of pasta, as well as the sauce. How many different combinations do the customers have to choose from?

27. What is the probability that a customer will order pasta made with spinach? $\frac{9}{27} = \frac{1}{3}$ *or 33%*

28. What is the probability that a customer will order pasta with the classic Italian sauce?

29. What is the probability that a customer will order egg fettuccine with Alfredo sauce? $\frac{1}{27}$ *or 4%*

24. 3 ice creams × 3 syrups × 4 toppings = 36 combinations

26. 3 pastas × 3 types × 3 sauces = 27 combinations

28. $\frac{9}{27} = \frac{1}{3}$ or 33%

32. P(glasses) = $\frac{6}{20}$
P(not glasses) = $\frac{14}{20}$ = 70%

Write the probability as a percent. Round to the nearest percent.

30. Mateo scored a field goal in 6 out of 10 games. What is the probability that he will score a field goal in today's game? $\frac{6}{10} = 60\%$

31. There are 11 boys and 14 girls in the school choir. What would be the probability of drawing a girl's name from a container having the names of all the choir members? $\frac{14}{25} = 56\%$

32. Mrs. Reed surveyed her class and found that 6 out of 20 students wear glasses. Find the probability of the complement, students *not* wearing glasses.

33. Lisa bought a bag of beads to make jewelry. All the beads are the same size and shape. There are 3 silver beads, 7 gold beads, and 5 white beads. What is the probability that the first bead Lisa pulls out of the bag will *not* be gold? $\frac{8}{15} \approx 53\%$

Use the survey results to find the probability of the event. Write the probability as a fraction and as a percent.

34. P(mystery)

35. P(historical fiction)

36. P(fantasy)

37. P(nonfiction)

Survey Results	
Book Type	**Results**
mystery	8
historical fiction	6
fantasy	4
nonfiction	2

Use the survey results to predict the type of book that would be chosen in a group of 100 people.

38. How many people would choose mysteries as their favorite type of book? $\frac{8}{20} = \frac{n}{100}$; *n = 40 people*

39. How many people would choose historical fiction? $\frac{6}{20} = \frac{n}{100}$; *n = 30 people*

40. How many people would choose nonfiction? $\frac{2}{20} = \frac{n}{100}$; *n = 10 people*

34. $\frac{8}{20}$ or $\frac{2}{5}$; 40% *36. $\frac{4}{20}$ or $\frac{1}{5}$; 20%*

35. $\frac{6}{20}$ or $\frac{3}{10}$; 30% *37. $\frac{2}{20}$ or $\frac{1}{10}$; 10%*

Chapter Review 361

If 1 player were to pick odd numbers and another player were to pick even numbers, and the player whose numbers were spun most often won, would "Even or Odd Spin" be a fair or an unfair game? fair; The probabilities of spinning an even or an odd number are equally likely: $P(\text{odd}) = \frac{3}{6} = \frac{1}{2}$, $P(\text{even}) = \frac{3}{6} = \frac{1}{2}$.

Finding the complement of the probability of an event

- Direct attention to the spinner on the page.

  What is the complement of spinning 2 on the spinner? not spinning 2

  Assist the students to conclude that complementary events cannot occur at the same time; either 2 will be spun or 2 will not be spun.

  If $P(2)$ is $\frac{1}{2}$, what is $P(\text{not } 2)$? $\frac{1}{2}$; $P(2) + P(\text{not } 2) = 1$, so $1 - P(2) = P(\text{not } 2)$; $1 - \frac{1}{2} = \frac{1}{2}$

  If $P(3)$ is $\frac{1}{3}$, what is $P(\text{not } 3)$? $\frac{2}{3}$

  If $P(1)$ is $\frac{1}{6}$, what is $P(\text{not } 1)$? $\frac{5}{6}$

- Direct the students to think about a number cube again.

  If $P(4)$ for rolling a 1–6 number cube is 17%, what is $P(\text{not } 4)$? 83%; $100\% - 17\% = 83\%$

Making a tree diagram to list the sample space; finding the number of possible outcomes

- Read aloud the following word problem.

  For dinner some students can choose either ham or beef for their meat; green beans, peas, or mashed potatoes for their vegetable; and cake or ice cream for dessert. How many different combinations of meals do the students have to choose from?

- Guide the students to conclude the number of choices that the students have for each part of their dinner: meat 2, vegetable 3, and dessert 2.

  What equation can you write to find the number of possible combinations of a meat, a vegetable, and a dessert? $2 × 3 × 2 = \underline{\quad}$

Are your actual results the same as your expected results? Answers will vary. Most likely the actual results will differ from the expected results; guide the students to conclude that every experiment has variables.

How can you use the actual results to calculate the experimental probability? I can divide the actual results (the number of favorable results or the frequency) for each event by the number of trials (12).

- Direct each student to calculate the experimental probability of spinning each number and then to record it on his table. Guide a discussion about the differences and similarities of the probabilities.

If 3 players were each to pick a different number on the spinner, and the player whose number was spun most often won, would "Spin 1, 2, or 3" be a fair or an unfair game? unfair; The probabilities of spinning each of the numbers are not equally likely: $P(1) = \frac{1}{6}$; $P(2) = \frac{3}{6} = \frac{1}{2}$; $P(3) = \frac{2}{6} = \frac{1}{3}$.

- Follow a similar procedure to guide the students as they complete the "Even or Odd Spin" table. $P(\text{even}) = \frac{3}{6} = \frac{1}{2}$, $P(\text{odd}) = \frac{3}{6} = \frac{1}{2}$; Each expected result is 10; the actual results and the experimental probabilities will vary.

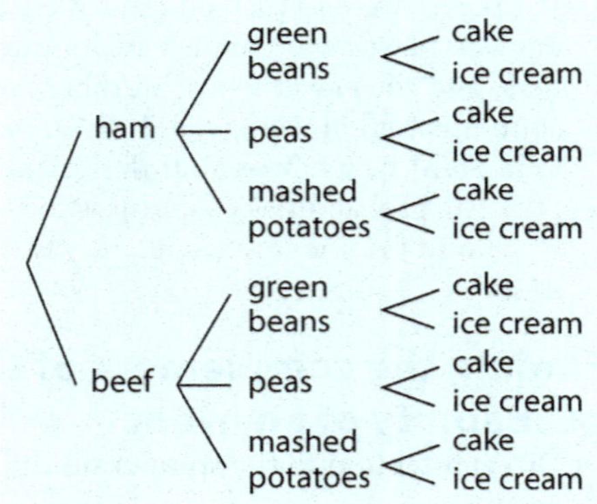

How many possible combinations are there? 12

What can help you to list the sample space when choices are given? a tree diagram

- Guide the students as they make a tree diagram and use it to list the sample space. {hgc, hgi, hpc, hpi, hmc, hmi, bgc, bgi, bpc, bpi, bmc, bmi}

What is the probability of a student choosing a meal with ice cream for dessert? P(ice cream) $= \frac{6}{12} = \frac{1}{2}$, 0.5, 50%; 6 out of the 12 possible meal combinations (outcomes) include ice cream for dessert.

What is the probability of a student choosing ham, peas, and cake for dinner? P(ham, peas, cake) $= \frac{1}{12}$, 0.08, 8%

Making predictions using probability

- Read aloud the following word problem.

Samuel is hoping his parents will let him use a spinner to determine what time he must go to bed at night. He divided the spinner into 3 equal sections and labeled the sections 9:00, 10:00, and 11:00. If his parents agree to use the spinner, will it give Samuel a greater opportunity to stay up until 11:00?

- Direct the students to draw the spinner on paper and determine the probability of each outcome.

P(9:00) $= \frac{1}{3}$, P(10:00) $= \frac{1}{3}$, P(11:00) $= \frac{1}{3}$
Will this spinner provide Samuel with a greater opportunity to stay up until 11:00 than 9:00 or 10:00? No; all the outcomes are equal.

- Instruct the students to draw a different spinner that includes the same times (9:00, 10:00, 11:00) but will give Samuel a greater opportunity to stay up until 11:00. Spinners will vary, but the size of the space for landing on 11:00 must indicate a greater probability than the probability of landing on 9:00 or 10:00.

- Write the following in a column for display: red ||, yellow |, blue ||||, and green |||. Explain that 10 students were asked to identify which of these 4 colors is their favorite. The tallies indicate the favorite color choices of the 10 students that were surveyed.

What is the sample space for the survey? {red, yellow, blue, green}

- Remind students that the probability calculated using the data gathered from a small sample of people can be used to make predictions about a larger population of the same people. Instruct the students to use the data from the survey to predict the expected results if 100 students were surveyed: red $\frac{1}{5} \times 100 = 20$; yellow $\frac{1}{10} \times 100 = 10$; blue $\frac{2}{5} \times 100 = 40$; green $\frac{3}{10} \times 100 = 30$.

- Direct the students to use the data from the survey results of 10 students to predict the expected results for each color if 40 students were surveyed: red $\frac{1}{5} \times 40 = 8$; yellow $\frac{1}{10} \times 40 = 4$; blue $\frac{2}{5} \times 40 = 16$; green $\frac{3}{10} \times 40 = 12$.

Student Edition pages 360–61

- Read and explain the directions for pages 360–61. Assist the students as they complete the pages independently.

Student Edition pages 362–64

CHAPTER 16 TEST

CUMULATIVE REVIEW

CONCEPT REVIEW

- **Identifying the standard form of a number written in exponent form (Lesson 14)**
- **Rounding a decimal to a given place (Lessons 68–69)**
- **Estimating the location of a fraction on a number line (Lesson 33)**
- **Classifying figures as congruent (Lesson 59)**
- **Identifying the type of triangle (Lesson 57)**
- **Solving measurement and money word problems (Lessons 5, 6, 84, Chapter 14)**
- **Finding the mean of data on a chart (Lessons 147–49)**
- **Finding the unknown measure for similar figures (Lesson 59)**
- **Finding an equivalent fraction or ratio (Lessons 35, 36, 127)**
- **Determining the least common multiple (Lessons 32, 36)**
- **Dividing fractions and whole numbers (Chapters 3, 8)**
- **Estimating the sum of mixed numbers (Lessons 42–44)**
- **Simplifying an expression using the order of operations (Lesson 94)**
- **Multiplying decimals (Lessons 68–69)**
- **Reading and interpreting a stem-and-leaf plot (Chapter 15)**

CUMULATIVE REVIEW

Choose the answer.

1. What is 10^6 in standard form?
- **A.** 10,000
- **B.** 100,000
- **C.** 1,000,000
- **D.** none of the above

2. Round 67.39788 to the nearest ten thousandth.
- **A.** 67.3979
- **B.** 67.3980
- **C.** 67.3988
- **D.** none of the above

3. The fraction $\frac{13}{20}$ is closest to what point on the number line?

- **A.** 0
- **B.** $\frac{1}{2}$
- **C.** 1
- **D.** $1\frac{1}{2}$

4. Which description best fits the figures?

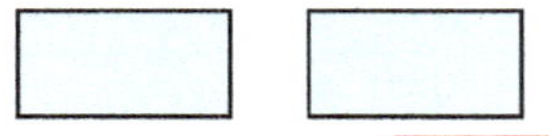

- **A.** obtuse angles
- **B.** similar figures
- **C.** congruent figures
- **D.** all of the above

5. What two words identify the type of triangle?
- **A.** acute, equilateral
- **B.** right, isosceles
- **C.** scalene, obtuse
- **D.** none of the above

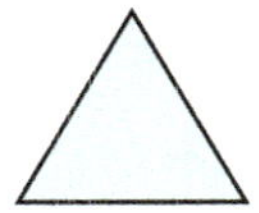

6. How many pieces of ribbon 12.5 cm long can Jaclyn cut from a piece 145 cm long?
- **A.** 11 pieces
- **B.** 13 pieces
- **C.** 12 pieces
- **D.** 14 pieces

7. Leah paid $25.95 for a new coat. If she gave the cashier forty dollars, what was her change?
- **A.** $12.05
- **B.** $14.05
- **C.** $13.05
- **D.** $11.05

8. Grandfather's house is $104\frac{7}{10}$ km from Finn's house. If Finn's family drove $60\frac{1}{2}$ km before lunch, how many more kilometers will they need to drive after lunch to reach Grandfather's house?
- **A.** $42\frac{1}{2}$ km
- **B.** $44\frac{1}{5}$ km
- **C.** $43\frac{1}{10}$ km
- **D.** $44\frac{4}{5}$ km

9. How many hours are in 420 min?
- **A.** 6 hr
- **B.** 8 hr
- **C.** 7 hr
- **D.** 9 hr

10. What is the mean temperature for the park for the given months?

Monthly Average High Temperature for Bryce Canyon National Park				
May	June	July	August	September
63°F	75°F	80°F	77°F	72°F

- **A.** 72.8°F
- **B.** 73.4°F
- **C.** 77.3°F
- **D.** 78°F

362 Chapter 16

- To prepare the students for the format of achievement tests, instruct them to work on a separate sheet of paper, if necessary, and to mark their answers on the *Cumulative Review Answer Sheet.*

Student Edition pages 362–64

The Cumulative Review provides additional practice of previously learned concepts. These pages may be completed during this lesson or anytime after this lesson, since they require limited or no teaching.

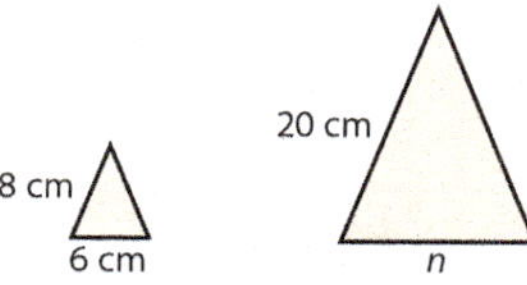

Choose the answer.

11. Use a proportion to find the unknown measure for the pair of similar figures.

8 cm / 6 cm · 20 cm / n

A. $n = 15$ cm   **C.** $n = 24$ cm
B. $n = 32$ cm   **D.** none of the above

12. Rename $\frac{32}{48}$ in lowest terms.

A. $\frac{1}{3}$ **C.** $\frac{2}{5}$
B. $\frac{3}{7}$ **D.** $\frac{2}{3}$

13. What is the least common multiple of 14 and 21?

A. 35   **C.** 42
B. 49   **D.** 60

14. $\frac{9}{10} \div 3$

A. $\frac{1}{10}$ **C.** $\frac{3}{5}$
B. $\frac{3}{10}$ **D.** $\frac{1}{2}$

15. Estimate the sum.

$1\frac{6}{14} + 2\frac{7}{12}$

A. 1   **C.** 4
B. 2   **D.** 5

16. Find an equivalent ratio for $\frac{11}{12}$.

A. $\frac{30}{36}$ **C.** $\frac{55}{56}$
B. $\frac{44}{48}$ **D.** $\frac{90}{100}$

17. $25 - 7 \times 2 + 44$

A. 35   **C.** 50
B. 40   **D.** 55

18. $34\overline{)6{,}908}$

A. $203\frac{6}{34}$ **C.** $213\frac{3}{34}$
B. $203\frac{3}{17}$ **D.** $230\frac{1}{34}$

19. 100×365.93

A. 36,593   **C.** 3.5993
B. 0.35993   **D.** none of the above

20. 85×6.051 g

A. 51,435 g   **C.** 514.005 g
B. 514.335 g   **D.** 5,143.35 g

Cumulative Review   363

Use the stem-and-leaf plot to find the answer.

Ages of the First 44 US Presidents at Inauguration	
Stem	**Leaf**
4	2 3 6 6 7 7 8 9 9
5	0 1 1 1 1 1 2 2 4 4 4 4 4 5 5 5 5 6 6 6 7 7 7 7 8
6	0 1 1 1 2 4 4 5 8 9

Key 4|2 = 42

21. What is the age of the youngest president?

    A. 40   **C.** 49

    B. 42   **D.** 50

22. What age is the median?

    A. 52   **C.** $54\frac{1}{2}$

    B. 54   **D.** 55

23. What is the range of ages?

    A. 20   **C.** 30

    B. 27   **D.** 32

24. What ages are the modes?

    A. 51 and 54   **C.** 51 and 61

    B. 56 and 57   **D.** 54 and 57

25. The oldest president served 8 yr as president. How old was he at the end of his term?

    A. 66   **C.** 75

    B. 68   **D.** 77

Lesson Plan Overview

<table>
<tr><th colspan="6">Chapter 17: Integers</th></tr>
<tr><th>Day</th><th>Lesson</th><th>Student Edition Pages</th><th>Teacher Edition Pages</th><th>Teacher Resources & Additional Materials</th><th>Topics, Skills & Biblical Worldview Shaping</th></tr>
<tr>
<td>170</td><td>167</td><td>366–67</td><td>366–67b</td>
<td>Teacher Resources
• 3 Positive & Negative Number Line</td>
<td>Topic: Integers
Skills: identifying integers on a number line, finding the absolute value of a number, comparing and ordering integers</td>
</tr>
<tr>
<td>171</td><td>168</td><td>365, 368–69</td><td>365, 368–69b</td>
<td>Teacher Resources
• 3 Positive & Negative Number Line
• 128 Algebra Mat
• 129 Positive & Negative Number Lines
Additional Materials
• counters: black and red</td>
<td>Topic: Adding Integers
Skills: adding integers, solving a word problem
BWS: Modeling</td>
</tr>
<tr>
<td>172</td><td>169</td><td>370–71</td><td>370–71b</td>
<td>Teacher Resources
• 3 Positive & Negative Number Line
• 128 Algebra Mat
• 129 Positive & Negative Number Lines
Additional Materials
• counters: black and red</td>
<td>Topic: Subtracting Integers
Skills: subtracting integers, solving a word problem
BWS: Modeling</td>
</tr>
<tr>
<td>173</td><td>170</td><td>372–73</td><td>372–73b</td>
<td>Teacher Resources
• 3 Positive & Negative Number Line
• 128 Algebra Mat
• 130 Subtraction Patterns
• 131 More Subtraction Patterns
Additional Materials
• counters: black and red
Assessments
• Chapter 17 Quiz 1</td>
<td>Topic: Adding & Subtracting Integers
Skills: adding and subtracting integers, solving a word problem</td>
</tr>
<tr>
<td>174</td><td>171</td><td>374–75</td><td>374–75b</td>
<td>Teacher Resources
• 3 Positive & Negative Number Line</td>
<td>Topic: More Integers
Skills: adding and subtracting integers, solving real-world problems</td>
</tr>
<tr>
<td>175</td><td>172</td><td>376–77</td><td>376–77b</td>
<td>Teacher Resources
• 128 Algebra Mat
• 132 Multiplication Patterns
Additional Materials
• counters: black and red</td>
<td>Topic: Multiplying Integers
Skills: multiplying integers, solving a word problem</td>
</tr>
<tr>
<td>176</td><td>173</td><td>378–79</td><td>378–79b</td>
<td>Teacher Resources
• 3 Positive & Negative Number Line
• 128 Algebra Mat
Additional Materials
• counters: black and red</td>
<td>Topic: Multiplying & Dividing Integers
Skills: multiplying and dividing integers, solving real-world problems
BWS: Modeling</td>
</tr>
</table>

Day	Lesson	Student Edition Pages	Teacher Edition Pages	Teacher Resources & Additional Materials	Topics, Skills & Biblical Worldview Shaping
177	174	380–81	380–81b	Teacher Resources • 3 *Positive & Negative Number Line* • 128 *Algebra Mat* • 133 *Order of Operations* Additional Materials • counters: black and red	**Topic:** Mixed Review **Skills:** adding, subtracting, multiplying, and dividing integers; subtracting integers by adding the opposite; solving problems by applying the order of operations
178	175	382–83	382–83b	Teacher Resources • 27 *Coordinate Plane* • 28 *Coordinate Planes* Assessments • Chapter 17 Quiz 2	**Topic:** Coordinate Planes **Skills:** graphing points on a coordinate plane, writing ordered pairs to identify points on a coordinate plane
179	176	384–85	384–85b	Teacher Resources • 3 *Positive & Negative Number Line* • 128 *Algebra Mat* Additional Materials • counters: black and red	**Topic:** Chapter Review **BWS:** Modeling
180	177	386–88	386–88		**Topic:** Test & Cumulative Review

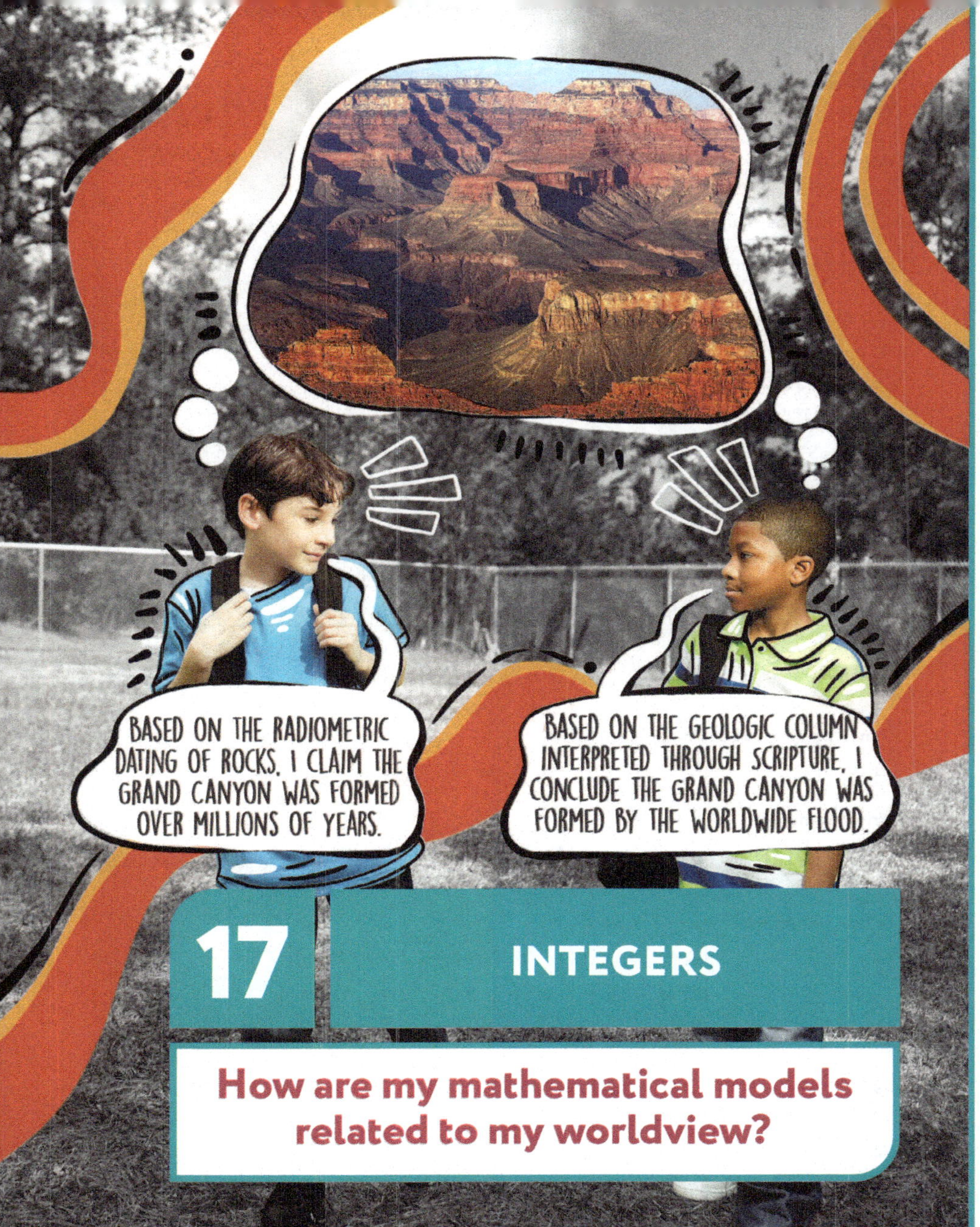

CHAPTER 17

CHAPTER OBJECTIVES

- Compare and order integers.
- Add, subtract, multiply, and divide integers.
- Solve word problems by writing equations.
- Relate points on a graph to their ordered pairs.
- Defend the claim that every model is influenced by a person's worldview.

To determine whether your students are prepared for the concepts taught in this chapter, you may wish to consult the preassessment checklist found on TeacherToolsOnline.com.

Throughout this chapter, select problems from the list of mental math problems provided on the 134–36 *Mental Math Strings* pages.

Make fact practice, both oral and written, part of your daily math routine to help the students with mastery.

Visit AfterSchoolHelp.com for math practice resources, or visit TeacherToolsOnline.com for additional resources to enhance the lessons.

Student Edition pages 366–67
Daily Review Chapter 17, section *a*

OBJECTIVES

- Identify integers on a number line.
- Find the absolute value of a number.
- Compare and order integers.

TEACHER RESOURCES

- 3 *Positive & Negative Number Line*

Preparation

Draw for display the number line pictured on page 367 of this Teacher Edition but do not draw the jumps.

Engage

- Direct the students to **brainstorm** to explore the essential question on Student Edition page 366, "Why do we have negative numbers?"

Instruct

Identifying integers; finding absolute value

- **Model** with a number line to help the students identify positive and negative numbers.

- Draw a vertical number line for display. Draw a short line across the middle and label it "0."

  What type of measuring unit does a thermometer use to measure temperature? degrees

- Read aloud the following scenario.

  The weather forecaster reported that the temperature was 10°F during the day, but the temperature will drop 20 degrees during the night.

  How could you show 10 degrees on the number line? I can draw 10 tick marks in a positive direction from 0 and label the number 10°.

  Invite a student to demonstrate.

  How far from 0 is 10°F? 10 degrees (units)

  In what direction is 10°F? up

  How can you show a drop in temperature of 20°? I can count down 10 degrees (units) to arrive at 0 and then draw 10 more tick marks in a negative direction from 0 and label the number ⁻10°.

  Choose a student to demonstrate.

  What is the temperature now? 10°F below 0

  How far from 0 is 10° below 0? 10 degrees (units)

  In what direction is 10° below 0? down

  Besides the direction, how does a thermometer distinguish between 10° above 0 and 10° below 0? A negative sign is used to indicate a temperature below 0.

  Choose a student to write the daytime temperature and the nighttime temperature for display. 10°F; ⁻10°F

- Explain that a thermometer is a type of vertical number line. Negative and positive numbers are represented on a number line. Point out that 0 is neither positive nor negative but that it separates the positive numbers from the negative numbers.

  Which numbers are positive on a vertical number line? all the numbers above 0

  Which numbers are negative? all the numbers below 0

- Two numbers are opposites if they are the same distance from 0 on a number line but in opposite directions. Point out that

---

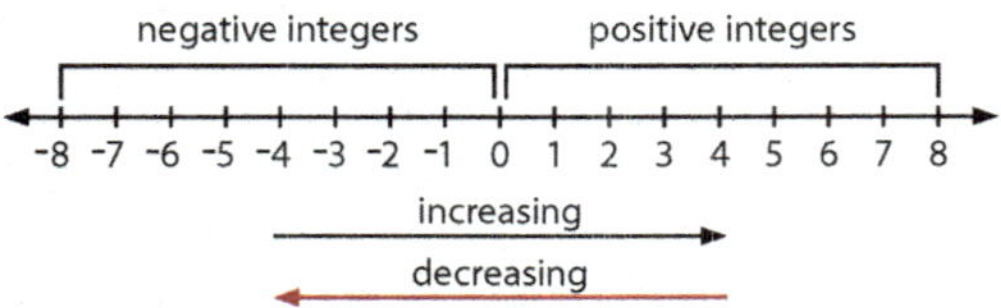

Integers

Why do we have negative numbers?

Integers consist of whole numbers and their opposites.

{. . ., -3, -2, -1, 0, 1, 2, 3, . . .}

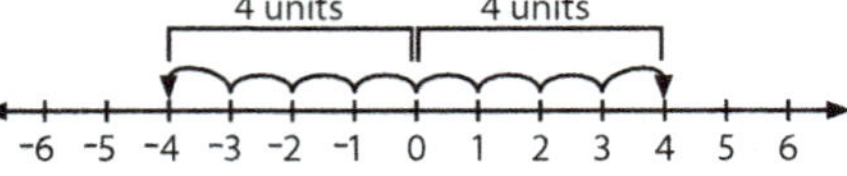

The values of negative numbers continue to *decrease* as you move left on a number line.

-6 < -3   -8 < -7   -2 > -4

The value of a negative number will always be less than the value of a positive number.

1 > -5   -3 < 2   7 > -7

The numbers 4 and ⁻4 are **opposites** because they are the same distance from 0 in opposite directions. **Absolute value** is the distance a number is from 0. Distance is expressed as a positive value. It is indicated by the symbol |*n*| and is read "the absolute value of *n*."

|4| The *absolute value* of 4 is 4.
|⁻4| The *absolute value* of ⁻4 is 4.
|4| = |⁻4|

Exercises

Draw a number line and mark the integer.

1. 5   **2.** -3   **3.** -5   **4.** 2   **5.** -1

Write the integer.

6. eighty feet below sea level   ⁻80 ft

7. twelve degrees above zero   12°

8. a golf score of 7 under par   ⁻7

Write a comparison sentence by using > or <.

9. -3 ≤ 1   **10.** -5 ≤ 5   **11.** 0 ≥ -4   **12.** 4 ≥ -4   **13.** 8 ≤ 11

14. 8 ≥ -9   **15.** -11 ≤ -10   **16.** -2 ≥ -6   **17.** -17 ≤ -12   **18.** 20 ≥ -25

Write the integers from least to greatest.

19. -9 12 10 -5 → -9 -5 10 12

20. 1 -2 -8 -3 → -8 -3 -2 1

21. 17 -15 -1 12 → -15 -1 12 17

22. 10 -5 4 -3 → -5 -3 4 10

23. -6 -8 5 1 → -8 -6 1 5

24. 14 -16 16 -15 → -16 -15 14 16

366   Chapter 17

Write the opposite of the given integer.

25. 17 *-17*   26. -8 *8*   27. -15 *15*   28. -9 *9*   29. 12 *-12*   30. -1 *1*

Draw a number line to show the given number and its opposite.

31. -8   32. 6   33. -9

Find the absolute value.

34. |-5| *5*   35. |12| *12*   36. |-1| *1*   37. |-7| *7*   38. |3| *3*

39. |10| *10*   40. |-21| *21*   41. |18| *18*   42. |-13| *13*   43. |-84| *84*

Practice & Application

44. Write the integer that is one greater than -8. *-7*

45. Write the integer that is one less than -4. *-5*

46. Write the absolute value of -14. *14*

47. Write the absolute value of 10. *10*

48. $\sqrt{144}$ *12*

49. 8^3 *512*

50. 100 × 3.42 *342*

51. 16.8 ÷ 1,000 *0.0168*

52. Explain the equation |-18| = 18.
The absolute value of -18 is 18.

53. Why do we have negative numbers?
Negative numbers represent the numbers below a standard, such as below sea level or below 0° temperatures.

Remind the students that the counting numbers (natural numbers) and 0 make the set of whole numbers. Explain that the set of whole numbers and their opposites makes the set of integers. Write "negative integers" above the numbers to the left of 0 and write "positive integers" above the numbers to the right of 0.

What is true about the numbers to the left of 0? They are the opposites of the numbers to the right of 0; they are all negative numbers; their values decrease as I move farther from 0 (to the left).

- Direct attention to the -200 to 200 number line previously prepared for display. Instruct students to listen to the next situation and determine how they could use this number line to picture the changing amount in the bank account.

- Read aloud the following scenario.

Joe had $115 in his checking account. He deposited $85 more. Joe wants to purchase a $225 lawnmower for his business.

Choose students to show the positive and negative actions as jumps on the number line.

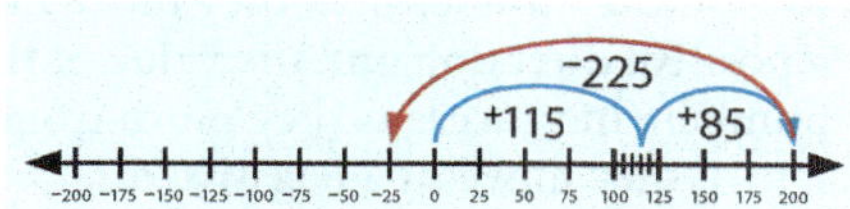

What will Joe's account balance be if he were to purchase the lawnmower? -$25; $115 + $85 − $225 = -$25

Does Joe have enough money in his checking account to purchase the lawn mower? No; a negative balance would be shown in his account.

- Write "absolute value" and explain that the absolute value of a number, x, is its distance from 0 on the number line. It is indicated by the symbol $|x|$ and is read "the absolute value of x." Point out 4 and -4 on the number line and explain that they have the same absolute value. They are both 4 units from 0. Absolute value will always be a positive number since distance from 0 is expressed as a positive number of units.

What type of integers always have the same absolute value? opposites

---

the opposite of a negative number is positive, and the opposite of a positive number is negative. Zero is its own opposite.

What measurement is the opposite of 4°C? -4°C; 4°C is 4 units above 0, and its opposite is 4 units below 0.

- Read aloud the following scenario.

Eric reported to the class that Death Valley is approximately 86 m below sea level and that the Dead Sea depression is approximately 413 m below sea level.

What numbers can you write to represent 86 m below sea level and 413 m below sea level? -86 m and -413 m

Explain that sea level is considered 0, and any distance below sea level is represented by a negative number.

What measurement is the opposite of -86 m? 86 m; Since -86 m is eighty-six meters (units) below sea level (0), its opposite is eighty-six meters (units) above sea level (0).

Why do we have negative numbers? Negative numbers represent the numbers below a standard, such as below sea level or below 0° temperatures.

- Display the *Positive & Negative Number Line* page.

What do the arrows at either end of a number line represent? They indicate that the number line extends infinitely in both directions with positive numbers to the right of 0 and negative numbers to the left of 0.

LESSON 167

Write "|6|" and "|⁻6|" for display.
What is the absolute value of 6? 6
What is the absolute value of ⁻6? 6

- Write the following for display and direct the students to find the absolute values.

|12| 12 |⁻8| 8 |⁻1| 1
|9| 9 |⁻25| 25

Comparing & ordering integers

- Use **direct instruction** to help the students compare and order integers.

- Explain that integers can be compared using the signs > and <. Redirect attention to the *Positive & Negative Number Line* page.
What do you know about the value of the numbers as you move to the right on the number line? The numbers increase in value.
What do you know about the value of the numbers as you move to the left on the number line? The numbers decrease in value.

- Point out that as students move from 4 to 7 on the number line, they move in a positive direction and the value of the numbers increases; as they move from 7 to 4, they move in a negative direction and the value of the numbers decreases. Choose a student to write a comparison of 4 and 7 using the signs > and <. 7 > 4 or 4 < 7

- Direct a student to find 6 and ⁻7 on the number line.
Does 6 or ⁻7 have the greater value? 6; The numbers increase in value as I move to the right on the number line; therefore, a positive number always has a greater value than a negative number.
Which number has the lesser value, ⁻6 or ⁻2? ⁻6; The negative numbers decrease in value as I move to the left on the number line.
Choose a student to write the comparisons using the signs > and <. ⁻2 > ⁻6 and ⁻6 < ⁻2

Use 647,325,689,038 to write the answer.

1. Write the value of the 5 in standard form. *5,000,000*

2. Write the digit in the hundred billions place. *6*

3. Round to the nearest one billion. *647,000,000,000*

4. Write the 3 digits in the thousands period. *6, 8, 9*

Write a comparison sentence by using >, <, or =.

5. 124 million $\leq$ 1 billion   6. 21.8 $\geq$ 21.09   7. twenty-one million $\geq$ 9,475,389

Write the numbers from least to greatest.

8. 784,983 7,840,983 7,850,983 7,849,983
   784,983 7,840,983 7,849,983 7,850,983

9. 3,721 3.721 372.1 37.21
   3.721 37.21 372.1 3,721

Round the number to the greatest place.

10. 453,279 *500,000*
11. 1,982,400 *2,000,000*
12. 820,761,398 *800,000,000*
13. 4.7 *5*

Write the number in standard form.

14. five hundred thirty-two billion, one million, four hundred twenty-seven thousand, ninety-six *532,001,427,096*

15. 200,000,000 + 40,000,000 + 8,000,000 + 300,000 + 60,000 + 9,000 + 100 + 50 + 7 *248,369,157*

16. 10 billions + 427 millions + 801 thousands + 119 ones *10,427,801,119*

17. (7 × 100,000) + (4 × 10,000) + (3 × 1,000) + (9 × 100) + (5 × 10) + (2 × 1) *743,952*

- Follow a similar procedure to complete the following number sentences.
  ⁻1 __ 1 <   ⁻3 __ ⁻4 >   ⁻6 __ 0 <
  7 __ ⁻7 >   0 __ 5 <   2 __ 1 >

- Write the following sets of numbers for display. Direct the students to use a number line to order each set of numbers from least to greatest.
  ⁻2, ⁻4, 7, ⁻1   *⁻4, ⁻2, ⁻1, 7*
  ⁻3, ⁻5, 5, 2   *⁻5, ⁻3, 2, 5*
  6, 4, ⁻1, 2   *⁻1, 2, 4, 6*

- Repeat the procedure without the number line.
  ⁻12, 8, ⁻9, 3   *⁻12, ⁻9, 3, 8*
  18, ⁻19, ⁻20, 20   *⁻20, ⁻19, 18, 20*
  25, ⁻26, 26, ⁻25   *⁻26, ⁻25, 25, 26*

Apply

Student Edition pages 366–67
• Read and explain the directions for pages 366–67. Assist the students as they complete the pages independently.

Daily Review
• Students should complete Chapter 17, section *a*.

DIFFERENTIATED INSTRUCTION

Use the following to provide extra help for students who experience difficulty with the concepts taught in Chapter 17.

Recognize an integer as being positive.
Allow the students who have difficulty recognizing that an integer without a symbol is a positive integer to write positive symbols (+) in front of positive integers. This will help them in comparing and ordering integers, in seeing that opposites cancel each other out within an equation (e.g., $5 + {}^-5 = 0$), and in solving equations (e.g., $3 + {}^-4 = {}^+3 + {}^-4 = {}^-1$ and ${}^-3 + 4 = {}^-3 + {}^+4 = {}^+1$).

Recognize plus (+) and minus (–) signs as operations (actions) or as integer symbols.
Solving addition and subtraction equations without the algebra mat may be helpful for the students who have difficulty in determining whether a plus (+) or minus (–) sign indicates an action in an equation or is an integer symbol. Write a positive symbol (+) on black counters, using a white marker, and write a negative symbol (–) on red counters, using a black marker. Guide the students in using the counters to solve addition and subtraction equations.

NOTES

LESSON 168

Student Edition pages 365, 368–69
Daily Review Chapter 17, section *b*

OBJECTIVES

- Add integers.
- Distinguish between consistency and reality in using a model. **BWS**
- Solve a word problem by writing an addition equation.

BIBLICAL WORLDVIEW SHAPING

- **Modeling (Explain):** Good math models represent reality well.

TEACHER RESOURCES

- 3 *Positive & Negative Number Line*
- 128 *Algebra Mat*
- 129 *Positive & Negative Number Lines* (for each student)

ADDITIONAL MATERIALS

- counters: black and red

As an alternative to using the provided *Algebra Mat* page, you may prepare a large algebra mat. Use white poster-board for the mat and cut circles from red and black paper to use as counters.

Preparation

On the *Algebra Mat* page, color the positive sign black and the negative sign red.

Engage

- Guide the students in a **visual analysis** of the chapter opener to explore the chapter essential question, "How are my mathematical models related to my worldview?"
- Direct attention to Student Edition page 365 and the picture of the Grand Canyon.
- You may use the following questions to guide the discussion.
  What contrasting views do the boys have regarding the formation of the Grand Canyon?
  On what basis does each boy form his belief?
- Point out that the boys come to different conclusions after viewing the same evidence. Throughout the chapter, encourage students to learn even more about models in order to answer the chapter essential question.

Instruct

Adding integers; models & reality

- **Model** with an algebra mat to help the students learn to add integers.
- Display the *Algebra Mat* page. Explain that an algebra mat can be used to illustrate the addition of integers.
  What are integers? whole numbers and their opposites such as -3, -2, -1, 0, 1, 2, 3
- Point out that the algebra mat has a "+" side for positive integers and a "–" side for negative integers. Display the counters and explain that the black counters represent positive integers, and the red counters represent negative integers.

You may explain that accountants often refer to a positive financial balance as being "in the black" and a negative financial balance as being "in the red."

Adding Integers

Does the model for adding on a number line represent reality well?

Adding Integers Using an Algebra Mat

1. Draw the first addend on the mat.
2. Draw the second addend on the mat.
3. A positive counter and a negative counter cancel each other out to make 0.

   1 + -1 = 0
4. The answer is the number of counters that have *not* been cancelled out.

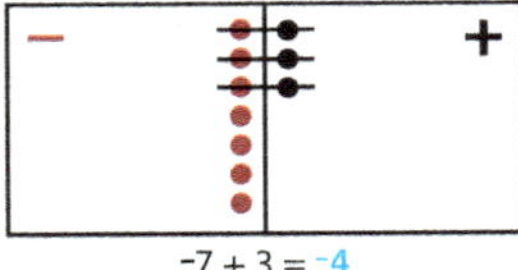

-7 + 3 = -4

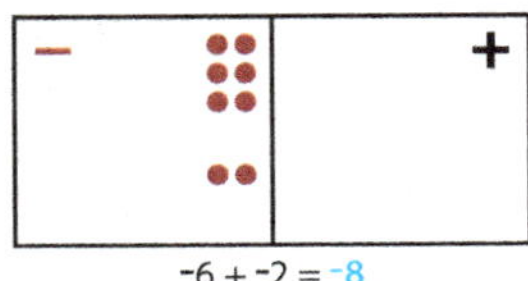

-6 + -2 = -8

Exercises

Draw an algebra mat to find the sum.

1. -5 + 2 *-3*
2. -7 + -2 *-9*
3. 7 + 4 *11*
4. -4 + 8 *4*

Add. Draw an algebra mat and counters if needed.

5. 8 + -3 *5*
6. 7 + -7 *0*
7. -2 + -4 *-6*
8. 4 + -2 *2*
9. -5 + 2 *-3*
10. 12 + -7 *5*
11. 2 + -9 *-7*
12. -15 + 8 *-7*
13. -8 + 5 *-3*
14. -8 + -6 *-14*
15. -6 + -5 *-11*
16. 7 + -12 *-5*

Write a comparison sentence by using >, <, or =.

17. 1 + 5 $\leq$ 12 + -4 *6 < 8*
18. -5 + -2 $\leq$ -1 + -4 *-7 < -5*
19. -4 + -1 $\geq$ -8 + 2 *-5 > -6*
20. 3 + 1 $=$ -2 + 6 *4 = 4*
21. 5 + 4 $\geq$ -14 + 6 *9 > -8*
22. 7 + -2 $\geq$ -1 + -4 *5 > -5*
23. 7 + -1 $=$ 9 + -3 *6 = 6*
24. -4 + 4 $=$ -2 + 2 *0 = 0*
25. -1 + 1 $\leq$ 2 + -1 *0 < 1*
26. |-6| + 2 $\geq$ 4 + -4 *6 + 2 > 0*
    8 > 0

Adding Integers Using a Number Line

1. Begin at 0.
2. Draw an arrow to the first addend.
3. From the first addend, draw a second arrow *right* to add a positive number or *left* to add a negative number.
4. The final stopping place is the sum.

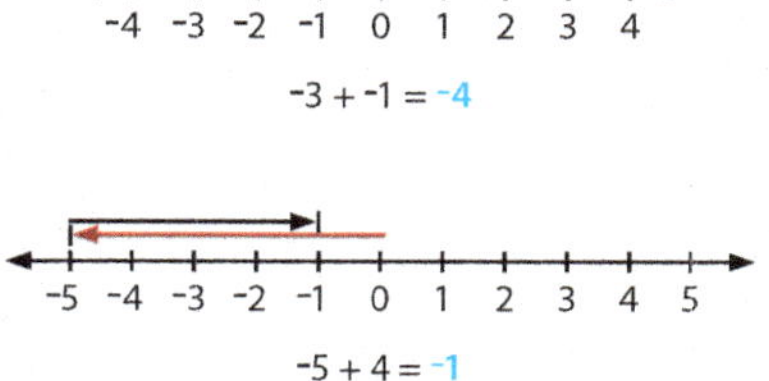

$$-3 + {}^-1 = {}^-4$$

$$-5 + 4 = {}^-1$$

Draw a number line to find the sum.

27. $4 + 1$ *5*
28. $^-4 + 2$ *-2*
29. $^-1 + {}^-2$ *-3*
30. $^-3 + 6$ *3*

Add. Draw a number line if needed.

31. $^-5 + {}^-3$ *-8*
32. $5 + 3$ *8*
33. $8 + {}^-6$ *2*
34. $^-6 + 1$ *-5*
35. $^-4 + 3$ *-1*
36. $^-7 + {}^-6$ *-13*
37. $7 + {}^-4$ *3*
38. $^-5 + 7$ *2*
39. $5 + {}^-5$ *0*

Write a positive or a negative integer to match the sentence.

40. Leslie scored 2 soccer goals. *2 goals*
41. The temperature was twelve degrees below zero. *-12°*
42. The city is seven feet below sea level. *-7 ft*
43. Marcy lost ten pounds last month. *-10 lb*

Practice & Application

44. The water in the lake was 10 ft below normal. A drought caused the level to drop 2 more feet. Write an addition equation using integers to show the water level in the lake. *-10 + ⁻2 = -12 ft*
45. Reed scored 2 strokes under par (-2) on his first game in the golf tournament. He scored 4 strokes under par (-4) on his second game. Write an addition equation using integers to show his total score for the two games. *-2 + ⁻4 = -6*
46. Chris and his friends played a game of charades, subtracting points from a team whenever someone spoke while acting. Chris spoke during both rounds of the game, losing 5 points in round 1 and 7 points in round 2. What addition equation shows the point loss? *-5 + ⁻7 = -12 points*
47. Jared's cell phone company automatically charges his bank account $45.00 each month. How much money has the company deducted after a 6-month period? *6 × $45.00 = $270.00*
48. Write the prime factorization for 132, using exponents. *2² • 3 • 11*
49. What is the value of $2^3 \cdot 3^2 \cdot 5$? *8 × 9 × 5 = 360*
50. Identify each number as prime or composite.

    51 78 97 *51: composite; 78: composite; 97: prime*
51. Find the sum of $8 + {}^-5$. What subtraction equation can be written to show the same solution? *8 + ⁻5 = 3; 8 − 5 = 3*
52. Does the model for adding on a number line represent reality well? Why or why not? *Yes, the model for adding on a number line represents reality well because the numbers and the answer correspond to real objects or values.*

How could you be sure that your answer (10) matches real life? sample answer: I could use objects to represent numbers. I could join 7 objects with 3 objects and count to find the sum, 10 objects. This enables me to check that the model (number line) gives the same answer.

If you repeated the procedure, what answer would you get? I would get the same answer.

Point out that the model is consistent because the answer is the same each time. It is also a good model because it gives an answer that is true to life.

- Direct attention to the essential question on Student Edition page 368.

  Does the model for adding on a number line represent reality well? Why or why not? Yes; the model for adding on a number line represents reality well because the numbers and the answer correspond to real objects or values.

- Write the equation "⁻3 + ⁻8 = __" for display.

  How could you show ⁻3 + ⁻8 on the mat? I could place 3 red counters on the negative side of the mat and then add 8 more red counters on the negative side for a total of 11 negative counters.

- Invite students to model the following problems on the mat: ⁻7 + ⁻2 = ⁻9; ⁻5 + ⁻3 = ⁻8; ⁻4 + ⁻6 = ⁻10

  What is true about the sum when the addends are all negative integers? The sum is a negative integer of lesser value but of greater absolute value.

  How can you find ⁻3 + ⁻8 on a number line? I can begin at 0 and move 3 units to the left, stopping at ⁻3. Then I can move 8 more units to the left, ending at ⁻11.

  Choose students to demonstrate ⁻3 + ⁻8 = ⁻11 and the other addition problems, using a number line.

- Write "1 + ⁻1 = __" for display.

  How do you solve this equation using the number line? I begin at 0 and draw an arrow 1 space (unit) to the right (in a positive direction) from 0 to 1; then from the first addend, I draw a second arrow 1 space (unit) to the left (in a negative direction), ending at 0.

- Write the equation "7 + 3 = __" for display.

  How could you show 7 + 3 on the mat? I could place 7 black counters on the positive side of the mat and then add 3 more black counters on the positive side for a total of 10 positive counters.

  Demonstrate the problem on the mat.

- Invite students to solve the following problems on the mat for display: 4 + 3 = 7; 6 + 2 = 8; 1 + 8 = 9

  What is true about the sum when the addends are all positive integers? The sum is a positive integer of greater value.

- Display the *Positive & Negative Number Line* page. Explain that a number line can be used to illustrate the addition of integers by drawing arrows to represent each addend. Demonstrate finding 7 + 3 on the number line. Begin at 0 and draw an arrow to the first addend (7 units to the right); then from the first addend, draw a second arrow (3 units to the right) ending at 10.

  You may draw red (negative) and black (positive) arrows as shown on Student Edition page 369.

- Remind the students that a good model is one that comes as close as possible to reality.

LESSON 168

What does $1 + {}^-1$ equal? **0**

Demonstrate $1 + {}^-1 = 0$ on the number line.

- Direct a student to demonstrate $1 + {}^-1 = 0$ on your algebra mat for display.

  What do you think is the value of the 1 positive counter and the 1 negative counter? **sample answer: 0; The counters cancel each other out.**

  How do you think you can show that the 1 positive counter and the 1 negative counter cancel each other out to equal 0? **sample answer: I can remove the pair of counters (1 red and 1 black) from the mat so that no counters remain on the mat.**

- Demonstrate removing 1 red and 1 black counter. Remind the students that negative 1 is the opposite of positive 1 and that every number has an opposite positive or negative number, with the exception of 0, which is neither negative nor positive. Repeat the procedure to demonstrate $3 + {}^-3 = 0$, $5 + {}^-5 = 0$, and $7 + {}^-7 = 0$ on the mat.

  What is true about the sum when the addends are a number and its opposite? **The sum is 0.**

- Remind the students that the Zero Principle allows them to add or subtract 0 without changing the value of an equation. Removing a pair of counters (1 red and 1 black) illustrates the Zero Principle.

- Write the equation "${}^-3 + 7 = \_\_$" for display.

  How can you show ${}^-3 + 7$ on the algebra mat? **I can place 3 red counters on the negative side and 7 black counters on the positive side.**

  Choose a student to demonstrate.

- Demonstrate removing the pairs of red and black counters $(1 + {}^-1 = 0)$ until only 4 black counters remain. Point out that the remaining counters indicate the sum, positive 4.

  What does ${}^-3 + 7$ equal? **4**

  How can you show ${}^-3 + 7$ on the number line? **I can begin at 0 and draw an arrow 3 units to the left, stopping at ${}^-3$. Then I can draw an arrow 7 units to the right from ${}^-3$ to 4.**

Solve. Write the answer in lowest terms. *Answer is shown using cancellation.*

1. $\frac{5}{6} \div \frac{1}{3}$
$\frac{5}{6} \times \frac{3}{1} = \frac{5}{2} = 2\frac{1}{2}$

2. $\frac{4}{8} \div 2$
$\frac{4}{8} \times \frac{1}{2} = \frac{2}{8} = \frac{1}{4}$

3. $3\frac{1}{2} \div 1\frac{1}{4}$
$\frac{7}{2} \times \frac{4}{5} = \frac{14}{5} = 2\frac{4}{5}$

4. $\frac{6}{8} \div \frac{1}{2}$
$\frac{6}{8} \times \frac{2}{1} = \frac{6}{4} = 1\frac{1}{2}$

5. $4 \times \frac{3}{4}$ **3**

6. $\frac{3}{6} \times \frac{2}{5}$ **$\frac{3}{15} = \frac{1}{5}$**

7. $5\frac{1}{3} \times 2\frac{1}{4}$ **$\frac{16}{3} \times \frac{9}{4} = 12$**

8. $\frac{3}{5} \times \frac{4}{9}$ **$\frac{4}{15}$**

9. $\frac{3}{9} = \frac{1}{3}$
$+ \frac{2}{3} = \frac{2}{3}$
$\frac{3}{3} = 1$

10. $6\frac{1}{2} = 6\frac{5}{10}$
$+ 2\frac{3}{5} = 2\frac{6}{10}$
$8\frac{11}{10} = 9\frac{1}{10}$

11. $9\frac{4}{5}$
$+ 2\frac{3}{5}$
$11\frac{7}{5} = 12\frac{2}{5}$

12. $8\frac{1}{5} = 8\frac{4}{20}$
$+ \frac{6}{20} = \frac{6}{20}$
$8\frac{10}{20} = 8\frac{1}{2}$

13. $\frac{9}{12}$
$- \frac{4}{12}$
$\frac{5}{12}$

14. $4\frac{7}{10} = 4\frac{7}{10}$
$- 2\frac{3}{5} = 2\frac{6}{10}$
$2\frac{1}{10}$

15. $7 = 6\frac{2}{2}$
$- 3\frac{1}{2} = 3\frac{1}{2}$
$3\frac{1}{2}$

16. $10\frac{3}{4} = 10\frac{9}{12}$
$- 5\frac{2}{3} = 5\frac{8}{12}$
$5\frac{1}{12}$

Determine whether the fraction is closest to 0, $\frac{1}{2}$, or 1.

17. $\frac{6}{10}$ **$\frac{1}{2}$**

18. $\frac{9}{10}$ **1**

19. $\frac{1}{10}$ **0**

20. $\frac{5}{10}$ **$\frac{1}{2}$**

Choose a student to demonstrate the equation on the number line.

- Follow a similar procedure for ${}^-5 + 2 = \_\_$, using the mat and the number line to illustrate the problem.

  What is the sum of ${}^-5 + 2$? **${}^-3$**

- Arrange the students in pairs. Distribute the *Positive & Negative Number Lines* page. Write the following equations for display. Direct students to use the number lines to solve the problems. Choose students to demonstrate solving the problems.

  $9 + {}^-8 =$ **1**

  ${}^-4 + 8 =$ **4**

  ${}^-4 + {}^-3 =$ **${}^-7$**

  ${}^-7 + 5 =$ **${}^-2$**

- Follow a similar procedure to complete the following number sentences using >, <, or =.

  $4 + {}^-2 > {}^-7 + 3$

  $3 + {}^-8 = {}^-7 + 2$

  ${}^-13 + {}^-5 < 4 + {}^-9$

  ${}^-6 + {}^-4 < {}^-12 + 3$

Solving a word problem

- Direct the students to **model** a word problem to help them write an addition equation.

- Read aloud the following word problem.

  John competes on a school quiz team. The quiz teams receive 1 point for each question answered correctly, and the teams lose 1 point for each question answered incorrectly. John's team answered

12 questions correctly and 4 questions incorrectly. What was the team's final score? 8 points

What addition equation can you write to solve this word problem? $12 + {}^-4 =$ __
Write the equation for display.

How can you solve the equation using the number line? I begin at 0 and move 12 units to the right (a positive direction) and then move 4 units to the left from 12 (a negative direction).

- Choose a student to demonstrate how to find the sum using a number line.
  What was the team's final score? 8 points
  Write "8" to complete the question.

- Read aloud the following word problem.
  The opposing quiz team answered 9 questions correctly and 2 questions incorrectly. What was the opposing team's final score? 7 points

  What addition equation can you write to solve this word problem? $9 + {}^-2 =$ __

- Choose a student to write the equation for display and use the algebra mat to solve the equation.

  How can you use the mat to solve this equation? I can place 9 black counters on the positive side of the mat and 2 red counters on the negative side of the mat. Then, I can remove 2 pairs of counters, leaving 7 positive counters.

  What was the opposing team's final score? 7 points

  Who won the competition? John's team won by 1 point.

Apply

Student Edition pages 368–69
- Read and explain the directions for pages 368–69. Assist the students as they complete the pages independently.

Daily Review
- Students should complete Chapter 17, section *b*.

LESSON 168

MATH TALK

Follow the General Procedure for Math Talks as outlined in Lesson 5 on Teacher Edition page 13b.

- Use the following digits and decimal point to build each indicated number.

  7 3 4 9 .

  sample answers:

  the largest number 9743.

  the smallest number .3479

  the number closest to 10^3 974.3

  the number whose absolute value is the same as $|{}^-493.7|$ 493.7

  the number that is closest to 100 97.43

  a number that rounds to 35 34.97 or 34.79

NOTES

LESSON 169

Student Edition pages 370–71
Daily Review Chapter 17, section c

OBJECTIVES

- Subtract integers.
- Explain the danger of trusting every model. **BWS**
- Solve a word problem by writing a subtraction equation.

BIBLICAL WORLDVIEW SHAPING

- **Modeling (Explain):** The validity of a model depends on right assumptions of the one who makes the model.

TEACHER RESOURCES

- 3 *Positive & Negative Number Line*
- 128 *Algebra Mat*
- 129 *Positive & Negative Number Lines* (for each student)

ADDITIONAL MATERIALS

- counters: black and red

Engage

- Direct the students to do a **Think-Pair-Write** to explore the essential question on Student Edition page 370, "How can I model subtracting a negative number on a number line?"

  You may initiate the thinking with the following question:

  How do you model subtracting a *positive* number (such as 6 − 2) on a number line? sample answer: I move in a negative direction (left) to subtract a positive number.

Instruct

Subtracting integers using manipulatives

- **Model** an equation with counters to help the students subtract integers.

- Display the *Algebra Mat* page and write "8 − 5 = 3" for display.

How can I model subtracting a negative number on a number line?

Key Terms
- Zero Principle

Subtracting Integers Using an Algebra Mat

1. Draw the minuend (the total) on the mat.
2. Cross out the subtrahend counters (the number being subtracted).
3. If there are not enough counters to subtract, draw pairs of positive and negative counters on the mat until the subtrahend can be subtracted (the Zero Principle).
4. The answer is the number of counters that have *not* been cancelled out.

Zero Principle
A number may be renamed by adding or subtracting 0 without changing the value of that number.

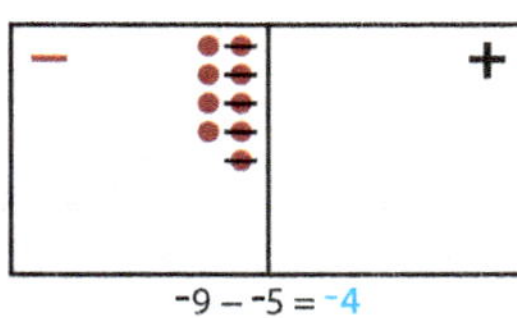

$$^-9 - {}^-5 = {}^-4$$

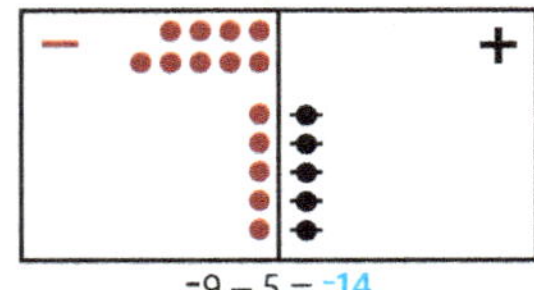

$$^-9 - 5 = {}^-14$$

Exercises

Draw an algebra mat to find the difference.

1. $3 - 7$ *-4*
2. $^-5 - 3$ *-8*
3. $8 - {}^-2$ *10*
4. $^-7 - {}^-2$ *-5*

Subtract. Draw an algebra mat and counters if needed.

5. $4 - 7$ *-3*
6. $^-2 - {}^-7$ *5*
7. $^-8 - {}^-4$ *-4*
8. $^-7 - 9$ *-16*
9. $^-6 - {}^-4$ *-2*
10. $^-10 - 2$ *-12*
11. $2 - 8$ *-6*
12. $3 - 6$ *-3*
13. $4 - {}^-2$ *6*
14. $10 - {}^-4$ *14*
15. $^-6 - 4$ *-10*
16. $^-8 - {}^-3$ *-5*

Solve.

17. Mr. Perry's golf score was 6 strokes under par. Jason's score was 6 strokes over par. Write a subtraction equation using integers to find the difference between Mr. Perry's score and Jason's score. *6 − ⁻6 = 12 strokes difference*

18. Trey shot his bow and arrow 7 times at the target. He missed the bull's eye 4 times. Write a subtraction equation to show how many times Trey hit the bull's eye. *7 − 4 = 3 times*

370 Chapter 17

How can you show this subtraction equation using the algebra mat? I can place 8 black counters on the positive side of the mat and then remove 5 black counters so that 3 black counters remain on the positive side of the mat.

- Guide students in demonstrating this equation on the mat by placing counters to represent the minuend (first number), and then removing counters to represent the subtrahend (second number).

  How many counters are left on the mat? 3

  Remind the students that the answer to a subtraction equation is called the *difference*.

- Write "⁻8 − ⁻5 = __" for display.

How could you solve this equation using an algebra mat? I could place 8 red counters on the negative side of the mat and then remove 5 red counters so that 3 red counters remain on the negative side of the mat.

Invite a student to solve the equation for display on the algebra mat.

What does ⁻8 − ⁻5 equal? ⁻3

Complete the equation.

- Point out that the procedure for subtracting a negative number from a negative number is the same as the procedure for subtracting a positive number from a positive number. First, you place the number of counters representing the minuend on the mat, and then you

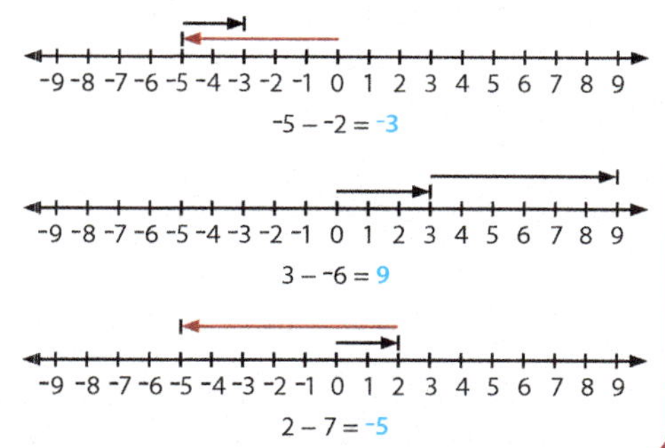

Use the number line to solve the equation.

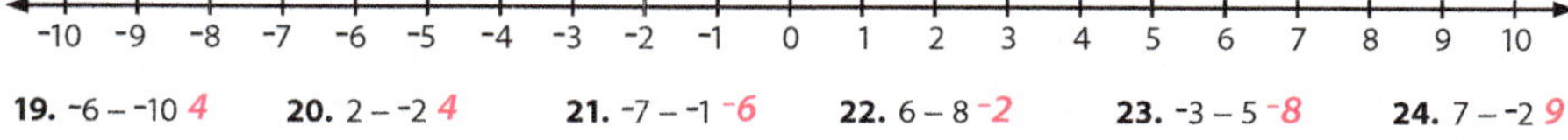

19. -6 − -10 *4*   20. 2 − -2 *4*   21. -7 − -1 *-6*   22. 6 − 8 *-2*   23. -3 − 5 *-8*   24. 7 − -2 *9*

25. 1 − 4 *-3*   26. -3 − 7 *-10*   27. 10 − 5 *5*   28. 4 − -4 *8*   29. -6 − 3 *-9*   30. 3 − 8 *-5*

Add. Draw an algebra mat and counters or a number line if needed.

31. 7 + -2 *5*   32. -5 + -3 *-8*   33. -2 + 4 *2*   34. 4 + -4 *0*   35. -6 + -1 *-7*

36. -7 + 2 *-5*   37. 8 + -4 *4*   38. -1 + -8 *-9*   39. 5 + -3 *2*

Practice & Application

40. How many inches are in $3\frac{3}{4}$ yd? $\frac{15}{4} \times 36 = 135$ *in.*

41. Write the numbers from least to greatest.

5.76 $\sqrt{25}$ $\frac{17}{3}$ $5\frac{3}{4}$ *$5\frac{3}{4}$ $\sqrt{25}$ $\frac{17}{3}$ 5.76*

42. Write a comparison sentence using < to compare 800.01 and 800.009. *800.009 < 800.01*

43. Use the order of operations to solve the equation.

$15 + 3 \times 8 \div 2^2 - 7 =$ __
$15 + 24 \div 4 - 7 =$
$15 + 6 - 7 = 14$

44. Write 17,301 in word form and in expanded form.
seventeen thousand, three hundred one
10,000 + 7,000 + 300 + 1

45. The *Milltown Pride* film is $2\frac{1}{4}$ hr long. How many minutes long is the film? *$\frac{9}{4} \times 60 = 135$ min*

46. The weather forecaster said that the temperature was 79°F. Write a subtraction equation and an addition equation to show what the temperature will be if it drops 15° as expected.
79° − 15° = 64°F; 79° + -15° = 64°F

47. How can I model subtracting a negative number on a number line? *First I draw an arrow to the minuend (total). Then I draw another arrow right of the minuend to represent the number I am subtracting.*

Repeat the procedure, adding 0 (1 negative counter and 1 positive counter) until 5 negative counters are on the mat.

- Invite a student to subtract (remove) the 5 negative counters from the mat.
What is 8 − -5? 13
Complete the equation.

- Write "-8 − 5 = __" for display.
How can you show this subtraction equation using the algebra mat? sample answer: I can place 8 red counters on the negative side of the mat. Since there are not enough black counters to subtract 5 from the positive side of the mat, I can place pairs of 1 negative counter and 1 positive counter on the mat until there are 5 counters on the positive side of the mat. Then I can remove 5 black counters from the positive side of the mat.
Demonstrate the subtraction equation on the algebra mat.
What is -8 − 5? -13
Complete the equation.

- Write "5 − 8 = __" for display.
How do you think you could solve this equation using your mat? sample answer: I can place 5 black counters on the positive side of the mat. Since there are not enough counters to subtract 8 from the positive side of the mat, I can add 0: I can place pairs of 1 positive and 1 negative counter on the mat until there are enough positive counters to remove 8.
Invite a student to solve the equation for display.
What is 5 − 8? -3
Complete the equation.

Subtracting integers using a number line; trusting a model

- Direct the students to **model** with a number line to help them subtract integers.

- Distribute the *Positive & Negative Number Lines* page to each student and display the *Positive & Negative Number Line* page. Write the equation "-7 − -3 = __" for display.

When subtracting a *positive* number, what direction do you move on the number line? to the left

remove the number of counters representing the subtrahend.

- Write "8 − -5 = __" for display.
How do you think you could solve this equation using your mat? I could place 8 black counters on the positive side of the mat and remove 5 red counters from the negative side.

Choose a student to place 8 black counters on the positive side of the mat for display.

Can you remove 5 negative counters from the mat? No; there are no red counters on the negative side of the mat.

- Remind the students that the Identity Property of Addition or the Zero Principle allows them to add 0 without changing the value of the equation.

If you add 0 to the 8 counters on the mat, will the value of the counters change? no

- Remind the students that they can "add 0" by adding the same number of counters to each side of the mat until there are enough counters to subtract the second number (subtrahend). Place 1 black counter on the positive side of the mat and 1 red counter on the negative side of the mat for display.

Has the value of the counters on the mat changed? No; the value has not changed since a positive and a negative counter together equal 0.

LESSON 169

What direction do you think you should move when subtracting a *negative* number? sample answer: Since subtraction is the inverse (opposite) operation of addition, I should move in a positive direction (to the right) to subtract a negative number.

Beginning at 0, how could you solve ⁻7 − ⁻3? I first move from 0 to ⁻7, and then I move 3 units to the right (positive direction) to subtract ⁻3, stopping on ⁻4.

Demonstrate on the displayed number line and complete the equation.

- Write "6 − ⁻4 = __" for display.
  How could you represent this equation on the number line model? sample answer: I could begin at 0 and draw a line to the minuend (6, the total).

- Direct the students to model the problem using a number line on the page. Choose a student to demonstrate solving the equation for display.
  How would you show subtracting a *positive* 4? I would draw a line from the minuend (6) four spaces to the left.
  Since this is not 4 but negative 4, how do you think you can model subtracting negative 4? I can draw another arrow from the minuend to the right 4 spaces to represent the number I am subtracting (⁻4).
  What is the result? 10

- Point out that in both these equations they moved in a positive direction (to the right) to subtract a negative number.

- Complete the equation. Remind the students that subtraction is finding the difference between 2 numbers.
  What is the difference in the number of units between 6 and ⁻4 on the number line? 10

Guide the students as they count the units.

How can you model subtracting a negative number on a number line? First I draw an arrow to the minuend (total). Then I draw another arrow from the minuend to the right to represent the number I am subtracting.

Write the measure of the angle.

1.

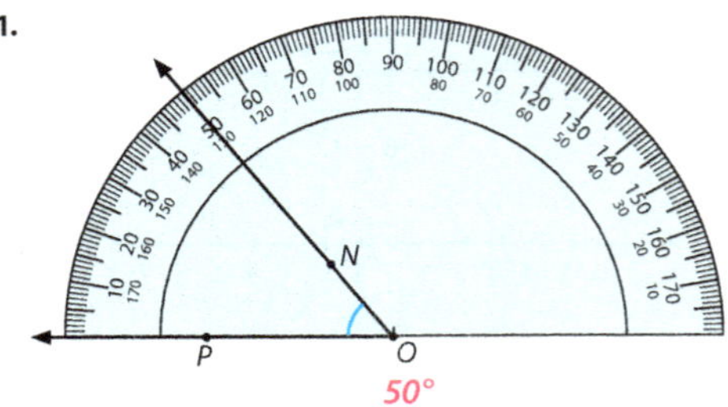

50°

2.

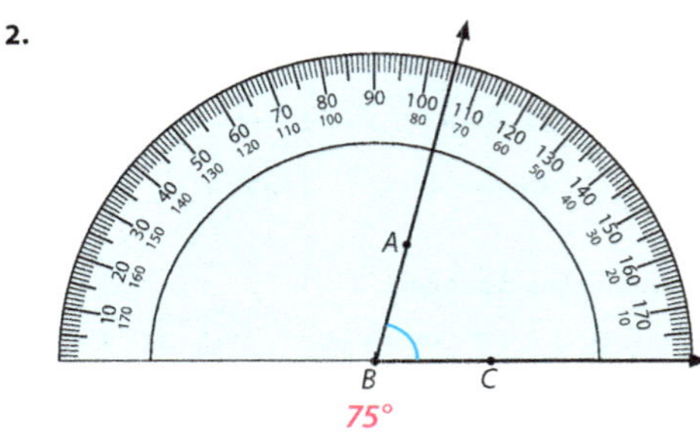

75°

Classify the angle as acute, obtuse, right, or straight.

3. 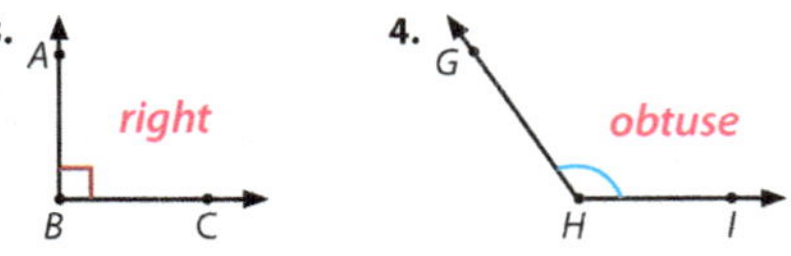 right

4. obtuse

5. 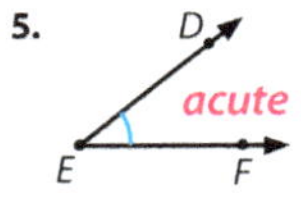 acute

6. 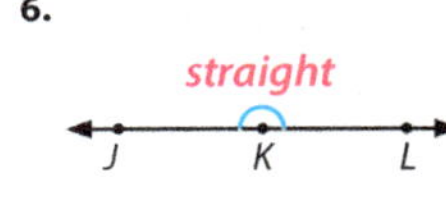 straight

Use the figure to find the answer.

7. Name the diameter.

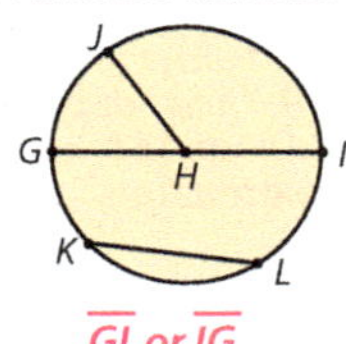

$\overline{GI}$ or $\overline{IG}$

8. Find the measure of the unknown angle.

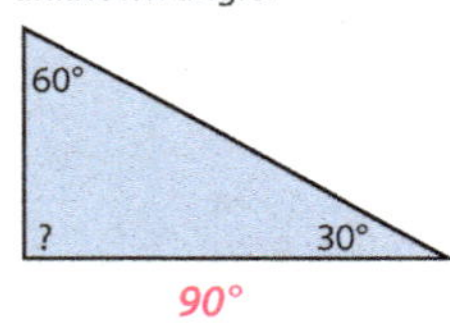

90°

9. Name the shape.

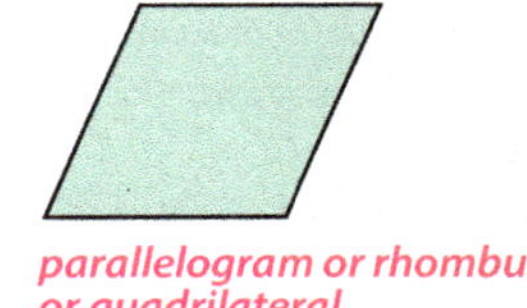

parallelogram or rhombus or quadrilateral

Daily Review 515

- Write "8 − 2 = __" for display. Direct the students to solve the problem using a number line on the page. Choose a student to demonstrate solving this equation on a number line. Instruct him to explain the process. I begin at 0, move 8 units to the right, and then move 2 units to the left, stopping on 6.
  Complete the equation.

- Write "4 − 7 = __" for display. Direct the students to solve the problem using a number line on the page.
  How can you solve this equation? I can begin at 0, move 4 units to the right, and then move 7 units to the left, stopping on ⁻3.

What is 4 − 7? ⁻3

Complete the equation.

Point out that in the equation 4 − 7 = ⁻3, they moved right from 0 to 4, and then they moved to the left (negative direction) to subtract a positive number, since subtracting a positive number is the inverse (opposite) of adding a positive number.

- Follow a similar procedure for ⁻3 − 6 = __. Begin at 0, move 3 units to the left (a negative direction) and then move 6 units to the left (a negative direction) to subtract a positive 6, stopping on ⁻9.

- Point out that a number line is a helpful model for subtracting integers. Remind the students that as reliable as number lines are, all models are formed by humans, who all have different assumptions and goals. (See Lesson 97 for the model of a child's growth that was based on wrong assumptions.)

  If a person's assumptions are wrong, what could you conclude about the model? The model will not be true to reality.

- Remind the students that all models should be based on reality, which God has revealed to us in nature (creation) and in Scripture.

Solving a word problem

- Guide the students in a **discussion** to help them solve a word problem.

- Read aloud the following word problem.

  Sarah's score was $^-5$. How many points must she score consecutively to reach a final score of 0?

  What subtraction equation can you write for this problem? $0 - {}^-5 = \_$; Subtraction is used to find the difference between 2 numbers.

  Write "$0 - {}^-5 = \_$" for display.

  How can you show $0 - {}^-5$ using the number line? I can begin at 0 and move 5 units to the right (positive direction) to subtract $^-5$.

  How many points must Sarah score to reach a final score of 0? 5 points

  Write "5" to complete the equation.

  What missing addend equation can you write for this problem? $^-5 + \_ = 0$

  Write "$^-5 + \_ = 0$" for display.

  How can you show this problem using the algebra mat? I can place 5 counters on the negative side of the mat; then I can determine that adding 5 counters to the positive side of the mat will make 0 ($^-5 + 5 = 0$).

  What is the missing addend? 5; An integer and its opposite will always equal 0.

  How can you show the problem using the number line? sample answer: I can begin at $^-5$ and determine that I need to move 5 units to the right (positive 5) to arrive at 0.

  What conclusion can you reach about subtracting a number and adding its opposite? Subtracting a number is the same as adding its opposite.

Apply

Student Edition pages 370–71

- Read and explain the directions for pages 370–71. Assist the students as they complete the pages independently.

Daily Review

- Students should complete Chapter 17, section *c*.

NOTES

LESSON 170

Student Edition pages 372–73
Daily Review Chapter 17, section *d*

OBJECTIVES
- Add and subtract integers.
- Solve a word problem by writing an equation.

TEACHER RESOURCES
- 3 *Positive & Negative Number Line*
- 128 *Algebra Mat*
- 130 *Subtraction Patterns* (for the teacher and for each student)
- 131 *More Subtraction Patterns*

ADDITIONAL MATERIALS
- counters: black and red

ASSESSMENTS
- Chapter 17 Quiz 1

Engage
- Choose a student to draw a simple number line and another student to draw a thermometer for display. Instruct each to draw tick marks and label the numbers $^-5$ to 5.
- Direct the students to do a **Turn and Talk** to answer the essential question on Student Edition page 372, "How is a thermometer like a number line?" sample answer: A thermometer measures positive and negative values like a number line does.

Instruct
Subtracting integers
- **Model** subtracting on a number line to help the students subtract integers.
- Distribute the *Subtraction Patterns* page to each student and display your copy. Review the steps at the top of the page on how to use a number line to subtract or add integers. Point out the first number line and the arrows above the number line.

  How do the arrows above the number line illustrate $6 - 4$? sample answer: I begin at 0 and move 6 units to the right (positive direction) stopping at 6; from there I move 4 units to the left (negative direction). The final stopping place of 2 illustrates that $6 - 4 = 2$.

- Direct the students to study the number line and arrows carefully since they also demonstrate another equation besides $6 - 4 = 2$.

  What addition equation do you think this number line illustrates? $6 + {}^-4 = 2$; I begin at 0 and move right 6 units, indicating a positive 6; then I move left 4 units, indicating that I am adding $^-4$ to the 6. The final stopping place of 2 illustrates that $6 + {}^-4 = 2$.

  Write the new equation below the number line and direct the students to do the same.

- Direct attention to the second number line and guide the students as they draw arrows to illustrate the equation $^-4 - {}^-8 = \underline{\quad}$.

  How can you illustrate the minuend on the number line? I begin at 0 and move 4 units to the left (negative direction) to $^-4$.

  How can you illustrate subtracting the subtrahend ($^-8$) from the minuend ($^-4$) on the number line? From $^-4$, I move 8 units to the right (the opposite of a negative direction) ending at 4.

  What is $^-4 - {}^-8$? 4

  Complete the equation and direct the students to do the same.

How is a thermometer like a number line?

1. Start at 0 and move 7 units to the right.
2. Move 3 units to the left to **subtract 3**.
   or
   Move 3 units to the left to **add** $^-3$.

$7 - 3 = 4$
$7 + {}^-3 = 4$

Subtracting a number is the same as adding its opposite. The opposite of 3 is $^-3$.

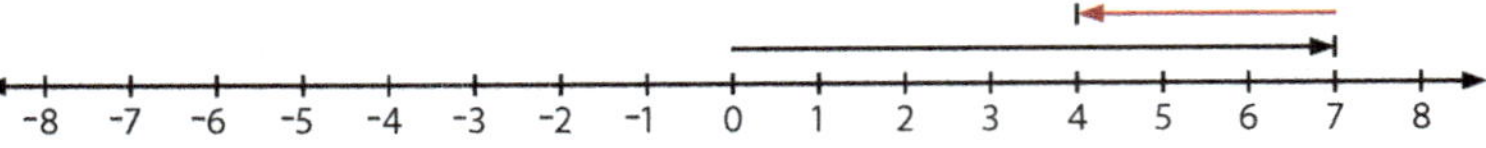

Exercises

Write the opposite of the given integer.

1. 8 $^-8$
2. $^-1$ *1*
3. $^-12$ *12*
4. 4 $^-4$
5. $^-7$ *7*

Solve.

6. $4 - 9$ $^-5$
7. $7 - {}^-2$ *9*
8. $1 + {}^-9$ $^-8$
9. $3 + {}^-3$ *0*

10. $^-1 - 3$ $^-4$
11. $^-13 - 8$ $^-21$
12. $^-5 + {}^-2$ $^-7$
13. $8 + 7$ *15*

14. $7 - 5$ *2*
15. $^-9 - {}^-11$ *2*
16. $1 + {}^-8$ $^-7$
17. $^-1 + {}^-3$ $^-4$

18. $^-11 - {}^-7$ $^-4$
19. $6 - 10$ $^-4$
20. $^-6 + 2$ $^-4$
21. $2 + {}^-1$ *1*

22. $^-8 - {}^-3$ $^-5$
23. $12 - {}^-7$ *19*
24. $^-2 + 11$ *9*
25. $^-9 + {}^-11$ $^-20$

Write an addition equation for the subtraction equation. Solve.

26. $^-8 - 4$ $^-8 + {}^-4 = {}^-12$
27. $3 - 5$ $3 + {}^-5 = {}^-2$
28. $^-12 - 5$ $^-12 + {}^-5 = {}^-17$

29. $^-10 - {}^-3$ $^-10 + 3 = {}^-7$
30. $2 - 7$ $2 + {}^-7 = {}^-5$
31. $9 - {}^-5$ $9 + 5 = 14$

Write a comparison sentence by using >, <, or =.

32. $3 + {}^-1 \underline{\geq} 2 - 3$ *2 > $^-1$*
33. $^-5 - 2 \underline{=} {}^-3 + {}^-4$ $^-7 = {}^-7$

34. $^-8 + {}^-2 \underline{\leq} 5 - 5$ $^-10 < 0$
35. $1 - 6 \underline{\geq} 5 + {}^-12$ $^-5 > {}^-7$

36. $3 - 4 \underline{\geq} 2 - 4$ $^-1 > {}^-2$
37. $^-1 - 1 \underline{\leq} 4 + {}^-4$ $^-2 < 0$

DID YOU KNOW?

The word *infinite* indicates something that goes on forever. The love of God is infinite. The symbol ∞ represents infinity. It can be used to show that a series of numbers continues forever. The set of integers can be written: ∞, . . ., -5, -4, -3, -2, -1, 0, 1, 2, 3, 4, 5, . . ., ∞.

The mercy [lovingkindness] of the LORD is from everlasting to everlasting upon them that fear him.

Psalm 103:17

Practice & Application

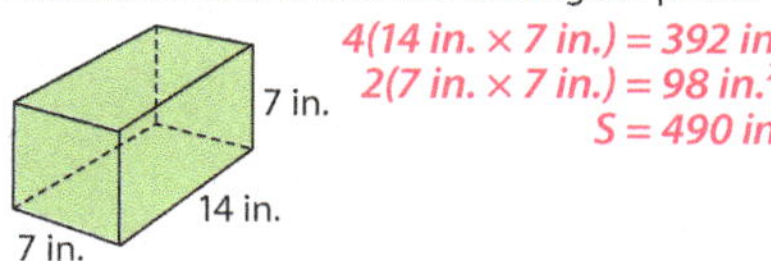

38. In 3 days a certain stock on the stock market lost 4 points, gained 2 points, and then lost 7 points. What was the total change in 3 days? *-4 + 2 − 7 = -9 points or -4 + 2 + -7 = -9 points*

39. Stacie attended a coin show. She sold 10 of her duplicate coins in order to buy 2 coins she did not have in her collection. If Stacie began with 25 coins, how many does she have now? *25 − 10 + 2 = 17 coins or 25 + -10 + 2 = 17 coins*

40. Find the surface area of the rectangular prism.

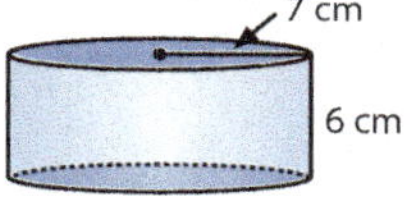

4(14 in. × 7 in.) = 392 in.²
2(7 in. × 7 in.) = 98 in.²
S = 490 in.²

41. Find the surface area of the cylinder.

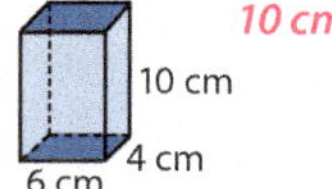

42. Find the volume of the rectangular prism.

10 cm × 4 cm × 6 cm = 240 cm³

43. Find the volume of the triangular prism.

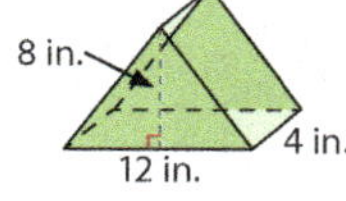

V = Bh
V = [½(12 in. × 8 in.)] × 4 in.
V = 48 in. × 4 in.
V = 192 in.³

44. Solve the equation $c \div 21 = 250$. *c = 5,250*

45. Find the product of 205 × 36. *7,380*

46. Write 9,843.067 in expanded form, using multiplication and powers of 10.

47. Find the quotient of $1\frac{7}{8} \div \frac{3}{4}$. *$\frac{15}{8} \times \frac{4}{3} = \frac{5}{2} = 2\frac{1}{2}$*

48. Use -10 − 6 = -16 to explain the following statement: Subtracting an integer is the same as adding its opposite. *-10 − 6 = -16 is the same as -10 + -6 = -16.*

49. How is a thermometer like a number line? *A thermometer measures positive and negative values like a number line does.*

41. *2 • 3.14 • (7 cm)² = 307.72 cm²*
(2 • 3.14 • 7 cm) • 6 cm = 263.76 cm²
S = 571.48 cm²

46. *(9 × 10³) + (8 × 10²) + (4 × 10¹) + (3 × 10⁰) + (6 × $\frac{1}{10^2}$) + (7 × $\frac{1}{10^3}$)*

Lesson 170   373

(where I began and where I finished). To get from one point on the number line to a second point, the same action had to be performed.

What action that appears different was the same in these 2 equations: 3 − -5 = 8 and 3 + 5 = 8? sample answer: Moving right to add an integer and moving right to subtract that integer's opposite are the same action.

- Instruct the students to use the fourth number line to illustrate the equation 2 − 6 = __.

How can you illustrate this equation on the number line? I move right from 0 to 2, and from there I move left 6 units to subtract a positive six, stopping on -4.

What is 2 − 6? -4

Complete the equation and direct the students to do the same.

What other equation does this number line illustrate? 2 + -6 = -4; I begin at 0 and move right, indicating a positive 2; then I move from there 6 units to the left to add a negative 6. The final stopping place of -4 illustrates that 2 + -6 = -4.

What pattern do you see in each pair of written equations? sample answer: The operation sign changed to its inverse, and the integer following the operation sign changed to its opposite.

How are the subtraction equations related to the addition equations? Subtracting an integer gives the same result as adding its opposite.

- Display the *More Subtraction Patterns* page. Direct the students to study the number patterns to find the missing numbers needed to complete the equations. 6 − -2 = 8 and 6 + 2 = 8; 5 − 6 = -1 and 5 + -6 = -1

What is true about subtracting an integer? Subtracting an integer is the same as adding its opposite.

Solving a word problem

- Guide a **discussion** to help the students write an equation for a word problem.

- Read aloud the following word problem.

Sam's team has a score of 5 points. Jason's team has a score of -3 points. What is the difference between the scores? How many points separate their scores?

What addition equation does this number line illustrate? sample answer: -4 + 8 = 4; I begin at 0 and move 4 units (spaces) to the left (negative direction) to -4; from there I move 8 units to the right (positive direction), ending at 4.

Write the new equation below the number line and direct the students to do the same.

- Instruct the students to use the third number line to illustrate the equation 3 − -5 = __. Guide the students as they draw arrows on their worksheet as you demonstrate on the page.

How can you illustrate this equation on the number line? I draw an arrow 3 units long to the right from 0 to 3, and from

there I draw another arrow 5 units long to the right, stopping on 8.

Why did you move 5 units to the right? I would move left to add -5, so I move the opposite direction to subtract -5.

What is 3 − -5? 8

Complete the equation and direct the students to do the same.

Do the arrows on this number line illustrate another equation? Which one? yes; 3 + 5 = 8

Direct the students to write this new equation below the number line as you do the same.

What is the same in these 2 equations: 3 − -5 = 8 and 3 + 5 = 8? The first number and the last number are the same

LESSON 170

- Display the *Positive & Negative Number Line* page. Choose a student to place dots on the number line to mark Sam's team score of 5 points and Jason's team score of $^-3$ points.

  What is the difference between the 2 teams' scores? 8 points; I can count the number of units that separate the two points on the number line.

- Remind students that the absolute value of a number is its distance from 0 on the number line. Since distances are always positive (or 0), the absolute value of any number is never negative.

  How many units is 5 from 0 on the number line? 5 units

  How many units is $^-3$ from 0? 3 units

  What equation can you write to find the difference between these 2 numbers? $5 - {}^-3 = $ __; I always subtract the lesser number from the greater number to find the difference.

  Write the equation for display.

  How can you solve this equation on a number line? I draw an arrow 5 units long to the right (positive direction) from 0 to 5, and from there I draw another arrow 3 units long from 5 to 8 (positive direction) to indicate subtracting a negative 3. The final stopping place of 8 illustrates that $5 - {}^-3 = 8$.

  Choose a student to demonstrate on the *Positive & Negative Number Line* page.

- Display the *Algebra Mat* page. Direct the students to draw a similar algebra mat on their own paper and then to draw counters to illustrate $5 - {}^-3 = 8$.

  How can you solve this problem using the algebra mat? I can place 5 black counters on the positive side of the mat. Since there are no negative counters to subtract, I must add pairs of 1 negative counter and 1 positive counter until there are 3 negative counters to subtract or take away.

  The students will need to draw lines through the pairs of counters, rather than remove them ($1 + {}^-1 = 0$).

  Choose a student to demonstrate on your mat.

  What is left on the mat? 8 positive counters

  What addition equation can you write for $5 - {}^-3 = $ __ using the new rule that you discovered in this lesson? $5 + 3 = 8$; add the opposite

  What is the difference between the scores in this word problem? 8 points

  Complete the equation.

- Follow a similar procedure for the following word problem.

  The temperature was 6°F at 8 p.m. By midnight the temperature had dropped 10 degrees. What was the temperature at midnight? $6°F - 10°F = {}^-4°F$; $6°F + {}^-10°F = {}^-4°F$

Use mental math to solve.

1. 10×15.3 *153*
2. 100×0.247 *24.7*
3. 10×4.5 *45*
4. 100×23 *2,300*
5. $89.5 \div 10$ *8.95*
6. $241.3 \div 100$ *2.413*
7. $894 \div 10$ *89.4*
8. $52.47 \div 100$ *0.5247*

Solve.

9.
$$\begin{array}{r} 2.45 \\ \times\ 3 \\ \hline 7.35 \end{array}$$

10.
$$\begin{array}{r} 398.01 \\ +\ 45.732 \\ \hline 443.742 \end{array}$$

11.
$$\begin{array}{r} 42.1 \\ -\ 3.87 \\ \hline 38.23 \end{array}$$

12. $8 - 3.804$ *4.196*

13. $50\overline{)6}$ *0.12*
14. $21\overline{)71.4}$ *3.4*
15. $12\overline{)6.48}$ *0.54*
16. $9\overline{)56.25}$ *6.25*

Write the fraction as a decimal.

17. $\frac{3}{4}$ *0.75*
18. $\frac{5}{10}$ *0.5*
19. $\frac{2}{5}$ *0.4*
20. $\frac{1}{4}$ *0.25*

Apply

Student Edition pages 372–73

- Read and explain the directions for pages 372–73. Assist the students as they complete the pages independently.

Daily Review

- Students should complete Chapter 17, section *d*.

Assess

Quiz 1

- Use the **summative assessment** to evaluate the students' progress at this point in the chapter.

NOTES

LESSON 171

Student Edition pages 374–75
Daily Review Chapter 17, section e

OBJECTIVES

- Add and subtract integers.
- Solve real-world problems by writing an equation.

TEACHER RESOURCES

- 3 *Positive & Negative Number Line*

Engage

- Direct the students to **brainstorm** to explore the essential question on Student Edition page 374, "How can integers help me represent elevation?"

Instruct

Adding & subtracting integers; solving real-world problems

- **Model** with a number line to help the students solve real-world problems with integers.

- Display the *Positive & Negative Number Line.*

  What is an integer? sample answer: Integers include the whole numbers (0, 1, 2, 3, . . .) and their opposites.

  What is the opposite of 3? ⁻3

  What is the opposite of 12? ⁻12

  What is the opposite of ⁻23? 23

  What does it mean that 3 is the opposite of ⁻3? The numbers are the same distance from 0 on the number line but in opposite directions. The opposite of a negative number is a positive number, and the opposite of a positive number is a negative number.

  Which integers are to the right of 0? positive integers

  Which integers are to the left of 0? negative integers

  Is 0 positive or negative? 0 is neither positive nor negative; it has no value.

  How do negative numbers compare to 0? They are less than 0.

  How do the positive numbers compare to 0? They are greater than 0.

  What do the arrows at the ends of the number line tell you about the set of integers? sample answer: As the set of whole numbers has infinite growing possibilities, the set of integers has infinite growing and decreasing possibilities.

- Draw a simple thermometer for display. Draw tick marks and label each number ⁻5 to 10. Following each problem, invite a student to demonstrate the equation for display.

  What do the integers on a thermometer represent? sample answer: Positive integers represent temperatures greater than or warmer than 0°, while negative integers represent temperatures less than or colder than 0°.

- Read aloud the following word problem.

  At 5 a.m. the temperature was ⁻1°F. By 10 a.m. the temperature had risen 6 degrees. What was the temperature at 10 a.m.? I begin at ⁻1°F then proceed up 6 degrees in a positive or warmer direction, stopping at 5°F.

  What was the temperature at 10 a.m.? 5°F

  What equation can you write for this problem? ⁻1°F + 6°F = 5°F

  Choose a student to write the equation.

- Read aloud the following word problem.

More Integers

How can integers help me represent elevation?

Exercises

Use the table to find the answer.

1. At 6:30 a.m. the temperature was ⁻10°F. Two hours later the temperature had risen 8°. What was the temperature at 8:30 a.m.?
   ⁻10° + 8° = ⁻2°F

2. From 6:30 a.m. to 9:30 a.m. there was a rise in temperature of 10°. What was the temperature at 9:30 a.m.? ⁻10° + 10° = 0°F

3. What was the difference in temperature from 12:30 p.m. to 6:30 p.m.? 10° − ⁻2° = 12°

4. At 2:30 p.m. the temperature was 12°F. How much did the temperature decrease by 6:30 p.m.? 12° − ⁻2° = 14°

5. Find the difference between the highest and the lowest temperatures recorded for January 5. 12° − ⁻10° = 22°

Solve.

6. Chad's golf score was 2 under par through the 16th hole. On the 17th hole, he shot 2 over par. What was Chad's net score? par, or 0; ⁻2 + 2 = 0

7. A football team lost 10 yd on their first down of the third quarter. They gained 5 yd on their next down. What was their total loss or gain? ⁻10 + 5 = ⁻5 yd; 5 yd loss

8. Ashley had 17 beads. She lost 3 beads and used 8 beads to make a bracelet. How many beads did Ashley have left? 17 − 3 − 8 = 6 beads

9. A scuba diver went 25 ft below the surface of the water. If he swims up 8 ft, what will his depth be? ⁻25 + 8 = ⁻17; 17 ft below sea level

Temperature Readings January 5	
Time	Temperature
6:30 a.m.	⁻10°F
7:30 a.m.	⁻7°F
8:30 a.m.	
9:30 a.m.	
10:30 a.m.	3°F
11:30 a.m.	8°F
12:30 p.m.	10°F
1:30 p.m.	12°F
2:30 p.m.	12°F
3:30 p.m.	10°F
4:30 p.m.	5°F
5:30 p.m.	0°F
6:30 p.m.	⁻2°F

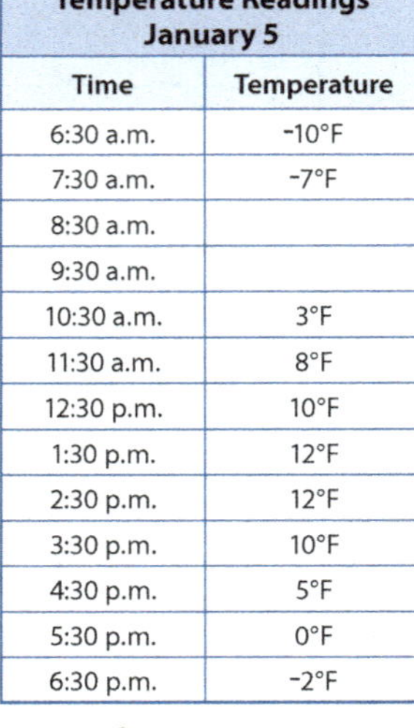

374   Chapter 17

Write a comparison sentence by using > or <.

10. $-3°F > -5°F$ 11. $-1°C > -3°C$ 12. $0°F < 1°F$

13. $-23°F < 23°F$ 14. $-18°C < 18°C$ 15. $5°C > 4°C$

16. $0°F > -1°F$ 17. $-2°C < -1°C$ 18. $-1°F < 1°F$

Practice & Application

19. Margo had $100 in her checking account. She made an error and wrote a check for $125. What was her balance after the check was written? What is her debt? *$100 − $125 = ⁻$25; Margo's debt is $25.*

20. If Margo's account requires a minimum balance of $25.00 at all times, how much money must she deposit to reach her minimum balance? *$25.00 − ⁻$25.00 = $50.00*

21. In one possession a football team gained 8 yd, lost 4 yd, lost 3 yd, and gained 2 yd. What was the total loss or gain? *8 − 4 − 3 + 2 = 3 yd gain or 8 + ⁻4 + ⁻3 + 2 = 3 yd gain*

22. Brett's team scored 15 points and then lost 35. Hudson's team lost 45 points and then gained 30. Whose team won? By how many points did they win?

23. Rachel's baby sister weighed 8 lb when she was born. When she came home from the hospital, she lost 2 lb. At her first week check-up, she had gained back 1 lb. How much does Rachel's sister weigh now? *8 − 2 + 1 = 7 lb or 8 + ⁻2 + 1 = 7 lb*

24. Evan played golf while on vacation. On his first game he scored 5 strokes over par, which is written as ⁺5 on his score card. On his second game he scored 1 stroke under par, which is written as ⁻1 on his score card. What is the difference between his two scores? *⁺5 − ⁻1 = 6 strokes*

25. If there is a 40% chance of rain tomorrow, what is the chance of it *not* raining? *60%*

26. A spinner is divided into 5 equal sections. There are 2 green sections, 2 red sections, and 1 yellow section. Write a ratio to show the probability of landing on a green section. *P(green) = $\frac{2}{5}$*

27. Using the spinner from problem 26, does *certain*, *equally likely*, or *impossible* best describe the chance of landing on a blue section? *impossible*

28. Find the value of *n* to complete the proportion.

$$\frac{15}{51} = \frac{n}{17} \quad n = 5$$

29. The ratio $\frac{12}{96}$ represents $\frac{tables}{chairs}$. Write the ratio in lowest terms and explain its terms. *$\frac{1}{8}$; There are 8 chairs for each table.*

30. Workmen were repairing the elevator at the building where Jesse makes deliveries. Jesse used the stairs to deliver a package to Mr. Samuels on floor 31. Mr. Samuels gave him a package to deliver to Miss Murphy. Jesse descended 18 floors to Miss Murphy's office. She gave him a package to deliver to Mr. Hays, whose office is 22 floors above hers. After delivering Mr. Hays's package, Jesse took a break in the employee lounge on floor 5. How many floors did he descend to get to the lounge? *31 − 18 = 13; 13 + 22 = 35; 35 − 5 = 30 floors*

31. How can integers help me represent elevation? *I can use integers to represent elevation above sea level (positive) and below sea level (negative).*

20.   *⁻$25.00 + n = $25.00*
⁻$25.00 + $25.00 + n = $25.00 + $25.00
n = $50.00

22. *Brett: 15 − 35 = ⁻20 points; Hudson: ⁻45 + 30 = ⁻15 points; Hudson; ⁻15 − ⁻20 = 5 points*

At 8:30 p.m. the temperature was 8°C. By 2 a.m. the temperature had dropped 11 degrees. What was the temperature at 2 a.m.? I begin at 8°C and then proceed down 11 degrees in a negative or colder direction, ending at ⁻3°C.

What was the temperature at 2 a.m.? ⁻3°C

What subtraction equation can you write for this problem? 8°C − 11°C = ⁻3°C

What addition equation can you write for this problem? 8°C + ⁻11°C = ⁻3°C

Choose a student to write the equations.

• Explain that the set of integers is used in everyday tasks for recording information that is above or below 0; these tasks include recording temperature, financial balances, golf scores, and depths of the ocean.

• Read aloud the following word problem.

Nolan's checking account balance is ⁻$5.00. If he deposits $10.00, what will his account balance be?

• Direct the students to draw a simple number line using arrows to show the change in the checking account balance. I begin at 0 and move left 5 units (dollars) in a negative direction to ⁻5. From ⁻5, I move right 10 units (dollars) in a positive direction, stopping at 5.

Some students may draw vertical number lines moving up and down rather than right and left.

What will Nolan's account balance be if he makes the $10.00 deposit? $5.00

What equation can you write for this problem? ⁻$5.00 + $10.00 = $5.00

Choose a student to write the equation.

Demonstrate on the *Positive & Negative Number Line* page. (Rotating the horizontal number line $\frac{1}{4}$ turn to the left will show the movement on a vertical number line.)

If Nolan's account balance is $5.00, can he afford to make a purchase greater than $5.00? no

What would his bank balance read if he made a $7.00 purchase with only $5.00 in his account? ⁻$2.00

What does a negative bank balance indicate? sample answer: A negative bank balance indicates that I have spent more money than I have, so I am now in debt or owe that amount to the bank.

• Read aloud the following word problem.

Jake went scuba diving. He was 20 ft below the surface of the water when he decided to dive down 6 ft more. What is Jake's elevation?

• Direct the students to draw a simple number line using arrows to show the change in Jake's depth. I begin at 0 and move left 20 units (feet) in a negative direction to ⁻20. From ⁻20, I move left 6 units (feet) in a negative (deeper) direction stopping at ⁻26.

What is Jake's elevation? 26 ft below sea level

Why does the number line read ⁻26 ft? The negative sign indicates that the 26 ft are below sea level (0).

• Direct the students to write a subtraction equation and an addition equation for this problem. Choose a student to write the equations for display. ⁻20 ft − 6 ft = ⁻26 ft; ⁻20 ft + ⁻6 ft = ⁻26 ft

How can integers help me represent elevation? I can use integers to represent elevation above sea level (positive) and below sea level (negative).

• Remind the students that it is possible to have a negative score in a game in which points are taken away.

LESSON 171

What would cause you to have a negative score in such a game? I would have had more points taken away than the number of points awarded to me.

- Read aloud the following word problem.

  Carl scored 21 points, and then he lost 23 points during the course of the game. What is his score?

- Direct the students to draw a simple number line using arrows to show the change in Carl's score. I begin at 0 and move right 21 units (points) in a positive direction to 21. From 21, I move left 23 units (points) in a negative direction, stopping at ⁻2.

  What is Carl's score? ⁻2 points

- Direct the students to write a subtraction equation and an addition equation for this problem. Choose a student to write the equations for display. 21 points – 23 points = ⁻2 points; 21 points + ⁻23 points = ⁻2 points

Apply

Student Edition pages 374–75
- Read and explain the directions for pages 374–75. Assist the students as they complete the pages independently.

Daily Review
- Students should complete Chapter 17, section *e*.

Write an algebraic expression for the word phrase.

1. seven times an unknown number *7n*

2. three more than a number *n + 3*

3. four less than five times *n* *5n – 4*

4. six more than 2 times a number *2n + 6*

Evaluate the expression if *n* = 5.

5. 3*n* *15*

6. 8 + *n* *13*

7. $\frac{15}{n}$ *3*

8. 20 – *n* *15*

Simplify the expression.

9. *a* + *a* *2a*

10. (2 + 4) + *n* *6 + n*

11. 3(4*x*) *12x*

12. 8 + *y* + 2 *10 + y*

Complete the table.

13.

x	3x
2	*6*
5	*15*
7	*21*

14.

a	a²
4	*16*
6	*36*
8	*64*

15.

n	2n + 3
7	*17*
9	*21*
10	*23*

LESSON 172

Student Edition pages 376–77
Daily Review Chapter 17, section *f*

OBJECTIVES
- Multiply integers.
- Solve a word problem by writing a multiplication equation.

TEACHER RESOURCES
- 128 *Algebra Mat*
- 132 *Multiplication Patterns*

ADDITIONAL MATERIALS
- counters: black and red

Engage

- Display the *Algebra Mat* page.
- Direct the students in a **discussion** to explore the essential question on Student Edition page 376, "Why can ⁻2 × 4 not be shown on an algebra mat?"
- You may use the following discussion prompts:
  In a multiplication equation, what does the first number represent?
  In a multiplication equation, what does the second number represent?
  Could you represent 2 sets?
  Could you represent ⁻2 sets?

Instruct

Multiplying positive & negative numbers

- **Model** with counters to help the students multiply integers.

- Write "4 + 4 + 4 = ___" for display. Choose a student to show this equation on the algebra mat. 3 rows of 4 black counters on the positive side of the mat
  What is 4 + 4 + 4? 12
  Write "12" to complete the equation.
  How many rows of 4 counters are shown? 3
  What multiplication equation can you write for 3 sets (rows) of 4? 3 × 4 = 12
  Write the equations "3 × 4 = 12" and "3(4) = 12" for display. Point out that the

---

Multiplying Integers

Why can ⁻2 × 4 not be shown on an algebra mat?

Multiplying Integers Using an Algebra Mat

4 × ⁻2 = ⁻8   ⁻2 × 4 = ⁻8

⁻2 sets of 4 counters cannot be shown on the algebra mat, but the Commutative Property states that the product for ⁻2 × 4 is the same as the product for 4 × ⁻2.

Commutative Property of Multiplication

The order of factors may be changed without changing the product.

4 × ⁻2 = ⁻2 × 4

4 sets of 2 negative counters can be shown on an algebra mat.

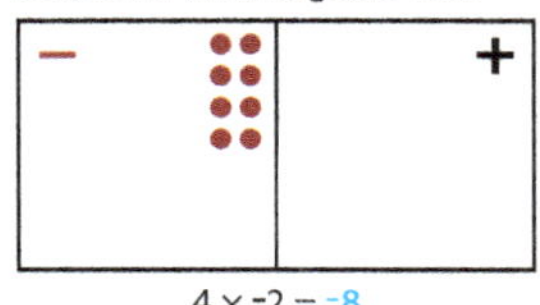

4 × ⁻2 = ⁻8

Rules for Multiplying Integers

When 2 factors have the same sign, the product is positive.

positive factor × positive factor = positive product   4 × 2 = 8
negative factor × negative factor = positive product   ⁻4 × ⁻2 = 8

When 2 factors have different signs, the product is negative.

positive factor × negative factor = negative product   4 × ⁻2 = ⁻8
negative factor × positive factor = negative product   ⁻4 × 2 = ⁻8

Exercises

Use the Commutative Property to write the related multiplication fact.

1. 5 × ⁻2 = ⁻10	**2.** ⁻10 × 15 = ⁻150	**3.** ⁻12 × ⁻11 = 132
⁻2 × 5 = ⁻10	15 × ⁻10 = ⁻150	⁻11 × ⁻12 = 132

Find the product by using the rules for multiplying integers. Notice the pattern when 1 factor remains the same.

4.		**5.**		**6.**	
4 × 2	8	4 × ⁻3	⁻12	⁻5 × 4	⁻20
3 × 2	6	3 × ⁻3	⁻9	⁻5 × 3	⁻15
2 × 2	4	2 × ⁻3	⁻6	⁻5 × 2	⁻10
1 × 2	2	1 × ⁻3	⁻3	⁻5 × 1	⁻5
0 × 2	0	0 × ⁻3	0	⁻5 × 0	0
⁻1 × 2	⁻2	⁻1 × ⁻3	3	⁻5 × ⁻1	5
⁻2 × 2	⁻4	⁻2 × ⁻3	6	⁻5 × ⁻2	10
⁻3 × 2	⁻6	⁻3 × ⁻3	9	⁻5 × ⁻3	15
⁻4 × 2	⁻8	⁻4 × ⁻3	12	⁻5 × ⁻4	20

376   Chapter 17

---

mat shows both the repeated addition equation and the multiplication equation.

What related fact does the Commutative Property of Multiplication allow you to write? 4 × 3 = 12; The order of the factors does not change the product.

How does changing the order of the factors change the picture? 4 rows of 3 counters

> You may rotate the pictured 3 × 4 array to show the 4 × 3 array. Write "4 × 3 = 12" and "4(3) = 12" for display.

- Write "⁻4 + ⁻4 + ⁻4 = ___" for display. Choose a student to demonstrate on the mat. 3 rows of 4 counters on the negative side of the mat

What is ⁻4 + ⁻4 + ⁻4? ⁻12
Write "⁻12" to complete the equation.
How many rows of ⁻4 counters are shown? 3

What multiplication equation can you write for 3 sets (rows) of ⁻4? 3 × ⁻4 = ⁻12

Write the equations "3 × ⁻4 = ⁻12" and "3(⁻4) = ⁻12" for display. Point out that the mat shows both the repeated addition equation and the multiplication equation.

What related fact does the Commutative Property of Multiplication allow you to write? ⁻4 × 3 = ⁻12; The order of the factors does not affect the product.

Write "⁻4 × 3 = ⁻12" and "⁻4(3) = ⁻12" for display.

Multiply.

7. $5 \times {}^-5$ *${}^-25$* 8. 4×12 *48* 9. ${}^-4 \times {}^-3$ *12* 10. ${}^-1 \times 8$ *${}^-8$*

11. $6 \times {}^-3$ *${}^-18$* 12. ${}^-6 \times {}^-3$ *18* 13. $5 \times {}^-10$ *${}^-50$* 14. $7 \times {}^-5$ *${}^-35$*

15. ${}^-6 \times 8$ *${}^-48$* 16. $2 \times {}^-5$ *${}^-10$* 17. ${}^-5 \times {}^-2$ *10* 18. ${}^-12 \times 2$ *${}^-24$*

Write a comparison sentence by using >, <, or =.

19. $9 \times {}^-2 \underline{\le} {}^-5 \times 3$ *${}^-18 < {}^-15$* 20. ${}^-5 \times {}^-8 \underline{\ge} 6 \times 6$ *$40 > 36$*

21. ${}^-2 \times 7 \underline{=} 7 \times {}^-2$ *${}^-14 = {}^-14$* 22. ${}^-4 \times {}^-6 \underline{=} 2 \times 12$ *$24 = 24$*

23. $0 \times 2 \underline{\ge} {}^-1 \times 2$ *$0 > {}^-2$* 24. $6 \times 3 \underline{\ge} {}^-9 \times 2$ *$18 > {}^-18$*

Practice & Application

25. Mackenzie borrowed $10 from her brother to purchase her rocket model. Then she borrowed $5 more for lunch. Write an equation to show Mackenzie's debt. *$10 + $5 = $15 debt or ${}^-$10 + ${}^-$5 = ${}^-$15*

26. Stock in the Davenport Sofa Company lost 12 points each day for 4 days. Write an equation to show the number of total points lost. *$4 \times {}^-12 = {}^-48$ points*

27. Show 4 sets of ${}^-3$ on a number line. Write a multiplication equation and solve. *$4 \times {}^-3 = {}^-12$*

28. Draw an algebra mat and show how 1 positive counter cancels out 1 negative counter. Solve the equation ${}^-11 + 15$. *${}^-11 + 15 = 4$*

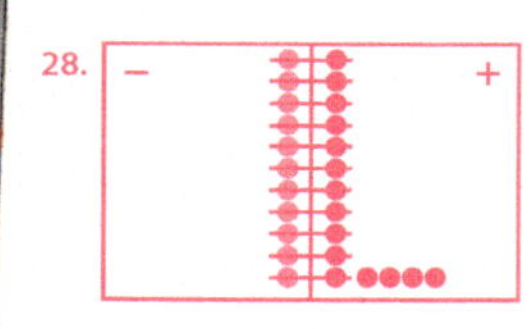

29. Write an addition equation for ${}^-4 - 7$. Solve. (Remember that subtracting an integer is the same as adding its opposite.) *${}^-4 + {}^-7 = {}^-11$*

30. The quarterback was sacked on 3 consecutive plays. He lost 10 yd on each sack. Write an addition equation and a multiplication equation to show how many yards he lost in 3 plays. *${}^-10 + {}^-10 + {}^-10 = {}^-30$ yd; $3 \times {}^-10 = {}^-30$ yd*

31. Explain how the number line helps you solve the problem $6 + {}^-12$. Write the sum. *${}^-6$*

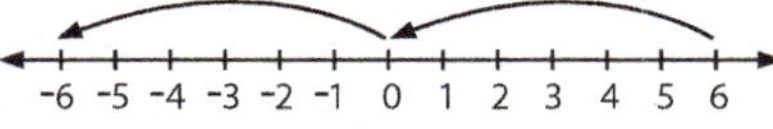

32. Why can ${}^-2 \times 4$ not be shown on an algebra mat? *It is not possible to show fewer than 0 sets of anything.*

28.

31. *12 jumps away from 6 are needed. Using the fact $6 + 6 = 12$ allows for 2 jumps instead of 12 individual jumps. 6 to 0 = 6 jumps 0 to ${}^-6$ = 6 jumps*

How could you represent this equation on the mat? Can you show ${}^-4$ sets of something? sample answer: A picture of negative sets of counters cannot be shown.

- Write "${}^-3 \times {}^-4 = \_\_$" and "${}^-4 \times {}^-3 = \_\_$."

What do you know about 3 and ${}^-3$? These numbers are opposites.

If 3 sets of ${}^-4 = {}^-12$, what do you think the opposite of 3 sets (${}^-3$ sets) of ${}^-4$ would be? 12; the opposite of ${}^-12$

If 4 sets of ${}^-3 = {}^-12$, what do you think the opposite of 4 sets (${}^-4$ sets) of ${}^-3$ would be? 12

Write "12" to complete the equations.

- Display the *Multiplication Patterns* page. Direct the students to study the first list of equations to find the pattern of multiplying a negative number and a positive number. Instruct them to use the pattern to complete the last 2 equations.

What is ${}^-3 \times 3$? ${}^-9$

What is ${}^-4 \times 3$? ${}^-12$

What pattern do you see in this list of equations? Decreasing the first factor by 1 decreases the product by 3. To continue the pattern, the last 2 products must be negative.

What property could also help you find ${}^-3$ sets of 3? Commutative; $3 \times {}^-3 = {}^-9$

LESSON 172

- Direct the students to study the second list of equations to find the pattern of multiplying a negative number by a negative number. Instruct them to use the pattern to complete the last 2 equations.

What is ${}^-4 \times {}^-3$? 12

What is ${}^-4 \times {}^-4$? 16

What pattern do you see in this list of equations? As the second factor decreases by 1, the product increases by 4. To continue the pattern, the last 2 products must be positive.

What do you notice about the product when the factors have the same sign? The product is positive.

What do you notice about the product when the factors have different signs? The product is negative.

- Direct attention to the essential question at the top of Student Edition page 376.

Why can ${}^-2 \times 4$ not be shown on an algebra mat? It is not possible to show fewer than 0 sets of anything.

You may point out that students can use the Commutative Property to show 4 sets of ${}^-2$.

Solving a word problem

- **Model** with integers to help the students write an equation for a word problem.

- Read aloud the following word problem.

The value of stock in Davis Lumber Company dropped 10 points each day this week. What was the change in stock points at the end of the 5-day week? ${}^-50$ points

What integer represents the change in stock points each day? ${}^-10$

How many times (days) was this same change in stock points repeated? 5; It was a 5-day week, and the stock fell 10 points each day this week.

What multiplication equation can you write for this problem? $5 \times {}^-10 = \_\_$; There were 5 days of a loss of 10 points (5 sets of ${}^-10$ points).

- Direct the students to draw a number line to show $5 \times {}^-10$. sample answer: I begin at 0 and move 10 units (points) to the left; then I repeat the movement of 10 units left 4 more times, stopping at ${}^-50$.

LESSON 172

- Follow a similar procedure for the following problems.

  Samantha lost 2 points on each of her first 4 turns of the game. What was her score after 4 turns? $4 \times {}^-2 = {}^-8$ points

  A marathon runner in training lost 2 lb a month for 6 months. What was the total change in his weight after the 6 months? $6 \times {}^-2 = {}^-12$ lb

Apply

Student Edition pages 376–77
- Read and explain the directions for pages 376–77. Assist the students as they complete the pages independently.

Daily Review
- Students should complete Chapter 17, section *f*.

Find the perimeter of the figure. *Equations may vary.*

1.
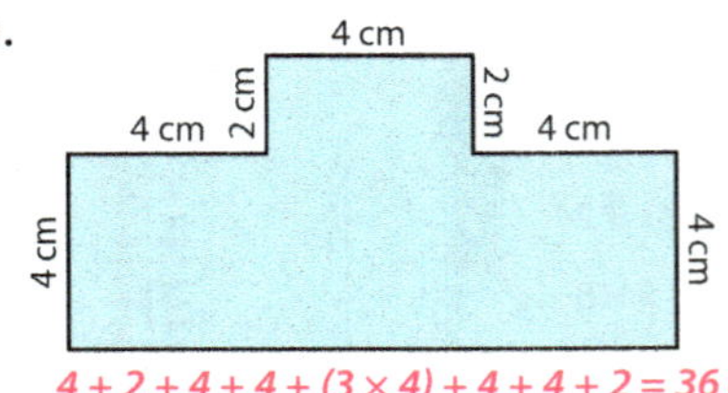

$4 + 2 + 4 + 4 + (3 \times 4) + 4 + 4 + 2 = 36$ cm

2.
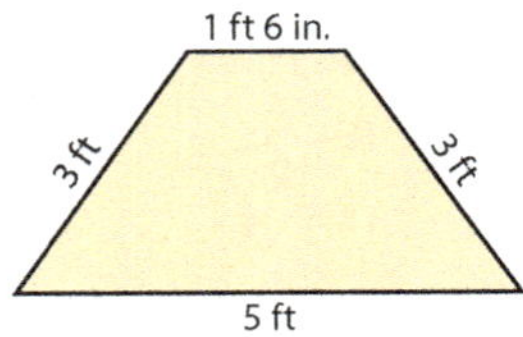

1 ft 6 in. $+ 3$ ft $+ 5$ ft $+ 3$ ft $= 12$ ft 6 in.

Write the formula for the circumference of a circle. Find the circumference.

3.
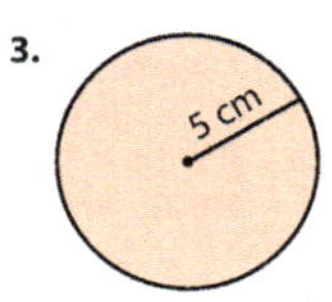

$C = 2\pi r$
$2 \times 3.14 \times 5 = 31.4$ cm

4.
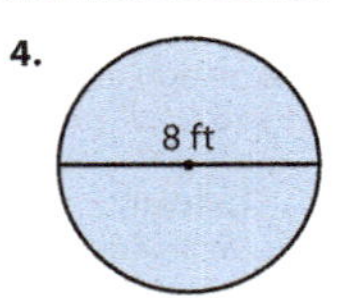

$C = \pi d$
$3.14 \times 8 = 25.12$ ft

Find the area of the figure.

5.
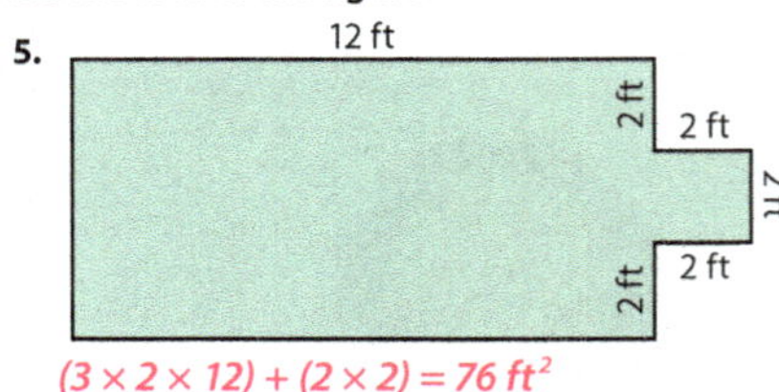

$(3 \times 2 \times 12) + (2 \times 2) = 76$ ft^2

Write the formula for the area of a circle. Find the area.

6.
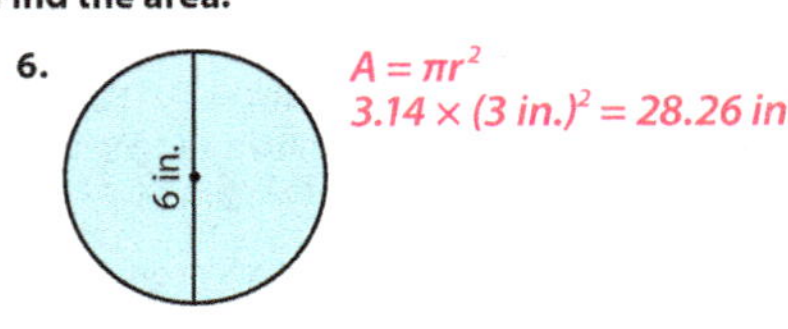

$A = \pi r^2$
$3.14 \times (3$ in.$)^2 = 28.26$ in.^{2}

OBJECTIVES

- Multiply and divide integers.
- Solve real-world problems by writing an equation.
- Describe the qualities of a good math model. **BWS**

BIBLICAL WORLDVIEW SHAPING

- **Modeling (Explain):** A good model represents God's world well.

TEACHER RESOURCES

- 3 _Positive & Negative Number Line_
- 128 _Algebra Mat_

ADDITIONAL MATERIALS

- counters: black and red

Engage

- Direct the students to do a **Think-Pair-Share** to explore the essential question on Student Edition page 378, "What makes a math model good?"

  You may use the following discussion prompts.

  What are some examples of math models that you have used in this chapter? number lines and algebra mats

  In what ways are the models good or helpful? They can be used to represent equations with integers.

Instruct

Multiplying & dividing integers

- **Model** equations to help the students multiply and divide integers.

- Display the _Algebra Mat_ page. Choose a student to place 2 red counters on the negative side of the mat. Repeat this procedure 3 more times leaving space between each group of 2 red counters.

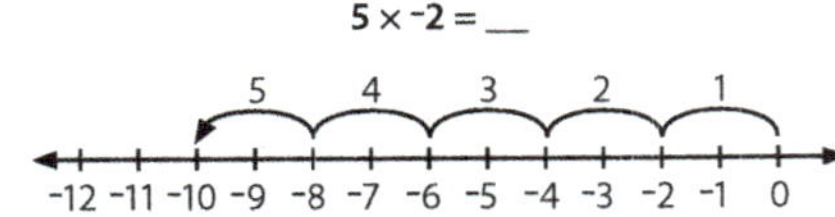

Multiplying & Dividing Integers

What makes a math model good?

Multiplying and Dividing Integers Using a Number Line

$5 \times {}^-2 = $ ___

1. Begin at 0.
2. Add 5 sets of ${}^-2$.
   $5 \times {}^-2 = {}^-10$
related division facts:
${}^-10 \div {}^-2 = 5$ and ${}^-10 \div 5 = {}^-2$

${}^-10 \div {}^-2 = $ ___

1. Begin at ${}^-10$.
2. Subtract sets of ${}^-2$ until you reach 0.
   ${}^-10 \div {}^-2 = 5$
related multiplication fact:
$5 \times {}^-2 = {}^-10$

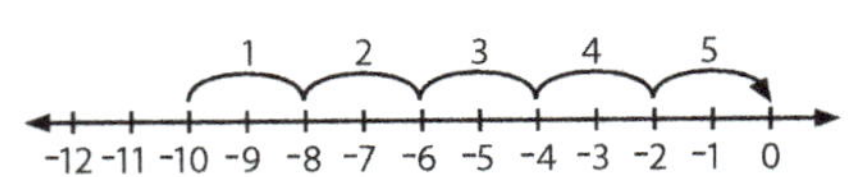

Exercises

Solve the related division facts.

1. ${}^-3 \times 5 = {}^-15$; therefore, ${}^-15 \div 5 = \underline{{}^-3}$ and ${}^-15 \div {}^-3 = \underline{5}$.
2. $6 \times {}^-4 = {}^-24$; therefore, ${}^-24 \div {}^-4 = \underline{6}$ and ${}^-24 \div 6 = \underline{{}^-4}$.
3. ${}^-2 \times 7 = {}^-14$; therefore, ${}^-14 \div 7 = \underline{{}^-2}$ and ${}^-14 \div {}^-2 = \underline{7}$.
4. ${}^-4 \times {}^-2 = 8$; therefore, $8 \div {}^-2 = \underline{{}^-4}$ and $8 \div {}^-4 = \underline{{}^-2}$.

Write a statement that explains the model for the division equation. Write the related multiplication equation.

5. 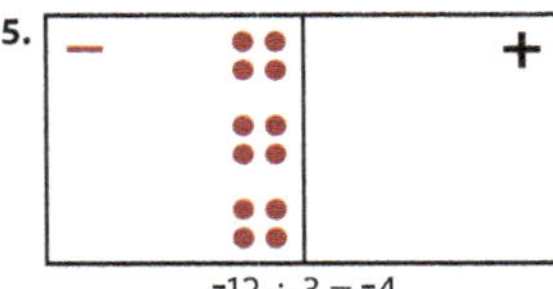
   ${}^-12 \div 3 = {}^-4$
   $3 \times {}^-4 = {}^-12$

6. 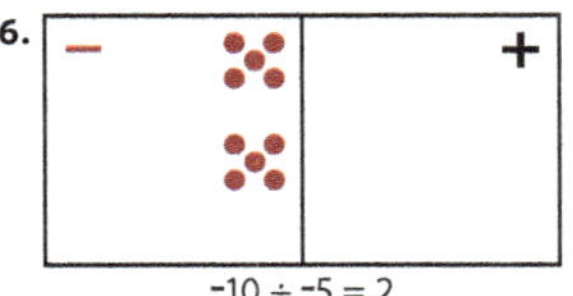
   ${}^-10 \div {}^-5 = 2$
   $2 \times {}^-5 = {}^-10$

Draw a mat to illustrate the sentence. Write an equation and solve.

7. Divide 21 negative counters into 7 equal sets.
   ${}^-21 \div 7 = {}^-3$
8. Divide 30 negative counters into sets of 6 negative counters. ${}^-30 \div {}^-6 = 5$
9. What makes a math model good? _A math model is good when it represents God's world well. It imitates nature as closely as possible._

378   Chapter 17

What equation can you write to represent these counters? ${}^-2 + {}^-2 + {}^-2 + {}^-2 = {}^-8$ or $4 \times {}^-2 = {}^-8$ since there are 4 sets of ${}^-2$ producing one large set of ${}^-8$

Write "${}^-2 + {}^-2 + {}^-2 + {}^-2 = {}^-8$" and "$4 \times {}^-2 = {}^-8$" for display.

- Display the _Positive & Negative Number Line_ page. Direct the students to draw a number line similar to the one displayed and to use arrows to show that $4 \times {}^-2 = {}^-8$. sample answer: I begin at 0; since I am adding 4 sets of ${}^-2$, I can draw an arrow to show 4 jumps of 2 in a negative direction (left) from 0: ${}^-2$, ${}^-4$, ${}^-6$ and ${}^-8$. Demonstrate each step.

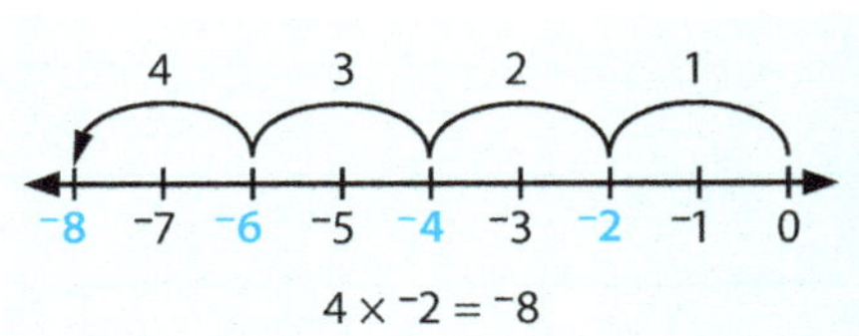

$4 \times {}^-2 = {}^-8$

- Remind the students that division is the inverse operation of multiplication. Write "$4 \times 2 = 8$" for display.

  What are the related division equations for $4 \times 2 = 8$? $8 \div 2 = 4$ and $8 \div 4 = 2$; When a product is divided by either of its factors, the quotient is the other factor.

  How can you show $8 \div 2$ (in each set) = 4 (sets)? 8 positive counters can be partitioned into sets of 2 positive counters in each set to make 4 sets.

Multiply.

10. $5 \times \text{-}7$ *-35* **11.** $\text{-}7 \times \text{-}9$ *63* **12.** $2 \times \text{-}12$ *-24* **13.** $\text{-}8 \times \text{-}1$ *8*

14. $\text{-}2 \times 6$ *-12* **15.** $\text{-}8 \times 6$ *-48* **16.** 4×10 *40* **17.** 4×7 *28*

18. $3 \times \text{-}12$ *-36* **19.** $\text{-}5 \times \text{-}15$ *75* **20.** $\text{-}6 \times 0$ *0* **21.** $16 \times \text{-}6$ *-96*

Divide.

22. $\text{-}15 \div 3$ *-5* **23.** $80 \div 4$ *20* **24.** $42 \div 2$ *21* **25.** $\text{-}8 \div 4$ *-2*

26. $\text{-}100 \div 2$ *-50* **27.** $18 \div \text{-}3$ *-6* **28.** $\text{-}2 \div \text{-}2$ *1* **29.** $\text{-}12 \div 4$ *-3*

30. $99 \div \text{-}9$ *-11* **31.** $\text{-}16 \div \text{-}2$ *8* **32.** $\text{-}12 \div 6$ *-2* **33.** $\text{-}20 \div \text{-}5$ *4*

Write a comparison sentence by using >, <, or =.

34. $\text{-}16 \div 4 \underline{=} 2 \times \text{-}2$ *-4 = -4* **35.** $4 \div \text{-}4 \underline{\leq} 3 \div 3$ *-1 < 1*

36. $0 \times 3 \underline{\geq} \text{-}1 \times 2$ *0 > -2* **37.** $\text{-}2 \times \text{-}3 \underline{\geq} \text{-}1 \times 5$ *6 > -5*

Practice & Application

38. Marc's golf team scored 6 strokes under par (-6) each day for 5 days. Write an equation to show the team's combined score. *$5 \times \text{-}6 = \text{-}30$*

39. At the end of a 3-day golf tournament, Mr. Davis had a final score of -12, or 12 strokes under par. Write an equation to show his average score each day. *$\text{-}12 \div 3 = \text{-}4$ strokes average per day*

40. Cody collected 36 blankets for the homeless. He gave away 6 blankets each day for 5 days. Write a multiplication and an addition equation to show how many blankets Cody has left.

41. Clark made 3 clay pots each day during 4 days at art camp. While in the kiln, 5 of the pots cracked. How many pots does Clark have left? *$4 \times 3 - 5 = 12 - 5 = 7$ pots*

42. Lexie's lap times at the swim meet were 12 sec slower than the best time. If her time decreases 2 sec each day, how many days will it take her to tie the best time? *$\text{-}12 \div \text{-}2 = 6$ days or $12 \div 2 = 6$ days*

43. Find the perimeter and the area of the figure.

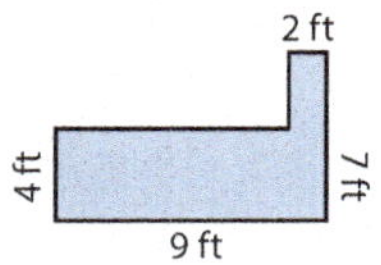

$P = 4\,ft + 7\,ft + 3\,ft + 2\,ft + 7\,ft + 9\,ft$
$P = 32\,ft$
$A = (4\,ft \times 9\,ft) + (2\,ft \times 3\,ft)$
$A = 42\,ft^2$

44. Find the perimeter and the area of the trapezoid.

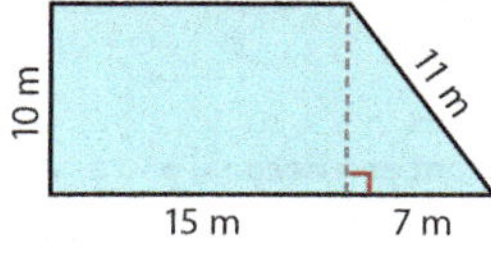

45. Find the area of the triangle.

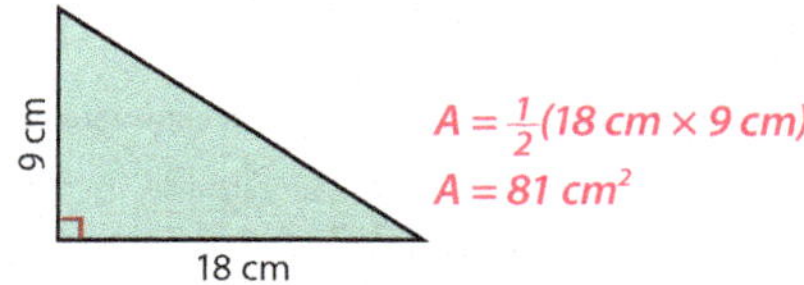

$A = \frac{1}{2}(18\,cm \times 9\,cm)$
$A = 81\,cm^2$

46. Use $10 \times 18 = 180$ to explain the following statement: The inverse operation of multiplication is division. *Division takes apart what multiplication puts together. 180 items can be put into 10 sets of 18 or 18 sets of 10.*

40. *$36 - (5 \times 6) = 6$ blankets; $6 + 6 + 6 + 6 + 6 = 30$; $36 - 30 = 6$ blankets*

44. *$P = 10\,m + 15\,m + 11\,m + 22\,m$*
$P = 58\,m$

$A = (10\,m \times 15\,m) + \frac{1}{2}(7\,m \times 10\,m)$
$A = 185\,m^2$

How can you show $8 \div 2$ (sets) = 4 (in each set)? 8 positive counters can be equally shared between 2 sets to find that there are 4 positive counters in each set.

Write the equations and demonstrate on the algebra mat.

What are the related division equations for $4 \times \text{-}2 = \text{-}8$? $\text{-}8 \div \text{-}2 = 4$ and $\text{-}8 \div 4 = \text{-}2$; When a product is divided by either of its factors, the quotient is the other factor.

How can you show $\text{-}8 \div \text{-}2$ (in each set) = 4 (sets)? 8 negative counters can be partitioned into sets of 2 negative counters in each set to make 4 sets

How can you show $\text{-}8 \div 4$ (sets) = $\text{-}2$ (in each set)? 8 negative counters can be equally shared among 4 sets to find that there are 2 negative counters in each set.

You may remind the students that it is not possible to show negative sets.

Write the equations and demonstrate on the algebra mat.

- Direct the students to draw another -10 to 10 number line and to use arrows to show that $\text{-}8 \div \text{-}2 = 4$. I begin at -8 and make jumps (sets) of -2 units until I reach 0.

Demonstrate each step.

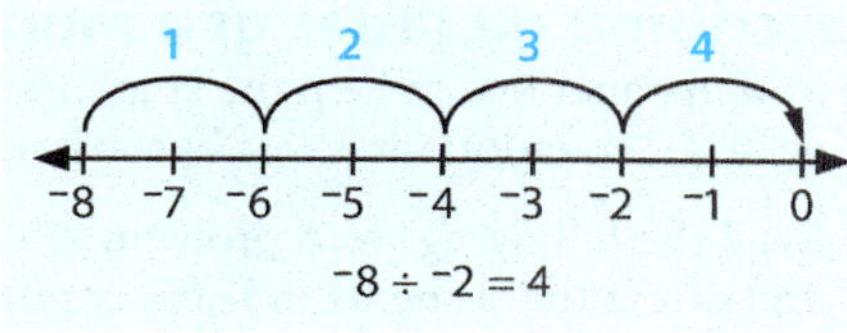

- Point out that repeated addition of integers is to multiplication as repeated subtraction of integers is to division: $\text{-}8 - \text{-}2 - \text{-}2 - \text{-}2 - \text{-}2 = 0$.

How many sets (jumps) of -2 did you make? 4

How many times did you subtract -2? 4

Explain that the quotient is the number of times you subtracted -2, so the quotient in the equation $\text{-}8 \div \text{-}2$ is 4. Display 8 red counters on the negative side of the mat. Remove sets of -2 (2 red counters) four times.

- Write the following equations for display: $4 \times \text{-}2 = \text{-}8$, $\text{-}8 \div \text{-}2 = 4$, and $\text{-}8 \div 4 = \text{-}2$; $\text{-}5 \times \text{-}3 = 15$, $15 \div \text{-}3 = \text{-}5$, and $\text{-}3 \times \text{-}5 = 15$. Choose students to provide the missing equation in each fact family. *$\text{-}2 \times 4 = \text{-}8$; $15 \div \text{-}5 = \text{-}3$*

What pattern do you see between the factors and products of these problems? Multiplication and division with like signs result in a positive answer, and multiplication and division with unlike signs result in a negative answer.

- Guide the students in identifying a mental picture to represent a problem. Write "$\text{-}24 \div \text{-}4 = \underline{\quad}$" for display.

What picture comes to your mind when you see this problem? sample answers: 24 red counters on the negative side of the mat can be partitioned into sets of 4 red counters in each set to make 6 sets; I begin at -24 on the number line and make jumps of -4, with a total of 6 jumps (sets).

- Direct the students to do a **Think-Pair-Model**. Instruct each pair of students to write a multiplication or division equation including negative integers. Direct them to trade equations and then picture the equation on a drawn algebra mat or number line.

LESSON 173

Solving a word problem; describing qualities of a model

- **Model** equations to help the students write an equation for a word problem.

- Read the following word problem aloud and direct the students to listen carefully in order to form a picture in their minds.

During a board game each player had 5 turns. On each turn Katelyn lost 2 points. What was her score at the end of the game?

What picture came to mind? Accept any reasonable answer including 5 sets of ⁻2.

- Explain that it is hard to picture a turn in a game; therefore, you may see only a loss of 2 points repeated 5 times in a row.

Write "⁻2" five times in 1 horizontal row, leaving space for an operation sign to be inserted between each integer.

Consider the 4 mathematical operations: addition, subtraction, multiplication, and division. Which of these 4 can be used to show this repetition? addition and/or multiplication; The ⁻2 was added 5 times in a row, or there were 5 sets of ⁻2.

The student who answers repeated subtraction is not incorrect, but he must establish that Katelyn began the game with a score of 0 and repeatedly subtracted 2. Division is not an option since the final score was not given, and ⁻10 cannot be obtained without first performing addition or multiplication.

Which operation sign (+, −, ×, or ÷) can be inserted between the five ⁻2s to form the needed equation? + sign

Write an addition sign (+) between the ⁻2s: ⁻2 + ⁻2 + ⁻2 + ⁻2 + ⁻2.

What integer represents the points scored each turn? ⁻2

How many times (turns) was this same score repeated? 5; Each player had 5 turns during the game.

What multiplication equation can you write for this problem? 5 × ⁻2 = __; There were 5 turns of a loss of 2 points (⁻10 points).

Write "5 × ⁻2 = __."

- Direct one of the students in each pair to model 5 × ⁻2 on the mat while the other student shows it on the number line. I place 5 sets of 2 counters on the negative side of the mat for a total of ⁻10; I begin at 0 and move 2 units (points) to the left 5 times repeatedly, ending at ⁻10.

What was Katelyn's score after her fifth turn? ⁻10 points

Complete the equation for display.

- Point out that using counters to represent numbers is a way to make sure the answer (⁻10) matches real life.

If you repeated this procedure 2–3 times, what do you think would happen? I'd get the same answer.

Guide a **discussion** to help the students answer the essential question.

What does it mean that the answer (⁻10) is consistent? I get the same answer each time.

Remind the students that a model is only as good as its representation of reality.

What is meant by reality? Something is true to life or God's world.

Point out that a model is useful as long as it corresponds to what God has revealed to us about His creation.

What makes a math model good? sample answer: A math model is good when it represents God's world well. It imitates nature as closely as possible.

Find the unit rate.

1. The Laphams drove 315 mi and used 15 gal of gas. *21 mi/gal*

2. Marcus earned $40.00 cleaning several cars. He worked 5 hr. *$8/hr*

3. Mrs. Bowers bought 8 lb of bananas for $4.72. *$0.59/lb*

4. The team traveled 1,450 mi in two days. *725 mi/d*

Find the distance traveled in the given time.

5. 4 days at 350 mi/d *1,400 mi*

6. 5 hr at 65 mph *325 mi*

Write a ratio. *Ratio form may vary.*

7. 3 cans for $2.00 *3:2*

8. 2 bags for $3.00 *2:3*

9. one computer for every 2 students *1:2*

Write the percent as a decimal and as a fraction in lowest terms.

10. 78% *0.78; $\frac{39}{50}$*

11. 50% *0.5; $\frac{1}{2}$*

12. 4% *0.04; $\frac{1}{25}$*

Write a proportion to find an equivalent ratio. Answer the question.

13. It takes Mrs. Snow 2 hr to grade 50 math pages. At this rate, how long would it take her to grade 100 math pages? *$\frac{2}{50} = \frac{4}{100}$; 4 hr*

14. It takes Brian 25 min to complete a math page. At this rate, how long would it take him to complete 4 math pages? *$\frac{25}{1} = \frac{100}{4}$; 100 min, or 1 hr and 40 min*

Daily Review   519

Math 6

- Direct the students to solve the following word problems.

Charlie lost 3 points on each of his first 3 turns of the game. On each of his last 2 turns he gained 5 points. What was his score at the end of 5 turns? $^-3 + {}^-3 + {}^-3 + 5 + 5 = {}^-9 + 10 = 1$; $(3 \times {}^-3) + (2 \times 5) = {}^-9 + 10 = 1$

The value of stock in the Hancock Steel Company gained a total of 15 points in 5 days. What was the average change each day? $15 \div 5 = 3$ points per day

Apply

Student Edition pages 378–79
- Read and explain the directions for pages 378–79. Assist the students as they complete the pages independently.

Daily Review
- Students should complete Chapter 17, section *g*.

NOTES

LESSON 174

Student Edition pages 380–81
Daily Review Chapter 17, section h

OBJECTIVES

- Add, subtract, multiply, and divide integers.
- Subtract integers by adding the opposite.
- Solve problems by applying the order of operations to integers.

TEACHER RESOURCES

- 3 *Positive & Negative Number Line*
- 128 *Algebra Mat*
- 133 *Order of Operations* (for the teacher and for each student)

ADDITIONAL MATERIALS

- counters: black and red

In this lesson the red (negative) and black (positive) counters may be used apart from the mat.

Engage

- Give the students a **preassessment** to explore the essential question on Student Edition page 380, "How can I subtract by adding?"
- Direct the students to complete the statement using >, <, or =.

$6 - 4 \_\_ 6 + {}^-4 =$ $2 - 3 \_\_ 2 + {}^-3 =$
$7 - 2 \_\_ 7 + {}^-2 =$

Instruct

Adding & subtracting integers

- Direct the students to **model** equations to help them add and subtract integers.
- Write "5 + 4 = __" and "$^-5 + {}^-4 =$ __" for display. Display the *Algebra Mat* page.

What is true about the sum when all the addends are positive integers? The sum will be a greater positive integer; positive integers plus positive integers make more positive integers.

What is true about the sum when all the addends are negative integers? The sum will be a lesser negative integer; negative integers plus negative integers make more negative integers.

- Direct each student to write his predicted answer to each equation. Then choose students to demonstrate with the mat and counters.

$5 + 4 = 9$

I place 5 positive counters, place 4 more positive counters, and then combine the counters for a total of 9 positive counters.

$^-5 + {}^-4 = {}^-9$

I place 5 negative counters, place 4 more negative counters, and then combine the counters for a total of 9 negative counters.

- Write "5 − 4 = __" and "4 − 5 = __" for display.

What is true about the difference of 2 positive integers? The difference is a positive integer if the minuend has the greater value; the difference is a negative integer if the subtrahend has the greater value.

Direct each student to write his predicted answer to each equation and then to show the equation using counters.

$5 - 4 = 1$

I place 5 positive counters and remove 4 positive counters, leaving a difference of 1 positive counter.

$4 - 5 = {}^-1$

I place 4 positive counters; since there is 1 fewer positive counter than the 5 needed to be taken away, I add 1 positive

---

How can I subtract by adding?

Order of Operations

1. Do all operations inside the parentheses first.
2. Do all multiplication and division in order from left to right.
3. Do all addition and subtraction in order from left to right.

$2(^-5 - 4) =$ $^-9 + {}^-8 \div 4 =$
$2 \times {}^-9 = {}^-18$ $^-9 + {}^-2 = {}^-11$

Exercises

Solve. Follow the order of operations.

1. $^-7 + {}^-8 - 3$ *$^-18$*
2. $^-4 - 5 + {}^-3$ *$^-12$*
3. $^-1 + 1 - 1$ *$^-1$*
4. $5 - (6 + {}^-5)$ *4*
5. $8 + (2 - 7)$ *3*
6. $9 - 7 + {}^-3$ *$^-1$*
7. $2(^-7 + 5)$ *$2(^-2) = {}^-4$*
8. $4 + {}^-6 \div 3$ *$4 + {}^-2 = 2$*
9. $8(^-1 + {}^-3)$ *$8(^-4) = {}^-32$*
10. $^-8 \div 2 - 3$ *$^-4 - 3 = {}^-7$*
11. $6 \div (6 - 9)$ *$6 \div {}^-3 = {}^-2$*
12. $^-5 - 2 \times 2$ *$^-5 - 4 = {}^-9$*

Add.

13. $^-8 + 3$ *$^-5$*
14. $6 + 8$ *14*
15. $^-10 + 9$ *$^-1$*
16. $1 + {}^-5$ *$^-4$*
17. $^-4 + 7$ *3*
18. $13 + {}^-2$ *11*
19. $^-7 + {}^-3$ *$^-10$*
20. $2 + {}^-2$ *0*
21. $^-15 + 9$ *$^-6$*
22. $^-2 + 7$ *5*
23. $14 + {}^-3$ *11*
24. $^-6 + 6$ *0*

Subtract.

25. $3 - 8$ *$^-5$*
26. $10 - {}^-2$ *12*
27. $5 - 7$ *$^-2$*
28. $^-4 - 10$ *$^-14$*
29. $^-6 - {}^-4$ *$^-2$*
30. $^-4 - {}^-4$ *0*
31. $^-7 - {}^-2$ *$^-5$*
32. $^-2 - 5$ *$^-7$*
33. $2 - 3$ *$^-1$*
34. $5 - {}^-4$ *9*
35. $3 - 9$ *$^-6$*
36. $6 - 5$ *1*

Multiply.

37. $6 \times {}^-5$ *$^-30$*
38. $^-1 \times 1$ *$^-1$*
39. $^-3 \times 6$ *$^-18$*
40. $^-8 \times {}^-7$ *56*
41. $2 \times {}^-3$ *$^-6$*
42. $^-1 \times {}^-9$ *9*
43. $^-4 \times 3$ *$^-12$*
44. $^-4 \times {}^-5$ *20*
45. $7 \times {}^-2$ *$^-14$*

Divide.

46. $^-15 \div 3$ *$^-5$*
47. $6 \div {}^-3$ *$^-2$*
48. $^-4 \div {}^-1$ *4*
49. $4 \div {}^-2$ *$^-2$*
50. $^-10 \div {}^-5$ *2*
51. $18 \div {}^-9$ *$^-2$*
52. $^-12 \div {}^-6$ *2*
53. $16 \div 4$ *4*
54. $39 \div 3$ *13*

380   Chapter 17

As of 2011, there were more than 1,500 search-and-rescue (SAR) teams in the United States. The histogram shows the number of teams that have their own website by state.

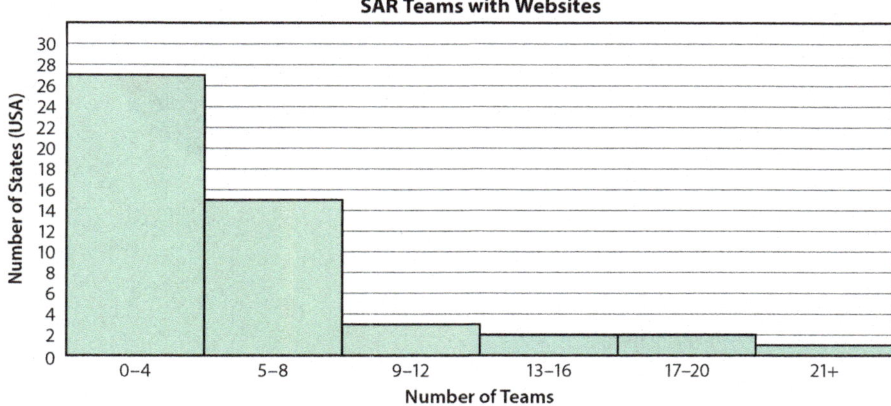

55. How many states list more than 20 search-and-rescue teams? *1 state*

56. What interval shows the mode? *0–4*

57. Pennsylvania has 14 teams listed. In which interval is Pennsylvania? *13–16*

58. How many states have 13 or more search-and-rescue teams listed? *5 states*

59. How many states have 12 or fewer search-and-rescue teams listed? *45 states*

60. What percentage of states is in the 0–4 interval?

61. Hannah is training for a 5K race. Her finish time has decreased 30 sec each week for 6 weeks. What is the total change in her finish time at the end of the 6 weeks? *$6 \times {}^-30 = {}^-180$ sec = $^-3$ min*

62. The weather forecaster predicted that the temperature would rise to 12°F. If the temperature is ⁻3°F, how many degrees must it rise to make the prediction true?

63. The youth group filled 15 trash bags with leaves while helping Mrs. Slocum with yard work. When they carried the bags to the street to be picked up, 2 of the bags broke, spilling the leaves. How many full bags of leaves did they still have? *$15 + {}^-2 = 13$ bags or $15 - 2 = 13$ bags*

64. How can I subtract by adding? *I can subtract an integer by adding its opposite.*

60. *$\dfrac{27}{50} = \dfrac{54}{100} = 54\%$*

62. *$^-3° + n = 12°$*
$^-3° - {}^-3° + n = 12° - {}^-3°$
$n = 15°$

$4 + {}^-2 = 2$

I place 4 positive counters, place 2 negative counters, and combine the counters to show $^-2 + 2 = 0$, with a difference of 2 positive counters.

- Write "$4 - 2 = 4 + {}^-2$" for display.

  What conclusion can you reach about subtracting an integer and adding its opposite? The answers are the same.

  How can I subtract by adding? I can subtract an integer by adding its opposite.

- Repeat a similar procedure to guide students in predicting and solving the following pairs of problems.

  $^-9 + 4 = {}^-5$ and $9 + {}^-4 = 5$

  $^-2 + 7 = 5$ and $2 + {}^-7 = {}^-5$

  $^-9 - {}^-5 = {}^-4$ and $^-5 - {}^-9 = 4$

- Display the *Positive & Negative Number Line*.

  When using a number line, how is adding 2 negative numbers different from adding 2 positive numbers? When adding 2 positive numbers, both moves are in a positive direction (to the right), but when adding 2 negative numbers, both moves are in a negative direction (to the left).

  Choose students to demonstrate solving $5 + 4 = 9$ and $^-2 + {}^-6 = {}^-8$ on the number line.

  When using a number line, how is adding $^-3 + 7$ different from $3 + {}^-7$? I begin at 0 and move in a negative direction or in a positive direction as each addend indicates. For these problems, I am moving in opposite directions for each addend.

  Choose students to demonstrate solving $^-3 + 7 = 4$ and $3 + {}^-7 = {}^-4$ on a number line.

Multiplying & dividing integers

- Direct the students to **model** equations to help them multiply and divide integers.

- Write the equation "$6 \times {}^-2 = \_\_$" for display. Direct students to demonstrate finding the product using counters or a number line. I place 6 sets of 2 negative counters; I begin at 0 and draw an arrow to the left to $^-2$, $^-4$, $^-6$, $^-8$, $^-10$, and $^-12$.

---

counter and 1 negative counter ($1 + {}^-1 = 0$); then I remove the 5 positive counters, leaving a difference of 1 negative counter.

- Repeat the procedure with the following problems:

  $7 - 5 = 2$

  I place 7 positive counters and remove 5 positive counters, leaving a difference of 2 positive counters.

  $5 - 7 = {}^-2$

  I place 5 positive counters; since there are 2 fewer positive counters than the 7 needed to be taken away, I add 2 positive counters and 2 negative counters ($2 + {}^-2 = 0$); then I remove the 7 positive counters, leaving a difference of 2 negative counters.

- Write "$4 - 2 = \_\_$" and "$4 + {}^-2 = \_\_$" for display. Arrange students in pairs. Direct each pair of students to write their predicted answers to each equation. Invite a student to demonstrate with the counters.

Guide the students in aligning positive and negative counters in a 1-to-1 correspondence so the canceling out can be seen.

$4 - 2 = 2$

I place 4 positive counters and remove 2 positive counters, leaving a difference of 2 positive counters.

LESSON 174

What repeated addition equation does this represent? $-2 + -2 + -2 + -2 + -2 + -2 = -12$

Write the repeated addition equation and complete the equation $6 \times -2 = -12$.

What is the inverse operation of multiplication? division

What division equations can you write from this multiplication equation?
$-12 \div -2 = 6$ and $-12 \div 6 = -2$

Write the equations.

- Direct students to show finding the quotient using counters or a number line. Remind the students that they can use repeated addition to multiply integers and repeated subtraction to divide integers. I place 12 negative counters (-12) and divide into sets of 2 negative counters (-2); I begin at -12 and subtract -2 until I reach 0.

  What does the equation $-12 \div -2 = 6$ tell you? There are 6 sets of -2 in -12.

- Write "$-5 \times 2 = $ __" for display. Remind students that they can use the Commutative Property to help them solve a problem with a negative multiplier: $-5 \times 2 = 2 \times -5$. -10

- Write "-10" to complete the answer.

- Write "$2 \times -3 = $ __" for display.

  What do you know about 2 and -2? They are opposites.

  What is 2×-3 equal to? -6

  If 2 sets of $-3 = -6$, what do you think the opposite of 2 sets (-2 sets) of -3 would be? 6; the opposite of -6

  Write "6" in the answer blank of the problem.

  What rules apply to both multiplication and division? Multiplication and division with like signs result in a positive answer, and multiplication and division with unlike signs result in a negative answer.

- Guide students in applying their knowledge of fact families and inverse operations to solve the following division problems.

  $-9 \div 3 = -3$ $8 \div -2 = -4$

Use the spinner to find the answer.

1. What color is the spinner most likely to land on? Write a fraction and a percent to show the probability. red; $\frac{4}{8}$, 50%

2. Find the probability of the spinner landing on blue. Write it as a fraction and a percent. $\frac{3}{8}$; 37.5%

3. Find the probability of the spinner landing on green. Write it as a fraction and a percent. $\frac{1}{8}$; 12.5%

Answer the question.

4. What are the possible combinations for a pizza with two different toppings? {pm, po, ps, mo, ms, os}

5. How many possible combinations are there for a pizza with two different toppings? 6

520   Daily Review

Applying the order of operations

- Guide the students in a **practice exercise** to help apply the order of operations.

- Display the *Order of Operations* page and review the explanation at the top of the page. Guide the students in following the order of operations to complete the problems. Remind them that if a step does not apply to an expression, students should proceed to the next step. (If a student struggles with performing the operations in the correct order, direct him to underline the operation that he is about to apply.) $-9, -5, -18, -8, -5, -7, 4, 7, 16, 41$

Apply

Student Edition pages 380–81

- Read and explain the directions for pages 380–81. Assist the students as they complete the pages independently.

Daily Review

- Students should complete Chapter 17, section *h*.

LESSON 175

Student Edition pages 382–83
Daily Review Chapter 17, section *i*

OBJECTIVES

- Graph points on a four-quadrant coordinate plane.
- Write ordered pairs to identify points on a four-quadrant coordinate plane.

TEACHER RESOURCES
- 27 *Coordinate Plane*
- 28 *Coordinate Planes* (for each student)

ASSESSMENTS
- Chapter 17 Quiz 2

Preparation

Write the following terms for display: coordinate plane, quadrants, *x*-axis, *y*-axis, origin (0, 0), ordered pair (*x*, *y*), *x*-coordinate, *y*-coordinate.

Engage

- Direct the students to **brainstorm** to explore the essential question on Student Edition page 382, "What is always true about the location of a point whose first coordinate is 0?"

Instruct

Graphing points on a four-quadrant coordinate plane

- Guide an **interactive activity** to help the students graph points.

- Distribute a *Coordinate Planes* page to each student and display the *Coordinate Plane* page. Remind the students that a coordinate plane is formed by 2 number lines intersecting at right angles and that the axes divide the coordinate plane into the 4 quadrants.

- Write the Roman numerals I–IV to label the quadrants and instruct the students to label their coordinate planes; begin with the top right quadrant and proceed in a counterclockwise direction.

Coordinate Planes

What is always true about the location of a point whose first coordinate is 0?

An **ordered pair** is used to describe the location of a point on a **coordinate plane** in relation to the **origin** (0, 0). Each ordered pair is located in a quadrant of the graph or on the ***x*-axis** or ***y*-axis**.

Quadrant II contains points with a negative *x*-coordinate and a positive *y*-coordinate.

$B = (-4, 3)$

Quadrant I contains points with positive *x*- and *y*-coordinates.

$A = (4, 2)$

Quadrant III contains points with negative *x*- and *y*-coordinates.

$C = (-3, -2)$

Quadrant IV contains points with a positive *x*-coordinate and a negative *y*-coordinate.

$D = (3, -4)$

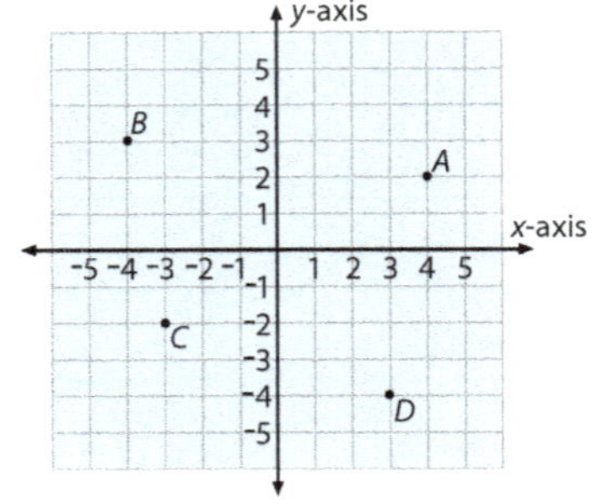

Exercises

Write the ordered pair for each point on the coordinate plane. Use the graph to find the answer.

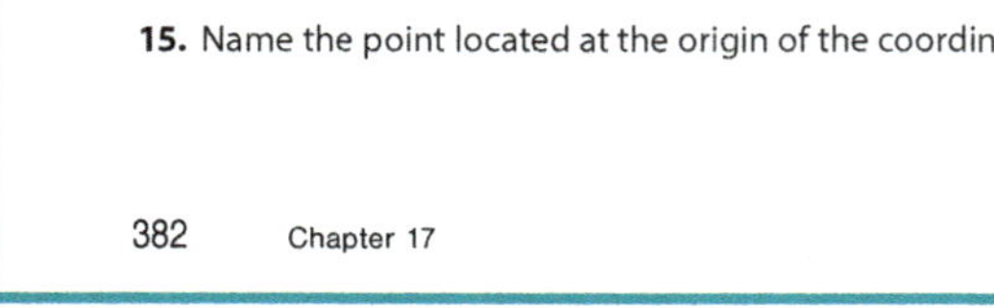

1. A *(0, 0)*
2. F *(-6, -4)*
3. B *(1, 4)*
4. G *(-5, 5)*
5. C *(3, 5)*
6. H *(2, -4)*
7. D *(-2, 2)*
8. I *(4, -3)*
9. E *(0, -2)*

10. How many points are in quadrant I? *2*

11. How many points are in quadrant II? *2*

12. What point is located in quadrant III? *F*

13. What points are located in quadrant IV? *H and I*

14. Where is point E located? *on the y-axis*

15. Name the point located at the origin of the coordinate plane. *point A*

382   Chapter 17

- Review with the students that the horizontal number line is the x-*axis* and the vertical number line is the y-*axis*. Remind students that the point where the *x*-axis and *y*-axis intersect is called the *origin*.

- Point out that an *ordered pair* describes the location of every point on a coordinate plane. The *x*-coordinate (first coordinate) tells the distance of the point along the *x*-axis—how far to move to the right or left from the origin. The *y*-coordinate (second coordinate) tells the distance of the point along the *y*-axis—how far to move up or down from the origin.

- Remind the students that the location of the origin is described by the ordered pair (0, 0). To find the location of any point, begin at the origin (0, 0) and move the number of units right or left as specified by the *x*-coordinate; then move the number of units up or down as specified by the *y*-coordinate (*x*, *y*).

  Where is the location of the point described by the ordered pair (2, 3)? It is 2 units right and 3 units up from the origin (0, 0) in quadrant I.

- Direct attention to the first coordinate plane (top left) on the *Coordinate Planes* page. Instruct the students to graph the point on this graph and to label it point A as you demonstrate.

  Where is the point described by the ordered pair (-2, 3) located? It is 2 units

Graph each ordered pair on a coordinate plane. Connect the points in order to form a figure.

16. (1, 1) ➔ (5, -3) ➔ (3, 3) ➔ (0, 0) ➔ (-3, -3)

17. (3, -3) ➔ (-3, 3) ➔ (-5, -3) ➔ (-1, 1)

Graph each ordered pair on a coordinate plane. Connect the points in order. Connect the last point to the first. Identify the quadrilateral.

18. (1, -2) ➔ (5, -2) ➔ (5, -5) ➔ (1, -5) *rectangle*

19. (-1, 0) ➔ (-5, 0) ➔ (-5, 3) ➔ (-3, 3) *trapezoid*

20. (0, 0) ➔ (2, 2) ➔ (5, 2) ➔ (3, 0) *parallelogram*

21. (-6, 6) ➔ (6, 6) ➔ (6, -6) ➔ (-6, -6) *square*

Write the ordered pair for each point on the coordinate plane.

22. A *(-5, -5)*   **23.** F *(-5, -4)*

24. B *(5, -5)*   **25.** G *(-2, 4)*

26. C *(5, 1)*   **27.** H *(-2, 5)*

28. D *(0, 6)*   **29.** I *(-4, 5)*

30. E *(-5, 1)*   **31.** J *(-4, 2)*

Write the letter of the point named by the ordered pair.

32. (-1, -5) *K*   **33.** (1, -2) *M*

34. (-1, -2) *L*   **35.** (1, -5) *N*

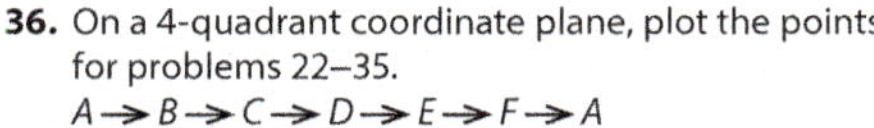

Connect the points in order. Identify the figure. Answer the question.

36. On a 4-quadrant coordinate plane, plot the points for problems 22–35.
A ➔ B ➔ C ➔ D ➔ E ➔ F ➔ A
G ➔ H ➔ I ➔ J
K ➔ L ➔ M ➔ N

37. Name the figure that is made. *a house*

38. What is always true about the location of a point whose first coordinate is 0? *A first coordinate of 0 means that I do not move left or right on the x-axis, so the point is always on the y-axis.*

Write "(-, -)" below the Roman numeral III in quadrant III.

What is true about the x- and y-coordinates of an ordered pair whose point is located in quadrant IV? The x-coordinate is a positive integer, and the y-coordinate is a negative integer.

Write "(+, -)" below the Roman numeral IV in quadrant IV.

In which of the 4 quadrants could a point with a positive x-coordinate possibly be located? quadrants I and IV

In which of the 4 quadrants could a point with a negative x-coordinate possibly be located? quadrants II and III

In which of the 4 quadrants could a point with a positive y-coordinate possibly be located? quadrants I and II

In which of the 4 quadrants could a point with a negative y-coordinate possibly be located? quadrants III and IV

- Follow a similar procedure to graph points J (0, 3) and K (0, -3) on the top right coordinate plane.

What do the ordered pairs of points J and K have in common, and how does it affect their location? They both have an x-coordinate of 0, causing them to be located somewhere along the y-axis; their y-coordinates are opposite integers with the same absolute value of 3, causing them to be 3 units from the origin in opposite directions.

What do you think is true of the x- and y-coordinates of the ordered pair of a point whose location is somewhere along the x-axis? The x-coordinate could be any integer, but the y-coordinate must be 0.

Ask students to give ordered pairs with 0 as their y-coordinate and guide the students in graphing them.

What is always true about the location of a point whose first coordinate is 0? sample answer: A first coordinate of 0 means that I do not move left or right on the x-axis, so the point is always on the y-axis.

- Invite students to give ordered pairs with 0 as the first coordinate and guide the students in graphing them.

left and 3 units up from the origin (0, 0) in quadrant II.

Graph and label point B (-2, 3).

Where is the point described by the ordered pair (-2, -3) located? It is 2 units left and 3 units down from the origin (0, 0) in quadrant III.

Graph and label point C (-2, -3).

Where is the point described by the ordered pair (2, -3) located? It is 2 units right and 3 units down from the origin (0, 0) in quadrant IV.

Graph and label point D (2, -3).

- Direct the students to graph points E (4, 6), F (4, -6), G (-4, -6), and H (-4, 6) on the same coordinate plane as points A, B, C, and D.

What is true about the x- and y-coordinates of an ordered pair whose point is located in quadrant I? Both coordinates are positive integers.

Write "(+, +)" below the Roman numeral I in quadrant I.

What is true about the x- and y-coordinates of an ordered pair whose point is located in quadrant II? The x-coordinate is a negative integer, and the y-coordinate is a positive integer.

Write "(-, +)" below the Roman numeral II in quadrant II.

What is true about the x- and y-coordinates of an ordered pair whose point is located in quadrant III? Both coordinates are negative integers.

LESSON 175

Writing ordered pairs to identify points

- Direct the students in a **collaborative exercise** to help them write ordered pairs.

- Direct the students to graph 8 points on the bottom left coordinate plane and to label them *A–H*. Encourage them to vary the location of the points among all 4 quadrants and the axes.

- Next, direct the students to write each letter (*A–H*) and the ordered pair that indicates its location on the backside of their page as an answer key.

- Direct the students to trade papers with a partner. Then tell them to write the ordered pair beside each point's letter name on the graph. After writing an ordered pair for each point, each partner can check his answers by comparing his ordered pairs with the answer key written on the back of the page.

You may also have students identify the quadrant number or axis name of the location.

- Write the ordered pairs "(4, 0)," "(0, 8)," and "(0, -4)" for display. Ask the students to write a prediction of the shape that would be formed if these 3 points were graphed and connected in order. Direct them to graph the points on the remaining coordinate plane to check their prediction. triangle

Direct the students to erase the points on the fourth coordinate plane.

- Write the following ordered pairs for display. Then direct the students to use the coordinates to find the length of a side joining the points with the same first coordinate or the same second coordinate.

polygon 1
(0, 5) (0, -3) 8 units
(5, 3) (5, 2) 1 unit
polygon 2
(5, 1) (9, 1) 4 units
(3, -5) (9, -5) 6 units

Write the numbers in order from least to greatest.

1. 0   -1   -3   4
-3   -1   0   4

2. 15   0   -12   -8
-12   -8   0   15

3. -15   15   13   -12
-15   -12   13   15

4. -8   -14   8   19
-14   -8   8   19

Write a comparison sentence by using >, <, or =.

5. -30 $\leq$ 29

6. -21 $\leq$ 0

7. 18 $\geq$ -45

8. 48 $\geq$ -48

9. 3 + -2 $\leq$ 5

10. -2 + -5 $\leq$ -4

11. -3 + 7 $\geq$ -3 + 4

12. 8 – 2 $=$ 10 + -4

Find the sum.

13. -9 + -1 *-10*

14. -8 + 5 *-3*

15. 7 + -4 *3*

16. -9 + -5 *-14*

Subtract.

17. 8 – -2 *10*

18. -3 – 8 *-11*

19. 9 – 15 *-6*

20. -3 – -1 *-2*

Apply

Student Edition pages 382–83

- Read and explain the directions for pages 382–83. Assist the students as they complete the pages independently.

Daily Review

- Students should complete Chapter 17, section *i*.

Assess

Quiz 2

- Use the **summative assessment** to evaluate the students' progress at this point in the chapter.

Student Edition pages 384–85

CHAPTER REVIEW

OBJECTIVES

- Compare and order integers.
- Add and subtract integers.
- Multiply and divide integers.
- Write an equation for a word problem.
- Defend the claim that every model is influenced by a person's worldview. **BWS**

BIBLICAL WORLDVIEW SHAPING

- **Modeling (Formulate):** The ideas and assumptions of the person making the model influence the design of the model.

TEACHER RESOURCES

- 3 *Positive & Negative Number Line*
- 128 *Algebra Mat*

ADDITIONAL MATERIALS

- counters: black and red

CHAPTER REVIEW

Complete the sentence, using the given choices.

> 0   integers   absolute value

1. The set of whole numbers and their opposites is called the set of ___. *integers*

2. The distance a number is from 0 is that number's ___. *absolute value*

3. The numbers 4 and ⁻4 are the same distance from ___. *0*

Write the integer described by the sentence.

4. The thermometer shows the temperature as 13° below zero. *⁻13*

5. Every leap year February has an extra day. *1*

6. The distance to school is 25 mi. *25*

7. Three inches of rain fell in the month of June. *3*

8. Part of the city of New Orleans is 7 ft below sea level. *⁻7*

9. David's miniature golf score was 4 strokes under par. *⁻4*

Write a comparison sentence by using > or <.

10. ⁻2 **>** ⁻27
11. ⁻16 **<** ⁻14
12. 3 **>** ⁻6
13. 6 **>** ⁻6
14. 8 **>** ⁻9
15. ⁻2 **<** ⁻1
16. 0 **>** ⁻1
17. ⁻3 **<** 1
18. 2 **<** 3

The Chapter Review offers an opportunity for students to discuss the concepts they have learned in the chapter. They may work collaboratively or independently as you review concepts. Circulate among the students, giving individual help as needed. Students who demonstrate proficiency with the discussion, the modeling, and the Student Edition pages are ready for the Chapter Test. Students who encounter difficulties with the review concepts would benefit from additional coaching and practice before testing.

Comparing & ordering integers

- Display the *Positive & Negative Number Line* page.

What number is in the center of the number line? 0

What kind of numbers are to the right of 0? positive numbers

What kind of numbers are to the left of 0? negative numbers

Remind the students that 0 is neither negative nor positive, but it separates the negative and positive numbers.

What name is given to the set of whole numbers (0, 1, 2, 3, . . .) and their opposites? integers

Write "integers" for display.

What integer is 3 units to the right of 0? 3

What integer is 3 units to the left of 0? ⁻3

What is the absolute value of 5? 5; The absolute value of a number is its distance from 0 on the number line.

What other integer has the same absolute value as 5? ⁻5; ⁻5 is 5 units from 0 on the number line.

- Write "|5| = 5" and "|⁻5| = 5" for display. Remind the students that these symbols indicate the absolute value of a number and are read "the absolute value of 5 is 5" and "the absolute value of negative 5 is 5." The absolute value of a number is

Write the integers from least to greatest.

19. ⁻6 3 ⁻8 1
⁻8 ⁻6 1 3

20. 16 ⁻17 ⁻13 12
⁻17 ⁻13 12 16

21. ⁻1 5 2 ⁻5
⁻5 ⁻1 2 5

22. ⁻4 ⁻1 ⁻2 1
⁻4 ⁻2 ⁻1 1

Add.

23. ⁻7 + 9 *2*

24. ⁻8 + 5 *⁻3*

25. ⁻4 + ⁻7 *⁻11*

26. ⁻6 + ⁻4 *⁻10*

27. ⁻12 + ⁻1 *⁻13*

28. 7 + ⁻4 *3*

29. ⁻3 + 3 *0*

30. 0 + ⁻7 *⁻7*

31. 8 + ⁻9 *⁻1*

Subtract.

32. 6 − ⁻3 *9*

33. ⁻12 − 8 *⁻20*

34. 0 − ⁻6 *6*

35. 5 − 13 *⁻8*

36. ⁻4 − ⁻1 *⁻3*

37. 8 − 17 *⁻9*

Solve.

38. 6 × ⁻2 *⁻12*

39. ⁻3 × ⁻2 *6*

40. 12 ÷ ⁻6 *⁻2*

41. ⁻16 ÷ ⁻4 *4*

42. 27 ÷ ⁻9 *⁻3*

43. ⁻3 × 7 *⁻21*

44. ⁻21 ÷ ⁻3 *7*

45. ⁻3 × ⁻8 *24*

46. 24 ÷ ⁻8 *⁻3*

Write a comparison sentence by using > or <.

47. 3 + ⁻5 ≤ 1 − 1 *⁻2 < 0*

48. 0 − ⁻3 ≥ 2 + ⁻2 *3 > 0*

49. ⁻6 + 3 ≤ 4 − 5 *⁻3 < ⁻1*

50. 4 − 7 ≥ 2 × ⁻3 *⁻3 > ⁻6*

51. ⁻2 − ⁻1 ≤ ⁻4 + 6 *⁻1 < 2*

52. 4 + 2 ≥ ⁻6 ÷ 1 *6 > ⁻6*

Use the order of operations to solve.

53. 3(⁻5 + ⁻1) *⁻18*

54. ⁻4 × 3 − 6 *⁻18*

55. 2 + ⁻12 ÷ 2 *⁻4*

56. How are my mathematical models related to my worldview? *My ideas and assumptions about the world influence the way I design a model.*

⁻4 + 5 = *1*

⁻3 + ⁻5 = *⁻8*

3 + 2 = *5*

3 − 7 = *⁻4*

9 − ⁻6 = *15*

⁻8 − ⁻2 = *⁻6*

3 − 5 = *⁻2*

⁻9 − ⁻2 = *⁻7*

⁻6 + 2 = *⁻4*

⁻7 − 3 = *⁻10*

6 − ⁻5 = *11*

⁻9 − 6 = *⁻15*

- Remind the students that subtracting a number is the same as adding its opposite. Write and solve an addition equation for each of the previous subtraction problems. (Refer to Lesson 170.)

3 + ⁻7 = ⁻4   *9 + 6 = 15*

⁻8 + 2 = ⁻6   *3 + ⁻5 = ⁻2*

⁻9 + 2 = ⁻7   *⁻7 + ⁻3 = ⁻10*

6 + 5 = 11   *⁻9 + ⁻6 = ⁻15*

- Write the following comparisons for display and choose students to write > or <.

⁻1°F *<* 1°F   ⁻6 + ⁻3 *>* ⁻1 − 10

⁻4 − 2 *<* 1 + 2   8°C *>* 2°C

Multiplying & dividing integers

- Write the following problems for display. Direct the students to draw a number line and use arrows to show the following multiplication and division problems. Choose students to solve each problem using an algebra mat and a number line.

4 × ⁻3 = *⁻12*   ⁻12 ÷ ⁻3 = *4*

7 × ⁻2 = *⁻14*   ⁻14 ÷ ⁻2 = *7*

- Write the following problems for display. Remind students that negative sets cannot be pictured. Direct the students to solve the problems using the Commutative Property.

⁻3 × 2 = *⁻6*   ⁻2 × 4 = *⁻8*   ⁻5 × 2 = *⁻10*

Guide the students in solving the related division equations.

⁻6 ÷ 2 = *⁻3*   ⁻8 ÷ 4 = *⁻2*   ⁻10 ÷ 2 = *⁻5*

What pattern do the integer signs follow in both multiplication and division? *Multiplication and division with like signs result in a positive answer, and multiplication and division with unlike signs result in a negative answer.*

- Write the following comparisons for display and choose students to write > or <.

⁻7 × 2 *<* ⁻3 × ⁻2   ⁻6 ÷ ⁻2 *>* ⁻16 ÷ 2

⁻10 ÷ 2 *>* ⁻3 × 2   ⁻3 × 2 *<* ⁻3 ÷ ⁻1

---

always a positive number since distance is expressed as a positive value.

What types of integers always have the same absolute value? *opposites*

In what direction are you moving on a number line if the values of the numbers are increasing? *in a positive direction; right on a horizontal number line and up on a vertical number line*

In what direction are you moving on a number line if the values of the numbers are decreasing? *in a negative direction; left on a horizontal number line and down on a vertical number line*

- Write the following number comparisons for display. Direct the students to copy and complete the number sentences using > or <.

⁻1 *>* ⁻2   6 *>* ⁻6   ⁻1 *<* 0   0 *<* 5

- Write the following sets of numbers for display and direct the students to order each set from least to greatest on their own paper.

8, ⁻1, ⁻8, 3 *⁻8, ⁻1, 3, 8*

⁻9, 5, ⁻4, ⁻1 *⁻9, ⁻4, ⁻1, 5*

Adding & subtracting integers

- Write the following equations for display. Direct the students to copy and complete the equations. Choose students to solve each problem using an algebra mat and a number line. (Refer to Lessons 168–70.)

- Ask students to recall the order of operations. parentheses, exponents, multiplication and division (left to right), addition and subtraction (left to right)

- Write the following equations for display. Guide the students in applying the order of operations to solve the equations.

$$10 \div {^-2} + {^-5} = {^-10} \qquad 4 + {^-6} \times 2 = {^-8}$$
$$5(2 - 3) = {^-5} \qquad {^-3} \times 4 - 5 = {^-17}$$

Writing an equation for a word problem

- Direct the students to write an equation for the following word problems. Choose students to explain the procedure they used to solve the problem.

  At 6:00 p.m. the temperature was $^-8°$F. By midnight the temperature had dropped 6 degrees. What was the temperature then? $^-8°F - 6°F = {^-14°}F$

  Ciara's checking account balance was $5.00. She wrote a check for $7.00. Then she discovered her error and hurried to the bank to deposit $12.00. What is her account balance now? $5.00 − $7.00 + $12.00 = $10.00 or $5.00 + $^-$7.00 + $12.00 = $10.00

  Jonathan's game score was 12. Madison's score was $^-8$. What is the difference between the 2 game scores? $12 - {^-8} =$ 20 points

Models' dependence on worldview

- Guide a discussion to answer the chapter essential question, "How are my mathematical models related to my worldview?"

  What mathematical models have we worked with this year? sample answers: pictured sets, arrays, bar models, part-whole models, number lines, algebra mats

  Are all of these models to be trusted just because they are models? No; they also must represent reality well.

- Remind students that a person's ideas and assumptions make up a person's worldview, which is the overall perspective they use to interpret the world.

  What is the only way we can learn the truth about the real world? from what God reveals to us in nature and in Scripture

- Redirect attention to the chapter opener picture on Student Edition page 365 and read the 2 worldviews presented about the formation of the Grand Canyon.

- Reinforce the idea that people make models based on what they believe about the world and that models alone cannot be trusted to represent reality. A model is useful as long as it corresponds to what God has revealed to us about His creation.

  How are my mathematical models related to my worldview? My ideas and assumptions about the world influence the way I design a model.

Student Edition pages 384–85

- Read and explain the directions for pages 384–85. Assist the students as they complete the pages independently.

LESSON 177

Student Edition pages 386–88

CHAPTER 17 TEST

CUMULATIVE REVIEW

CONCEPT REVIEW

- **Interpreting a double line graph (Lessons 148, 154–55)**
- **Identifying the geometric figure (Lesson 56)**
- **Identifying a geometric figure by its net (Lesson 109)**
- **Finding the area of a parallelogram (Lesson 106)**
- **Finding the surface area and volume of a rectangular prism (Lessons 114, 117)**
- **Identifying planes and angles (Lessons 52, 54)**
- **Calculating the unknown measure of an angle (Lesson 58)**
- **Estimating an answer (Lessons 41, 67, 76)**
- **Simplifying a fraction (Lesson 35)**
- **Expressing numbers in different forms (Lesson 1)**
- **Writing an expression (Lesson 94)**

- To prepare the students for the format of achievement tests, instruct them to work on a separate sheet of paper, if necessary, and to mark their answers on the *Cumulative Review Answer Sheet*.

Student Edition pages 386–88

The Cumulative Review provides additional practice of previously learned concepts. These pages may be completed during this lesson or anytime after this lesson, since they require limited or no teaching.

Use the double line graph to find the answer.

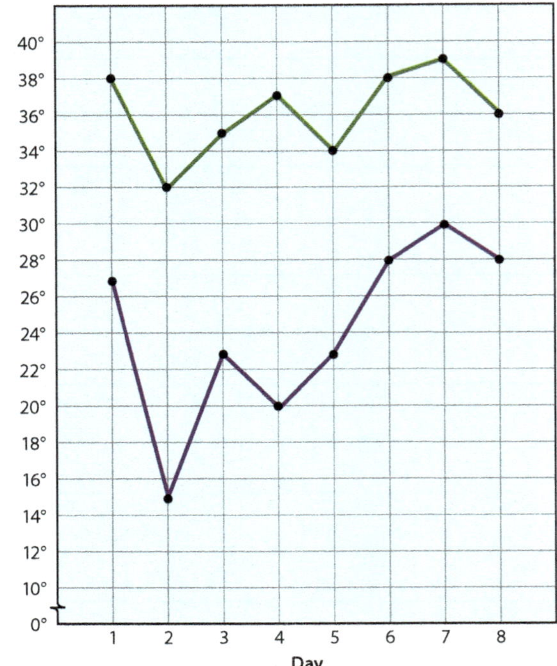

1. What is the range in the temperature?
- **A.** 24°
- **B.** 27°
- **C.** 30°
- **D.** 38°

2. What is the range of high temperatures?
- **A.** 5°
- **B.** 6°
- **C.** 7°
- **D.** 8°

3. What is the average high temperature?
- **A.** 35.75°F
- **B.** 36.125°F
- **C.** 36.5°F
- **D.** 37°F

4. What day shows the least difference between the high and low temperatures?
- **A.** day 1
- **B.** day 2
- **C.** day 4
- **D.** day 8

5. What is the mode for the high temperature?
- **A.** 32°F
- **B.** 34°F
- **C.** 38°F
- **D.** 40°F

6. What was the change in low temperature from day 1 to day 2?
- **A.** −12°
- **B.** 12°
- **C.** 10°
- **D.** −10°

386   Chapter 17

Choose the answer.

7. Which polygon has 7 sides and 7 angles?

 A. pentagon   **C.** heptagon

 B. hexagon   **D.** quadrilateral

8. Match the figure to its net.

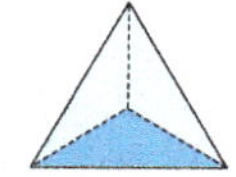

 A. 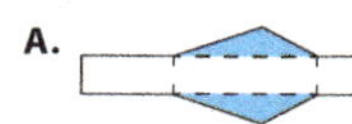   **C.** 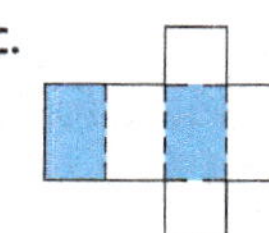

 B. 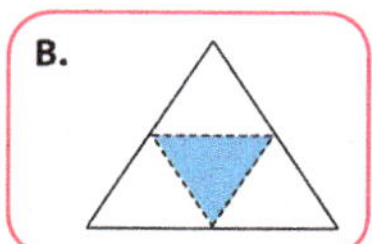   **D.** 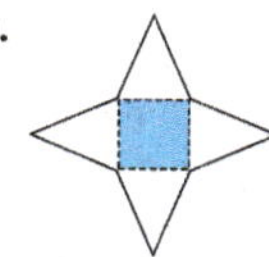

9. Find the area of the figure.

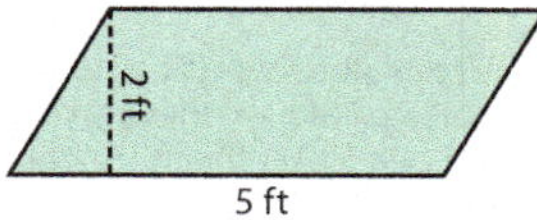

 A. 7 ft   **C.** 10 ft²

 B. 10 ft³   **D.** 12 ft

10. Find the surface area.

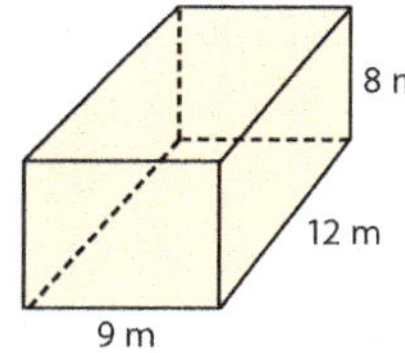

 A. 204 m³   **C.** 864 m³

 B. 552 m²   **D.** 900 m²

11. Find the volume.

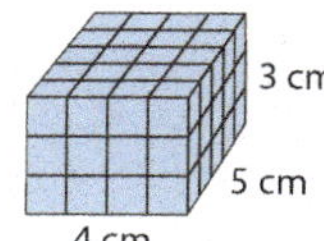

 A. 20 cm³   **C.** 35 cm³

 B. 60 cm³   **D.** 70 cm³

12. Which figure represents a plane?

 A.    **C.** 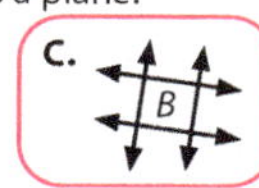

 B.    **D.** •Z

13. Which angle is obtuse?

 A.    **C.** 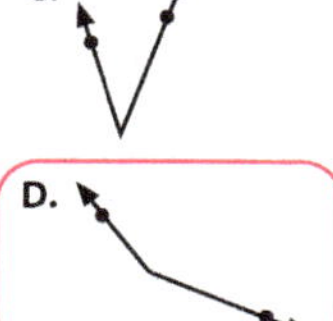

 B.   **D.**

14. What is the measure of the unknown angle?

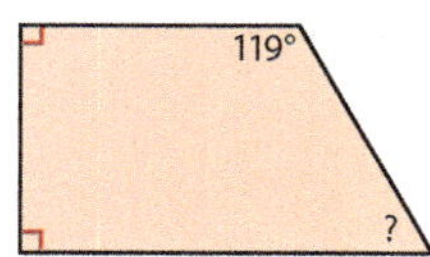

 A. 21°   **C.** 50°

 B. 39°   **D.** 61°

Choose the answer.

15. Use rounding to estimate.

$$28.36$$
$$- 12.52$$

A. 10   **C.** 30

B. 20   **D.** 40

16. Round to the nearest cent.

$26)\overline{\$86.00}$

A. \$2.07   **C.** \$3.08

B. \$3.07   **D.** \$3.31

17. Estimate the answer.

$\frac{3}{8} + \frac{5}{6} \approx$ ___

A. $1\frac{1}{2}$ **C.** $2\frac{1}{2}$

B. 2   **D.** 3

18. Solve for n.

$n \times 132 = 528$

A. $n = 1$ **C.** $n = 3$

B. $n = 2$ **D.** $n = 4$

19.   4 pt 3 c
    − 3 pt 1 c

A. 1 pt 7 c   **C.** 2 pt 5 c

B. 1 pt 2 c   **D.** 4 pt

20. $3\frac{1}{8} - \frac{3}{4} =$ ___

A. $1\frac{5}{8}$ **C.** $2\frac{1}{4}$

B. $2\frac{3}{8}$ **D.** $3\frac{1}{4}$

21. $4\frac{1}{2} \times 2\frac{1}{6} =$ ___

A. $8\frac{1}{6}$ **C.** $9\frac{1}{4}$

B. $8\frac{1}{2}$ **D.** $9\frac{3}{4}$

22. $2\frac{3}{4} \div \frac{1}{2} =$ ___

A. $1\frac{3}{8}$ **C.** $5\frac{1}{2}$

B. $3\frac{1}{8}$ **D.** 6

23. Devin has 60 min to do his chores. If he has worked for 36 min, what part of his chore time has passed?

A. $\frac{2}{5}$ **C.** $\frac{3}{5}$

B. $\frac{4}{5}$ **D.** $\frac{5}{5}$

24. What is three hundred one and six thousand, four hundred two ten thousandths in standard form?

A. 301.6402   **C.** 3,001.6042

B. 310.6042   **D.** 301.642

25. A photographer took student pictures for 3 schools. He took 125 pictures at the first school, 230 at the second, and 560 at the third. Which expression could be used to determine the number of pictures taken at the 3 schools?

A. $560 - (230 + 125)$ **C.** $(125 + 230) \times 560$

B. $125 + 230 + 560$ **D.** $(230 - 125) + 560$

Chapter 10

page 205

33. **36:** 1, 2, 3, 4, 6, 9, 12, 18, 36
    72: 1, 2, 3, 4, 6, 8, 9, 12, 18, 24, 36, 72
    GCF = 36

36. The fractions $\frac{6}{16}$, $\frac{9}{24}$, and $\frac{12}{32}$ are equivalent to $\frac{3}{8}$.

page 209

22. $8x + (4x + 2) =$ Commutative
    $(8x + 4x) + 2 =$ Associative
    $12x + 2$

23. $(3 \cdot 7)x$ Associative
    $21x$

24. $6 + (5x + 7)$ Associative
    $6 + (7 + 5x)$ Commutative
    $(6 + 7) + 5x$ Associative
    $5x + 13$ Commutative

26. $(8 + 2) + x$ Associative
    $x + 10$ Commutative

27. $x \cdot (5 \cdot 6)$ Associative
    $30x$ Commutative

28. $(2 + 3) + x$ Associative
    $x + 5$ Commutative

29. $x \cdot (8 \cdot 9)$ Associative
    $72x$ Commutative

30. $(12 \cdot 2)x$ Associative
    $24x$

page 210

1. $n + 5 - 5 = 21 - 5$
   $n = 16$
   $16 + 5 = 21$

2. $d + 45 - 45 = 90 - 45$
   $d = 45$
   $45 + 45 = 90$

3. $d + 60 = 85$
   $d + 60 - 60 = 85 - 60$
   $d = 25$
   $25 + 43 + 17 = 85$

4. $x + 12 - 12 = 40 - 12$
   $x = 28$
   $28 + 12 = 40$

5. $16 - 16 + f = 35 - 16$
   $f = 19$
   $16 + 19 = 35$

6. $s + 11.5 = 20$
   $s + 11.5 - 11.5 = 20 - 11.5$
   $s = 8.5$
   $8.5 + 14 - 3 + 0.5 = 20$

7. $c - 6 + 6 = 17 + 6$
   $c = 23$
   $23 - 6 = 17$

8. $s - 39 + 39 = 61 + 39$
   $s = 100$
   $100 - 39 = 61$

9. $19.8 + b = 29$
   $19.8 - 19.8 + b = 29 - 19.8$
   $b = 9.2$
   $3.8 + 16 + 9.2 = 29$

10. $a - 4 + 4 = 36 + 4$
    $a = 40$
    $40 - 4 = 36$

11. $24 - 24 + n = 100 - 24$
    $n = 76$
    $24 + 76 = 100$

12. $1\frac{1}{4} + f = 1\frac{1}{2}$
    $1\frac{1}{4} - 1\frac{1}{4} + f = 1\frac{1}{2} - 1\frac{1}{4}$
    $f = \frac{1}{4}$
    $\frac{3}{4} + \frac{1}{2} + \frac{1}{4} = 1\frac{1}{2}$

page 211

13. $14 - 8 = 6$; yes

14. $0.8 + 0.8 \neq 1.7$; no
    $n + 0.8 - 0.8 = 1.7 - 0.8$
    $n = 0.9$

15. $17 + 13 \neq 40$; no
    $x + 13 - 13 = 40 - 13$
    $x = 27$

16. $8 - 1\frac{1}{2} = 6\frac{1}{2}$; yes

17. $2.6 - 17 \neq 9$; no
    $f - 17 + 17 = 9 + 17$
    $f = 26$

18. $20.2 + 17 + 3.5 = 40.7$; yes

28. $n + 5 = 12$
    $n + 5 - 5 = 12 - 5$
    $n = 7$
    $7 + 5 = 12$

29. $n - 6 = 12$
    $n - 6 + 6 = 12 + 6$
    $n = 18$
    $18 - 6 = 12$

30. $n - 8 = 3$
    $n - 8 + 8 = 3 + 8$
    $n = 11$
    $11 - 8 = 3$

31. $n + 3 = 16$
    $n + 3 - 3 = 16 - 3$
    $n = 13$
    $13 + 3 = 16$

32. $10 + n = 17$
    $10 - 10 + n = 17 - 10$
    $n = 7$
    $10 + 7 = 17$

33. $n - 7 = 9$
    $n - 7 + 7 = 9 + 7$
    $n = 16$
    $16 - 7 = 9$

34. $n - 2 = 5$
    $n - 2 + 2 = 5 + 2$
    $n = 7$
    $7 - 2 = 5$

35. $7 + n = 10$
    $7 - 7 + n = 10 - 7$
    $n = 3$
    $7 + 3 = 10$

36.
$$\begin{array}{r} 113 \\ \times 609 \\ \hline 1017 \\ + 67800 \\ \hline 68{,}817 \end{array}$$

42.
$$\begin{array}{r} 17 \\ 3\overline{)51} \\ -3 \\ \hline 21 \\ -21 \\ \hline 0 \end{array}$$

page 212

1. $a \times \frac{5}{5} = \frac{30}{5}$
   $a = 6$
   $6 \times 5 = 30$

2. $\frac{9p}{9} = \frac{54}{9}$
   $p = 6$
   $9(6) = 54$

3. $c \bullet \frac{7}{7} = \frac{49}{7}$
   $c = 7$
   $7 \bullet 7 = 49$

4. $\frac{2m}{2} = \frac{18}{2}$
   $m = 9$
   $2(9) = 18$

5. $\frac{8}{8} \times p = \frac{64}{8}$
   $p = 8$
   $8 \times 8 = 64$

6. $d \times 2^2 = 48$
   $d \times 4 = 48$
   $d \times \frac{4}{4} = \frac{48}{4}$
   $d = 12$
   $12 \times 2^2 = 48$

7. $f \bullet \frac{3}{3} = \frac{24}{3}$
   $f = 8$
   $8 \bullet 3 = 24$

8. $\frac{4n}{4} = \frac{32}{4}$
   $n = 8$
   $4(8) = 32$

9. $a \bullet \frac{0.9}{0.9} = \frac{3.6}{0.9}$
   $a = 4$
   $4 \bullet 0.9 = 3.6$

10. $\frac{12}{12} \bullet b = \frac{36}{12}$
    $b = 3$
    $12 \bullet 3 = 36$

11. $\frac{7n}{7} = \frac{56}{7}$
    $n = 8$
    $7(8) = 56$

12. $\frac{1.8}{1.8} \times a = \frac{36}{1.8}$
    $a = 20$
    $1.8 \times 20 = 36$

13. $a \bullet 7 = 21$ or $7a = 21$
    $a \bullet \frac{7}{7} = \frac{21}{7}$
    $a = 3$

14. $y \bullet 9 = 54$ or $9y = 54$
    $y \bullet \frac{9}{9} = \frac{54}{9}$
    $y = 6$

15. $8 \bullet b = 48$ or $8b = 48$
    $\frac{8}{8} \bullet b = \frac{48}{8}$
    $b = 6$

16. $3 \bullet n = 36$ or $3n = 36$
    $\frac{3}{3} \bullet n = \frac{36}{3}$
    $n = 12$

17. $x \bullet 7 = 35$ or $7x = 35$
    $x \bullet \frac{7}{7} = \frac{35}{7}$
    $x = 5$

page 214

1. $n \div 7 \bullet 7 = 63 \bullet 7$
   $n = 441$
   $441 \div 7 = 63$

2. $\frac{m}{4} \bullet 4 = 3 \bullet 4$
   $m = 12$
   $\frac{12}{4} = 3$

3. $x \div 8 \bullet 8 = 14.5 \bullet 8$
   $x = 116$
   $116 \div 8 = 14.5$

4. $\frac{3}{3} \bullet x = \frac{15}{3}$
   $x = 5$
   $3 \bullet 5 = 15$

5. $n \bullet \frac{1}{5} \div \frac{1}{5} = 9 \div \frac{1}{5}$
   $n = 9 \times \frac{5}{1}$
   $n = 45$
   $45 \bullet \frac{1}{5} = 9$

6. $r \div 0.3 \bullet 0.3 = 57.5 \bullet 0.3$
   $r = 17.25$
   $17.25 \div 0.3 = 57.5$

7. $\frac{2}{3} \div \frac{2}{3} \bullet x = 15 \div \frac{2}{3}$
   $x = 15 \bullet \frac{3}{2}$
   $x = \frac{45}{2} = 22\frac{1}{2}$
   $\frac{2}{3} \bullet \frac{45}{2} = 15$

8. $\frac{x}{9} \bullet 9 = 18 \bullet 9$
   $x = 162$
   $\frac{162}{9} = 18$

9. $\frac{4}{4} \bullet c = \frac{32}{4}$
   $c = 8$
   $4 \bullet 8 = 32$

10. $\frac{1}{2} \div \frac{1}{2} \bullet p = 45 \div \frac{1}{2}$
    $p = 45 \times \frac{2}{1}$
    $p = 90$
    $\frac{1}{2}(90) = 45$

11. $\frac{3n}{3} = \frac{78.12}{3}$
    $n = 26.04$
    $3(26.04) = 78.12$

12. $p \bullet \frac{3}{5} \div \frac{3}{5} = 5 \div \frac{3}{5}$
    $p = 5 \bullet \frac{5}{3}$
    $p = \frac{25}{3} = 8\frac{1}{3}$
    $\frac{25}{3} \bullet \frac{3}{5} = 5$

13. $n \div 12 = 3$
    $\frac{n}{12} \bullet 12 = 3 \bullet 12$
    $n = 36$
    $36 \div 12 = 3$

14. $y \div 7 = 2$
    $\frac{y}{7} \bullet 7 = 2 \bullet 7$
    $y = 14$
    $14 \div 7 = 2$

15. $x \bullet 7 = 35$
    $x \bullet \frac{7}{7} = \frac{35}{7}$
    $x = 5$
    $5 \bullet 7 = 35$

16. $n \div 20 = 3$
    $\frac{n}{20} \bullet 20 = 3 \bullet 20$
    $n = 60$
    $60 \div 20 = 3$

17. $n \div 5 = 11$
    $\frac{n}{5} \bullet 5 = 11 \bullet 5$
    $n = 55$
    $55 \div 5 = 11$

18. $2n = 18$
    $\frac{2n}{2} = \frac{18}{2}$
    $n = 9$
    $2(9) = 18$

19. $6n = 54$
    $\frac{6n}{6} = \frac{54}{6}$
    $n = 9$
    $6(9) = 54$

20. $n \bullet 5 = 10$
    $n \bullet \frac{5}{5} = \frac{10}{5}$
    $n = 2$
    $2 \bullet 5 = 10$

page 217

21. $a + 5 - 5 = 33 - 5$
    $a = 28$
    $28 + 5 = 33$

22. $\frac{8x}{8} = \frac{480}{8}$
    $x = 60$
    $8(60) = 480$

23. $\frac{1.5w}{1.5} = \frac{30}{1.5}$
    $w = 20$
    $1.5(20) = 30$

24. $x - 1.2 + 1.2 = 10 + 1.2$
    $x = 11.2$
    $11.2 - 1.2 = 10$

25. $y - 43 + 43 = 129 + 43$
    $y = 172$
    $172 - 43 = 129$

26. $\frac{3.8p}{3.8} = \frac{64.6}{3.8}$
    $p = 17$
    $3.8(17) = 64.6$

27. $\frac{a}{12} \bullet 12 = 3 \bullet 12$
    $a = 36$
    $\frac{36}{12} = 3$

28. $\frac{3}{4} \div \frac{3}{4} \bullet x = 6 \div \frac{3}{4}$
    $x = 6 \bullet \frac{4}{3}$
    $x = \frac{24}{3} = 8$
    $\frac{3}{4}(8) = \frac{24}{4} = 6$

Solutions

29. $\frac{x}{9} \cdot 9 = 4 \cdot 9$
$x = 36$
$\frac{36}{9} = 4$

30. $\frac{8x}{8} = \frac{1}{8}$
$x = \frac{1}{8}$
$8 \cdot \frac{1}{8} = 1$

31. $\frac{2x}{2} = \frac{14.8}{2}$
$x = 7.4$
$2(7.4) = 14.8$

32. $a + 1.7 - 1.7 = 1.9 - 1.7$
$a = 0.2$
$0.2 + 1.7 = 1.9$

33. $n - 16 + 16 = 140 + 16$
$n = 156$
$156 - 16 = 140$

34. $x - 6 + 6 = 1.4 + 6$
$x = 7.4$
$7.4 - 6 = 1.4$

35. $x \div 12 \cdot 12 = 62 \cdot 12$
$x = 744$
$744 \div 12 = 62$

39. $p \cdot 8 = 200$
$p \cdot \frac{8}{8} = \frac{200}{8}$
$p = 25$ pizzas

40. $x + 9 = 20$
$x + 9 - 9 = 20 - 9$
$x = 11$ insects

44. $\frac{1}{2}n = 118$
$\frac{1}{2}n \div \frac{1}{2} = 118 \div \frac{1}{2}$
$n = 118 \cdot \frac{2}{1}$
$n = 236$

page 218

1. $d = r \cdot t$
$11{,}247 = r \cdot 102$
$\frac{11{,}247}{102} = r \cdot \frac{102}{102}$
$r \approx 110$ km/hr

2. $d = r \cdot t$
$d = 85 \cdot 4$
$d = 340$ km

3. $d = r \cdot t$
$96 = r \cdot 4$
$\frac{96}{4} = r \cdot \frac{4}{4}$
$r = 24$ mi/day

4. $d = r \cdot t$
$33 = 11 \cdot t$
$\frac{33}{11} = \frac{11}{11} \cdot t$
$t = 3$ hr

5. $d = r \cdot t$
$d = 93 \cdot 52$
$d = 4{,}836$ km

6. $d = r \cdot t$
$d = 22 \cdot 15$
$d = 330$ yd

7. $d = r \cdot t$
$380 = 95 \cdot t$
$\frac{380}{95} = \frac{95}{95} \cdot t$
$t = 4$ hr

8. $d = r \cdot t$
$1{,}760 = 22 \cdot t$
$\frac{1{,}760}{22} = \frac{22}{22} \cdot t$
$t = 80$ min

page 221

34. $\frac{x}{4} \cdot 4 = 9 \cdot 4$
$x = 36$
$\frac{36}{4} = 9$

35. $n - 8 + 8 = 61 + 8$
$n = 69$
$69 - 8 = 61$

36. $\frac{4a}{4} = \frac{56}{4}$
$a = 14$
$4(14) = 56$

37. $n + 3 - 3 = 32 - 3$
$n = 29$
$29 + 3 = 32$

38. $x - 1.6 + 1.6 = 1.4 + 1.6$
$x = 3$
$3 - 1.6 = 1.4$

39. $y \div 5 \cdot 5 = 25 \cdot 5$
$y = 125$
$125 \div 5 = 25$

40. $b + 5 - 5 = 48 - 5$
$b = 43$
$43 + 5 = 48$

41. $\frac{7n}{7} = \frac{85.4}{7}$
$n = 12.2$
$7(12.2) = 85.4$

Chapter 11
page 226

1. $P = (2 \cdot l) + (2 \cdot w)$
$P = (2 \cdot 5 \text{ cm}) + (2 \cdot 2 \text{ cm})$
$P = 10 \text{ cm} + 4 \text{ cm} = 14 \text{ cm}$

2. $P = a + b + c + d$
$P = 2\frac{1}{2} \text{ cm} + 5 \text{ cm} + 2\frac{1}{2} \text{ cm} + 1 \text{ cm}$
$P = 11$ cm

3. $P = n \cdot s$
$P = 3 \cdot 3$ cm
$P = 9$ cm

page 227

16. $12 \text{ m} = (3 \text{ m} + 4 \text{ m}) + n$
$12 \text{ m} = 7 \text{ m} + n$
$12 \text{ m} - 7 \text{ m} = 7 \text{ m} - 7 \text{ m} + n$
$n = 5$ m

17. $24 \text{ yd} = (2 \cdot 4 \text{ yd}) + 6 \text{ yd} + 8 \text{ yd} + n$
$24 \text{ yd} = 8 \text{ yd} + 6 \text{ yd} + 8 \text{ yd} + n$
$24 \text{ yd} = 22 \text{ yd} + n$
$24 \text{ yd} - 22 \text{ yd} = 22 \text{ yd} - 22 \text{ yd} + n$
$n = 2$ yd

18. $P = (2 \cdot l) + (2 \cdot w)$
$P = 2(2 \cdot 4 \text{ cm}) + (2 \cdot 4 \text{ cm})$
$P = 2(8 \text{ cm}) + 8 \text{ cm}$
$P = 16 \text{ cm} + 8 \text{ cm}$
$P = 24$ cm

19. $P = (2 \cdot l) + (2 \cdot w)$
$P = 2(4 \text{ cm} + 3 \text{ cm}) + (2 \cdot 4 \text{ cm})$
$P = 2(7 \text{ cm}) + 8 \text{ cm}$
$P = 14 \text{ cm} + 8 \text{ cm}$
$P = 22$ cm

23. $P = (2 \cdot l) + (2 \cdot w)$
$P = (2 \cdot 100 \text{ yd}) + (2 \cdot 50 \text{ yd})$
$P = 200 \text{ yd} + 100 \text{ yd}$
$P = 300$ yd

24. $P = (2 \cdot l) + (2 \cdot w)$
$P = (2 \cdot 50 \text{ yd}) + (2 \cdot 20 \text{ yd})$
$P = 100 \text{ yd} + 40 \text{ yd}$
$P = 140$ yd

25. $P = (2 \cdot l) + (2 \cdot w)$
$P = (2 \cdot 50 \text{ yd}) + (2 \cdot 55 \text{ yd})$
$P = 100 \text{ yd} + 110 \text{ yd}$
$P = 210$ yd

26. $P = (2 \cdot l) + (2 \cdot w)$
$P = (2 \cdot 40 \text{ yd}) + (2 \cdot 50 \text{ yd})$
$P = 80 \text{ yd} + 100 \text{ yd}$
$P = 180$ yd

27. $P = (2 \cdot l) + (2 \cdot w)$
$P = (2 \cdot 200 \text{ yd}) + (2 \cdot 75 \text{ yd})$
$P = 400 \text{ yd} + 150 \text{ yd}$
$P = 550$ yd

page 229

9. $C = \pi d$
$C = 3.14 \cdot 8$ yd
$C = 25.12$ yd

10. $C = \pi d$
$C = 3.14 \cdot 3.5$ m
$C = 10.99$ m

11. $C = 2\pi r$
$C = 2 \cdot 3.14 \cdot 2.7$ cm
$C = 16.956$ cm ≈ 16.96 cm

12. $C = \pi d$
$C = 3.14 \cdot 1\frac{3}{4}$ ft
$C = 3.14 \cdot 1.75$ ft
$C = 5.495$ ft ≈ 5.50 ft or 5.5 ft

13. $C = 2\pi r$
$C = 2 \cdot 3.14 \cdot 13$ m
$C = 81.64$ m

14. $P = n \cdot s$
$P = 6 \cdot 4$ ft
$P = 24$ ft

15. $P = (2 \cdot l) + (2 \cdot w)$
$P = (2 \cdot 6$ cm$) + (2 \cdot 2$ cm$)$
$P = 12$ cm $+ 4$ cm
$P = 16$ cm

16. $P = (2 \cdot 20$ m$) + (2 \cdot 15$ m$) +$
$\qquad 25$ m $+ 30$ m
$P = 40$ m $+ 30$ m $+ 25$ m $+ 30$ m
$P = 125$ m

page 230

19. $C = 2\pi r$
$C = 2 \cdot 3.14 \cdot 4$ ft
$C = 25.12$ ft

20. $C = \pi d$
$C = 3.14 \cdot 60$ in.
$C = 188.4$ in.

21. $P = (2 \cdot l) + (2 \cdot w)$
$P = (2 \cdot 10) + (2 \cdot 5)$
$P = 20 + 10$
$P = 30$ ft
$30 \cdot \$4.50 = \$135.00$

22. $C = \pi d$
$C = 3.14 \cdot 212$ ft
$C = 665.68$ ft

23. $C = \pi d$
$C = 3.14 \cdot 20$ in.
$C = 62.8$ in.

24. $P = (2 \cdot l) + (2 \cdot w)$
$P = (2 \cdot 12) + (2 \cdot 8)$
$P = 24 + 16$
$P = 40$ ft
$40 \div 15 = 2.\overline{6}$; 3 rolls

page 232

5. $A = b \cdot h$
$A = 2.3$ m $\cdot 2.7$ m
$A = 6.21$ m^2

6. $A = b \cdot h$
$A = 5$ in. $\cdot 12$ in.
$A = 60$ in.^{2}

7. $A = l \cdot w$
$A = (10$ yd $\cdot 12$ yd$) + (18$ yd $\cdot 20$ yd$)$
$A = 120$ yd$^2 + 360$ yd^2
$A = 480$ yd^2

8. $A = l \cdot w$
$A = (2$ cm $\cdot 2$ cm$) + (2$ cm $\cdot 1$ cm$) +$
$\qquad (1$ cm $\cdot 1$ cm$)$
$A = 4$ cm$^2 + 2$ cm$^2 + 1$ cm^2
$A = 7$ cm^2

9. $n = A \div s$
$n = 25$ ft$^2 \div 5$ ft
$n = 5$ ft

10. $n = A \div s$
$n = 136$ yd$^2 \div 17$ yd
$n = 8$ yd

11. $n = A \div s$
$n = 70$ m$^2 \div 5.6$ m
$n = 12.5$ m

12. $n = A \div s$
$n = 140$ cm$^2 \div 10$ cm
$n = 14$ cm

19. $A = l \cdot w$
$A = (5$ cm $\cdot 5$ cm$) + (6$ cm $\cdot 10$ cm$)$
$A = 25$ cm$^2 + 60$ cm^2
$A = 85$ cm^2

20. $A = l \cdot w$
$A = 2(3$ m $\cdot 3$ m$) + (13$ m $\cdot 6$ m$)$
$A = 18$ m$^2 + 78$ m^2
$A = 96$ m^2

page 234

1. $A = b \cdot h$
$A = 8$ m $\cdot 6$ m
$A = 48$ m^2

2. $A = \frac{1}{2}(b \cdot h)$
$A = \frac{1}{2}(8$ m $\cdot 6$ m$)$
$A = \frac{1}{2}(48$ m$^2)$
$A = 24$ m^2

3. $A = b \cdot h$
$A = 7$ in. $\cdot 5$ in.
$A = 35$ in.^{2}

4. $A = \frac{1}{2}(b \cdot h)$
$A = \frac{1}{2}(7$ in. $\cdot 5$ in.$)$
$A = \frac{1}{2}(35$ in.$^2)$
$A = 17.5$ in.^{2}

5. $A = \frac{1}{2}(b \cdot h)$
$A = \frac{1}{2}(5$ in. $\cdot 3$ in.$)$
$A = \frac{1}{2}(15$ in.$^2)$
$A = 7.5$ in.^{2}

6. $A = \frac{1}{2}(b \cdot h)$
$A = \frac{1}{2}(2.5$ cm $\cdot 2.5$ cm$)$
$A = \frac{1}{2}(6.25$ cm$^2)$
$A = 3.125$ cm^2

7. $A = \frac{1}{2}(b \cdot h)$
$A = \frac{1}{2}(1\frac{1}{4}$ in. $\cdot 2$ in.$)$
$A = \frac{1}{2}(\frac{5}{4}$ in. $\cdot 2$ in.$)$
$A = \frac{1}{2}(\frac{10}{4}$ in.$^2)$
$A = \frac{10}{8}$ in.$^2 = 1\frac{1}{4}$ in.^{2}

page 235

12. $A = \frac{1}{2}(b \cdot h)$
$30 = \frac{1}{2}(10 \cdot h)$
$30 = \frac{1}{2}(10h)$
$\frac{30}{5} = \frac{5h}{5}$
$h = 6$ ft

13. $A = \frac{1}{2}(b \cdot h)$
$15 = \frac{1}{2}(b \cdot 10)$
$15 = \frac{1}{2}(10b)$
$\frac{15}{5} = \frac{5b}{5}$
$b = 3$ m

14. $A = \frac{1}{2}(b \cdot h)$
$24 = \frac{1}{2}(6 \cdot h)$
$24 = \frac{1}{2}(6h)$
$\frac{24}{3} = \frac{3h}{3}$
$h = 8$ yd

18. $A = l \cdot w$
$A = (2 \cdot w) \cdot w$
$A = (2 \cdot 6$ ft$) \cdot 6$ ft
$A = 12$ ft $\cdot 6$ ft
$A = 72$ ft^2

20. $C = \pi d$
$C = 3.14 \cdot 8.6$ in.
$C = 27.004$ in.
27 beads

page 236

1. $A = \pi r^2$
   $A = 3.14(6^2)$
   $A = 3.14(36)$
   $A = 113.04 \text{ cm}^2$

2. $A = \pi r^2$
   $A = 3.14(9^2)$
   $A = 3.14(81)$
   $A = 254.34 \text{ yd}^2$

3. $A = \pi r^2; d = 42; r = 21$
   $A = 3.14(21^2)$
   $A = 3.14(441)$
   $A = 1{,}384.74 \text{ in.}^2$

4. $A = \pi r^2; d = 20; r = 10$
   $A = 3.14(10^2)$
   $A = 3.14(100)$
   $A = 314 \text{ m}^2$

5. $3 \cdot 15^2$
   $3 \cdot 225 = 675 \text{ m}^2$

6. $3 \cdot 2^2$
   $3 \cdot 4 = 12 \text{ cm}^2$

7. $3 \cdot 6^2$
   $3 \cdot 36 = 108 \text{ ft}^2$

8. $3 \cdot 13^2$
   $3 \cdot 169 = 507 \text{ ft}^2$

page 237

21. $A = \pi r^2; d = 4; r = 2$
    $A = \frac{1}{2}(\pi r^2)$
    $A = \frac{1}{2}(3.14 \cdot 2^2)$
    $A = \frac{1}{2}(3.14 \cdot 4)$
    $A = \frac{1}{2}(12.56)$
    $A = 6.28 \text{ cm}^2$

22. $A = \pi r^2; d = 6; r = 3$
    $A = \frac{1}{4}(\pi r^2)$
    $A = \frac{1}{4}(3.14 \cdot 3^2)$
    $A = \frac{1}{4}(3.14 \cdot 9)$
    $A = \frac{1}{4}(28.26)$
    $A = 7.065 \text{ cm}^2$

23. $A = l \cdot w$ and $A = \pi r^2$
    $A = (10 \cdot 10) - (3.14 \cdot 5^2)$
    $A = 100 - (3.14 \cdot 25)$
    $A = 100 - 78.5$
    $A = 21.5 \text{ ft}^2$

page 239

17. $2(10 \cdot 5) = 100 \text{ in.}^2$
    $2(10 \cdot 5) = 100 \text{ in.}^2$
    $2(5 \cdot 5) = 50 \text{ in.}^2$
    total surface area $= 250 \text{ in.}^2$ or
    $2(10 \cdot 5) + 2(10 \cdot 5) + 2(5 \cdot 5) =$
    $100 + 100 + 50 = 250 \text{ in.}^2$

18. $6(5 \cdot 5) = 150 \text{ in.}^2$
    total surface area $= 150 \text{ in.}^2$

19. $16 \cdot 3 = 48 \text{ cm}^2$
    $2(10 \cdot 3) = 60 \text{ cm}^2$
    $2 \cdot \frac{1}{2}(16 \cdot 6) = 96 \text{ cm}^2$
    total surface area $= 204 \text{ cm}^2$ or
    $(16 \cdot 3) + 2(10 \cdot 3) + 2 \cdot \frac{1}{2}(16 \cdot 6) =$
    $48 + 60 + 96 = 204 \text{ cm}^2$

20. $2(4 \cdot 7) = 56 \text{ cm}^2$
    $2(4 \cdot 2) = 16 \text{ cm}^2$
    $2(2 \cdot 7) = 28 \text{ cm}^2$
    total surface area $= 100 \text{ cm}^2$ or
    $2(4 \cdot 7) + 2(4 \cdot 2) + 2(2 \cdot 7) =$
    $56 + 16 + 28 = 100 \text{ cm}^2$

page 240

1. $A = \pi r^2$ $A = l \cdot w; l = 2\pi r$
   $A = 2(3.14 \cdot 5^2)$ $A = (2 \cdot 3.14 \cdot 5) \cdot 6$
   $A = 2(3.14 \cdot 25)$ $A = 188.4 \text{ cm}^2$
   $A = 2(78.5)$
   $A = 157 \text{ cm}^2$
   total surface area $= 157 + 188.4 =$
   345.4 cm^2

2. $A = \pi r^2$ $A = l \cdot w; l = 2\pi r$
   $A = 2(3.14 \cdot 2^2)$ $A = (2 \cdot 3.14 \cdot 2) \cdot 10$
   $A = 2(3.14 \cdot 4)$ $A = 125.6 \text{ cm}^2$
   $A = 2(12.56)$
   $A = 25.12 \text{ cm}^2$
   total surface area $= 25.12 + 125.6 =$
   150.72 cm^2

3. $A = \pi r^2$ $A = l \cdot w; l = 2\pi r$
   $A = 2(3.14 \cdot 2^2)$ $A = (2 \cdot 3.14 \cdot 2) \cdot 6$
   $A = 2(3.14 \cdot 4)$ $A = 75.36 \text{ in.}^2$
   $A = 2(12.56)$
   $A = 25.12 \text{ in.}^2$
   total surface area $= 25.12 + 75.36 =$
   100.48 in.^2

4. $A = \pi r^2$ $A = l \cdot w; l = 2\pi r$
   $A = 2(3.14 \cdot 3^2)$ $A = (2 \cdot 3.14 \cdot 3) \cdot 7$
   $A = 2(3.14 \cdot 9)$ $A = 131.88 \text{ in.}^2$
   $A = 2(28.26)$
   $A = 56.52 \text{ in.}^2$
   total surface area $= 56.52 + 131.88 =$
   188.4 in.^2

5. $A = \pi r^2$ $A = l \cdot w; l = 2\pi r$
   $A = 2(3.14 \cdot 3^2)$ $A = (2 \cdot 3.14 \cdot 3) \cdot 4$
   $A = 2(3.14 \cdot 9)$ $A = 75.36 \text{ in.}^2$
   $A = 2(28.26)$
   $A = 56.52 \text{ in.}^2$
   total surface area $= 56.52 + 75.36 =$
   131.88 in.^2

6. $A = \pi r^2$ $A = l \cdot w; l = 2\pi r$
   $A = 2(3.14 \cdot 4^2)$ $A = (2 \cdot 3.14 \cdot 4) \cdot 5$
   $A = 2(3.14 \cdot 16)$ $A = 125.6 \text{ cm}^2$
   $A = 2(50.24)$
   $A = 100.48 \text{ cm}^2$
   total surface area $= 100.48 + 125.6 =$
   226.08 cm^2

7. $A = \pi r^2$ $A = l \cdot w; l = 2\pi r$
   $A = 2(3.14 \cdot 3^2)$ $A = (2 \cdot 3.14 \cdot 3) \cdot 6$
   $A = 2(3.14 \cdot 9)$ $A = 113.04 \text{ in.}^2$
   $A = 2(28.26)$
   $A = 56.52 \text{ in.}^2$
   total surface area $= 56.52 + 113.04 =$
   169.56 in.^2

8. $A = \pi r^2$ $A = l \cdot w; l = 2\pi r$
   $A = 2(3.14 \cdot 6^2)$ $A = (2 \cdot 3.14 \cdot 6) \cdot 5$
   $A = 2(3.14 \cdot 36)$ $A = 188.4 \text{ cm}^2$
   $A = 2(113.04)$
   $A = 226.08 \text{ cm}^2$
   total surface area $= 226.08 + 188.4 =$
   414.48 cm^2

page 241

21. Perimeter is the distance around a figure; area is the number of square units needed to cover a figure or a surface.

page 242

1. $A = 1 \text{ m} \cdot 6 \text{ m}$
   $P = 2(1 \text{ m}) + 2(6 \text{ m}) = 14 \text{ m}$
   or
   $A = 2 \text{ m} \cdot 3 \text{ m}$
   $P = 2(2 \text{ m}) + 2(3 \text{ m}) = 10 \text{ m}$

2. $A = 1 \text{ in.} \cdot 10 \text{ in.}$
   $P = 2(1 \text{ in.}) + 2(10 \text{ in.}) = 22 \text{ in.}$
   or
   $A = 2 \text{ in.} \cdot 5 \text{ in.}$
   $P = 2(2 \text{ in.}) + 2(5 \text{ in.}) = 14 \text{ in.}$

3. $A = 1 \text{ ft} \cdot 16 \text{ ft}$
   $P = 2(1 \text{ ft}) + 2(16 \text{ ft}) = 34 \text{ ft}$
   or
   $A = 2 \text{ ft} \cdot 8 \text{ ft}$
   $P = 2(2 \text{ ft}) + 2(8 \text{ ft}) = 20 \text{ ft}$
   or
   $A = 4 \text{ ft} \cdot 4 \text{ ft}$
   $P = 2(4 \text{ ft}) + 2(4 \text{ ft}) = 16 \text{ ft}$

Chapter 12

page 250

1. $V = (l \bullet w) \bullet h$
$V = (7 \bullet 4) \bullet 2$
$V = 28 \bullet 2$
$V = 56 \text{ units}^3$

2. $V = (l \bullet w) \bullet h$
$V = (5 \bullet 3) \bullet 3$
$V = 15 \bullet 3$
$V = 45 \text{ units}^3$

3. $V = (l \bullet w) \bullet h$
$V = (6 \bullet 5) \bullet 4$
$V = 30 \bullet 4$
$V = 120 \text{ units}^3$

4. $V = (l \bullet w) \bullet h$
$V = (4 \bullet 3) \bullet 2$
$V = 12 \bullet 2$
$V = 24 \text{ units}^3$

5. $V = (l \bullet w) \bullet h$
$V = (8 \bullet 5) \bullet 1$
$V = 40 \bullet 1$
$V = 40 \text{ units}^3$

6. $V = (l \bullet w) \bullet h$
$V = (7 \bullet 2) \bullet 3$
$V = 14 \bullet 3$
$V = 42 \text{ units}^3$

7. $V = (l \bullet w) \bullet h$
$V = (3 \text{ cm} \bullet 2 \text{ cm}) \bullet 3 \text{ cm}$
$V = 6 \text{ cm}^2 \bullet 3 \text{ cm}$
$V = 18 \text{ cm}^3$

8. $V = (l \bullet w) \bullet h$
$V = (4 \text{ cm} \bullet 1 \text{ cm}) \bullet 2 \text{ cm}$
$V = 4 \text{ cm}^2 \bullet 2 \text{ cm}$
$V = 8 \text{ cm}^3$

9. $V = (l \bullet w) \bullet h$
$V = (4 \text{ cm} \bullet 3 \text{ cm}) \bullet 4 \text{ cm}$
$V = 12 \text{ cm}^2 \bullet 4 \text{ cm}$
$V = 48 \text{ cm}^3$

page 251

10. $V = (l \bullet w) \bullet h$
$V = (4 \text{ cm} \bullet 3 \text{ cm}) \bullet 9 \text{ cm}$
$V = 12 \text{ cm}^2 \bullet 9 \text{ cm}$
$V = 108 \text{ cm}^3$

11. $V = (l \bullet w) \bullet h$
$V = (5.2 \text{ m} \bullet 2.4 \text{ m}) \bullet 3.5 \text{ m}$
$V = 12.48 \text{ m}^2 \bullet 3.5 \text{ m}$
$V = 43.68 \text{ m}^3 \approx 43.7 \text{ m}^3$

12. $V = (l \bullet w) \bullet h$
$V = (\frac{3}{4} \text{ in.} \bullet \frac{1}{2} \text{ in.}) \bullet 4 \text{ in.}$
$V = \frac{3}{8} \text{ in.}^2 \bullet 4 \text{ in.}$
$V = \frac{3}{2} \text{ in.}^3 = 1\frac{1}{2} \text{ in.}^3$

13. $V = (l \bullet w) \bullet h$
$V = (8 \text{ m} \bullet 4 \text{ m}) \bullet 7 \text{ m}$
$V = 32 \text{ m}^2 \bullet 7 \text{ m}$
$V = 224 \text{ m}^3$

14. $V = (l \bullet w) \bullet h$
$V = (4.2 \text{ cm} \bullet 3.5 \text{ cm}) \bullet 6 \text{ cm}$
$V = 14.7 \text{ cm}^2 \bullet 6 \text{ cm}$
$V = 88.2 \text{ cm}^3$

15. $V = (l \bullet w) \bullet h$
$V = (8 \text{ in.} \bullet 2 \text{ in.}) \bullet 2 \text{ in.}$
$V = 16 \text{ in.}^2 \bullet 2 \text{ in.}$
$V = 32 \text{ in.}^3$

16. $V = (l \bullet w) \bullet h$
$V = (5 \text{ in.} \bullet 7 \text{ in.}) \bullet 2 \text{ in.}$
$V = 35 \text{ in.}^2 \bullet 2 \text{ in.}$
$V = 70 \text{ in.}^3$

17. $V = (l \bullet w) \bullet h$
$V = (12 \text{ cm} \bullet 8 \text{ cm}) \bullet 10 \text{ cm}$
$V = 96 \text{ cm}^2 \bullet 10 \text{ cm}$
$V = 960 \text{ cm}^3$

18. $V = (l \bullet w) \bullet h$
$V = (\frac{3}{8} \text{ in.} \bullet \frac{2}{3} \text{ in.}) \bullet \frac{1}{2} \text{ in.}$
$V = \frac{1}{4} \text{ in.}^2 \bullet \frac{1}{2} \text{ in.}$
$V = \frac{1}{8} \text{ in.}^3$

19. $V = Bh$
$V = 15 \text{ ft}^2 \bullet 2 \text{ ft}$
$V = 30 \text{ ft}^3$

20. $V = Bh$
$V = 25 \text{ m}^2 \bullet 8 \text{ m}$
$V = 200 \text{ m}^3$

21. $V = Bh$
$V = 46 \text{ cm}^2 \bullet 5.2 \text{ cm}$
$V = 239.2 \text{ cm}^3$

22. $V = (l \bullet w) \bullet h$
$V = (24 \text{ in.} \bullet 18 \text{ in.}) \bullet 12 \text{ in.}$
$V = 432 \text{ in.}^2 \bullet 12 \text{ in.}$
$V = 5,184 \text{ in.}^3$

23. $S = 2(24 \text{ in.} \bullet 18 \text{ in.}) + 2(24 \text{ in.} \bullet 12 \text{ in.}) + 2(18 \text{ in.} \bullet 12 \text{ in.}) = 2(432 \text{ in.}^2) + 2(288 \text{ in.}^2) + 2(216 \text{ in.}^2) = 864 \text{ in.}^2 + 576 \text{ in.}^2 + 432 \text{ in.}^2 = 1,872 \text{ in.}^2$

page 252

1. $V = s^3$
$V = 16 \text{ cm}^2 \bullet 4 \text{ cm}$
$V = 64 \text{ cm}^3$

2. $V = s^3$
$V = (10 \text{ in.})^3$
$V = 1,000 \text{ in.}^3$

3. $V = (s \bullet s) \bullet s$
$V = (3.2 \text{ m} \bullet 3.2 \text{ m}) \bullet 3.2 \text{ m}$
$V = 10.24 \text{ m}^2 \bullet 3.2 \text{ m}$
$V = 32.768 \text{ m}^3 \approx 32.8 \text{ m}^3$

4. $V = s^3$
$V = 36 \text{ ft}^2 \bullet 6 \text{ ft}$
$V = 216 \text{ ft}^3$

5. $V = (s \bullet s) \bullet s$
$V = (1\frac{1}{2} \text{ yd} \bullet 1\frac{1}{2} \text{ yd}) \bullet 1\frac{1}{2} \text{ yd}$
$V = (\frac{3}{2} \text{ yd} \bullet \frac{3}{2} \text{ yd}) \bullet \frac{3}{2} \text{ yd}$
$V = \frac{9}{4} \text{ yd}^2 \bullet \frac{3}{2} \text{ yd}$
$V = \frac{27}{8} \text{ yd}^3 = 3\frac{3}{8} \text{ yd}^3$

6. $V = (s \bullet s) \bullet s$
$V = (4.5 \text{ m} \bullet 4.5 \text{ m}) \bullet 4.5 \text{ m}$
$V = 20.25 \text{ m}^2 \bullet 4.5 \text{ m}$
$V = 91.125 \text{ m}^3 \approx 91.1 \text{ m}^3$

11. $V = (l \bullet w) \bullet h$
$V = (2 \text{ cm} \bullet 2 \text{ cm}) \bullet 2 \text{ cm}$
$V = 4 \text{ cm}^2 \bullet 2 \text{ cm}$
$V = 8 \text{ cm}^3$
or
$V = s^3$
$V = (2 \text{ cm})^3$
$V = 8 \text{ cm}^3$

12. $V = (l \bullet w) \bullet h$
$V = (3 \text{ cm} \bullet 3 \text{ cm}) \bullet 3 \text{ cm}$
$V = 9 \text{ cm}^2 \bullet 3 \text{ cm}$
$V = 27 \text{ cm}^3$
or
$V = s^3$
$V = (3 \text{ cm})^3$
$V = 27 \text{ cm}^3$

13. $V = (l \bullet w) \bullet h$
$V = (4 \text{ cm} \bullet 3 \text{ cm}) \bullet 5 \text{ cm}$
$V = 12 \text{ cm}^2 \bullet 5 \text{ cm}$
$V = 60 \text{ cm}^3$

page 253

14. $V = (l \bullet w) \bullet h$
$24 \text{ ft}^3 = (2 \text{ ft} \bullet 3 \text{ ft}) \bullet h$
$\frac{24 \text{ ft}^3}{6 \text{ ft}^2} = \frac{6 \text{ ft}^2 \bullet h}{6 \text{ ft}^2}$
$h = 4 \text{ ft}$

15. $V = (l \cdot w) \cdot h$
$248 \text{ m}^3 = (l \cdot 4 \text{ m}) \cdot 6.2 \text{ m}$
$248 \text{ m}^3 = l(4 \text{ m} \cdot 6.2 \text{ m})$
$\frac{248 \text{ m}^3}{24.8 \text{ m}^2} = \frac{l \cdot 24.8 \text{ m}^2}{24.8 \text{ m}^2}$
$l = 10 \text{ m}$

16. $V = (l \cdot w) \cdot h$
$144 \text{ cm}^3 = (6 \text{ cm} \cdot w) \cdot 8 \text{ cm}$
$144 \text{ cm}^3 = w(6 \text{ cm} \cdot 8 \text{ cm})$
$\frac{144 \text{ cm}^3}{48 \text{ cm}^2} = \frac{w \cdot 48 \text{ cm}^2}{48 \text{ cm}^2}$
$w = 3 \text{ cm}$

17. $V = (l \cdot w) \cdot h$
$75 \text{ in.}^3 = (5 \text{ in.} \cdot 3 \text{ in.}) \cdot h$
$\frac{75 \text{ in.}^3}{15 \text{ in.}^2} = \frac{15 \text{ in.}^2 \cdot h}{15 \text{ in.}^2}$
$h = 5 \text{ in.}$

18. $V = (l \cdot w) \cdot h$
$48 \text{ cm}^3 = (2 \text{ cm} \cdot w) \cdot 4 \text{ cm}$
$48 \text{ cm}^3 = w(2 \text{ cm} \cdot 4 \text{ cm})$
$\frac{48 \text{ cm}^3}{8 \text{ cm}^2} = \frac{w \cdot 8 \text{ cm}^2}{8 \text{ cm}^2}$
$w = 6 \text{ cm}$

19. $V = Bh$
$90 \text{ in.}^3 = 6 \text{ in.}^2 \cdot h$
$\frac{90 \text{ in.}^3}{6 \text{ in.}^2} = \frac{6 \text{ in.}^2 \cdot h}{6 \text{ in.}^2}$
$h = 15 \text{ in.}$

20. $V = Bh$
$51.3 \text{ cm}^3 = 5.7 \text{ cm}^2 \cdot h$
$\frac{51.3 \text{ cm}^3}{5.7 \text{ cm}^2} = \frac{5.7 \text{ cm}^2 \cdot h}{5.7 \text{ cm}^2}$
$h = 9 \text{ cm}$

25. white and orange box:
$S = 2(3 \text{ in.} \cdot 4 \text{ in.}) + 2(1 \text{ in.} \cdot 4 \text{ in.}) + 2(1 \text{ in.} \cdot 3 \text{ in.}) = 24 \text{ in.}^2 + 8 \text{ in.}^2 + 6 \text{ in.}^2 = 38 \text{ in.}^2$
$V = (l \cdot w) \cdot h$
$V = (3 \text{ in.} \cdot 1 \text{ in.}) \cdot 4 \text{ in.}$
$V = 3 \text{ in.}^2 \cdot 4 \text{ in.}$
$V = 12 \text{ in.}^3$

green and red box:
$S = 6(8 \text{ cm} \cdot 8 \text{ cm}) = 6 \cdot 64 \text{ cm}^2 = 384 \text{ cm}^2$
$V = s^3$
$V = 64 \text{ cm}^2 \cdot 8 \text{ cm}$
$V = 512 \text{ cm}^3$

gold box:
$S = 2(3 \text{ in.} \cdot 10 \text{ in.}) + 2(1\frac{1}{2} \text{ in.} \cdot 10 \text{ in.}) + 2(3 \text{ in.} \cdot 1\frac{1}{2} \text{ in.}) = 60 \text{ in.}^2 + 30 \text{ in.}^2 + 9 \text{ in.}^2 = 99 \text{ in.}^2$
$V = (l \cdot w) \cdot h$
$V = (3 \text{ in.} \cdot 1\frac{1}{2} \text{ in.}) \cdot 10 \text{ in.}$
$V = (3 \text{ in.} \cdot \frac{3}{2} \text{ in.}) \cdot 10 \text{ in.}$
$V = \frac{9}{2} \text{ in.}^2 \cdot 10 \text{ in.}$
$V = 45 \text{ in.}^3$

page 254

1. $V = Bh$ or $(\frac{1}{2}bh_1)h_2$
$V = (\frac{1}{2} \cdot 4 \text{ cm} \cdot 3 \text{ cm}) \cdot 10 \text{ cm}$
$V = 6 \text{ cm}^2 \cdot 10 \text{ cm}$
$V = 60 \text{ cm}^3$

2. $V = Bh$ or $(\frac{1}{2}bh_1)h_2$
$V = (\frac{1}{2} \cdot 6 \text{ cm} \cdot 2 \text{ cm}) \cdot 8 \text{ cm}$
$V = 6 \text{ cm}^2 \cdot 8 \text{ cm}$
$V = 48 \text{ cm}^3$

3. $V = Bh$ or $(\frac{1}{2}bh_1)h_2$
$V = (\frac{1}{2} \cdot 4 \text{ cm} \cdot 2 \text{ cm}) \cdot 7 \text{ cm}$
$V = 4 \text{ cm}^2 \cdot 7 \text{ cm}$
$V = 28 \text{ cm}^3$

page 255

4. $V = Bh$ or $(\pi r^2)h$
$V = 3.14 \cdot (3 \text{ cm})^2 \cdot 9 \text{ cm}$
$V = (3.14 \cdot 9 \text{ cm}^2) \cdot 9 \text{ cm}$
$V = 28.26 \text{ cm}^2 \cdot 9 \text{ cm}$
$V = 254.34 \text{ cm}^3 \approx 254.3 \text{ cm}^3$

5. $V = Bh$ or $(\pi r^2)h$
$V = 3.14 \cdot (5 \text{ cm})^2 \cdot 6 \text{ cm}$
$V = (3.14 \cdot 25 \text{ cm}^2) \cdot 6 \text{ cm}$
$V = 78.5 \text{ cm}^2 \cdot 6 \text{ cm}$
$V = 471 \text{ cm}^3$

6. $V = Bh$ or $(\pi r^2)h$
$V = 3.14 \cdot (4 \text{ cm})^2 \cdot 10 \text{ cm}$
$V = (3.14 \cdot 16 \text{ cm}^2) \cdot 10 \text{ cm}$
$V = 50.24 \text{ cm}^2 \cdot 10 \text{ cm}$
$V = 502.4 \text{ cm}^3$

18. $V = (l \cdot w) \cdot h$
$36 \text{ ft}^3 = (2 \text{ ft} \cdot 3 \text{ ft}) \cdot h$
$\frac{36 \text{ ft}^3}{6 \text{ ft}^2} = \frac{6 \text{ ft}^2 \cdot h}{6 \text{ ft}^2}$
$h = 6 \text{ ft}$

19. $V = (l \cdot w) \cdot h$
$162 \text{ m}^3 = (l \cdot 4 \text{ m}) \cdot 4.5 \text{ m}$
$162 \text{ m}^3 = l(4 \text{ m} \cdot 4.5 \text{ m})$
$\frac{162 \text{ m}^3}{18 \text{ m}^2} = \frac{l \cdot 18 \text{ m}^2}{18 \text{ m}^2}$
$l = 9 \text{ m}$

20. $V = (l \cdot w) \cdot h$
$240 \text{ cm}^3 = (5 \text{ cm} \cdot w) \cdot 12 \text{ cm}$
$240 \text{ cm}^3 = w(5 \text{ cm} \cdot 12 \text{ cm})$
$\frac{240 \text{ cm}^3}{60 \text{ cm}^2} = \frac{w \cdot 60 \text{ cm}^2}{60 \text{ cm}^2}$
$w = 4 \text{ cm}$

24. $V = Bh$ or $(\pi r^2)h$
$V = 3.14 \cdot (9 \text{ ft})^2 \cdot 4 \text{ ft}$
$V = (3.14 \cdot 81 \text{ ft}^2) \cdot 4 \text{ ft}$
$V = 254.34 \text{ ft}^2 \cdot 4 \text{ ft}$
$V = 1,017.36 \text{ ft}^3$

25. Students may solve for the exact answer or they may estimate.
$1,017.36 \cdot 7.5 \text{ gal} = 7,630.2 \text{ gal}$ or
$1,017 \cdot 8 \text{ gal} = 8,136 \text{ gal}$

page 256

1. top & bottom: $2(7 \text{ cm} \cdot 1 \text{ cm}) = 14 \text{ cm}^2$
front & back: $2(7 \text{ cm} \cdot 1 \text{ cm}) = 14 \text{ cm}^2$
left & right sides: $2(1 \text{ cm} \cdot 1 \text{ cm}) = 2 \text{ cm}^2$
total surface area $= 30 \text{ cm}^2$

2. top & bottom: $2(10 \text{ cm} \cdot 1 \text{ cm}) = 20 \text{ cm}^2$
front & back: $2(10 \text{ cm} \cdot 1 \text{ cm}) = 20 \text{ cm}^2$
left & right sides: $2(1 \text{ cm} \cdot 1 \text{ cm}) = 2 \text{ cm}^2$
total surface area $= 42 \text{ cm}^2$

top & bottom: $2(5 \text{ cm} \cdot 2 \text{ cm}) = 20 \text{ cm}^2$
front & back: $2(5 \text{ cm} \cdot 1 \text{ cm}) = 10 \text{ cm}^2$
left & right sides: $2(2 \text{ cm} \cdot 1 \text{ cm}) = 4 \text{ cm}^2$
total surface area $= 34 \text{ cm}^2$

3. top & bottom: $2(12 \text{ cm} \cdot 1 \text{ cm}) = 24 \text{ cm}^2$
front & back: $2(12 \text{ cm} \cdot 1 \text{ cm}) = 24 \text{ cm}^2$
left & right sides: $2(1 \text{ cm} \cdot 1 \text{ cm}) = 2 \text{ cm}^2$
total surface area $= 50 \text{ cm}^2$

top & bottom: $2(6 \text{ cm} \cdot 2 \text{ cm}) = 24 \text{ cm}^2$
front & back: $2(6 \text{ cm} \cdot 1 \text{ cm}) = 12 \text{ cm}^2$
left & right sides: $2(2 \text{ cm} \cdot 1 \text{ cm}) = 4 \text{ cm}^2$
total surface area $= 40 \text{ cm}^2$

top & bottom: $2(4 \text{ cm} \cdot 3 \text{ cm}) = 24 \text{ cm}^2$
front & back: $2(4 \text{ cm} \cdot 1 \text{ cm}) = 8 \text{ cm}^2$
left & right sides: $2(3 \text{ cm} \cdot 1 \text{ cm}) = 6 \text{ cm}^2$
total surface area $= 38 \text{ cm}^2$

top & bottom: $2(3 \text{ cm} \cdot 2 \text{ cm}) = 12 \text{ cm}^2$
front & back: $2(3 \text{ cm} \cdot 2 \text{ cm}) = 12 \text{ cm}^2$
left & right sides: $2(2 \text{ cm} \cdot 2 \text{ cm}) = 8 \text{ cm}^2$
total surface area $= 32 \text{ cm}^2$

4. top & bottom: $2(16 \text{ cm} \cdot 1 \text{ cm}) = 32 \text{ cm}^2$
front & back: $2(16 \text{ cm} \cdot 1 \text{ cm}) = 32 \text{ cm}^2$
left & right sides: $2(1 \text{ cm} \cdot 1 \text{ cm}) = 2 \text{ cm}^2$
total surface area $= 66 \text{ cm}^2$

top & bottom: $2(8 \text{ cm} \cdot 2 \text{ cm}) = 32 \text{ cm}^2$
front & back: $2(8 \text{ cm} \cdot 1 \text{ cm}) = 16 \text{ cm}^2$
left & right sides: $2(2 \text{ cm} \cdot 1 \text{ cm}) = 4 \text{ cm}^2$
total surface area $= 52 \text{ cm}^2$

top & bottom: $2(4 \text{ cm} \cdot 2 \text{ cm}) = 16 \text{ cm}^2$
front & back: $2(4 \text{ cm} \cdot 2 \text{ cm}) = 16 \text{ cm}^2$
left & right sides: $2(2 \text{ cm} \cdot 2 \text{ cm}) = 8 \text{ cm}^2$
total surface area $= 40 \text{ cm}^2$

5. top & bottom: 2(24 cm • 1 cm) = 48 cm²
front & back: 2(24 cm • 1 cm) = 48 cm²
left & right sides: 2(1 cm • 1 cm) = 2 cm²
total surface area = 98 cm²

top & bottom: 2(12 cm • 2 cm) = 48 cm²
front & back: 2(12 cm • 1 cm) = 24 cm²
left & right sides: 2(2 cm • 1 cm) = 4 cm²
total surface area = 76 cm²

top & bottom: 2(8 cm • 3 cm) = 48 cm²
front & back: 2(8 cm • 1 cm) = 16 cm²
left & right sides: 2(3 cm • 1 cm) = 6 cm²
total surface area = 70 cm²

top & bottom: 2(6 cm • 4 cm) = 48 cm²
front & back: 2(6 cm • 1 cm) = 12 cm²
left & right sides: 2(4 cm • 1 cm) = 8 cm²
total surface area = 68 cm²

top & bottom: 2(6 cm • 2 cm) = 24 cm²
front & back: 2(6 cm • 2 cm) = 24 cm²
left & right sides: 2(2 cm • 2 cm) = 8 cm²
total surface area = 56 cm²

top & bottom: 2(4 cm • 3 cm) = 24 cm²
front & back: 2(4 cm • 2 cm) = 16 cm²
left & right sides: 2(3 cm • 2 cm) = 12 cm²
total surface area = 52 cm²

page 257

6. Lateral Surface Area
$LS = (2\pi r)h$
$LS = 2(3.14 • 3 \text{ cm}) • 4 \text{ cm}$
$LS = 2(9.42 \text{ cm}) • 4 \text{ cm}$
$LS = 18.84 \text{ cm} • 4 \text{ cm}$
$LS = 75.36 \text{ cm}^2 \approx 75 \text{ cm}^2$

$LS = (2\pi r)h$
$LS = 2(3.14 • 1 \text{ cm}) • 12 \text{ cm}$
$LS = 2(3.14 \text{ cm}) • 12 \text{ cm}$
$LS = 6.28 \text{ cm} • 12 \text{ cm}$
$LS = 75.36 \text{ cm}^2 \approx 75 \text{ cm}^2$

Volume
$V = (\pi r^2)h$
$V = 3.14 • (3 \text{ cm})^2 • 4 \text{ cm}$
$V = (3.14 • 9 \text{ cm}^2) • 4 \text{ cm}$
$V = 28.26 \text{ cm}^2 • 4 \text{ cm}$
$V = 113.04 \text{ cm}^3 \approx 113 \text{ cm}^3$

$V = (\pi r^2)h$
$V = 3.14 • (1 \text{ cm})^2 • 12 \text{ cm}$
$V = (3.14 • 1 \text{ cm}^2) • 12 \text{ cm}$
$V = 3.14 \text{ cm}^2 • 12 \text{ cm}$
$V = 37.68 \text{ cm}^3 \approx 38 \text{ cm}^3$

7. Lateral Surface Area
$LS = (2\pi r)h$
$LS = 2(3.14 • 1.5 \text{ cm}) • 14 \text{ cm}$
$LS = 2(4.71 \text{ cm}) • 14 \text{ cm}$
$LS = 9.42 \text{ cm} • 14 \text{ cm}$
$LS = 131.88 \text{ cm}^2 \approx 132 \text{ cm}^2$

$LS = (2\pi r)h$
$LS = 2(3.14 • 1 \text{ cm}) • 21 \text{ cm}$
$LS = 2(3.14 \text{ cm}) • 21 \text{ cm}$
$LS = 6.28 \text{ cm} • 21 \text{ cm}$
$LS = 131.88 \text{ cm}^2 \approx 132 \text{ cm}^2$

Volume
$V = (\pi r^2)h$
$V = 3.14 • (1.5 \text{ cm})^2 • 14 \text{ cm}$
$V = (3.14 • 2.25 \text{ cm}^2) • 14 \text{ cm}$
$V = 7.065 \text{ cm}^2 • 14 \text{ cm}$
$V = 98.91 \text{ cm}^3 \approx 99 \text{ cm}^3$

$V = (\pi r^2)h$
$V = 3.14 • (1 \text{ cm})^2 • 21 \text{ cm}$
$V = (3.14 • 1 \text{ cm}^2) • 21 \text{ cm}$
$V = 3.14 \text{ cm}^2 • 21 \text{ cm}$
$V = 65.94 \text{ cm}^3 \approx 66 \text{ cm}^3$

8. Lateral Surface Area
$LS = 2(6 \text{ cm} • 2 \text{ cm}) = 24 \text{ cm}^2$
$LS = 2(4 \text{ cm} • 2 \text{ cm}) = 16 \text{ cm}^2$
$LS = 40 \text{ cm}^2$

$LS = 2(5 \text{ cm} • 2 \text{ cm}) = 20 \text{ cm}^2$
$LS = 2(5 \text{ cm} • 2 \text{ cm}) = 20 \text{ cm}^2$
$LS = 40 \text{ cm}^2$

Volume
$V = (l • w) • h$
$V = (6 \text{ cm} • 4 \text{ cm}) • 2 \text{ cm}$
$V = 24 \text{ cm}^2 • 2 \text{ cm}$
$V = 48 \text{ cm}^3$

$V = (l • w) • h$
$V = (5 \text{ cm} • 5 \text{ cm}) • 2 \text{ cm}$
$V = 25 \text{ cm}^2 • 2 \text{ cm}$
$V = 50 \text{ cm}^3$

9. Students may refer to the chart on page 256 to find the least amount of surface area for 16 and 24 blocks. From that chart they should be able to follow the pattern to find the least surface area for 36 blocks.

surface area for containers holding 36 blocks:
36 in. • 1 in. • 1 in.; 146 in.²
18 in. • 2 in. • 1 in.; 112 in.²
12 in. • 3 in. • 1 in.; 102 in.²
9 in. • 4 in. • 1 in.; 98 in.²
6 in. • 6 in. • 1 in.; 96 in.²
9 in. • 2 in. • 2 in.; 80 in.²
6 in. • 3 in. • 2 in.; 72 in.²
4 in. • 3 in. • 3 in.; 66 in.²; the least surface area

page 258

5. $V = (l • w) • h$
$V = (3 \text{ cm} • 3 \text{ cm}) • 3 \text{ cm}$
$V = 9 \text{ cm}^2 • 3 \text{ cm}$
$V = 27 \text{ cm}^3$

6. $V = (l • w) • h$
$V = (4 \text{ cm} • 1 \text{ cm}) • 5 \text{ cm}$
$V = 4 \text{ cm}^2 • 5 \text{ cm}$
$V = 20 \text{ cm}^3$

7. $V = Bh$
$V = 6 \text{ cm}^2 • 2 \text{ cm}$
$V = 12 \text{ cm}^3$

8. $V = (l • w) • h$
$V = (3 \text{ cm} • 2 \text{ cm}) • 5 \text{ cm}$
$V = 6 \text{ cm}^2 • 5 \text{ cm}$
$V = 30 \text{ cm}^3$

9. $V = (l • w) • h$
$V = (4.2 \text{ m} • 1.5 \text{ m}) • 2.8 \text{ m}$
$V = 6.3 \text{ m}^2 • 2.8 \text{ m}$
$V = 17.64 \text{ m}^3 \approx 17.6 \text{ m}^3$

10. $V = (l • w) • h$
$V = (\frac{2}{3} \text{ m} • \frac{1}{2} \text{ m}) • 6 \text{ m}$
$V = \frac{1}{3} \text{ m}^2 • 6 \text{ m}$
$V = 2 \text{ m}^3$

11. $V = s^3$
$V = (4 \text{ ft})^3$
$V = 4 \text{ ft} • 4 \text{ ft} • 4 \text{ ft}$
$V = 64 \text{ ft}^3$

12. $V = s^3$
$V = (2 \text{ yd})^3$
$V = 2 \text{ yd} • 2 \text{ yd} • 2 \text{ yd}$
$V = 8 \text{ yd}^3$

13. $V = s^3$
$V = (3.2 \text{ m})^3$
$V = 3.2 \text{ m} • 3.2 \text{ m} • 3.2 \text{ m}$
$V = 32.768 \text{ m}^3 \approx 32.8 \text{ m}^3$

14. $V = (\frac{1}{2} bh_1)h_2$
$V = \frac{1}{2}(5 \text{ cm} • 2 \text{ cm}) • 3 \text{ cm}$
$V = \frac{1}{2}(10 \text{ cm}^2) • 3 \text{ cm}$
$V = 5 \text{ cm}^2 • 3 \text{ cm}$
$V = 15 \text{ cm}^3$

15. $V = (\frac{1}{2} bh_1)h_2$
$V = \frac{1}{2}(4 \text{ cm} • 6 \text{ cm}) • 5 \text{ cm}$
$V = \frac{1}{2}(24 \text{ cm}^2) • 5 \text{ cm}$
$V = 12 \text{ cm}^2 • 5 \text{ cm}$
$V = 60 \text{ cm}^3$

16. $V = (\frac{1}{2}bh_1)h_2$
$V = \frac{1}{2}(6 \text{ in.} \bullet 4 \text{ in.}) \bullet 2 \text{ in.}$
$V = \frac{1}{2}(24 \text{ in.}^2) \bullet 2 \text{ in.}$
$V = 12 \text{ in.}^2 \bullet 2 \text{ in.}$
$V = 24 \text{ in.}^3$

17. $V = (\pi r^2)h$
$V = 3.14 \bullet (2 \text{ cm})^2 \bullet 7 \text{ cm}$
$V = (3.14 \bullet 4 \text{ cm}^2) \bullet 7 \text{ cm}$
$V = 12.56 \text{ cm}^2 \bullet 7 \text{ cm}$
$V = 87.92 \text{ cm}^3 \approx 87.9 \text{ cm}^3$

18. $V = (\pi r^2)h$
$V = 3.14 \bullet (6 \text{ cm})^2 \bullet 5 \text{ cm}$
$V = (3.14 \bullet 36 \text{ cm}^2) \bullet 5 \text{ cm}$
$V = 113.04 \text{ cm}^2 \bullet 5 \text{ cm}$
$V = 565.2 \text{ cm}^3$

19. $V = (\pi r^2)h$
$V = 3.14 \bullet (3 \text{ in.})^2 \bullet 6 \text{ in.}$
$V = (3.14 \bullet 9 \text{ in.}^2) \bullet 6 \text{ in.}$
$V = 28.26 \text{ in.}^2 \bullet 6 \text{ in.}$
$V = 169.56 \text{ in.}^3 \approx 169.6 \text{ in.}^3$

page 259

24. $V = (l \bullet w) \bullet h$
$36 \text{ ft}^3 = (3 \text{ ft} \bullet 3 \text{ ft}) \bullet h$
$\frac{36 \text{ ft}^3}{9 \text{ ft}^2} = \frac{9 \text{ ft}^2 \bullet h}{9 \text{ ft}^2}$
$h = 4 \text{ ft}$

25. $V = (l \bullet w) \bullet h$
$70 \text{ m}^3 = (l \bullet 2 \text{ m}) \bullet 3.5 \text{ m}$
$70 \text{ m}^3 = l(2 \text{ m} \bullet 3.5 \text{ m})$
$\frac{70 \text{ m}^3}{7 \text{ m}^2} = \frac{l \bullet 7 \text{ m}^2}{7 \text{ m}^2}$
$l = 10 \text{ m}$

26. $V = (l \bullet w) \bullet h$
$216 \text{ cm}^3 = (6 \text{ cm} \bullet w) \bullet 9 \text{ cm}$
$216 \text{ cm}^3 = w(6 \text{ cm} \bullet 9 \text{ cm})$
$\frac{216 \text{ cm}^3}{54 \text{ cm}^2} = \frac{w \bullet 54 \text{ cm}^2}{54 \text{ cm}^2}$
$w = 4 \text{ cm}$

27. $V = (l \bullet w) \bullet h$
$80 \text{ in.}^3 = (5 \text{ in.} \bullet 2 \text{ in.}) \bullet h$
$\frac{80 \text{ in.}^3}{10 \text{ in.}^2} = \frac{10 \text{ in.}^2 \bullet h}{10 \text{ in.}^2}$
$h = 8 \text{ in.}$

28. $V = (l \bullet w) \bullet h$
$144 \text{ cm}^3 = (6 \text{ cm} \bullet w) \bullet 4 \text{ cm}$
$144 \text{ cm}^3 = w(6 \text{ cm} \bullet 4 \text{ cm})$
$\frac{144 \text{ cm}^3}{24 \text{ cm}^2} = \frac{w \bullet 24 \text{ cm}^2}{24 \text{ cm}^2}$
$w = 6 \text{ cm}$

29. $V = (l \bullet w) \bullet h$
$V = (9 \text{ ft} \bullet 4 \text{ ft}) \bullet 1.5 \text{ ft}$
$V = 36 \text{ ft}^2 \bullet 1.5 \text{ ft}$
$V = 54 \text{ ft}^3$

30. $V = (l \bullet w) \bullet h$
$V = (4 \text{ ft} \bullet 4 \text{ ft}) \bullet 1.5 \text{ ft}$
$V = 16 \text{ ft}^2 \bullet 1.5 \text{ ft}$
$V = 24 \text{ ft}^3$

31. $V = (\pi r^2)h$
$V = 3.14 \bullet (10 \text{ in.})^2 \bullet 37 \text{ in.}$
$V = (3.14 \bullet 100 \text{ in.}^2) \bullet 37 \text{ in.}$
$V = 314 \text{ in.}^2 \bullet 37 \text{ in.}$
$V = 11{,}618 \text{ in.}^3$

Chapter 13
page 267

24. $\frac{480 \text{ mi}}{16 \text{ gal}} = \frac{n}{1 \text{ gal}}$
$n = 480 \text{ mi} \div 16 \text{ gal}$
$n = 30 \text{ mi/gal}$

25. $\frac{\$48}{8 \text{ hr}} = \frac{n}{1 \text{ hr}}$
$n = \$48 \div 8 \text{ hr}$
$n = \$6/\text{hr}$

26. $\frac{2{,}250 \text{ mi}}{3 \text{ days}} = \frac{n}{1 \text{ day}}$
$n = 2{,}250 \text{ mi} \div 3 \text{ days}$
$n = 750 \text{ mi/day}$

27. $\frac{30 \text{ pg}}{60 \text{ min}} = \frac{n}{1 \text{ min}}$
$n = 30 \text{ pg} \div 60 \text{ min}$
$n = \frac{30 \text{ pg}}{60 \text{ min}} = \frac{1}{2} \text{ pg/min}$

28. $\frac{\$3.16}{4 \text{ lb}} = \frac{n}{1 \text{ lb}}$
$n = \$3.16 \div 4 \text{ lb}$
$n = \$0.79/\text{lb}$

29. $\frac{165 \text{ words}}{3 \text{ min}} = \frac{n}{1 \text{ min}}$
$n = 165 \text{ words} \div 3 \text{ min}$
$n = 55 \text{ words/min}$

30. $\frac{\$3.15}{1 \text{ gal}} = \frac{n}{5 \text{ gal}}$
$n = 5 \times \$3.15$
$n = \$15.75$

31. $\frac{\$7}{1 \text{ hr}} = \frac{n}{4 \text{ hr}}$
$n = 4 \times \$7$
$n = \$28$

32. $\frac{60 \text{ mi}}{1 \text{ hr}} = \frac{n}{6 \text{ hr}}$
$n = 6 \times 60 \text{ mi}$
$n = 360 \text{ mi}$

33. $\frac{230 \text{ mi}}{1 \text{ day}} = \frac{n}{9 \text{ days}}$
$n = 9 \times 230 \text{ mi}$
$n = 2{,}070 \text{ mi}$

34. $\frac{7 \text{ km}}{1 \text{ hr}} = \frac{n}{0.5 \text{ hr}}$
$n = 0.5 \times 7 \text{ km}$
$n = 3.5 \text{ km}$

35. $\frac{\$10}{1 \text{ hr}} = \frac{n}{3.5 \text{ hr}}$
$n = 3.5 \times \$10$
$n = \$35$

36. $\frac{180 \text{ mi}}{2 \text{ hr}} = \frac{n}{6 \text{ hr}}$
$n = 3 \times 180 \text{ mi}$
$n = 540 \text{ mi}$

37. $\frac{420 \text{ mi}}{2 \text{ hr}} = \frac{n}{4 \text{ hr}}$
$n = 2 \times 420 \text{ mi}$
$n = 840 \text{ mi}$

38. Note: Finding the unit rate first makes this problem easier to solve.
$\frac{300 \text{ mi}}{12 \text{ gal}} = \frac{1{,}000 \text{ mi}}{n}$
$\frac{300 \text{ mi}}{12 \text{ gal}} = \frac{25 \text{ mi}}{1 \text{ gal}}$
$\frac{25 \text{ mi}}{1 \text{ gal}} = \frac{1{,}000 \text{ mi}}{n}$
$n = 1{,}000 \div 25 = 40$
$n = 40 \times 1 \text{ gal}$
$n = 40 \text{ gal}$

page 268

10–12.

ft	5,280	10,560	15,840	21,120	26,400
mi	1	2	3	4	5

page 269

21. $4 \times 5 = 20$
$4 \times \$32.50 = \$130.00$

22. $2 \times 12 = 24$
$2 \times \$78 = \$156.00$

23. $9 \times 3 = 27$
$9 \times \$19.50 = \$175.50$

24. $2 \times 20 = 40$
$2 \times 560 \text{ mi} = 1{,}120 \text{ mi}$

25. $20 + 30 = 50$
$560 \text{ mi} + 840 \text{ mi} = 1{,}400 \text{ mi}$

26. $20 \div 5 = 4$
$560 \div 5 = 112 \text{ mi}$

29. 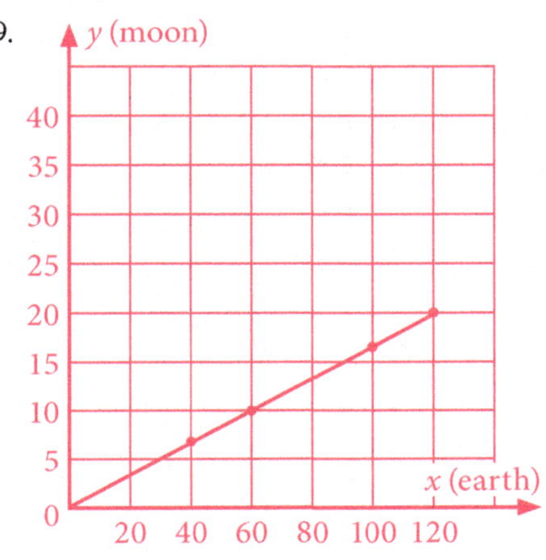

page 270

13.
$$\begin{array}{r} 0.5 \\ 4\overline{)2.0} \\ -2\,0 \\ \hline 0 \end{array} \qquad \begin{array}{r} 0.5 \\ 30\overline{)15.0} \\ -15\,0 \\ \hline 0 \end{array}$$
$$0.5 = 0.5$$

14.
$$\begin{array}{r} 0.3\,7\,5 \\ 8\overline{)3.0\,0\,0} \\ -2\,4 \\ \hline 6\,0 \\ -5\,6 \\ \hline 4\,0 \\ -4\,0 \\ \hline 0 \end{array} \qquad \begin{array}{r} 0.2\,5 \\ 16\overline{)4.0\,0} \\ -3\,2 \\ \hline 8\,0 \\ -8\,0 \\ \hline 0 \end{array}$$
$$0.375 \neq 0.25$$

15.
$$\begin{array}{r} 0.\overline{6} \\ 3\overline{)2.0\,0} \\ -1\,8 \\ \hline 2\,0 \end{array} \qquad \begin{array}{r} 0.6 \\ 5\overline{)3.0} \\ -3\,0 \\ \hline 0 \end{array}$$
$$0.\overline{6} \neq 0.6$$

16.
$$\begin{array}{r} 0.1\,2\,5 \\ 72\overline{)9.0\,0\,0} \\ -7\,2 \\ \hline 1\,8\,0 \\ -1\,4\,4 \\ \hline 3\,6\,0 \\ -3\,6\,0 \\ \hline 0 \end{array} \qquad \begin{array}{r} 0.1\,2\,5 \\ 48\overline{)6.0\,0\,0} \\ -4\,8 \\ \hline 1\,2\,0 \\ -9\,6 \\ \hline 2\,4\,0 \\ -2\,4\,0 \\ \hline 0 \end{array}$$
$$0.125 = 0.125$$

page 271

23. $\frac{\$6}{4 \text{ lb}} = \frac{\$15}{10 \text{ lb}}$
$\$6 \div 4 = \$1.50/\text{lb}$
$\$15 \div 10 = \$1.50/\text{lb}$

24. $\frac{55¢}{12 \text{ oz}} \neq \frac{95¢}{20 \text{ oz}}$
$\$0.55 \div 12 \approx \$0.045/\text{oz}$
$\$0.95 \div 20 \approx \$0.047/\text{oz}$

25. $\frac{\$2}{12 \text{ eggs}} = \frac{\$3}{18 \text{ eggs}}$
$\$2 \div 12 = \$0.1\overline{6}/\text{egg}$
$\$3 \div 18 = \$0.1\overline{6}/\text{egg}$

26. $\frac{\$1}{1 \text{ qt}} = \frac{\$4}{\cancel{1 \text{ gal}} \atop 4 \text{ qt}}$
$\$1 \div 1 = \$1/\text{qt}$
$\$4 \div 4 = \$1/\text{qt}$

27. $\frac{\$7}{5 \text{ bottles}} \neq \frac{\$8}{6 \text{ bottles}}$
$\$7 \div 5 = \$1.40/\text{bottle}$
$\$8 \div 6 \approx \$1.33/\text{bottle}$

28. $\frac{50¢}{5 \text{ oz}} \neq \frac{75¢}{8 \text{ oz}}$
$\$0.50 \div 5 = \$0.10/\text{oz}$
$\$0.75 \div 8 \approx \$0.09/\text{oz}$

29. $\frac{3 \text{ hr}}{2 \text{ driveways}} = \frac{n}{5 \text{ driveways}}$
$3 \text{ hr} \div 2 = 1.5 \text{ hr/driveway};$
$n = 5 \cdot 1.5$
$n = 7.5 \text{ hr}$

30. $\frac{3}{1} = \frac{n}{5}$
$5 \times 3 = n$
$n = 15 \text{ votes}$

31. $\frac{480 \text{ mi}}{3 \text{ hr}} = \frac{n}{5 \text{ hr}}$
$480 \text{ mi} \div 3 \text{ hr} = 160 \text{ mi/hr}$
$n = 5 \cdot 160 \text{ mi}$
$n = 800 \text{ mi}$

32. $\frac{24}{3} = \frac{n}{4}$
$24 \times 4 = 3n$
$\frac{96}{3} = \frac{3n}{3}$
$n = 32 \text{ oz}$

33. $\frac{\$15}{2 \text{ pizzas}} = \frac{n}{7 \text{ pizzas}}$
$\$15 \div 2 = \$7.50/\text{pizza}$
$n = 7 \cdot \$7.50$
$n = \$52.50$

34. $\frac{36}{120} = \frac{n}{10}$
$10 \cdot 36 = 120n$
$\frac{360}{120} = \frac{120n}{120}$
$n = 3 \text{ students}$

35. $\frac{30 \text{ hr}}{25 \text{ items}} = \frac{n}{60 \text{ items}}$
$30 \div 25 = 1.2 \text{ hr/item}$
$n = 60 \times 1.2$
$n = 72 \text{ hr}$

page 272

1. $\frac{10}{5} = \frac{70}{n}$ or $\frac{10}{70} = \frac{5}{n}$
$10n = 5 \cdot 70$
$\frac{10n}{10} = \frac{350}{10}$
$n = 35 \text{ cm}$

2. $\frac{8}{12} = \frac{6}{n}$ or $\frac{8}{6} = \frac{12}{n}$
$8n = 12 \cdot 6$
$\frac{8n}{8} = \frac{72}{8}$
$n = 9 \text{ m}$

3. $\frac{10}{n} = \frac{75}{60}$ or $\frac{10}{75} = \frac{n}{60}$
$60 \cdot 10 = 75n$
$\frac{600}{75} = \frac{75n}{75}$
$n = 8 \text{ cm}$

4. $\frac{12}{n} = \frac{9}{15}$ or $\frac{12}{9} = \frac{n}{15}$
$15 \cdot 12 = 9n$
$\frac{180}{9} = \frac{9n}{9}$
$n = 20 \text{ m}$

5. $\frac{15}{n} = \frac{30}{60}$ or $\frac{15}{30} = \frac{n}{60}$
$60 \cdot 15 = 30n$
$\frac{900}{30} = \frac{30n}{30}$
$n = 30 \text{ cm}$

6. $\frac{n}{60} = \frac{25}{125}$ or $\frac{n}{25} = \frac{60}{125}$
$125n = 60 \cdot 25$
$\frac{125n}{125} = \frac{1,500}{125}$
$n = 12 \text{ cm}$

page 273

7. $\frac{2}{4} = \frac{6}{n}$
$2n = 4 \cdot 6$
$\frac{2n}{2} = \frac{24}{2}$
$n = 12 \text{ in.}$

8. $\frac{126}{4.5} = \frac{n}{8}$
$8 \cdot 126 = 4.5n$
$\frac{1,008}{4.5} = \frac{4.5n}{4.5}$
$n = 224 \text{ mi}$

9. $\frac{9}{12} = \frac{n}{10}$
$10 \cdot 9 = 12n$
$\frac{90}{12} = \frac{12n}{12}$
$n = 7.5 \text{ in.}$

10. $\frac{9}{16.5} \neq \frac{10.5}{24}$
$24 \cdot 9 = 216$
$16.5 \cdot 10.5 = 173.25$
$216 \text{ cm} \neq 173.25 \text{ cm}$

11. $\frac{\$0.48}{3} = \frac{n}{20}$
$20 \cdot \$0.48 = 3n$
$\frac{\$9.60}{3} = \frac{3n}{3}$
$n = \$3.20$

12. $\frac{50}{25} = \frac{20}{n}$
$50n = 25 \cdot 20$
$\frac{50n}{50} = \frac{500}{50}$
$n = 10 \text{ m}$

page 275

1. $\frac{1}{12} = \frac{1.2}{n}$
$n = 12 \cdot 1.2$
$n = 14.4 \text{ ft}$

2. $\frac{1}{12} = \frac{3.5}{n}$
$n = 12 \cdot 3.5$
$n = 42 \text{ ft}$

3. $\frac{1}{12} = \frac{2.3}{n}$
$n = 12 \cdot 2.3$
$n = 27.6 \text{ ft}$

4. $\frac{1}{12} = \frac{0.5}{n}$ $\qquad \frac{1}{12} = \frac{1}{n}$
$n = 12 \cdot 0.5$ $\qquad n = 12 \cdot 1$
$n = 6$ $\qquad n = 12$
$n = 6 \text{ ft long}$ $\qquad n = 12 \text{ ft wide}$

page 276

5. $\frac{1}{32} = \frac{3}{n}$ $\qquad$ 6. $\frac{1}{32} = \frac{4.5}{n}$
$n = 32 \cdot 3$ $\qquad\qquad n = 32 \cdot 4.5$
$n = 96 \text{ km}$ $\qquad\quad n = 144 \text{ km}$

7. $\frac{1}{32} = \frac{7}{n}$ $\qquad$ 8. $\frac{1}{32} = \frac{5.5}{n}$
$n = 32 \cdot 7$ $\qquad\qquad n = 32 \cdot 5.5$
$n = 224 \text{ km}$ $\qquad\; n = 176 \text{ km}$

9. $\frac{1}{32} = \frac{3.5}{n}$
$n = 32 \bullet 3.5$
$n = 112$ km

10. $\frac{1}{32} = \frac{6}{n}$
$n = 32 \bullet 6$
$n = 192$ km

11. $\frac{1}{32} = \frac{5}{n}$
$n = 32 \bullet 5$
$n = 160$ km

page 277

15. $\frac{1}{150} = \frac{4}{n}$
$n = 150 \bullet 4$
$n = 600$ mi

16. $\frac{1}{150} = \frac{9}{n}$
$n = 150 \bullet 9$
$n = 1{,}350$ mi

17. $\frac{1}{150} = \frac{3.6}{n}$
$n = 150 \bullet 3.6$
$n = 540$ mi

18. $\frac{1}{150} = \frac{0.6}{n}$
$n = 150 \bullet 0.6$
$n = 90$ mi

19. $\frac{1}{150} = \frac{5.8}{n}$
$n = 150 \bullet 5.8$
$n = 870$ mi

20. $\frac{1}{150} = \frac{13}{n}$
$n = 150 \bullet 13$
$n = 1{,}950$ mi

21. $\frac{1}{16} = \frac{n}{80}$
$\frac{80}{16} = \frac{16n}{16}$
$n = 5$ in.

22. $\frac{1}{16} = \frac{n}{160}$
$\frac{160}{16} = \frac{16n}{16}$
$n = 10$ in.

23. $\frac{1}{16} = \frac{n}{120}$
$\frac{120}{16} = \frac{16n}{16}$
$n = 7.5$ in.

24. $\frac{1}{16} = \frac{n}{48}$
$\frac{48}{16} = \frac{16n}{16}$
$n = 3$ in.

25. $\frac{1}{16} = \frac{n}{8}$
$\frac{8}{16} = \frac{16n}{16}$
$n = 0.5$ in.

26. $\frac{1}{16} = \frac{n}{67.2}$
$\frac{67.2}{16} = \frac{16n}{16}$
$n = 4.2$ in.

page 278

1. 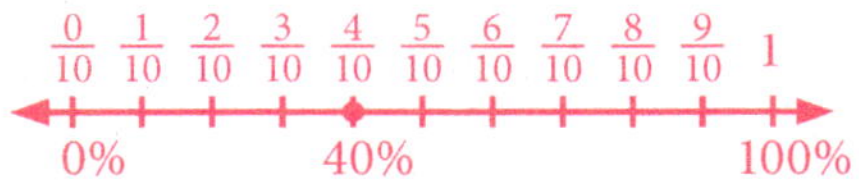

2. 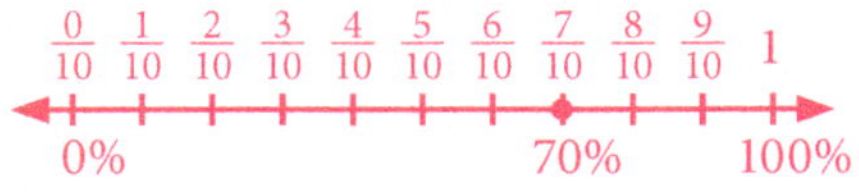

3. 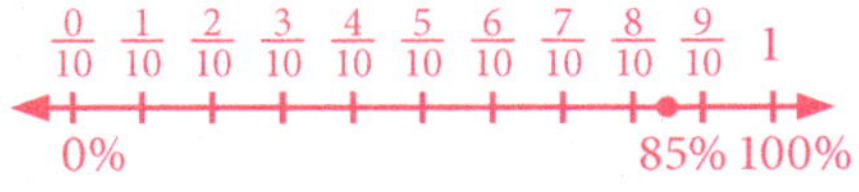

13. 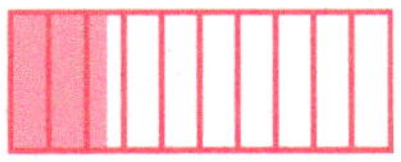

14. 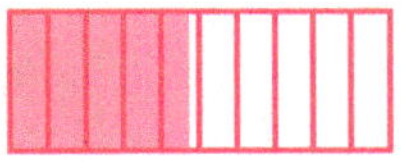

15. 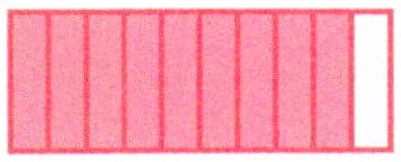

page 280

1. $\frac{50}{100} \times 78 = 39$
2. $\frac{30}{100} \times 80 = \frac{120}{5} = 24$
3. $\frac{40}{100} \times 200 = 80$
4. $\frac{25}{100} \times 48 = 12$
5. $\frac{60}{100} \times 25 = 15$
6. $\frac{75}{100} \times 52 = 39$
7. $\frac{20}{100} \times 85 = 17$
8. $\frac{33}{100} \times 100 = 33$
9. $\frac{10}{100} \times 250 = 25$
10. $\frac{70}{100} \times 15 = 10.5$ or $10\frac{1}{2}$
11. $0.15 \times 80 = 12$
12. $0.35 \times 120 = 42$
13. $0.24 \times 400 = 96$
14. $0.05 \times 64 = 3.2$
15. $1.0 \times 25 = 25$
16. $0.18 \times 65 = 11.7$
17. $0.52 \times 65 = 33.8$
18. $0.39 \times 200 = 78$
19. $0.99 \times 50 = 49.5$
20. $0.45 \times 20 = 9$
21. 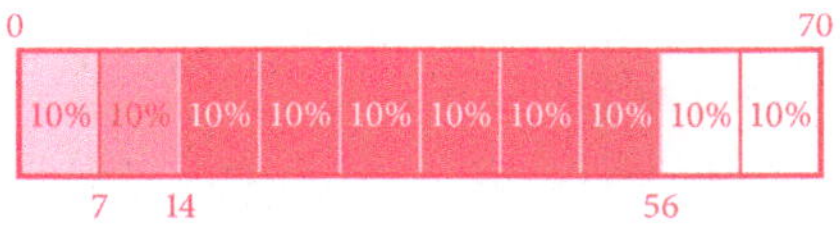

10% of 70 = 7
$20\% = 2 \times 7 = 14$
$80\% = 8 \times 7 = 56$

22. 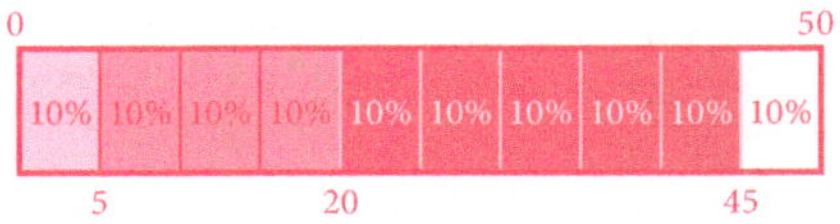

10% of 50 = 5
$40\% = 4 \times 5 = 20$
$90\% = 9 \times 5 = 45$

23. 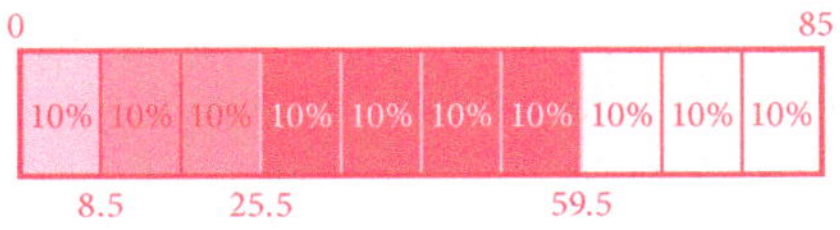

10% of 85 = 8.5
$30\% = 3 \times 8.5 = 25.5$
$70\% = 7 \times 8.5 = 59.5$

page 281

33. $\frac{5}{100} = \frac{n}{24}$; $n = 1.2$ hr
34. $\frac{35}{100} = \frac{n}{40}$; $n = 14$ shots
35. $\frac{25}{100} = \frac{n}{\$64}$; $n = \$16$; $\$64 - \$16 = \$48$
36. $\frac{8}{100} = \frac{n}{\$37.50}$; $n = \$3.00$
37. $\frac{20}{100} = \frac{n}{\$60}$; $n = \$12$
38. $\frac{80}{100} = \frac{n}{6}$; $n = 4.8$ ft
39. $\frac{10}{100} = \frac{n}{\$80}$; $n = \$8$; $\frac{40}{100} = \frac{n}{\$80}$; $n = \$32$
40. $\frac{2}{100} = \frac{n}{\$150}$; $n = \$3$
41. $\frac{65}{100} = \frac{n}{200}$; $n = 130$ people

page 282

1. $15\% \bullet n = 12$
$\frac{0.15n}{0.15} = \frac{12}{0.15}$
$n = 80$

2. $75\% \bullet n = 9$
$\frac{0.75n}{0.75} = \frac{9}{0.75}$
$n = 12$

3. $20\% \bullet n = 50$
$\frac{0.20n}{0.20} = \frac{50}{0.20}$
$n = 250$

4. $16 = 25\% \bullet n$
$\frac{16}{0.25} = \frac{0.25n}{0.25}$
$n = 64$

5. $60\% \bullet n = 15$
$\frac{0.60n}{0.60} = \frac{15}{0.60}$
$n = 25$

6. $14 = 35\% \bullet n$
$\frac{14}{0.35} = \frac{0.35n}{0.35}$
$n = 40$

7. $\frac{35}{100} = \frac{42}{n}$
$\frac{35n}{35} = \frac{4{,}200}{35}$
$n = 120$

8. $\frac{3}{100} = \frac{6}{n}$
$\frac{3n}{3} = \frac{600}{3}$
$n = 200$

9. $\frac{60}{100} = \frac{24}{n}$
$\frac{60n}{60} = \frac{2{,}400}{60}$
$n = 40$

10. $\frac{14}{100} = \frac{7}{n}$
$\frac{14n}{14} = \frac{700}{14}$
$n = 50$

11. $\frac{52}{100} = \frac{78}{n}$
$\frac{52n}{52} = \frac{7{,}800}{52}$
$n = 150$

12. $\frac{45}{100} = \frac{36}{n}$
$\frac{45n}{45} = \frac{3{,}600}{45}$
$n = 80$

page 283

13. 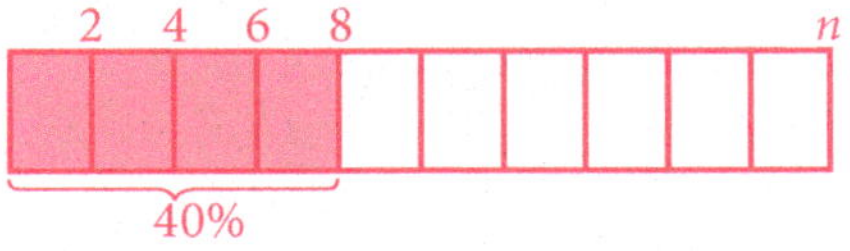

$40\% \bullet n = 8$
$8 \div 4 = 2$; Each part is 2.
$100\% = 10 \bullet 2 = 20$
$n = 20$
$40\% \bullet 20 = 8$

14. 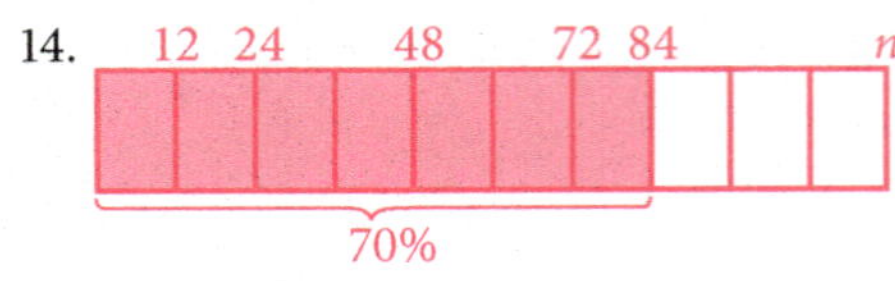

$70\% \cdot n = 84$
$84 \div 7 = 12$; Each part is 12.
$100\% = 10 \cdot 12 = 120$
$n = 120$
$70\% \cdot 120 = 84$

15. 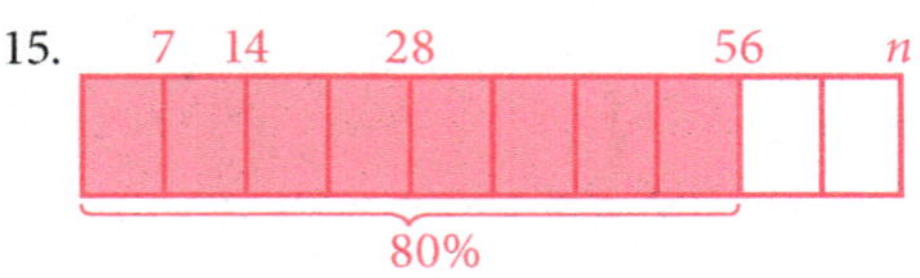

$80\% \cdot n = 56$
$56 \div 8 = 7$; Each part is 7.
$100\% = 10 \cdot 7 = 70$
$n = 70$
$80\% \cdot 70 = 56$

16. 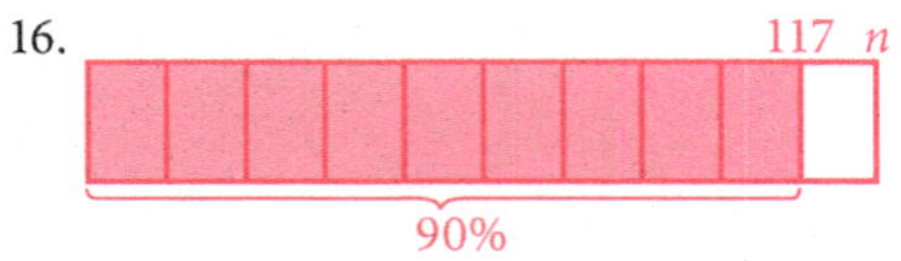

$117 = 90\% \cdot n$
$117 \div 9 = 13$; Each part is 13.
$100\% = 10 \cdot 13 = 130$
$n = 130$
$90\% \cdot 130 = 117$

17. 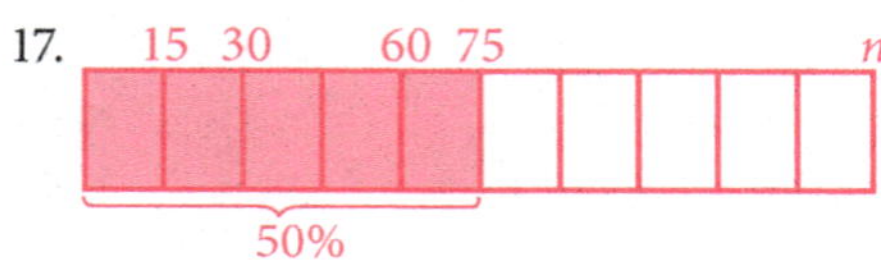

$50\% \cdot n = 75$
$75 \div 5 = 15$; Each part is 15.
$100\% = 10 \cdot 15 = 150$
$n = 150$
$50\% \cdot 150 = 75$

18. 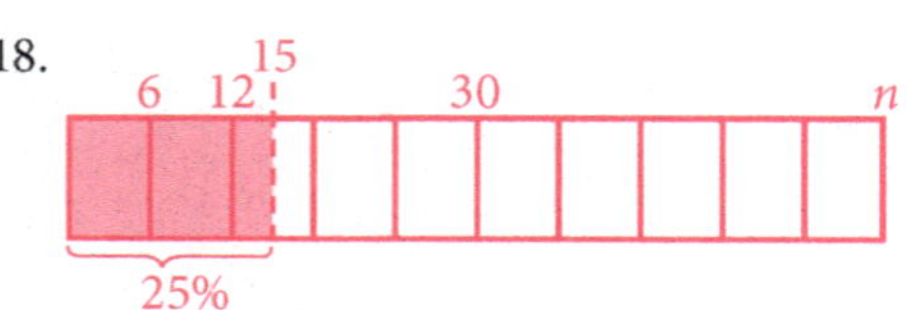

$15 = 25\% \cdot n$
$15 \div 2\frac{1}{2} = 15 \cdot \frac{2}{5} = 6$; Each part is 6.
$100\% = 10 \cdot 6 = 60$
$n = 60$
$25\% \cdot 60 = 15$

34. $\frac{4}{6} = \frac{5}{d}$; $d = 7.5$ hr

35. $\frac{240}{2} = \frac{m}{5}$; $m = 600$ mi

36. $\frac{1}{3} \cdot \$45 = \frac{45}{3} = \$15$; $\$45 - \$15 = \$30$

37. $\frac{8}{100} = \frac{t}{\$6.50}$; $t = \$0.52$

38. $\frac{35}{100} = \frac{d}{24}$; $d = 8.4$ hr

39. $\frac{30}{100} = \frac{\$12}{b}$; $b = \$40$

40. $\frac{20}{100} = \frac{2}{g}$; $g = 10$ goals

41. $\frac{2}{1} = \frac{v}{8}$; $v = 16$ votes

42. $\frac{6}{8} = \frac{n}{28}$; $n = 21$ ft

43. $\frac{1}{3} = \frac{n}{12}$; $n = 4$ m

44. $\frac{29}{33} = \frac{n}{100}$; $n \approx 88\%$

45. $\frac{20}{100} = \frac{n}{\$65}$; $n = \$13.00$;
$\frac{45}{100} = \frac{s}{\$65}$; $s = \$29.25$;
$\$65.00 - \$29.25 = \$35.75$

page 285

1. $\frac{4\text{ mi}}{1\text{ hr}} = \frac{n}{2\text{ hr}}$
$n = 2 \cdot 4$ mi
$n = 8$ mi

2. $\frac{52\text{ km}}{1\text{ hr}} = \frac{n}{5\text{ hr}}$
$n = 5 \cdot 52$ km
$n = 260$ km

3. $\frac{7\text{ ft}}{1\text{ sec}} = \frac{n}{12\text{ sec}}$
$n = 12 \cdot 7$ ft
$n = 84$ ft

4. $\frac{3\text{ mi}}{1\text{ hr}} = \frac{5\text{ mi}}{n}$
$\frac{3n}{3} = \frac{5}{3}$
$n \approx 1.67$ hr

5. $\frac{50\text{ mi}}{1\text{ hr}} = \frac{200\text{ mi}}{n}$
$\frac{50n}{50} = \frac{200}{50}$
$n = 4$ hr

6. $\frac{330\text{ mi}}{1\text{ hr}} = \frac{165\text{ mi}}{n}$
$\frac{330n}{330} = \frac{165}{330}$
$n = 0.5$ hr

page 286

7. $\frac{224\text{ km}}{3.5\text{ hr}} = \frac{n}{1\text{ hr}}$
$n = 64$ km/hr

8. $\frac{140\text{ mi}}{4\text{ hr}} = \frac{n}{1\text{ hr}}$
$n = 35$ mi/hr

9. $\frac{270\text{ mi}}{3\text{ hr}} = \frac{n}{1\text{ hr}}$
$n = 90$ mi/hr

10. $\frac{350\text{ mi}}{1\text{ hr}} = \frac{d}{2.25\text{ hr}}$
$d = 787.5$ mi

11. $\frac{340\text{ mi}}{1\text{ hr}} = \frac{1{,}190\text{ mi}}{t}$
$\frac{340t}{340} = \frac{1{,}190\text{ mi}}{340}$
$t = 3.5$ hr

12. $\frac{12\text{ mi}}{1\text{ hr}} = \frac{24\text{ mi}}{t}$
$\frac{12t}{12} = \frac{24}{12}$
$t = 2$ hr

13. $\frac{50\text{ mi}}{1\text{ hr}} = \frac{600\text{ mi}}{t}$
$\frac{50t}{50} = \frac{600}{50}$
$t = 12$ hr

$\frac{60\text{ mi}}{1\text{ hr}} = \frac{600\text{ mi}}{t}$
$\frac{60t}{60} = \frac{600}{60}$
$t = 10$ hr

$12 - 10 = 2$ hr

14. 1 hr = 60 min
$\frac{30\text{ mi}}{60\text{ min}} = \frac{5\text{ mi}}{t}$
$30t = 5 \cdot 60$
$\frac{30t}{30} = \frac{300}{30}$
$t = 10$ min

$\frac{25\text{ mi}}{60\text{ min}} = \frac{5\text{ mi}}{t}$
$25t = 5 \cdot 60$
$\frac{25t}{25} = \frac{300}{25}$
$t = 12$ min

$12 - 10 = 2$ min

page 287

15. 1 hr = 60 min
$\frac{50\text{ mi}}{60\text{ min}} = \frac{d}{60\text{ min}}$
$60 \cdot 50 = 60d$
$\frac{3{,}000}{60} = \frac{60d}{60}$
$d = 50$ mi

16. 20 yd = 60 ft
$\frac{5}{1} = \frac{60\text{ ft}}{t}$
$\frac{5t}{5} = \frac{60}{5}$
$t = 12$ min

17. $\frac{\frac{1}{2}\text{ mi}}{1\text{ hr}} = \frac{d}{5\text{ hr}}$
$5 \cdot \frac{1}{2} = d$
$d = \frac{5}{2} = 2\frac{1}{2}$ mi

18. $\frac{0.5\text{ mi}}{5\text{ min}} = \frac{r}{1\text{ min}}$
$\frac{0.5}{5} = \frac{5r}{5}$
$r = 0.1$ mi/min

19. 3 km = 3,000 m
$\frac{250\text{ m}}{1\text{ min}} = \frac{3{,}000\text{ m}}{t}$
$\frac{250t}{250} = \frac{3{,}000}{250}$
$t = 12$ min

20. $\frac{1}{2}$ day = 12 hr
$\frac{540\text{ mi}}{12\text{ hr}} = \frac{r}{1\text{ hr}}$
$\frac{540\text{ mi}}{12} = \frac{12r}{12}$
$r = 45$ mi/hr

21. $\frac{21\text{ ft}}{1\text{ sec}} = \frac{d}{15\text{ sec}}$
$21 \cdot 15 = d$
$d = 315$ ft

22. 1 day = 24 hr
$\frac{600\text{ mi}}{24\text{ hr}} = \frac{r}{1\text{ hr}}$
$\frac{600}{24} = \frac{24r}{24}$
$r = 25$ mi/hr

23. 1 mi = 5,280 ft
$\frac{40\text{ ft}}{1\text{ sec}} = \frac{5{,}280\text{ ft}}{t}$
$\frac{40t}{40} = \frac{5{,}280}{40}$
$t = 132$ sec

24. 1 hr = 60 min
$\frac{70\text{ km}}{60\text{ min}} = \frac{d}{30\text{ min}}$
$70 \cdot 30 = 60d$
$\frac{2{,}100}{60} = \frac{60d}{60}$
$d = 35$ km

25. $\frac{66\text{ ft}}{1\text{ min}} = \frac{d}{15\text{ min}}$
$66 \cdot 15 = d$
$d = 990$ ft
$\frac{66\text{ ft}}{1\text{ min}} = \frac{5{,}280\text{ ft}}{t}$
$\frac{66t}{66} = \frac{5{,}280}{66}$
$t = 80$ min

Chapter 14
page 296

8. $5 \times 36 = 180$

9. $117 \div 36 = 3\frac{9}{36} = 3\frac{1}{4}$

10. $\frac{3}{4} \cdot 36 = 27$

11. $\frac{1}{9} \cdot 36 = 4$

12. $12 \div 3 = 4$

13. $30 \div 12 = 2\frac{6}{12} = 2\frac{1}{2}$

14. $\frac{2}{3} \cdot 12 = 8$

15. $\frac{1}{2} \cdot 36 = 18$

16. $2 \times 5{,}280 = 10{,}560$

17. $7 \div 3 = 2$ r1

18. $\frac{1}{4} \cdot 5{,}280 = 1{,}320$

19. $\frac{1}{2} \cdot 5{,}280 = 2{,}640$

20. $108 \div 36 = 3$

21. $(6 \cdot 12) + 6 = 78$

22. $\frac{3}{4} \cdot 12 = 9$

23. $\frac{2{,}640}{1{,}760} = 1\frac{1}{2}$

24. $3 \times \frac{5}{3} = \frac{15}{3} = 5$ or 3 ft + 2 ft = 5 ft

25. $5{,}290 \div 5{,}280 = 1$ r10

26. $\frac{2}{3} \cdot 5{,}280 = 3{,}520$

27. $82 \div 12 = 6\frac{5}{6}$

page 298

9. $6 \cdot 16 = 96$

10. $(6 \cdot 16) + 9 =$
$96 + 9 = 105$

11. $(2 \cdot 2{,}000) + 25 =$
$4{,}000 + 25 = 4{,}025$

12. $3 \cdot 2{,}000 = 6{,}000$

13. $(12 \cdot 16) + 9 =$
$192 + 9 = 201$

14. $9 \div 2 = 4$ r1

15. $(4 \cdot 16) + 2 =$
$64 + 2 = 66$

16. $10 \div 4 = 2$ r2

page 299

17. $\frac{1}{2} \cdot 2{,}000 = 1{,}000$

18. $\frac{5}{8} \cdot 16 = 10$

19. $\frac{3}{4} \cdot 4 = 3$

20. $\frac{3}{8} \cdot 8 = 3$

21. $\frac{1}{2} \cdot 16 = 8$

22. $\frac{1}{4} \cdot 2{,}000 = 500$

23. $\frac{3}{4} \cdot 8 = 6$

24. $\frac{1}{2} \cdot 4 = 2$

25. $3\frac{1}{2} \cdot 2 = \frac{7}{2} \cdot 2 = 7$

26. $1\frac{1}{2} \cdot 16 = \frac{3}{2} \cdot 16 = 24$

27. $10\frac{1}{4} \cdot 2{,}000 = \frac{41}{4} \cdot 2{,}000 = 20{,}500$

28. $3\frac{1}{2} \cdot 4 = \frac{7}{2} \cdot 4 = 14$

44. $\frac{3}{4} \cdot 16 = 12$ oz

45. $(2\frac{1}{4} \cdot 12) - 21 =$
$(\frac{9}{4} \cdot 12) - 21 =$
$27 - 21 = 6$ in. or $\frac{1}{2}$ ft

46. Though the mass of my body would not change, the pull of gravity on the moon is less than the pull of gravity on the earth.

page 304

17.
$$3\overline{)48.6\text{ m}} = 16.2\text{ m}$$
$$-3$$
$$18$$
$$-18$$
$$06$$
$$-6$$
$$0$$

18.
$$4\overline{)16\text{ ft }8\text{ in.}} = 4\text{ ft }2\text{ in.}$$
$$-16\text{ ft}$$
$$0\text{ ft }8\text{ in.}$$
$$-8\text{ in.}$$
$$0$$

19.
$$2\overline{)6\text{ tn }210\text{ lb}} = 3\text{ tn }105\text{ lb}$$
$$-6\text{ tn}$$
$$0\text{ tn }2$$
$$-2$$
$$010$$
$$-10$$
$$0$$

20.
$$3\overline{)18\text{ gal }9\text{ pt}} = 6\text{ gal }3\text{ pt}$$
$$-18\text{ gal}$$
$$0\text{ gal }9\text{ pt}$$
$$-9\text{ pt}$$
$$0$$

page 306

15. $C = \frac{5}{9} \cdot (F - 32)$
$C = \frac{5}{9} \cdot (59 - 32)$
$C = \frac{5}{9} \cdot 27$
$C = \frac{5}{9} \cdot \frac{27}{1}$
$C = 15°$

16. $F = (\frac{9}{5} \cdot C) + 32$
$F = (\frac{9}{5} \cdot 10) + 32$
$F = (\frac{9}{5} \cdot \frac{10}{1}) + 32$
$F = 18 + 32$
$F = 50°$

17. $C = \frac{5}{9} \cdot (F - 32)$
$C = \frac{5}{9} \cdot (41 - 32)$
$C = \frac{5}{9} \cdot 9$
$C = \frac{5}{9} \cdot \frac{9}{1}$
$C = 5°$

18. $F = (\frac{9}{5} \cdot C) + 32$
$F = (\frac{9}{5} \cdot 20) + 32$
$F = (\frac{9}{5} \cdot \frac{20}{1}) + 32$
$F = 36 + 32$
$F = 68°$

19. $C = \frac{5}{9} \cdot (F - 32)$
$C = \frac{5}{9} \cdot (95 - 32)$
$C = \frac{5}{9} \cdot 63$
$C = \frac{5}{9} \cdot \frac{63}{1}$
$C = 35°$

20. $F = (\frac{9}{5} \cdot C) + 32$
$F = (\frac{9}{5} \cdot 50) + 32$
$F = (\frac{9}{5} \cdot \frac{50}{1}) + 32$
$F = 90 + 32$
$F = 122°$

page 307

21. $F = (\frac{9}{5} \cdot C) + 32$
$F = (\frac{9}{5} \cdot 15) + 32$
$F = (\frac{9}{5} \cdot \frac{15}{1}) + 32$
$F = 27 + 32$
$F = 59°$

page 317

17. $\frac{36\text{ in.}}{1\text{ yd}}$ or $\frac{1\text{ yd}}{36\text{ in.}}$

$\frac{7\text{ yd}}{1} \cdot \frac{36\text{ in.}}{1\text{ yd}} = 252$ in.

18. $\frac{16\text{ oz}}{1\text{ lb}}$ or $\frac{1\text{ lb}}{16\text{ oz}}$

$\frac{64\text{ oz}}{1} \cdot \frac{1\text{ lb}}{16\text{ oz}} = 4$ lb

19. $\frac{4\text{ qt}}{1\text{ gal}}$ or $\frac{1\text{ gal}}{4\text{ qt}}$

$\frac{8\text{ gal}}{1} \cdot \frac{4\text{ qt}}{1\text{ gal}} = 32$ qt

20. $\frac{60\text{ min}}{1\text{ hr}}$ or $\frac{1\text{ hr}}{60\text{ min}}$

$\frac{3\text{ hr}}{1} \cdot \frac{60\text{ min}}{1\text{ hr}} = 180$ min

21. $\frac{1{,}000\text{ g}}{1\text{ kg}}$ or $\frac{1\text{ kg}}{1{,}000\text{ g}}$

$\frac{3\text{ kg}}{1} \cdot \frac{1{,}000\text{ g}}{1\text{ kg}} = 3{,}000$ g

22. $\frac{1{,}000\text{ mL}}{1\text{ L}}$ or $\frac{1\text{ L}}{1{,}000\text{ mL}}$

$\frac{2{,}500\text{ mL}}{1} \cdot \frac{1\text{ L}}{1{,}000\text{ mL}} = 2.5$ L

23. $\dfrac{100 \text{ cm}}{1 \text{ m}}$ or $\dfrac{1 \text{ m}}{100 \text{ cm}}$

$\dfrac{450 \text{ cm}}{1} \cdot \dfrac{1 \text{ m}}{100 \text{ cm}} = 4.5$ m

24. $\dfrac{8 \text{ fl oz}}{1 \text{ c}}$ or $\dfrac{1 \text{ c}}{8 \text{ fl oz}}$

$\dfrac{3 \text{ c}}{1} \cdot \dfrac{8 \text{ fl oz}}{1 \text{ c}} = 24$ fl oz

25. $\dfrac{5 \text{ tn}}{1} \cdot \dfrac{2,000 \text{ lb}}{1 \text{ tn}} = 10,000$ lb

26. $\dfrac{13 \text{ gal}}{1} \cdot \dfrac{4 \text{ qt}}{1 \text{ gal}} = 52$ qt

27. $\dfrac{20 \text{ c}}{1} \cdot \dfrac{1 \text{ pt}}{2 \text{ c}} = 10$ pt

28. $\dfrac{8 \text{ mi}}{1} \cdot \dfrac{1,760 \text{ yd}}{1 \text{ mi}} = 14,080$ yd

29. $\dfrac{108 \text{ in.}}{1} \cdot \dfrac{1 \text{ ft}}{12 \text{ in.}} = 9$ ft

30. $\dfrac{120 \text{ fl oz}}{1} \cdot \dfrac{1 \text{ c}}{8 \text{ fl oz}} = 15$ c

31. $\dfrac{4 \text{ yd}}{1} \cdot \dfrac{36 \text{ in.}}{1 \text{ yd}} = 144$ in.

32. $\dfrac{128 \text{ oz}}{1} \cdot \dfrac{1 \text{ lb}}{16 \text{ oz}} = 8$ lb

Chapter 15

Graphs and diagrams not included in the Solutions section are available at TeacherToolsOnline.com.

page 326

6. mean: $[(2 \cdot 10) + (4 \cdot 11) + (3 \cdot 12) + (2 \cdot 13) + (2 \cdot 14)] \div 13 = 154 \div 13 \approx 11.8$
median: ~~10~~ ~~10~~ ~~11~~ ~~11~~ ~~11~~ ~~11~~ (12) ~~12~~ ~~12~~ ~~13~~ ~~13~~ ~~14~~ ~~14~~

7–10. See frequency table at Teacher Tools Online.

8. mean: $(83 + 81 + 86 + 88 + 80 + 82 + 85) \div 7 = 585 \div 7 \approx 83.6$

9. median: ~~80~~ ~~81~~ ~~82~~ (83) ~~85~~ ~~86~~ ~~88~~;
$83° < 83.6°$

page 327

14. mean: The sum of the data is 499;
$499 \div 6 \approx 83.2$.
median: ~~79~~ ~~81~~ (82 85) ~~85~~ ~~87~~

15. mean: The sum of the data is 232;
$232 \div 5 = 46.4$.
median: ~~40~~ ~~41~~ (48) ~~51~~ ~~52~~

16. mean: The sum of the data is 115;
$115 \div 8 \approx 14.4$.
median: ~~10~~ ~~12~~ ~~12~~ (12 15) ~~17~~ ~~18~~ ~~19~~

17. $14 + 11 + 10 + 24 = 59$ chairs

page 328

25. $4(6 \text{ ft} \cdot 3 \text{ ft}) + 2(3 \text{ ft} \cdot 3 \text{ ft}) =$
$4 \cdot 18 \text{ ft}^2 + 2 \cdot 9 \text{ ft}^2 =$
$72 \text{ ft}^2 + 18 \text{ ft}^2 = 90 \text{ ft}^2$

26. $6 \text{ ft} \cdot 3 \text{ ft} \cdot 3 \text{ ft} = 54 \text{ ft}^3$

page 328

4. mean for Maria:
$(30 + 90 + 15 + 75 + 45) \div 5 =$
$255 \div 5 = 51$ min
mean for Mitchell:
$(15 + 90 + 30 + 60 + 60) \div 5 =$
$255 \div 5 = 51$ min

8. median for Mitchell:
~~15~~ ~~30~~ (60) ~~60~~ ~~90~~
median for Maria:
~~15~~ ~~30~~ (45) ~~75~~ ~~90~~

page 329

16. range: $5 - 1.5 = 3.5$ km
mean: $(1.5 + 2 + 3 + 3 + 4 + 5) \div 6 =$
$18.5 \div 6 \approx 3.1$ km

18. range: $4 - 1 = 3$ km
mean: $(1 + 2 + 2 + 1.5 + 3 + 4) \div 6 =$
$13.5 \div 6 \approx 2.3$ km

19. See double line graph at Teacher Tools Online.

21. mean: The sum of the data is 1,120;
$1,120 \div 5 = 224$.

24. See double bar graph at Teacher Tools Online.

page 330

1–7. See stem-and-leaf plot at Teacher Tools Online.

8–15. See stem-and-leaf plot at Teacher Tools Online.

page 331

22. See double line graph at Teacher Tools Online.

23. mean: The sum of the data is 2,645;
$2,645 \div 5 = 529$ calls

24. mean: The sum of the data is 2,650;
$2,650 \div 5 = 530$ calls

33. mean (average): the sum of the data divided by the number of addends
median: the middle value or an average of the two middle values of a set of data when ordered from least to greatest
mode: the value that occurs most often or has the greatest frequency. Some sets may have more than one mode, and some sets may not have a mode.

page 333

14–22. See line plot at Teacher Tools Online.

17. The sum of the data is 370; $370 \div 19 \approx 19.5$.

26. $12 \div 19 \approx 0.63$; $0.63 \times 100 = 63\%$

28. See line plot at Teacher Tools Online.
See line graph at Teacher Tools Online.

page 335

11–19. See histogram at Teacher Tools Online.

28. See double bar graph at Teacher Tools Online.

page 336

1–7. See box-and-whisker plot at Teacher Tools Online.

8–15. See box-and-whisker plot at Teacher Tools Online.

page 337

22. See box-and-whisker plot at Teacher Tools Online.

23. See box-and-whisker plot at Teacher Tools Online.

28. See histogram at Teacher Tools Online.

page 341

10. See box-and-whisker plot at Teacher Tools Online.

11. See box-and-whisker plot at Teacher Tools Online.

16. See histogram at Teacher Tools Online.

Chapter 16

page 351

17. $\dfrac{4}{6} = 0.\overline{6} \approx 0.67 = 67\%$
$\dfrac{2}{6} = 0.\overline{3} \approx 0.33 = 33\%$

18. $\dfrac{2}{6} = 0.\overline{3} \approx 0.33 = 33\%$
$\dfrac{4}{6} = 0.\overline{6} \approx 0.67 = 67\%$

19. $\dfrac{3}{6} = 0.5 = 50\%$
$\dfrac{3}{6} = 0.5 = 50\%$

20. $\frac{1}{6} = 0.1\overline{6} \approx 0.17 = 17\%$

$\frac{5}{6} = 0.8\overline{3} \approx 0.83 = 83\%$

21. $\frac{1}{6} = 0.1\overline{6} \approx 0.17 = 17\%$

22. $\frac{3}{6} = \frac{1}{2} = 0.50 = 50\%$

23. $\frac{2}{6} = \frac{1}{3} = 0.\overline{3} \approx 0.33 = 33\%$

24. $\frac{0}{6} = 0 = 0\%$

25. $\frac{5}{6} = 0.8\overline{3} \approx 0.83 = 83\%$

26. $\frac{4}{6} = \frac{2}{3} = 0.\overline{6} \approx 0.67 = 67\%$

27. $\frac{1}{6} = 0.1\overline{6} \approx 0.17 = 17\%$

28. $\frac{5}{6} = 0.8\overline{3} \approx 0.83 = 83\%$

29. $\frac{3}{6} = \frac{1}{2} = 0.50 = 50\%$

30. $\frac{2}{6} = \frac{1}{3} = 0.\overline{3} \approx 0.33 = 33\%$

page 353

15. 2 sleeve lengths × 8 colors = 16 shirt choices

16. 9 digits × 9 digits × 9 digits × 9 digits = 6,561 combinations

17. 2 breads × 3 meats × 2 cheeses = 12 sandwich choices

18. 4 beds × 2 nightstands × 3 desks = 24 bedroom sets

19. 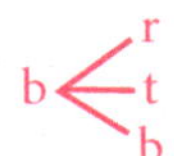

sample space: {s2, s4, a2, a4}
P(automatic transmission, 4-door) = $\frac{1}{4}$; 25%

20. 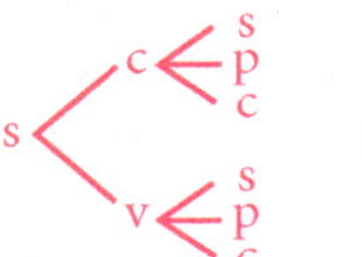

sample space: {wr, wt, wb, br, bt, bb}
P(white or black car, black interior) = $\frac{2}{6} = \frac{1}{3}$; 33%

21. 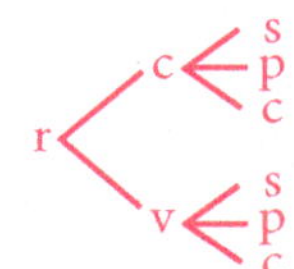

sample space: {scs, scp, scc, svs, svp, svc, rcs, rcp, rcc, rvs, rvp, rvc}
P(cone with chocolate ice cream) = $\frac{6}{12} = \frac{1}{2}$; 50%

22. 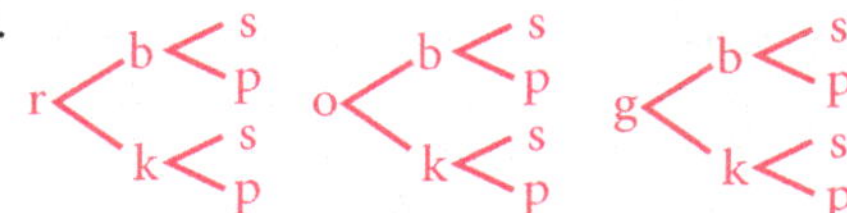

sample space: {rbs, rbp, rks, rkp, obs, obp, oks, okp, gbs, gbp, gks, gkp}
P(red shirt, blue pants, solid sweatshirt) = $\frac{1}{12}$; 8%

23. (Note: T = thick; t = thin)

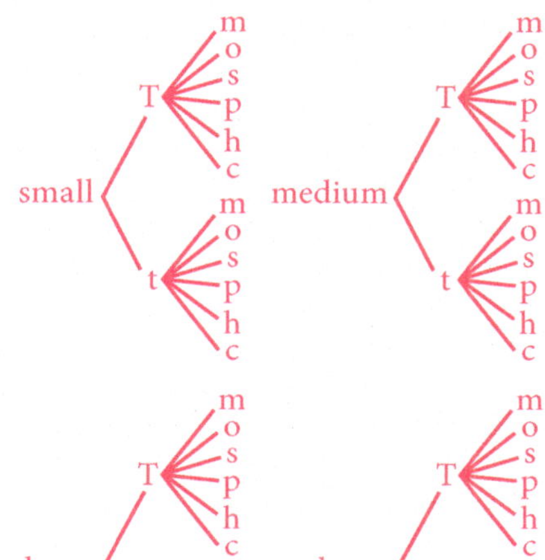

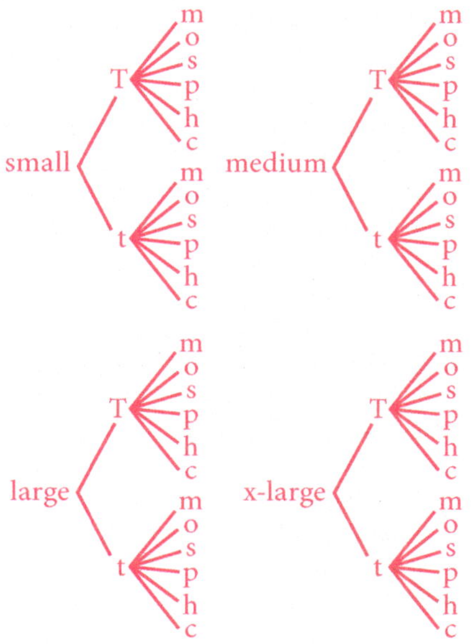

24. sample space: {sTm, sTo, sTs, sTp, sTh, sTc, stm, sto, sts, stp, sth, stc, mTm, mTo, mTs, mTp, mTh, mTc, mtm, mto, mts, mtp, mth, mtc, lTm, lTo, lTs, lTp, lTh, lTc, ltm, lto, lts, ltp, lth, ltc, xTm, xTo, xTs, xTp, xTh, xTc, xtm, xto, xts, xtp, xth, xtc}

25. $\frac{6}{48} = \frac{1}{8}$ or 13%

26. $\frac{1}{48}$ or 2%

page 356

6. sample space: {HHH, HHT, HTH, HTT, THH, THT, TTH, TTT}
P(at least 2 heads) = $\frac{4}{8} = \frac{1}{2}$;
P(at least 2 tails) = $\frac{4}{8} = \frac{1}{2}$;
P(at least 2 heads) = P(at least 2 tails); fair

page 357

29.

	Genetic Survey Results (sample = 200 students)			
Trait	Dimples	Straight Hair	Attached Earlobes	Widow's Peak
yes	80	100	60	100
no	120	100	140	100

page 359

14. P(red) = {r, r, b, b, y, g} = $\frac{2}{6}$

P(yellow) = {r, b, b, y, g} = $\frac{1}{5}$

$\frac{2}{6} \times \frac{1}{5} = \frac{2}{30} = \frac{1}{15}$

15. P(yellow) = {r, r, b, b, y, g} = $\frac{1}{6}$

P(green) = {r, r, b, b, g} = $\frac{1}{5}$

$\frac{1}{6} \times \frac{1}{5} = \frac{1}{30}$

16. P(red) = {r, r, b, b, y, g} = $\frac{2}{6}$

P(*not* yellow) = {r, b, b, y, g} = $\frac{4}{5}$

$\frac{2}{6} \times \frac{4}{5} = \frac{8}{30} = \frac{4}{15}$

17. P(red) = {r, r, b, b, y, g} = $\frac{2}{6}$

P(blue) = {r, b, b, y, g} = $\frac{2}{5}$

$\frac{2}{6} \times \frac{2}{5} = \frac{4}{30} = \frac{2}{15}$

18. P(yellow) = {r, r, b, b, y, g} = $\frac{1}{6}$

P(blue) = {r, r, b, b, g} = $\frac{2}{5}$

$\frac{1}{6} \times \frac{2}{5} = \frac{2}{30} = \frac{1}{15}$

19. P(green) = {r, r, b, b, y, g} = $\frac{1}{6}$

P(*not* blue) = {r, r, b, b, y} = $\frac{3}{5}$

$\frac{1}{6} \times \frac{3}{5} = \frac{3}{30} = \frac{1}{10}$

Chapter 17
page 366

1. 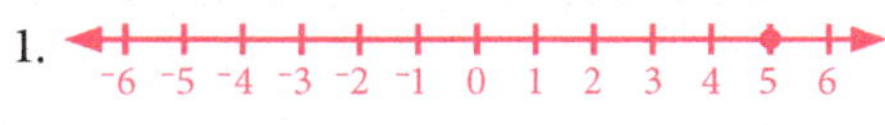

2. 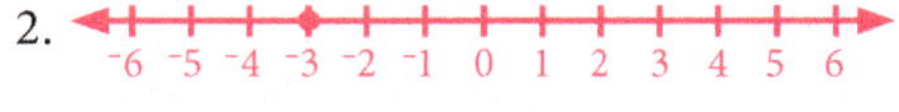

3.

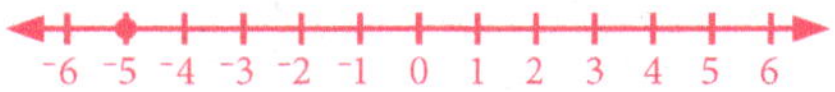

4.

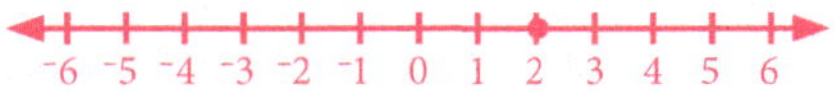

5.

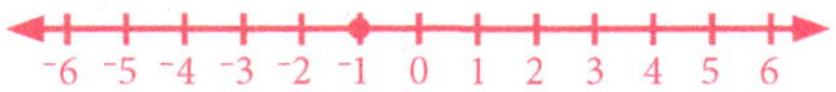

page 367

31. 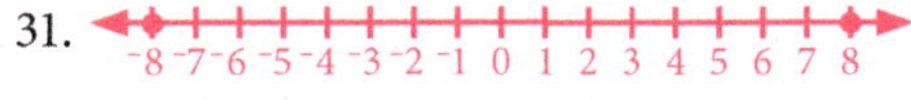

32. 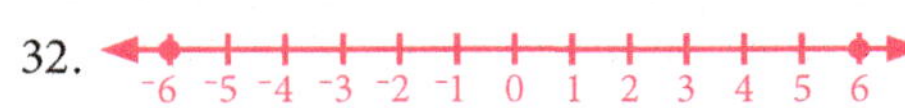

33.

48. 12 • 12 = 144

49. 8 • 8 • 8 = 512

1. 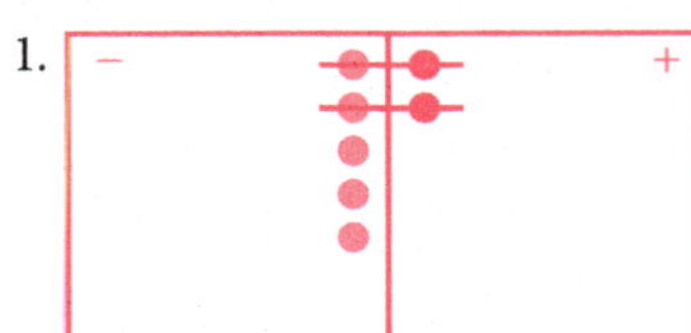
2. 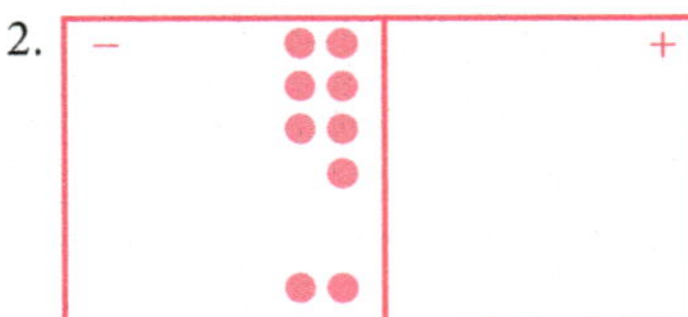
3. 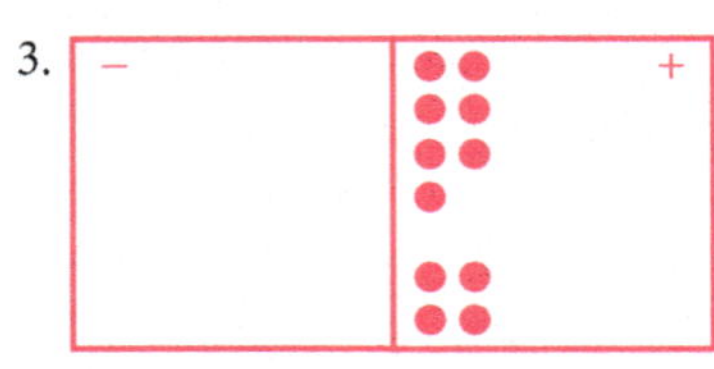
4. 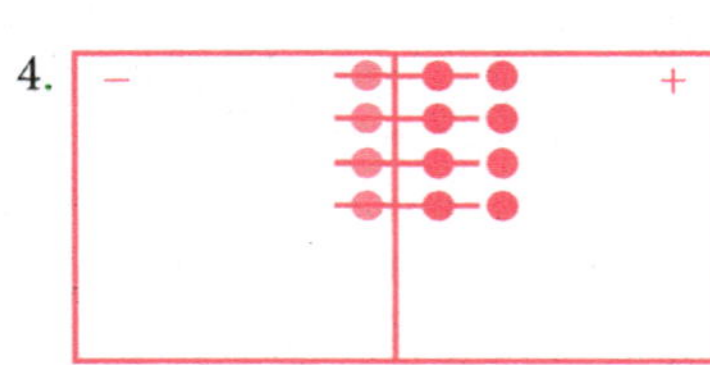

27. 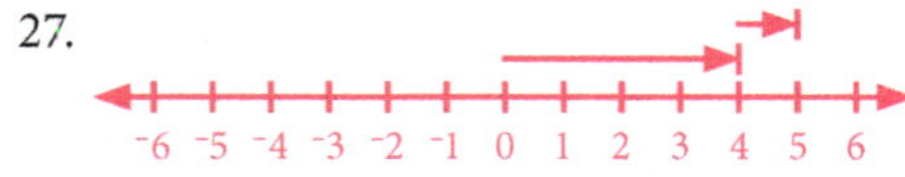
28. 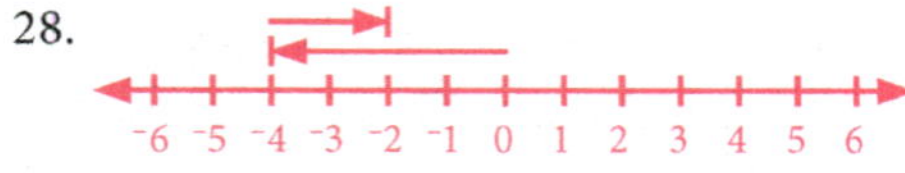
29. 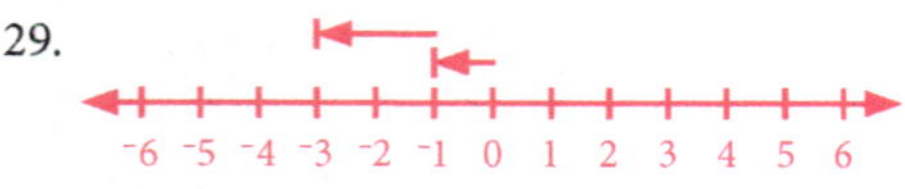
30. 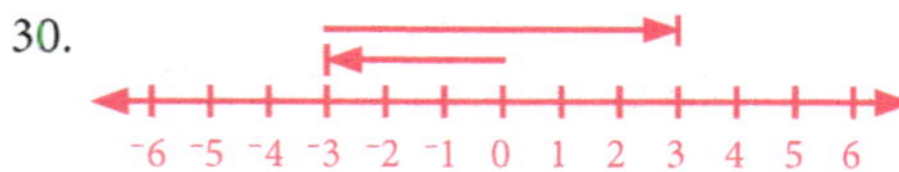

1. 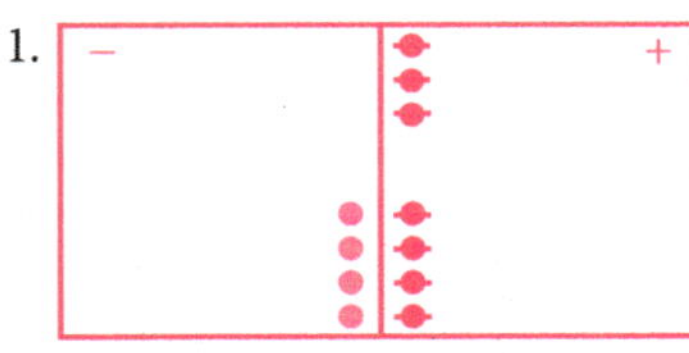
2. 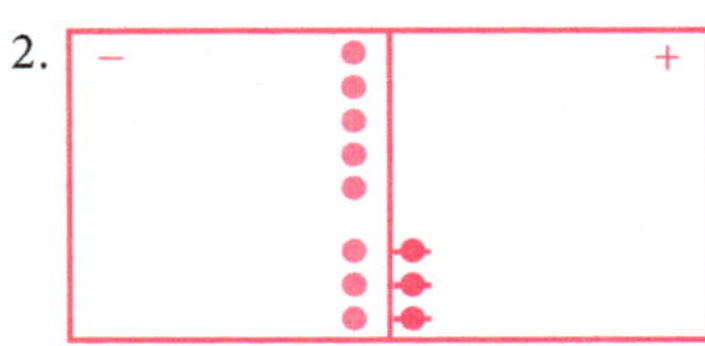
3. 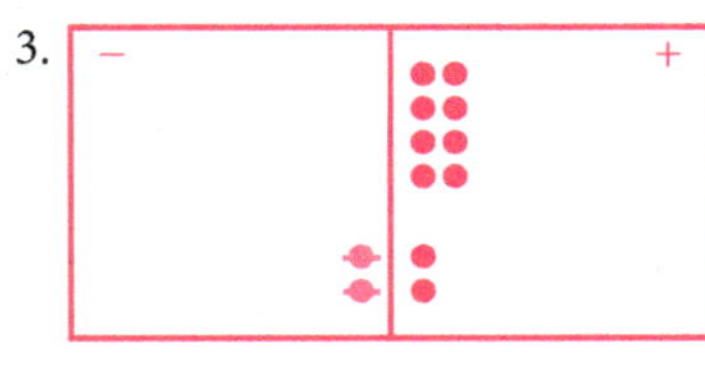
4. 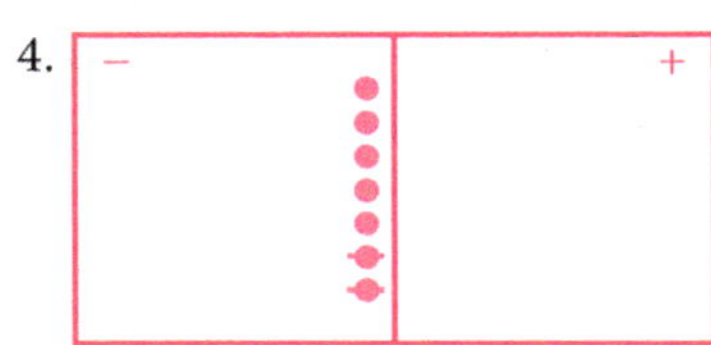

27. 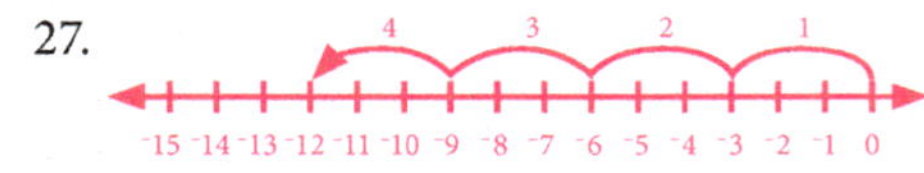

5. 12 negative counters divided into 3 sets equals 4 negative counters in each set.

6. 10 negative counters divided into equal sets of 5 negative counters will make 2 sets.

7. 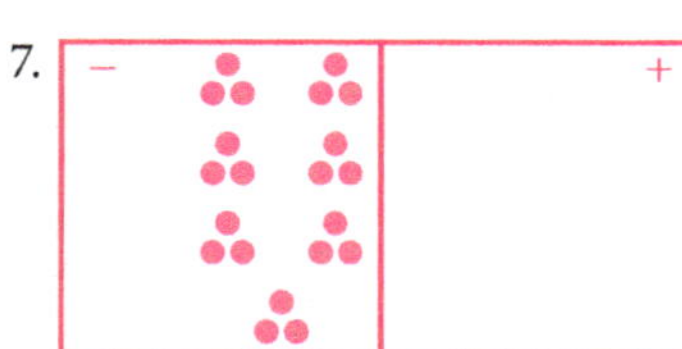
8. 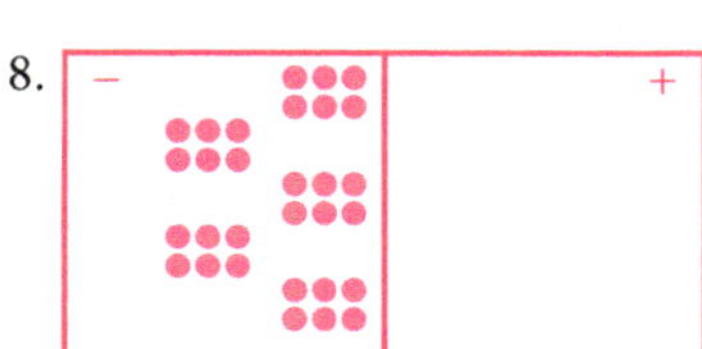

16–17. 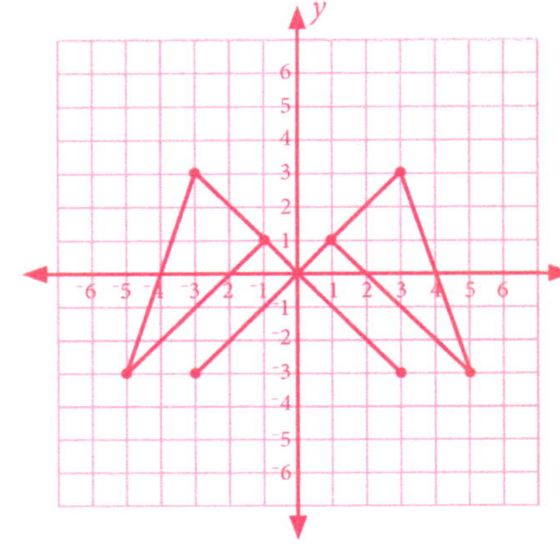

18–21. 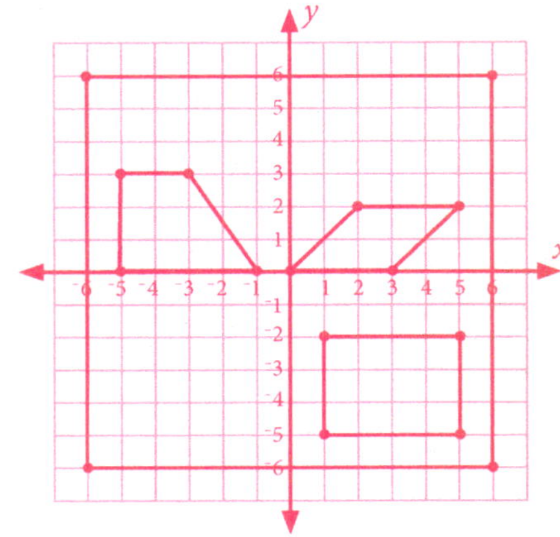

36. 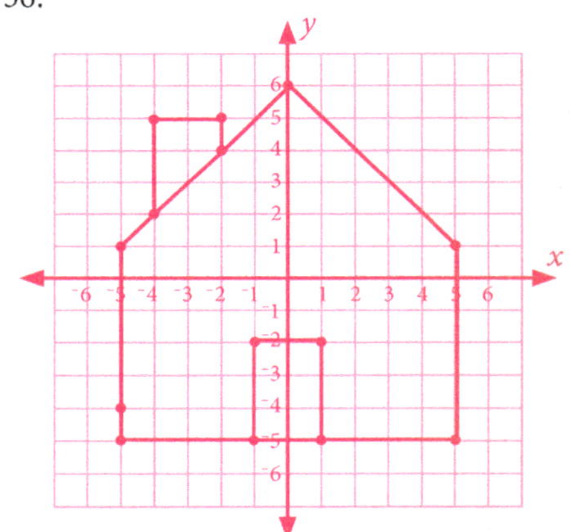

Solutions

EXPLAINING THE GOSPEL

One of the greatest desires of Christian teachers is to lead students to faith in the Savior. God has called you to present the gospel to your students so that they may repent and trust Christ, thereby being acceptable to God through Christ.

Relying on the Holy Spirit, you should take advantage of the opportunities that arise for presenting the good news of Jesus Christ. Ask questions to personally apply the Ten Commandments to your students (e.g., What is sin? Have you ever told a lie or taken something that wasn't yours? Are you a sinner?). You may also ask questions to discern the student's sincerity or to reveal any misunderstanding he or she might have (e.g., What is the gospel? What does it mean to repent? Can you do anything to save yourself?). Read verses from your Bible. You may find the following outline helpful, especially when dealing individually with a student.

1. I have sinned (Romans 3:23).

- Sin is disobeying God's Word (1 John 3:4). I break the Ten Commandments (Exodus 20:2–17) by loving other people or things more than I love God, worshiping other things or people, using God's name lightly, disobeying and dishonoring my parents, lying, stealing, cheating, thinking harmful and sinful thoughts, or wanting something that belongs to somebody else.
- Therefore, I am a sinner (Psalm 51:5; 58:3; Jeremiah 17:9).
- God is holy and must punish me for my sin (Isaiah 6:3; Romans 6:23).
- God hates sin, and there is nothing that I can do to get rid of my sin by myself (Titus 3:5; Romans 3:20, 28). I cannot make myself become a good person.

2. Jesus died for me (Romans 5:8).

- God loves me even though I am a sinner.
- He sent His Son, Jesus Christ, to die on the cross for me. Christ is sinless and did not deserve death. Because of His love for me, Christ took my sin on Himself and was punished in my place (1 Peter 2:24*a*; 1 Corinthians 15:3; John 1:29).
- God accepted Christ's death as the perfect substitution for the punishment of my sin (2 Corinthians 5:21).
- Three days later, God raised Jesus from the dead. Jesus is alive today and offers salvation to all. This is the gospel of Jesus Christ: He died on the cross for our sins according to the Scriptures, and He rose again the third day according to the Scriptures (1 Corinthians 15:1–4; 2 Peter 3:9; 1 Timothy 2:4).

3. I need to put my trust in Jesus (Romans 10:9–10, 13–14*a*).

- I must repent (turn away from my sin) and trust only Jesus Christ for salvation (Mark 1:15).
- If I repent and believe in what Jesus has done, I am putting my trust in Jesus.
- Everyone who trusts in Jesus is forgiven of sin (Acts 2:21) and will live forever with God (John 3:16). I am given Christ's righteousness and become a new creation, with Christ living in me (2 Corinthians 5:21; Colossians 1:27).

If a student shows genuine interest and readiness, ask, "Are you ready to put your trust in Jesus and depend on only Christ for salvation?" If the student says yes, then ask him or her to talk to God about this. Perhaps he or she will pray something like the following:

> God, I know that I've sinned against You and that You hate sin, but that You also love me. I believe that Jesus died to pay for my sin and that He rose from the dead, so I put my trust in Jesus to forgive me and give me a home with You forever. In Jesus' name I pray. Amen.

Show the student how to know from God's Word whether he or she is forgiven and in God's family (1 John 5:12–13; John 3:18). Encourage the student to follow Jesus by obeying Him each day. Tell the student that whenever he or she sins, God will grant forgiveness as he or she confesses those sins to God (1 John 1:9).

n	n	n
n	n	n

X	*X*	*X*
X	*X*	*X*
X	*X*	*X*
X	*X*	*X*
X	*X*	*X*

Graphing an Equation

$$distance = rate \times time$$
$$d = r \times t$$

Rate (mph)	10	20	30	40	50	60	70	80	90	100
Time (hours)										

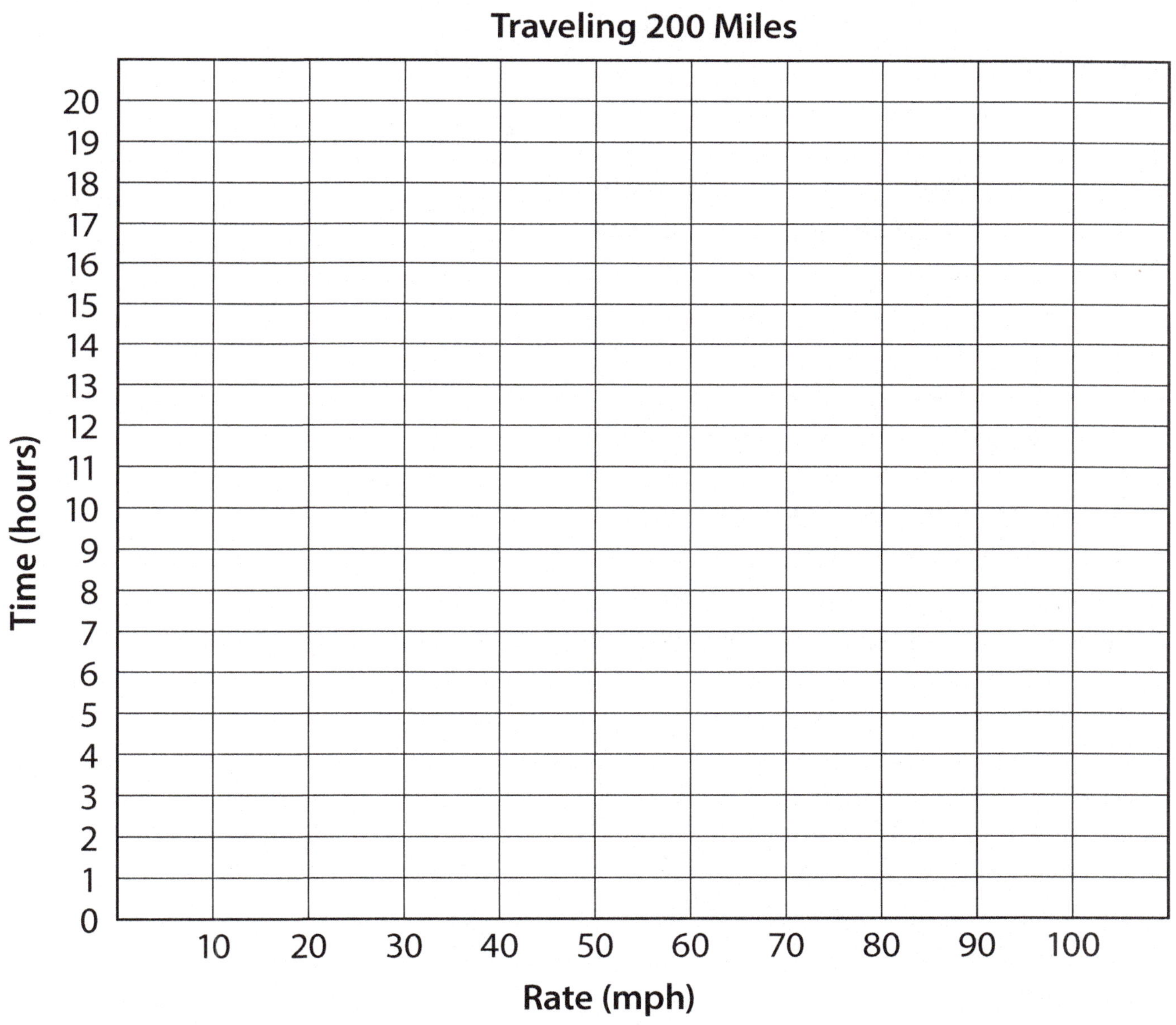

Traveling 200 Miles

Finding the Circumference

1. Write the number of the object in the first column.

2. Place a string around the object to find the circumference. Measure the string to the nearest millimeter. Record the circumference in the second column.

3. Use the ruler to measure the diameter of the object to the nearest millimeter. Write the diameter in the third column.

4. Divide the circumference by the diameter. Round your answer to the nearest hundredth and write it in the fourth column.

5. Multiply the diameter of the object by π (3.14) to find the circumference. Write the circumference in the last column.

Object	Measured Circumference	Measured Diameter	Circumference/ Diameter Relationship $\frac{c}{d} = \pi$	Circumference Formula $C = \pi d$

Multiply the diameter of each circle by π ($\frac{22}{7}$) to find the circumference.

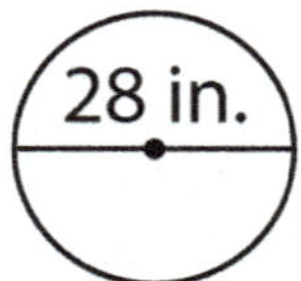

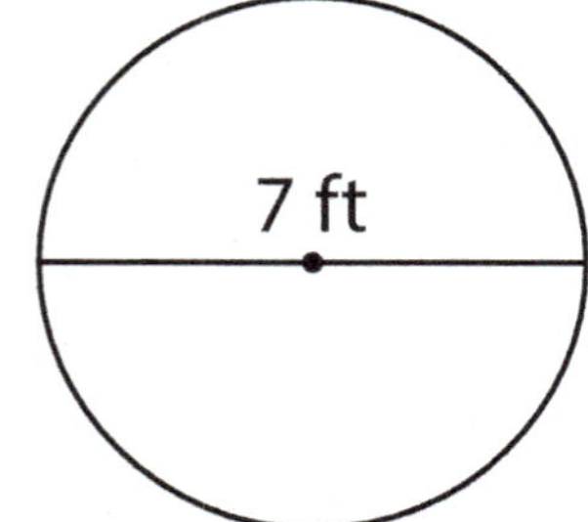

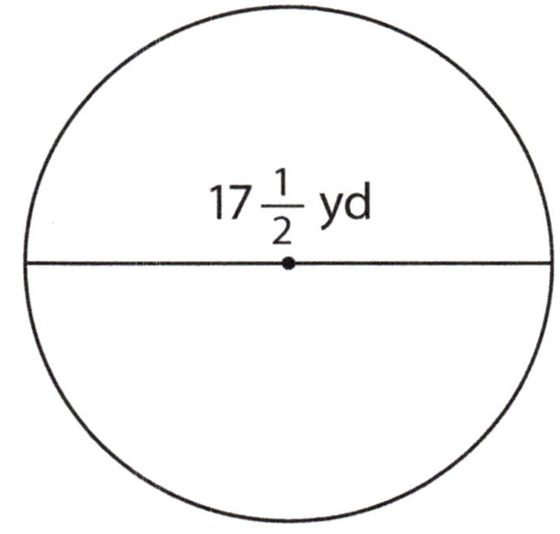

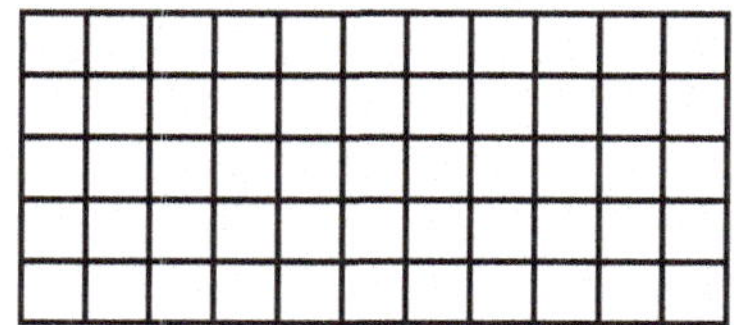

rectangle

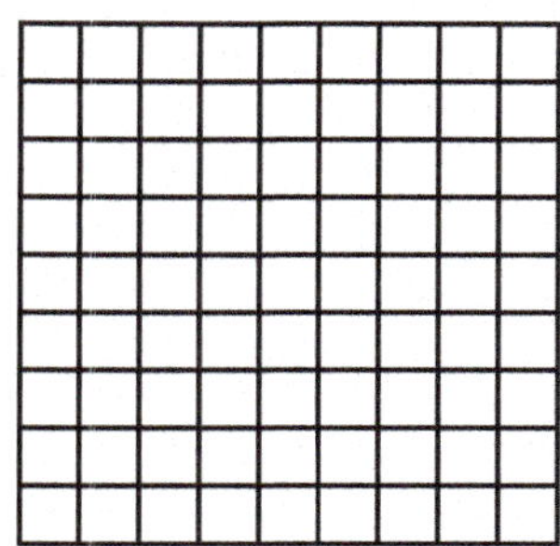

square

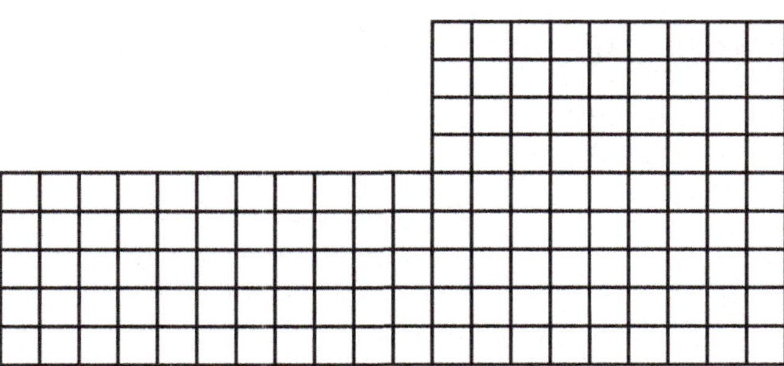

complex figure

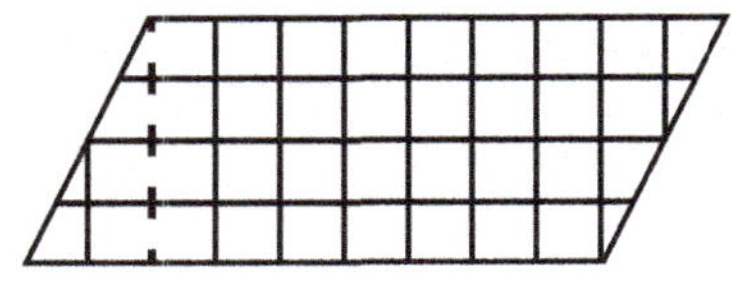

parallelogram

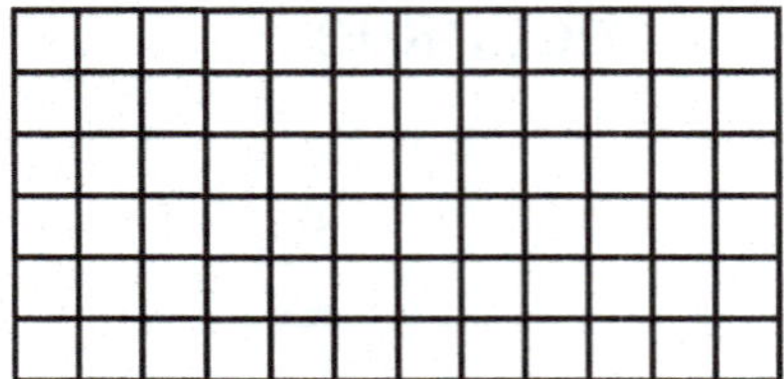

rectangle

triangle

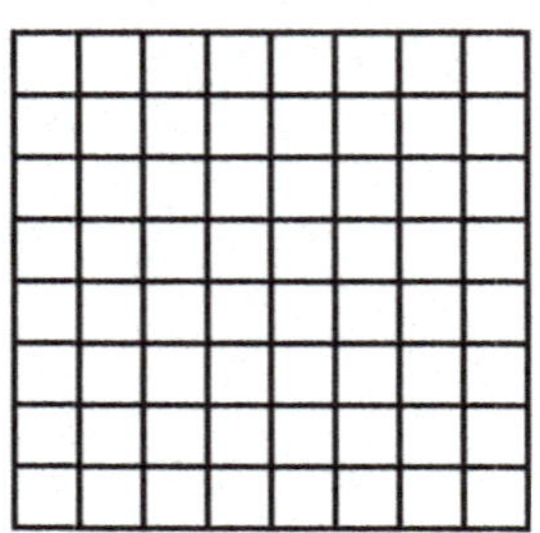

square

triangle

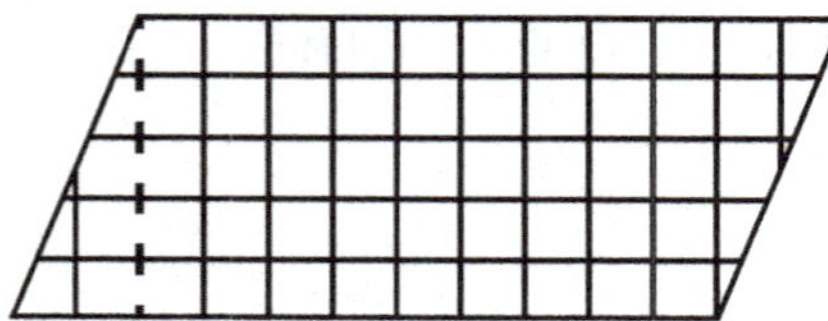

parallelogram

triangle

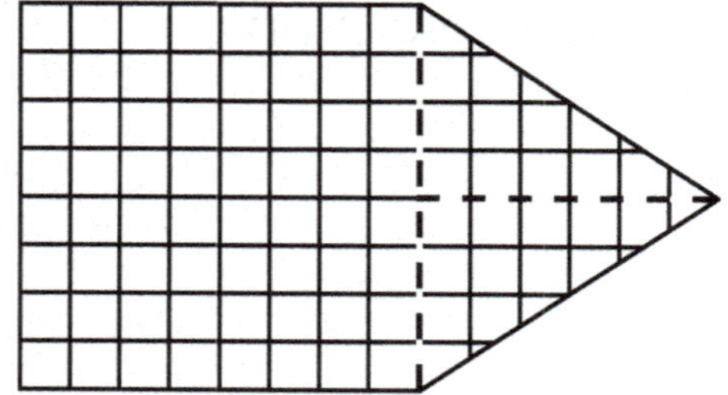

complex figure

Area: Circles

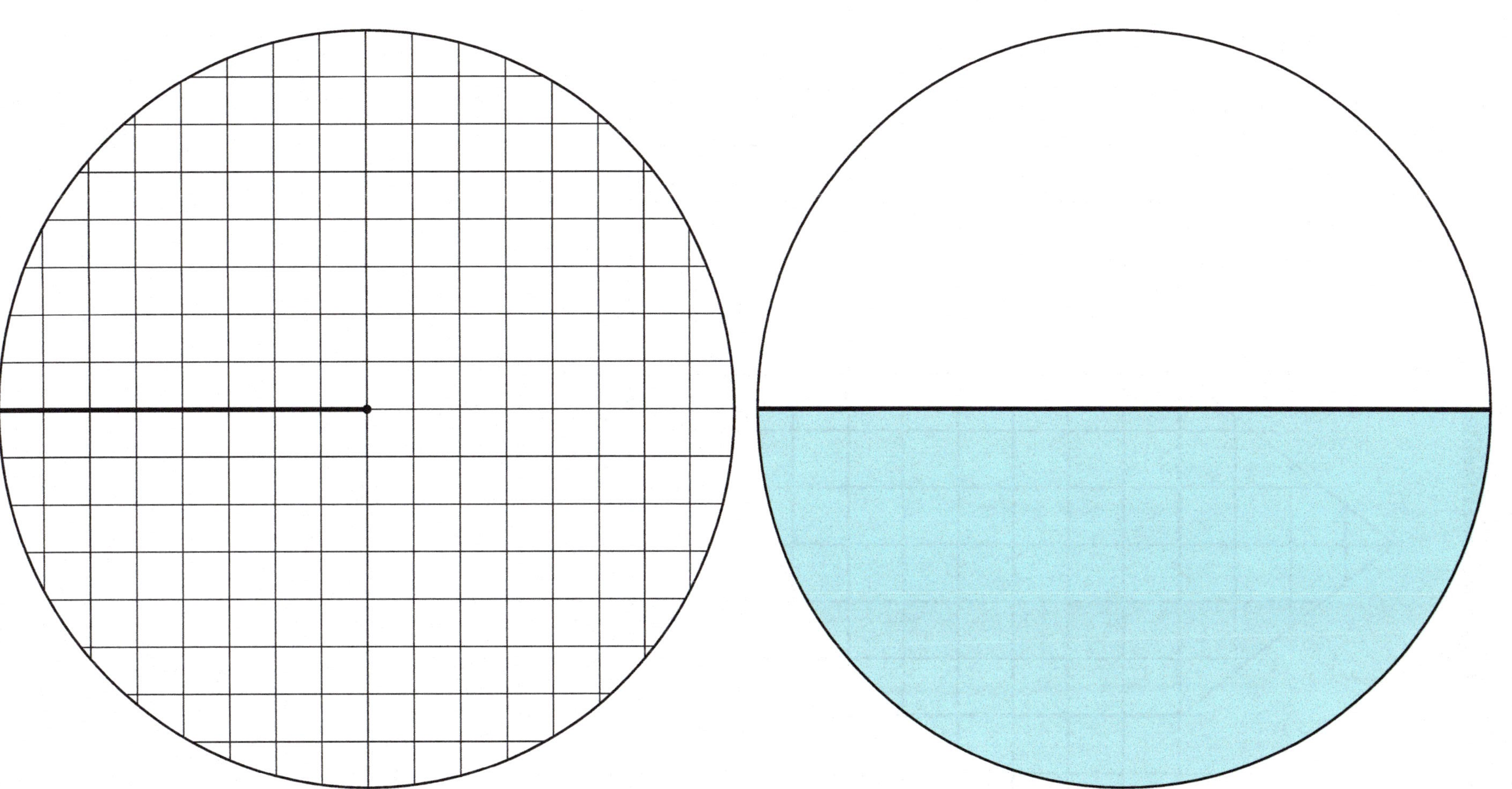

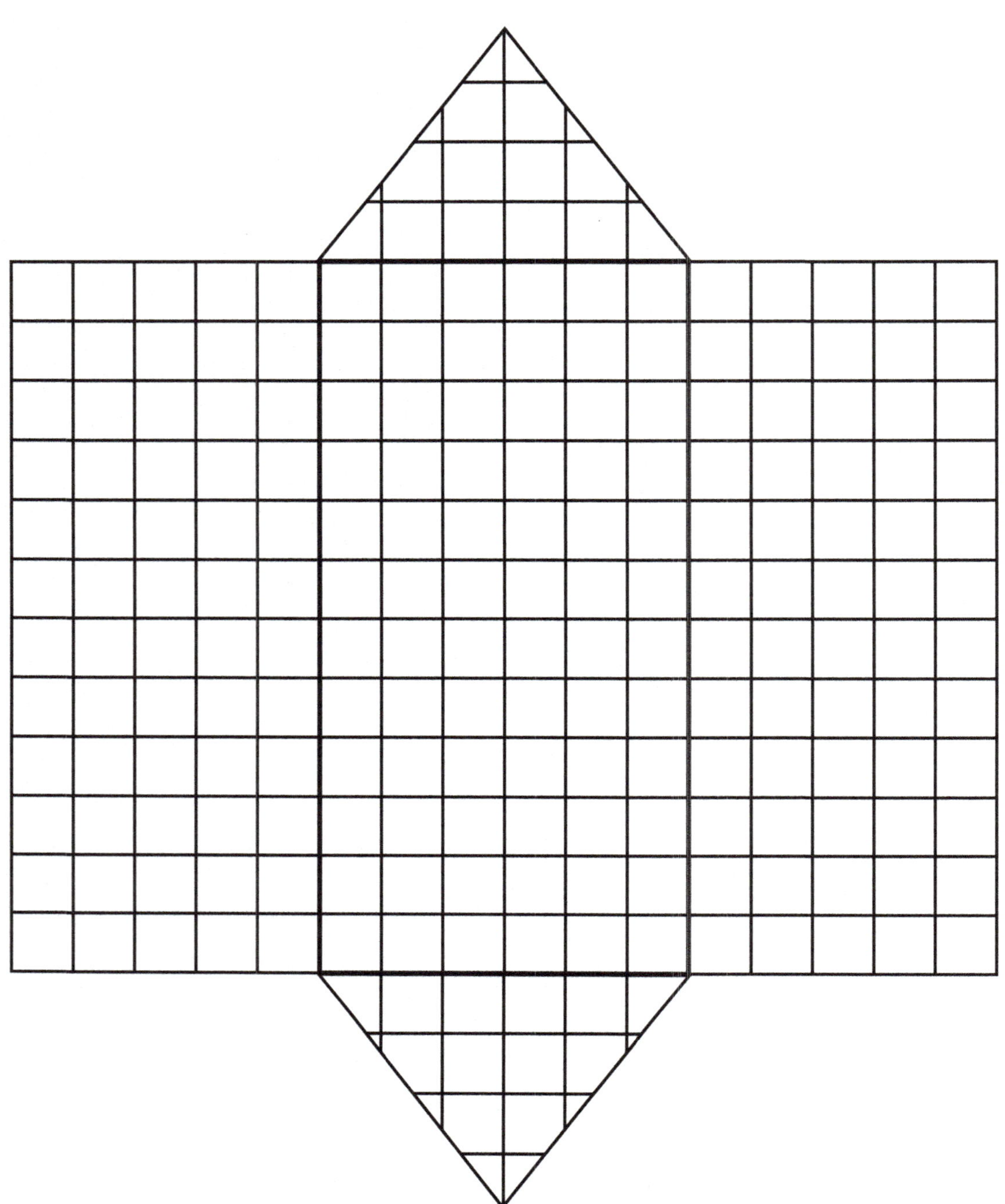

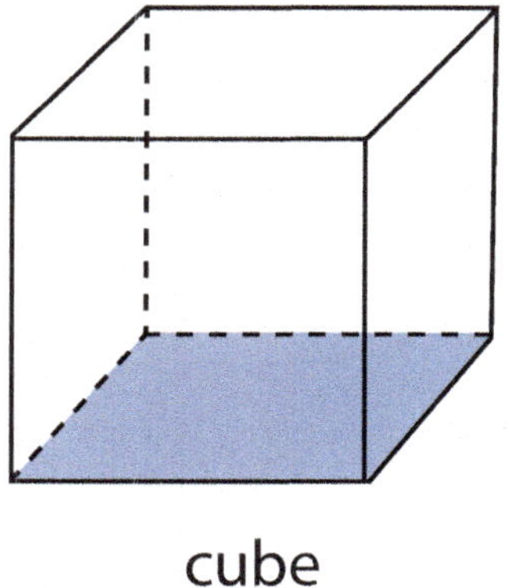

cube

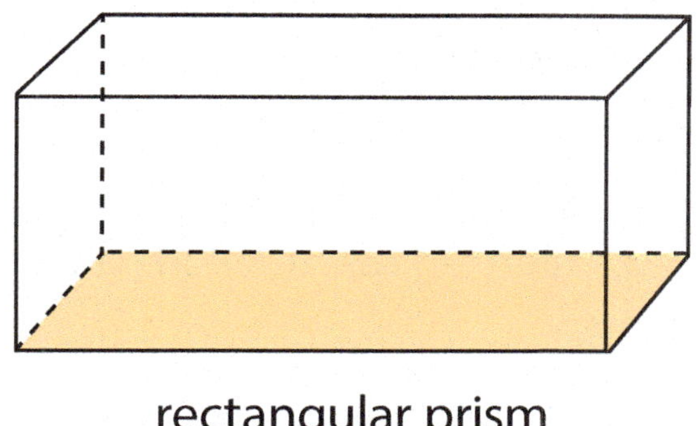

rectangular prism

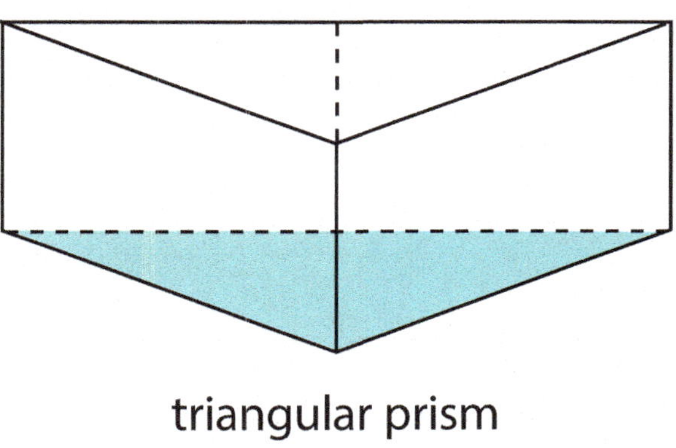

triangular prism

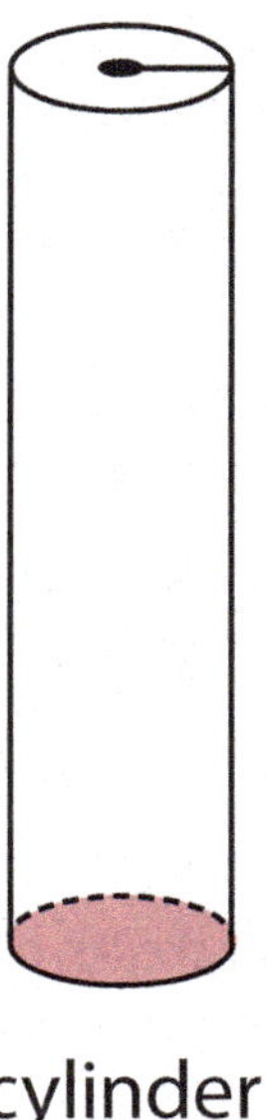

cylinder

Floor Plan Activity

Assignment:

Create a floor plan for a house with an area of 1,500 ft^2.

Materials:

- *Floor Plan Grid* (or graph paper, 30 squares × 50 squares)
- 9" × 12" sheet of construction paper
- scissors
- glue stick

Requirements:

1. Use all 1,500 squares in your floor plan. The fixed area of the house is 1,500 ft^2. The area of a house includes only the heated living areas. Your floor plan does not include a carport, a garage, or additional outside buildings.
2. Include a minimum of 6 rooms: 3 bedrooms, 1 bathroom, a kitchen, and a living room.
3. Include hallways where needed.
4. Use only full squares in your floor plan.

Procedures:

1. Sketch the general layout of the house.
2. Consider the area you want in a room. Draw an outline of the room on the *Floor Plan Grid* and then cut out the area.
3. Place all the rooms on the construction paper.
4. Analyze the layout. Do you need to add a hallway? Do you want to make a room larger or smaller? If you have leftover squares, what extra room will you add to the house, or will you enlarge some rooms?
5. After you have used all your squares and have checked the layout of the rooms, glue the rooms to the construction paper.

Extra Details:

You may choose to draw the doorways leading into rooms. (A doorway should be at least 3 ft wide.) You may choose to draw closets in the bedrooms. (A closet can be any size.) You may choose to draw the cabinets in the kitchen and bathroom. (Because you are working with full squares only, cabinets should be 2 ft deep.)

Floor Plan Grid

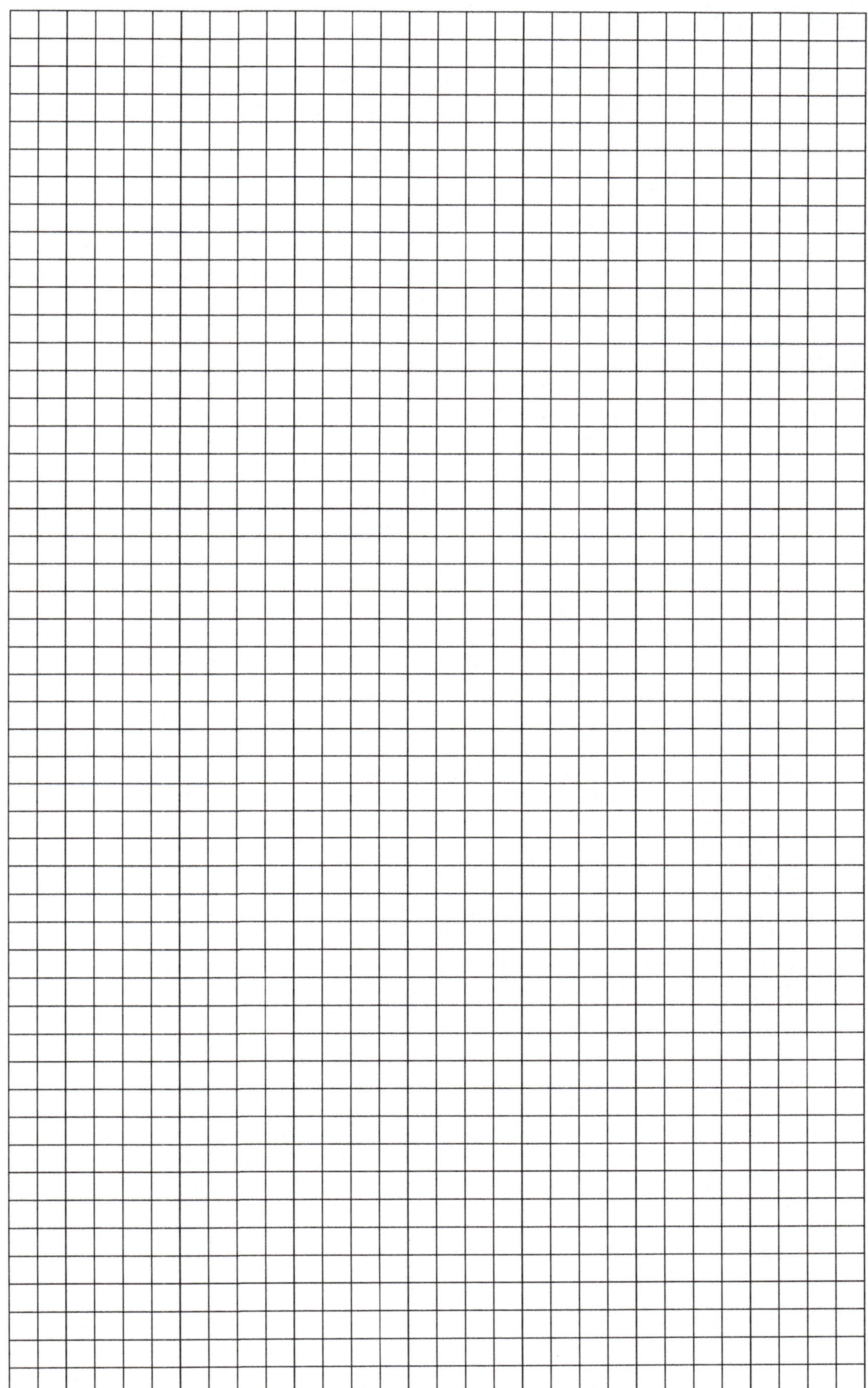

Instructional Aid 62 • For use with Lesson 111

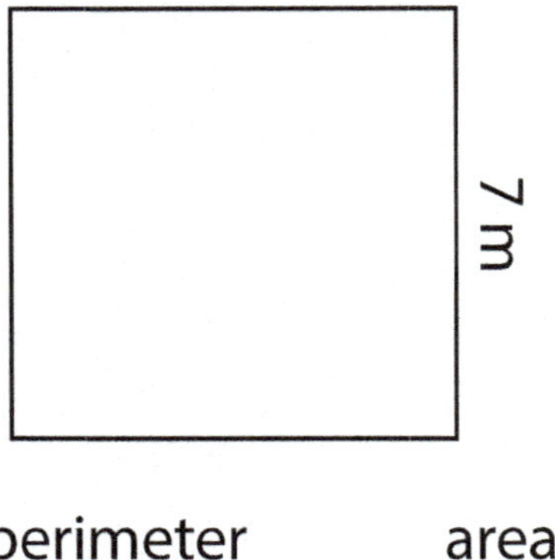

perimeter   area

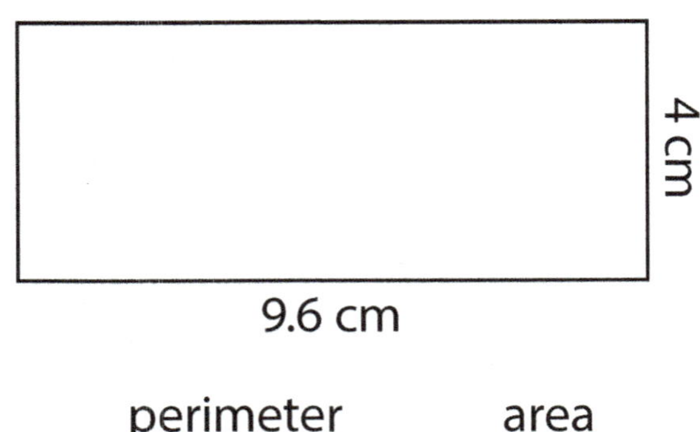

perimeter   area

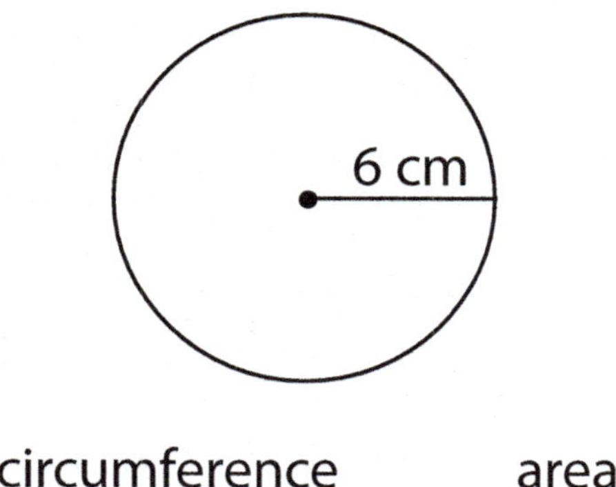

circumference   area

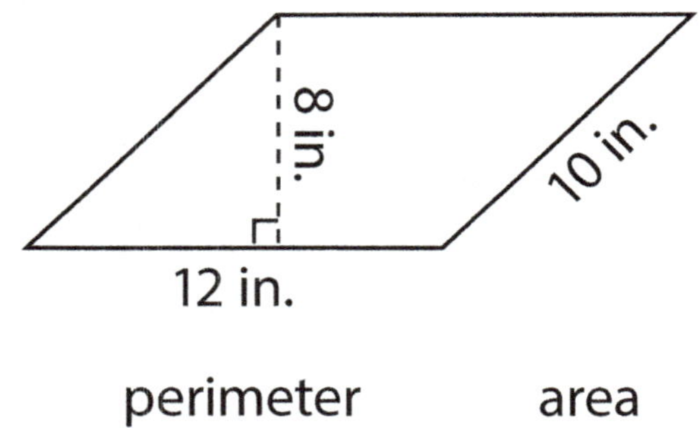

perimeter   area

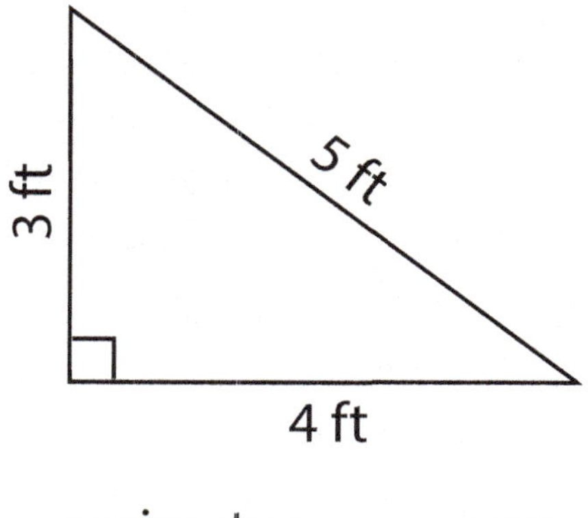

perimeter   area

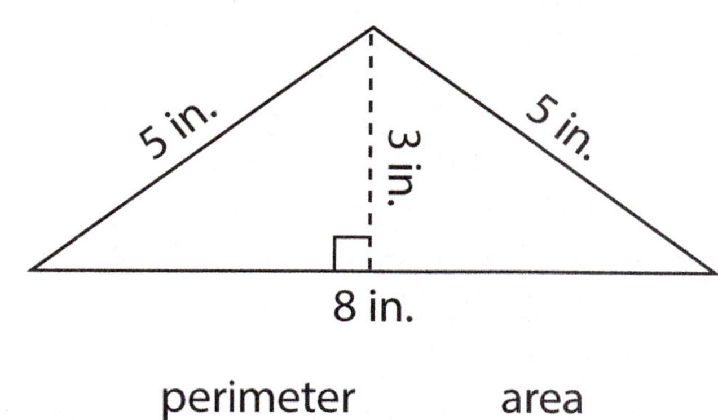

perimeter   area

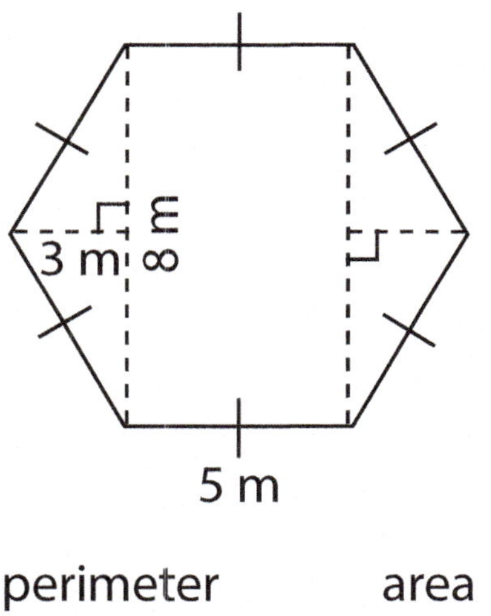

perimeter   area

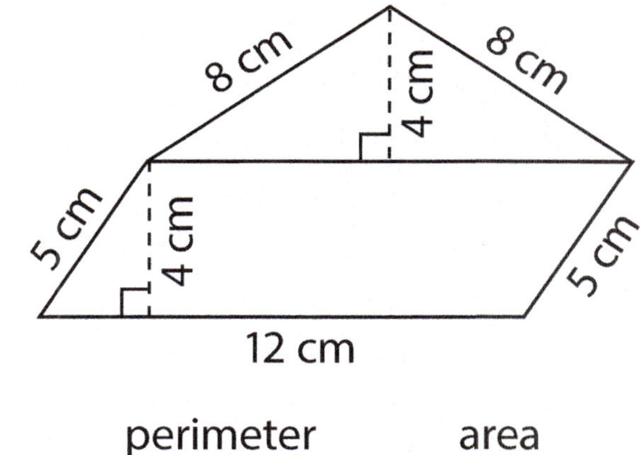

perimeter   area

Nets

1. 

____

2. 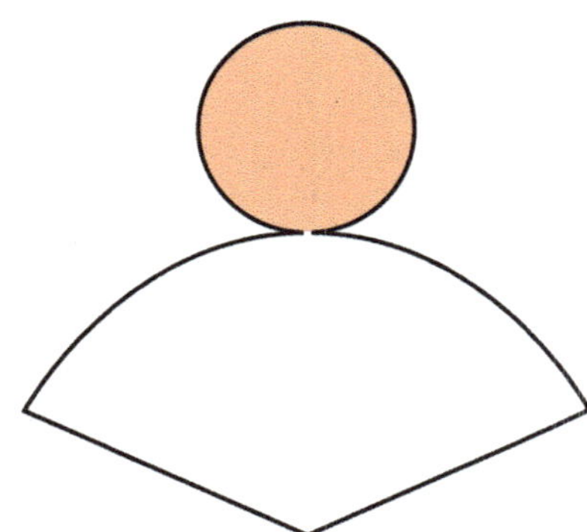

____

3. 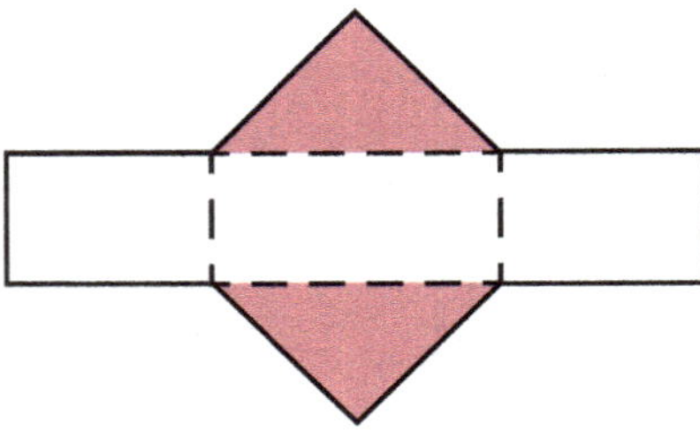

____

4. 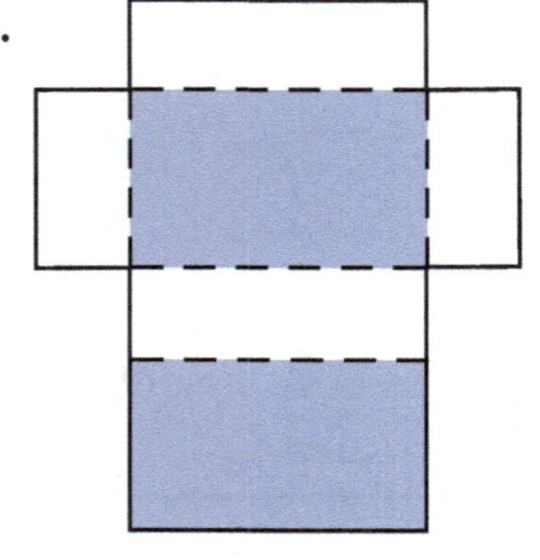

____

5. 

____

6. 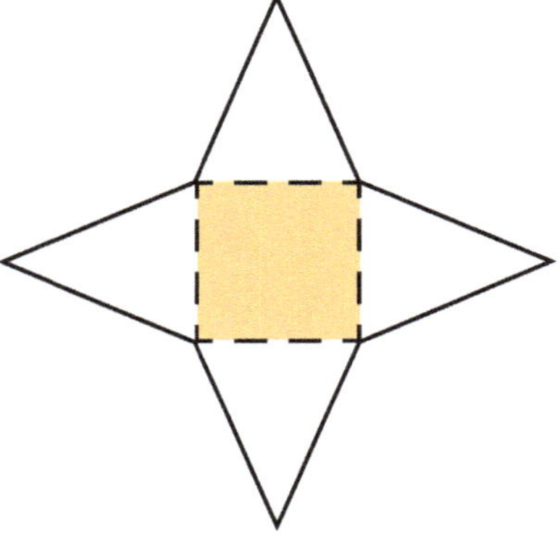

____

7. 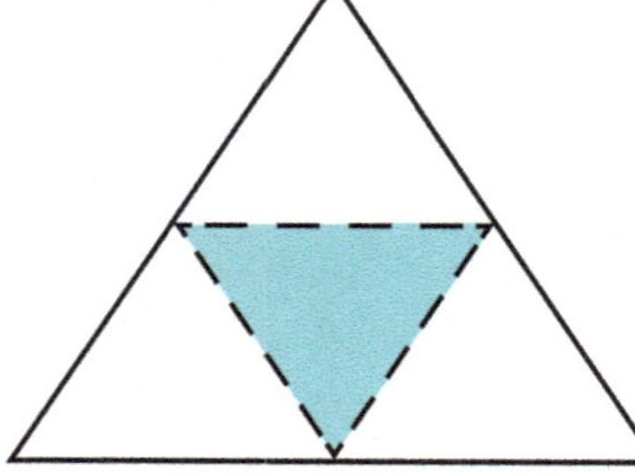

____

8. 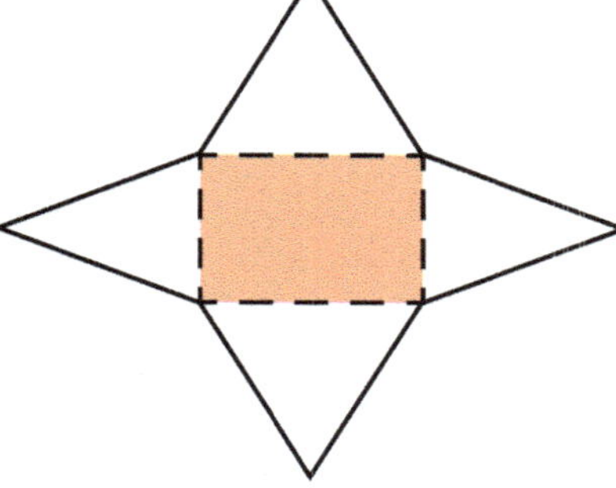

____

Shipping Volume

Shipping companies transport *freight*, the items being shipped, in large enclosed containers called *trailers* that are hauled by powerful trucks called *tractors*. Shippers calculate the volume of their trailers to help them load their freight efficiently and provide better service to their customers.

Freight is secured onto a square platform to make a unit called a *skid cube* for shipping. A trailer's volume can be measured in skid cubes.

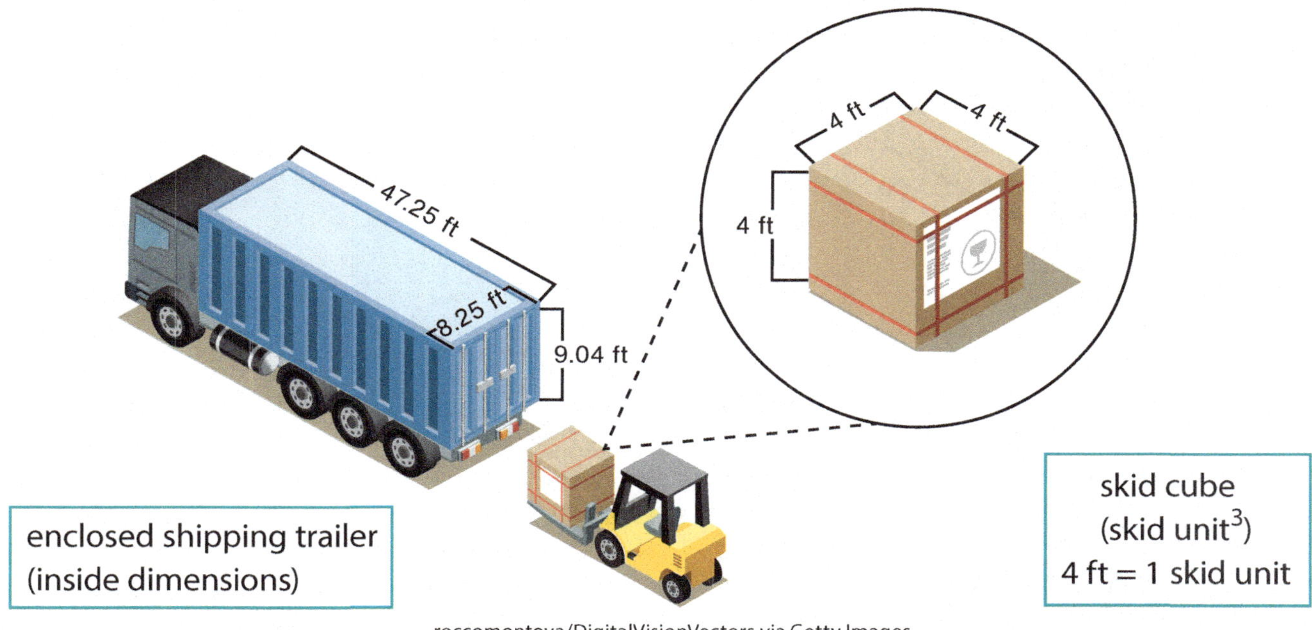

roccomontoya/DigitalVisionVectors via Getty Images

How many skid cubes will fit in the trailer?
Round a decimal answer to the nearest hundredth.

l = 47.25 ft ÷ _______ ft = _______ skid units; _______ skid units will fit

w = 8.25 ÷ _______ ft = _______ skid units; _______ skid units will fit

h = 9.04 ft ÷ _______ ft = _______ skid units; _______ skid units will fit

$V = Bh$

$V = (l \times w) \times h$

$V = ($_______ skid units $\times$ _______ skid units$) \times$ _______ skid units

$V =$ _______ skid units^3 (skid cubes)

Triangular Prisms & Cylinders

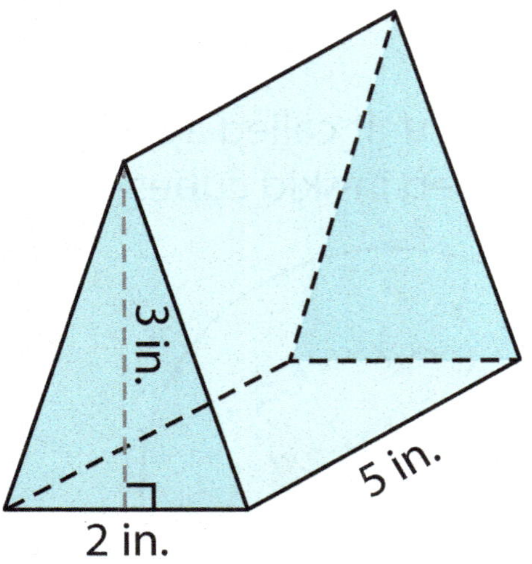

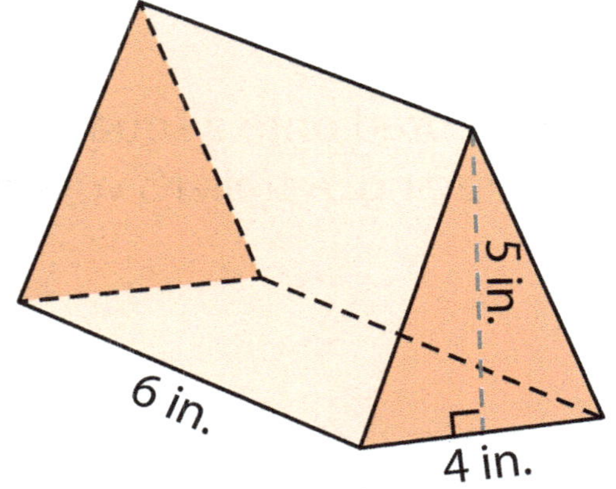

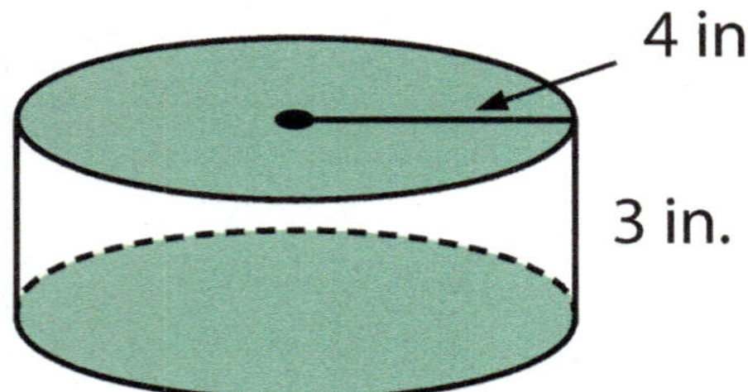

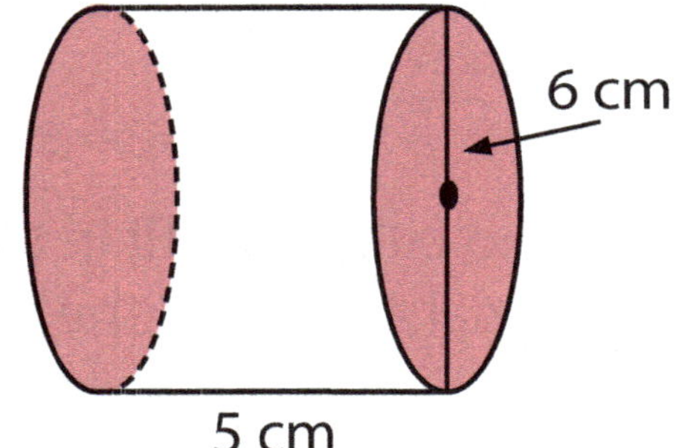

Fixed Volume

Volume units^3	Length unit	Width unit	Height unit	Surface Area units^2

Volume Review

1. 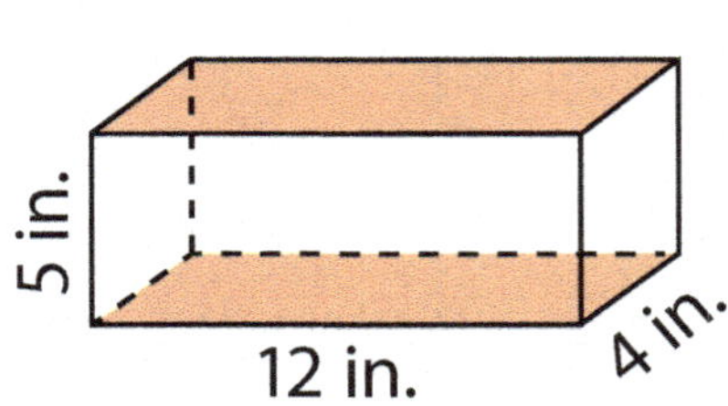

5 in. · 12 in. · 4 in.

2. 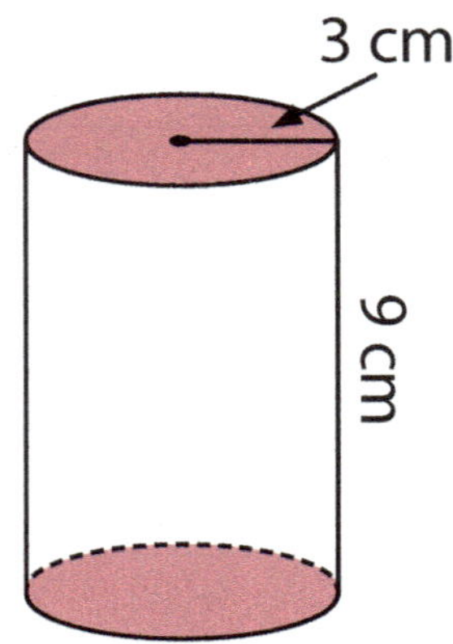

3 cm · 9 cm

3. 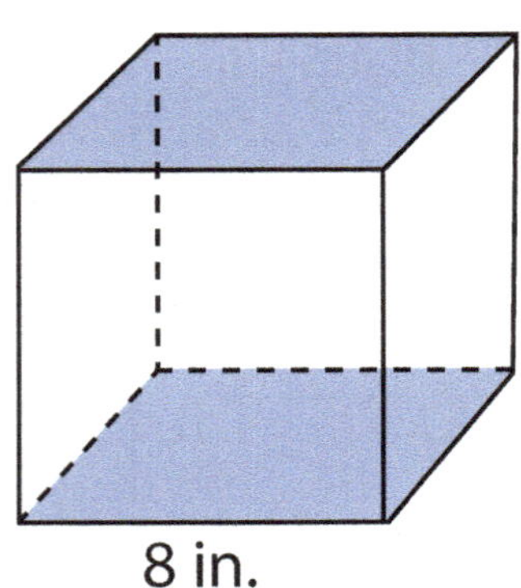

8 in.

4. 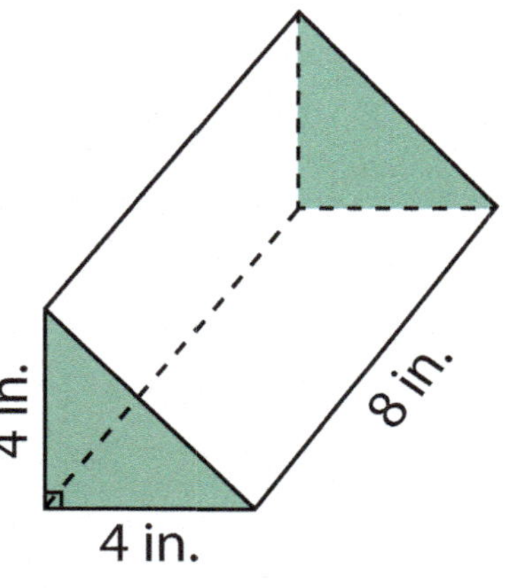

4 in. · 4 in. · 8 in. · 4 in.

5. 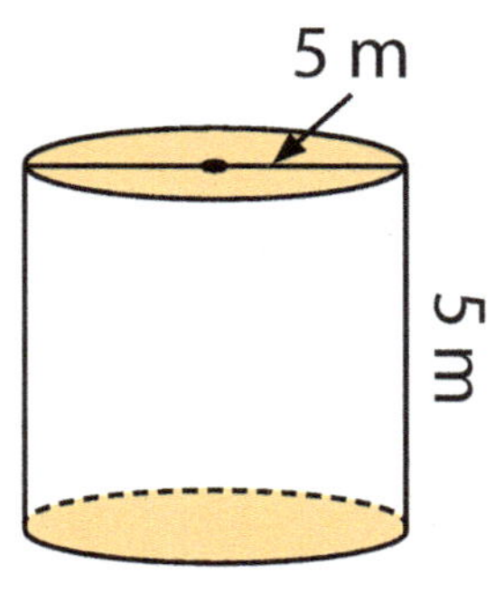

5 m · 5 m

6. 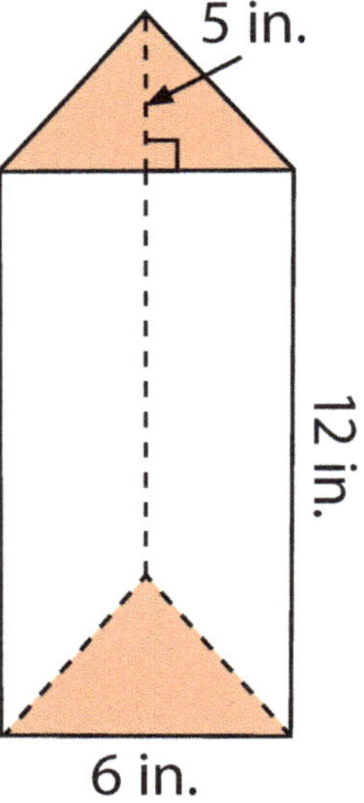

5 in. · 12 in. · 6 in.

7. 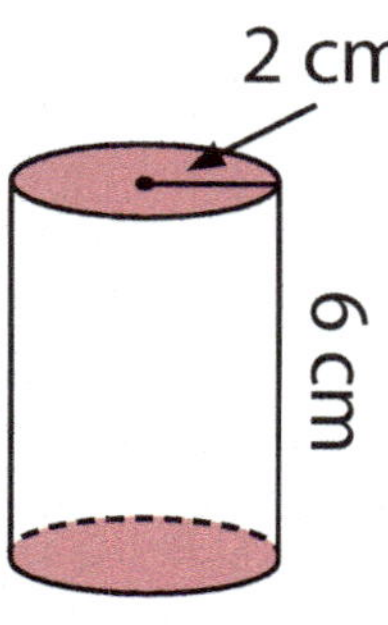

2 cm · 6 cm

8. 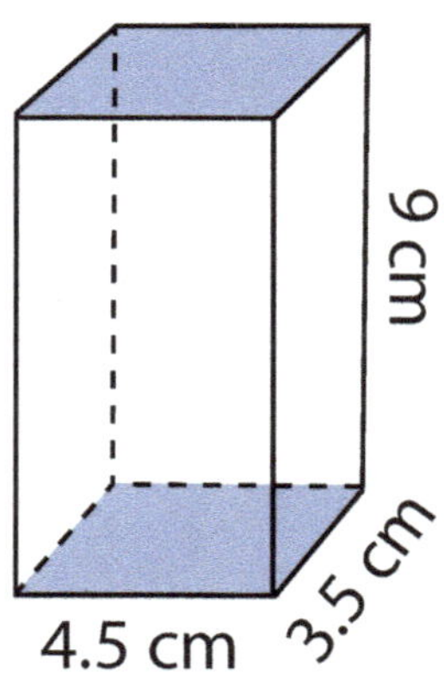

9 cm · 4.5 cm · 3.5 cm

9. 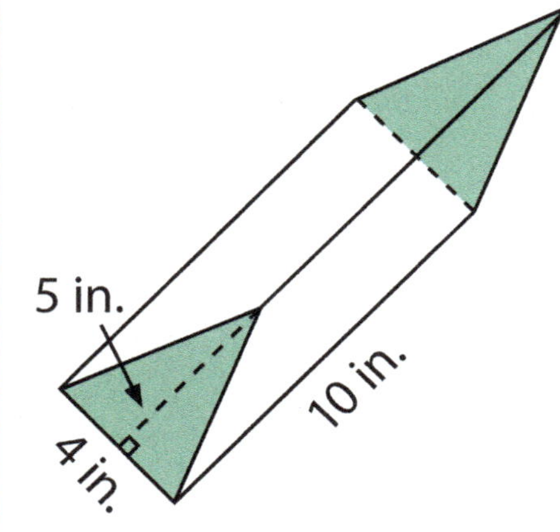

5 in. · 4 in. · 10 in.

Volume Word Problems

1. The Miller family purchased 4 planters to put on their new deck. All the edges of the gray planter are 2 ft in length. How much potting soil did it take to fill the gray planter?

2. The Millers purchased a terra cotta planter with a diameter of 2 ft and a height of 2 ft. How much potting soil did it take to fill the terra cotta planter?

3. The Millers bought a brown planter that is 4 ft long, 2 ft wide, and 1.5 ft high. How much potting soil did it take to fill the brown planter?

4. The fourth planter is blue with a triangular base and 3 rectangular sides. The triangle's base is 4 ft long, and its height is 4 ft. The height of the planter is 2 ft. How much potting soil did it take to fill the blue planter?

Dear Parent:

____
(date)

While studying measurement in math class, the students will be estimating and measuring the length, weight/mass, and capacity of real-world objects. To help with these lessons, please send one or more of the following items to class by __________________.

(date)

Please write your child's name on any item that you would like returned after we have completed the measurement chapter. Items without a name will either be kept in the classroom or be donated to a needy family or a food bank.

Thank you for the objects that you are able to provide.   Sincerely,

- a bag of candy, dried beans, apples, or other weighable items
- a weighable item that fits in your hand
- a boxed food item
- a canned food item
- a 1-liter bottle of water for each student
- waterproof containers of various shapes and sizes: 1 cup, 1 pint, 1 quart, or 1 gallon

---

Dear Parent:

____
(date)

While studying measurement in math class, the students will be estimating and measuring the length, weight/mass, and capacity of real-world objects. To help with these lessons, please send one or more of the following items to class by __________________.

(date)

Please write your child's name on any item that you would like returned after we have completed the measurement chapter. Items without a name will either be kept in the classroom or be donated to a needy family or a food bank.

Thank you for the objects that you are able to provide.   Sincerely,

- a bag of candy, dried beans, apples, or other weighable items
- a weighable item that fits in your hand
- a boxed food item
- a canned food item
- a 1-liter bottle of water for each student
- waterproof containers of various shapes and sizes: 1 cup, 1 pint, 1 quart, or 1 gallon

Instructional Aid 71 • For use with Chapter 13   Math 6 • Teacher Resources

Pictured Ratios

1. 

2. 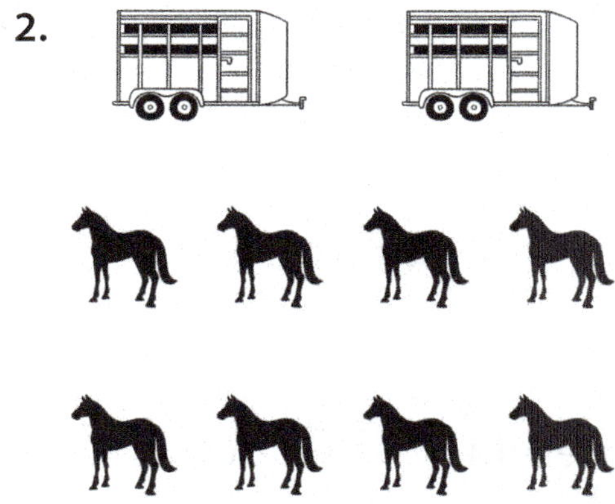

3. 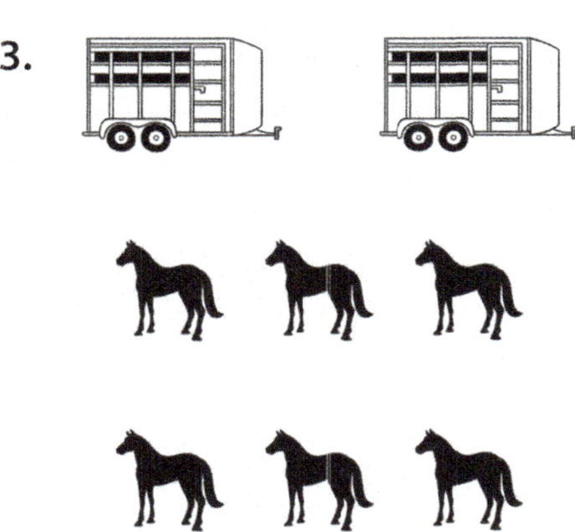

4. 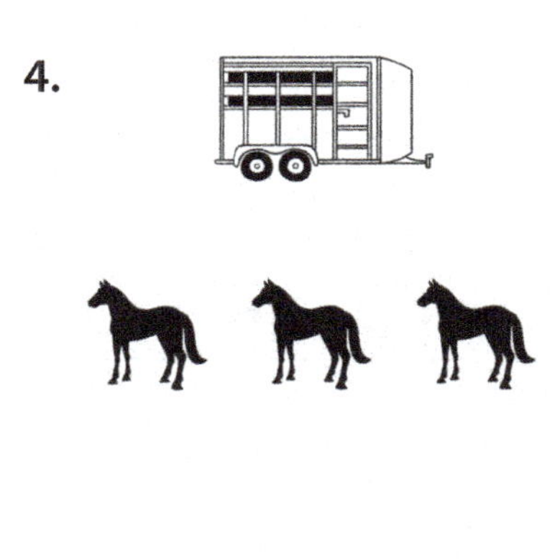

5. 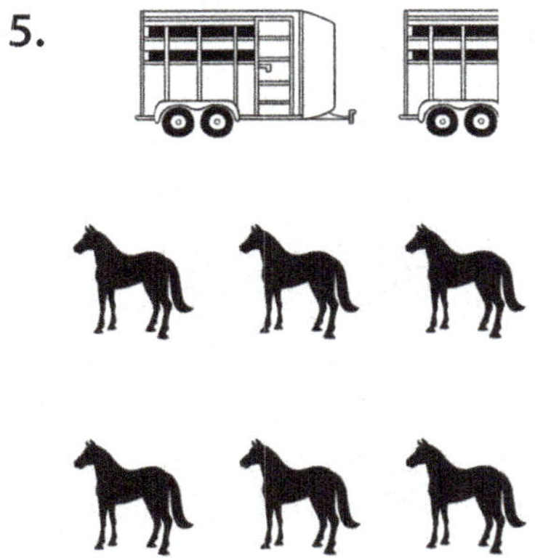

6. 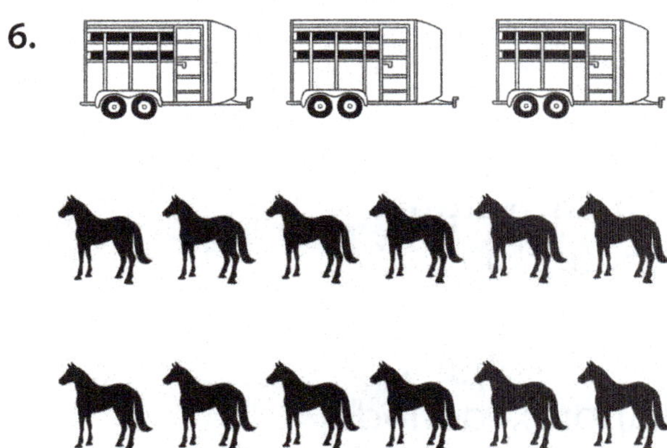

Ratio Tables

Horses	4	6	8		
Trailers	1	1.5	2	3	4

Centimeters	100				
Meters	1	2	3	4	5

Ounces	16				
Pounds	1	2	4	8	16

Tax					
Purchase					

Earth Weight (pounds)	100	60	40	20	10
Mars Weight (approx. pounds)	40	24	16	8	4

Missing Measurements

Similar Figures A and B

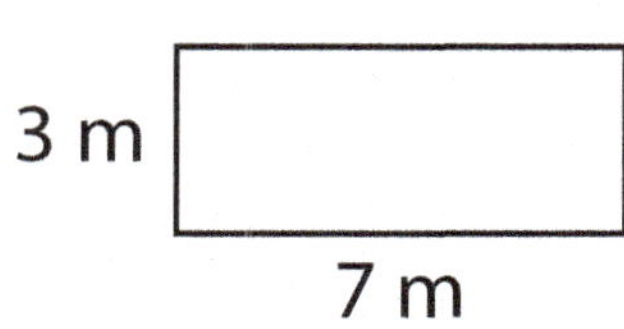

Figure A

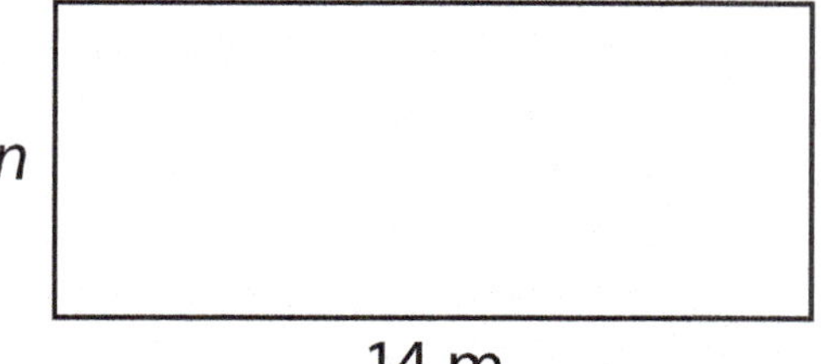

Figure B

Similar Figures C and D

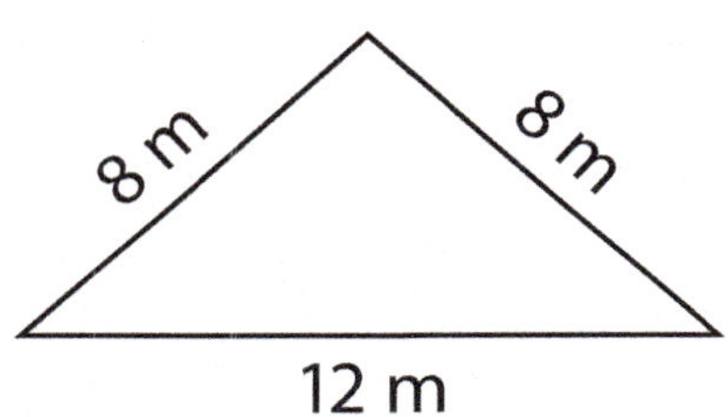

Figure C

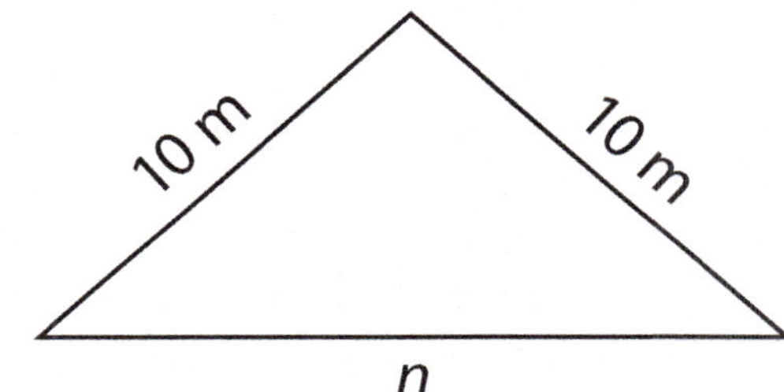

Figure D

Similar Figures E and F

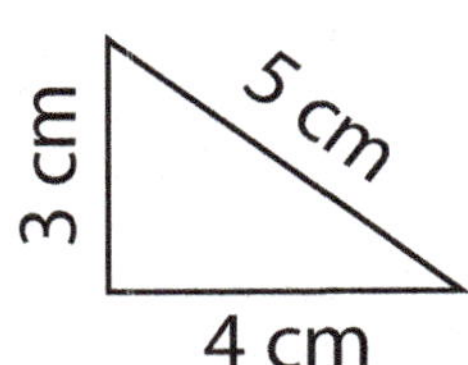

Figure E

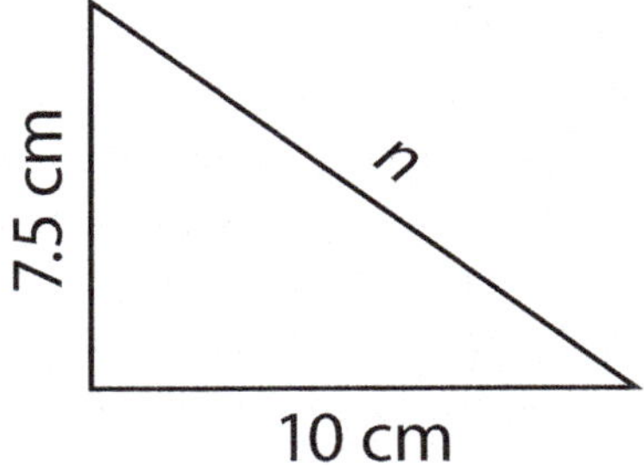

Figure F

Percent

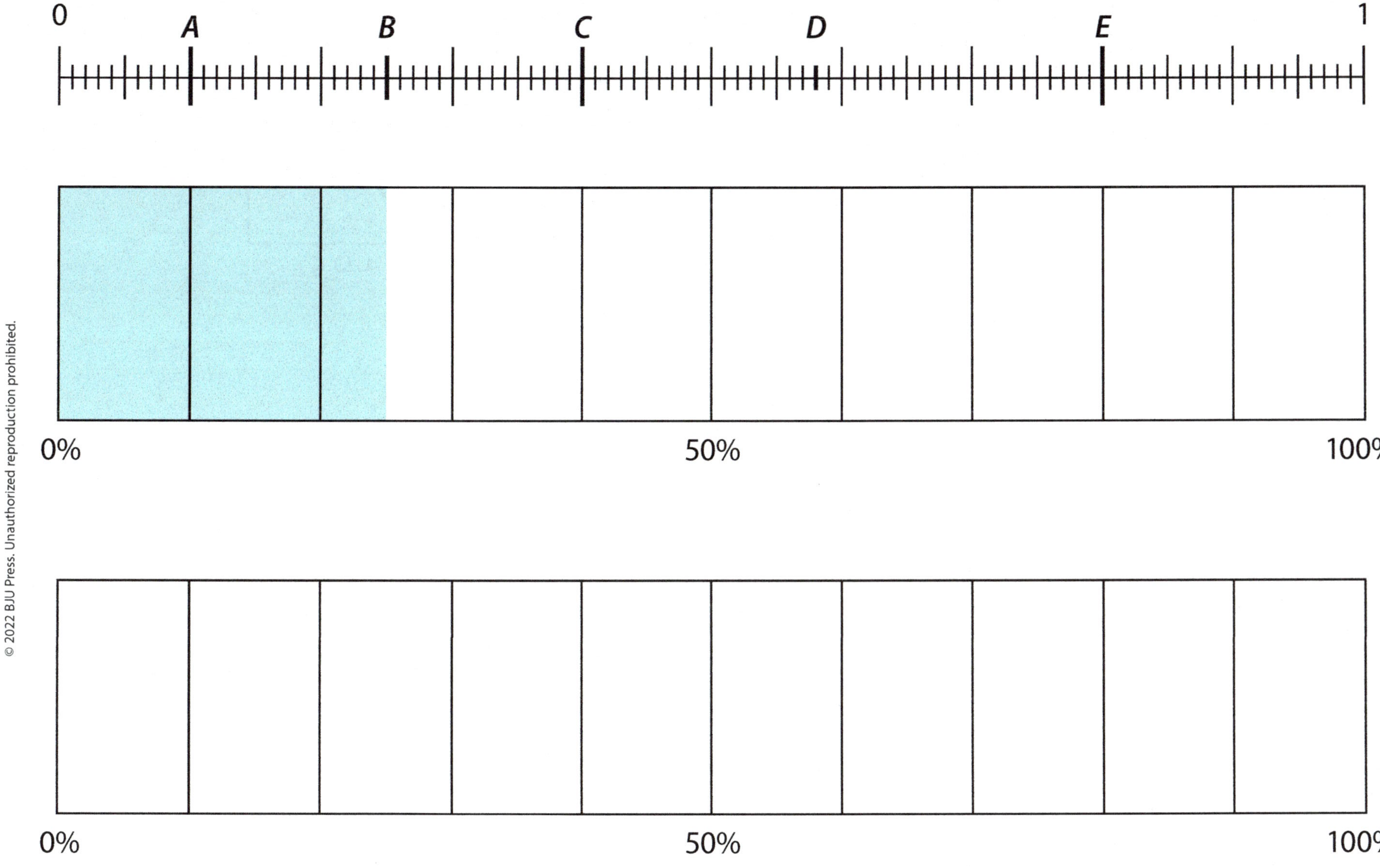

Instructional Aid 75 • For use with Lesson 125

Math 6 • Teacher Resources

Percent Models: Finding the Part

Brad earned $60. He gave 10% to the church and saved 40%. How much money did Brad give to the church and save?

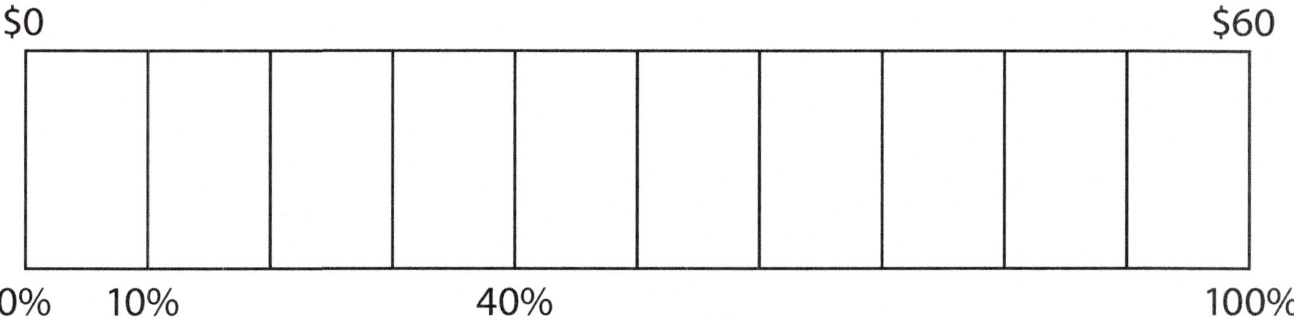

Shannon saved 60% of the $75 she had earned. How much money did Shannon save?

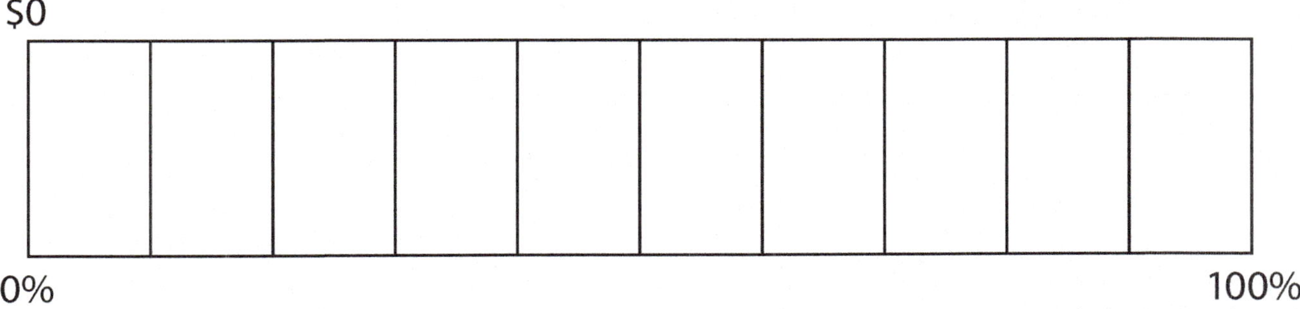

John purchased a bike during a 30% off sale at the store. The original price of the bike was $180. What was the amount of the discount? What was the sale price?

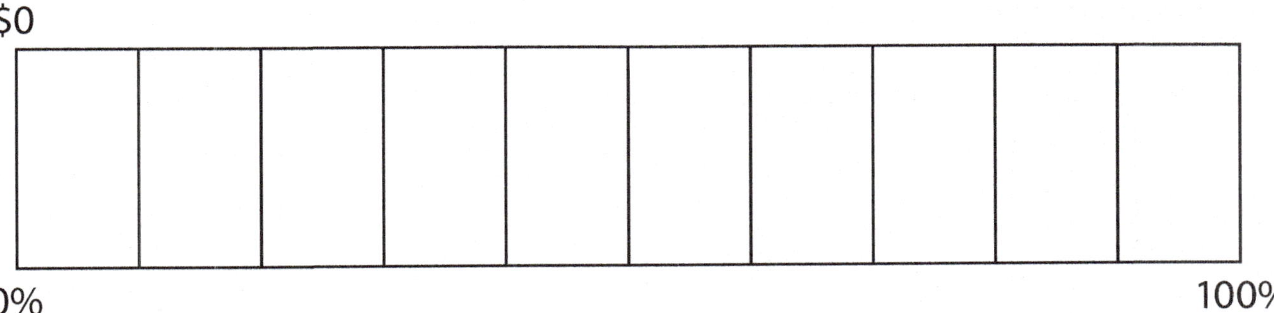

Percent Models: Finding the Whole

Dori saved $120 this week. This amount is 60% of the money she earned. What is the total amount that Dori earned this week?

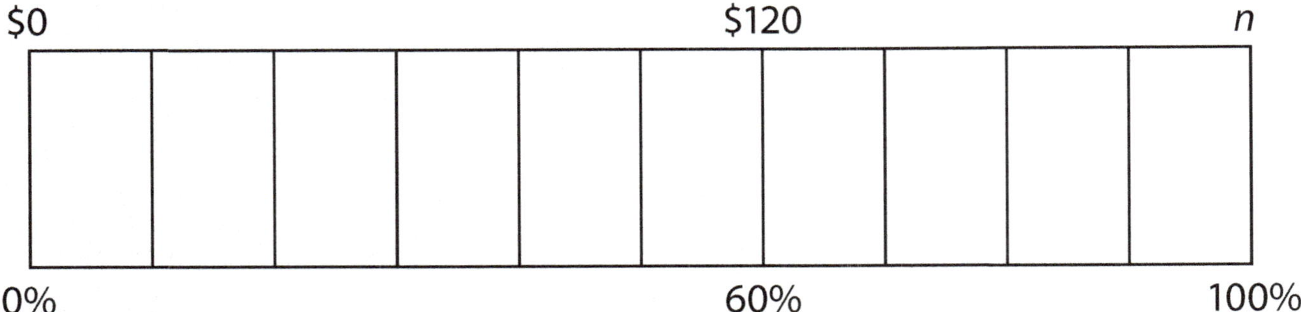

Board games were on sale for 40% off the regular price. Eric received a discount of $14 on the game he purchased. What was the original price of the board game?

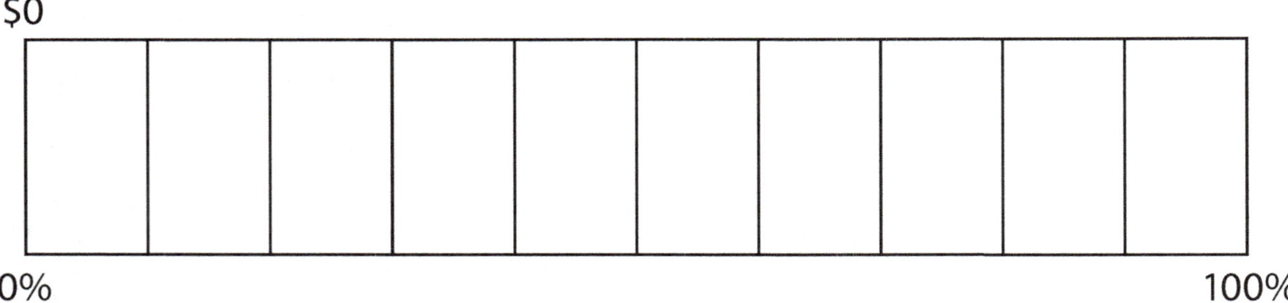

Caleb has saved $72 for summer camp. He has saved 30% of the amount he needs. What is the total amount of money he needs for camp?

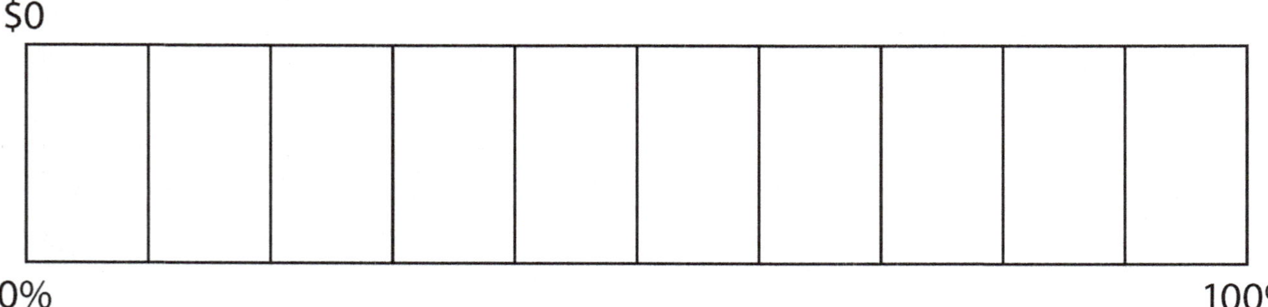

Elements	Percent	Ratio (fraction form)	Fraction (lowest terms)	Decimal
oxygen	47%			
silicon	28%			
aluminum	8%			
iron	5%			
calcium	4%			
other	8%			

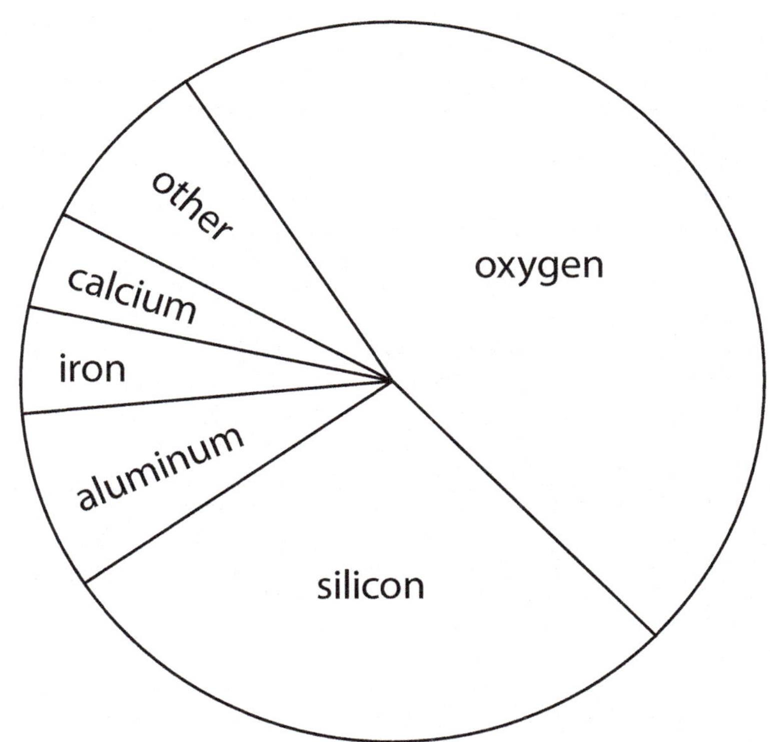

Bridges bear loads by balancing the forces acting on them.

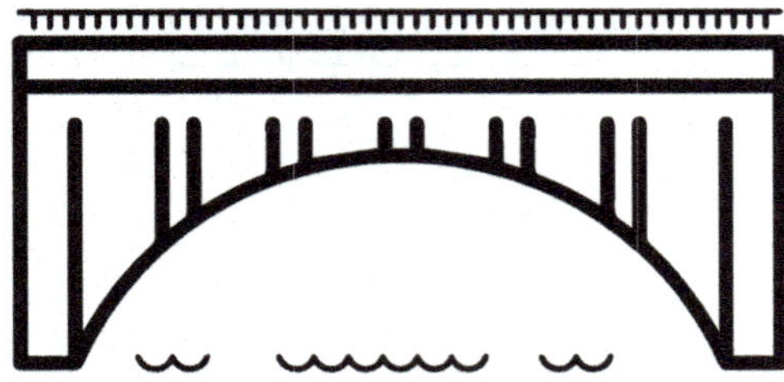

The **arch bridge** relies on the naturally strong structure of an arch.

A **beam bridge's** strong horizontal roadway is supported underneath by pillars.

Strong cables fan out from one or more towers in a **cable-stayed bridge** to support the bridge deck.

The strong top cables and the smaller vertically suspended cables of a **suspension bridge** help support the weight of the bridge's deck.

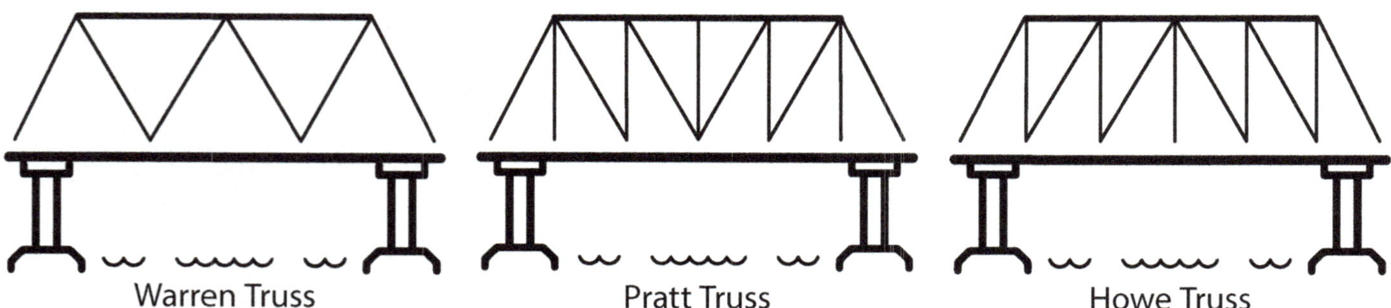

A system of triangles spread out the load and support the roadway of a **truss bridge**.

shopplaywood/iStock/Getty Images Plus via Getty Images

STEM Rubric: Building a Bridge

Name_______________________

	Excellent (3)	Good (2)	Needs Improvement (1)	Points Earned
The Engineering Design Process	Uses all the steps of the Engineering Design Process to solve the problem of designing and building an efficient bridge.	Uses some of the steps of the Engineering Design Process to solve the problem of designing and building an efficient bridge.	Uses few or does not use the steps of the Engineering Design Process to solve the problem of designing and building an efficient bridge.	
Model	Bridge design shows understanding of concepts.	Bridge design shows understanding of most concepts.	Bridge design shows lack of understanding of concepts, or model is missing.	
Data Record	Data record is complete, well organized, and easy to follow.	Data record is mostly complete, organized, and easy to follow.	Data record is incomplete, lacks organization, or is hard to follow.	
Collaboration with Peers	Always listens carefully to others and offers detailed, constructive feedback. Participates fully and shares the workload fairly.	Sometimes listens to others; occasionally offers constructive feedback. Participates but sometimes does not share the workload fairly.	Does not listen to others and often interrupts them. Does not offer constructive feedback. Does not participate and relies on others to carry the workload most of the time.	

Comments **Total**

Measurement Flashcards

1 lb = ______ oz	16 oz = ______ lb
1 tn = ______ lb	2,000 lb = ______ tn
1 c = ______ fl oz	8 fl oz = ______ c
2 c = ______ pt	1 pt = ______ c
2 pt = ______ qt	1 qt = ______ pt
1 qt = ______ c	4 c = ______ qt

16 oz = 1 lb

1 lb = 16 oz

2,000 lb = 1 tn

1 tn = 2,000 lb

8 fl oz = 1 c

1 c = 8 fl oz

1 pt = 2 c

2 c = 1 pt

1 qt = 2 pt

2 pt = 1 qt

4 c = 1 qt

1 qt = 4 c

4 qt = _______ gal

1 gal = _______ qt

12 in. = _______ ft

1 ft = _______ in.

36 in. = _______ yd

1 yd = _______ in.

3 ft = _______ yd

1 yd = _______ ft

5,280 ft = _______ mi

1 mi = _______ ft

1,760 yd = _______ mi

1 mi = _______ yd

1 gal = 4 qt

4 qt = 1 gal

1 ft = 12 in.

12 in. = 1 ft

1 yd = 36 in.

36 in. = 1 yd

1 yd = 3 ft

3 ft = 1 yd

1 mi = 5,280 ft

5,280 ft = 1 mi

1 mi = 1,760 yd

1,760 yd = 1 mi

1 m = _______ cm

100 cm = _______ m

1 m = _______ mm

1,000 mm = _______ m

1 km = _______ m

1,000 m = _______ km

1 L = _______ mL

1,000 mL = _______ L

1 kg = _______ g

1,000 g = _______ kg

100 cm = 1 m

1 m = 100 cm

1,000 mm = 1 m

1 m = 1,000 mm

1,000 m = 1 km

1 km = 1,000 m

1,000 mL = 1 L

1 L = 1,000 mL

1,000 g = 1 kg

1 kg = 1,000 g

Customary Measurement Craze

Measurable Item	Estimate	Measurement

Metric Measurement

	kilo- 1,000	hecto- 100	deka- 10	1	deci- 0.1	centi- 0.01	milli- 0.001
Length	kilometer km	hectometer hm	dekameter dkm	meter m	decimeter dm	centimeter cm	millimeter mm
Capacity	kiloliter kL	hectoliter hL	dekaliter dkL	liter L	deciliter dL	centiliter cL	milliliter mL
Mass	kilogram kg	hectogram hg	dekagram dkg	gram g	decigram dg	centigram cg	milligram mg

Metric Measure Mania

Item		Meters	Centimeters	Millimeters
	estimate			
	measurement			
	estimate			
	measurement			
	estimate			
	measurement			
	estimate			
	measurement			
	estimate			
	measurement			
	estimate			
	measurement			

Customary Measurement Word Problems

1. The Poole Family raises broiler chickens. They feed the baby chicks starter feed.

   Week 1   4 oz of starter feed per chick

   Week 2   9 oz of starter feed per chick

   Week 3   1 lb 1 oz of starter feed per chick

   How much starter feed is needed for 1 chick for the first 3 weeks?

2. One of the Poole's broiler chickens weighs 5 lb 4 oz after seven weeks of intense feeding. One of their laying hens that is the same age weighs 4 lb 8 oz. What is the difference in the weights of the two chickens?

3. The Pooles had 12 baby chicks to raise. How many pounds of starter feed were needed to feed the 12 chicks for the first 3 weeks?

4. The baby chicks were kept in a small square area that is easy to keep warm. The length of the area is 5 ft 8 in. What is its perimeter?

5. At the end of seven weeks, the twelve broiler chickens had a total weight of 60 lb 12 oz. What was the average weight of one broiler chicken?

6. The chicks ate half of a 5 lb bag of starter feed. How many pounds of starter feed were eaten? How many ounces of starter feed were eaten?

Metric Measurement Word Problems

1. The Poole family raises laying hens. Each chick consumes 5.74 kg of chicken feed in seven weeks. How many kilograms of feed should the Pooles purchase for 12 chickens?

2. The Pooles have 24 laying hens. Each laying hen needs 0.5 L of water each day. How much water do the Pooles need to supply for the hens each day?

3. The Poole family also raises broiler chickens. One of their broiler chickens has a mass of 2.4 kg after seven weeks of intense feeding. One of their laying hens that is the same age has a mass of 1.9 kg. What is the difference in the masses of the two chickens?

4. At the end of seven weeks, 12 of the Poole's broiler chickens had a total mass of 37.2 kg. What was the average mass of each broiler chicken?

5. The broiler chicks ate two-thirds of a 3.27 kg bag of starter feed. How many kilograms of the starter feed were left? How many grams of the starter feed were left?

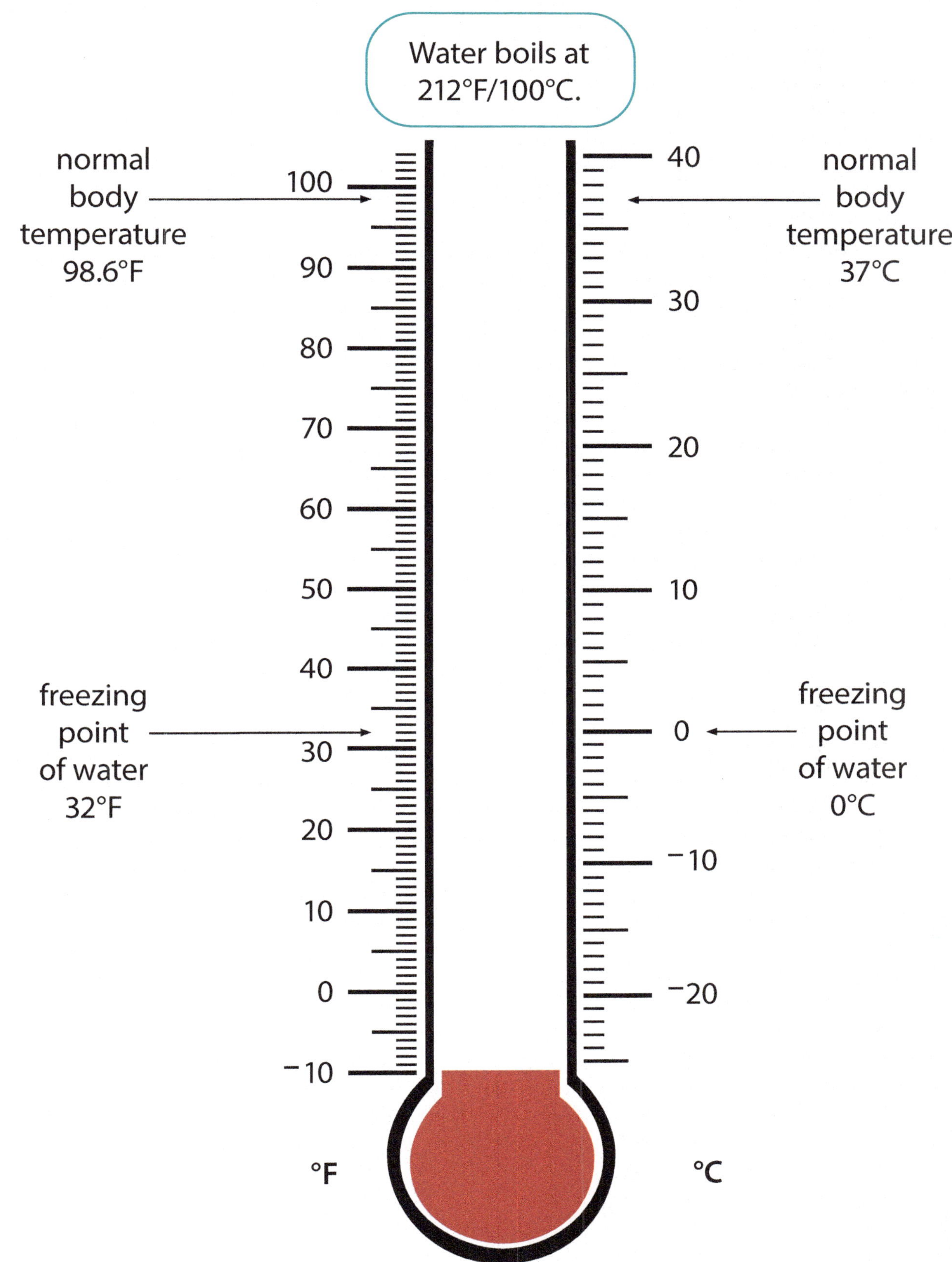

Water boils at 212°F/100°C.
normal body temperature 98.6°F
normal body temperature 37°C
freezing point of water 32°F
freezing point of water 0°C
100
90
80
70
60
50
40
30
20
10
0
−10
40
30
20
10
0
−10
−20
°F
°C

Temperature Hunt

Item	Type of Thermometer	Temperature (°F or °C)	Time

Double-Scale Thermometer

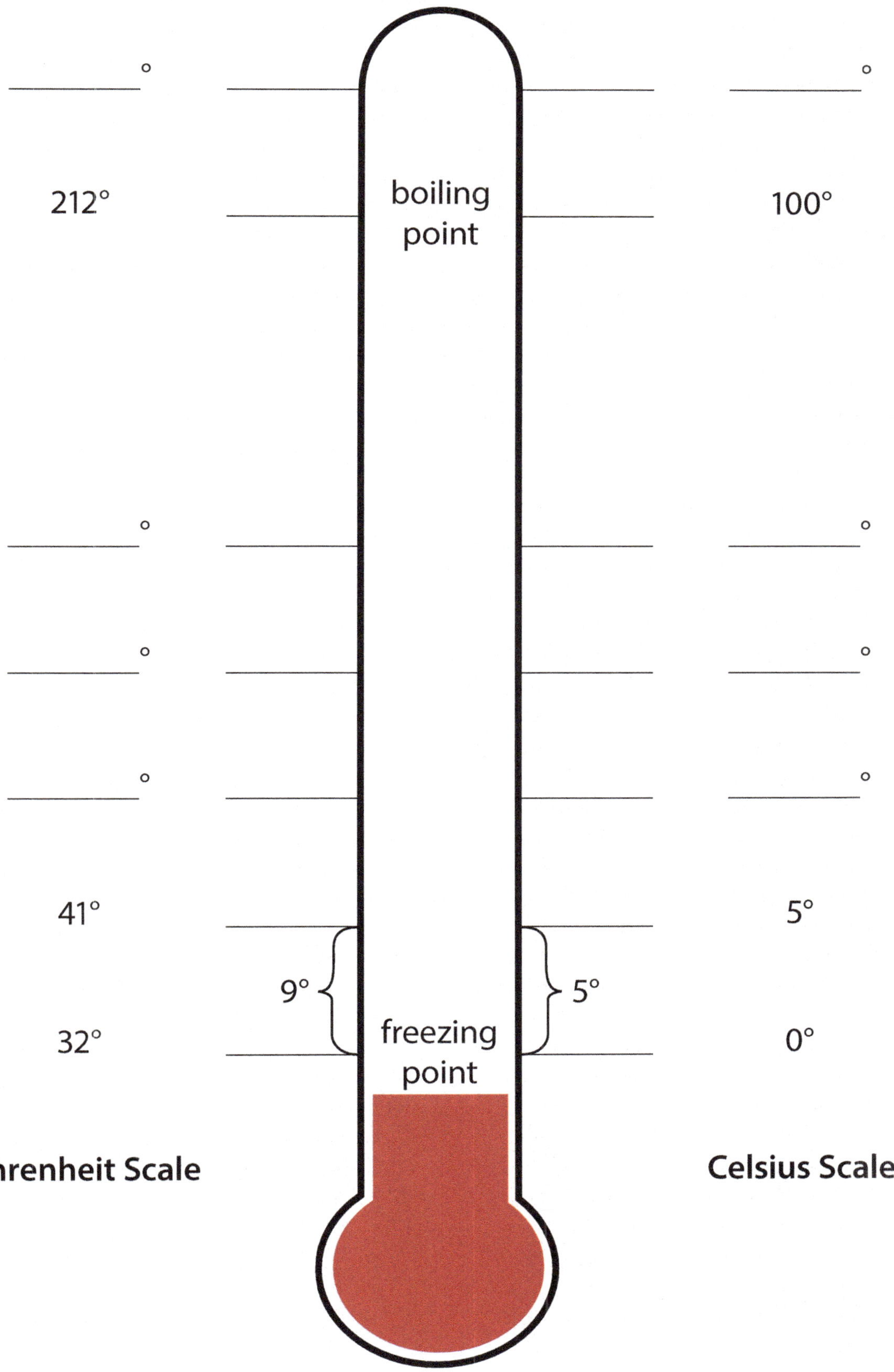

Customary & Metric Conversions

Length	Capacity	Weight or Mass
1 in. ≈ 2.5 cm	1 fl oz ≈ 30 mL	1 oz ≈ 30 g
1 ft ≈ 30 cm	1 L ≈ 1 qt	1 kg ≈ 2.2 lb
1 mi ≈ 1.6 km		
1 m ≈ 39 in.		

Time Measurement

1 minute (min) = 60 seconds (sec)

1 hour (hr) = 60 minutes

1 day (d) = 24 hours

1 week (wk) = 7 days

1 month (mo) = 28–31 days

1 year (yr) = 12 months

1 year = 52 weeks

1 year = 365 days

1 leap year = 366 days

1 decade = 10 years

1 century = 100 years

1 millennium = 1,000 years

Time Zones of the World

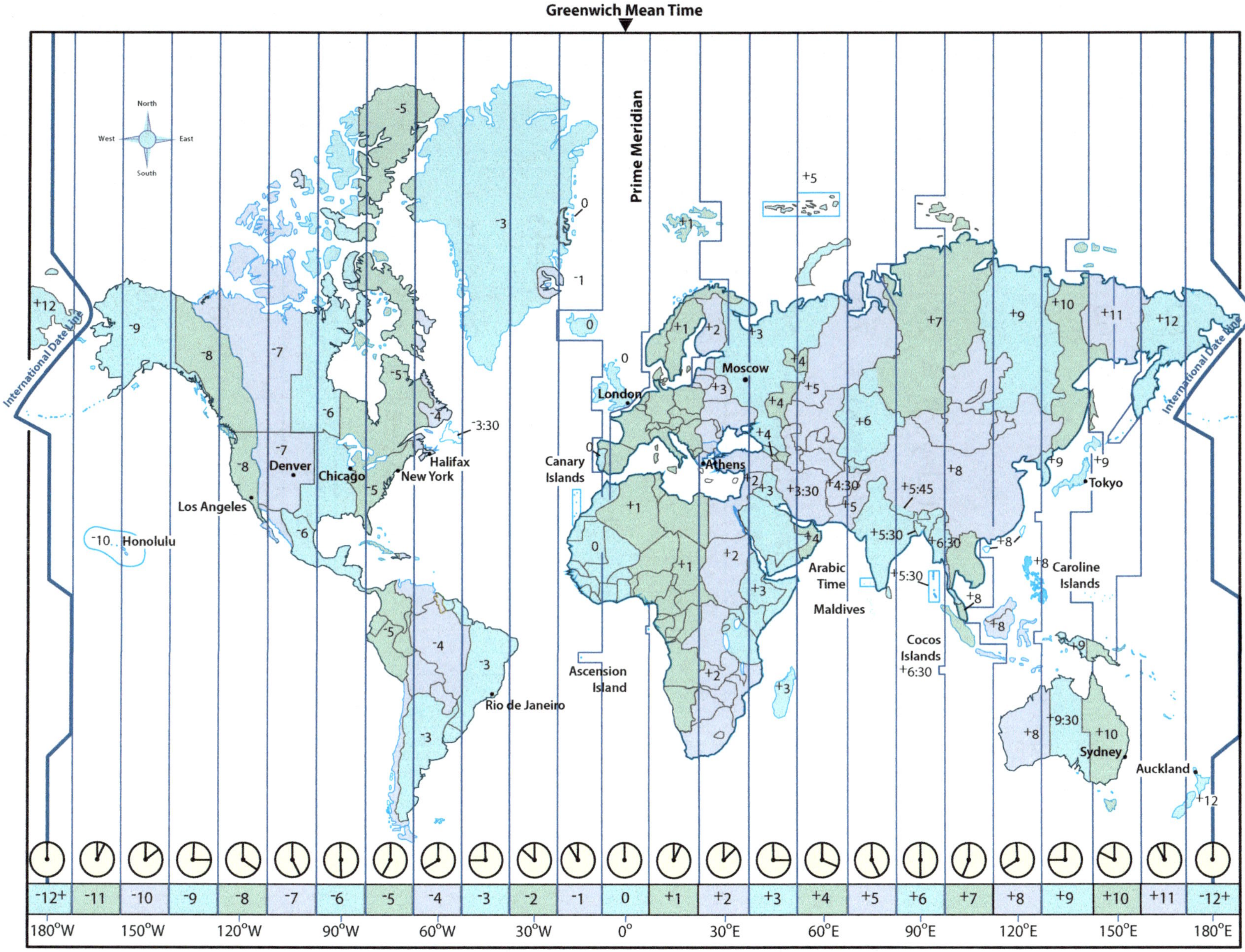

Time Zones of Major Cities

Greenwich Mean Time	City	Zone	Time
	London		
	Moscow		
	Tokyo		
	New York		
	Los Angeles		

Time Zones of Major Cities

Greenwich Mean Time	City	Zone	Time
	London		
	Moscow		
	Tokyo		
	New York		
	Los Angeles		

Map Key

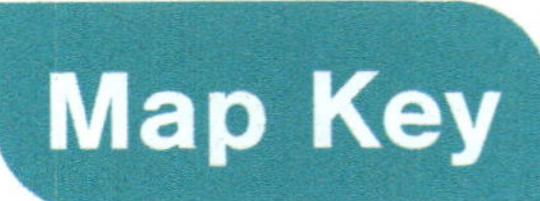

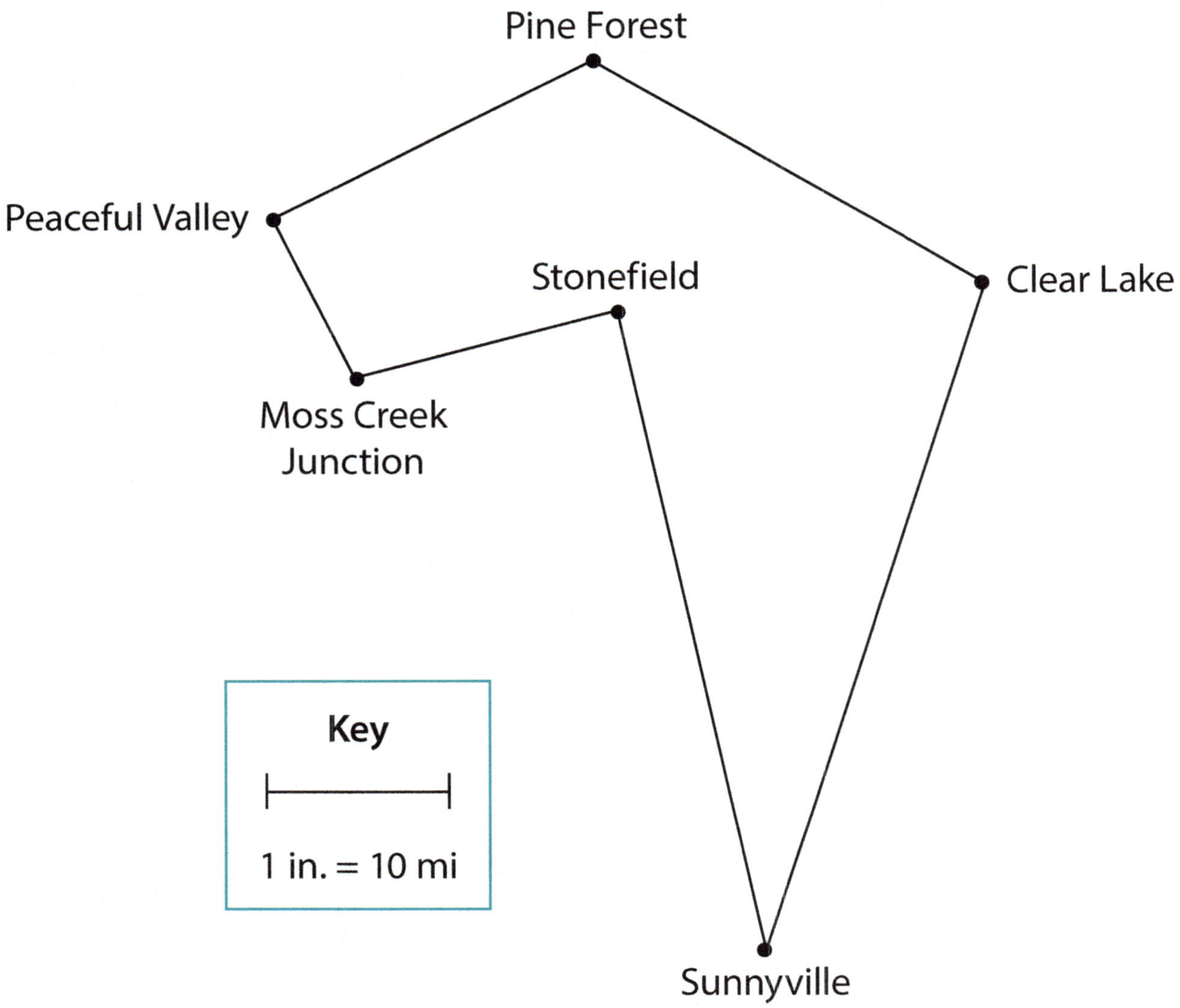

Distance	Inches	Miles
Peaceful Valley to Stonefield		

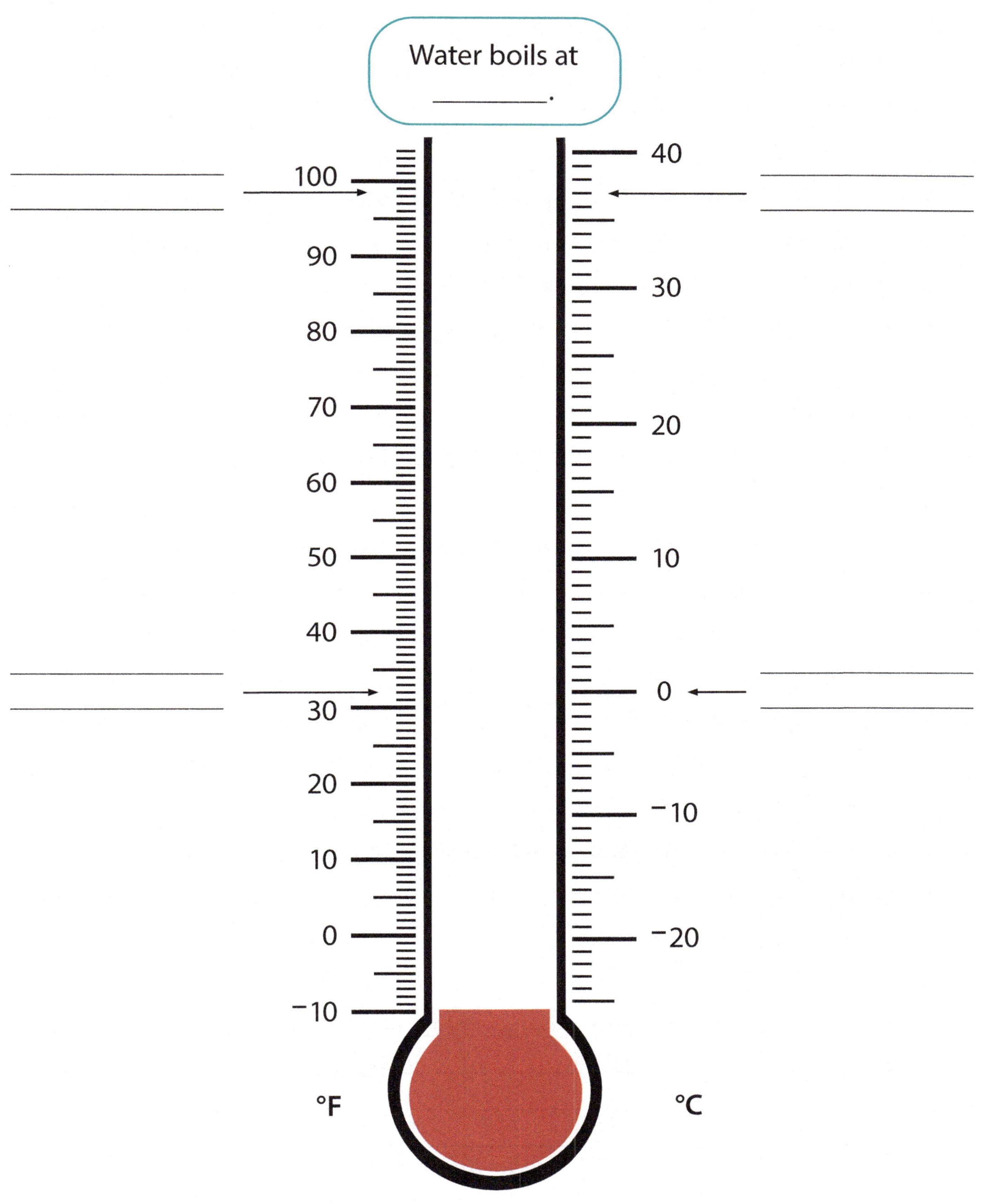

Water boils at
__________.
100
90
80
70
60
50
40
30
20
10
0
−10
°F
40
30
20
10
0
−10
−20
°C

Spelling Test Scores										
95	93	89	95	100	86	82	98	90	89	95

Spelling Test Scores		
Score	**Tally**	**Frequency**

range: mean: median: mode:

Daily Low Temperatures							
71°	69°	78°	74°	69°	73°	74°	72°

Daily Low Temperatures		
Temperature	**Tally**	**Frequency**

range: mean: median: mode:

Double Bar Graph

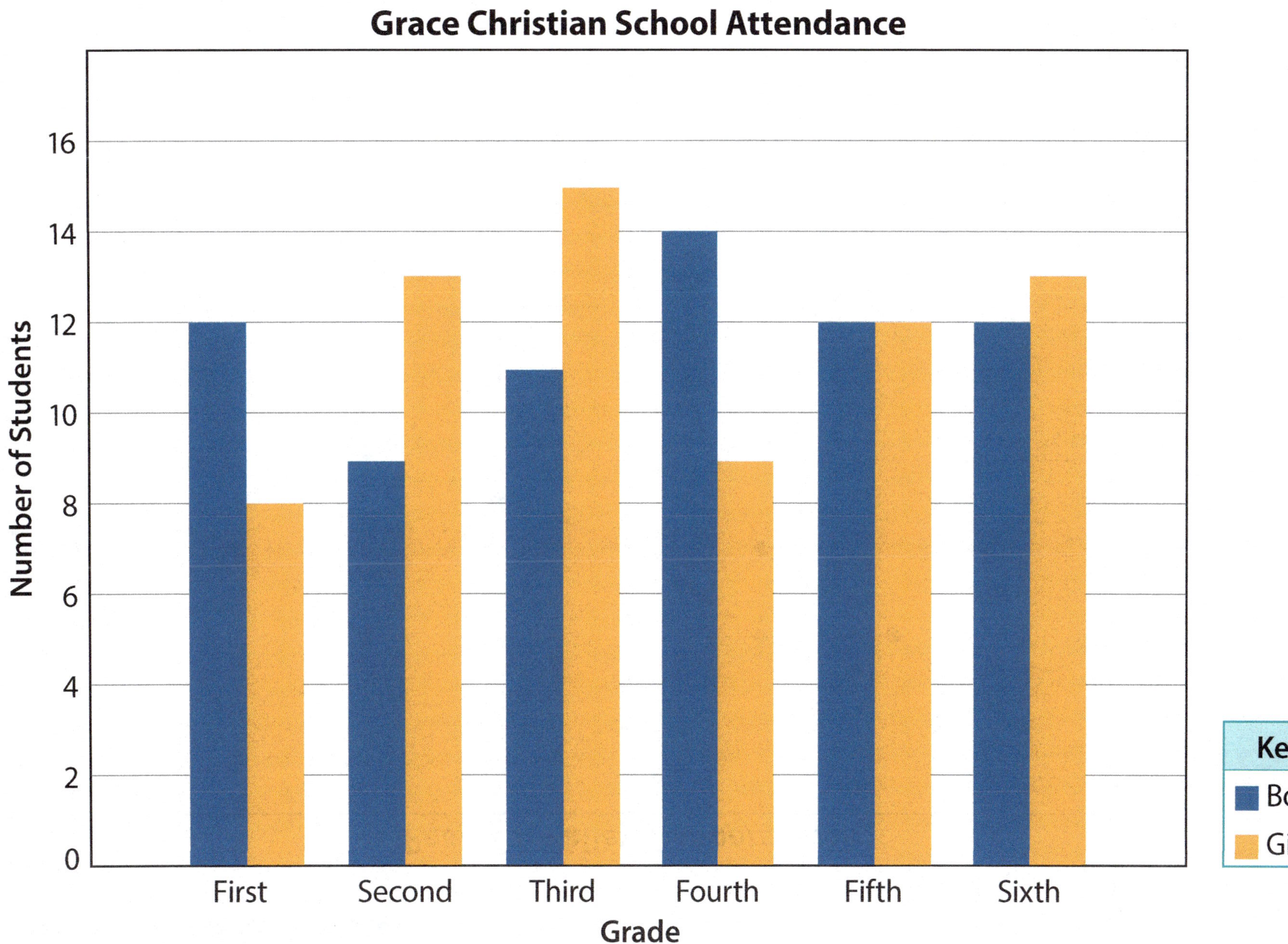

Instructional Aid 97 • For use with Lesson 148

Math 6 • Teacher Resources

Double Line Graph

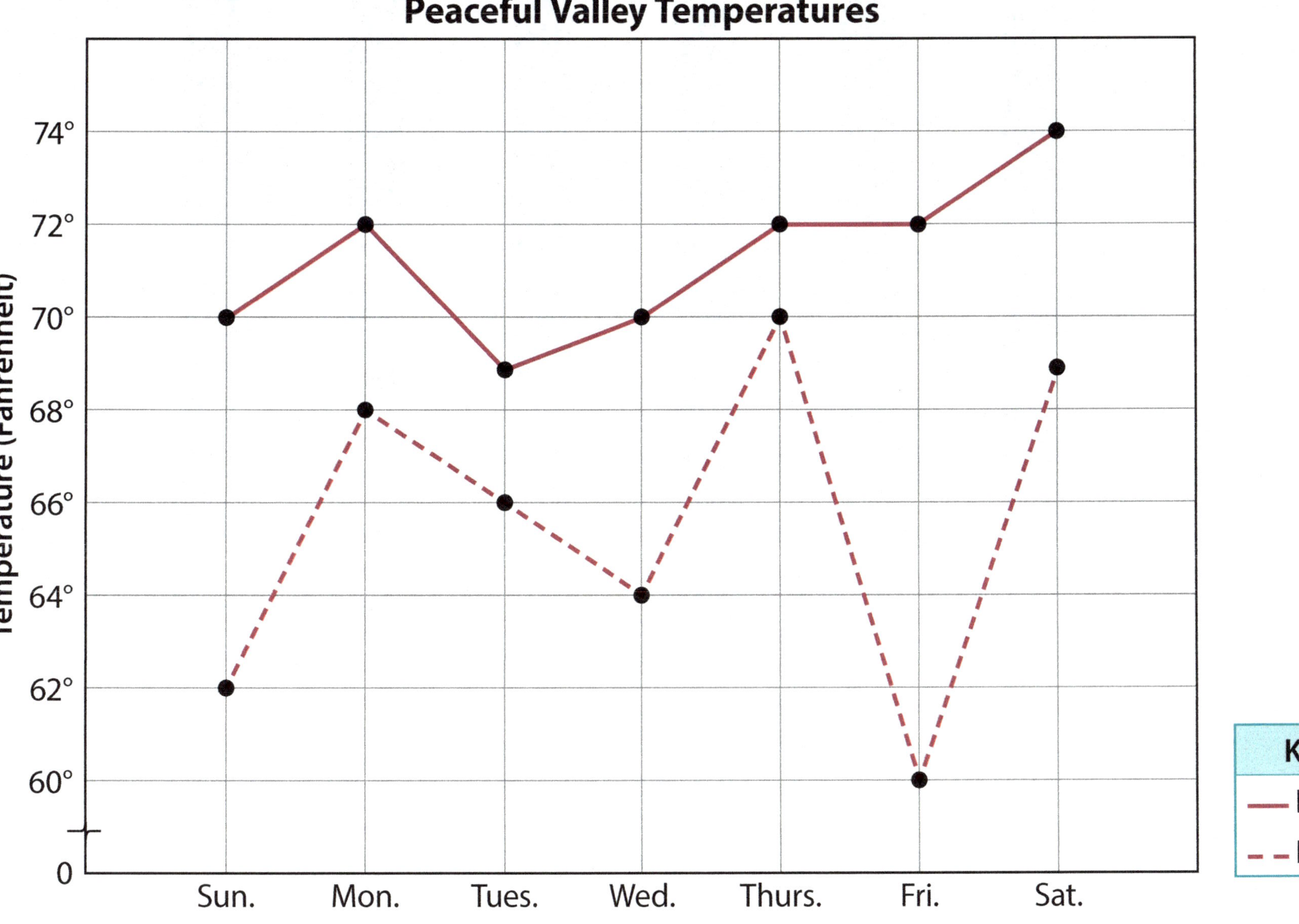

bar graph

double bar graph

used to compare data

compares two sets of
related or similar data
on the same graph

line graph	**double line graph**
shows changes over a period of time for a set of data	shows changes over a period of time for two sets of related data

stem-and-leaf plot

line plot

used to display the frequency of data by using the actual digits in each piece of data to record the results

used to display the frequency of data by creating columns of Xs above numbers in the scale as each piece of data is recorded

histogram

box-and-whisker plot

used to display a large
amount of data by
separating the data
into equal intervals

used to summarize
data on a number line
by finding the median
of a set of data that
has been ordered
from least to greatest
and then finding the
median of each half of
the set of data

Enrollment at Carter's Christian School

Stem	Leaf
0	9
1	7 8 9
2	0 0 1 2

Key | 1|7 = 17 students

range: __________   median: __________

mode: __________   mean: __________

Stem	Leaf

range: __________   median: __________

mode: __________   mean: __________

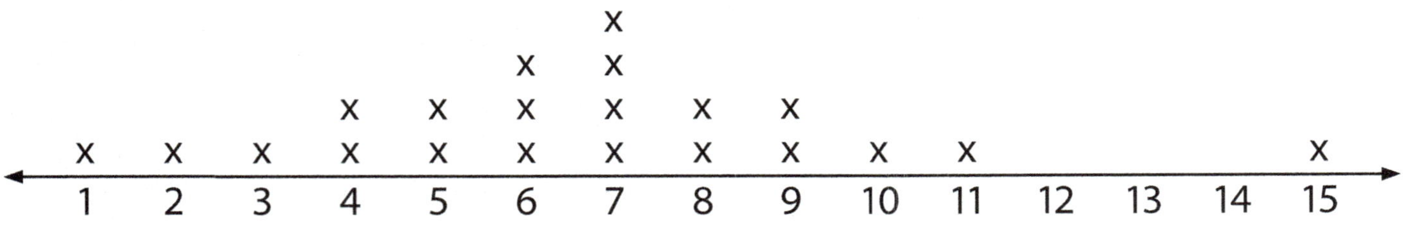

Number of Verses Memorized

range: __________   median: __________

mode: __________   mean: __________

without outlier

range: __________   median: __________

mode: __________   mean: __________

Science Test Scores

range: __________   median: __________

mode: __________   mean: __________

without outlier

range: __________   median: __________

mode: __________   mean: __________

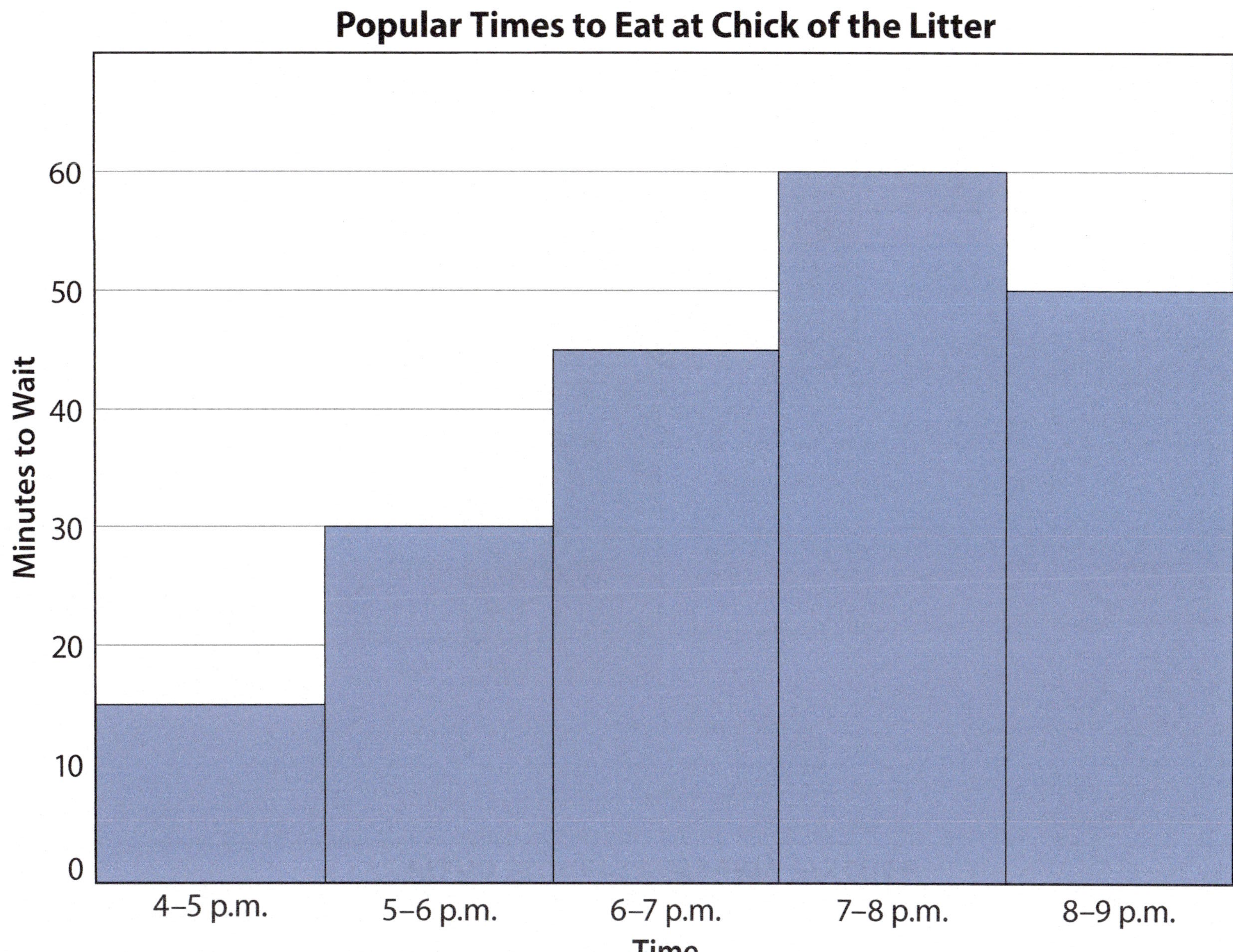

Popular Times to Eat at Chick of the Litter
Minutes to Wait
60
50
40
30
20
10
0
4–5 p.m.
5–6 p.m.
6–7 p.m.
7–8 p.m.
8–9 p.m.
Time

Histogram: Piano Practice Statistics

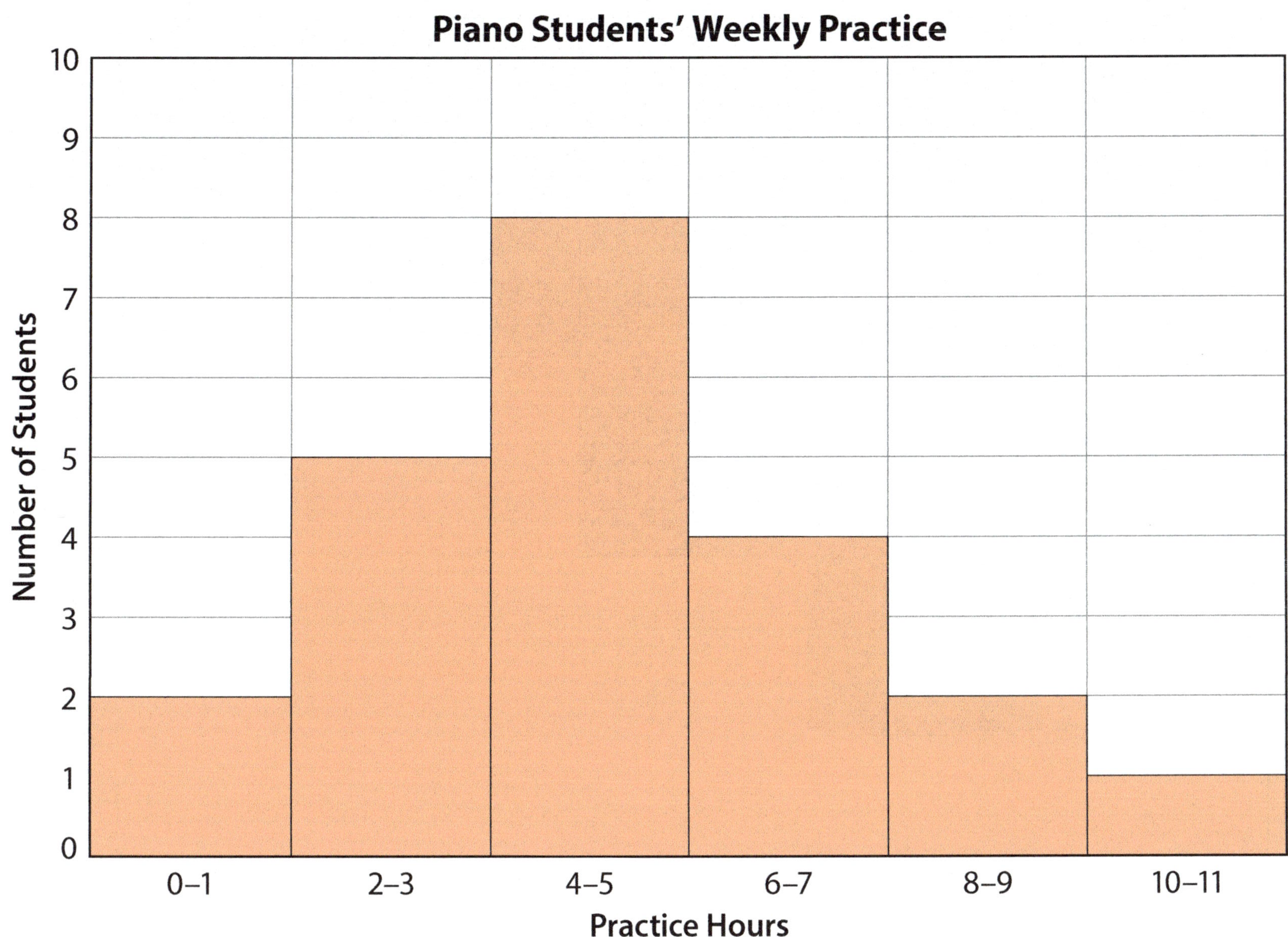

Histogram

Ages of Campers: Week 1

11, 10, 12, 10, 11, 12, 15, 8, 9,
12, 11, 12, 13, 11, 15, 8, 9, 9,
8, 12, 9, 12, 14, 12, 13

Ages of Campers: Week 1		
Age	Tally	Frequency

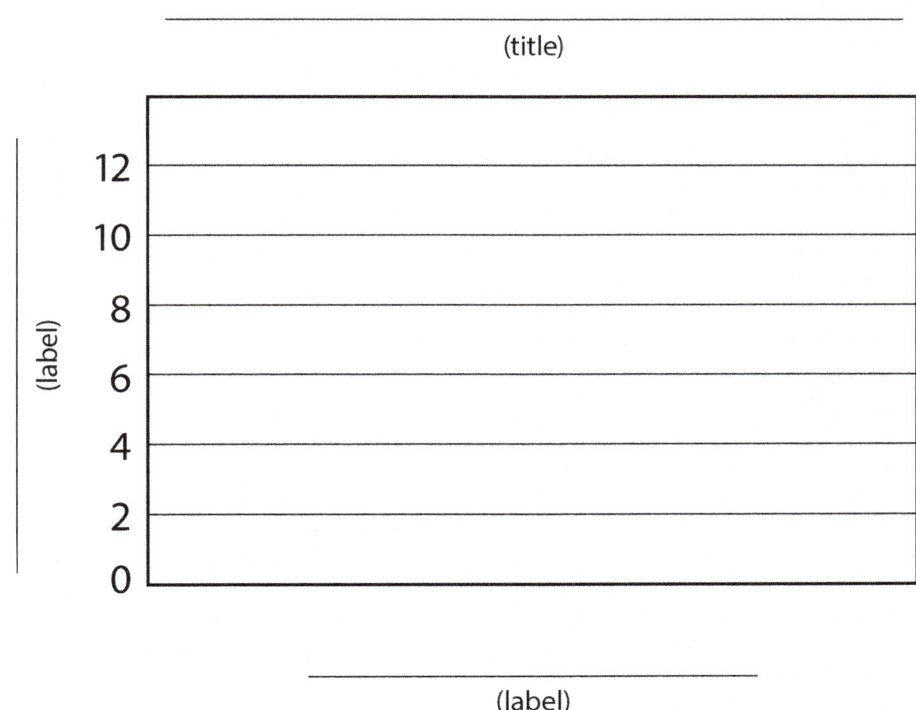

number of points Joel scored per basketball game: 7, 10, 12, 16, 4, 11, 2

ordered list of data: 2, 4, 7, 10, 11, 12, 16

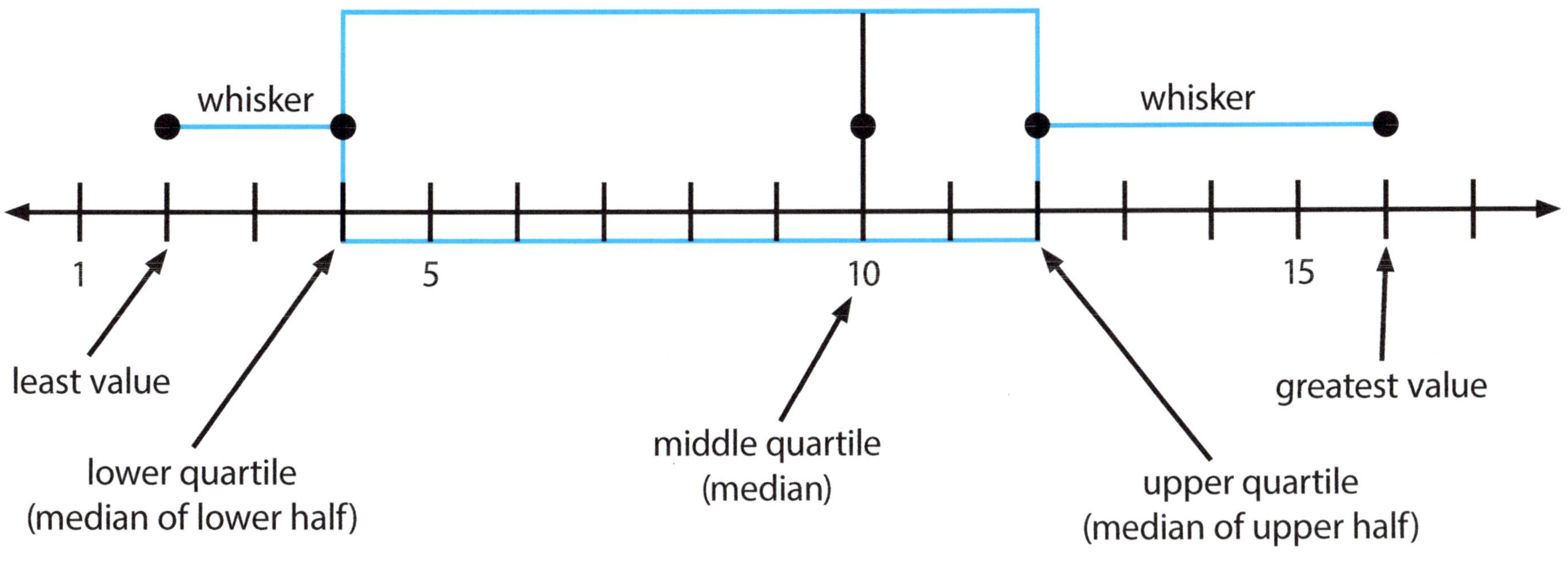

Box-and-Whisker Plot

Test Scores

data: 89, 76, 91, 82, 95, 98, 85, 80

ordered data: 76, 80, 82, 85, 89, 91, 95, 98

middle quartile	lower quartile	upper quartile
(median)	(median of lower half)	(median of upper half)

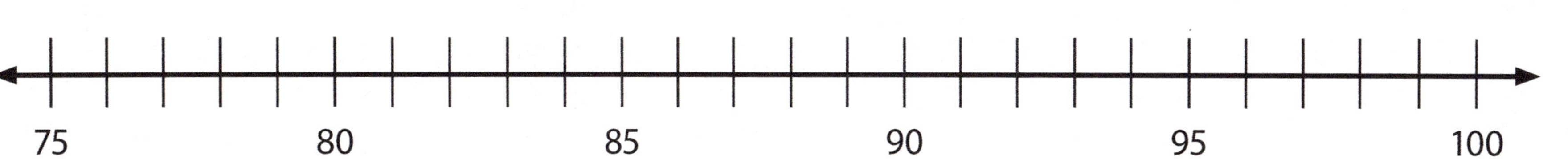

Graph: Double Bar Graph

Population Distribution

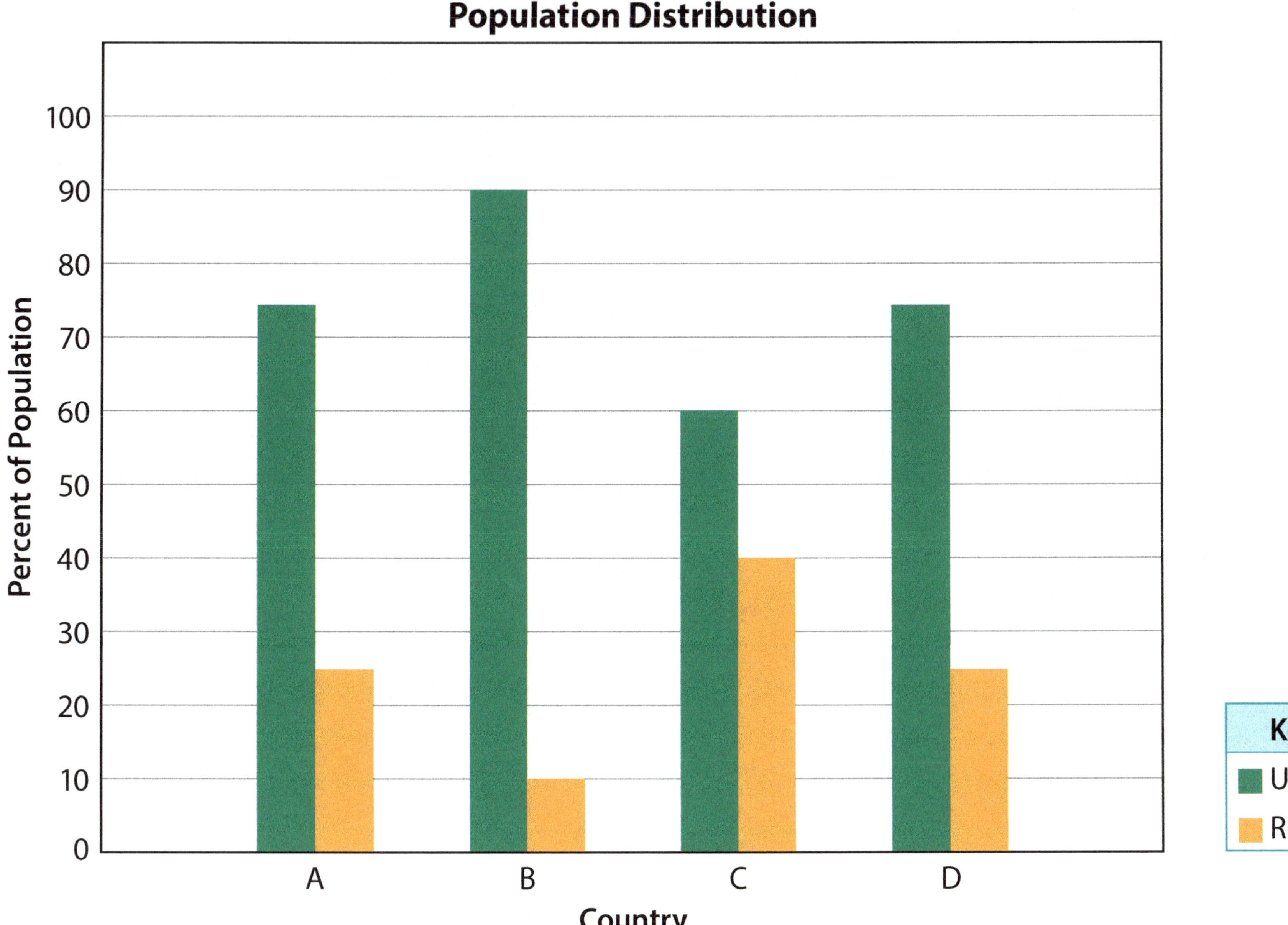

Graph: Double Line Graph

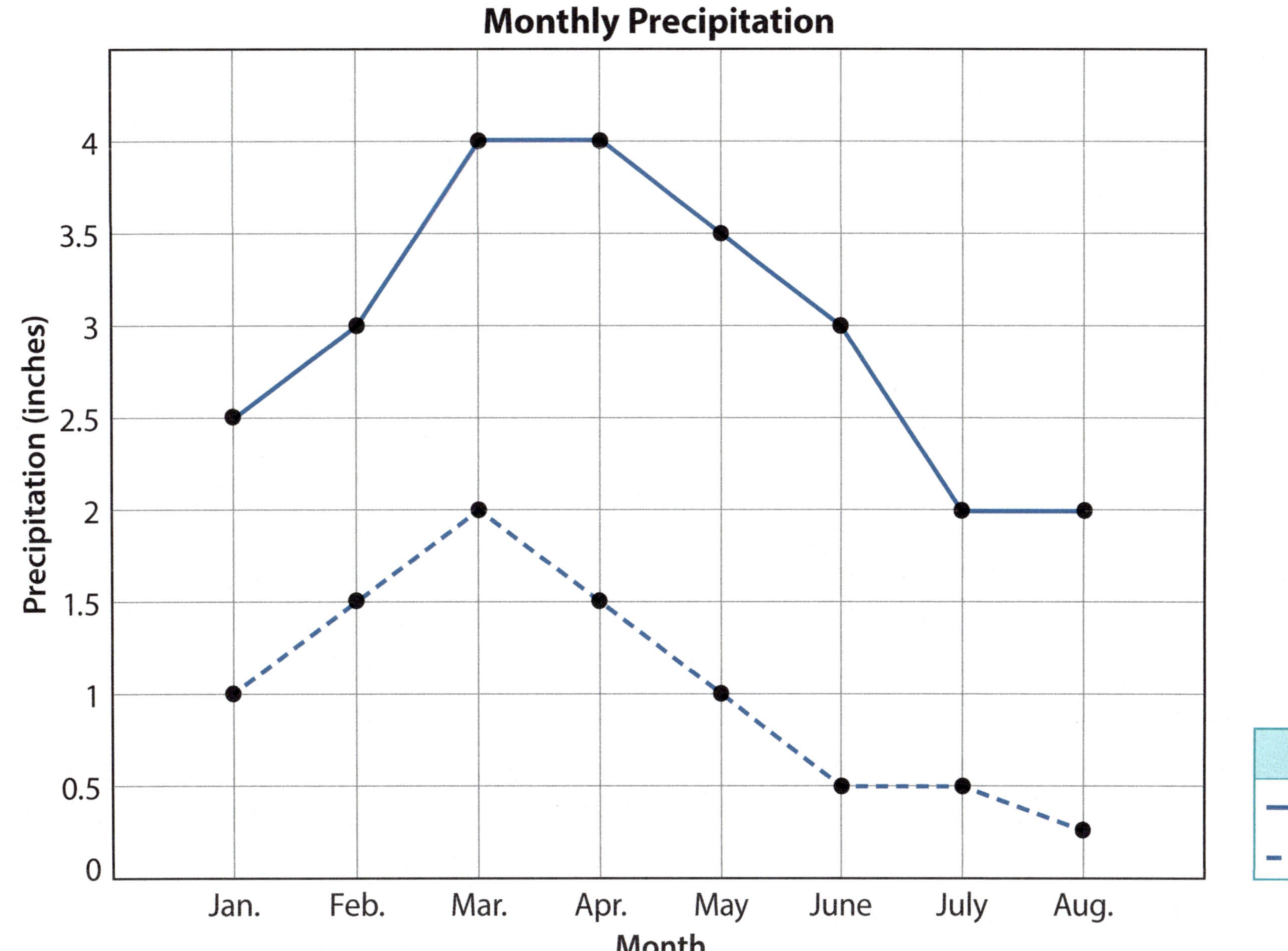

Stem-and-Leaf Plot & Line Plot

Employee Ages	
Stem	**Leaf**
1	9
2	0 1 2 3 5 7 9 9
3	1 4 4 5 7 7 9
4	2 8 9
5	3 7
6	2

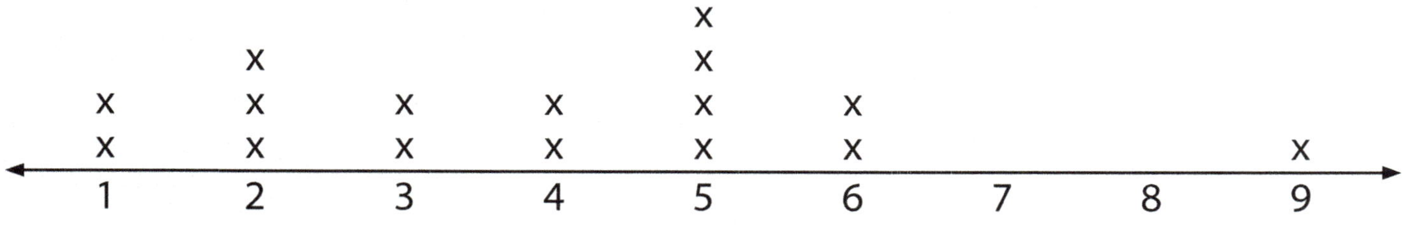

Graph: Histogram

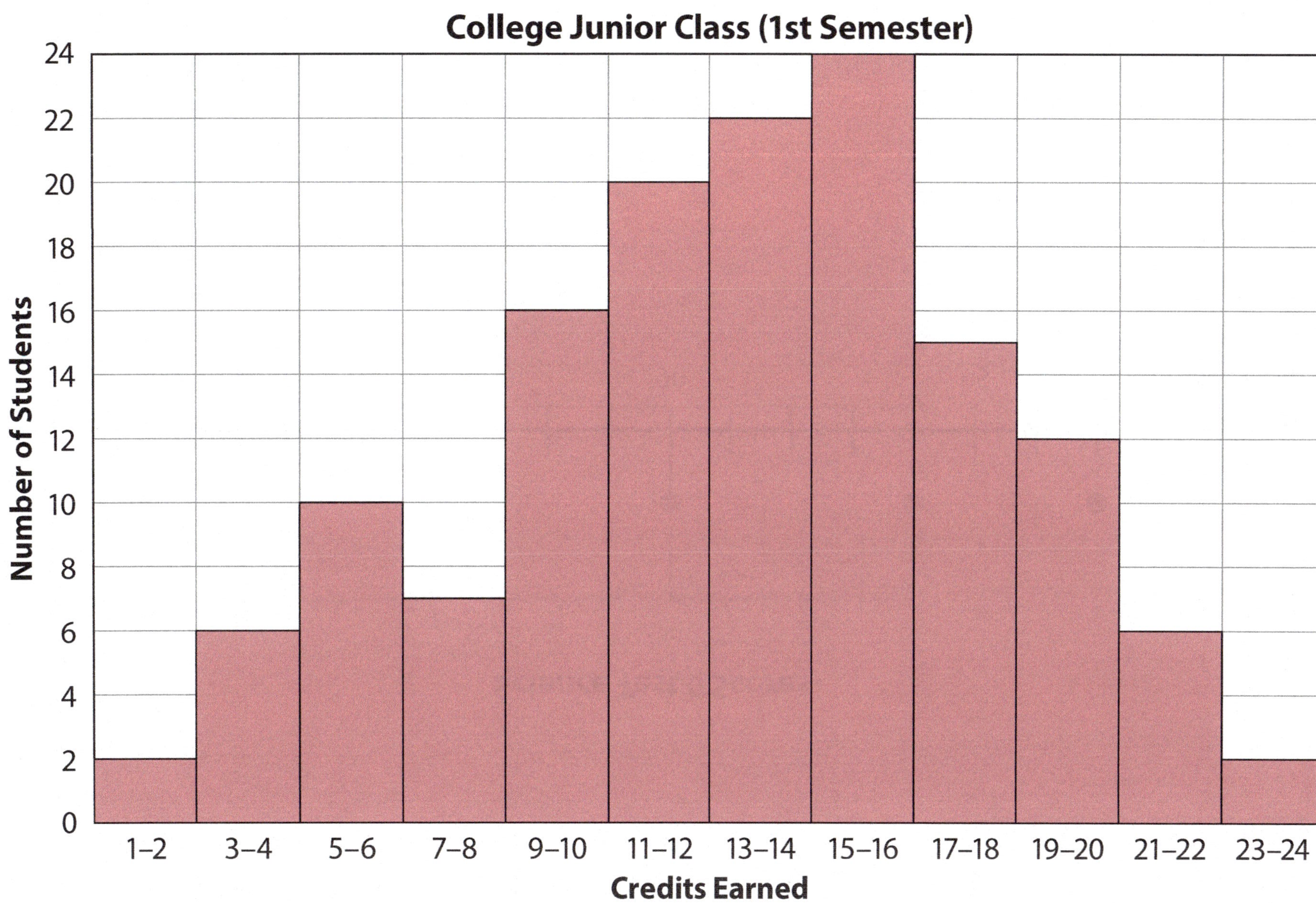

Instructional Aid 110 • For use with Lesson 153

Math 6 • Teacher Resources

Graph: Box-and-Whisker Plot

Science Test 6 Scores

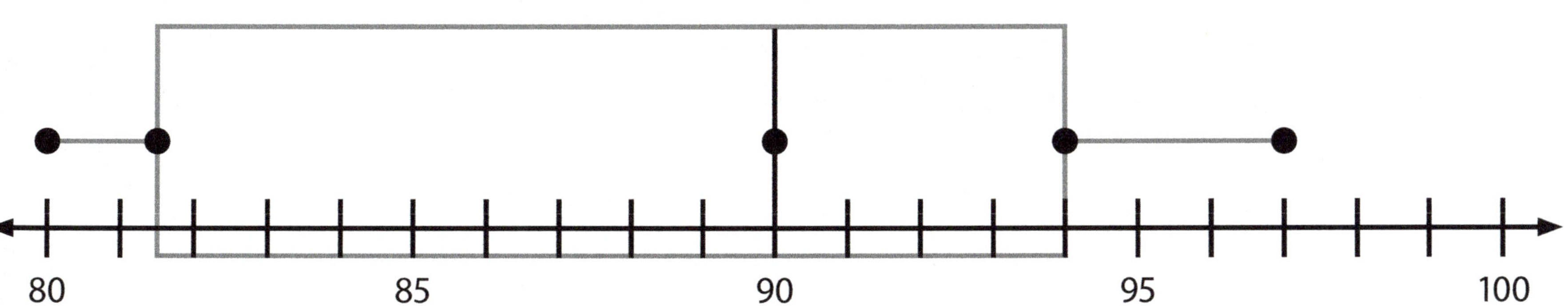

Hours Worked

Worker	Week 1	Week 2	Week 3	Week 4
Ethan	8	6	10	6
Logan	16	16	10	16
Samuel	10	10	5	6
Zachary	14	10	12	8

Hours	Tally	Frequency

Problem to Solve	Gifts/Abilities/Interests	Business Idea

best business idea: ____

Business Name: Rover's Romps

What we will do (purpose of business) Rover's Romps provides a dog-walking service for treasured pets.	**Problem (problem the business addresses)** Pet owners want to exercise their dogs for the pets' best health and temperament, but the owners are often too busy to walk them.
Solution/wow factor (solution to the problem; what sets your business apart) Our team of loving and careful pet guardians walks dogs according to the owners' schedules and gives excellent customer service by providing water, treats, and clean-up service during the walk.	**Customers (who will buy)** Targeted customers are pet owners in nearby neighborhoods within walking distance of our homes.
Competition (similar businesses) Three other dog-walking services are offered nearby, but none offers the level of service we will offer.	**Where/how sold (ways or places you will sell)** Rover's Romps will sell directly to customers in nearby neighborhoods.
Marketing activities (how to reach customers) We will leave printed fliers on doors and businesses and ask for referrals from satisfied customers. As we expand, we will use social media and an email list to connect with customers and prospects.	**Expenses (what you will spend money on)** dog leashes water dishes and bottled water dog treats scoop and disposable bags for doggy clean up backpack for carrying supplies advertising materials
Team and responsibilities (who does what) Jonathan will create and print fliers. Thomas will handle the records and money. Russ will purchase supplies. Jonathan, Thomas, and Russ will distribute fliers and walk dogs.	**Goals for growth (plans for the future)** As it grows, we would like to add more teams and expand into more neighborhoods. We would advertise more widely and add Rover's brand pet products (collars, leashes, treats, etc.).

Business Name:	
What we will do (purpose of business)	**Problem (problem the business addresses)**
Solution/wow factor (solution to the problem; what sets your business apart)	**Customers (who will buy)**
Competition (similar businesses)	**Where/how sold (ways or places you will sell)**
Marketing activities (how to reach customers)	**Expenses (what you will spend money on)**
Team and responsibilities (who does what)	**Goals for growth (plans for the future)**

Income − Expenses = Profit

After researching "dog walking services near me" to learn competitors' prices, Rover's Romps determined that they could charge $10 for a 20 min walk and that each person could walk 2 dogs each hour. Each person plans to work for 2 hr each day, 5 d/wk. They will share the duty of passing out 25 fliers before each week begins.

Table 1

Projected Income	Income/ Walk	Walks/Day 3 Walkers	Income/ Day	Days/ Week	Income/ Week
Income per week @ 12 walks/d, 60/wk	$10.00	12	$120.00	5	$600.00
Income per week @ 1 walk/person/d, 3 walks/d, 15/wk	$10.00	3	$ 30.00	5	$ 150.00

Table 2

Estimated expense for 1st week @ 60 walks	Estimated Cost w/Tax	# of Units Needed	Total Cost
dog leashes @ 1 per walker; 3 needed	$ 7.00	3	$ 21.00
collapsible water dishes @ 1 per walker; 3 needed; purchased in sets of 2	$13.00	2	$ 26.00
dog waste scoops @ 1 per walker; 3 needed	$13.00	3	$ 39.00
drawstring backpack @ 1 per walker; 3 needed	$ 5.00	3	$ 15.00
advertising: black & white posters; 25 posters @ $0.13 ea	$ 4.00	1	$ 4.00
dog treats; 150 treats/package, 1/dog/walk	$ 8.00	1	$ 8.00
disposable bags; 80 bags/package, 1/dog/walk	$ 5.00	1	$ 5.00
Total expenses for 1st week @ 60 walks			**$118.00**

Table 3

Estimated expense for 1st week @ 15 walks	Estimated Cost w/Tax	# of Units Needed	Total Cost
dog leashes @ 1 per walker; 3 needed	$ 7.00	3	$ 21.00
collapsible water dishes @ 1 per walker; 3 needed; purchased in sets of 2	$13.00	2	$ 26.00
dog waste scoops @ 1 per walker; 3 needed	$13.00	3	$ 39.00
drawstring backpack @ 1 per walker; 3 needed	$ 5.00	3	$ 15.00
advertising: black & white posters; 25 posters @ $0.13 ea	$ 4.00	1	$ 4.00
dog treats; 150 treats/package, 1/dog/walk	$ 8.00	1	$ 8.00
disposable bags; 80 bags/package, 1/dog/walk	$ 5.00	1	$ 5.00
Total expenses for 1st week @ 15 walks			$118.00

Table 4

*The Rover's Romps team decided to lower their expenses by purchasing some items secondhand at thrift stores or garage sales and using items that they had on hand in place of the backpacks.

Adjusted expense for 1st week @ 15–60 walks	Estimated Cost w/Tax	# of Units Needed	Total Cost
dog leashes @ 1 per walker; 3 needed	$ 7.00	3	$21.00
*water dishes @ 1 per walker; 3 needed	$ 1.00	3	$ 3.00
*dog waste scoops @ 1 per walker; 3 needed	$4.00	3	$12.00
*drawstring backpack @ 1 per walker; 3 needed	$ —	3	$ —
advertising: black & white posters; 25 posters @ $0.13 ea	$4.00	1	$ 4.00
dog treats; 150 treats/package, 1/dog/walk	$8.00	1	$ 8.00
disposable bags; 80 bags/package, 1/dog/walk	$5.00	1	$ 5.00
Total expenses for 1st week @ 15–60 walks			$53.00

Table 5

Profit @ 3 walks/d, 5 d/wk; *Income − Expenses = Profit*

Week 1		Week 2	
Expenses (start up)	$ (53.00)	Profit from previous week	$ 97.00
Income @ $30/d	$ 150.00	Income @ $30/d	$ 150.00
Profit	$ 97.00	**Subtotal**	$ 247.00
		Expenses (advertising)	$ (4.00)
		Profit	$ 243.00
Week 3		**Week 4**	
Profit from previous week	$ 243.00	Profit from previous week	$ 389.00
Income @ $30/d	$ 150.00	Income @ $30/d	$ 150.00
Subtotal	$ 393.00	**Subtotal**	$ 539.00
Expenses (advertising)	$ (4.00)	Expenses (advertising)	$ (4.00)
Profit	$ 389.00	**Profit**	$ 535.00

Table 6

Profit Summary	
Week 1	$ 97.00
Week 2	$ 243.00
Week 3	$ 389.00
Week 4	$ 535.00

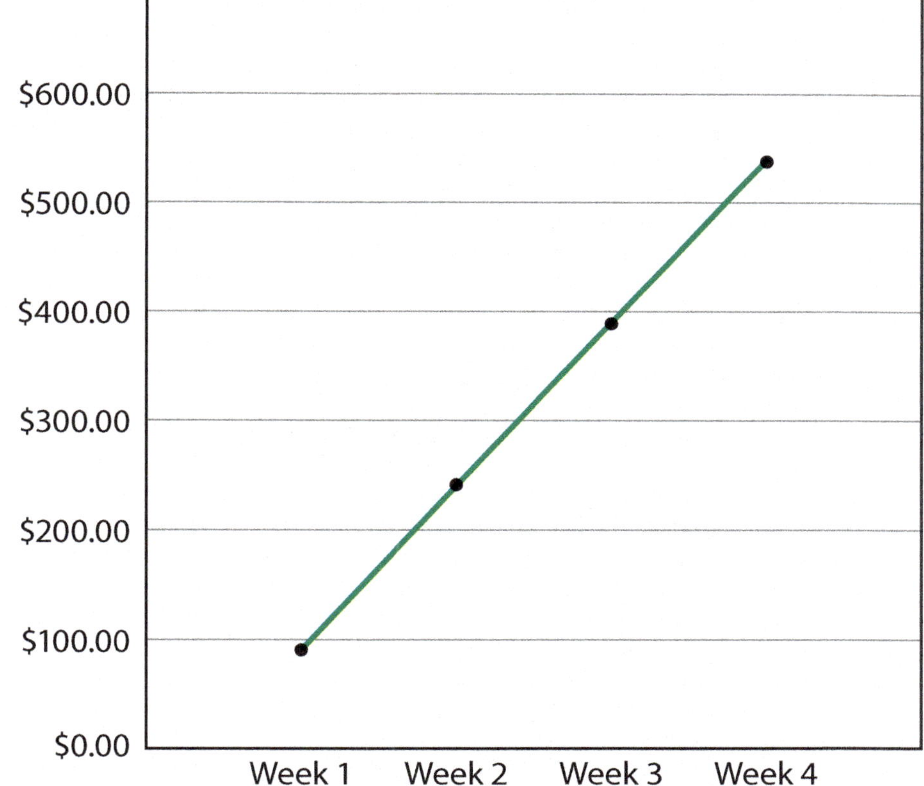

Profit Projection Month 1

Business Profit Projection

Directions: Figure your profit for the first 4 weeks of your business.

Week 1

Income $ _______________

Expenses   − $ _______________

Profit $ _______________

Week 2

Profit (loss) from Week 1 $ _______________

Income   + $ _______________

Subtotal $ _______________

Expenses   − $ _______________

Profit $ _______________

Week 3

Profit (loss) from Week 2 $ _______________

Income   + $ _______________

Subtotal $ _______________

Expenses   − $ _______________

Profit $ _______________

Week 4

Profit (loss) from Week 3 $ _______________

Income   + $ _______________

Subtotal $ _______________

Expenses   − $ _______________

Profit $ _______________

Profit Summary

Total at end of Week 1 $ _______________

Total at end of Week 2 $ _______________

Total at end of Week 3 $ _______________

Total at end of Week 4 $ _______________

STEM Rubric: Entrepreneurship

Name________________________________

	Excellent (3)	Good (2)	Needs Improvement (1)	Points Earned
The Engineering Design Process	Uses all the steps of the Engineering Design Process to solve the problem of identifying a good business to start.	Uses some of the steps of the Engineering Design Process to solve the problem of identifying a good business to start.	Uses few or does not use the steps of the Engineering Design Process to solve the problem of identifying a good business to start.	
Model	Plan shows understanding of concepts.	Plan shows understanding of most concepts.	Plan shows lack of understanding of concepts, or plan is missing.	
Organization/ Appearance	Presentation is organized, neat, and easy to follow.	Presentation is mostly organized, neat, and easy to follow.	Presentation lacks organization or neatness or is hard to follow.	
Collaboration with Peers	Always listens carefully to others and offers detailed, constructive feedback. Participates fully and shares the workload fairly.	Sometimes listens to others; occasionally offers constructive feedback. Participates but sometimes does not share the workload fairly.	Does not listen to others and often interrupts them. Does not offer constructive feedback. Does not participate and relies on others to carry the workload most of the time.	

Comments   **Total**

impossible	unlikely	equally likely	likely	certain
0	$\frac{1}{4}$	$\frac{1}{2}$	$\frac{3}{4}$	$\frac{4}{4}$
0	0.25	0.5	0.75	1
0%	25%	50%	75%	100%

Theoretical Probability

$$P(\text{event}) = \frac{\text{number of favorable outcomes}}{\text{number of possible outcomes}}$$

$P(\text{event}) + P(\textit{not } \text{event}) = 1 \text{ or } 100\%$

P(event)	Probability (fraction)	Probability (decimal)	Probability (percent)
P(1)			
P(not 1)			
P(2)			
P(not 2)			
P(3)			
P(not 3)			

Sample Spaces

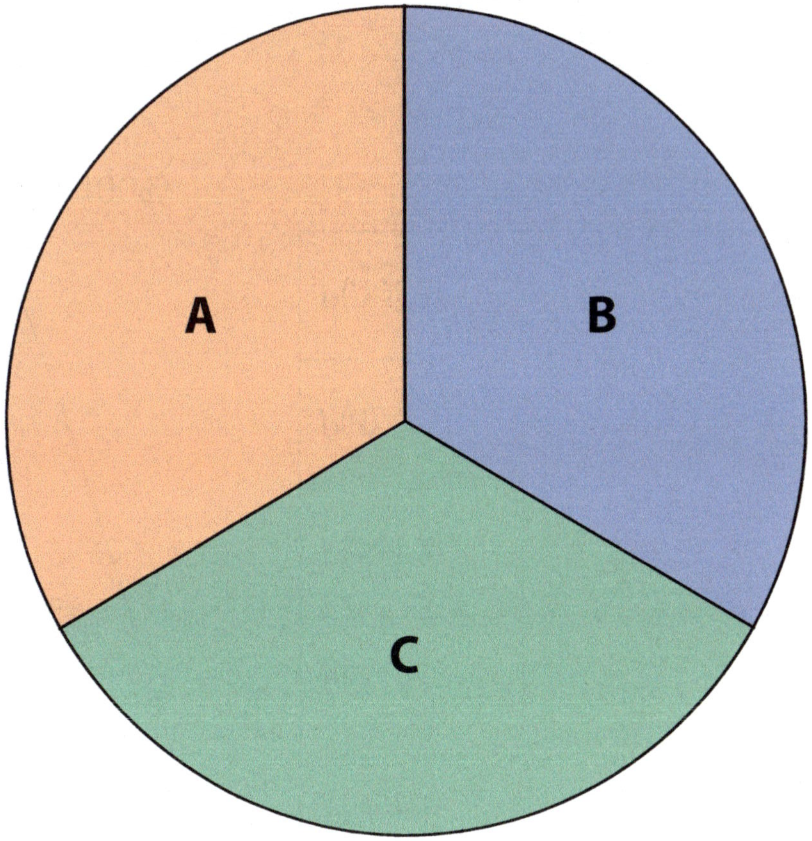

	Sample Space	Number of Possible Outcomes
1 spin	{A, B, C}	3
2 spins	{AA,	
3 spins	{AAA,	

Tree Diagram

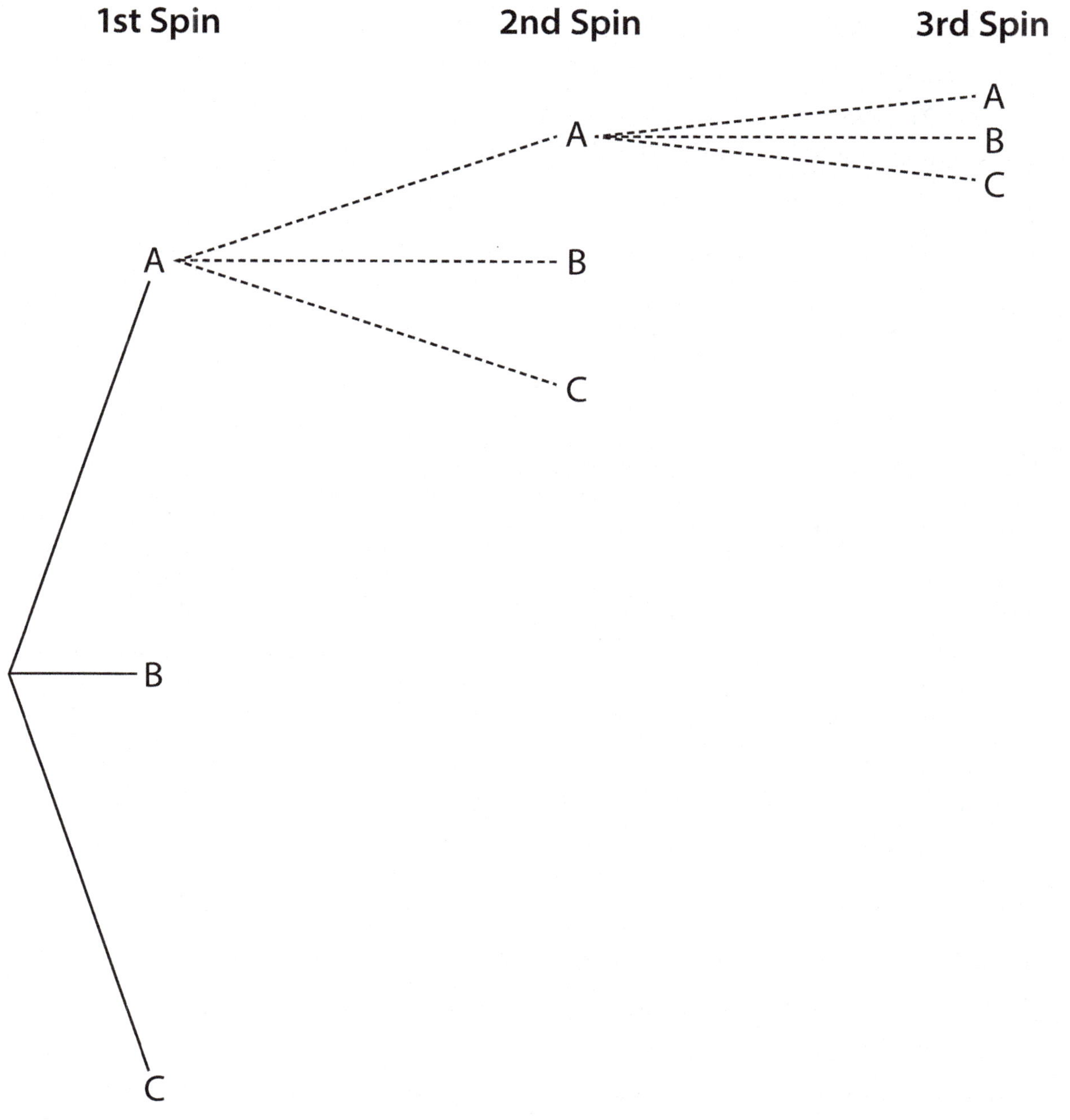

Number of Possible Outcomes		
3		

Multiplication Counting Principle

1. Mrs. Fraley is preparing peanut butter sandwiches with grape jelly, strawberry jelly, or honey on either white or multigrain bread. How many possible combinations are there? What is the probability of having a peanut butter sandwich with jelly?

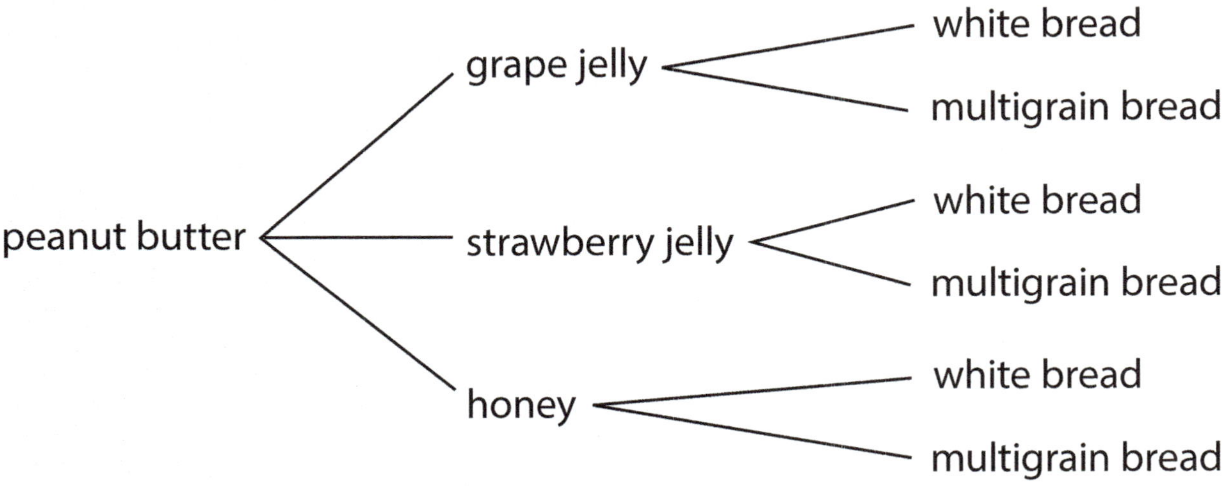

3 spread choices × 2 bread choices = 6 sandwich combinations

2. Mrs. Parker is making large and small fruit baskets. Each basket has yellow or clear cellophane and has a red, yellow, or blue bow. How many possible combinations are there? What is the probability of having a fruit basket with a red bow?

Experimental Probability

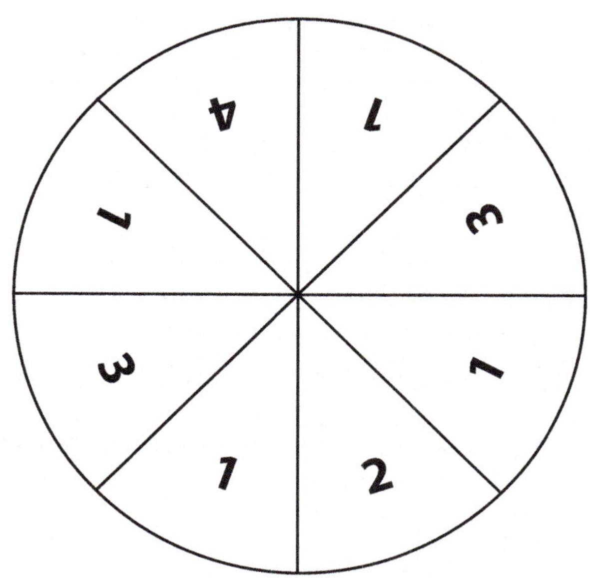

Theoretical Probability	Number of Trials	Expected Results	Actual Results	Experimental Probability
$P(1) =$	16			
$P(2) =$	16			
$P(3) =$	16			
$P(4) =$	16			

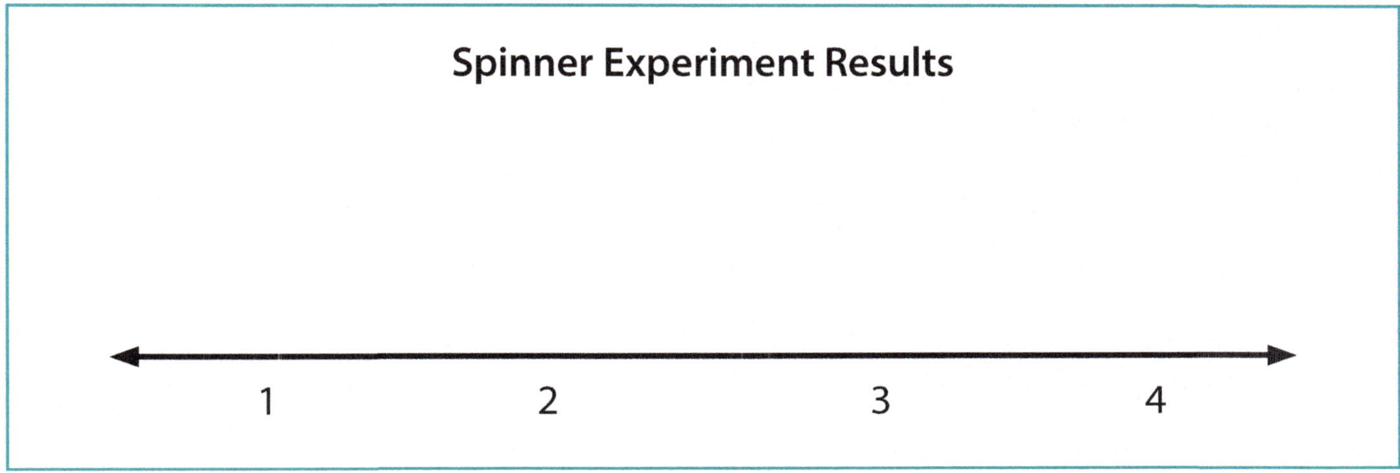

Spinning Penny Experiment

1. Determine P(heads) and P(tails) for spinning a penny. Write the theoretical probabilities in the table.

2. If you spin the penny 20 times, how many times do you expect to spin heads? to spin tails? Write the answers in the Expected Results column of the table.

3. Conduct the experiment: Stand a penny on its side. Holding the top of the penny with a finger on one hand, flick the side of the penny with a finger on your other hand so that the penny spins. When the penny stops spinning, record the result (heads or tails) by placing a tally in the Actual Results column of the table. Repeat the procedure, spinning the penny 20 times.

4. Use the frequency of your actual results to determine the experimental probability for P(heads) and P(tails).

Theoretical Probability	Number of Trials	Expected Results	Actual Results	Experimental Probability
P(heads) =	20			P(heads) =
P(tails) =	20			P(tails) =

Spinning Penny Experiment

1. Determine P(heads) and P(tails) for spinning a penny. Write the theoretical probabilities in the table.

2. If you spin the penny 20 times, how many times do you expect to spin heads? to spin tails? Write the answers in the Expected Results column of the table.

3. Conduct the experiment: Stand a penny on its side. Holding the top of the penny with a finger on one hand, flick the side of the penny with a finger on your other hand so that the penny spins. When the penny stops spinning, record the result (heads or tails) by placing a tally in the Actual Results column of the table. Repeat the procedure, spinning the penny 20 times.

4. Use the frequency of your actual results to determine the experimental probability for P(heads) and P(tails).

Theoretical Probability	Number of Trials	Expected Results	Actual Results	Experimental Probability
P(heads) =	20			P(heads) =
P(tails) =	20			P(tails) =

Fair or Unfair Games

Two-Cube Sum

1. Predict the sum that will occur most frequently.

2. Roll 2 number cubes. Add the numbers.

3. Draw a tally in the column indicating the sum.

4. Continue rolling the cubes until a number has 10 tallies in a column. The player(s) who predicted the number with 10 tallies wins.

2	3	4	5	6	7	8	9	10	11	12

Is this game fair or unfair? Why?

Two-Cube Even or Odd Sum

1. Predict the number of even sums and the number of odd sums for rolling 2 number cubes 20 times.

2. Roll 2 number cubes. Add the numbers.

3. Draw a tally in the column indicating the sum.

4. Roll the cubes and record the tally 19 more times.

5. Total the tallies for even sums and odd sums. The player(s) who predicted most correctly wins.

Even Sum	Odd Sum

Is this game fair or unfair? Why?

Spin 1, 2, or 3

1. Find the theoretical probability of spinning each number and write it as a fraction in lowest terms. Use the theoretical probability to predict the results you would expect if you spin the paper clip 12 times and record it.

2. Spin the paper clip 12 times. Draw tallies to record the frequency in the Actual Results column.

3. Write as a fraction in lowest terms the experimental probability of spinning each number.

Theoretical Probability	Number of Trials	Expected Results	Actual Results	Experimental Probability
$P(1) =$	12			
$P(2) =$	12			
$P(3) =$	12			

Even or Odd Spin

1. Find the theoretical probability and predict the expected results for spinning an odd or an even number.

2. Spin the paper clip.

3. Draw a tally in the Actual Results column.

4. Spin the spinner and tally the result 19 more times.

5. Write the experimental probability to complete the chart.

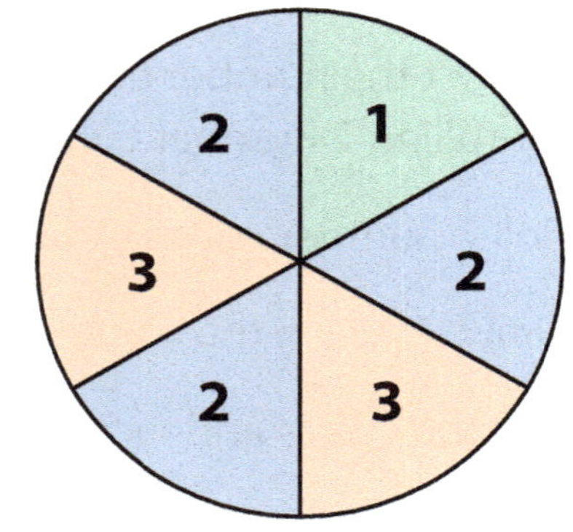

Theoretical Probability	Number of Trials	Expected Results	Actual Results	Experimental Probability
$P(\text{even}) =$	20			
$P(\text{odd}) =$	20			

Would Spin 1, 2, or 3 and Even or Odd Spin make fair or unfair games? Why?

Algebra Mat

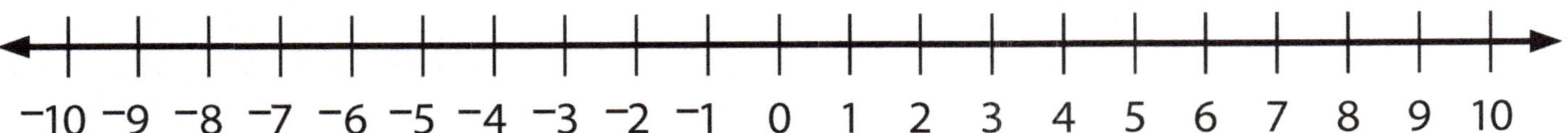

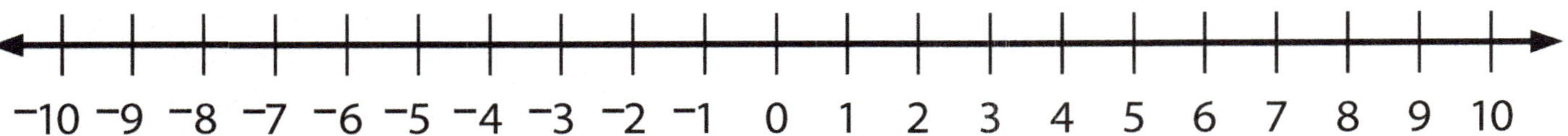

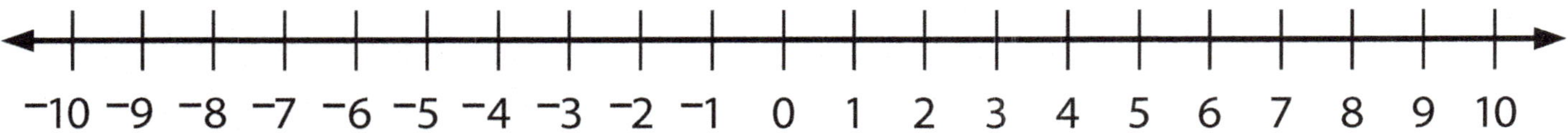

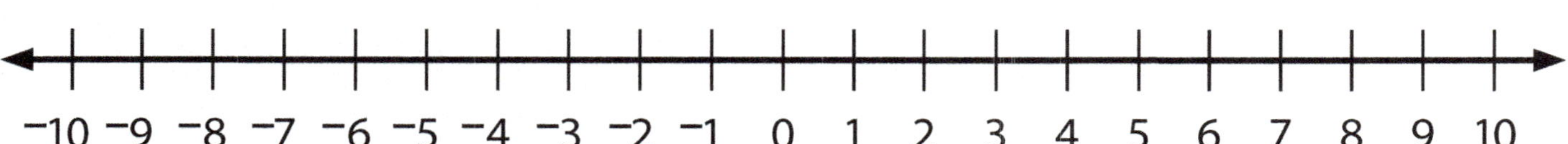

Subtracting Integers

1. Begin at 0 and draw an arrow to the minuend (first number).
2. Draw a second arrow: left to subtract a positive number, right to subtract a negative number.
3. The final stopping place is the difference.

Adding Integers

1. Begin at 0 and draw an arrow to the first addend.
2. Draw a second arrow: right to add a positive number, left to add a negative number.
3. The final stopping place is the sum.

1.
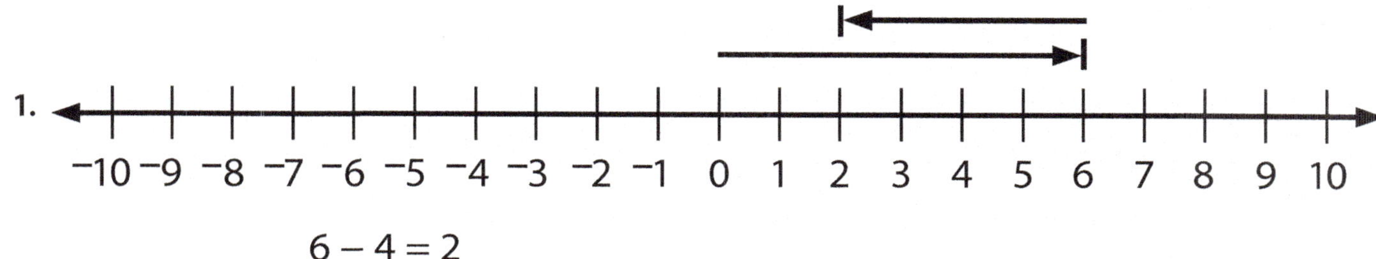

$$6 - 4 = 2$$

2.
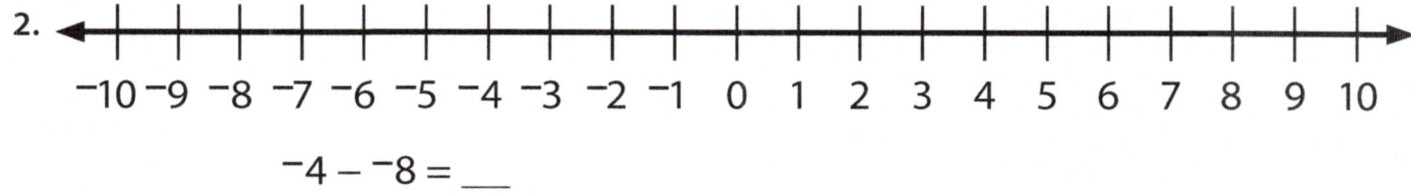

$$^-4 - {}^-8 = \underline{\quad}$$

3.
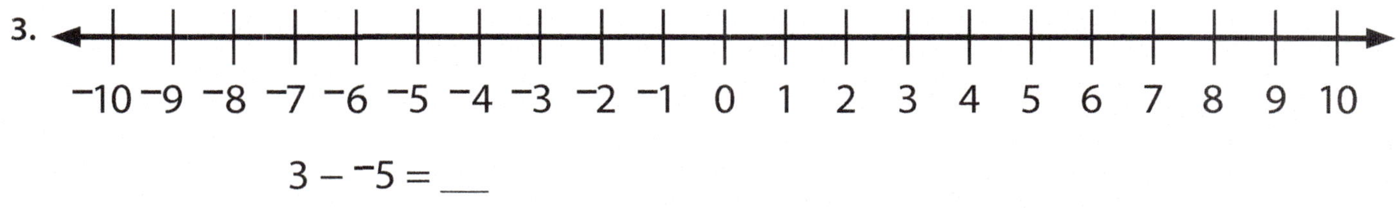

$$3 - {}^-5 = \underline{\quad}$$

4.
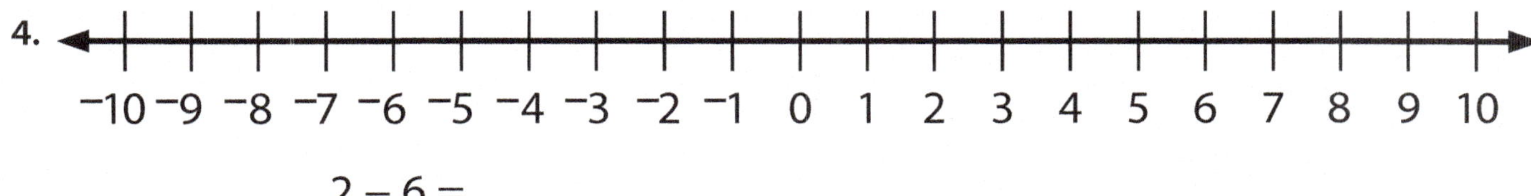

$$2 - 6 = \underline{\quad}$$

More Subtraction Patterns

Subtracting	Adding the Opposite
$6 - 3 = 3$	$6 + {}^-3 = 3$
$6 - 2 = 4$	$6 + {}^-2 = 4$
$6 - 1 = 5$	$6 + {}^-1 = 5$
$6 - 0 = 6$	$6 + 0 = 6$
$6 - {}^-1 = 7$	$6 + 1 = 7$
$6 - {}^-2 = \underline{}$	$6 + 2 = \underline{}$

Subtracting	Adding the Opposite
$5 - 1 = 4$	$5 + {}^-1 = 4$
$5 - 2 = 3$	$5 + {}^-2 = 3$
$5 - 3 = 2$	$5 + {}^-3 = 2$
$5 - 4 = 1$	$5 + {}^-4 = 1$
$5 - 5 = 0$	$5 + {}^-5 = 0$
$5 - 6 = \underline{}$	$5 + {}^-6 = \underline{}$

$$3 \times 3 = 9$$

$$2 \times 3 = 6$$

$$1 \times 3 = 3$$

$$0 \times 3 = 0$$

$$^-1 \times 3 = {}^-3$$

$$^-2 \times 3 = {}^-6$$

$$^-3 \times 3 = \underline{}$$

$$^-4 \times 3 = \underline{}$$

$$^-4 \times 4 = {}^-16$$

$$^-4 \times 3 = {}^-12$$

$$^-4 \times 2 = {}^-8$$

$$^-4 \times 1 = {}^-4$$

$$^-4 \times 0 = 0$$

$$^-4 \times {}^-1 = 4$$

$$^-4 \times {}^-2 = 8$$

$$^-4 \times {}^-3 = \underline{}$$

$$^-4 \times {}^-4 = \underline{}$$

Order of Operations

> 1. Do operations in parentheses first.
> 2. Find the value of exponents.
> 3. Multiply and divide from left to right.
> 4. Add and subtract from left to right.

1. $^{-}8 + 2 - 3 =$

2. $(^{-}1 + 16) \div {}^{-}3 =$

3. $6(2 + {}^{-}5) =$

4. $^{-}4 - 1 + {}^{-}3 =$

5. $(3 - 2) + (^{-}4 + {}^{-}2) =$

6. $^{-}15 \div 3 + {}^{-}2 =$

7. $2(^{-}12 \div {}^{-}6) =$

8. $3 \times 3 + {}^{-}2 =$

9. $14 - (^{-}4 + 2) =$

10. $7^2 + (^{-}16 \div 2) =$

Mental Math Strings

4-number strings

$12 + 2 \div 7 + 4 = 6$

$5 \times 8 \div 10 + 6 = 10$

$26 + 2 \div 4 - 3 = 4$

$16 + 4 \div 5 + 1 = 5$

$3 \times 5 + 10 \div 5 = 5$

$35 + 10 \div 5 \times 3 = 27$

$4 \times 11 - 4 \div 5 = 8$

$7 \times 9 - 9 - 4 = 50$

$2 \times 4 - 8 + 3 = 3$

$6 + 2 \times 4 + 3 = 35$

$5 \times 6 \div 3 + 7 = 17$

$8 \div 2 \times 6 \div 3 = 8$

$10 - 3 \times 3 + 4 = 25$

$30 - 10 \div 5 + 6 = 10$

$66 \div 11 \times 7 - 2 = 40$

$18 \div 3 + 7 + 3 = 16$

$9 + 8 + 4 \div 7 = 3$

$2 + 20 \div 11 \times 3 = 6$

$4 \times 9 \div 6 \times 8 = 48$

$5 \times 9 + 3 \div 6 = 8$

4-number strings

$3 \times 8 \div 4 + 7 = 13$

$8 \div 1 \times 6 \div 4 = 12$

$9 - 3 \times 7 + 4 = 46$

$75 - 3 \div 8 + 1 = 10$

$36 \div 12 \times 4 - 4 = 8$

$30 \div 3 + 4 + 8 = 22$

$7 + 8 - 4 + 7 = 18$

$12 + 8 \times 3 \div 6 = 10$

$4 \times 7 - 6 \div 11 = 2$

$3 \times 9 + 3 \div 6 = 5$

$12 + 10 \div 2 + 4 = 15$

$5 \times 5 + 10 \div 7 = 5$

$28 + 2 \div 6 - 3 = 2$

$7 + 4 + 5 - 1 = 15$

$3 \times 7 + 9 \div 5 = 6$

$45 + 10 \div 5 \times 3 = 33$

$4 \times 8 - 4 \div 4 = 7$

$7 \times 7 - 9 \div 4 = 10$

$2 \times 9 - 8 + 3 = 13$

$8 + 2 \times 4 \div 8 = 5$

5-number strings

$2 \times 7 + 8 + 3 \div 5 = 5$

$6 + 8 + 10 \div 3 + 5 = 13$

$9 \times 6 - 9 \div 5 - 2 = 7$

$56 \div 7 \times 8 \div 8 + 0 = 8$

$23 - 5 \div 3 + 4 \div 1 = 10$

$16 \div 4 \times 9 + 5 - 4 = 37$

$3 \times 9 + 20 - 9 + 5 = 43$

$7 \times 6 + 9 - 1 \div 10 = 5$

$4 \times 7 - 4 \div 6 + 9 = 13$

$7 \times 9 - 8 \div 5 - 4 = 7$

$9 + 2 \times 7 - 14 \div 9 = 7$

$9 \times 8 - 5 + 3 \div 7 = 10$

$43 + 7 \div 10 - 3 \times 4 = 8$

$16 + 9 \div 5 + 7 \div 2 = 6$

$29 - 8 + 7 \div 4 + 5 = 12$

$9 + 7 + 4 \div 2 \times 6 = 60$

$20 + 13 \div 11 \times 6 \div 9 = 2$

$4 \times 8 + 4 \div 6 \times 8 = 48$

$56 \div 8 \times 9 + 3 \div 6 = 11$

$10 \times 10 - 10 \div 10 \times 7 = 63$

5-number strings

$6 \times 6 - 7 + 5 + 4 = 38$

$9 + 7 \div 4 + 6 \div 5 = 2$

$9 \times 9 - 6 - 3 \div 9 = 8$

$34 + 8 \div 7 \times 3 + 4 = 22$

$9 + 9 \div 6 + 12 \div 5 = 3$

$9 - 8 + 6 \times 2 + 5 = 19$

$4 + 7 + 4 \div 3 \times 7 = 35$

$20 + 8 \div 4 \times 6 - 6 = 36$

$4 \times 8 + 10 \div 6 \times 8 = 56$

$64 \div 8 \times 9 + 3 - 15 = 60$

$10 \times 3 - 10 \div 10 \times 9 = 18$

$3 \times 3 + 8 + 1 \div 2 = 9$

$6 + 7 + 2 \div 3 + 5 = 10$

$4 \times 6 - 9 \div 5 - 2 = 1$

$54 \div 9 \times 2 \div 2 + 9 = 15$

$37 - 5 \div 8 + 7 \div 11 = 1$

$16 \div 2 \times 9 - 2 \div 7 = 10$

$3 \times 4 + 2 - 9 + 8 = 13$

$7 \times 1 + 9 \div 4 \times 9 = 36$

$4 \times 4 - 8 \div 2 + 9 = 13$

More Mental Math Strings

6-number strings

$12 - 7 + 2 \times 4 \div 4 + 3 = 10$

$5 \times 5 \times 4 \div 10 + 6 \div 4 = 4$

$30 + 2 \div 4 - 3 \times 9 + 0 = 45$

$11 + 4 \div 5 + 10 - 1 \times 3 = 36$

$1 + 5 \times 10 \div 12 \times 2 - 1 = 9$

$53 + 10 \div 7 \times 3 + 3 \div 3 = 10$

$3 \times 11 - 3 \div 6 + 2 \times 4 = 28$

$7 + 9 \div 4 + 6 \times 6 - 12 = 48$

$2 \times 4 + 9 + 3 \times 4 - 3 = 77$

$6 \div 2 \times 11 + 3 \div 9 + 2 = 6$

$5 \times 8 \div 2 + 7 \div 3 \times 2 = 18$

$10 \div 2 \times 1 + 13 \div 9 + 7 = 9$

$10 - 7 \times 3 + 1 \div 2 \times 7 = 35$

$70 - 10 \div 6 + 4 \div 2 \times 8 = 56$

$56 \div 8 \times 7 + 1 \div 5 + 3 = 13$

$18 \div 6 + 7 \div 2 + 8 - 2 = 11$

$6 + 8 - 4 \times 7 + 2 \div 9 = 8$

$4 + 20 \div 6 \times 5 + 1 \div 3 = 7$

$3 \times 12 \div 6 \times 8 + 2 \div 5 = 10$

$6 \times 9 + 6 \div 6 \times 2 \div 10 = 2$

6-number strings

$6 - 3 \times 8 \div 2 + 3 \div 5 = 3$

$8 \div 4 \times 5 + 32 \div 7 \times 9 = 54$

$9 + 3 \times 2 + 4 \div 7 - 2 = 2$

$24 - 3 \div 7 \times 9 + 3 \div 1 = 30$

$48 \div 12 \times 6 - 4 \div 5 \times 2 = 8$

$90 \div 9 + 12 \div 11 \times 8 - 1 = 15$

$7 \times 8 - 7 \div 7 + 5 \div 6 = 2$

$24 \div 12 + 8 \times 3 \div 6 \times 2 = 10$

$9 \times 7 + 8 + 10 \div 9 - 3 = 6$

$3 \times 3 + 2 \times 6 - 2 \div 8 = 8$

$11 + 10 \div 3 + 4 \times 9 + 1 = 100$

$35 \div 5 - 1 \times 8 \div 12 + 13 = 17$

$40 \div 8 \times 2 + 6 \div 4 + 3 = 7$

$5 + 4 \div 3 \times 12 - 4 \div 8 = 4$

$8 \times 7 + 9 - 1 \div 8 \times 2 = 16$

$44 + 10 \div 6 \times 3 + 3 \div 10 = 3$

$12 \div 4 \times 11 + 7 \div 5 + 9 = 17$

$6 \times 5 - 9 \div 3 \times 9 + 5 = 68$

$12 \div 6 \times 7 - 8 \times 8 + 3 = 51$

$18 + 2 \div 4 \times 5 + 8 - 4 = 29$

8-number strings

$2 \times 9 + 8 + 1 \div 3 \times 4 + 2 - 9 = 29$

$16 + 8 - 10 \div 7 + 5 \times 8 + 4 \div 6 = 10$

$5 \times 7 + 9 \div 11 + 8 \div 3 \times 7 - 2 = 26$

$42 \div 7 \times 8 \div 4 + 9 \div 7 \div 3 \times 2 = 2$

$45 \div 5 \div 3 \times 4 \div 2 \times 9 + 8 - 1 = 61$

$28 \div 4 \times 0 + 8 \div 4 \times 11 + 3 \div 5 = 5$

$6 \div 3 \times 9 + 10 - 10 \div 6 + 8 \times 5 = 55$

$8 \times 6 + 9 - 1 \div 7 \times 4 + 3 \div 5 = 7$

$26 + 2 \div 7 - 4 \times 7 + 9 \div 3 + 18 = 21$

$17 - 9 \div 8 \times 5 + 4 \times 2 \div 6 + 31 = 34$

$9 + 5 - 2 \times 7 - 3 \div 9 + 6 \div 5 = 3$

$8 \times 9 - 4 - 2 \div 6 + 3 \div 7 \times 12 = 24$

$37 + 7 \div 11 \times 10 - 4 \div 9 + 8 - 5 = 7$

$2 \times 6 \times 5 \div 6 \div 2 \times 10 + 4 \div 9 = 6$

$63 \div 7 \times 8 - 42 \div 5 + 3 \times 8 - 4 = 68$

$29 + 7 \div 6 \div 2 \times 4 \div 1 + 0 \times 5 = 60$

$31 + 2 \div 11 + 4 \times 8 - 7 \div 7 \times 9 = 63$

$4 \times 2 + 4 \div 6 \times 9 + 6 \div 8 \times 9 = 27$

$48 \div 8 \times 9 + 3 + 7 \div 8 \times 6 \div 12 = 4$

$100 \div 10 \times 10 - 10 \div 10 \times 7 - 3 \div 10 = 6$

8-number strings

$3 \times 6 - 7 + 5 \div 4 \times 7 - 3 \div 5 = 5$

$9 + 9 \div 3 + 6 \div 4 \times 9 + 5 \div 8 = 4$

$7 \times 7 - 4 \div 5 - 3 \div 2 \times 12 + 5 = 41$

$45 + 10 \div 11 \times 8 + 4 \div 11 + 6 \times 9 = 90$

$10 + 9 - 4 \div 3 + 6 \times 7 - 5 \div 8 = 9$

$11 - 8 + 9 \times 2 + 8 \div 4 \div 2 + 0 = 4$

$24 \div 4 + 7 + 5 \div 3 \times 8 \div 6 + 9 = 17$

$20 + 7 \div 9 \times 4 - 6 + 5 \times 10 - 11 = 99$

$6 - 4 \times 8 + 9 \div 5 \times 1 + 4 \div 3 = 3$

$1 \times 9 \div 3 \times 11 + 3 \div 9 + 7 - 9 = 2$

$6 \times 10 \div 12 \times 3 - 10 \times 10 - 8 \div 7 = 6$

$3 \times 7 + 8 + 1 \div 3 \times 10 - 12 \div 11 = 8$

$9 + 7 \div 2 \times 3 + 8 \div 4 + 3 \times 2 = 22$

$4 \times 6 - 3 \div 7 - 2 \times 9 \times 7 - 8 = 55$

$54 \div 6 \times 4 \div 9 + 9 - 1 \times 7 + 7 = 91$

$42 - 5 + 3 \div 8 + 7 \div 3 \times 8 - 5 = 27$

$18 \div 2 \times 9 - 11 \div 7 \times 5 - 7 + 9 = 52$

$7 \times 4 + 2 - 9 \div 3 + 8 \div 5 \times 12 = 36$

$4 \times 1 + 7 \times 5 - 6 \div 7 \times 9 + 9 = 72$

$9 \times 11 - 9 \div 9 + 11 \div 3 + 9 \div 4 = 4$

Mental Math Strings with Fractions

Adding & Subtracting Fractions

$$\frac{1}{4} + \frac{3}{4} + \frac{2}{4} - \frac{1}{4} + \frac{2}{4} - \frac{3}{4} = \frac{4}{4} \text{ or } 1$$

$$\frac{2}{2} + \frac{3}{2} - \frac{1}{2} + \frac{7}{2} - \frac{8}{2} + \frac{1}{2} = \frac{4}{2} \text{ or } 2$$

$$\frac{2}{3} + \frac{1}{3} + \frac{7}{3} - \frac{4}{3} + \frac{2}{3} - \frac{2}{3} = \frac{6}{3} \text{ or } 2$$

$$\frac{11}{5} + \frac{4}{5} - \frac{2}{5} + \frac{7}{5} - \frac{3}{5} + \frac{4}{5} = \frac{21}{5} \text{ or } 4\frac{1}{5}$$

$$\frac{1}{6} + \frac{5}{6} + \frac{3}{6} + \frac{8}{6} - \frac{10}{6} + \frac{2}{6} = \frac{9}{6} \text{ or } 1\frac{1}{2}$$

$$\frac{4}{7} - \frac{2}{7} + \frac{6}{7} - \frac{1}{7} + \frac{9}{7} - \frac{5}{7} = \frac{11}{7} \text{ or } 1\frac{4}{7}$$

$$\frac{3}{8} + \frac{2}{8} + \frac{5}{8} - \frac{4}{8} + \frac{7}{8} - \frac{1}{8} = \frac{12}{8} \text{ or } 1\frac{1}{2}$$

$$\frac{3}{9} + \frac{6}{9} - \frac{1}{9} + \frac{2}{9} - \frac{1}{9} - 1 = \frac{0}{9} \text{ or } 0$$

$$\frac{4}{5} + \frac{2}{5} - \frac{1}{5} + \frac{3}{5} + \frac{1}{5} - 1 = \frac{4}{5}$$

$$\frac{3}{12} + \frac{8}{12} + \frac{4}{12} + \frac{2}{12} + \frac{1}{12} - 1 = \frac{6}{12} \text{ or } \frac{1}{2}$$

$$\frac{1}{3} + \frac{5}{3} + \frac{2}{3} + \frac{5}{3} - \frac{1}{3} - 1 = \frac{9}{3} \text{ or } 3$$

$$\frac{5}{4} + \frac{3}{4} - \frac{2}{4} + \frac{4}{4} + \frac{1}{4} - 1 = \frac{7}{4} \text{ or } 1\frac{3}{4}$$

$$\frac{2}{6} + \frac{3}{6} - \frac{1}{6} + \frac{3}{6} + \frac{2}{6} - 1 = \frac{3}{6} \text{ or } \frac{1}{2}$$

$$\frac{1}{8} + \frac{6}{8} - \frac{2}{8} + \frac{4}{8} - \frac{1}{8} - \frac{1}{2} = \frac{4}{8} \text{ or } \frac{1}{2}$$

$$\frac{4}{10} - \frac{2}{10} - \frac{1}{10} + \frac{3}{10} + \frac{1}{10} - \frac{1}{2} = \frac{0}{10} \text{ or } 0$$

$$\frac{1}{12} + \frac{9}{12} - \frac{2}{12} + \frac{4}{12} - \frac{2}{12} - \frac{1}{2} = \frac{4}{12} \text{ or } \frac{1}{3}$$

$$1 - \frac{2}{3} + \frac{5}{3} + \frac{1}{3} - \frac{3}{3} + \frac{1}{3} = \frac{5}{3} \text{ or } 1\frac{2}{3}$$

$$\frac{1}{2} + \frac{5}{2} - \frac{4}{2} + \frac{7}{2} - \frac{1}{2} - 1 = \frac{6}{2} \text{ or } 3$$

$$1 - \frac{2}{5} + \frac{3}{5} + \frac{1}{5} + \frac{4}{5} - \frac{1}{5} = \frac{10}{5} \text{ or } 2$$

$$\frac{1}{4} + \frac{6}{4} + \frac{2}{4} - \frac{1}{4} + \frac{2}{4} - \frac{1}{2} = \frac{8}{4} \text{ or } 2$$

$$\frac{7}{9} + \frac{3}{9} - \frac{1}{9} - \frac{2}{9} + \frac{11}{9} - 1 = \frac{9}{9} \text{ or } 1$$

Multiplying & Dividing Fractions

$$2 \times \frac{2}{4} + \frac{1}{4} - \frac{3}{4} + \frac{3}{4} - \frac{1}{4} = \frac{4}{4} \text{ or } 1$$

$$\frac{1}{2} \times 2 + \frac{1}{2} \times 2 - \frac{1}{2} \times 1 = \frac{5}{2} \text{ or } 2\frac{1}{2}$$

$$\frac{1}{3} + \frac{2}{3} + \frac{2}{3} + \frac{1}{3} - \frac{2}{3} \div 2 = \frac{2}{3}$$

$$\frac{3}{9} - \frac{2}{9} + \frac{5}{9} + \frac{4}{9} - \frac{1}{9} \div 3 = \frac{3}{9} \text{ or } \frac{1}{3}$$

$$\frac{8}{2} \div 2 - \frac{2}{2} + \frac{3}{2} \times 2 \div 5 = \frac{2}{2} \text{ or } 1$$

$$\frac{2}{3} \times 3 \div 2 + \frac{9}{3} + \frac{2}{3} - \frac{1}{3} = \frac{13}{3} \text{ or } 4\frac{1}{3}$$

$$3 \times \frac{4}{5} - \frac{2}{5} \div 5 + \frac{1}{5} + 1 = \frac{8}{5} \text{ or } 1\frac{3}{5}$$

$$2 \times \frac{3}{4} + \frac{2}{4} - \frac{1}{4} + \frac{1}{4} \div 4 = \frac{2}{4} \text{ or } \frac{1}{2}$$

$$4 \times \frac{2}{3} \div 2 \times 7 \div 4 + \frac{1}{3} = \frac{8}{3} \text{ or } 2\frac{2}{3}$$

$$7 \times \frac{1}{7} + \frac{2}{7} \div 3 \times 2 - \frac{1}{7} = \frac{5}{7}$$

$$\frac{5}{6} \times 2 - \frac{1}{6} \div 3 + \frac{5}{6} \div 2 = \frac{4}{6} \text{ or } \frac{2}{3}$$

$$\frac{6}{8} \div 3 \times 1 + \frac{1}{8} \times 6 + \frac{6}{8} = \frac{24}{8} \text{ or } 3$$

$$\frac{3}{4} \times 6 \div 6 + \frac{2}{4} - \frac{1}{4} \times 2 = \frac{8}{4} \text{ or } 2$$

$$\frac{2}{10} \div 2 \times 9 + \frac{3}{10} - 1 \times 2 = \frac{4}{10} \text{ or } \frac{2}{5}$$

$$5 \times \frac{1}{9} \times 2 \div 5 + \frac{7}{9} \times 2 = \frac{18}{9} \text{ or } 2$$

$$\frac{5}{3} \div \frac{1}{3} \times \frac{1}{2} + \frac{3}{2} \div 2 \div 2 = \frac{2}{2} \text{ or } 1$$

$$\frac{2}{5} \times 6 \div 3 \times 2 \div \frac{2}{5} \times \frac{1}{9} = \frac{4}{9}$$

$$3 \times \frac{5}{8} \div 5 \div \frac{1}{8} \times \frac{2}{3} \div 2 = \frac{3}{3} \text{ or } 1$$

$$8 \times \frac{1}{2} \div 4 \times \frac{1}{3} \times 6 \div 3 = \frac{2}{3}$$

$$10 \times \frac{2}{5} \times \frac{2}{3} \div 4 \times 3 \div 2 = \frac{3}{3} \text{ or } 1$$

$$\frac{5}{6} \times 6 \div 5 \times 1 \times 7 + \frac{3}{6} = \frac{45}{6} \text{ or } 7\frac{1}{2}$$

Shipping Volume

Shipping companies transport *freight*, the items being shipped, in large enclosed containers called *trailers* that are hauled by powerful trucks called *tractors*. Shippers calculate the volume of their trailers to help them load their freight efficiently and provide better service to their customers.

Freight is secured onto a square platform to make a unit called a *skid cube* for shipping. A trailer's volume can be measured in skid cubes.

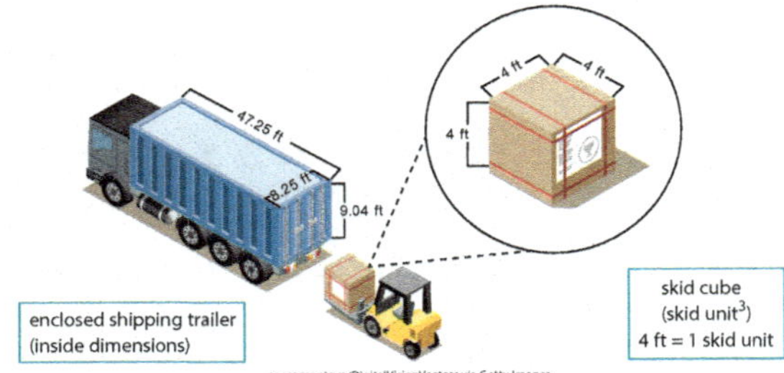

roccomontoya/DigitalVisionVectors via Getty Images

How many skid cubes will fit in the trailer?
Round a decimal answer to the nearest hundredth.

l = 47.25 ft ÷ __4__ ft = __11.81__ skid units; __11__ skid units will fit

w = 8.25 ÷ __4__ ft = __2.06__ skid units; __2__ skid units will fit

h = 9.04 ft ÷ __4__ ft = __2.26__ skid units; __2__ skid units will fit

$V = Bh$

$V = (l \times w) \times h$

$V = ($ __11__ skid units × __2__ skid units$) \times$ __2__ skid units

$V =$ __44__ skid units³ (skid cubes)

Business Idea Organizer

sample answers

Problem to Solve	Gifts/Abilities/Interests	Business Idea
pet sitter needed	*loves animals*	*pet sitting service*
people want to try new foods	*loves cooking, inherited a vintage cookbook*	*food blog*
demand for fresh vegetables	*enjoys gardening*	*vegetable stand*
babysitters needed	*great with kids, took CPR class*	*babysitting service*
neighbors need help with yard work	*loves being outdoors*	*lawn mowing business*
demand for farm fresh eggs	*raises chickens*	*selling eggs*
demand for unique jewelry	*enjoys design, owns a jewelry-making kit*	*designing & making jewelry to sell*

best business idea: ____

INDEX

Numbers listed indicate the lessons in which each entry appears.

PHOTO CREDITS

Key: (t) top; (c) center; (b) bottom; (l) left; (r) right; (bg) background; (i) inset

Teacher Edition

x SDI Productions/E+/via Getty Images; **xii** Tyler Olson/Shutterstock.com; **xvii** Matt Bird/Stone/Getty Images

Student Edition

Chapter 1

1 donald_gruener/iStock/Getty Images Plus/Getty Images; **4** andresr/E+/Getty Images; **5** Map Resources; **11tli** Bettmann/Bettmann/Getty Images; **11bg** Robert Kneschke/Eye Em/Getty Images; **11bci** Danlie Cheng/Eye Em/Getty Images; **11bri** takenobu/iStock/Getty Images Plus/Getty Images; **18r** mmac72/iStock/Getty Images Plus/Getty Images; **18li** Digital Light Source/Universal Images Group/Getty Images

Chapter 2

27 Enjoyyourlife/iStock/Getty Images Plus/Getty Images; **36** (one and twenty dollar bills) Unusual Films; **36** (quarter) Pete Spiro/Shutterstock.com; **39** fdastudillo/iStock Unreleased/Getty Images; **43** stellalevi/E+/Getty Images

Chapter 3

47 kali9/E+/Getty Images; **49** SDI Productions/E+/Getty Images; **58** Enrique Diaz/7cero/Moment/Getty Images; **62** fotografixx/iStock/Getty Images Plus/Getty Images

Chapter 4

69 ©Mint Images/Media Bakery; **75** Tierfotoagentur/Alamy Stock Photo

Chapter 5

90 Mayur Kakade/Moment/Getty Images; **91bg** Solstock/iStock/Getty Images Plus/Getty Images; **91br** ©2018 Memorial Sloan Kettering Cancer Center; **91bl** mediaphotos/E+/Getty Images; **95** Adobe Stock/ExQuisine; **102** Andrew Johnson/E+/Getty Images; **105** tmarvin/E+/Getty Images; **106** Philip Scalia/Alamy

Chapter 6

111 Rubberball/Mike Kemp/Getty Images; **111bg** Spooh/E+/Getty Images; **111bli** H_Yasui/iStock/Getty Images Plus/Getty Images; **129** Darrell Gulin/The Image Bank/Getty Images

Chapter 7

139 Johner Images/Getty Images; **143** kali9/E+/Getty Images; **151** Emma Farrer/Moment/Getty Images

Chapter 8

157 NattapolStudiO/Shutterstock.com; **157i** Erik Isakson/Getty Images; **157bg** francescoch/iStock/Getty Images Plus/Getty Images; **167** JudiParkinson/iStock/Getty Images Plus/Getty Images; **169** GlobalP/iStock/Getty Images Plus/Getty Images; **170** CASEZY/iStock/Getty Images Plus/Getty Images; **173** Smith Collection/Gado/Archive Photos/Getty Images; **177tr** Hi-Story/Alamy Stock Photo; **177cr** MediaPunch Inc/Alamy Stock Photo; **177bl** Rospotte Photography/Alamy Stock Photo; **177tl** Gado Images/Alamy Stock Photo; **177br** bill belknap/Alamy Stock Photo; **177cl** Everett Collection/Shutterstock.com

Chapter 9

178 Jose Luis Pelaez Inc/Digital Vision/Getty Images Plus/Getty Images; **179bg** vgajic/E+/Getty Images; **179br** ©Kent Gaylor; **179bl** Handout/Getty Images News/Getty Images; **182** Johner Images/Getty Images; **184** Difught/Shutterstock.com; **192** James D Morgan/Getty Images News/Getty Images

Chapter 10

203 Marc Dufresne/E+/Getty Images; **203ti** spooh/E+/Getty Images; **207br** THEPALMER/iStock/Getty Images Plus/Getty Images; **207bl** Thomas-Soellner/iStock/Getty Images Plus/Getty Images; **209** abilityriddle/Moment/Getty Images; **215** vernonwiley/iStock/Getty Images Plus/Getty Images; **221** Filip Micovic/500px/Getty Images

Chapter 11

225 digitalskillet/iStock/Getty Images; **225bg** Map Resources; **241** 3DMI/Shutterstock.com

Chapter 12

249 valeninrussanov/E+/Getty Images; **259** Elenathewise/iStock/Getty Images Plus/Getty Images; **263br** Murat Deniz/iStock/Getty Images Plus/Getty Images; **263bcr** CoffeeAndMilk/E+/Getty Images; **263bc** mustafagull/iStock/Getty Images Plus/Getty Images; **263bcl** Kuzmichstudio/iStock/Getty Images Plus/Getty Images; **263bl** Yobro10/iStock/Getty Images Plus/Getty Images

Chapter 13

264 skynesher/E+/Getty Images; **264bg** Dmytro Aksonov/E+/Getty Images; **265bg** bluejayphoto/iStock/Getty Images Plus/Getty Images; **265ci** ©Susan R Symonds of Infinity Portrait Design, courtesy of Rosales + Partners; **265bli** Kruck20/iStock/Getty Images Plus/Getty Images; **275l** Map Resources; **276l** GlobalP/iStock/Getty Images Plus/Getty Images; **276br** Barbara Rich/Moment Open/Getty Images; **276tr** Ashva/iStock/Getty Images Plus/Getty Images; **294** Matt Bird/Stone/Getty Images